Concise Encyclopaedia of BIOINFORMATICS and COMPUTATIONAL BIOLOGY

Concise Encyclopaedia of BIOINFORMATICS and COMPUTATIONAL BIOLOGY

SECOND EDITION

Edited by

John M. Hancock
Department of Physiology,
Development & Neuroscience
University of Cambridge
Cambridge, UK

Marketa J. Zvelebil
Breakthrough Breast Cancer Research
Institute of Cancer Research
London, UK

WILEY Blackwell

Registered office
John Wiley & Sons, Ltd, The Atrium, Southern Gate, Chichester, West Sussex, PO19 8SQ, UK

Editorial offices
9600 Garsington Road, Oxford, OX4 2DQ, UK
The Atrium, Southern Gate, Chichester, West Sussex, PO19 8SQ, UK
111 River Street, Hoboken, NJ 07030-5774, USA

For details of our global editorial offices, for customer services and for information about how to apply for permission to reuse the copyright material in this book please see our website at www.wiley.com/wiley-blackwell.

Library of Congress Cataloging-in-Publication Data

Dictionary of bioinformatics and computational biology
Concise encyclopaedia of bioinformatics and computational biology 2e / edited by John M. Hancock, MRC Mammalian Genetics Unit, Harwell, Oxfordshire, United Kingdom, Marketa J. Zvelebil, University College London, Ludwig Institute for Cancer Research, London, United Kingdom. -- 2e.
pages cm
The Concise Encyclopaedia of Bioinformatics and Computational Biology is a follow-up edition to the Dictionary of Bioinformatics and Computational Biology.
Includes bibliographical references and index.
ISBN 978-0-470-97871-9 (pbk. : alk. paper) – ISBN 978-0-470-97872-6 (cloth : alk. paper) – ISBN 978-1-118-59814-6 (emobi) – ISBN 978-1-118-59815-3 (epub) – ISBN 978-1-118-59816-0 (epdf) – ISBN 978-1-118-77297-3 1. Bioinformatics–Dictionaries. 2. Computational biology–Dictionaries. I. Hancock, John M., editor of compilation. II. Zvelebil, Marketa J., editor of compilation. III. Title.
QH324.2.D53 2014
572′.330285–dc23

2013029874

A catalogue record for this book is available from the British Library.

Wiley also publishes its books in a variety of electronic formats. Some content that appears in print may not be available in electronic books.

Cover images: Background image - iStock File #5630705 © nicolas_. Circular images left to right - iStock File #2286469 © ChristianAnthony, iStock File #6273558 © alanphillips, iStock File #21506562 © Andrey Prokhorov
Cover design by Steve Thompson

Set in 10/12pt Garamond by Laserwords Private Limited, Chennai, India
Printed in Singapore by Ho Printing Singapore Pte Ltd

1 2014

John M. Hancock
would like to thank his wife, Liz, for all her support
and his parents for everything they have done.

Marketa J. Zvelebil
would like to dedicate this book to the memory of her father, Professor K.V. Zvelebil, and brother, Professor Marek Zvelebil.

CONTENTS

List of Contributors ix
Preface xiii
Entries A to Z 1
Author Index 791
Colour plate section facing p 210

LIST OF CONTRIBUTORS

JOSEP F. ABRIL, Departament de Genètica/Institut de Biomedicina (IBUB), Universitat de Barcelona, Barcelona, Spain

BISSAN AL-LAZIKANI, Institute of Cancer Research, London, UK

TERESA K. ATTWOOD, Faculty of Life Sciences, University of Manchester, Manchester, UK

CONCHA BIELZA, Departamento de Inteligencia Artificial, Facultad de Informática, Universidad Politécnica de Madrid, Madrid, Spain

ENRIQUE BLANCO, Departament de Genètica/ Institut de Biomedicina (IBUB), Universitat de Barcelona (UB), Barcelona, Spain

DAN BOLSER, EMBL Outstation – Hinxton, European Bioinformatics Institute, Wellcome Trust Genome Campus, Cambridge, UK

STUART BROWN, Cell Biology, New York, NY, USA

AIDAN BUDD, EMBL Heidelberg, Heidelberg, Germany

JAMIE J. CANNONE, Integrative Biology, The University of Texas at Austin, Austin, TX, USA

FENG CHEN, Department of Computer Science and Computer Engineering, La Trobe University, Melbourne, VA, Australia

YI-PING PHOEBE CHEN, Department of Computer Science and Computer Engineering, La Trobe University, Melbourne, VA, Australia

ANDREW COLLINS, Human Genetics Research Division, University of Southampton, Southampton General Hospital, Southampton, UK

DARREN CREEK, Department of Biochemistry and Molecular Biology, University of Melbourne, Melbourne, VIC, Australia

ALISON CUFF, Institute of Structural and Molecular Biology, University College London, London, UK

MICHAEL P. CUMMINGS, Center for Bioinformatics and Computational Biology, University of Maryland, College Park, MD, USA

TJAART DE BEER, EMBL Outstation, European Bioinformatics Institute, Wellcome Trust Genome Campus, Cambridge, UK

ROLAND DUNBRACK, Institute for Cancer Research, Philadelphia, PA, USA

ANTON FEENSTRA, BIVU/Bioinformatics, Free University Amsterdam, Amsterdam, The Netherlands

PEDRO FERNANDES, Instituto Gulbenkian de Ciência, Oeiras, Portugal

JUAN ANTONIO GARCIA RANEA, Department of Biology Molecular and Biochemistry, Faculty of Sciences, Campus of Teatinos, Malaga, Spain

CAROLE GOBLE, Department of Computer Science, University of Manchester, Manchester, UK

DOV GREENBAUM, Molecular Biophysics & Biochemistry, Yale University, New Haven, CT, USA

MALACHI GRIFFITH, The Genome Institute, Washington University School of Medicine, Louis, MO, USA

OBI L. GRIFFITH, The Genome Institute, Washington University School of Medicine, Louis, MO, USA

SAM GRIFFITHS-JONES, Faculty of Life Sciences, University of Manchester, Manchester, UK

Roderic Guigó, Bioinformatics and Genomics Group, Centre for Genomic Regulation (CRG), Universitat Pompeu Fabra, Barcelona, Spain

Robin R. Gutell, Integrative Biology, The University of Texas at Austin, Austin, TX, USA

John M. Hancock, Department of Physiology, Development & Neuroscience, University of Cambridge, Cambridge, UK

Andrew Harrison, Department of Mathematical Sciences, University of Essex, Colchester, UK

Matthew He, Division of Math, Science, and Technology, Farquhar College of Arts and Sciences, Nova Southeastern University, Fort Lauderdale, FL, USA

Jaap Heringa, Centre for Integrative Bioinformatics VU, Department of Computer Science, Faculty of Sciences, Vrije Universiteit, Amsterdam, The Netherlands

A.R. Hoelzel, School of Biological and Biomedical Sciences, Durham University, Durham, UK

Simon Hubbard, Faculty of Life Sciences, University of Manchester, Manchester, UK

Austin L. Hughes, Department of Biological Sciences, University of South Carolina, Columbia, SC, USA

Pascal Kahlem, EMBL Outstation, European Bioinformatics Institute, Wellcome Trust Genome Campus, Cambridge, UK

Ana Kozomara, Faculty of Life Sciences, University of Manchester, Manchester, UK

Pedro Larrañaga, Departamento de Inteligencia Artificial, Facultad de Informática, Universidad Politécnica de Madrid, Madrid, España

Antonio Marco, School of Biological Sciences, University of Essex, Colchester, UK

James Marsh, School of Computer Science, University of Manchester, Manchester, UK

Erick Matsen, Computational Biology Program, Fred Hutchinson Cancer Research Center, Seattle, WA, USA

Luis Mendoza, Departamento de Biología Molecular y Biotecnología, Instituto de Investigaciones Biomédicas, Universidad Nacional Autónoma de México, Mexico City, México

Christine Orengo, Research Department of Structural and Molecular Biology, Institute of Structural and Molecular Biology, University College London, London, UK

Laszlo Patthy, Institute of Enzymology, Biological Research Center, Hungarian Academy of Sciences, Budapest, Hungary

Hedi Peterson, Institute of Computer Science, University of Tartu, Tartu, Estonia

Steve Pettifer, School of Computer Science, University of Manchester, Manchester, UK

Richard Scheltema, Department of Proteomics and Signal Transduction, Max-Planck Institute for Biochemistry, Martinsried, Germany

Thomas D. Schneider, National Institutes of Health, National Cancer Institute (NCI-Frederick), Gene Regulation and Chromosome Biology Laboratory, Molecular Information Theory Group, Frederick, MD, USA

Alexandros Stamatakis, Scientific Computing, HITS gGmbH, Heidelberg, Germany

Neil Swainston, School of Computer Science, University of Manchester, Manchester, UK

Denis Thieffry, Département de Biologie, École Normale Supérieure, Paris, France

David Thorne, School of Computer Science, University of Manchester, Manchester, UK

Jacques van Helden, Aix-Marseille Université, Inserm Unit UMR_S 1090, Technologie Avancée pour le Génome et la Clinique (TAGC), Marseille, France

JUAN ANTONIO VIZCANIO, EMBL Outstation, European Bioinformatics Institute, Wellcome Trust Genome Campus, Cambridge, UK

KATY WOLSTENCROFT, Department of Computer Science, University of Manchester, Manchester, UK

MARKETA J. ZVELEBIL, Breakthrough Breast Cancer Research, Institute of Cancer Research, London, UK

Non-active contributors whose contributions were carried over or modified in this edition

PATRICK ALOY, Institute for Research in Biomedicine (IRB Barcelona), Parc Científic de Barcelona, Barcelona, Spain

ROLF APWEILER, EMBL Outstation, European Bioinformatics Institute, Wellcome Trust Genome Campus, Cambridge, UK

JEREMY BAUM, Not Available

M.J.BISHOP, Not Available

LIZ CARPENTER, SGC, University of Oxford, Oxford, UK

JEAN-MICHEL CLAVERIE, Mediterranean Institute of Microbiology (IMM, FR3479), Scientific Parc of Luminy, Marseille, France

NELLO CRISTIANINI, Faculty of Engineering, University of Bristol, Bristol, UK

NIALL DILLON, MRC Clinical Sciences Centre, Faculty of Medicine, Imperial College London, London, UK

JAMES FICKETT, Not Available

ALAN FILIPSKI, Center for Evolutionary Medicine and Informatics, The Biodesign Institute at Arizona State University, Tempe, AZ, USA

KATHELEEN GARDINER, Colorado IDDRC, University of Colorado Denver, Aurora, CO, USA

DAVID JONES, Department of Computer Science, University College London, London, UK

SUDHIR KUMAR, Center for Evolutionary Medicine and Informatics, The Biodesign Institute at Arizona State University, Tempe, AZ, USA

ROMAN LASKOWSKI, European Bioinformatics Institute, Wellcome Trust Genome Campus, Cambridge, UK

ERIC MARTZ, Department of Microbiology, University of Massachusetts, Amherst, USA

MARK MCCARTHY, OCDEM, University of Oxford, Oxford, UK

IRMTRAUD MEYER, Centre for High-Throughput Biology Bioinformatics Laboratories, University of British Columbia, Vancouver, BC, Canada

RODGER STADEN, Not Available

ROBERT STEVENS, School of Computer Science, University of Manchester, Manchester, UK

GUENTER STOESSER, Not Available

STEVEN WILTSHIRE, Not Available

PREFACE

In 2004 we compiled the *Dictionary of Bioinformatics and Computational Biology* to 'provide clear definitions of the fundamental concepts of bioinformatics and computational biology'. We wrote then, 'The entries were written and edited to enhance the book's utility for newcomers to the field, particularly undergraduate and postgraduate students. Those already working in the field should also find it handy as a source for quick introductions to topics with which they are not too familiar' and this applies equally to this new edition, the *Concise Encyclopaedia of Bioinformatics and Computational Biology*, which is a follow-up edition to the Dictionary.

Over the last decade bioinformatics has undergone largely evolutionary change. This is reflected in the *Concise Encyclopaedia* by turnover in the description of popular software programs, many of which have changed although the most popular remain. Probably the area of most revolutionary change has been in the technology, and associated bioinformatics, of DNA sequencing with the advent of High-Throughput (Next-Generation) sequencing (NGS). This is reflected in this edition by a number of new entries relating to NGS. Although NGS is replacing microarray technology in many applications, we have retained entries on microarray techniques as they are still current and highly used.

Apart from the changes, the *Concise Encyclopaedia* has been enhanced by expanding coverage in some of the more computer science-derived areas of the subject, as these are becoming increasingly important. To keep the book's length within reasonable limits, we have removed some entries that apply to what on reflection we consider to be purely biological concepts that are not in themselves necessary for an understanding of bioinformatics applications.

We hope that this new edition will be a welcome addition to bookshelves and electronic libraries and that it will continue to perform a useful function for the community.

As always, we thank our contributors for their efforts in writing the vast majority of the entries in this book, although as editors we accept responsibility for their accuracy.

John M. Hancock
Marketa J. Zvelebil

Ab Initio

Ab Initio Gene Prediction, *see* Gene Prediction, ab initio.

ABNR, *see* Energy Minimization.

Accuracy (of Protein Structure Prediction)

Accuracy Measures, *see* Error Measures.

Adjacent Group

Admixture Mapping (Mapping by Admixture Linkage Disequilibrium)

Adopted-basis Newton–Raphson Minimization (ABNR), *see* Energy Minimization.

Affine Gap Penalty, *see* Gap Penalty.

Affinity Propagation-based Clustering

Affymetrix GeneChip™ Oligonucleotide Microarray

Affymetrix Probe Level Analysis

After Sphere, *see* After State.

After State (After Sphere)

AIC, *see* Akaike Information Criterion.

Akaike Information Criterion

Algorithm

Alignment (Domain Alignment, Repeats Alignment)

Alignment Score

Allele-Sharing Methods (Non-parametric Linkage Analysis)

Allelic Association

Allen Brain Atlas

Allopatric Evolution (Allopatric Speciation)

Allopatric Speciation, *see* Allopatric Evolution.

AlogP

Alpha carbon, *see* Cα (C-Alpha).

Alpha Helix

Alternative Splicing

Alternative Splicing Gene Prediction, *see* Gene Prediction, alternative splicing.

Amide Bond (Peptide Bond)

Amino Acid (Residue)

Amino Acid Abbreviations, *see* IUPAC-IUB Codes.

Amino Acid Composition

Amino Acid Exchange Matrix (Dayhoff Matrix, Log Odds Score, PAM (Matrix), BLOSUM Matrix)

AMINO Acid Substitution Matrix, *see* Amino Acid Exchange Matrix.

Amino-terminus, *see* N-terminus.

Amphipathic

Analog (Analogue)

Ancestral Lineage, *see* Offspring Lineage.

Ancestral State Reconstruction

Anchor Points

Concise Encyclopaedia of Bioinformatics and Computational Biology, Second Edition.
Edited by John M. Hancock and Marketa J. Zvelebil.

A

Annotation Refinement Pipelines, *see* Gene Prediction.
Annotation Transfer (Guilt by Association Annotation)
APBIONET (Asia-Pacific Bioinformatics Network)
Apomorphy
APOLLO, *see* Gene Annotation, visualization tools.
Arc, *see* Branch (of a Phylogenetic Tree).
Are We There Yet?, *see* AWTY.
Aromatic
Array, *see* Data Structure.
Artificial Neural Networks, *see* Neural Networks.
ASBCB (The African Society for Bioinformatics and Computational Biology)
Association Analysis (Linkage Disequilibrium Analysis)
Association Rule, *see* Association Rule Mining.
Association Rule Mining (Frequent Itemset, Association Rule, Support, Confidence, Correlation Analysis)
Associative Array, *see* Data Structure.
Asymmetric Unit
Atomic Coordinate File (PDB file)
Autapomorphy
Autozygosity, *see* Homozygosity, Homozygosity Mapping.
AWTY (Are We There Yet?)
Axiom

AB INITIO

Roland Dunbrack and Marketa J. Zvelebil

In quantum mechanics, calculations of physical characteristics of molecules based on first principles such as the Schrödinger equation. In protein structure prediction, calculations made without reference to a known structure homologous to the target to be predicted. In other words these methods attempt to predict protein structure essentially from first principles (i.e. from physics and chemistry). The main advantage of this type of method is that no homologous structure is required to predict the fold of the target protein. However the accuracy of *ab initio* methods is not as high as threading or homology modeling.

There are a number of often used *ab initio* methods: lattice folding, FragFOld, Rosetta and Unres.

Relevant website

Rosetta	http://robetta.bakerlab.org/

Further reading

Defay T, Cohen FE (1995) Evaluation of current techniques for *ab initio* protein structure prediction. *Proteins*, 23: 431–445.

Hinchliffe A (1995) *Modelling Molecular Structures*. Wiley, New York.

Jones DT (2001) Predicting novel protein folds by using FRAGFOLD. *Proteins* Suppl. 5: 127–132.

AB INITIO GENE PREDICTION, *SEE* GENE PREDICTION, AB INITIO.

ABNR, *SEE* ENERGY MINIMIZATION.

ACCURACY (OF PROTEIN STRUCTURE PREDICTION)

David Jones

The measurement of the agreement between a predicted structure and the true native conformation of the target protein.

Depending on the type of prediction, a variety of metrics can be defined to measure the accuracy of a prediction experiment. For predictions of protein secondary structure (i.e. assigning residues to helix, strand or coil states), the percentage of residues correctly assigned (the Q_3 score) is the simplest and most widely used metric. For predictions of 3-D structure, there are many different metrics of accuracy with a variety of advantages and disadvantages. Probably the most widely used metric is the root mean square deviation (RMSD) between the model and the native structure. Unfortunately, small errors in

A

the model, particularly those which accrue from errors in the alignment used to build the model, result in very large RMSD values, and so this metric is less useful for low quality models. For low quality models, it is more common to measure prediction accuracy in terms of the percentage of residues which have been correctly aligned when compared against a structural alignment of the target and template proteins, or the percentage of residues which have been correctly positioned to within a certain distance cut-off (e.g. 3 Angstroms).

Relevant website

Accuracy estimators	http://predictioncenter.llnl.gov/local/local.html

See also Q_{index}, Secondary Structure Prediction, RMSD.

ACCURACY MEASURES, *SEE* ERROR MEASURES.

ADJACENT GROUP

Aidan Budd and Alexandros Stamatakis

Subtrees of an unrooted (mostly binary) tree that are attached to the same internal node of the unrooted tree have been described as 'adjacent groups'. The term was first used in

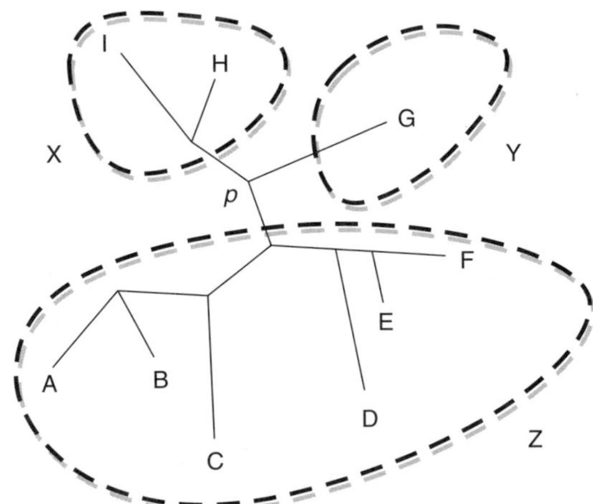

Figure A.1 An unrooted tree with nine operational taxonomic unit (OTU) labels A through I and one of the internal nodes labeled as p. The three adjacent groups associated with the internal node p are denoted by X (subtree containing OTUs I and H), Y (subtree containing OTU G), and Z (subtree containing OTUs A–F).

2007 to provide an unrooted tree equivalent to the term 'sister group', which should only be used in the context of rooted trees. See Figure A.1.

Further reading

Wilkinson M, et al. (2007) Of clades and clans: terms for phylogenetic relationships in unrooted trees. *Trends Ecol Evol* 22: 114–115.

ADMIXTURE MAPPING (MAPPING BY ADMIXTURE LINKAGE DISEQUILIBRIUM)

Andrew Collins, Mark McCarthy and Steven Wiltshire

A powerful method for identifying genes that underlie ethnic differences in disease risk: it focuses on recently-admixed populations and detects linkage by testing for association of the disease with ancestry at each typed marker locus.

Several major multifactorial traits (such as diabetes, hypertension) show marked ethnic differences in disease frequency, which may, at least in part, reflect differences in the prevalence of major susceptibility variants. Admixture mapping makes use of populations which have arisen through recent admixture of ancestral populations with widely-differing disease prevalences. In such admixed populations, the location of disease-susceptibility genes responsible for the prevalence difference between ancestral populations can be revealed by identifying chromosomal regions in which affected individuals show increased representation of the high-prevalence ancestral line.

Suitable populations include previously genetically separate populations which have shown recent admixture such as African Americans. A genome scan for disease association in such a recently admixed population can be achieved with only 1500–2500 'ancestry-informative markers'. The first genome-wide scan for admixture appeared in 2005 and enabled the identification of genes underlying a number of common diseases.

Related website

MALDsoft	http://pritch.bsd.uchicago.edu/software/maldsoft_download.html

Further reading

Collins-Schramm HE, et al. (2002) Ethnic-difference markers for use in mapping by admixture linkage disequilibrium. *Am J Hum Genet* 70: 737–750.

Freedman ML, Haiman CA, Patterson N, et al. (2006) Admixture mapping identifies 8q24 as a prostate cancer risk locus in African-American men. *Proc. Natl. Acad. Sci. USA* 103: 14068–14073.

Hoggart CJ, Shriver MD, Kittles RA, Clayton DG, McKeigue PM (2004) Design and analysis of admixture mapping studies. *Am J Hum Genet* 74: 965–978.

McKeigue PM (1998) Mapping genes that underlie ethnic differences in disease risk: methods for detecting linkage in admixed populations, by conditioning on parental admixture. *Am J Hum Genet* 63: 241–251.

Reich D, Patterson N, De Jager PL, et al. (2005) A whole-genome admixture scan finds a candidate locus for multiple sclerosis susceptibility. *Nat Genet* 37: 1113–1118.

Shriver MD, et al. (1998) Ethnic-affiliation estimation by use of population-specific DNA markers. *Am J Hum Genet* 60: 957–964.

Stephens JC, et al. (1994) Mapping by admixture linkage disequilibrium in human populations: limits and guidelines. *Am J Hum Genet* 55: 809–824.

Winkler CA, Nelson GW, Smith MW (2010) Admixture mapping comes of age. *Annual Review of Genomics and Human Genetics* 11: 65–89.

See also Linkage Analysis, Genome Scans for Linkage.

ADOPTED-BASIS NEWTON–RAPHSON MINIMIZATION (ABNR), *SEE* ENERGY MINIMIZATION.

AFFINE GAP PENALTY, *SEE* GAP PENALTY.

AFFINITY PROPAGATION-BASED CLUSTERING

Pedro Larrañaga and Concha Bielza

Affinity propagation (Frey and Dueck, 2007) is a clustering algorithm that, given a set of points and a set of similarity values between the points, finds clusters of similar points, and for each cluster gives a representative example or exemplar. A characteristic that makes affinity propagation different to other clustering algorithms is that the points directly exchange information between them regarding the suitability of each point to serve as an exemplar for a subset of other points. The algorithm takes as input a matrix of similarity measures between each pair of points $s(i,k)$. Affinity propagation works by exchanging messages between the points until a stop condition, which reflects an agreement between all the points with respect to the current assignment of the exemplars, is satisfied. These messages can be seen as the way the points share local information in the gradual determination of the exemplars.

There are two types of messages to be exchanged between data points. The responsibility $r(i,k)$, sent from data point i to candidate exemplar point k, reflects the accumulated evidence for how well-suited point k is to serve as the exemplar for point i, taking into account other potential exemplars for point i. The availability $a(i,k)$, sent from candidate exemplar point k to point i, reflects the accumulated evidence for how appropriate it would be for point i to choose point k as its exemplar, taking into account the support from other points that point k should be an exemplar.

The message-passing procedure may be terminated after a fixed number of iterations, when changes in the messages fall below a threshold, or after the local decisions stay constant for some number of iterations.

Affinity propagation has been applied in several Bioinformatics problems, such as structural biology (Bodenhofer et al., 2011), and the identification of subspecies among clonal organisms (Borile et al., 2011).

Further reading

Bodenhofer U, et al. (2011) APCluster: an R package for affinity propagation clustering. *Bioinformatics* 27 (17): 2463–2464.

Borile C, et al. (2011) Using affinity propagation for identifying subspecies among clonal organisms: lessons from *M. tuberculosis*. *BMC Bioinformatics* 12: 224.

Frey BJ, Dueck D (2007) Clustering by passing messages between data points. *Science* 315: 972–976.

AFFYMETRIX GENECHIP™ OLIGONUCLEOTIDE MICROARRAY

Stuart Brown and Dov Greenbaum

Technology for measuring expression levels of large numbers of genes simultaneously.

An oligonucleotide microarray technology, known as *GeneChip™* arrays, has been developed by Fodor, Lockhart, and colleagues at Affymetrix Inc. This involves the *in situ* synthesis of short DNA oligonucleotide probes directly on glass slides using photolithography techniques (similar to the methods used for the fabrication of computer chips). This method allows more than 400,000 individual features to be placed on a single 1.28 cm^2 slide. GeneChips produced in a single lot are identical within rigorous quality control standards. In the GeneChip system, a single mRNA sample is fluorescently labeled and hybridized to a chip, then scanned. Instead of having actual genes or large gene fragments on a chip, GeneChips typically have tens of oligonucleotides, each 25 bases long, corresponding to sequences from a single gene. Usually, 11 of the oligonucleotides are single base mismatches. The other 11 are overlapping matches corresponding to the correct sequence. In this way the technology aims to provide a higher degree of specificity than other microarray methodologies.

The entire laboratory process for using the GeneChip system is highly standardized with reagent kits for fluorescent labeling, automated equipment for hybridization and washing, and a fluorescent scanner with integrated software to control data collection and primary analysis. The system also provides indicators of sample integrity, assay execution, and hybridization performance through the assessment of control hybridizations of cDNA sequences spiked into the labeling reagents. As a result, Affymetrix arrays produce highly reproducible results when the same RNA is labelled and hybridized to duplicate GeneChips. The Affymetrix microarray software automatically calculates local background values for regions of each chip and provides a quality score as a flag (Present, Absent or Marginal) together with each gene's measured signal strength. This allows for the easy identification of genes with expression levels below the reliable detection threshold of the system, an important step in the data analysis process.

Affymetrix GeneChip expression experiments generate large amounts of data, 72 MB for a typical array, so it is critical to establish robust bioinformatics systems for data storage and management.

Related website

Affymetrix website	http://www.affymetrix.com

A

Further reading

Fodor SP, et al. (1991) Light-directed, spatially addressable parallel chemical synthesis. *Science* 251, 767–773.

Lockhart DJ, et al. (1996) Expression monitoring by hybridization to high-density oligonucleotide arrays. *Nat. Biotechnol.* 14: 1675–1680.

AFFYMETRIX PROBE LEVEL ANALYSIS

Stuart Brown

Nomalization of Affymetrix chip data.

In the Affymetrix GeneChip™ system, each gene is represented by a set of 11 to 20 25-base oligonucleotide probes that match positions distributed across the known cDNA sequence (a perfect match or PM probe). For each of these probes that match the cDNA sequence, an additional probe is created with a single base mismatch in the center (MM probe). The Affymetrix system uses the Microarray Analysis Suite (MAS) software to drive a fluorescent scanner that captures data from labeled target cDNA hybridized to each feature (probe sequence) of the GeneChip. The observed intensities of fluorescent signal show huge variability among the probes in a probe set due to sequence-specific differences in optimal hybridization temperatures as well as variable amounts of cross-hybridization between individual probes and cDNAs from other genes. In the MAS version 4 software, the signal for each probe set was calculated as a simple average of the differences between each PM probe and its corresponding MM probe (average difference). In the MAS version 5 software, a median weighted average (Tukey's Biweight Estimate) of the fluorescent intensity differences between paired PM and MM probes is used to compute a signal for each gene, but MM intensities that are greater than the corresponding PM are discarded so that negative signals are not produced for any genes.

The Affymetrix MAS software also assigns a quality flag (Present, Absent or Marginal) to each signal based on the overall signal strengths of PM vs. MM probes. In the MAS 4 software, each gene was called Present if a majority of probe pairs had stronger PM than MM signals. In MAS 5 software, this determination is made using a statistical approach. A 'discrimination score' of $(PM - MM)/(PM + MM)$ is calculated for each probe pair, which is then compared to a cutoff value (a small positive number between 0 and 0.1) and an overall p-value for each probe set is computed using the One-sided Wilcoxon's Signed Rank test. By adjusting the cutoff value, it is possible to change the probability that genes with low signal levels (in both PM and MM probes) will be flagged as Absent. In any case, if a probe set has many MM probes with stronger signals than their paired PM probes, then the gene will be flagged as Absent.

Several alternative techniques have been proposed to integrate probe set signals into a number that represents the expression level of a single gene. The Model Based Expression Index (MBEI) developed by Li and Wong (made available in the free dChip software) uses a model fitting technique for PM and MM values for a probe set across all chips in an experiment to achieve improved results. Speed and co-workers have developed a system called Robust Multi-Chip Analysis (RMA) which calculates expression values for each gene using $\log_2$ of quantile normalized, background adjusted, signals from only the PM probes across all of the chips in an experiment. The MBEI and RMA methods reduce the observed variability of low expressing genes allowing for the reliable detection of differential expression at lower mRNA concentrations.

Related websites

Bioconductor web site	http://www.bioconductor.org
dCHIP web site	http://biosun1.harvard.edu/~cli/complab/dchip/

Further reading

Affymetrix (1999) *Affymetrix Microarray Suite User Guide*. Affymetrix, Santa Clara, CA, version 4.

Affymetrix (2001a) *Affymetrix Microarray Suite User Guide*. Affymetrix, Santa Clara, CA, version 5.

Affymetrix (2001b) *Statistical Algorithms Reference Guide*. Affymetrix, Santa Clara, CA.

Bolstad BM, Irizarry RA, Astrand M, Speed TP (2003) A comparison of normalization methods for high density oligonucleotide array data based on variance and bias. *Bioinformatics* 19 (2): 185–193.

Li C, Wong WH (2001) Model-based analysis of oligonucleotide arrays: expression index computation and outlier detection. *Proc Natl Acad Sci USA* 98: 31–36.

Liu WM, et al. (2001) Rank-based algorithms for analysis of microarrays. *Proceedings SPIE* 4266: 56–67.

See also Microarray Normalization

AFTER SPHERE, *SEE* AFTER STATE.

AFTER STATE (AFTER SPHERE)

Thomas D. Schneider

The low energy state of a molecular machine after it has made a choice while dissipating energy is called its after state. This corresponds to the state of a receiver in a communications system after it has selected a symbol from the incoming message while dissipating the energy of the message symbol. The state can be represented as a sphere in a high dimensional space.

Further reading

Schneider TD (1991) Theory of molecular machines. I. Channel capacity of molecular machines. *J. Theor. Biol.* 148: 83–123. http://alum.mit.edu/www/toms/papers/ccmm/

See also Shannon Sphere, Channel Capacity, Gumball Machine

AIC, *SEE* AKAIKE INFORMATION CRITERION.

A

AKAIKE INFORMATION CRITERION

Pedro Larrañaga and Concha Bielza

The Akaike information criterion (AIC) (Akaike, 1974) is a measure of the relative goodness of fit of a statistical model with respect to a data set. It is based on the likelihood of the data given a model and on the complexity of this model. AIC is able to trade off between accuracy (likelihood) and complexity (number of parameters) of the model.

A general expression of AIC is:

$$\text{AIC} = k - \ln(L)$$

where k is the number of parameters in the statistical model, and L is the maximized value of the likelihood function for the estimated model. The selected model is the one with the minimum AIC value. Hence AIC not only rewards goodness of fit, but also includes a penalty term that discourages overfitting.

For example, AIC has been applied in Bioinformatics as a criterion for guiding a clustering algorithm in time-course microarray experiments (Liu et al., 2009); the analysis of the association of single SNPs (Solé et al., 2006); and in model selection in phylogenetics (Keane et al., 2006).

Further reading

Akaike H (1974) A new look at the statistical model identification. *IEEE Transactions on Automatic Control* 19 (6): 716–723.

Keane TM, et al. (2006) Assessment of methods for amino acid matrix selection and their use on empirical data shows that ad hoc assumptions for choice of matrix are not justified. *BMC Evolutionary Biology* 6: 29.

Liu T, et al. (2009) Information criterion-based clustering with order-restricted candidate profiles in short time-course microarray experiments. *BMC Bioinformatics* 10 (1): 146.

Solé X (2006) SNPStats: a web tool for the analysis of association studies. *Bioinformatics* 22 (15): 1928–1929.

See also Bayesian Information Criterion

ALGORITHM

Matthew He

An algorithm is a series of steps defining a procedure or formula for solving a problem that can be coded into a programming language and executed. Bioinformatics algorithms typically are used to process, store, analyse, visualize and make predictions from biological data.

ALIGNMENT (DOMAIN ALIGNMENT, REPEATS ALIGNMENT)

Jaap Heringa

Sequence alignment is the most common task in bioinformatics. It involves matching the amino acids or nucleotides of two or more sequences in such a way that their similarity can be determined best. Although many properties of nucleotide or protein sequences can be used to derive a similarity score; *e.g.*, nucleotide or amino acid composition, isoelectric point or molecular weight, the vast majority of sequence similarity calculations presuppose an alignment between two sequences from which a similarity score is inferred. Ideally, the alignment of two sequences should be in agreement with their evolution; i.e. the patterns of descent as well as the molecular structural and functional development. Unfortunately, the evolutionary traces are often very difficult to detect; for example, in divergent evolution of two protein sequences from a common ancestor where amino acid mutations, insertions and deletions of residues, gene doubling, transposed gene segments, repeats, domain structures and the like can blur the ancestral tie beyond recognition. The outcome of an alignment operation can be studied by eye for conservation patterns or be given as input to a variety of bioinformatics methods, ranging from sequence database searching to tertiary structure prediction.

Although very many different alignments can be created between two nucleotide or protein sequences, there has been only a single evolutionary pathway from one sequence to another. In the absence of observed evolutionary traces, the matching of two sequences is regarded as mimicking evolution best when the minimum number of mutations is used to arrive at one sequence from the other. An approximation of this is finding the highest similarity value determined from summing substitution scores along matched residue pairs minus any insertion/deletion penalties. Such an alignment is generally called the optimal alignment. Unfortunately, testing all possible alignments including the insertion of a gap at each position of each sequence, is unfeasible. For example, there are about 1088 possible alignments of two sequences of 300 amino acids (Waterman, 1989), a number clearly beyond all computing capabilities. However, when introductions of gaps are also assigned scoring values such that they can be treated in the same manner as the mutation of one residue into another, the number of calculations is greatly reduced and becomes readily computable. The technique to calculate the highest scoring or optimal alignment is generally known as the dynamic programming (DP) technique, introduced by Needleman and Wunsch (1970) to the biological community. The DP algorithm is an optimisation technique which guarantees, given an amino acid exchange matrix and gap penalty values, that the best scoring or optimal alignment is found.

There are two basic types of sequence alignment: global alignment and local alignment. Global alignment implies the matching of sequences over their complete lengths, whereas with local alignment the sequences are aligned only over the most similar parts of the sequences, carrying the clearest memory of evolutionary relatedness. In cases of distant sequences, global alignment techniques might fail to recognize such highly similar internal regions because these may be overshadowed by dissimilar regions and strong gap penalties required to achieve their proper matching. Moreover, many biological sequences are modular and show shuffled domains, which can render a global alignment of two complete sequences meaningless. For example, a two-domain sequence with domains A and B consecutive in sequence (AB) will be aligned incorrectly against a sequence with domain organisation BA. The occurrence of varying numbers of internal sequence repeats is known to limit the applicability of global alignment methods, as the necessity to insert gaps spanning excess domains and confusion due to recurring strong motifs is likely to lead to erroneous alignment. In general, when there is a large difference in the lengths of two sequences to be compared, global alignment should be avoided.

A special case of global alignment is semi-global alignment, where two sequences are aligned in their entirety but with gaps occurring at either flank of the sequences remaining unpenalized (zero gap penalties). Semi-global alignment leads to proper matching of a short sequence against a longer sequence if the short sequence is suspected to span a local region of the longer sequence; e.g. aligning a gene against an evolutionary related genome, or aligning a single homologous domain against a mosaic multi-domain sequence.

In searches for homology, local alignment methods should be included in the analysis. The first pairwise algorithm for local alignment was developed by Smith and Waterman (1981) as an adaptation of the algorithm of Needleman and Wunsch (1970). The Smith-Waterman technique selects the most similar region in each of two sequences, which are then aligned. Waterman and Eggert (1987) generalized the local alignment routine by devising an algorithm that allows the calculation of a user-defined number of top scoring local alignments instead of only the optimal local alignment. Various implementations of the Waterman-Eggert technique exist with memory requirements reduced from quadratic to linear, thereby allowing very long sequences to be searched at the expense of only a small increase in computational time (e.g. Huang et al., 1990; Huang and Miller, 1991). A number of fast implementations of local alignment have emerged to perform a homology search with a given query sequence against curated and annotated sequence databases, including SSEARCH (Pearson, 2000) and the popular heuristic methods BLAST and PSI-BLAST (Altschul et al., 1997).

It is not always clear whether the top scoring (local or global) alignment of two sequences is biologically the most meaningful. There may be biologically plausible alignments that score close to the top value. For example, alignments of proteins based on their structures or of DNA sequences based on evolutionary changes are often different from their associated optimal sequence alignments. Since for most sequences the true alignment is unknown, methods that either assess the significance of the optimal alignment, or provide few 'close' alternatives to the optimal one, are of great importance.

A suboptimal alignment is an alignment whose score lies within the neighbourhood of the optimal score. It can therefore be a natural candidate for an alternative to the optimal alignment. Enumeration of suboptimal alignments is not very useful since there are many such alignments. Other approaches that use only partial information about suboptimal alignments are more successful in practice. Vingron and Argos (1990) and Zuker (1991) approached this issue by constructing an algorithm, which determines all optimal and suboptimal alignments and depicts them in a dot plot. Reliably aligned regions can be defined as those for which alternative local alignments do not exist.

Pairwise alignments can also be made based on tertiary structures, if these are available for a target set of protein sequences. Methods have been devised for optimal superpositioning of two protein structures represented by their main-chain-trace (Cα atoms). The matched residues in a so-called structural alignment then correspond to equivalenced Cα atoms in the tertiary structure superpositioning (Taylor and Orengo, 1989). Following the notion that in divergent evolution, protein structure is more conserved than sequence, structural alignments are often used as a standard of truth to evaluate the performance of different sequence alignment methods.

Further reading
Altschul SF, Madden TL, Schaffer AA, et al. (1997) Gapped BLAST and PSI-BLAST: A new generation of protein database search programs. *Nucl Acids Res* 25: 3389–3402.

Huang X, Hardison RC, Miller W (1990) A space-efficient algorithm for local similarities. *CABIOS* 6: 373–381.

Huang X, Miller W (1991) A time-efficient, linear-space local similarity algorithm. *Adv. Appl. Math.* 12: 337–357.

Karlin S, Altschul SF (1990) Methods for assessing the statistical significance of molecular sequence features by using general scoring schemes. *Proc. Natl. Acad. Sci. USA* 87: 2264–2268.

Needleman SB, Wunsch CD (1970) A general method applicable to the search for similarities in the amino acid sequence of two proteins. *J. Mol. Biol.* 48: 443–453.

Pearson WR (2000) Flexible sequence similarity searching with the FASTA3 program package. *Methods Mol Bio.* 132: 185–219.

Smith TF, Waterman MS (1981) Identification of common molecular subsequences. *J. Mol. Biol.* 147: 195–197.

Taylor WR, Orengo CA (1989) Protein structure alignment. *J. Mol. Biol.* 208: 1–22.

Vingron M, Argos P (1990) Determination of reliable regions in protein sequence alignments. *Prot. Eng.* 3: 565–569.

Waterman MS (1989) Sequence alignments. In: *Mathematical methods for DNA sequences* (Waterman MS, ed.). CRC Press, Boca Raton, FL.

Waterman MS and Eggert M (1987) A new algorithm for best subsequences alignment with applications to the tRNA-rRNA comparisons. *J. Mol Biol.* 197: 723–728.

Zuker M (1991) Suboptimal sequence alignment in molecular biology. Alignment with error analysis. *J. Mol. Biol.* 221: 403–420.

See also Global Alignment, Local Alignment, Dynamic Programming, Alignment, multiple, Gap Penalty

ALIGNMENT SCORE

Laszlo Patthy

Alignment of the sequences of homologous genes or proteins requires that the nucleotides (or amino acids) from 'equivalent positions' (i.e. of common ancestry) are brought into vertical register. The procedures that attempt to align sites of common ancestry are referred to as sequence alignment procedures.

The correct alignment of two sequences is the one in which only sites of common ancestry are aligned, and all sites of common ancestry are aligned. The true alignment is most likely to be found as the one in which matches are maximized and mismatches and gaps are minimized. This may be achieved by using some type of scoring system that rewards similarity and penalizes dissimilarity and gaps. The alignment score is calculated as the sum of the scores assigned to the different types of matches, mismatches and gaps. Most of the commonly used sequence-alignment programs are based on the Needleman-Wunsch Algorithm, which identifies the best alignment as the one with maximum alignment score among all possible alignments.

Further reading

Needleman SB, Wunsch CD (1970) A general method applicable to the search for similarities in the amino acid sequence of two proteins. *J Mol Biol* 48: 443–453.

See also Alignment

ALLELE-SHARING METHODS (NON-PARAMETRIC LINKAGE ANALYSIS)

Mark McCarthy, Steven Wiltshire and Andrew Collins

Basis for one type of linkage analysis methodology which has been employed extensively in the analysis of multifactorial traits. The method relies on the detection of an excess of allele-sharing amongst related individuals who share a phenotype of interest (or equivalently, a deficit of allele-sharing amongst related individuals with differing phenotypes).

The simplest, and most frequently used, substrates for the application of allele-sharing methods are collections of affected sibling pairs (or sibpairs). The expectation, under the null hypothesis, is that any pair of full siblings selected at random will, on average, share 50% of their genomes identity-by-descent; and that a large collection of such sibpairs will show approximately 50% sharing at all chromosomal locations. If, however, sibpairs are not selected at random, but instead are ascertained on the basis of similarity for some phenotype of interest (for example, sharing a disease known to have some inherited component), then the location of genes contributing to variation in that phenotype can be revealed through detecting regions where those affected sibpairs share significantly more than 50% identity-by-descent. In other words, when relatives are correlated for phenotype, they will tend to show correlations of genotype around variants contributing to phenotype development. Some of the earliest programs for allele-sharing linkage analysis relied on sharing defined identity-by-state (IBS), but these have been superseded by identity-by-descent (IBD) sharing methods. A variety of methods have been developed to extend this methodology to include: other combinations of relatives; the analysis of quantitative as well as qualitative traits; and, relative pairs with dissimilar as well as similar phenotypes. Allele-sharing methods are, essentially, equivalent to non-parametric linkage analysis. This strategy for gene mapping in complex traits was largely superseded from ~2005 onwards by more powerful association mapping approaches, and particularly genome-wide association, as a route to identifying common disease variants.

The first example of a genome scan for a multifactorial trait using allele-sharing methods was published by Davies et al. (1994) using approximately 100 sibpair families ascertained for type 1 diabetes. This scan revealed evidence for linkage in several chromosomal regions, though the largest signal by far (as expected) fell in the HLA region of chromosome 6. Wiltshire et al. (2001) describe a scan of over 500 sibpair families ascertained for type 2 diabetes, which illustrates some of the methodological issues associated with this approach.

Related websites
Programs for allele-sharing analyses

GENEHUNTER	http://www-genome.wi.mit.edu/ftp/distribution/software/genehunter
MERLIN	http://www.sph.umich.edu/csg/abecasis/Merlin/
SPLINK	https://www-gene.cimr.cam.ac.uk/staff/clayton/software/

Further reading

Abecasis GR, et al. (2002) Merlin – rapid analysis of dense genetic maps using sparse gene flow trees. *Nat Genet* 30: 97–101.

Davies JL, et al. (1994) A genome-wide search for human type 1 diabetes susceptibility genes. *Nature* 371: 130–136.

Kruglyak L, et al. (1996) Parametric and non-parametric linkage analysis: a unified multipoint approach. *Am J Hum Genet* 58: 1347–1363.

Lander ES, Schork NJ (1994) Genetic dissection of complex traits. *Science* 265: 2037–2048.

Ott J (1999) *Analysis of Human Genetic Linkage*, 3rd Edition. Baltimore: Johns Hopkins University Press, pp 272–296.

Risch N, Zhang H (1995) Extreme discordant sib pairs for mapping quantitative trait loci in humans. *Science* 268: 1584–1589.

Terwilliger J, Ott J (1994) *Handbook of Human Genetic Linkage*. Baltimore, Johns Hopkins University Press.

Wiltshire S (2001) A genome-wide scan for loci predisposing to type 2 diabetes in a UK population (The Diabetes (UK) Warren 2 Repository): analysis of 573 pedigrees provides independent replication of a susceptibility locus on chromosome 1q. *Am J Hum Genet* 69: 553–569.

See also Linkage Analysis, Multifactorial Trait, Haseman-Elston Regression, Sibpair Methods

ALLELIC ASSOCIATION

Mark McCarthy, Steven Wiltshire and Andrew Collins

Non-independent distribution of alleles at different variant loci.

In the most general sense, allelic association refers to any observed association, whatever its basis. However, most associations of interest (at least from the gene mapping point of view) are those between loci that are linked, in which case, allelic association equates with linkage disequilibrium (in the stricter definition of that term).

Allelic association is the basis of association analysis (either case–control or family-based association) where the aim is to demonstrate allelic association between a marker locus (of known location, and genotyped) and a disease-susceptibility variant (the position of which is unknown, and which is represented in the analysis by the disease phenotype).

Examples: The HLA region on chromosome 6 comprises a number of highly polymorphic loci and displays strong linkage disequilibrium. Consequently, alleles at neighbouring HLA loci are in very strong allelic association in all studied populations. Since this region is known to harbour disease-susceptibility genes (particularly for diseases associated with disturbed immune regulation), there are many examples of

A

strong associations between certain HLA alleles and a variety of disease states. For example, possession of the HLA-B27 allele has long been known to be associated with ankylosing spondylitis; and individuals who carry DR3 and/or DR4 (now termed HLA-DQA1*0501-DQB1*0201/DQA1*0301-DQB1*0302) alleles are at increased risk of type 1 diabetes.

Related website

Repository for haplotype map data:

HapMap	http://hapmap.ncbi.nlm.nih.gov/

Further reading

Cardon LR, Bell JI (2001) Association study designs for complex disease. *Nat Rev Genet* 2: 91–99.

Herr M, et al (2000) Evaluation of fine mapping strategies for a multifactorial disease locus: systematic linkage and association analysis of IDDM1 in the HLA region on chromosome 6p21. *Hum Mol Genet* 9: 1291–1301.

Jeffreys AJ, et al (2001) Intensely punctate meiotic recombination in the class II region of the major histocompatibility complex. *Nat Genet* 29: 217–222.

Lonjou C, et al (1999) Allelic association between marker loci. *Proc Natl Acad Sci USA* 96: 1621–1626.

Owen MJ, McGuffin P (1993) Association and linkage: complementary strategies for complex disorders. *J Med Genet* 30: 638–639.

Sham P (1998) *Statistics in Human Genetics*. London, Arnold, pp 145–186.

The International HapMap Consortium (2005) A haplotype map of the human genome. *Nature* 437: 1299–1320.

The International HapMap Consortium (2007) A second generation human haplotype map of over 3.1 million SNPs. *Nature* 449: 851–861.

The International HapMap Consortium (2010) Integrating common and rare genetic variation in diverse human populations. *Nature* 467: 52–58.

See also Linkage Disequilibrium, Linkage Disequilibrium Map, Association Analysis, Family-Based Association Analysis, Phase, Haplotype

ALLEN BRAIN ATLAS

Dan Bolser

An integrated database of mouse gene expression and neuroanatomical data with integrated links to human and primate resources. Data sets include: a genome-wide atlas of gene expression in the adult and developing mouse and human brain, neural connections in the mouse brain, expression in the mouse brain across different genetic backgrounds, sexes, and sleep conditions, and gene expression data for postnatal primate brain development and the mouse spinal cord.

Related websites

ALLEN BRAIN ATLAS	http://www.brain-map.org/
Wikipedia	http://en.wikipedia.org/wiki/Allen_Brain_Atlas

Further reading

Jones AR, Overly CC, Sunkin SM. (2009) The Allen Brain Atlas: 5 years and beyond. *Nat Rev Neurosci* 10: 821–828.

See also FLYBRAIN

ALLOPATRIC EVOLUTION (ALLOPATRIC SPECIATION)

A.R. Hoelzel

Allopatric evolution occurs when populations of the same species diverge genetically in geographically isolated locations (having non-overlapping geographic ranges).

Such differentiation, together with the development of pre- or post-zygotic reproductive isolation, can result in allopatric speciation. In general terms this is the most widely accepted mechanism for speciation in animal species. Both natural selection and genetic drift can be factors generating divergence in allopatry, with the relative importance likely depending on the nature of the organism and its environment. Obvious examples of allopatric evolution include differentiation between populations on different continents, different oceans, or separated by physical barriers to migration (e.g. a mountain range or river).

The generation of natural biogeographical barriers to gene flow over time is referred to as a vicariance event, such as the incursion of a glacier, and this can lead to differentiation in allopatry, sometimes followed by reconvergence (e.g. after a glacier recedes). Populations may also be established in allopatry, for example by long-range dispersal.

Further reading

Hartl DL (2000) *A Primer of Population Genetics*. Sinauer Associates Inc.

See also Evolution

ALLOPATRIC SPECIATION, *SEE* ALLOPATRIC EVOLUTION.

ALOGP

Bissan Al-Lazikani

A theoretical calculation of the Partition Coefficient LogP using the individual atoms in the structure of a compound to estimate the overall LogP value of the structure.

See also LogP

A

ALPHA CARBON, *SEE* Cα (C-ALPHA).

ALPHA HELIX

Roman Laskowski and Tjaart de Beer

The most common of the regular secondary structure in globular proteins. It is characterized by the helical path traced by the protein's backbone, a complete turn of the helix taking 3.6 residues and involving a translation of 5.41Å. The rod-like structure of the helix is maintained by hydrogen bonds from the backbone carbonyl oxygen of each residue, *i*, in the helix to the backbone –NH of residue $i+4$.

In proteins, most alpha-helices are right-handed; that is, when viewed down the axis of the helix, the backbone traces a clockwise path. The right-handed form is more stable than the left-handed as the latter involves steric clashes between side chains and backbone.

Certain types of amino acids (residues), such as alanine, arginine and leucine, favour formation of alpha-helices, whereas others tend to occur less often in this type of secondary structure, the most significant being proline which breaks the hydrogen-bonding pattern necessary for alpha-helix formation and stability and can introduce a kink in the alpha-helix.

Other types of helix found in globular proteins, albeit very rarely, are the 3_{10} and *pi*-helix, which differ in that their hydrogen bonds are to residues $i+3$ and $i+5$, respectively.

Further reading

Branden C, Tooze J (1998) *Introduction to Protein Structure.* Garland Science, New York.

Lesk AM. (2000) *Introduction to Protein Architecture.* Oxford University Press, Oxford.

See also Secondary Structure, Globular, Backbone, Side Chains, Residue

ALTERNATIVE SPLICING

Enrique Blanco and Josep F. Abril

Splicing of the same pre-mRNA molecule (primary transcript) that can produce different mature transcripts.

Alternative splicing is a way in which several mRNAs can be synthesized from a single gene in function of certain splicing regulators, which influence the use of a particular arrangement of splice sites. By switching the set of acceptor and donor splice sites that is employed in each circumstance, resulting transcripts may exclude/include alternative exons, alter their length or even retain certain introns.

These modifications can affect untranslated regions reconfiguring the stability and the localization of the mRNA or introduce changes in the coding region of the gene, increasing the protein diversity. It is thought that alternative splicing is a mechanism to express different proteins from the same gene in different tissues or developmental stages.

However, it is unclear whether or not such isoforms are functional and structurally viable in the cell. Some alternative splicing transcripts may lead to truncated proteins. Most of those may be non-functional and will be degraded, although other may contain complete functional domains.

In any case, it is accepted that alternative splicing is a common mechanism that acts on many genes. For instance, about 95% of human genes are reported to produce alternatively spliced mRNAs. In addition, alternative splicing events give rise to sequence conservation levels among vertebrates, emphasising their relevance, although species-specific isoforms are documented as well.

Related websites

ASPicDB	http://t.caspur.it/ASPicDB/
A general definition and nomenclature for alternative splicing events	http://www.scivee.tv/node/6837
Alternative splicing transcriptional landscape visualization tool	http://genome.crg.cat/astalavista/index.html
Alternative splicing and transcript diversity databases	http://www.ebi.ac.uk/asd/index.html
Website for alternative splicing prediction	http://genes.toronto.edu/wasp/

Further reading

Barash Y, Calarco JA, Gao W, et al. (2010) Deciphering the splicing code. *Nature* 465: 53–59.

Mudge JM, Frankish A, Fernandez-Banet J, et al. (2011) The origins, evolution, and functional potential of alternative splicing in vertebrates. *Mol. Biol. Evol.* 28: 2949–2959.

Pan Q, Shai O, Lee LJ, Frey BJ, Blencowe BJ (2008) Deep surveying of alternative splicing complexity in the human transcriptome by high-throughput sequencing. *Nat. Genet.* 40: 1413–1415.

Tress ML, et al. (2007) The implications of alternative splicing in the ENCODE protein complement. *PNAS* 104: 5495–5500.

See also Splicing, Coding Region, Gene Prediction, alternative splicing, ENCODE

ALTERNATIVE SPLICING GENE PREDICTION, *SEE* GENE PREDICTION, ALTERNATIVE SPLICING.

AMIDE BOND (PEPTIDE BOND)

Roman Laskowski and Tjaart de Beer

The covalent bond joining two amino acids, between the carboxylic (–COOH) group of one and the amino group ($-NH_2$) of the other, to form a peptide.

The bond has a partial double bond character and so the atoms shown below tend to lie in a plane and act as a rigid unit.

Because of this, the dihedral angle ω (omega), defined about the bond, tends to be close to 180° degrees.

Further reading

Branden C, Tooze J. (1998) *Introduction to Protein Structure.* Garland Science, New York.

Lesk AM. *Introduction to Protein Architecture.* (2000) Oxford University Press, Oxford.

See also Amino Acid, Peptide, Dihedral Angle

AMINO ACID (RESIDUE)

Roman Laskowski, Jeremy Baum and Tjaart de Beer

An organic compound containing one or more amino groups ($-NH_2$) and one or more carboxyl groups ($-COOH$). Twenty alpha-amino acids, which have the form $RCH(NH_2)COOH$ where R is either hydrogen or an organic group, are the component molecules of proteins. The different amino acids are distinguished by the side chain substituent R (see Figure A.2) and the stereochemistry of the alpha carbon.

The sequence of amino acids that make up each protein is encoded in the DNA using the genetic code wherein each amino acid is represented by one or more codons in the DNA.

The central carbon atom, to which are bonded an amino group (NH_2), a carboxyl group (COOH), a hydrogen atom and a side chain group R, is called the Cα (C-alpha). The atoms of the side chain are designated β,γ,δ and ε in order from the Cα.

Except for glycine, the central Cα is asymmetric and can adopt one of two isomers, identified as L and D. In proteins, amino acids predominantly adopt the L-configuration.

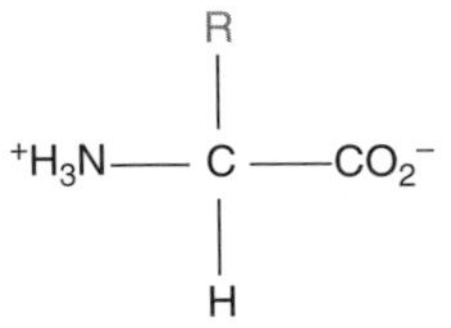

R
$^{+}H_3N$—C—CO_2^-
H

Figure A.2 Chemical structure of an amino acid.

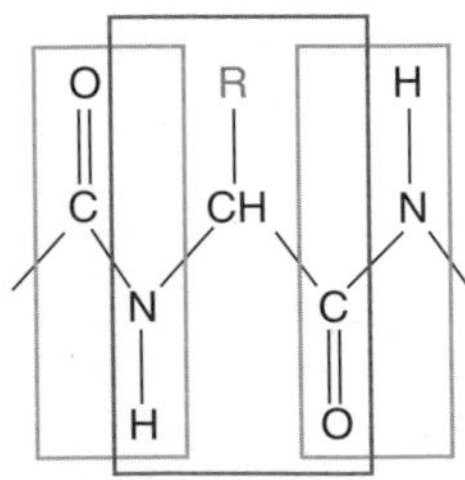

Figure A.3 An amino acid structure placed in the context of a polypeptide chain.

When amino acids polymerize they do so in linear chains with neighbouring residues connected together by (approximately) planar peptide bonds (See Figure A.3.). The basic structure of an amino acid in a polymer is shown in Figure A.3. They are then often referred to as amino acid residues. The -NH-CH-CO- component in common to all residues is called the backbone.

Related websites

Amino acid nomenclature	http://www.chem.qmw.ac.uk/iupac/AminoAcid
Amino acid information	http://www.imb-jena.de/IMAGE_AA.html
Amino acid viewer	http://www.bio.cmu.edu/courses/BiochemMols/AAViewer/AAVFrameset.htm

See also Protein, Side Chain, Sequence, Genetic Code, Codon, C-α.

AMINO ACID ABBREVIATIONS, *SEE* IUPAC-IUB CODES.

AMINO ACID COMPOSITION

Jeremy Baum

The relative abundance of the twenty amino acids (usually) in a protein. This measure can be used to make predictions of the protein class, as for example integral membrane proteins tend to have more hydrophobic residues than do cytoplasmic proteins.

See also Amino Acid

AMINO ACID EXCHANGE MATRIX (DAYHOFF MATRIX, LOG ODDS SCORE, PAM (MATRIX), BLOSUM MATRIX)

Jaap Heringa

An amino acid exchange matrix is a 20×20 matrix which contains probabilities for each possible mutation between 20 amino acids. The matrix is symmetric so that a mutation from amino acid X into Y is assigned an identical probability as the mutation Y into X. The matrix diagonal contains the odds for self-conservation.

Amino acid exchange matrices constitute a model for protein evolution and are a prerequisite for sequence alignment methods. For example, the dynamic programming technique relies on an amino acid exchange weight matrix and gap penalty values. For alignment of similar sequences (>35% in sequence identity), the scoring system utilized is not critical (Feng et al., 1985). In more divergent comparisons with residue identity fractions in the so-called twilight-zone (15 to 25%) (Doolittle, 1987), different scoring regimes can lead to dramatically deviating alignments.

Many different substitution matrices have been devised over more than three decades, each trying to approximate divergent evolution in order to optimise the signal-to-noise ratio in the detection of homologies among sequences. A combination of physical-chemical characteristics of amino acids can be used to derive a substitution matrix which then basically contains pairwise amino acid similarity values. Other data from which residue exchange matrices have been computed include sequence alignments, structure-based alignments and common sequence motifs.

Popular exchange matrices derived from multiple sequence alignments include the PAM series (Dayhoff et al., 1978, 1983) and BLOSUM series (Henikoff and Henikoff, 1992) of matrices, each series containing a range of matrices adapted to different evolutionary distances. Typically, amino acid exchange matrices contain log odds scores, allowing scoring of residues matched in an alignment by adding the corresponding exchange scores. This is often preferable for computational reasons relative to multiplication of the original (non-logarithmic) probabilities.

Further reading

Dayhoff MO, Barker WC, Hunt LT (1983) Establishing homologies in protein sequences. *Methods Enzymol* 91: 524–545.

Dayhoff MO, Schwartz RM, Orcutt BC (1978) A model of evolutionary change in proteins. In: *Atlas of Protein Sequence and Structure*, Dayhoff MO (ed.), vol.5, suppl. 3, Washington, DC, Natl. Biomed. Res. Found.), pp. 345–352.

Doolittle RF (1987) *Of URFS and ORFS. A Primer on How to Analyze Derived Amino Acid Sequences*. University Science Books, Mill Valley, CA.

Eisenberg D, Lüthy R, McLachlan AD (1991) Secondary structure-based profiles: Use of structure-conserving scoring tables in searching protein sequence databases for structural similaritiesimilarities. *Proteins* 10: 229–239.

Feng D-F, Johnson MS, Doolittle RF (1985) Aligning amino acid sequences: comparison of commonly used methods. *J. Mol. Evol.* 25: 351–360.

Gonnet GH, Cohen MA, Benner SA (1992) Exhaustive matching of the entire protein sequence database. *Science* 256: 1443–1445.

Henikoff S, Henikoff JG (1992) Amino acid substitution matrices from protein blocks. *Proc. Natl. Acad. Sci. USA* 89: 10915–10919.

Henikoff S, Henikoff JG (1993) Performance evaluation of amino acid substitution matrices. *Proteins* 17: 49–61.

Jones DT, Taylor WR, Thornton JM (1992) The rapid generation of mutation matrices from protein sequences. *CABIOS* 8: 275–282.

Koshi JM, Goldstein RA (1998) Models of natural mutations including site heterogeneity. *Proteins*, 32: 289–295.

Thompson MJ, Goldstein RA (1996) Constructing amino acid residue substitution classes maximally indicative of local protein structure. *Proteins* 25: 28–37.

See also Gap Penalty, Global Alignment, Local Alignment

AMINO ACID SUBSTITUTION MATRIX, *SEE* AMINO ACID EXCHANGE MATRIX.

AMINO-TERMINUS, *SEE* N-TERMINUS.

A

AMPHIPATHIC

Roman Laskowski and Tjaart de Beer

A molecule having both a hydrophilic ('water loving' or polar) end and a hydrophobic ('water hating' or non-polar) end. Examples are the phospholipids that form cell membranes. These have polar head groups, which form the outer surface of the membrane, and two hydrophobic hydrocarbon tails, which extend into the interior.

In proteins amphipathicity may be observed in surface β-helices and α-strands. Residues (amino acids) on the surface of proteins tend to be polar as they are in contact with the surrounding solvent, whereas those on the inside of the protein tend to be hydrophobic.

Thus an amphipathic helix is one which lies on, or close to, the protein's surface and hence has one side consisting largely of polar residues, facing out into the solvent, and the opposite side consisting largely of hydrophobic residues, facing the protein's hydrophobic core. Because of the helix's periodicity of 3.6 residues per turn, this gives rise to a characteristic pattern of polar and hydrophobic residues in the protein's sequence. For example, the pattern of hydrophobic residue might be of the form: i, $i+3$, $i+4$, $i+7$, with polar residues in between.

Similarly, an amphipathic strand has one side hydrophobic and the other side polar. From the geometry of strand residues, this gives rise to a simple pattern of alternating hydrophobic and polar residues.

Further reading

Branden C, Tooze J. (1998) *Introduction to Protein Structure*. Garland Science, New York.

Lesk AM. (2000) *Introduction to Protein Architecture*. Oxford University Press, Oxford.

See also Amino Acids, Polar, Hydrophobic, Sequence

ANALOG (ANALOGUE)

Dov Greenbaum

Genes or proteins that are similar in some way but show no signs of any common ancestry.

Examples of analogs are structural analogs: those proteins that share the same fold. Functional analogs are proteins that share the same function.

Analogs are interesting in that they are the exception to the rule in annotation transfer, such that a protein which shares no evolutionary history and does not have any similar structure can still have the same function as its analogue. Structural similarities are thought to be the result of convergence to a favourable fold, although an extremely distant divergence cannot be ruled out.

Determining the difference in sequences between analogs and distant homologs is important; evolutionary theory can only be used to analyse proteins that are similar through descent from a common ancestor and not due to random mutations that result in either similar fold or function. Conversely a protein that is misidentified as an analog rather than a distant homolog may not be analysed in its proper evolutionary context. Either error will be deleterious to a comparative genomic study.

Further reading

Fitch W (1970) Distinguishing homologous from analogous proteins. *Syst. Zool.* 19: 99–113.

ANCESTRAL LINEAGE, *SEE* OFFSPRING LINEAGE.

ANCESTRAL STATE RECONSTRUCTION

Sudhir Kumar and Alan Filipski

The process of inferring the nucleotide or amino acid state or the respective state probabilities at a specific site of an ancestral sequence located at an internal node p of a phylogenetic tree.

Maximum parsimony and likelihood-based (Maximum Likelihood and Bayesian) inference methods are available for ancestral state reconstruction. Parsimony methods can work reasonably well when the sequences are closely related (such that the expected number of substitutions per site is small) and when branches are relatively similar in length.

Likelihood-based methods can be deployed to compute the posterior probabilities or marginal likelihoods of observing (e.g., for DNA data) A, C, G, and T at a site i of an ancestral node p. To obtain discrete ancestral state sequences one may choose the state with the highest posterior probability as the best estimate.

Software

Maddison W P, Maddison DR (1992) *MacClade: Analysis of Phylogeny and Character Evolution*. Sunderland, Mass., Sinauer Associates.

Yang Z. (1997) PAML: a program package for phylogenetic analysis by maximum likelihood. *Comput Appl Biosci* 13: 555–556.

Related websites

Stamatakis, A. (2012) standard RAxML version:	https://github.com/stamatak
Zwickl, D. (2012) GARLI software:	http://code.google.com/p/garli/

Further reading

Nei M, Kumar S (2000) *Molecular Evolution and Phylogenetics*. Oxford, Oxford University Press.

Pupko T, Pe I, Shamir R, Graur DA (2000) A fast algorithm for joint reconstruction of ancestral amino acid sequences. *Molecular Biology and Evolution* 17(6): 890–896.

Yang Z, et al. (1995) A new method of inference of ancestral nucleotide and amino acid sequences. *Genetics* 141: 1641–1650.

Yang Z (2006) *Computational Molecular Evolution*. New York, Oxford University Press.

See also Maximum Parsimony Principle, Bayesian Phylogenetic Analysis

ANCHOR POINTS

Roland Dunbrack

In protein loop modeling, usually one or two residues from the parent structure before and after a loop that is being constructed to fit onto the template structure. These residues are usually kept fixed, and the constructed loop must be connected to them to produce a viable model for the loop. Therefore if a five residue loop is being modeled with a two residue anchor point at either end, a fragment of nine residues long will have to be found from witch the two residues at either end 'fit' well onto the parent structure.

See also Loop Prediction/Modeling, Parent

ANNOTATION REFINEMENT PIPELINES, *SEE* GENE PREDICTION.

ANNOTATION TRANSFER (GUILT BY ASSOCIATION ANNOTATION)

Dov Greenbaum

The transfer of information on the known function, or other property, of a gene or protein to another gene or protein based on pre-defined criteria.

Annotation transfer is often termed guilt by association. That is, given some knowledge of some other gene or protein that has some similarity to an unknown gene, it is possible to transfer the known annotation to the novel gene. For example, in the case of homologous proteins, it is often assumed that proteins with a high degree of sequence similarity share the same function. Another example could be that proteins that interact with each other are assumed to have similar functions, thus one can transfer annotation information from one gene to its interaction partner.

One has to be careful in utilizing annotation transfer, as it may not always be the case that the two proteins have similar function, even though they may share an interaction, localization, expression or sequence similarity. One pitfall to be avoided is the incorporation of potentially misleading information, resulting from an annotation transfer, into a database, as this will only propagate the error when additional annotation transfers are made from the formerly novel protein.

Further reading

Hegyi H, Gerstein M. (2001) Annotation transfer for genomics: measuring functional divergence in multi-domain proteins. *Genome Res* 11: 1632–1640.

Punta M, Ofran Y (2008) The Rough Guide to In Silico Function Prediction, or How To Use Sequence and Structure Information To Predict Protein Function. *PLoS Comput Biol* 4(10): e1000160.

A

APBIONET (ASIA-PACIFIC BIOINFORMATICS NETWORK)

Pedro Fernandes

APBioNet was founded in 1998.

Its main areas of activity include:

- organization and coordination of the International Conference on Bioinformatics (InCoB) since 2002
- development of bioinformatics network infrastructure for the distribution and mirroring of bioinformatics databases (Bio-Mirror)
- exchange of data and information and promotion of their standards: MIABi standards, author identification, biodb100 database reinstantiation repository, etc.
- development of training programs, workshops and symposia
- encouragement of collaboration through programms such as sponsorship of travelling fellowships for students.

It is affiliated to the International Society for Computational Biology (ISCB) and works closely with various other international organizations such as the ASEAN SubCommittee on Biotechnology (SCB), Federation of Asian and Oceanian Biochemists and Molecular Biologists (FAOBMB), International Society for BioCuration (ISB). To promote the quality of bioinformatics research in the region, it collaborates with various journals: BioMed Central (BMC) Bioinformatics, BMC Genomics, Immunome Research, Bioinformation. It has members in Australia, Canada, China, Hong Kong, India, Japan, S. Korea, Malaysia, Russia, Singapore and the USA.

Related websites

APBioNet	http://www.apbionet.org/
InCoB	http://incob.apbionet.org/
Bio-Mirror	http://bio-mirror.net
BioDB100	http://biodb100.apbionet.org

APOMORPHY

A.R. Hoelzel

A derived or newly-evolved state of an evolutionary character.

As a hypothesis about the pattern of evolution among operational taxonomic units, a phylogenetic reconstruction can inform about the relationship between character states. We can identify ancestral and descendent states, and therefore identify primitive and derived states. An Apomorphy (meaning 'from-shape' in Greek) is a derived state (shown as filled circles in Figure A.4).

Changes have occurred compared to the common ancestor, and these are now represented by the derived or apomorphic states. In Figure A.4, there are two apomorphies, each derived independently from the ancestral state, which is an example of homoplasy.

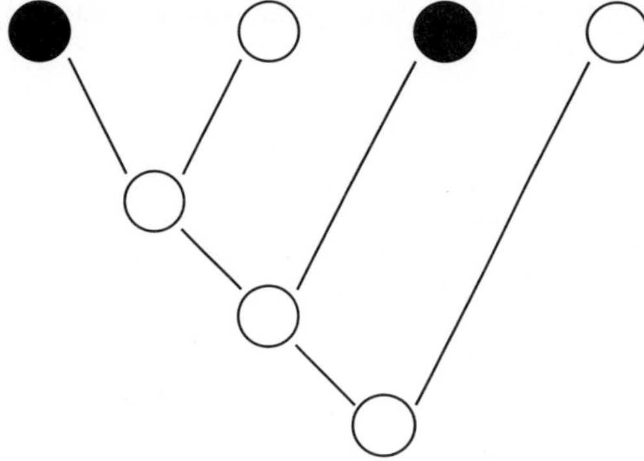

Figure A.4 Illustration of apomorphy. The two black circles represent examples of apomorphy.

Further reading
Maddison DR, Maddison WP (2001) *MacClade 4: Analysis of Phylogeny and Character Evolution* Sinauer Associates.

See also Plesiomorphy, Synapomorphy, Autapomorphy

APOLLO, *SEE* GENE ANNOTATION, VISUALIZATION TOOLS.

ARC, *SEE* BRANCH (OF A PHYLOGENETIC TREE).

ARE WE THERE YET?, *SEE* AWTY.

AROMATIC

Roman Laskowski and Tjaart de Beer

An aromatic molecule, or compound, is one that contains a planar, or near-planar, closed ring with particularly strong stability. The enhanced stability results from the overlap of p-orbital electrons to constitute a delocalized *Pi* molecular orbital system. Huckel's rule for identifying aromatic cyclic rings states that the total number of electrons in the *Pi* system must equal $(4n + 2)$, where n is any integer.

The archetypal aromatic molecule is benzene, which satisfies the Huckel rule with $n = 1$.

In proteins, the amino acid residues phenylalanine, tyrosine, tryptophan and histidine have aromatic side chains. Interactions between aromatic side chains in the cores of

A

protein structures often play an important part in conferring stability on the structure as a whole.

Relevant website

Atlas of protein side chain interaction	http://www.biochem.ucl.ac.uk/bsm/sidechains/index.html

Further reading

Burley SK, Petsko GA (1985) Aromatic-aromatic interaction: a mechanism of protein structure stabilization. *Science* 229: 23–28.

Singh J, Thornton JM (1985) The interaction between phenylalanine rings in proteins. *FEBS Lett* 191: 1–6.

Singh J, Thornton JM (1992) *Atlas of Protein Side-Chain Interactions*, Vols. I and II. IRL Press., Oxford

See also Amino Acid, Side Chain

ARRAY, *SEE* DATA STRUCTURE.

ARTIFICIAL NEURAL NETWORKS, *SEE* NEURAL NETWORKS.

ASBCB (THE AFRICAN SOCIETY FOR BIOINFORMATICS AND COMPUTATIONAL BIOLOGY)

Pedro Fernandes

The ASBCB is a non-profit professional association dedicated to the advancement of bioinformatics and computational biology in Africa. The Society serves as an international forum and resource devoted to developing competence and expertise in bioinformatics and computational biology in Africa. It complements its activities with those of other international and national societies, associations and institutions, public and private, that have similar aims. It also promotes the standing of African bioinformatics and computational biology in the global arena through liaison and cooperation with other international bodies.

Related website

ASBCB	http://www.asbcb.org/

ASSOCIATION ANALYSIS (LINKAGE DISEQUILIBRIUM ANALYSIS)

Andrew Collins, Mark McCarthy and Steven Wiltshire

Genetic analysis which aims to detect associations between alleles at different loci, and a central tool in the identification and characterisation of disease-susceptibility genes.

Typically association studies are undertaken as large scale case-control analyses in which a comparison is made of allele (or haplotype) frequencies between a sample of cases (individuals with the disease or phenotype of interest, and therefore enriched for disease-susceptibility alleles) and controls (either a population-based sample, or individuals selected not to have the disease). The interest here is in identifying associations which reflect linkage disequilibrium between the typed marker and disease (and hence with disease-susceptibility variants). Such associations indicate either that the typed marker is the disease susceptibility variant, or that it is in linkage disequilibrium (LD) with it (and, in turn, given the limited genomic extent of linkage disequilibrium in most populations, that the disease-susceptibility variant must be nearby). Case-control studies may generate false positives due to latent population stratification (i.e. reveal associations which do not reflect linkage disequilibrium). For this reason techniques are employed to correct for population stratification. These include genomic control which aims to correct the inflation of association statistics and structured association which determines sub-population clusters and stratifies association evidence by cluster. Family-based association tests in which transmissions of alleles to affected children in families are analogous to the 'case' sample, and untransmitted alleles represent controls, avoid population structure problems. However, genome-wide association studies have focused on very large case and control samples to detect low penetrance common disease alleles. In this context extensive replication samples have proven essential to confirm disease roles for specific alleles or genomic regions.

Examples: Altshuler and colleagues (2000) combined a variety of association analysis methods to demonstrate that the Pro12Ala variant in the PPARG gene is associated with type 2 diabetes. Dahlman and colleagues (2002) sought to replicate a previously-reported association between a variant in the 3'UTR region of the interleukin 12 p40 gene (IL12B) and type 1 diabetes: despite typing a much larger data set they were unable to find evidence to support the previous association.

Related websites

EIGENSTRAT stratification correction method:	http://www.hsph.harvard.edu/faculty/alkes-price/software/
PLINK software for genome association analysis	http://pngu.mgh.harvard.edu/~purcell/plink/

Further reading

Altshuler D, et al. (2000) The common PPARgamma Pro12Ala polymorphism is associated with decreased risk of type 2 diabetes. *Nat Genet* 26: 76–80.

Cardon LR, Bell JI (2001) Association study designs for complex disease. *Nat Rev Genet* 2: 91–99.

Dahlman I, et al. (2002) Parameters for reliable results in genetic association studies in common disease. *Nat Genet* 30: 149–150.

Hästbacka J, et al. (1993) Linkage disequilibrium mapping in isolated founder populations: diastrophic dysplasia in Finland. *Nat Genet* 2: 204–211.

Hirschhorn JN, et al. (2002) A comprehensive review of genetic association studies. *Genet Med* 4: 45–61.

Lander ES, Schork NJ (1994) Genetic dissection of complex traits. *Science* 265: 2037–2048.

Price AL, Zaitlen NA, Reich D, Patterson N (2010) New approaches to population stratification in genome-wide association studies. *Nature Reviews Genetics* 11: 459–463.

Risch N (2001) Implications of multilocus inheritance for gene-disease association studies. *Theor Popul Biol* 60: 215–220.

Risch N, Merikangas K (1996) The future of genetic studies of complex human diseases. *Science* 273: 1516–1517.

See also Genome-Wide Association Study, HapMap Project, Linkage Disequilibrium, Allelic Association, Multifactorial Trait, Family-based Association Analysis

ASSOCIATION RULE, *SEE* ASSOCIATION RULE MINING.

ASSOCIATION RULE MINING (FREQUENT ITEMSET, ASSOCIATION RULE, SUPPORT, CONFIDENCE, CORRELATION ANALYSIS)

Feng Chen and Yi-Ping Phoebe Chen

Association rule learning is a popular technique for discovering interesting relations between variables in large databases. It has originated from how to discover regularities between products in large-scale transaction data in supermarkets. The rule detected would indicate that for example if a customer buys onions and potatoes together, he or she is likely to also buy hamburger meat. Such information can be used as the basis for decisions about marketing activities. In addition to the above example from market basket analysis, association rules are also employed in many application areas such as web mining and bioinformatics.

The definition of association rule mining is as follows: Let $I=\{i_1,i_2,\ldots, i_n\}$ be a set of n binary attributes called items. Let $D=\{t_1,t_2,\ldots, t_m\}$ be a set of transactions called the database. Each transaction in D has a unique ID and contains a subset of the items in I. A rule is defined as $X \Rightarrow Y$ where X, Y ⊂ I and X ∩ Y = Ø. The sets of items X and Y are called antecedent and consequent respectively. For example, in a very small supermarket database including 4 products and 4 transactions, as shown in Table A.1, {bread, butter} = > {milk} could be meaningful because these three products occurs in one transaction, which signifies that if one buys bread and butter, he or she might also buy milk.

Table A.1 An example of association rule mining in a small database

ID	Beer	Milk	Butter	Bread
1	0	1	0	0
2	0	0	1	0
3	0	0	0	0
4	0	1	1	1

Before explaining how to detect association rules, there are some important concepts that have to be defined. First, itemset is a set of items. A frequent itemset means this itemset occurs much more frequently than others in the transactions. Additionally, support, confidence and lift are popular parameters in mining association rules. The former two came from the original association rule mining algorithm. The support (supp(X)) of an itemset X is the proportion of transactions in the data set which contain the itesssssmset. Confidence is defined as

$$conf\,(X \Rightarrow Y) = \frac{supp(X \cup Y)}{supp(X)}$$

which indicates how likely it is that Y is purchased when X is already purchased in a supermarket database.

However support and confidence are not efficient for reflecting the relations of the itemsets in one rule. That is why correlation analysis is introduced.

$$lif\,(X \Rightarrow Y) = \frac{supp(X \cup Y)}{supp(Y) \times supp(X)}$$

represents how correlated X and Y are. Lift > 1 means X and Y are positively related, and vice versa. X and Y are independent if this value is 1.

Association rule mining is usually required to satisfy a user-specified minimum support and a user-specified minimum confidence simultaneously. The mining process can be partitioned into two steps. First, minimum support is applied to find all frequent itemsets in a database. Second, these frequent itemsets and the minimum confidence constraint are used to form rules.

Association rule may be used for the discovery of co-working genes, or to speculate how one gene regulates another. For example, when we think of every tumor as a record and all the miRNAs in it as items, the miRNAs in one frequent itemset can be regarded as ones that cooperate together in a variety of tumors.

Further reading

Agrawal R, Imielinski T, Swami A (1993) *Mining Association Rules Between Sets of Items in Large Databases*, SIGMOD Conference, 207–216: 1993.

Hájek P, Feglar T, Rauch J, Coufal D (2003) *The GUHA Method, Data Preprocessing and Mining, Database Support for Data Mining* Applications. Springer.

Omiecinski ER (2003) Alternative interest measures for mining associations in databases. *IEEE Transactions on Knowledge and Data Engineering* 15(1): 57–69.

A

Tan P-N, Kumar V, Srivastava J (2004) Selecting the right objective measure for association analysis. *Information Systems* 29(4): 293–313.

Wu X, Jiang R, Zhang MQ, Li S (2008) Network-based global inference of human disease genes. *Molecular Systems Biology* 4: 189.

ASSOCIATIVE ARRAY, *SEE* DATA STRUCTURE.

ASYMMETRIC UNIT

Liz Carpenter

The asymmetric unit of a crystal is the smallest building block from which a crystal can be created by applying the crystallographic symmetry operators followed by translation by multiples of the unit cell vectors. The asymmetric unit may contain one or several copies of the molecules under study. See X-ray diffraction for more details.

see X-ray Diffraction, X-ray Crystallography

See also Crystal, Macromolecular, Space Group, Unit Cell

ATOMIC COORDINATE FILE (PDB FILE)

Eric Martz

A data file containing a list of the coordinates in three-dimensional space for a group of atoms. Typically the atoms constitute a molecule or a complex of molecules. An atomic coordinate file is required in order to look at the three-dimensional structure of a molecule using visualization software. Sources of coordinates, in order of decreasing reliability, are empirical (X-ray crystallography or nuclear magnetic resonance), homology modeling, or *ab initio* theoretical modeling. There are over a dozen popular formats for atomic coordinate files. One of the most popular for macromolecules is the original one used by the Protein Data Bank, commonly called a 'PDB file'. The Protein Data Bank has recently adopted a new standard format, the macromolecular crystallographic information format, or mmCIF, but some popular software remains unable to process this newer format. Therefore the PDB format continues to be supported at the Protein Data Bank as well as by most other sources of atomic coordinate files. When atomic coordinate files are transmitted through the internet, their formats are identified with MIME types.

Relevant website

Protein Data Bank	http://www.rcsb.org/pdb/home/home.do

See also MIME types, Protein Data Bank, Molecular Visualization

Eric Martz is grateful for help from Eric Francoeur, Peter Murray-Rust, Byron Rubin and Henry Rzepa.

AUTAPOMORPHY

A.R. Hoelzel

A unique derived state of an evolutionary character.

As a hypothesis about the pattern of evolution among operational taxonomic units, a phylogenetic reconstruction can inform us about the relationship between character states. We can identify ancestral and descendent states, and therefore identify primitive and derived states. An apomorphy (meaning 'from-shape' in Greek) is a derived state (shown as the filled circle in Figure A.5).

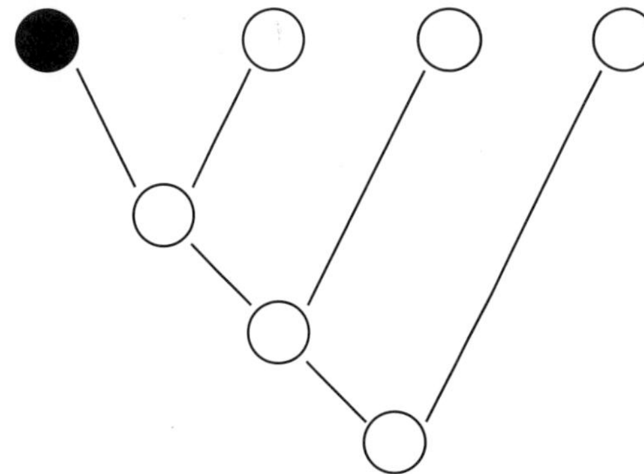

Figure A.5 Illustration of autapomorphy. The black circle is in a derived state not shared by any other OTU.

A unique derived state (as shown in Figure A.5) is known as an autapomorphy ('aut' means alone).

Further reading

Maddison DR, Maddison WP (2001) *MacClade 4: Analysis of Phylogeny and Character Evolution*. Sinauer Associates, Sunderland MA.

See also Plesiomorphy, Synapomorphy, Apomorphy

AUTOZYGOSITY, *SEE* HOMOZYGOSITY, HOMOZYGOSITY MAPPING.

AWTY (ARE WE THERE YET?)

Michael P. Cummings

An online system for graphically exploring convergence of Markov Chain Monte Carlo (MCMC) chains in Bayesian phylogenetic inference.

The graphics are designed to help assess whether an MCMC analysis has run sufficiently long such that tree topologies are being sampled in proportion to their posterior probabilities. Input consists of a NEWICK or NEXUS formatted tree file generated by programs such as MrBayes or BAMBE representing a set of trees sampled over a MCMC run. Graphics results can also be downloaded and analysed using other plotting packages.

Related website

AWTY online home page	http://king2.scs.fsu.edu/CEBProjects/awty/awty_start.php

Further reading

Nylander JA, Wilgenbusch JC, Warren DL, Swofford DL (2008) AWTY (are we there yet?): a system for graphical exploration of MCMC convergence in Bayesian phylogenetics. *Bioinformatics* 24: 581–583.

See also BAMBE, MrBayes

AXIOM

Robert Stevens

An axiom is a statement asserted into a logical system without proof.

Some knowledge representation languages, such as Description Logics, make use of axioms to make statements adding information to ontologies. For example, I can assert that the class *Nucleic Acid* has two children *DNA* and *RNA*. A disjointness axiom can then be made stating that *DNA* and *RNA* are disjoint, that is, it is not possible to be both a DNA and an RNA; that is, the classes do not overlap. Some systems include other axiom types that can assert the equivalence of two categories; that a set of sub-categories covers the meaning of a super-category exhaustively; or to assert additional necessary properties of a category or individual.

See also Description Logic

B

Backbone (Main Chain)
Backbone Models
Backpropagation Networks, *see* Neural Networks.
Bagging
Ball and Stick Models
BAMBE (Bayesian Analysis in Molecular Biology and Evolution)
Base-Call Confidence Values
Base Composition (GC Richness, GC Composition)
Bayes' Theorem
Bayesian Classifier (NaÏve Bayes)
Bayesian Evolutionary Analysis Utility, *see* BEAUti.
Bayesian Information Criterion (BIC)
Bayesian Network (Belief Network, Probabilistic Network, Causal Network, Knowledge Map)
Bayesian Phylogenetic Analysis
BEAGLE
Beam Search
BEAST (Bayesian Evolutionary Analysis by Sampling Trees)
BEAUti (Bayesian Evolutionary Analysis Utility)
Before State (Before Sphere)
Belief Network, *see* Bayesian Network.
Bemis and Murcko Framework (Murcko Framework)
Best-First Search
Beta Barrel
Beta Breaker
Beta Sheet
Beta Strand
BIC, *see* Bayesian Information Criterion.
Biclustering Methods
Bifurcation (in a Phylogenetic Tree)
Binary Numerals
Binary Relation
Binary Tree, *see* Data Structure.
Binding Affinity (K_d, K_i, IC_{50})
Binding Site
Binding Site Symmetry
Bio++
Bioactivity Database
Bioinformatics (Computational Biology)
The Bioinformatics Organization, Inc (formerly bioinformatics.org)
Bioinformatics Training Network, *see* BTN.
Biological Identifiers
BioMart
Bipartition, *see* Split
Bit

Concise Encyclopaedia of Bioinformatics and Computational Biology, Second Edition.
Edited by John M. Hancock and Marketa J. Zvelebil.

BLAST (Maximal Segment Pair, MSP)
BLASTX
BLAT (BLAST-like Alignment Tool)
BLOSUM (BLOSUM Matrix)
Boltzmann Factor
Boolean Logic
Boosting
Bootstrap Analysis, *see* Bootstrapping.
Bootstrapping (Bootstrap Analysis)
Bottleneck, *see* Population Bottleneck.
Box
Branch (of a Phylogenetic Tree) (Edge, Arc)
Branch Length, *see* Branch, Branch-length Estimation.
Branch-length Estimation
BTN (Bioinformatics Training Network)

BACKBONE (MAIN CHAIN)

Roman Laskowski and Tjaart de Beer

When amino acids link together to form a peptide or protein chain, the atoms that constitute the continuous link running the length of the chain are referred to as the backbone atoms. Each amino acid contributes its -N-Cα-C- atoms to this backbone. Also included as part of the backbone is the carbonyl oxygen attached to the backbone carbon, C- , and any hydrogens attached to these backbone atoms. All other atoms are termed side chain atoms. For most of the amino acids, the side chain atoms spring off from the main chain Cα, the exception being glycine, which has no side chain (other than a single hydrogen atom), and proline whose side chain links back onto its main chain nitrogen.

Further reading

Branden C, Tooze J (1998) *Introduction to Protein Structure*. Garland Science, New York.

Lesk AM (2000) *I*ntroduction to Protein Architecture. Oxford University Press, Oxford.

See also Amino Acid, Side Chain, Glycine.

BACKBONE MODELS

Eric Martz

Simplified depictions of proteins or nucleic acids in three dimensions that enable the polymer chain structures to be seen. (See illustration at Models, molecular.) Lines are drawn between the positions of alpha carbon atoms (for proteins) or phosphorus atoms (for nucleic acids), thus allowing the backbone chains to be visualized. These lines do not lie in the positions of any of the covalent bonds. Another type of backbone model is a line following the covalent bonds of the main-chain atoms, rendering the polypeptide as polyglycine. For nucleic acids, an alternative backbone consists of lines connecting the centers of the pentose rings. Most macromolecular visualization software packages can display backbone models. Sometimes there is an option to smooth the backbone trace; smoothed backbones may be rendered as wires, ribbons or schematic models. In the early 1970s, before computers capable of displaying backbone models were readily available, physical backbone models were made of metal wire. One apparatus for constructing wire backbone models was Byron's Bender (see Websites, Further reading). It was popular in the 1970s and early 1980s, and was then superseded for most purposes by computer visualization methods.

Related website

Martz, E and Francoeur E. *History of Visualization of Biological Macromolecules*	http://www.umass.edu/microbio/rasmol/history.htm

B

Further reading

Rubin B, Richardson JS (1972) The simple construction of protein alpha-carbon models. *Biopolymers* 11: 2381–2385.

Rubin, B (1985. Macromolecule backbone models. *Methods Enzymol.* 115: 391–397.

See also Visualization, molecular, Schematic ('Cartoon') Model, Models, molecular (illustrated)

Eric Martz is grateful for help from Eric Francoeur, Peter Murray-Rust, Byron Rubin and Henry Rzepa.

BACKPROPAGATION NETWORKS, *SEE* NEURAL NETWORKS.

BAGGING

Concha Bielza and Pedro Larrañaga

Bagging is short for Bootstrap AGGregatING, an ensemble learning algorithm. Different learning sets are first generated by bootstrap, i.e. from an original set of n learning examples, and each new learning set is generated by n-times randomly selecting (with replacement) an example. Some examples will occur more than once and some others may not occur (on average, 36.8% of examples). Each newly generated set is used as an input to a learning algorithm. All the algorithms form the ensemble. The class (predicted) label with more votes will be the final prediction of a new example.

Bagging is recommended when the learning set has a small size and when using unstable base algorithms, such as decision trees and regression trees, with high variance (slight changes in the training data may easily cause a large difference in the generated classifier, thereby producing diversity). Bagging is robust in the sense that increasing the number of base classifiers does not lead to overfitting.

Further reading

Breiman L (1996) Bagging predictors. *Machine Learning* 24(2): 123–140.

Dudoit S and Fridlyand J (2003) Bagging to improve the accuracy of a clustering procedure. *Bioinformatics* 19(9): 1090–1099.

Hanczar B, Nadif M (2011) Using the bagging approach for biclustering of gene expression data. *Neurocomputing* 74(10): 1595–1605.

He Z, Yang C, Yu W (2008) Peak bagging for peptide mass fingerprinting. *Bioinformatics* 24(10): 1293–1299.

See also Random Forests

BALL AND STICK MODELS

Eric Martz

Three-dimensional molecular models in which atoms are represented by balls, and covalent bonds by sticks connecting the balls. Also called (in the UK) 'ball and spoke models'. (See illustration at Models, molecular.) Ball and stick models can be traced back to John Dalton in the early 19th century. Because they are so detailed, ball and stick models are generally more useful for small molecules than for macromolecules. For macromolecules, hydrogen atoms are often omitted to simplify the model. Ball and stick models originated in physical models, and later were implemented in computer visualization programs. One of the early and widely-used packages is ORTEP (see Websites). In order to see details beneath the outer surface of the model, the balls are typically considerably smaller than the van der Waals radii of the atoms. In RasMol and its derivatives (see Visualization), the balls have a fixed radius of 0.45 Å, and the cylindrical sticks, 0.15 Å. Some software offers scaled ball and stick models, in which the balls, while still much smaller than the van der Waals radii, vary in size in proportion to the van der Waals radii. When Kendrew and coworkers solved the structure of myoglobin at atomic resolution in the early 1960s, they first built a wire-frame model. Shortly thereafter, they built ball and spoke models, twenty-nine of which were sold to interested research groups in the late 1960s (see Websites). Due to the complexity of proteins, it is difficult to discern major structural features from a ball and stick model. Therefore, physical models of entire proteins or protein domains were later more commonly backbone or schematic.

Related websites

Martz, E and Francoeur E. *History of Visualization of Biological Macromolecules*	http://www.umass.edu/microbio/rasmol/history.htm
ORTEP – Oak Ridge Thermal Ellipsoid Plot Program for Crystal Structure Illustrations	http://www.ornl.gov/ortep/ortep.html

Further reading

Brode W, Boord CE (1932) Molecular models in the elementary organic laboratory. *I. J. Chem. Ed.* 9: 177–782.

Corey RB, Pauling L (1953) Molecular models of amino acids, peptides and proteins. *Rev. Sci. Instr.* 24: 62–27.

Kendrew JC, Dickerson RE, Strandberg RG, et al. (1960) Structure of myoglobin. A three-dimensional fourier synthesis at 2 Å resolution. *Nature* 185: 42–47.

See also Wire-Frame Models, Models, molecular (illustrated), Visualization, molecular

Eric Martz is grateful for help from Eric Francoeur, Peter Murray-Rust, Byron Rubin and Henry Rzepa.

B

BAMBE (BAYESIAN ANALYSIS IN MOLECULAR BIOLOGY AND EVOLUTION)

Michael P. Cummings

Program for phylogenetic analysis nucleotide of sequence data using a Bayesian approach.

A Metropolis-Hastings algorithm for Markov Chain Monte Carlo (MCMC) is used to sample model space through a process of parameter modification proposal and acceptance/rejection steps (also called cycles or generations). After the process becomes stationary the frequency with which parameter values are visited in the process represents an estimate of their underlying posterior probability. Model parameters include substitution rates and specific taxa partitions. A choice of several commonly used likelihood models is available as are choices for starting tree of the Markov Chain including user-defined, UPGMA (Unweighted Pair Group Method with Arithmetic Mean), neighbor-joining, and random. An accessory program, Summarize, is used to process the output file of tree topology information and reports on topological features.

The programs are written in C++ and are available as source code and binaries for some systems.

Related website

BAMBE home page	http://www.mathcs.duq.edu/larget/bambe.html

Further reading

Larget B, Simon D (1999) Markov Chain Monte Carlo algorithms for the Bayesian analysis of phylogenetic trees. *Mol. Biol. Evol.* 16: 750–759.

Mau B, et al. (1999) Bayesian phylogenetic inference via Markov Chain Monte Carlo methods. *Biometrics* 55: 11–12.

Newton MA, et al. (1999) Markov Chain Monte Carlo for the Bayesian analysis of evolutionary trees from aligned molecular sequences. In: F. Seillier-Moseiwitch (ed.), *Statistics in Molecular Biology and Genetics. IMS Lecture Notes-Monograph Series*, 33: 1431–62.

See also MrBayes

BASE-CALL CONFIDENCE VALUES

Rodger Staden and John M. Hancock

Numerical values assigned to each base in a sequence reading to predict its reliability.

In Sanger DNA sequencing, the values are assigned by analysis of the fluorescence signals from which the basecalls where derived. The most widely used scale of values is named after the program PHRED which defines the confidence $C = -10\log(P_{error})$ where P_{error} is the probability that the base-call is erroneous.

Next-generation sequencing machines typically provide base-call confidence values on the PHRED scale as part of a broader suite of quality control measures.

Related website

PHRED	www.phrap.org/phredphrapconsed.html

BASE COMPOSITION (GC RICHNESS, GC COMPOSITION)

Katheleen Gardiner

The percent of G+C or A+T bases in a DNA sequence or genome.

For example, the average base composition of the human genome is ~38%GC. This is not a uniform value however, and is related to chromosome bands. G bands are consistently AT rich, ~344–0% GC. R bands are variable, with some segments, especially many telomeric regions, >60% GC.

Further reading

Nekrutenko A, Li WH (2000) Assessment of compositional heterogeneity within and between eukaryotic genomes Genome Res 10: 1986–1995.

See also Chromosome Band, Isochore

BAYES' THEOREM

Feng Chen, Yi-Ping Phoebe Chen and Matthew He

A mathematical formula used for calculating conditional probabilities. Bayes' Theorem simplifies the calculation of conditional probabilities and clarifies significant features of a subjectivist position. Bayes' theorem may be applied to discrete or continuous probability distributions in bioinformatics. It figures prominently in subjectivist or Bayesian approaches to epistemology, statistics, and inductive logic.

Statistically, Bayes' Theorem is:

$$P(A|B) = \frac{P(B|A)P(A)}{P(B)} = P(A)\frac{P(B|A)}{P(B)}$$

where P(A) and P(B) are the probabilities of A and B; P(A|B) and P(B|A) indicate the conditional probabilities of A given B and B given A respectively. Bayes' Theorem is always used for classification or prediction, such as the naive Bayesian classifier.

Bayes' Theorem is applied very widely in bioinformatics, for example in microarray analysis, protein informatics, quantitative network modeling.

Further reading

Dey DK, Ghosh S, Mallick BK (2010) *Bayesian Modeling in Bioinformatics*. Chapman and Hall/CRC.

Wilkinson DJ (2007) *Bayesian Methods in Bioinformatics and Computational Systems Biology*. Briefings in Bioinformatics, Vol 8, No 2, 101–1016.

B

BAYESIAN CLASSIFIER (NAÏVE BAYES)

Pedro Larrañaga and Concha Bielza

A Bayesian classifier assigns the most probable a posteriori class, c*, to a given new case to be classified, $\mathbf{x} = (x_1, x_2, \ldots, x_n)$, i.e.

$$c^* = arg\ max_c\ p(c|\boldsymbol{x}) = arg\ max_c\ p(c)\ p(\boldsymbol{x}|c) = arg\ max_c\ p(c)\ p(x_1, x_2, \ldots, x_n|c)$$

As the estimation of $p(x_1, x_2, \ldots, x_n\ |c)$ requires a number of parameters that is exponential in the number of features, n, assumptions avoiding the estimation of this huge number of parameters have been proposed in the literature.

Naive Bayes (Minsky, 1961) is the simplest Bayesian network classifier (see Figure B.1), since the conditional independence of the predictive variables given the class is assumed, computing the posterior probability of the class variable given the features as $p(c|\boldsymbol{x}) = p(c) \prod p(x_i|c)$.

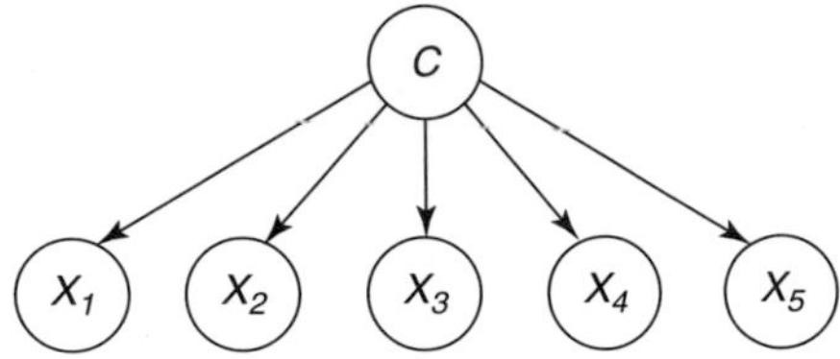

Figure B.1 Example of a naïve Bayes structure with 5 predictor variables. From it $p(c|\boldsymbol{x})$ is computed as proportional to $p(c)\ p(x_1|c)\ p(x_2|c)\ p(x_3|c)\ p(x_4|c)\ p(x_5|c)$.

Although this strong assumption of conditional independence is far away from real situations, the naïve Bayes model provides competitive results with other state of the art classification algorithms. A way for incorporating dependency relationships among predictor variables is by learning a tree structure among them and after that imposing a naïve Bayes classifier, getting the so called *tree augmented naïve Bayes classifier* (Friedman et al., 1997). See Figure B.2 for an example of tree augmented naïve Bayes structure.

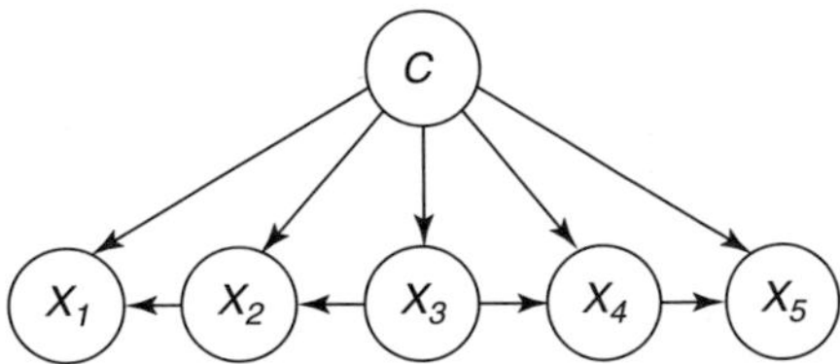

Figure B.2 Example of a tree augmented naïve Bayes structure with 5 predictor variables. From it $p(c|\mathbf{x})$ is computed as proportional to $p(c)\ p(x_1|c, x_2)\ p(x_2|c, x_3)\ p(x_3|c)\ p(x_4|c, x_3)\ p(x_5|c, x_4)$.

Examples of application of Bayesian classifiers include the prediction of the protein secondary structure (Robles et al. 2004) and the selection of genes for cancer classification from microarray data (Wang et al., 2005).

Further reading

Friedman N, et al. (1997) Bayesian network classifiers. *Machine Learning* 29: 131–163.

Minsky M (1961) Steps toward artificial intelligence. *Transactions on Institute of Radio Engineers* 49: 8–30.

Robles V, et al. (2004) Bayesian networks as consensed voting system in the construction of a multi-classifier for protein secondary structure prediction. *Artificial Intelligence in Medicine* 31: 117–136.

Wang Y, et al. (2005) Gene selection from microarray data for cancer classification – a machine learning approach. *Computational Biology and Chemistry* 29(1): 37–46.

BAYESIAN EVOLUTIONARY ANALYSIS UTILITY, *SEE* BEAUTI.

BAYESIAN INFORMATION CRITERION (BIC)

Concha Bielza and Pedro Larrañaga

The Bayesian Information Criterion (BIC) or Schwarz criterion (Schwarz, 1978) is a criterion for model selection, applicable in settings where the fitting is carried out by maximizing the (log)likelihood. The general form is

$$\mathrm{BIC} = -\mathrm{loglik} + (\log N)\cdot k$$

where N is the number of learning examples and k is the number of free parameters to be estimated. If the estimated model is a linear regression, k is the number of regressors, including the intercept.

The rationale for BIC lies in the observation that when fitting models, it is possible to increase the likelihood by adding parameters, but this may result in overfitting. The BIC solves this problem by penalizing the complexity of the model, where complexity refers to the number of parameters in the model. Given any two estimated models, the model with the lower value of BIC is the one to be preferred.

BIC is closely related to AIC, where the penalty term is larger in BIC than in AIC. It is approximately equal to the minimum description length criterion but with negative sign.

Further reading

Liu T, Lin N, Shi N, Zhang B (2009) Information criterion-based clustering with order-restricted candidate profiles in short time-course microarray experiments. *BMC Bioinformatics* 10:146.

Schwarz GE (1978) Estimating the dimension of a model. *Annals of Statistics* 6(2): 461–464.

Xi R, Hadjipanayis AG, Luquette LJ, et al. (2011) Copy number variation detection in whole-genome sequencing data using the Bayesian information criterion. *Proceedings of the National Academy of Sciences* 108(46): 18583–18584.

See also Regression Analysis, Akaike Information Criterion

B

BAYESIAN NETWORK (BELIEF NETWORK, PROBABILISTIC NETWORK, CAUSAL NETWORK, KNOWLEDGE MAP)

John M Hancock

A modeling tool that combines directed acyclic graphs with Bayesian probability.

Figure B.3 shows a simple example of a Bayesian network, which consists of a causal graph combined with an underlying probability distribution. Each node of the network in the figure corresponds to a variable and the edges represent causality between these events, which is directional. The other element of a Bayesian network are the probability distributions (e.g. P(Y|X), P(X|V,W)) associated with each node. With this information the network can model probabilities of complex causal relationships.

Bayesian networks are widely used modeling tools. Techniques also exist for inferring or estimating nework parameters. Such an approach is applicable, for example, to modeling gene networks from microarray data.

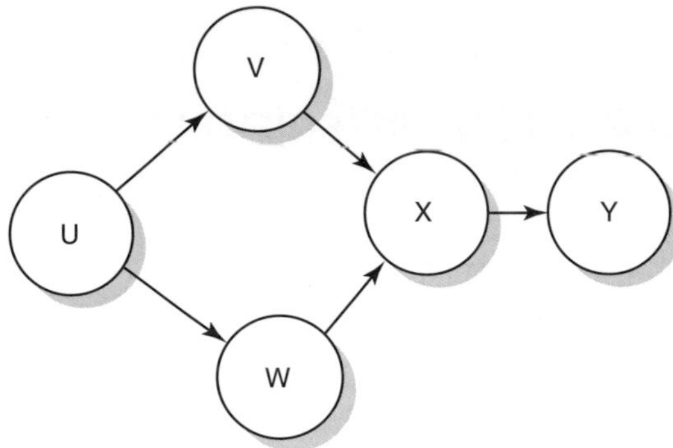

Figure B.3 Representation of a Bayesian Network. The nodes U-Y are linked causally as represented by the arrows.

Further reading

Friedman N, et al. (2000) Using Bayesian networks to analyze expression data. *J Comput Biol* 7: 601–620.

Neapolitan RE (2003) *Learning Bayesian Network*, Prentice Hall.

See also Directed Acyclic Graph, Bayesian Analysis, MCMC

BAYESIAN PHYLOGENETIC ANALYSIS

Sudhir Kumar and Alan Filipski

A probabilistic method based on the Bayes theorem. It is often used to infer distributions of phylogenetic trees and to estimate evolutionary parameters.

Given a substitution model, an assumed prior distribution for all parameters (typically including the tree topologies), and a set of sequences (mostly a multiple sequence alignment), Bayesian statistics can be used to compute a posterior distribution for the unknown quantities (e.g., tree topology, branch lengths, ancestral states, divergence times, and more recently also a multiple sequence alignment). Estimates for these quantities are obtained by sampling from these posterior distributions. Directly applying Bayes' rule to compute the posterior distributions for parameters of interest is mostly not practical because of the need to evaluate high-dimensional integrals. This difficulty is almost always addressed by using Markov Chain Monte Carlo (MCMC) methods or more advanced versions thereof such as Metropolis-Coupled MCMC (MC^3). MCMC provides a computationally tractable method for sampling from the posterior distribution of the parameters (e.g. topology, rates, alignments, or branch lengths) being estimated. One key challenge is the assessment of MCMC chain convergence. Chain convergence assessment software can only be deployed to show that chains have not converged and not to assess if chains have converged. Another difficulty is the specification of the prior distributions (e.g., should all trees, even the most unreasonable ones, have the same prior probability?).

Software

Felsenstein J (1993) PHYLIP: Phylogenetic Inference Package. Seattle,WA, University of Washington.

Ronquist F, Huelsenbeck JP (2003) MRBAYES 3: Bayesian phylogenetic inference under mixed models. *Bioinformatics* 19(12): 1571–1573.

Yang Z (1997). PAML: a program package for phylogenetic analysis by maximum likelihood. *Comput Appl Biosci* 13: 5555–5556.

Related websites

PhyloBayes:	http://www.phylobayes.org/
Beast:	http://beast.bio.ed.ac.uk/
RevBayes	http://sourceforge.net/projects/revbayes/
AWTY:	http://ceb.csit.fsu.edu/awty/
BaliPhy:	http://www.biomath.ucla.edu/msuchard/bali-phy/

Further reading

Metropolis, N, et al. (1953) Equation of state calculations by fast computing machines. *Journal of Chemical Physics* 21(6): 1087–1092.

Murphy WJ, et al. (2001) Resolution of the early placental mammal radiation using Bayesian phylogenetics. *Science* 294: 2342–2351.

Yang Z, Rannala B (1997) Bayesian phylogenetic inference using DNA sequences: a Markov Chain Monte Carlo Method. *Mol Biol Evol* 14: 717–7124.

See also Phylogenetic Reconstruction, Markov Chain Monte Carlo

BEAGLE

Michael P. Cummings

An application programming interface (API) and library for high-performance statistical phylogenetic inference.

The API provides a uniform interface for performing phylogenetic likelihood calculations on a variety of computing hardware platforms. The library includes a set of efficient implementations and can currently exploit hardware including graphical processing units (GPUs) using NVIDIA CUDA, central processing units (CPUs) with Streaming SIMD Extensions (SSE) and related processor supplementary instruction sets, and multicore CPUs via OpenMP. Proper use of the library can result in dramatic performance increases for likelihood and Bayesian analyses, particularly for data sets with long sequences and parameter-rich models. The library has been incorporated into several popular phylogenetic software packages including BEAST, GARLI and MrBayes.

The core BEAGLE library is implemented in C++ with C and Java JNI interfaces, and these are all available as free open source software licensed under the Lesser GPL.

Related website

BEAGLE home page	https://code.google.com/p/beagle-lib/

Further reading

Ayres DL, Darling A, Zwickl DJ, et al. (2012) BEAGLE: an application programming interface and high-performance computing library for statistical phylogenetics. *Syst. Biol.* 61: 171–173.

See also BEAST, GARLI, MrBayes

BEAM SEARCH

Concha Bielza and Pedro Larrañaga

Beam search is similar to best-first search but truncates its list of candidate subsets at each state so that it only contains a fixed number – the beam width – of most promising candidates. We stop when we cannot get a better subset.

If the beam width is 1, then we get the basic greedy search. Beam search has the potential of avoiding the myopia associated with the greedy search, although if the beam width is not large enough, myopia may still occur.

Further reading

Carlson JM, Chakravarty A, Gross RH (2006). BEAM: A beam search algorithm for the identification of cis-regulatory elements in groups of genes. *Journal of Computational Biology* 13(3): 686–701.

Sun Y, Buhler J (2005) Designing multiple simultaneous seeds for DNA similarity search. *Journal of Computational Biology* 12(6): 847–861.

See also Feature Subset Selection, Best-First Search

BEAST (BAYESIAN EVOLUTIONARY ANALYSIS BY SAMPLING TREES)

Michael P. Cummings

A program for Bayesian MCMC analysis of molecular sequences to infer rooted, time-measured phylogenies using strict or relaxed molecular clock models.

Among the most sophisticated programs for phylogenetic analysis, it can be used for reconstructing phylogenies and testing evolutionary hypotheses including phylographical hypotheses. The program BEAUti can be used for setting up standard analyses, and programs Tracer and FigTree can be used for analysing and displaying results.

Related website

BEAST home page	http://beast.bio.ed.ac.uk/Main_Page

Further reading

Drummond AJ, Ho SYW, Phillips MJ, Rambaut A (2006) Relaxed phylogenetics and dating with confidence. *PLoS Biol.* 4: e88.

Drummond AJ, Rambaut A (2007) BEAST: Bayesian evolutionary analysis by sampling trees. *BMC Evol. Biol.* 7: 214.

Drummond AJ, Suchard MA (2010) Bayesian random local clocks, or one rate to rule them all. *BMC Biology* 8: 114.

Lemey P, Rambaut A, Drummond AJ, Suchard MA (2009) Bayesian phylogeography finds its roots. *PLoS Comput. Biol.* 5: e1000520.

Lemey P, Rambaut A, Welch JJ, Suchard MA (2010) Phylogeography takes a relaxed random walk in continuous space and time. *Mol. Biol. Evol.* 27: 1877–1885.

See also BEAGLE, BEAUti, FigTree, Tracer

BEAUTI (BAYESIAN EVOLUTIONARY ANALYSIS UTILITY)

Michael P. Cummings

A program to select various models and options, import data in NEXUS format, and generate an XML file ready for use in BEAST.

Related website

BEAUti home page	http://beast.bio.ed.ac.uk/BEAUti

See also BEAST

BEFORE STATE (BEFORE SPHERE)

Thomas D. Schneider

The high energy state of a molecular machine before it makes a choice is called its before state. This corresponds to the state of a receiver in a communications system before it has selected a symbol from the incoming message. The state can be represented as a sphere in a high dimensional space.

Further reading

Schneider TC (1991) Theory of molecular machines. I. Channel capacity of molecular machines. *J. Theor. Biol.* 148: 83–123.

See also Shannon Sphere, Channel Capacity, Gumball Machine

BELIEF NETWORK, *SEE* BAYESIAN NETWORK.

BEMIS AND MURCKO FRAMEWORK (MURCKO FRAMEWORK)

Bissan Al-Lazikani

Sometimes referred to only as the Murcko framework, is a representation of the chemical core of a compound. This is defined using an algorithm developed by Bemis and Murcko and includes all rings in a compound as well as all atoms and bonds that link these rings, but excludes all other atoms. See Figure B.4. Bemis and Murcko frameworks are widely used as a useful means to cluster chemically similar compounds.

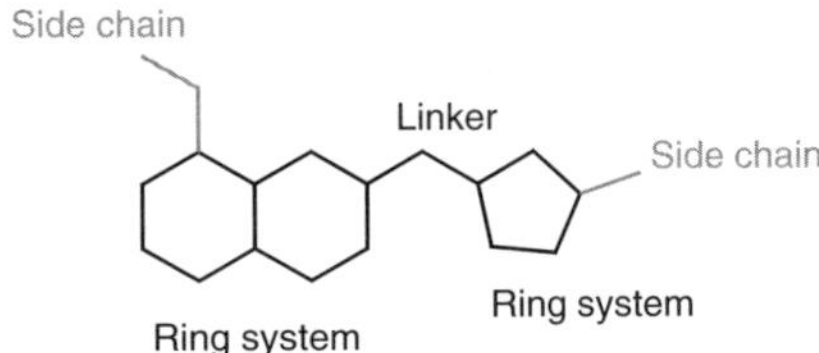

Figure B.4 An organic molecule composed of multiple ring systems and aliphatic linkers and side-chains. The Bemis and Murcko framework is defined by including all ring systems in the structure together with connecting bonds. All 'side-chains' are excluded.

Further reading

Bemis GW, Murcko MA. The properties of known drugs. Molecular frameworks. *J Med Chem.* 39(15):2887–2893. PMID: 8709122.

Shelat AA, Guy RK. Scaffold composition and biological relevance of screening libraries. *Nat Chem Biol.* 2007, 3(8): 44.

See also Scaffold

BEST-FIRST SEARCH

Concha Bielza and Pedro Larrañaga

This is a search method that does not terminate when the performance starts to drop but keeps a list of all candidate subsets evaluated so far, ranked by their performance measure, so that it can revisit an earlier configuration instead. As an example of search, feature subset selection involves searching the space of variables for the subset that predicts best the class. With many variables an exhaustive search is impractical and this is why other more efficient (not necessarily optimal) search methods exist. Given enough time, best-first search will explore the entire space, unless this is prevented by some stopping criterion. It is in general more efficient than the branch and bound method. However, the space complexity remains as a problem.

Further reading

Gopalakrishnan V, Lustgarten JL, Visweswaran S, Cooper GF (2009) Bayesian rule learning for biomedical data mining. *Bioinformatics* 26(5): 668–675.

Noy E, Goldblum A (2010) Flexible protein-protein docking based on best-first search algorithm. *Journal of Computational Chemistry* 31(9): 1929–1943.

See also Feature Subset Selection, Beam Search

BETA BARREL

Roman Laskowski

A beta sheet that curves around to close up into a single, cylindrical structure. Barrels most commonly consist of purely parallel or antiparallel sheets of six or eight beta strands.

Related websites

3Dee database	http://www.compbio.dundee.ac.uk/3Dee/
CATH server	http://www.cathdb.info/

Further reading

Buchanan SK, Smith BS, Venkatramani L, et al (1999) *Crystal structure of the outer membrane active transporter FepA from Escherichia coli: Nature Struc Biol* 6: 56–63.

See also Beta Sheet, Beta Strands

BETA BREAKER

Patrick Aloy

Beta breakers are amino acid residues that have been found to disrupt and terminate beta-strands. They were first split into two categories by Chou and Fasman: *breakers* (Lys, Ser, His, Asn, Pro) and *strong breakers* (Glu), depending on the frequency in which these residues were found in extended structures. More recent analyses performed on larger datasets of protein structures have changed the classification, the *breakers* now being Gly, Lys and Ser and the *strong breakers* Asp, Glu and Pro. It is worth noting that the structural parameters empirically derived do not always agree with those determined experimentally for individual proteins.

Further reading

Chou PY, Fasman GD (1974) Prediction of protein conformation. *Biochemistry* 13: 222–245.

Munoz V, Serrano L (1994) Intrinsic secondary structure propensities of the amino acids, using statistical phi-psi matrices: comparison with experimental scales. *Proteins* 20: 301–311.

BETA SHEET

Roman Laskowski and Tjaart de Beer

The second most commonly found regular secondary structure in globular proteins (after the alpha helix). The sheet is formed from a number of beta strands, lying side by side, and held together by hydrogen bonds between the backbone NH and CO groups of adjacent strands. Each strand is in a fully extended conformation. The sheet can be made from beta strands running parallel to one another (*i.e* in the same N-terminal to C-terminal sense), anti-parallel, or a mixture of the two, the most common being purely anti-parallel sheets.

Further reading

Branden C, Tooze J (1998) *Introduction to Protein Structure*. Garland Science, New York.

Lesk M (2000) *Introduction to Protein Architecture*. Oxford University Press, Oxford.

See also Secondary Structure, Globular, Alpha Helix, Beta Strands, Backbone, N-terminal, C-terminal

BETA STRAND

Roman Laskowski and Tjaart de Beer

Part of the polypeptide chain of a protein which, together with other beta strands, forms a beta sheet. The backbone in a beta strand is almost fully extended (in contrast to the alpha helix structure in which the backbone is tightly coiled).

Further reading

Branden C, Tooze J (1998) *Introduction to Protein Structure*. Garland Science, New York.

Lesk AM (2000) *Introduction to Protein Architecture*. Oxford University Press, Oxford.

See also Polypeptide, Beta Sheet, Backbone, Alpha Helix.

BIC, *SEE* BAYESIAN INFORMATION CRITERION.

BICLUSTERING METHODS

Pedro Larrañaga and Concha Bielza

Clustering methods can be applied to either the rows or the columns of a data matrix, separately. Biclustering methods (Hartigan, 1972) perform clustering in the two dimensions simultaneously. This means that clustering methods derive a global model for the whole matrix, while biclustering algorithms produce several local models.

For example in genomics, when clustering algorithms are used, each gene in a given gene cluster is defined using all the conditions. Similarly, each condition in a cluster is characterized by the activity of all the genes. However, each gene in a bicluster is selected using only a subset of the conditions and each condition in a bicluster is selected using only a subset of the genes. The goal of biclustering techniques is thus to identify subgroups of genes and subgroups of conditions, by performing simultaneous clustering of both rows and columns of the gene expression matrix, instead of clustering these two dimensions separately. We can then conclude that, unlike clustering algorithms, biclustering algorithms identify groups of genes that show similar activity patterns under a specific subset of the experimental conditions.

Biclustering algorithms can be classified (Madeira and Oliveira, 2004) according to the type of biclusters the algorithm is able to find, the bicluster structure provided by the algorithm, and the type of algorithms used for searching for the optimal biclusters.

Examples in Bioinformatics include works by Sheng et al. (2003), Tanay et al. (2002) and van Uitert et al. (2008).

Further reading

Hartigan JA (1972) Direct clustering of a data matrix. *Journal of the American Statistical Association* 67(337): 123–129.

Madeira SC, Oliveira AL (2004) Biclustering algorithms for biological data analysis: A survey. *IEEE/ACM Transactions on Computational Biology and Bioinformatics* 1(1): 24–45.

Sheng Q, Moreau Y, De Moor B (2003) Biclustering microarray data by Gibbs sampling. *Bioinformatics* 19 (Suppl.2): ii196–ii205.

Tanay A, Sharan R, Shamir R (2002) Discovering statistically significant biclusters in gene expression data. *Bioinformatics* 18 (Suppl. 1): S136–S144.

van Uitert M, Meuleman W, Wessels L (2008). Biclustering sparse binary genomic data. *Journal of Computational Biology* 15(10): 1329–1345.

BIFURCATION (IN A PHYLOGENETIC TREE)

Aidan Budd and Alexandros Stamatakis

An internal node of a phylogenetic tree at which an ancestral lineage diverges to yield only two offspring (or 'daughter') lineages. A tree that contains only bifurcations can be described as a bifurcating (or 'fully binary') tree. In unrooted binary trees, inner nodes are denoted as trifurcations, since they do not have a direction.

See also Multifurcation, Node

B

BINARY NUMERALS

John M. Hancock

A system for representing numbers using only two digits, typically 0 and 1. As examples, Table B.1 shows the numbers 1–10 in binary notation.

Table B.1 Binary Numeral Equivalents of the Base-10 Numbers 1 to 10

Number	Binary	Number	Binary
1	1	6	110
2	10	7	111
3	11	8	1000
4	100	9	1001
5	101	10	1010

Related website

Binary numeral system	http://en.wikipedia.org/wiki/Binary_numeral_system

BINARY RELATION

Robert Stevens

A relation between exactly two objects or concepts.

Most Knowledge Representation Languages are based on binary relationships. Binary relationships are easy to visualise as trees or graphs. (Systems of binary relationships are sometimes said to consist of Object-Attribute-Value – OAV – triples, where the attribute represents the relationship between the object and its value).

BINARY TREE, *SEE* DATA STRUCTURE.

BINDING AFFINITY (K_D, K_I, IC_{50})

Bissan Al-Lazikani and John M. Hancock

A measure of how 'tightly' a ligand (e.g. a drug or transcription factor) binds its receptor (e.g. enzyme active site or binding site). This can be derived experimentally by measuring the dissociation or inhibitory constants, K_d or K_i.

For any complex, K_d is the product of the concentrations of the free components divided by the concentration of the complex. K_i is the equivalent measure for an enzyme-inhibitor complex.

Binding affinity is also approximated through other experimentally derived measurements such as IC_{50} (concentration of compound required to inhibit 50% of enzyme activity).

Related websites

Dissociation constant	http://en.wikipedia.org/wiki/Dissociation_constant
Enzyme inhibition and K_i	http://en.wikipedia.org/wiki/Enzyme_inhibitor

BINDING SITE

Thomas D. Schneider

A binding site is a place on a nucleic acid that a recognizer (protein or macromolecular complex) binds. A classic example is the set of binding sites for the bacteriophage Lambda Repressor (cI) protein on DNA Ptashne et al. (1980). These happen to be the same as the binding sites for the Lambda *cro* protein.

Related websites

Hawaii figure	http://alum.mit.edu/www/toms/gallery/hawaii.fig1.gif
Hawaii	http://alum.mit.edu/www/toms/papers/hawaii/

Further reading

Ptashne M, et al. (1980) How the λ repressor and cro work. *Cell*, 19: 11–21.

Shaner MC (1993) Sequence logos: A powerful, yet simple, tool. In: TN Mudge, V Milutinovic, L. Hunter (eds) *Proceedings of the Twenty-Sixth Annual Hawaii International Conference on System Sciences, Volume 1: Architecture and Biotechnology Computing*, pp. 813–8121, Los Alamitos, CA. IEEE Computer Society Press. http://alum.mit.edu/www/toms/toms/paper/hawaii/

See also Binding Site Symmetry, Molecular Efficiency, Sequence Logo, Sequence Walker

BINDING SITE SYMMETRY

Thomas D. Schneider

Binding sites on nucleic acids have three kinds of symmetry:

Asymmetric: All sites on RNA and probably most if not all sites on DNA bound by a single polypeptide will be asymmetric. Examples: RNA: splice sites (Figure B.5); DNA: T7 RNA polymerase (Figure B.6).

Symmetric: Sites on DNA bound by a dimeric protein usually (there are exceptions) have a two-fold dyad axis of symmetry. This means that there is a line passing through the DNA, perpendicular to its long axis, about which a 180 degree rotation will bring the DNA helix phosphates back into register with their original positions. There are two places that the dyad axis can be set:

Odd symmetric: The axis is on a single base so that the site contains an odd number of bases. Examples: lambda cI and cro, and Lambda O (Figure B.7).

Even symmetric: The axis is between two bases, so that the site contains an even number of bases. Examples: 434 cI and cro, ArgR, CRP, TrpR, FNR, LexA (Figure B.7).

See also Binding Site

B

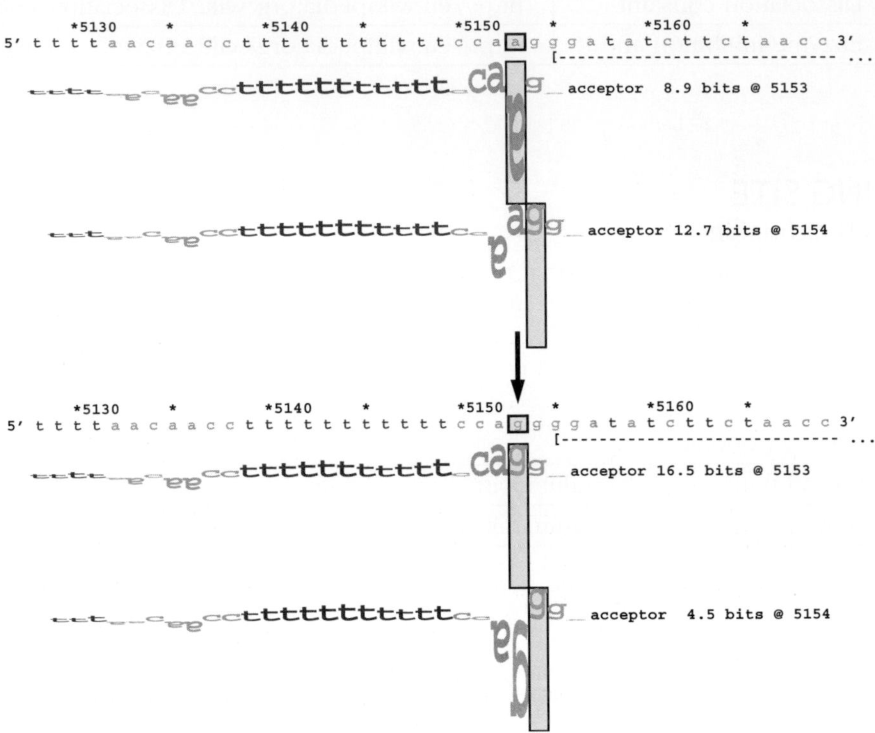

Figure B.5 Illustrations of asymmetrical binding sites.

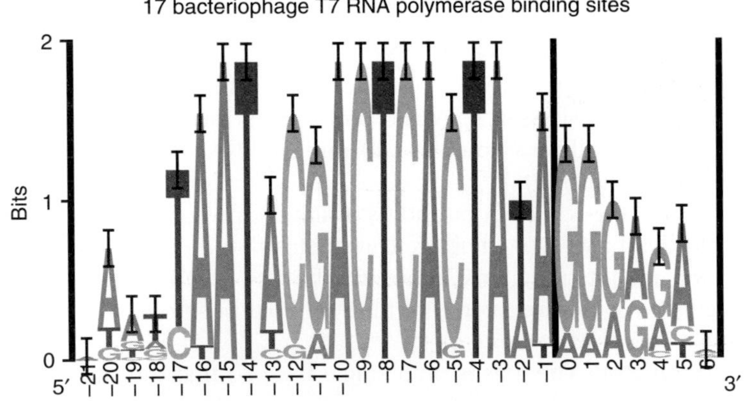

Figure B.6 Illustrations of a symmetrical binding site.

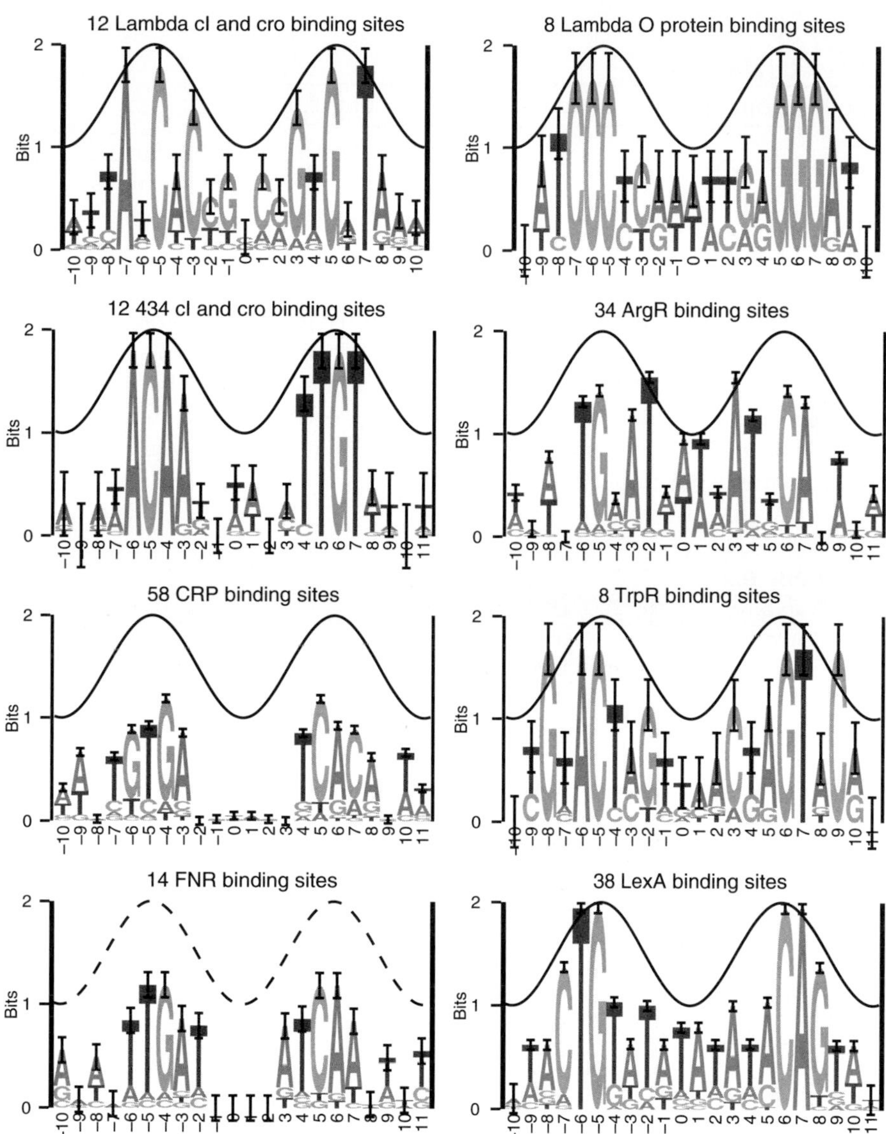

Figure B.7 Illustrations of odd binding sites.

B

BIO++

Michael P. Cummings

A set of C++ libraries for sequence analysis, phylogenetics, molecular evolution and population genetics.

Related website

Bio++ home page	http://biopp.univ-montp2.fr/

BIOACTIVITY DATABASE

Bissan Al-Lazikani

In the context of cheminformatics, these are databases that store chemical structures of compounds together with any measured biological activity. This can be, for example, a dissassociation constant of the compound from a protein, the % inhibition of enzymatic activity caused by the compound, the effect on cell growth of the compound or the affect on an entire organism. There are several open public and commercial databases. Table B.2 summarizes the major public bioactivity databases.

Table B.2

Database	Ref	URL, description
ChEMBL	1	https://www.ebi.ac.uk/chembldb/ Largest curated bioactivity data from the literature, contains dose-response in vitro measurements as well as functional and in vivo data
PubChem BioAssay	2	http://www.ncbi.nlm.nih.gov/pcassay Largely high throughput screening data. Both from literature and by online submission of user data.
ChemBank	3	http://chembank.broadinstitute.org/ Chemical biology data generated at the Broad Institute. Primarily cell-based/functional measurements.
BindingDB	4	http://www.bindingdb.org/ Dose response data from the literature
DrugBank	5	http://drugbank.ca/ Drug and clinical candidate data
PDBbind	6	http://sw16.im.med.umich.edu/databases/pdbbind/index.jsp Binding affinities of PDB ligands-protein complexes

These databases are useful resources for generating annotated virtual libraries for screening as well as considering chemical biology questions (e.g. what compound can I use to test the effect of my protein in cells). Care should be taken if downloading any data to avoid redundancy, as many of these databases now contain data from each other.

Further reading

Gaulton A, Bellis LJ, Bento AP, et al. (2012) ChEMBL: a large-scale bioactivity database for drug discovery. *Nucleic Acids Res 40* (Database issue): D110.

Liu T, Lin Y, Wen X, Jorissen RN, Gilson M (2007). BindingDB: a web-accessible database of experimentally determined protein-ligand binding affinities. *Nucleic Acids Res 35*(Database issue): D1980–1981.

Reymond JL, Awale M. (2012) Exploring chemical space for drug discovery using the chemical universe database. *ACS Chem Neurosci* 19;3(9): 6495–6497.

Wang R, Fang X, Lu Y, Wang S. (2004) The PDBbind database: collection of binding affinities for protein-ligand complexes with known three-dimensional structures. *J Med Chem* 3; 47(12): 2978–29790.

Wang Y, Xiao J, Suzek TO, et al. (2012) *PubChem's BioAssay Database. Nucleic Acids Res 40* (Database issue): D4001–4002.

See also Virtual Library

BIOINFORMATICS (COMPUTATIONAL BIOLOGY)

John M. Hancock

In general terms, the application of computers and computational techniques to biological data.

The field covers a wide range of applications from the databasing of fundamental datasets such as protein and DNA sequences, and even laboratory processes, to sophisticated analyses such as evolutionary modeling, to the modeling of protein structures and cellular networks. Areas related to bioinformatics are Neuroinformatics, the modelling of nervous systems, and Medical Informatics, the application of computational techniques to medical datasets. While the boundaries of these related disciplines are reasonably clearly drawn it is not unambiguously clear whether they form part of Bioinformatics however they are not included in this Dictionary. Systems Biology, the modeling of biological systems in general, is included here as a sub-discipline of Bioinformatics.

Bioinformatics may be regarded as a synonym for Computational Biology. It is the more often-used synonym in the UK and Europe, whereas Computational Biology is more commonly used in the USA, although these differences are not exclusive. A case could be made for defining Bioinformatics more narrowly in terms of the computational storage and manipulation (but not analysis) of biological information and Computational Biology as a more biology-oriented discipline aimed at learning new knowledge about biological systems. However given the difficulty of drawing this distinction at the margins, particularly as researchers may move seamlessly between the two areas, we have not adopted it here.

Bioinformatics draws on a range of disciplines including biochemistry, molecular biology, genomics, molecular evolution, computer science and mathematics.

Related websites

A Definition of Bioinformatics	http://bioinformatics.org/wiki/Bioinformatics
A Definition of Computational Biology	http://bioinformatics.org/wiki/Computational_biology

B

Further reading

Luscombe NM, et al. (2001) What is bioinformatics? A proposed definition and overview of the field. *Methods Inf Med* 40: 346–358.

THE BIOINFORMATICS ORGANIZATION, INC (FORMERLY BIOINFORMATICS.ORG)

Pedro Fernandes

Established in 1988, the Bioinformatics Organization, Inc. serves the scientific and educational needs of bioinformatic practitioners and the general public. It develops and maintains computational resources to facilitate world-wide communications and collaborations between people of all educational and professional levels. It provides and promotes open access to the materials and methods required for, and derived from, research, development and education.

It offers:

- Life sciences R&D: The Bioinformatics Organization provides group hosting services and a number of online tools, databases and forums, all free to use. Through its Bioinformatics Core Facility, it provides fee-based data analysis and software development services to cater to the needs of scientists and biomedical researchers.
- Education: In addition to the hosted resources, which are also available for educational use, the Bioinformatics Organization offers professional development courses for continuing scientific education. These are short, graduate- or postgraduate-level courses on computational topics of interest to prospective and practicing researchers in the life sciences.

Related website

Bioinformatics.org web site	http://bioinformatics.org/

BIOINFORMATICS TRAINING NETWORK, *SEE* BTN.

BIOLOGICAL IDENTIFIERS

Carole Goble and Katy Wolstencroft

Information about biological entities (e.g. genes, proteins, pathways, metabolites etc) exists in over 1000 different Life Science databases. These databases may have overlapping information, but none are identical. Despite attempts to introduce such schemes, there is no universal agreement on naming and identifying biological entities, so each database has its own naming scheme and set of identifiers for each biological entity it contains. Even databases that exchange content regularly (e.g. EMBL, Genbank and DDBJ) have different identifiers for entities in their own systems.

An identifier is a unique primary key for a database entry, which is typically a collection of letters and numbers in a recognisable format. For example, an identifier from GenBank is in the format GI:11265141 and an identifier from ENSEMBL is in the form ESNG: ENSG00000139618. Researchers can link entities across databases together by using cross references, so identifiers for each database must be stable and unchanging through the life of the database.

Identifiers.org is a system providing resolvable persistent URIs used to identify data for the scientific community, with a current focus on the Life Sciences domain. The provision of resolvable identifiers prepares biological datasets for the Linked Data initiative.

Related website

Identifiers.org	http://identifiers.org

See also Cross Reference, Linked Data

BIOMART

Obi L. Griffith and Malachi Griffith

BioMart is a freely available, open source, federated database system that provides unified access to disparate, geographically distributed data sources. It is designed to be data agnostic and platform independent, such that existing databases can easily be incorporated into the BioMart framework. As of version 0.8 BioMart includes three main portals. The 'Bio Portal' allows retrieval and search from COSMIC, Ensembl, MGI, VEGA, HapMap, HGNC, InterPro, UniProt, Reactome, Wormbase, and more. A 'Cancer Portal' provides access to ICGC/TCGA cancer data and a 'Mouse Portal' provides access to data from the International Knockout Mouse Consortium. Probably its most popular use in bioinformatics is as a user-friendly interface for accessing and interrogating Ensembl. A typical workflow involves choosing a database (e.g., Ensembl), selecting a dataset (e.g., Homo sapiens genes), restricting the search using filters (e.g., to a specific chromosome, region, or pre-defined list of genes) and finally defining attributes for output in the final results table (e.g., Gene ID, GO Terms, and protein domains). Queries can also be specified in XML format, SPARQL, or Java for programmatic access and output is typically provided as simple tab-delimited text files.

Related websites

BioMart	http://www.biomart.org/
BioMart – Ensembl	www.ensembl.org/biomart/martview

Further reading

Baker M (2011) Quantitative data: learning to share. *Nat Methods* 9(1): 39–41.

Kasprzyk A (2011) BioMart: driving a paradigm change in biological data management. *Database (Oxford). bar049.*

BIPARTITION, *SEE* SPLIT

B

BIT

Thomas D. Schneider

A bit is a binary digit, or the amount of information required to distinguish between two equally likely possibilities or choices. If I tell you that a coin is 'heads' then you learn one bit of information. It's like a knife slice between the possibilities. Likewise, if a protein picks one of the four bases, then it makes a two bit choice. For 8 things it takes 3 bits. In simple cases the number of bits is the log base 2 of the number of choices or messages M:

$$\text{bits} = \log_2 M$$

Claude Shannon figured out how to compute the average information when the choices are not equally likely. The reason for using this measure is that when two communication systems are independent, the number of bits is additive. The log is the only mathematical measure that has this property. Both of the properties of averaging and additivity are important for sequence logos and sequence walkers.

Even in the early days of computers and information theory people recognized that there were already two definitions of bit and that nothing could be done about it. The most common definition is 'binary digit', usually a 0 or a 1 in a computer. This definition allows only for two integer values. The definition that Shannon came up with is an average number of bits that describes an entire communication message (or, in molecular biology, a set of aligned protein sequences or nucleic-acid binding sites). This latter definition allows for real numbers. Fortunately the two definitions can be distinguished by context.

The idea that DNA can carry 2 bits per base goes back a long way. It was implied by Seeman, Rosenberg and Rich's (1976) famous paper that the major groove of DNA can support up to 2 bits of sequence conservation, while the minor groove can only support 1 bit, but practical application of this idea to molecular biology only came when it was discovered that RepA binding sites are strangely anomalous in this regard (Papp et al., 1993). More recent experiments by Lyakhov et al. (2001) have confirmed this prediction.

A byte is a binary string consisting of 8 bits.

Related websites

Primer	http://alum.mit.edu/www/toms/papers/primer/
Hawaii	http://alum.mit.edu/www/toms/papers/hawaii/
Nano2	http://alum.mit.edu/www/toms/papers/nano2/

Further reading

Lyakhov IG, et al. (2001) The P1 phage replication protein RepA contacts an otherwise inaccessible thymine N3 proton by DNA distortion or base flipping. *Nucl. Acid Res* 29: 489–490. http://alum.mit.edu/www/toms/papers/repan3/

Papp PP, et al. (1993) Information analysis of sequences that bind the replication initiator RepA. *J. Mol. Biol.*, 233: 21–30.

Schneider TD (1994) Sequence logos, machine/channel capacity, Maxwell's demon, and molecular computers: a review of the theory of molecular machines. *Nanotechnology*, 5: 11–8. http://alum.mit.edu/www/toms/papers/nano2/

Seeman NC, et al. (1976) Sequence-specific recognition of double helical nucleic acids by proteins. *Proc. Natl Acad. Sci. USA* 73: 804–808.

See also Byte, Digit, Molecular Efficiency, Nit

BLAST (MAXIMAL SEGMENT PAIR, MSP)

Jaap Heringa

BLAST (Basic Local Alignment Search Tool; Altschul *et al*., 1990) is a fast heuristic homology search algorithm which comprises five basic routines to search with a query sequence against a sequence database, including all combinations of nucleotide and protein sequences: (i) BLASTP compares an amino acid query sequence against a protein sequence database; (ii) BLASTN compares a nucleotide query sequence against a nucleotide sequence database; (iii) BLASTX compares the six-frame conceptual protein translation products of a nucleotide query sequence against a protein sequence database; (iv) TBLASTN compares a protein query sequence against a nucleotide sequence database translated in six reading frames, and (v) TBLASTX compares the six-frame translations of a nucleotide query sequence against the six-frame translations of a nucleotide sequence database.

The BLAST suite is the most widely used technique for sequence database searching that maintains sensitivity based on an exhaustive statistical analysis of ungapped alignments (Karlin and Altschul, 1990). The basic idea behind BLAST is the generation of all tripeptides from a query sequence and for each of those the derivation of a table of tripeptides deemed similar; the number of which is only a fraction of the total number possible. The BLAST program quickly scans a database of protein sequences for ungapped regions showing high similarity, which are called high-scoring segment pairs (HSP), using the tables of similar peptides. The initial search is done for a word of length W that scores at least the threshold value T when compared to the query using a substitution matrix.

Word hits are then extended in either direction in an attempt to generate an alignment with a score exceeding the threshold of S, and as far as the cumulative alignment score can be increased. The original BLAST program (Altschul et al., 1990) only detects local alignments without gaps, and therefore might miss some divergent but significant similarities. The current implementation, Gapped BLAST (Altschul et al., 1997), employs the dynamic programming technique to extend high-scoring segment pairs, leading to increased sensitivity. Word hits are first filtered using the so-called Two-Hit method, which selects pairs of HSPs separated by the same number of residues in either sequence, and this number should not exceed a preset threshold. The two-hit method typically filters out a large number of HSPs. For pairs of HSPs passing the two-hit method, a single matched residue pair is then selected as the optimal seed from which extension in each direction is performed using Dynamic Programming, thus allowing gaps in the local alignment. Extension is halted when the cumulative alignment score falls off by a quantity X from its maximum achieved value; (ii) the cumulative score goes to zero or below due to the accumulation of one or more negative-scoring residue alignments; or (iii) upon reaching the end of either sequence. A maximal-scoring segment pair (MSP) is defined as the highest scoring of all thus aligned segment pairs. The *T* parameter is the most important for the speed and sensitivity of the search resulting in the high-scoring segment pairs, from which the gapped local alignments are derived.

The BLAST algorithm provides a rigorous statistical framework (Karlin and Altschul, 1990) based on the extreme value-theorem to estimate the statistical significance of tentative homologues. The E value given for each database sequence found indicates the randomly expected number of sequences with an alignment score equal or greater than that of the sequence considered. Only if the value is lower than the user-selectable threshold (E value threshold) will the hit be reported to the user.

The most significant development for the BLAST suite is Position Specific Iterated BLAST (PSI-BLAST) (Altschul et al., 1997) which exploits increased sensitivity offered by multiple alignments and derived profiles in an iterative fashion. The program initially operates on a single query sequence by performing a gapped BLAST search. Then, the program takes the significant local alignments found, constructs a multiple alignment and abstracts a position specific scoring matrix (PSSM) from this alignment. This is a type of profile which is used to rescan the database in a subsequent round to find more homologous sequences. The scenario is iterated until the user decides to stop or the search has converged; i.e. no more significantly scoring sequences can be found in subsequent iterations. To optimize meaningful searching, query sequences are first scanned for the presence of so-called low-complexity regions (Wooton and Federhen, 1996), i.e. regions with a biased composition likely to lead to spurious hits, which are excluded from alignment. The web server for PSI-BLAST, located at http://www.ncbi.nlm.nih.gov/BLAST, http://www.ncbi.nlm.nih.gov/BLAST, enables the user to specify at each iteration round which sequences should be included in the profile, while by default all sequences are included that score below a user-set expectation value (E value). However, the user needs to activate every subsequent iteration. An alternative to the PSI-BLAST web server is a stand-alone version of the program, downloadable from the aforementioned WWW address, which allows the user to specify beforehand the desired number of iterations.

Although a consistent and very powerful tool, a limitation of the PSI-BLAST engine is the fact that no safeguards are in place to control the number of sequences added to the PSSM at each iterative step, i.e. all sequences having an E-value lower than a preset threshold are selected. It is clear that permissive E-value settings and/or erroneous alignments are likely to progressively drive the method into inclusion of false positives, a scenario dubbed profile wander. Also, the various BLAST programs are not generally useful for motif searching, for which specialized software has been developed.

For more specific searches, pattern hit initiated Blast (PHI-BLAST) can be used. In addition to a uery sequence, the user has to provide a pattern representing a sequence motif. The syntax for specifying the sequence motif is taken from the PROSITE database. PHI-BLAST will then only report hits that contain one or more copies of the sequence motif specified by the user. Another alternative to PSI-BLAST is reverse position specific BLAST (RPS-BLAST), also called reverse PSI-BLAST. This method searches a query protein sequence against a database of position specific scoring matrices (PSSMs, profiles) representing known conserved protein domains, thus revealing the conserved domain the query protein is the most similar to.

Related website

BLAST	http://www.ncbi.nlm.nih.gov/BLAST/

Further reading

Altschul SF, Gish W, Miller W, Meyers EW, Lipman DJ (1990) Basic local alignment search tool. *J. Mol. Biol.* 215: 40–10.

Altschul SF, Gish W (1996) Local alignment statistics. In: *Methods in Enzymology*, RF Doolittle (ed.), Vol. 266 pp. 46–80, Academic Press, San Diego.

Altschul SF, Madden TL, Schäffer AA., et al. (1997) Gapped BLAST and PSI-BLAST: a new generation of protein database search programs. *Nucl. Acids. Res.* 25: 338–402.

Karlin S, Altschul SF (1990) Methods for assessing the statistical significance of molecular sequence features by using general scoring schemes. *Proc. Natl. Acad. Sci. USA* 87: 226–268.

Wooton JC, Federhen S (1996) Analysis of compositionally biased regions in sequence databases. In: *Methods in Enzymology*, RF Doolittle (ed.), *Vol. 266*, pp. 55–71. Academic Press, San Diego.

See also Homology Search, High-Scoring Segment Pair, Maximal-Scoring Segment Pair, E Value

BLASTX

Roderic Guigó

A program of the BLAST suite for sequence comparison, BLASTX compares the translation of the nucleotide query sequence to a protein database.

Because BLASTX translates the query sequence in all six reading frames and provides combined significance statistics for hits to different frames, it is particularly useful when the reading frame of the query sequence is unknown or it contains errors that may lead to frame shifts or other coding errors. Thus a BLASTX search is often the first analysis performed when analysing an anonymous genomic sequence. Sequence conservation with known proteins may indicate the existence (and location) of coding exons on the genomic sequence.

BLASTX is particularly useful when analysing predicted ORFs in bacterial genomes, or ESTs. It is also useful for analyzing short anonymous eukaryotic sequences, although BLASTX searches do not resolve exon splice boundaries well. For large sequences (such as eukaryotic chromosomes), however, the BLASTX search may become computationally prohibitive.

Related website

NCBI BLAST Home Page	http://www.ncbi.nlm.nih.gov/BLAST/

Further reading

Gish W, States D (1993) Identification of protein coding regions by database similarity search. *Nature Genetics*, 3: 26–72

See also BLAST, Alignment, Gene Prediction, homology-based

B

BLAT (BLAST-LIKE ALIGNMENT TOOL)

J.M. Hancock

Program to rapidly align protein or nucleic acid sequences to a genome.

BLAT uses a heuristic similar to that used by BLAST to carry out rapid alignment of input sequences (protein or nucleic acid) to a genomic sequence. The program makes use of a catalogue of 11-mer oligonucleotide sequences that is small enough to be held in memory, making searches rapid. The program identifies matches between 11-mers in the probe sequence and 11-mers in the genomes and retrieves the genomic region likely to match the input sequence. It then aligns the sequence to the region and links together nearby alignments into a longer alignment. Finally it attempts to find small exons that might have been missed and attempts to adjust gap boundaries to correspond to splice sites. For protein searches BLAT uses 4-mers rather than 11-mers. The program is designed to find matches of better than 95% identity and length of 40 or more bases, or for proteins >80% similarity and length 20 or more amino acids.

Related website

BLAT search page	http://genome.ucsc.edu/cgi-bin/hgBlat

Further reading

Kent WJ. (2002) BLAT-the BLAST-like alignment tool. *Genome Res.* 12: 65–64.

See also BLAST

BLOSUM (BLOSUM MATRIX)

Laszlo Patthy

Whereas Dayhoff amino acid substitution matrices are based on substitution rates derived from global alignments of protein sequences that are at least 85% identical, the BLOSUM (BLOcks amino acid Substitution Matrix) matrices are derived from local alignments of conserved blocks of aligned amino acid sequence segments of more distantly related proteins. The primary difference is that in the case of distantly related proteins the Dayhoff matrices make predictions based on observations of closely related sequences, whereas the BLOSUM approach makes direct observations on blocks of distantly related proteins. The matrices derived from a database of blocks in which sequence segments are identical at 45% and 80% of aligned residues are referred to as BLOSUM 45, BLOSUM 80, and so on.

The BLOSUM matrices show some consistent differences when compared with Dayhoff matrices. Since the blocks were derived from the most highly conserved regions of proteins, the differences between BLOSUM and Dayhoff matrices arise from the more significant structural and functional constraint on conserved regions. The differences between BLOSUM and Dayhoff matrices are primarily due to the fact that the most variable, surface exposed regions of proteins (loops, β-turns) are underrepresented, and the highly conserved regions (secondary structure elements) that form the conserved core of protein folds are overrepresented in BLOSUM matrices. The relatively weak conservation of polar residues in the Dayhoff matrices is due to the fact that in the case of surface residues it is the hydrophilicity rather than the actual residue that is conserved.

BLOSUM matrices are used in the alignment of amino acid sequences to calculate alignment scores and in sequence similarity searches and are especially useful for the detection of distant homologies.

Related website

BLOCKS	http://blocks.fhcrc.org/blocks/

Further reading

Henikoff JG, Greene EA, Pietrokovski S, Henikoff S. (2000) Increased coverage of protein families with the blocks database servers. *Nucleic Acids Res.* 1;28(1):2282–2230.

Henikoff S, Henikoff TG (1992) Amino acid substitution matrix from protein blocks. *Proc. Natl. Acad Sci USA* 89: 1091–10919.

See also Amino Acid Exchange Matrix, Dayhoff Amino Acid Substitution Matrices, Global Alignment, Local Alignment, Alignment Scores, Sequence Similarity

BOLTZMANN FACTOR

Roman Laskowski and Tjaart de Beer

In any system of molecules at equilibrium, the number possessing an energy E is proportional to the Boltzmann factor $\exp^{(-E/kT)}$ where k is Boltzmann's constant and T is the temperature of the system.

It can be used in various dynamic studies of proteins.

Further reading

Bromberg S (2002) *Molecular Driving Forces: Statistical Thermodynamics in Chemistry and Biology.* Garland Science, New York.

BOOLEAN LOGIC

James Marsh, David Thorne and Steve Pettifer

Boolean logic is a form of algebra in which all values are either 1 (true) or 0 (false). Operators are defined, primarily AND (conjunction), OR (disjunction), and NOT (negation) that compare input values and produce an output, which in turn can be used as an input to further operators. Such logic is at the heart of modern digital electronics and most computer programming languages.

In programming languages boolean operators are often used in two distinct contexts: the first is to implement boolean propositional logic used to evaluate the truth of a statement combining other true or false values in order to decide whether to perform some further operation, for example 'if a and not (b or c) then perform x'. The other use is as operators on integer values in which case the operator applies separately to each corresponding pair of bits in two binary numbers to produce another binary number.

See also Binary, Propositional Logic

B

BOOSTING

Concha Bielza and Pedro Larrañaga

Boosting is a theoretically well-founded ensemble learning algorithm. The learning examples are weighted according to how difficult they are for being classified. Unlike bagging, the probability of selecting an example is not uniform, but proportional to its weight. Starting from equally weighted learning examples, boosting builds its first classifier. Depending on how this classifier predicts the class label of each learning example the weights are adjusted. The weights of examples correctly classified are decreased, whereas the weights of misclassified examples are increased. The usual weight adjustment is by multiplying the weight times $e/(1-e)$, where e is the classifier error. After adjusting all the weights, they are normalized. The process is repeated until the overall error becomes sufficiently small or large, the latter meaning that the remaining difficult examples are not solvable. These models are rejected as they tend, respectively, to overfit the learning data and to not contribute any useful knowledge. The remaining models form the ensemble in an ordered chain. For a new example, the prediction is given by a weighted voting, where the weight of each model is proportional to its performance (e.g. classification accuracy).

Boosting frequently yields better performance than bagging. Unlike bagging, it can be used with stable classifiers, with small variance.

AdaBoost is very popular and perhaps the most significant boosting algorithm. However, there are many more recent algorithms such as *LPBoost, TotalBoost, BrownBoost, MadaBoost, LogitBoost*, and others.

Further reading

Binder H, Schumacher M (2009) Incorporating pathway information into boosting estimation of high-dimensional risk prediction models. *BMC Bioinformatics* 10(1): 18.

Dettling M, Bühlmann P (2003) Boosting for tumor classification with gene expression data. *Bioinformatics* 19(9): 1061–1069.

Freund Y, Schapire RE (1997) A decision-theoretic generalization of on-line learning and an application to boosting. *Journal of Computer and System Sciences* 55(1): 119–139.

See also Bagging, Ensemble of Classifiers

BOOTSTRAP ANALYSIS, *SEE* BOOTSTRAPPING.

BOOTSTRAPPING (BOOTSTRAP ANALYSIS)

Aidan Budd and Alexandros Stamatakis

General bootstrap technique: A method used in statistics to computationally assess the uncertainty with which a parameter is estimated from a sample of data. Bootstrapping uses the original data sample as the basis for the generation of a certain number ('many') of pseudoreplicate (or 'pseudo-random') data samples on which the parameter of interest is estimated again. These samples are described as 'pseudoreplicates' as they are not obtained using a larger number of distinct samples, but are based on re-sampling from the original data sample.

When the original sample contains many data points, and these individual measurements are independent of each other (this is an important condition), then the distribution of values in the sampled data is expected to represent a good approximation of the distribution of values in a larger number of distinct samples from which the data at hand was obtained. In this case, the variability or repeatability of parameter estimates obtained from psuedoreplicate datasets represents a good approximation of the variability or repeatability with which these parameters are estimated in the original sample.

In phylogenetics, the parameter of interest is usually the tree topology and the sampled data are usually the sites of the multiple sequence alignment representing the set of operational taxonomic units under study. However, the approach can also be used to estimate the variability of other phylogenetic model parameters (e.g, the rate parameters of a nucleotide substitution matrix).

The most common application of the bootstrap in phylogenetics is the 'nonparametric bootstrap'. In this method, the original sequence data set (usually a multiple sequence alignment) is used to generate a large number (typically hundreds or thousands) of replicates by pseudo-randomly sampling columns/sites from the original dataset with replacement. Each replicate is a complete dataset containing the same number of sites and taxa as the original dataset. These replicates are then subject to the same analysis (e.g., tree search algorithm) and the estimates obtained are used to generate confidence intervals, variances, or other measures of the robustness of the initial inference. For instance, in tree reconstruction, each pseudo-random replicate is used to infer a phylogeny (using, e.g., a Maximum Likelihood algorithm) and the frequency of occurrence in the replicate tree set of bipartitions that are also contained in the original tree (e.g., the Maximum Likelihood tree inferred on the original dataset) is referred to as the bootstrap value for that branch (the bipartition induced by that branch). Hence, bootstrap values always refer to branches (bipartitions) of the tree and not to nodes. Alternatively, the information contained in a pseudoreplicate tree set can also be summarized directly by building consensus trees on the tree set. The question of how many pseudoreplicates need to be generated for obtaining stable estimates is still open. Empirical studies suggest that the number of replicates depends on the dataset at hand.

'Parametric bootstrapping', a less commonly used approach in phylogenetics, deploys the initial data set to obtain estimates of the parameters of a phylogenetic model, typically including the topology and branch lengths of a phylogenetic tree, along with other parameters of a statistical evolutionary model. Random samples are then drawn from a phylogenetic model that is parametrized using the values obtained from the analysis of the initial, original, dataset. The model parameters are used to create/simulate a large number of pseudoreplicate multiple sequence alignments. Parameters of interest can then be estimated from the pseudoreplicate data sets to analyze the variation with which they were estimated from the initial data set.

Software

Kumar S, et al. (2001) MEGA2: molecular evolutionary genetics analysis software. *Bioinformatics* 17: 124–245.

B

Stamatakis A (2006) RAxML-VI-HPC: Maximum Likelihood-based Phylogenetic Analyses with Thousands of Taxa and Mixed Models. *Bioinformatics* 22: 268–690.

Swofford DL (1998) *PAUP*: Phylogenetic Analysis Using Parsimony (and other methods)*. Sunderland, MA., Sinauer Associates.

Stamatakis A, Hoover P, Rougemont J (2008) A Fast Bootstrapping Algorithm for the RAxML Web-Servers. *Systematic Biology* 57(5): 75–7.

Further reading

Dopazo J (1994) Estimating errors and confidence intervals for branch lengths in phylogenetic trees by a bootstrap approach. *J Mol Evol* 38: 3003–3004.

Efron B (1982) *The Jackknife, the Bootstrap, and other Resampling Plans*. Society for Industrial and Applied Mathematics, Philadelphia, PA.

Efron B, Tibshirani R (1993) *An Introduction to the Bootstrap*. New York, Chapman & Hall.

Felsenstein J (1985) Confidence limits on phylogenies: an approach using the bootstrap. *Evolution* 39: 78–91.

Nei M, Kumar S (2000) *Molecular Evolution and Phylogenetics*. Oxford University Press, Oxford.

Pattengale ND, Alipour, Bininda-Emonds M, Moret ORP, Stamatakis, BME, (2010) How many bootstrap replicates are necessary? *Journal of Computational Biology* 17(3), 33–54.

Zhaxybayeva O, Gogarten JP. (2002) Bootstrap, Bayesian probability and maximum likelihood mapping: exploring new tools for comparative genome analyses. *BMC Genomics* 3: 4.

See also Phylogenetic Reconstruction, Jackknife, Consensus Tree, Split

BOTTLENECK, *SEE* POPULATION BOTTLENECK.

BOX

Thomas D. Schneider

Commonly used term for region of sequence with a particular function.

A sequence logo of a binding site will often reveal that there is significant sequence conservation outside such a 'box'. The term 'core' is sometimes used to acknowledge this, but sequence logos reveal that the division is arbitrary. A better usage is to replace this concept with binding site for nucleic acids or 'motif' for proteins. For example, in a paper by Margulies and Kaguni (1996), the authors use the conventional model that DnaA binds to 9 bases and they call the sites 'boxes'. However, in the paper they demonstrate that there are effects of the sequence outside the 'box'.

Further reading

Margulies C, Kaguni JM (1996) Ordered and sequential binding of DnaA protein to oriC, the chromosomal origin of *Escherichia coli*. *J. Biol. Chem.* 271: 17035–17040.

B

BRANCH (OF A PHYLOGENETIC TREE) (EDGE, ARC)

Aidan Budd and Alexandros Stamatakis

Lineages of taxonomic units that link nodes within a phylogenetic tree.

In a rooted tree, branches indicate direct transmission of genetic information from the taxonomic unit located at one end (parent) of the branch to the other (child). To determine the direction of transmission, consider removing the branch of interest from the tree to yield two unconnected subtrees, each of which contains only one of the nodes directly linked by the branch of interest. The subtree containing the root node contains the more ancestral of the two nodes linked by the branch of interest; hence, the direction of transfer of genetic information is from this more ancestral node to the other node linked by the branch of interest. For an unrooted tree, it is unknown which of the subtrees contains the root node, hence in unrooted trees the direction of transmission of genetic information is not specified.

If the branch is part of a scaled phylogenetic tree, then a value is associated with the branch that indicates some measure of the difference between the two taxonomic units directly linked by the branch; this value is often referred to as the 'branch length'. If the tree is 'unscaled', no such value is associated with the branch, that is, no branch length is specified or defined.

A branch that links two internal nodes is known as an internal, inner, or interior branch. Branches linking an internal and an external node are referred to as external branches (also terminal branches).

See also Phylogenetic Tree

BRANCH LENGTH, *SEE* BRANCH, BRANCH-LENGTH ESTIMATION.

BRANCH-LENGTH ESTIMATION

Sudhir Kumar and Alan Filipski

The process of estimating branch lengths in a given, fixed phylogenetic tree topology.

Given a set of sequences, their pairwise distances, and a phylogenetic tree, the branch lengths can be estimated by an ordinary least squares method, which chooses branch lengths so as to minimize the sum of squared differences between observed distances and patristic distances. The *patristic distance* between two sequences is the sum of branch lengths connecting them. A computationally efficient analytical formula to obtain this estimate was developed by Rzetsky and Nei (1993).

Alternatively, branch lengths may be estimated using the Maximum Likelihood method, in which the probability of observing the given sequences on a given, fixed tree topology is maximized. There may exist multiple local branch length optima on a fixed tree. With Bayesian methods one can sample the posterior distribution of the branches in a tree. Under the maximum parsimony model, algorithms are available to compute branch lengths by reconstructing the ancestral states and comparing the ancestral and descendant sequences for a given branch. However, this latter method can lead to underestimation because of multiple substitutions at individual sites.

Software

Kumar S, et al. (2001) *MEGA2*: molecular evolutionary genetics analysis software. *Bioinformatics* 17: 124–245.

Swofford DL (1998) *PAUP*: Phylogenetic Analysis Using Parsimony (and other methods).* Sinauer Associates, Sunderland, MA.

Maddison WP, Maddison DR (1992) *MacClade : analysis of phylogeny and character evolution.* Sinauer Associates, Sunderland, Mass.

Further reading

Chor B, Hendy MD, Holland BR, Penny D (2000) Multiple maxima of likelihood in phylogenetic trees: an analytic approach. *Molecular Biology and Evolution* 17(10): 152–541.

Felsenstein J (1981) Evolutionary trees from DNA sequences: a maximum likelihood approach. *J Mol Evol* 17: 36–87.

Nei M, Kumar S (2000) *Molecular Evolution and Phylogenetics.* Oxford University Press, Oxford.

Rzhetsky A, Nei M (1993) Theoretical foundation of the minimum-evolution method of phylogenetic inference. *Mol Biol Evol* 10: 107–195.

See also Phylogenetic Reconstruction

BTN (BIOINFORMATICS TRAINING NETWORK)

Pedro Fernandes

The Bioinformatics Training Network (BTN) is an independent organization of individuals that are involved in the provision of bioinformatics training in several ways. The goal is to tackle training challenges in bioinformatics, build pragmatic solutions to them, and provide mutual support. It seeks contribution from the bioinformatics community in the form of participation in its studies.

The BTN, as a community-based project, aims at to providing a centralised facility to share materials under the Creative Commons Attribution-Share Alike 2.5 License, to list training events (including course contents and trainers), and to share and discuss training experiences.

Participants in BTN have co authored several publications, mostly in the form of review papers. The BTN also prepares working documents such as white papers, guidelines and reports.

Related website

BTN	http://www.biotnet.org/

C

C-α (C-alpha)
Cancer Gene Census (CGC)
Candidate Gene (Candidate Gene-based Analysis)
Carboxy-Terminus, *see* C-Terminus.
CASP
Catalogue of Somatic Mutations in Cancer, *see* COSMIC
Catalytic Triad
Category
Causal Network, *see* Bayesian Network
CCDS (Consensus Coding Sequence Database)
CDS, *see* Coding Region.
Centimorgan
Centromere (Primary Constriction)
CGC, *see* Cancer Gene Census.
Channel Capacity (Channel Capacity Theorem)
Character (Site)
CHARMM
Chemical Biology
Chemical Hashed Fingerprint
Chemoinformatics
ChIP-seq
Chou & Fasman Prediction Method
Chromatin
Chromosomal Deletion
Chromosomal Inversion
Chromosomal Translocation
Chromosome
Chromosome Band
Circos, *see* Gene Annotation, visualization tools.
Cis-regulatory Element
Cis-regulatory Module (CRM)
Cis-regulatory Module Prediction
Clade (Monophyletic Group)
Cladistics
Clan
Classification
Classification in Machine Learning (Discriminant Analysis)
Classifier (Reasoner)
Classifiers, Comparison
ClogP
CLUSTAL
Clustal Omega, *see* Clustal.
ClustalW, *see* Clustal
ClustalX, *see* Clustal.
Cluster
Cluster Analysis, *see* Clustering.
Cluster of Orthologous Groups (COG, COGnitor)
Clustering (Cluster Analysis)
Clustering Analysis, *see* Clustering.

Concise Encyclopaedia of Bioinformatics and Computational Biology, Second Edition.
Edited by John M. Hancock and Marketa J. Zvelebil.

CNS

Code, Coding, *see* Coding Theory.

Coding Region (CDS)

Coding Region Prediction

Coding Statistics (Coding Measure, Coding Potential, Search by Content)

Coding Theory (Code, Coding)

Codon

Codon Usage Bias

Coevolution (Molecular Coevolution)

Coevolution of Protein Residues

Cofactor

COG, *see* Cluster of Orthologous Groups.

COGnitor, *see* Cluster of Orthologous Groups.

Coil (Random Coil)

Coiled-Coil

Coincidental Evolution, *see* Concerted Evolution.

Comparative Gene Prediction, *see* Gene Prediction, comparative.

Comparative Genomics

Comparative Modeling (Homology Modeling, Knowledge-based Modeling)

Complement

Complex Trait, *see* Multifactorial Trait.

Complexity

Complexity Regularization, *see* Model Selection.

Components of Variance, *see* Variance Components.

Compound Similarity and Similarity Searching

Computational Biology, *see* Bioinformatics.

Computational Gene Annotation, *see* Gene Prediction.

Concept

Conceptual Graph

Concerted Evolution (Coincidental Evolution, Molecular Drive)

Confidence, *see* Association Rule Mining.

Conformation

Conformational Energy

Conjugate Gradient Minimization, *see* Energy Minimization.

Connectionist Networks, *see* Neural Networks.

Consensus, *see* Consensus Sequence, Consensus Pattern, Consensus Tree.

Consensus Coding Sequence Database, *see* CCDS.

Consensus Pattern (Regular Expression, Regex)

Consensus Pattern Rule

Consensus Sequence

Consensus Tree (Strict Consensus, Majority-Rule Consensus, Supertree)

Conservation (Percentage Conservation)

Constraint-based Modeling (Flux Balance Analysis)

Contact Map

Contig

Continuous Trait, *see* Quantitative Trait.

Convergence

Coordinate System of Sequences

Copy Number Variation

Core Consensus

Correlation Analysis, *see* Regression Analysis, Association Rule Mining, Lift

COSMIC (Catalogue of Somatic Mutations in Cancer)

Covariation Analysis

CpG Island
CRM, *see* Cis-regulatory Module.
Cross-Reference (Xref)
Cross-Validation (K-Fold Cross-Validation, Leave-One-Out, Jackknife, Bootstrap)
C-Value, *see* Genome Size.

C

C-α (C-ALPHA)

Roman Laskowski and Tjaart de Beer

The central carbon atom, Cα, common to all amino acids to which is attached an amino group (NH_2), a carboxyl group (COOH), a hydrogen atom and a side chain. Each amino acid is distinguished from every other by its specific side chain.

Further reading

Branden C, Tooze J (1998) *Introduction to Protein Structure.* Garland Science, New York.

Lesk AM (2000) *Introduction to Protein Architecture.* Oxford University Press, Oxford.

See also Amino Acids, Side Chain

CANCER GENE CENSUS (CGC)

Malachi Griffith and Obi L. Griffith

The Cancer Gene Census is an ongoing effort to catalogue the subset of all genes that have been demonstrated to have acquired (somatic) or inherited (germline) mutations that lead to cancer. Currently approximately 1% of human genes have been implicated in cancer by observation of such mutations. Of these genes, 90% have somatic mutations, 20% have germline mutations, and 10% have both germline and somatic. The database includes a variety of mutation types including point mutations, small insertions and deletions, translocations and amplifications. The gene census list is freely available and at the time of writing contains 473 distinct genes. Each cancer gene is annotated with information regarding the types of mutations involved, the type of genetics (dominant, recessive), the cancer types and subtypes that arise by mutations of the gene, gene fusion partners, and more.

Related website

CGP homepage	http://www.sanger.ac.uk/genetics/CGP/Census/

Further reading

Futreal PA, et al. (2004) A census of human cancer genes. *Nat Rev Cancer.* 4(3):177–83.

CANDIDATE GENE (CANDIDATE GENE-BASED ANALYSIS)

Mark McCarthy, Steven Wiltshire and Andrew Collins

In the context of disease-gene mapping, used to denote a gene with a strong prior claim for involvement in determining trait susceptibility, usually based on a perceived match between the known (or presumed) function of the gene and/or its product, and the biology of the disease under study.

By extension, the term is often used (as in 'candidate-gene-based analysis'), to represent a particular strategy for susceptibility gene identification, which focuses on association analysis of biological candidates. This is often contrasted with approaches which rely on an unbiased genome-wide analysis (such as genome scans for linkage and association). Since, for many diseases, the basic pathophysiological mechanisms are unclear, the candidate approach has obvious limitations: most obviously, it will be unlikely to uncover susceptibility variants when they lie in pathways not previously suspected of a role in disease involvement. Increasingly, the boundaries between the 'candidate-gene' and 'genome-wide' approaches are becoming blurred as many gene-mapping efforts start from an initial genome scan, followed by detailed examination of all the positional candidates within linked/associated regions of interest. That is, they seek to identify the strongest candidates based on both chromosomal position and biology.

Examples: Multifactorial trait susceptibility loci which were initially identified through biological candidacy include the HLA region and the insulin gene (in type 1 diabetes) and Factor V Leiden (in deep vein thrombosis)

Related websites

Online Mendelian Inheritance in Man	http://www.ncbi.nlm.nih.gov/sites/entrez?db=omim
Genecanvas (candidate genes for cardiovascular disease)	http://genecanvas.idf.inserm.fr/news.php

Further reading

Collins FS (1995) Positional cloning moves from the perditional to traditional. *Nat Genet* 9: 347–350.

Halushka MK, et al. (1999) Patterns of single-nucleotide polymorphisms in candidate genes for blood-pressure homeostasis. *Nat Genet* 22: 239–247.

Hirschhorn JN, et al. (2002) A comprehensive review of genetic association studies. *Genet Med* 4: 45–61.

Tabor HK, et al. (2002) Candidate-gene approaches for studying complex genetic traits: practical considerations. *Nat Rev Genet* 3: 1–7.

See also Positional Candidate Approach, Genome Scan for Linkage, Genome-Wide Association, Linkage

CARBOXY-TERMINUS, *SEE* C-TERMINUS.

CASP

David Jones

CASP stands for Critical Assessment of Techniques for Protein Structure Prediction.

An ongoing international experiment, run every two years, to assess the state-of-the-art in protein structure prediction by blind testing.

One problem with benchmarking protein structure prediction methods is that if predictions are made on protein whose structure is already known then it is hard to be certain that prior knowledge of the correct answer did not influence the prediction. Published benchmarking results for prediction methods might not therefore be representative of the results that might be expected on proteins of entirely unknown structure.

To tackle this problem, John Moult and colleagues initiated an international experiment to evaluate the accuracy of protein structure prediction methods by blind testing. The first Critical Assessment in Structure Prediction (CASP) Experiment was in 1994, and has since been run every two years. In CASP, crystallographers and NMR spectroscopists release the sequences of as yet unpublished protein structures to the prediction community. Each research group attempts to predict the structures of these proteins and e-mail their predictions in before the structures are make available to the public. At the end of the prediction season, all of the predictions are collated and evaluated by a small group of independent assessors. In this way, the current state-of-the-art in protein structure prediction can be determined without any chance of bias or cheating. CASP has proven to be an extremely important development in the field of protein structure prediction, and has not only allowed progress in the field to be accurately estimated, but has also greatly stimulated interest in the field itself.

Related website

CASP	http://predictioncenter.org/

Further reading

Fischer D, Barrett C, Bryson K, et al. (1999) *Proteins: Structure, Function, and Bioinformatics* 79(S10): 1–207.

CATALOGUE OF SOMATIC MUTATIONS IN CANCER, *SEE* COSMIC

CATALYTIC TRIAD

Roman Laskowski and Tjaart de Beer

A group of three amino acid residues in an enzyme structure responsible for its catalytic activity. The three residues may be far apart in the amino acid sequence of the protein yet in the final, folded structure come together in a specific conformation in the active site to perform the enzyme's function on its substrate.

The best-known catalytic triad is the Ser-His-Asp triad of the serine proteinases and lipases, first identified in 1969. Its role is to cleave a given peptide substrate at a specific peptide bond. Specificity is governed by the substrate residue that fits into the P-subsite, or specificity pocket, immediately adjacent to the scissile bond. For example, in the serine proteinase trypsin, the peptide bond that is cut is the one downstream from an arginine or lysine residue, either of these fitting neatly into the specificity pocket.

The enzymes that use the Ser-His-Asp triad are a ubiquitous group responsible for a range of physiological responses such as the onset of blood clotting and digestion, as well as playing a major role in the tissue destruction associated with arthritis, pancreatitis and pulmonary emphysema.

A number of different protein families possess the Ser-His-Asp triad and, although their overall folds and topologies may differ substantially, the 3D conformation of the triad itself is remarkably well conserved, illustrating the importance of the precise 3D arrangement of these residues for catalytic activity to take place. What is more, because the protein structures differ so greatly, it is though the triad may have been arrived at independently as a consequence of convergent evolution.

Further reading

Barth A, Wahab M, Brandt W, Frost K (1993) Classification of serine proteases derived from steric comparisons of their active sites. *Drug Design and Discovery*, 10: 297–317.

Blow DM (1990) More of the catalytic triad. *Nature*, 343: 694–695.

Wallace AC, Laskowski RA, Thornton JM (1996) Derivation of 3D coordinate templates for searching structural databases: application to Ser-His-Asp catalytic triads in the serine proteinases and lipases. *Prot. Sci.* 5: 1001–1013.

See also Residue, Sequence, Peptide, Peptide Bond, Fold, Topology, Conformation

CATEGORY

Robert Stevens

Types of things, which are defined either through a collection of shared properties (an intensional definition), or enumeration (an extensional definition).

Country, *Person*, *Bacterium*, are all examples of categories. Categories can be thought of as having conditions that are necessary for an individual to be a member of that category. For example, a *Bachelor* is a man who has never married (strictly speaking). The conditions for category membership fall into two kinds:

1. Necessary condition: A condition that must be satisfied for an **individual** to be a member of that category. For example, a *Protein enzyme* is a *polymer-of Amino acids*, but not all *polymer- of Amino acids* are *Protein enzymes*.
2. Sufficient condition: A condition that must be satisfied for an **individual** to be a member of that category; in addition, fulfilling this condition is enough for category membership. For example, a *Protein enzyme* is a kind of *Protein* that *catalyses a Chemical reaction*. A Protein Enzyme must do this and what is more doing this is sufficient for a thing to be a *Protein enzyme*.

Categories that have such sufficiency conditions are known as *defined*. It is not possible to determine sufficiency conditions for all categories. Such categories are known as *natural kinds*. It is often true that no matter how many conditions are made or how carefully they are constructed, a member of the category will not fulfil those conditions. For example, an *Elephant* without tusks, a trunk and coloured pink is still an *Elephant*.

CAUSAL NETWORK, *SEE* BAYESIAN NETWORK

CCDS (CONSENSUS CODING SEQUENCE DATABASE)

Obi L. Griffith and Malachi Griffith

The Consensus CDS project is a collaborative effort to converge towards a consistent set of human and mouse gene annotations. This is necessary because there exist similar but not identical representations of genes, transcripts and proteins in the major genome browsers: NCBI, Ensembl and UCSC. CCDS assigns stable identifiers, tracks identical protein annotations between these systems, ensures that they are consistently represented, and coordinates manual review of any inconsistencies.

Related website

Consensus CDS database	homepage http://www.ncbi.nlm.nih.gov/CCDS/

Further reading

Pruitt KD, et al. (2009) The consensus coding sequence (CCDS) project: Identifying a common protein-coding gene set for the human and mouse genomes. *Genome Res.* 19(7):1316–1323.

CDS, *SEE* CODING REGION.

CENTIMORGAN

Katheleen Gardiner

Unit of distance between two genes in a genetic linkage map; based on recombination frequency.

One CM equals a 1% recombination frequency. The correlation between CM and physical distance in base pairs is not uniform within or between organisms. Where recombination frequency is high, one CM corresponds to fewer base pairs. In the human genome, one CM averages 1 Mb; in mouse, one CM averages 2 Mb.

Further reading

Liu B-H (1997) *Statistical Genomics: Linkage, Mapping and QTL analysis.* CRC Press.

Ott J (1999) *Analysis of human genetic linkage*. Johns Hopkins University Press, Baltimore.

Yu, et al. (2001) Comparison of human genetic and sequence-based physical maps. *Nature* 409: 951–953.

CENTROMERE (PRIMARY CONSTRICTION)

Katheleen Gardiner

The chromosomal region containing the kinetochore, where the spindle fibers attach during meiosis and mitosis. A centromere is essential for stable chromosome maintenance and for separation of the chromatids to daughter cells during cell division. In metaphase chromosomes, where the DNA is highly condensed, the centromere is seen as a narrowed or constricted region. The centromere defines the two chromosome arms, the chromatin (DNA plus associated proteins) above or below the centromere.

Relative centromere location categorizes a chromosome as acrocentric, the centromere at or near one end producing one very short and one long arm; metacentric, the centromere near the middle of the chromosome, and two arms of approximately equal size; or submetacentric, the centromere between the middle and one end, producing one long and one short arm. In human chromosomes, the short arm is referred to as the p arm, and the long, as the q arm.

Human centromeres are comprised of a large block of a complex family of repetitive DNA, called alphoid sequences. The basic alphoid sequences of ~170 basepairs are similar among chromosomes. They are arranged into tandem arrays and successively larger blocks of arrays to generate segments ranging from several hundred kilobases to several megabases in size that are chromosome specific.

Further reading

Miller OJ, Therman E (2000) Human Chromosomes. Springer.

Schueler MG, Higgens AW, Rudd MK, Gustashaw K, Willard HF (2001) Genomic and genetic definition of a functional human centromere. *Science* 294: 109–115.

Scriver CR, Beaudet AL, Sly WS, Valle D (eds) (2002) *The Metabolic and Molecular Basis of Inherited Disease*. McGraw-Hill, Maidenhead,UK.

Sullivan BA, Schwartz S, Willard HF (1996) Centromeres of human chromosomes. *Environ Mol Mutagen* 28: 182–191.

Wagner RP, Maguire MP, Stallings RL (1993) *Chromosomes – A Synthesis*. Wiley-Liss.

See also Kinetochore, Satellite DNA

CGC, *SEE* CANCER GENE CENSUS.

CHANNEL CAPACITY (CHANNEL CAPACITY THEOREM)

Thomas D. Schneider

The maximum information, in bits per second, that a communications channel can handle is the channel capacity:

$$C = W \log_2 \left(\frac{P}{N} + 1\right)_{\text{bits/second,}}$$

where W is the bandwidth (cycles per second, or hertz), P is the received power (joules per second) and N is the noise (joules per second). Shannon derived this formula by realizing that the received message can be represented as a sphere in a high dimensional space. The maximum number of messages is determined by the diameter of these spheres and the available space. The diameter of the spheres is determined by the noise and the available space is a sphere determined by the total power and the noise. Shannon realized that by dividing the volume of the larger sphere by the volume of the smaller message spheres, one would obtain the maximum number of messages. The logarithm (base 2) of this number is the channel capacity. In the formula, the signal-to-noise ratio is P/N. Shannon's channel capacity theorem states that if one attempts to transmit information at a rate R greater than C only at best C will be received. On the other hand if R is less than or equal to C then one may have as few errors as desired, so long as the channel is properly coded.

Further reading

Shannon CE (1948) A mathematical theory of communication. *Bell System Tech. J.*, 27: 379–423, 623–656. http://cm.bell-labs.com/cm/ms/what/shannonday/paper.html

Shannon CE (1949) Communication in the presence of noise. *Proc. IRE*, 37: 10–21.

See also Molecular Efficiency, Molecular Machine Capacity, Message, Shannon Sphere

CHARACTER (SITE)

Sudhir Kumar and Alan Filipski

A feature, or an attribute, of an organism or genetic sequence.

For instance, a backbone is an attribute of animals, which takes a value of 1 for vertebrates and 0 for all other animals. In a pair-wise or multiple molecular sequence alignment, each column (position or site in the sequence) corresponds to a character. The observed value of a character is called a character state. In molecular sequence analysis, a character can take either 4 (for DNA or RNA) or 20 (for amino acid) possible states, or can be absent because of an insertion or a deletion (indel) event for a particular sequence in an alignment. Modern phylogeny programs can also handle binary or multi-state morphological (where morphology is not described by simple presence/absence of features) data alignments as input. Also, in RNA alignments, pairs of RNA sites that form a stem can be grouped together into a single RNA secondary structure state to model that their evolution is linked. Also, three adjacent DNA nucleotides can be represented or condensed into a Codon character.

Further reading

Schuh, RT (2000) *Biological Systematics: Principles and Applications*. Ithaca, NY, Cornell University Press.

See also Sequence Alignment

CHARMM

Roland Dunbrack

A molecular dynamics simulation and modeling computer program developed by Martin Karplus and colleagues at Harvard ('Chemistry at HARvard Macromolecular Mechanics'). CHARMM uses an empirical molecular mechanics potential energy function for studying the structures and dynamics of proteins, DNA, and other molecules.

Related website

Alexander MacKerell webpage (CHARMM parameterization)	http://www.pharmacy.umaryland.edu/faculty/amackere/

Further reading

Brooks BR, et al. (1983) CHARMM: A program for macromolecular energy, minimization, and dynamics calculations. *J. Comput. Chem.* 4: 187–217.

Karplus M, McCammon JA (2002) Molecular dynamics simulations of biomolecules. *Nat Struct Biol* 9: 646–652.

MacKerell AD Jr., et al. (1998) All-atom empirical potential for molecular modeling and dynamics studies of proteins. *J. Phys. Chem.* B102: 3586–3616.

See also Comparative Modeling

CHEMICAL BIOLOGY

Bissan Al-Lazikani

The discipline of exploring biological systems using chemical probes. For example, analysing the phenotypic, genomic and proteomic effects on cancer cell lines before and after treatment with a series of compounds to decipher disease mechanisms.

Related website

Wikipedia	http://en.wikipedia.org/wiki/Chemical_biology

CHEMICAL HASHED FINGERPRINT

Bissan Al-Lazikani

A numerical representation of a chemical structure. This is typically a bit-array, which can be thousands of bits in length, where each bit is set corresponding to the existence of a chemical group, bond or element, e.g. a tetrahedral carbon, an ester bond etc. This allows rapid comparison of compounds using established profile or array comparison methods, such as Euclidean or Tanimoto distances, can be calculated rapidly. There are several fingerprints in use, most of which are proprietary, including MDL-keys, Unity2D and Daylight.

Illustration of a fingerprint (Figure C.1):

C	C≡N	-O-	=O	$-NH_2$	-OH	 ⬡	...
1	0	0	1	1	1	0	...

Figure C.1 Simple illustration of a binary fingerprint: the top row represents a series of chemical groups varying in complexity from single atoms to complex charge or ring systems. The bottom row represents the binary signature where 1 denotes that the chemical group exists in the compound being represented and 0 denotes that the group does not exist in the compound. Using these fingerprints, a large library of compounds can be represented in a consistent and easily comparable manner.

See also Compound Similarity and Similarity Searching, Euclidean Distance, Tanimoro Distance

CHEMOINFORMATICS

Bissan Al-Lazikani

The discipline of applying computational and mathematical techniques to chemistry. Typically, this is applied to organic chemistry in the context of biological research and drug discovery. It is a broad field that encompasses chemical library design, drug design, virtual screening, molecular modeling, QSAR and predictive ADME.

CHIP-SEQ

Stuart Brown

Next Generation DNA Sequencing (NGS) has been applied to the study of protein-DNA interactions. Chromatin-immunoprecipitation (ChIP) captures proteins by covalently cross-linking them to DNA at their sites of interaction using reagents such as formaldehyde applied to living cells. DNA can then be extracted from the cells and sheared into small fragments (200-500 bp). An antibody specific to a protein of interest is then used to immunoprecipitate protein-DNA complexes and the DNA fragments are released by reversing the cross-links. In earlier work, DNA fragments captured by ChIP were identified by hybridization to a microarray containing probes that correspond to known positions on the genome. ChIP-seq involves sequencing the captured DNA fragments on a NGS machine. The sequence reads are then mapped to a reference genome using alignment software optimized for NGS data. Positions on the genome where protein binding occurs are identified by software that detects clusters or 'peaks' in the mapping positions of the reads.

It has been shown that direct whole genome NGS of sheared chromatin does not yield a perfectly random distribution of fragments. If these DNA sequence reads are mapped to the genome and then screened with peak-finding software, clusters of reads are often detected at the promoters of transcribed genes and other non-random sites on the genome. Therefore, the accurate detection of ChIP-seq peaks (and the suppression of false-positives) requires a comparison between the distribution of mapped reads from the ChIP sample and an untreated 'input DNA' sample as a control for bias in the background distribution. This comparison between ChIP and input sample may be computed as a fold change, direct subtraction, or a more complex function of the two distributions.

The bioinformatics analysis becomes even more complex when the ChIP-seq method is used to detect changes in protein-DNA interactions as the result of biological treatments. Experiments of this design require biological replicates of each condition and a statistical method to detect significant changes in detected peaks due to treatment that is outside of the variance observed among replicates with the same treatment.

ChIP-seq has been applied to identify the binding sites of transcription factors (empirically identifying sites of promoters and enhancers), sites of RNA-polymerase activity, and epigenetic modifications of histone proteins. With some modifications ChIP-seq methdos have also been used to study DNAse hypersensitive sites (FAIRE-seq) and DNA-DNA interactions as a result of 3-dimensional chromatin conformation (4C).

Further reading

Johnson DS, Mortazavi A, et al. (2007) Genome-wide mapping of in vivo protein–DNA interactions. *Science* 316: 1497–1502.

Jothi, et al. (2008) Genome-wide identification of in vivo protein–DNA binding sites from ChIP-seq data. *Nucl Acids Res* 36(16): 5221–5231.

Zhang, et al. (2008) Model-based Analysis of ChIP-Seq (MACS). *Genome Biol* 9(9): R137.

CHOU & FASMAN PREDICTION METHOD

Patrick Aloy

The Chou-Fasman is a method for protein secondary structure prediction from sequence information. It was one of the first automated implementations and falls into the categories of 'first generation' structure prediction methods. It is based on single residue preferences derived from a very limited database (15 proteins). According to these preferences to adopt a given conformation, residues were classified as strong formers, formers, weak formers, indifferent formers, breakers and strong breakers.

The method uses empirical rules for predicting the initiation and termination of helical and beta regions in proteins (i.e. four helix formers out of six residues; three beta formers out of five) and then extends these segments in both directions until a tetra-peptide of breakers is found.

The authors claimed a three-state-per-residue accuracy of 77% when evaluated on the same set used to derive the parameters. Ulterior analyses on realistic benchmarks have shown the Chou-Fasman accuracy to be around 48%.

Related website

Chou & Fasman Program	http://www.biogem.org/tool/chou-fasman/

Further reading

Chou PY, Fasman GD (1974) Prediction of protein conformation. *Biochemistry* 13: 222–245.

See also Secondary Structure Prediction

CHROMATIN

Katheleen Gardiner

DNA within a chromosome complexed with proteins.

Euchromatin, comprising most of the chromosome material, contains transcribed genes, both active and inactive, and is in an expanded conformation during interphase. Heterochromatin, within and adjacent to centromeres, contains repetitive DNA and remains condensed during interphase.

Further reading

Miller OJ, Therman E (2000) *Human Chromosomes* (Springer).

Schueler MG et al. (2001) Genomic and genetic definition of a functional human centromere. *Science* 294:109–115.

Wagner RP et al. (1993) *Chromosomes – a synthesis* (Wiley-Liss)

CHROMOSOMAL DELETION

Katheleen Gardiner

Breakage and loss of a segment of a chromosome.

A deletion may be telomeric, where material is lost from the end of a chromosome, or internal. Large deletions (megabase in size) can be detected as a loss of, or and change in, the chromosome band pattern. Some specific deletions are associated with human genetic disease, for example Charcot Marie Tooth (chromosome 4) and DiGeorge syndrome (chromosome 22).

Further reading

Borgaonkar DS (1994) *Chromosomal Variation in Man*. Wiley-Liss.

Miller OJ, Therman E (2000) *Human Chromosomes*. Springer.

Scriver CR, Beaudet AL, Sly WS, Valle D (eds) (2002) *The Metabolic and Molecular Basis of Inherited Disease*. McGraw-Hill, Maidenhead, UK.

Wagner RP, Maguire MP, Stallings RL (1993) *Chromosomes – A Synthesis*. Wiley-Liss.

CHROMOSOMAL INVERSION

Katheleen Gardiner

Breakage within a chromosome at two positions, followed by rejoining of the ends after the internal segment reverses orientation.

Even if no material is lost, gene order is changed. A pericentric inversion occurs when the two breakpoints are located on opposite sides of the centromere; a paracentric inversion, when they are located on the same side of the centromere.

C

Further reading

Borgaonkar DS (1994) *Chromosomal Variation in Man*. Wiley-Liss

Miller OJ, Therman E (2000) *Human Chromosomes*. Springer.

Scriver CR, et al. (eds) (1995) *The Metabolic and Molecular Basis of Inherited Disease*. OMBID.

Wagner RP, et al. (1993) *Chromosomes – A Synthesis*. Wiley-Liss.

CHROMOSOMAL TRANSLOCATION

Katheleen Gardiner

Breakage of two nonhomologous chromosomes with exchange and rejoining of broken fragments.

A reciprocal translocation occurs when there is no loss of genetic material and each product contains a single centromere. Chromosome translocations can be identified by an abnormal karyotype, where the chromosome number is correct, but two chromosomes show abnormal sizes and banding patterns. Specific translocations are associated with some forms of malignancy.

Further reading

Borgaonkar DS (1994) *Chromosomal Variation in Man*. Wiley-Liss.

Miller OJ, Therman E (2000) *Human Chromosomes*. Springer.

Scriver CR, et al. (eds) (1995) *The Metabolic and Molecular Basis of Inherited Disease*.

Wagner RP, et al. (1993) *Chromosomes – A Synthesis*. Wiley-Liss.

CHROMOSOME

Katheleen Gardiner

A DNA molecule containing genes in an ordered linear sequence and complexed with protein.

In prokaryotes, one or more circular chromosomes generally contain the complete set of genes of the organism. In eukaryotes, the complete set of genes is divided among multiple linear chromosomes. The number, size and gene content of each chromosome is species specific and invariant among individuals of a species.

Structural features describing eukaryotic chromosomes include the centromere, the telomere and the long and short arms.

The DNA of the human genome is divided among 23 pairs of chromosomes, the autosomes, numbered in order of decreasing size from 1 to 22, plus the sex chromosomes, X and Y.

Individual chromosomes in a cell can be identified by a light/dark band pattern produced by various staining procedures. The number, size and chromosome band pattern of an organism defines its karyotype.

Further reading

Miller OJ, Therman E (2000) *Human Chromosomes*. Springer.

Wagner RP, et al. (1993) *Chromosomes – A Synthesis*. Wiley-Liss.

See also Centromere, Telomere, Chromosome Band

CHROMOSOME BAND

Katheleen Gardiner

Metaphase human chromosomes can be stained with a variety of reagents to produce a banding pattern visible under the light microscope.

Banding pattern plus chromosome size usually allows the unambiguous identification of individual normal chromosomes. For human chromosomes, a total of 450 bands is commonly achieved; 850 bands can be identified in longer, prophase chromosomes. Bands are numbered from the centromere.

When chromosomes are treated with trypsin or hot salt solutions and stained with Giemsa darkly staining regions are referred to as G bands, or Giemsa dark bands. R bands are produced when chromosomes are first heated in a phosphate buffer, and then stained with Giemsa. The pattern obtained is the reverse (hence R) of the G band pattern. T bands are the very bright R bands, resistant to heat denaturation, found most often, but not exclusively at the telomeres. Q bands are obtained by staining with quinacrine, with bright bands corresponding to those seen with Giemsa. C banding highlights centromeres by treatment with alkali and controlled hydrolysis.

The exact molecular basis of chromosomal band patterns is not clear, however, there is thought to be a relationship with regional base composition. G bands are regions that are high in AT content. R bands are variable in base composition, but the brightest R bands are the regions that are highest in GC content, and include most telomeric regions of human chromosomes. G bands are also gene-poor, LINE-rich, Alu-poor, CpG island-poor and late replicating. Opposite features are found in a subset of R bands.

Further reading

Holmquist GP (1992) Chromosome bands, their chromatin flavors and their functional features *Am J Hum Genet* 51: 17–37.

Miller OJ, Therman E (2000) *Human Chromosomes.* Springer.

Scriver CR. et al. (eds) (1995) *The Metabolic and Molecular Basis of Inherited Disease. OMBID.*

Standing Committee on Human Cytogenetic Nomenclature (1985) *An international System for Human Cytogenetic Nomenclature.* Karger.

Wagner RP, et al. (1993) *Chromosomes – a synthesis.* Wiley-Liss.

CIRCOS, *SEE* GENE ANNOTATION, VISUALIZATION TOOLS.

CIS-REGULATORY ELEMENT

Jacques van Helden

A cis-regulatory element is a short genomic region that exerts a positive or negative effect on the level of transcription of a neighbouring gene by mediating the interaction between sequence-specific DNA-binding transcription factors and the promoter of their target gene.

The prefix 'cis' refers to the genetics concept of a 'cis-trans' test: two genetic loci are said to interact '*in cis*' if the phenotype differs when they are separated by a chromosomal rearrangement (e.g. translocation). A typical example of a 'cis' effect is the action of a transcription factor binding site (TFBS) on the target gene. By contrast, two loci are said to interact '*in trans*' if their interaction does not depend on their respective chromosomal location. A typical example of 'trans' effect is the interaction between the transcription factor and its target gene.

See also Regulatory Region, Transcription Factor Binding Site, Cis-regulatory Module

CIS-REGULATORY MODULE (CRM)

Jacques van Helden

A cis-regulatory module is a genomic region that combines multiple cis-regulatory elements, mediating the interaction between several transcription factors and a promoter. CRMs are qualified of *homotypic* when they are essentially composed of multiple binding sites for a single transcription factor, or *heterotypic* if they combine sites bound by several distinct transcription factors. A CRM typically covers a few tens to a few hundreds base pairs.

The modularity of a CRM (i.e. its ability to act separately from its native region) can be established experimentally by measuring its capability to drive the expression of a reporter gene. A CRM is qualified of *enhancer* or *silencer* depending on whether it increases or decreases the transcription level of the target gene. The enhancing/silencing effect is typically measured by deletion analysis: a CRM will be qualified of enhancer (resp. silencer) if its deletion provokes a reduction (resp. increase) of the target gene. Enhancers can also be characterized by reporter gene experiments. The distinction between enhancers and silencers seems somewhat simplistic, since the same cis-regulatory module can enhance the expression of a target gene in some conditions (tissue, developmental time), and silence it in other conditions.

Related websites

TRANSCompel: Database of composite elements	http://www.gene-regulation.com/pub/databases/transcompel/compel.html
FlyReg: A database of cis-regulatory modules and cis-regulatory elements in the fruit fly Drosophila melanogaster	http://www.flyreg.org/

See also Cis-regulatory Element, Cis-regulatory Module Prediction

CIS-REGULATORY MODULE PREDICTION

Jacques van Helden

Computer-based analysis of DNA sequences to detect regions putatively acting as cis-regulatory modules (CRM).

Sequences are scanned with a user-provided set of transcription factor binding motifs (position-specific scoring matrices or consensus sequences). Predicted CRMs are the regions showing a higher density of predicted binding sites than expected by chance.

Related websites

Cluster-Buster (Scans DNA sequences to detect homotypic or heterotypic clusters of user-specified motifs)	http://zlab.bu.edu/cluster-buster/
matrix-scan (RSAT suite) (Scans DNA sequences for TFBS and predicts CRMs by detecting regions enriched in predicted cis-regulatory elements)	http://www.rsat.eu/
MCAST (Motif cluster alignment and search tool, part of the MEME suite)	http://meme.nbcr.net/meme/cgi-bin/mcast.cgi
MSCAN	http://cisreg.cmmt.ubc.ca/cgi-bin/mscan/MSCAN

Further reading

Su J, Teichmann S, Down TA (2010) Assessing computational methods of cis-regulatory module prediction. *PLoS Comput Biol* 6: e1001020.

See also Cis-regulatory Module (CRM), Transcription Factor Binding Site

CLADE (MONOPHYLETIC GROUP)

Aidan Budd and Alexandros Stamatakis

Clade or Monophyletic Group are terms that have their origins in the field of cladistics. Most bioinformatics-related uses of the term deploy it to refer to any of the subtrees associated with a given internal node of a phylogenetic tree that do not contain the root of the tree. Thus, for an unrooted tree (in which the position of the root is unspecified) it is not possible to distinguish which of the subtrees linked to a node are clades and which are not. Hence the use of the term 'clan' to refer to subtrees in the context of unrooted trees.

Further reading

Schuh RT (2000) *Biological Systematics: Principles and Applications*. Ithaca, NY, Cornell University Press.

See also Cladistics

CLADISTICS

Aidan Budd and Alexandros Stamatakis

An approach and set of associated methods used in the classification of taxonomic units, typically at the species level. These approaches are either based on the identification of taxonomic units that are more closely related to each other than they are to any other units and/or on the basis of the identification of shared common characters within a group of units that are derived with respect to other units. Such characters are described as synapomorphies. Many of the terms associated with cladistic classification have different definitions depending on whether the focus is on patterns of taxonomic relatedness or on the distribution of characters amongst taxa.

For example, the most common definition of 'monophyletic group' or 'clade' is a group of taxonomic units (both operational and hypothetical) that include an ancestral organism, all the descendant taxonomic units of that ancestor, and no others; however there is also another definition based on the distribution of characters within a group of taxonomic units, where the characters are believed to be shared due to their being derived with respect to other taxonomic units in the analysis. Non-monophyletic groups, depending on the shape of the tree and/or on the distribution of specific characters within the tree may be described as paraphyletic or polyphyletic.

An interesting discussion of cladistics is given by Felsenstein in the his book *Inferring Phylogenies* (Chapter 10 'A digression on history and philosophy').

Further reading

Felsenstein J (2004) A digression on history and philosophy. *Inferring Phylogenies* 123–146.

See also Clade, Synapomorphy

CLAN

Aidan Budd and Alexandros Stamatakis

A subtree of an unrooted tree.

The term is used to provide an analogous concept to a clade in the context of an unrooted tree; 'clade' only makes sense in the context of rooted phylogenies, as it requires knowledge or at least assertion of the position of the root in the phylogeny of interest.

Further reading

Wilkinson M, et al. (2007) Of clades and clans: terms for phylogenetic relationships in unrooted trees. *Trends Ecol Evol*, 22 114–115.

See also Adjacent Group, Cladistics, Clade

CLASSIFICATION

Robert Stevens

1. A grouping of concepts into useful categories or classes, not necessarily in a hierarchical fashion. For most purposes, a 'class' is the same as a category. In biology, *Nucleic acid*, *Protein*, *Gene*, would all be classes to which other concepts (classes or individuals) may belong. Often classes are arranged into a hierarchy of progressively more specific subclasses, but the primary nature of classification is the grouping of concepts.

2. The process of arranging individuals into classes or classes into a hierarchy of sub- and super-classes based on their definitions, often with the help of software (a reasoner or classifier) using formal logical criteria.

See also Classification in Machine Learning

CLASSIFICATION IN MACHINE LEARNING (DISCRIMINANT ANALYSIS)

Nello Cristianini

The statistical and computational task of assigning data points to one of k classes (for finite k), specified in advance.

This process is also known as Discriminant Analysis. The function that maps data into classes is called a 'classifier' or a 'discriminant function'. Particularly important is the special case of binary classification ($k = 2$), to which the general case $k > 2$ (multiclass) is often reduced.

The task of inferring a classifier function from a labeled data set is one of the standard problems in machine learning and in statistics.

Discriminant Analysis is also used for determining which variables are more relevant in forming the decision function, and hence for obtaining information about the 'reasons' for a certain classification rule.

Common approaches include linear Fisher Discriminant Analysis and nonlinear methods such as Neural Networks or Support Vector Machines.

Further reading

Cristianini N, Shawe-Taylor J (2000) *An Introduction to Support Vector Machines* Cambridge University Press, Cambridge.

Duda RO, Hart PE, Stork DG (2001 *Pattern Classification*, John Wiley & Sons, USA.

See also Machine Learning, Label, Clustering, Regression, Neural Networks, Support Vector Machine

CLASSIFIER (REASONER)

Robert Stevens

A mechanism, usually a software program, for performing classification.

See also Classification, Classifiers, Comparison

CLASSIFIERS, COMPARISON

Pedro Larrañaga and Concha Bielza

The tests used to assess the difference between two classifiers on a single data set and the differences over multiple data sets are not similar. When testing on a single data set, we usually compute the mean performance and its variance over repetitive training and testing on random samples of examples. Since these samples are usually related, much care is needed in designing the statistical procedures and tests that avoid problems with biased estimations of variance (Nadeau and Bengio, 2003).

Statistical tests for comparisons of two or more supervised classification algorithms, evaluated using classification accuracy, Area Under Curve or some other measure, on multiple data sets were reviewed by Demšar (2006). He recommends the use of robust non-parametric tests: the Wilcoxon signed ranks test (Wilcoxon, 1945) for comparison of two classifiers and the Friedman test (Friedman, 1940) for comparing more than two classifiers, this one with the corresponding post-hoc test (Nemenyi, 1963) if the null hypothesis is rejected.

When all classifiers are compared with a control classifier, we can instead of the Nemenyi test use one of the general procedures for controlling the family-wise error (the probability of making at least one Type I error in any of the comparisons) in multiple hypothesis testing, such as the Bonferroni correction (Dunn, 1961) or similar procedures, like the Holm (1979) and Hochberg (1988) tests.

Examples of applications in Bioinformatics include comparisons in microarray area (Hanczar and Dougherty, 2010) and in spectral domains (Menze et al., 2009).

Further reading

Demšar J (2006) Statistical comparisons of classifiers over multiple data sets. *Journal of Machine Learning Research* 7: 1–30.

Dunn OJ (1961) Multiple comparisons among means. *Journal of the American Statistical Association*. 56: 52–64.

Friedman M (1940) A comparison of alternative tests of significance for the problem of m rankings. *Annals of Mathematical Statistics* 11: 86–92.

Hanczar B, Dougherty ER (2010) On the comparison of classifiers for microarray data. *Current Bioinformatics* 5: 29–39.

Hochberg Y (1988) A sharper Bonferroni procedure for multiple tests of significance. *Biometrika* 75: 800–803.

Holm S (1979) A simple sequentially rejective multiple test procedure. *Scandinavian Journal of Statistics* 6: 65–70.

Menze BJ, et al. (2009) A comparison of random forest and its Gini importance with standard chemometric methods for the feature selection and classification of spectral data. *BMC Bioinformatics* 10: 213.

Nadeau C, Bengio Y (2003) Inference for the generalization error. *Machine Learning*, 52: 239–281.

Wilcoxon F (1945) Individual comparisons by ranking methods. *Biometrics* 1: 80–83.

CLOGP

Bissan Al-Lazikani

A theoretical calculation of LogP using the fragmented structure of a compound.

See also LogP

CLUSTAL

Jaap Heringa

Higgins and Sharp (1988) constructed the first implementation of the fast and widely used method CLUSTAL for multiple sequence alignment, which was especially designed for use on small workstations. The method follows the heuristic of the tree-based progressive alignment protocol, which implies repeated use of a pairwise alignment algorithm to construct the multiple alignment in *N*-1 steps (with *N* the number of sequences) following the order dictated by a guide tree. Speed of the early version of CLUSTAL was obtained during the pairwise alignments of the sequences through the Wilbur and Lipman (1983, 1984) algorithm. From the resulting pairwise similarities, a tree was constructed using the UPGMA clustering criterion, after which the sequences were aligned following the branching order of this tree. For the comparison of groups of sequences, Higgins and Sharp (1988) used consensus sequences to represent aligned subgroups of sequences and also employed the Wilbur-Lipman technique to match these.

Since its early inception, the CLUSTAL package has been subjected to a number of revision cycles. Higgins et al., (1992) implemented an updated version CLUSTAL V in which the memory-efficient dynamic programming routine of Myers and Miller (1988) is used, enabling the alignment of large sets of sequences using little memory. Further, two alignment positions, each from a different alignment, are compared in CLUSTAL V using the average alignment similarity score of Corpet (1988). The largely extended version CLUSTAL W (Thompson et al., 1994) uses the alternative Neighbour-Joining (NJ) algorithm (Saitou and Nei, 1984), which is widely used in phylogenetic analysis, to construct a guide tree. Sequence blocks are represented by a profile, in which the individual sequences are additionally weighted according to the branch lengths in the NJ tree. Further carefully crafted heuristics to optimise the exploitation of sequence information include: (i) local gap penalties, (ii) automatic selection of the amino acid substitution matrix, (iii) automatic gap penalty adjustment and (iv) a mechanism to delay the alignment of sequences that appear to be distant at the time they are considered.

The method CLUSTAL W does not provide the possibility to iterate the procedure as do several other multiple sequence alignment algorithms (e.g. Hogeweg and Hesper, 1984; Corpet, 1988, Gotoh, 1996; Heringa, 1999, 2002). An integrated user interface for the CLUSTAL W method has been implemented in CLUSTAL X (Thompson *et al.*, 1997), which is freely available and includes accessory programs for tree depiction. The CLUSTAL W method is the most popular method for multiple sequence alignment, and is reasonably robust for a wide variety of alignment cases. A thoroughly updated implementation is CLUSTAL Omega (Sievers et al., 2011), which has a much improved speed and very good alignment accuracy for large sequence sets.

Further reading

Corpet F (1988) Multiple sequence alignment with hierarchical clustering. *Nucl. Acids Res.* 16: 10881–10890.

Gotoh O. Sigificant improvement in accuracy of multiple protein sequence alignments by iterative refinement as assessed by reference to structural alignments. *J. Mol. Biol.* 264: 823–838.

Heringa J (1999) Two strategies for sequence comparison: Profile-preprocessed and secondary structure-induced multiple alignment. *Comp. Chem.* 23: 341–364.

C

Heringa J (2002) Local weighting schemes for protein multiple sequence alignment. *Comput. Chem.*, 26(5), 459–477.

Higgins DG, Bleasby AJ, Fuchs R (1992) CLUSTAL V: improved software for multiple sequnce alignment. *CABIOS* 8: 189–191.

Higgins DG, Sharp PM (1988) CLUSTAL: a package for performing multiple sequence alignment on a microcomputer. *Gene* 73: 237–244.

Hogeweg P, Hesper B (1984) The alignment of sets of sequences and the construction of phyletic trees: an integrated method. *J. Mol. Evol.* 20: 175–186.

Myers EW, Miller W (1988) Optimal alignment in linear space. *CABIOS* 4: 11–17.

Saitou N, Nei M (1987) The neighbor-joining method: a new method for reconstructing phylogenetic trees. *Mol. Biol. Evol* 4: 406–425.

Sievers F, Wilm A, Dineen DG, et al. (2011) Fast, scalable generation of high-quality protein multiple sequence alignments using Clustal Omega. *Molecular Systems Biology* 7: 539–544.

Thompson JD, Gibson TJ, et al. (1997). The ClustalX windows interface: flexible strategies for multiple sequence alignment aided by quality analysis tools. *Nucleic Acids Res* 25: 4876–4882.

Thompson JD, Higgins DG, Gibson TJ (1994) CLUSTAL W: improving the sensitivity of progressive multiple sequence alignment through sequence weighting, positions-specific gap penalties and weight matrix choice. *Nucleic Acids Res.* 22: 4673–4680.

Wilbur WJ, Lipman DJ (1983) Rapid similarity searches of nucleic acid and protein data banks. *Proc. Natl. Acad. Sci. USA* 80: 726–730.

Wilbur WJ, Lipman DJ (1984) The context dependent comparison of biological sequences. *SIAM J. Appl. Math.* 44: 557–567.

See also Tree-based Progressive Alignment, Sequence Similarity, Alignment, multiple

CLUSTAL OMEGA, *SEE* CLUSTAL.

CLUSTALW, *SEE* CLUSTAL

CLUSTALX, *SEE* CLUSTAL.

CLUSTER

Dov Greenbaum

A group of related objects such as sequences, genes or gene products.

The relatedness of a cluster is established through the use of different clustering methods and distance metrics; their distance from a seed sequence or gene determines a member's membership within a cluster. Clustering methods, many extracted from other disciplines, include hierarchical, self-organizing maps (SOM), self-organizing tree algorithm (SOTA) or K-means clustering. Distance methods include Euclidean distance or Pearson correlations. Clustering has become a basic tool of bioinformatics in the studying of microarray and other high throughput data.

Clustering has many uses. Most commonly it associates genes with other genes based on a specific feature such as mRNA expression. Based on this clustering, i.e. that genes are coregulated because they have the same expression patterns, one can then assign function to novel protein based on the functions of other proteins in the cluster: guilt by association.

Clustering can unsupervised, that is there is no additional information regarding the genes, or supervised, where information extraneous to the data being clustered is used to help cluster.

Further reading

Eisen MB, et al. (1998) Cluster analysis and display of genome-wide expression patterns. *Proc. Natl. Acad. Sci. USA* 95: 14863–14868.

Michaels GS, et al. (1998) Cluster analysis and data visualization of large-scale gene expression data. *Pac Symp Biocomput* 1998: 42–53.

CLUSTER ANALYSIS, *SEE* CLUSTERING.

CLUSTER OF ORTHOLOGOUS GROUPS (COG, COGNITOR)

Dov Greenbaum

The NCBI database of Clusters of Orthologous Groups (COGs) of proteins, contains groups of proteins clustered together on the basis of their purported evolutionary similarity.

Using an all-against-all sequence comparison of all the sequenced genomes, the authors found and clustered together all proteins that are more similar to each of the proteins in the group than any of the other proteins in their respective genomes. This technique was used to account for both slow and rapid evolution of proteins. The major disadvantage in this method is the clustering of proteins that may have only minimal similarity. Each cluster contains at least one protein from archea, prokaryotes and eukaryotes.

COGs in the database can be selected on the basis of one or any of the following criteria; function, text, species, or groups of species. COGs are named within the database on the basis of known functions (similar or otherwise) of the proteins in the group. These functions also include putative and predicted. Tools are also provided to search through the database using a gene/protein name, text string or organism.

The COG database provides three distinct forms of information for the user: (i) Functional and structural (two and/or three-dimensional structures) annotation of the proteins in the COG, (ii) Multiple alignments of the proteins in the COG allowing the user to identify and describe conserved regions in proteins of similar function, (iii) Phylogenetic patterns of each COG, allowing for the determination of whether a given organism is present or absent from a COG, thus possibly identifying missing genes in a particular pathway.

The database also provides the user with a tool to assign new proteins to a COG. The COGnitor compares the inputted sequence (FASTA or flat file format) with all the underlying sequence information in the database, and provides a suggested cluster based on these sequences.

Additional information is provided for many of the COGs including, (i) classification schemes (e.g. Transport Classification or Enzyme Commission numbers) (ii) gene names (iii) basis for the cluster – (i.e. motif, experimental data, operon structure, similarity based on PSI-BLAST (iii) domains contained in the clustered proteins (iv) COG structure information (e.g. the existence of sub-COGs) (v) protein notes (vi) background information (vii) predications (viii) references and (ix) modified or new protein sequences

Related website

COG	http://www.ncbi.nlm.nih.gov/COG/

Further reading

Tatusov RL, et al. (2001) The COG database: new developments in phylogenetic classification of proteins from complete genomes. *Nucleic Acids Res* 29: 22–28.

Tatusov RL, et al. (2003) The COG database: an updated version includes eukaryotes. *BMC Bioinformatics* 114(1),41.

Tatusov RL, Koonin EV, Lipman DJ (1997) A genomic perspective on protein families. *Science* 278: 631–637.

CLUSTERING (CLUSTER ANALYSIS)

Nello Cristianini

The statistical and computational problem of partitioning a set of data into 'clusters' of similar items, or in other words of assigning each data item to one of a finite set of classes not known *a priori*.

No external information about class membership is provided, hence making this an instance of an unsupervised learning algorithm. Often no assumption is made about the number of classes existing in the dataset.

This methodology is often used in exploratory data analysis, and in visualization of complex multivariate datasets, and is equally a topic of statistics and of machine learning.

Clustering is done on the basis of a similarity measure (or a distance) between data items, which needs to be provided to the algorithm. The criteria with which performance should be measured are less standardized than in other forms of machine learning (*see* Error Measures) but overall the goal is typically to produce a clustering of the data that maximizes some notion of within-cluster similarity and minimizes the between-clusters similarity (over all possible partitions of the data).

Since examining all possible organizations of the data into clusters is computationally too expensive, a number of algorithms have been devised to find 'reasonable' clusterings, without having to examine all configurations. Typical methods include hierarchical clustering techniques, that proceed by a series of successive mergers. At the beginning there are as many clusters as data items, then the most similar items are grouped, and these initial groups are merged according to their similarity, and so on. Eventually, this method would produce a single cluster, so that some stopping criterion needs to be specified.

Another popular heuristic is the one known as k-means clustering. In such a method, a set of k prototypes (or centers, or means) is found, and each point is assigned to the class of the nearest prototype. The method is iterative: first some partitioning into k-classes is found (possibly by choosing random 'seeds' as temporary centers). Then the means of the k classes so defined are computed. Then each data point is assigned to the nearest center, so finding another partitioning, that has new means, and so on, iterating until convergence. Notice that the final solution is dependent on the initial choice of centers.

See also Phylogeny Reconstruction Methods, Microarray

CLUSTERING ANALYSIS, *SEE* CLUSTERING.

CNS

Roland Dunbrack

A software CLUSTERING (CLUSTER ANALYSIS)CNS suite for determination and refinement of X-ray crystallographic and nuclear magnetic resonance structures developed by Axel Brünger and colleagues at Yale and Stanford. CNS (Crystallography and NMR System) has a hierarchical structure with an HTML interface, a structure determination language, and low-level source code that allow for new algorithms to be easily integrated.

Related website

CNS	http://cns-online.org/v1.3/

Further reading

Brunger AT, et al. (1998) Crystallography & NMR system: A new software suite for macromolecular structure determination. *Acta Crystallogr D Biol Crystallogr* 54 (Pt 5), 905–921.

Brunger AT, Adams PD (2002) Molecular dynamics applied to X-ray structure refinement. *Acc Chem Res* 35: 404–412.

See also X-ray Crystallography, NMR

CODE, CODING, *SEE* CODING THEORY.

C

CODING REGION (CDS)

Roderic Guigó

Region in genomic DNA or cDNA that codes for protein--either a fraction of the protein (coding exon) or the whole protein.

The term denotes only the regions in genomic DNA or cDNA that end up translated into amino acid sequences. That is, in spliced genes, the coding regions are the coding fraction of coding exons, and do not include untranslated exons, or the untranslated segments of the coding exons.

CDS (coding sequence) is synonym of coding region, and it is a feature key in the DDBJ/EMBL/GENBANK nucleotide sequence data banks. It is defined as 'sequence of nucleotides that corresponds with the sequence of amino acids in a protein (location includes stop codon); feature includes amino acid conceptual translation.' The CDS feature is used to obtain automatic amino acid translations of the sequences in the DNA databases, such as DAD from DDBJ, TREMBL from EMBL or GENPEPT from GENBANK.

Related websites

DDBJ/EMBL/GENBANK Feature Table Definition	http://www.ebi.ac.uk/embl/Documentation/FT_definitions/feature_table.html#components
TREMBL	http://www.ebi.ac.uk/trembl/

Further reading

Guigó R (1998) Assembling genes from predicted exons in linear time with dynamic programming. *J Comp Biol* 5: 681–702.

See also Coding Region Prediction

CODING REGION PREDICTION

Roderic Guigó

Identification of regions in genomic DNA or cDNA sequences that code for proteins.

Protein coding regions exhibit characteristic sequence composition bias. The bias is a consequence of the unequal representation of amino acids in real proteins, and of the unequal usage of synonymous codons (*see* Codon Bias). The bias can be measured (see Coding Statistics), and used to discriminate protein coding regions from non-coding ones.

Protein coding regions are usually delimited by sequence patterns (translation initiation and termination codons in prokaryotic organisms, acceptor and donor splice sites in addition in eukaryotes) that may also contribute to their identification. The identification of protein coding regions is the first, essential step once the genome sequence of an organism has been obtained – even if only partially.

See also Coding Region, Codon Bias, Coding Statistic, Gene Prediction, *ab initio*

CODING STATISTICS (CODING MEASURE, CODING POTENTIAL, SEARCH BY CONTENT)

Roderic Guigó

A coding statistic can be defined as a function that computes a real number related to the likelihood that a given DNA sequence codes for a protein (or a fragment of a protein).

Protein coding regions exhibit characteristic DNA sequence composition bias, which is absent on non-coding regions. The bias is a consequence of (1) the uneven usage of the amino acids in real proteins, and (2) of the uneven usage of synonymous codons (*see* Codon Bias). Coding statistics measure this bias, and thus help to discriminate between protein coding and non-coding regions.

Most coding statistics measure directly or indirectly either codon (or di-codon) usage bias, base compositional bias between codon positions, or periodicity (correlation) in base occurrence (or a mixture of them all). Since the early eighties, a great number of coding statistics have been published in the literature (*see,* for example, Gelfand, (1995) and references therein). Fickett and Tung, (1992) showed that all these measures reduce essentially to a few independent ones: the Fourier spectrum of the DNA sequence, the length of the longest ORFs, the number of runs (repeats) of a single base or of any set of bases, and in-phase hexamer frequencies.

Guigó (1998) distinguishes between measures dependent on a model of coding DNA and measures independent of such a model. The model of coding DNA is usually probabilistic, i.e, the probability distribution of codons, or di-codons in coding sequences, or more complex Markov models. To estimate the model, a set of previously known coding sequences from the genome of the species under consideration is required. This model is often very species-specific. Under the model one can compute the probability of a DNA sequence, given that the sequence codes for a protein. The probability of the DNA sequence can also be computed under the alternative non-coding model. Often the log-likelihood ratio of these two probabilities is taken as the coding measure. Examples of model dependent coding measures are Codon Usage (Staden and McLachlan, 1982), Amino-Acid Usage (McCaldon and Argos, 1988), Codon Preference (Gribskov et al., 1984), Hexamer Usage (Claverie et al., 1990), Codon Prototype (Mural et al., 1991), and Heterogeneous Markov Models (Borodovsky and McIninch, 1993).

In contrast, model independent measures do not require a set of previously known coding sequences from the species genome under study. They capture intrinsic bias in coding regions, which is very general, but they do not measure the direction of this bias, which is very species specific. Examples of model independent measures are Position Asymmetry (Fickett, 1982), Periodic Asymmetry (Konopka, 1990), Average Mutual Information (Herzel and Grosse, 1995) and the Fourier spectrum (Silverman and Linsker, 1986). Model independent coding measures are useful when no previous coding sequences are known for a given genome. Because the signal they produce is weaker than that produced by model dependent measures, usually longer sequences are required to obtain discrimination. This limits their utility mostly to prokaryotic genomes where genes are continuous ORFs.

Typically, coding statistics are computed on a sliding window along the query genomic sequence. This generates a profile in which peaks tend to correspond to coding regions and valleys to non-coding ones. Nowadays, coding statistics are used within *ab initio* gene prediction programs that resolve the limits between peaks and valleys at legal splice junctions. 5th order Markov models are among the most popular coding statistics used within gene prediction programs.

Two popular coding region identification programs are TestCode (Fickett, 1982) and GRAIL (Uberbacher and Mural, 1991).

Further reading

Borodovsky M, McIninch J (1993) GenMark: Parallel gene recognition for both DNA strands. *Comput Chem* 17: 123–134.

Claverie J-M, et al. (1990) k-Tuple frequency analysis: From intron/exon discrimination to T-cell epitope mapping. *Methods Enzymol* 183,: 237–252.

Fickett JW (1982) Recognition of protein coding regions in DNA sequences. *Nucleic Acids Res* 10: 5303–5318.

Fickett JW, Tung C-S (1992) Assessment of protein coding measures. *Nucleic Acids Res* 20: 6441–6450.

Gelfand MS (1995) Prediction of function in DNA sequence analysis. *J Comput Biol* 1: 87–115.

Gribskov M, et al, (1984) The codon preference plot: graphic analysis of protein coding sequences and prediction of gene expression. *Nucleic Acids Res* 12: 539–549.

Guigó R (1998) Assembling genes from predicted exons in linear time with dynamic programming. *J Comput Biol* 5: 681–702.

Guigó R (1999) DNA composition, codon usage and exon prediction. In: M. Bishop (ed.) *Genetic Databases*, pp 53–80, Academic Press.

Herzel H, Grosse I (1995) Measuring correlations in symbol sequences. *Physica A* 216: 518–542.

Konopka AK (1990) Towards mapping functional domains in indiscriminantly sequenced nucleic acids: A computational approach. In: R Sarma, M Sarma (eds) *Structure and Methods: VI. Human Genome Initiative and DNA Recombination*, pp. 113–125. Adenine Press, Guilderland, New York.

Konopka AK (1999) Theoretical molecular biology. In: Meyers RA (ed.) *Encyclopedia of Molecular Biology and Molecular Medicine*, pp 37–53, John Wiley and Sons Inc, New York.

McCaldon P, Argos P (1988) Oligopeptide biases in protein sequences and their use in predicting protein coding regions in nucleotide sequences. *Proteins* 4: 99–122.

Mural RJ, et al. (1991) Pattern recognition in DNA sequences: the intron-exon junction problem. In: CC Cantor, HA Lim (eds) *Proceedings of the First International Conference on Electrophoresis, Supercomputing and the Human Genome*, pp 164–172. World Scientific Co, Singapore.

Silverman BD, Linsker R (1986) A measure of DNA periodicity. *J Theor Biol* 118: 295–300.

Staden R, McLachlan AD (1982) Codon preference and its use in identifying protein coding regions in long DNA sequences. *Nucleic Acids Res* 10: 141–156.

Uberbacher EC, Mural RJ (1991) Locating protein-coding regions in human DNA sequences by a multiple sensor-neural network approach. *Proc Natl Acad Sci USA* 88: 11261–11265.

See also Coding Region, Coding Region Prediction, Codon Usage Bias, GRAIL, Gene Prediction, *ab initio*, GENSCAN, GENEID

CODING THEORY (CODE, CODING)

Thomas D. Schneider

Coding is the transformation of a message into a form suitable for transmission over a communications line.

This protects the message from noise. Since messages can be represented by points in a high dimensional space (*see* Message), the coding corresponds to the placement of the messages relative to each other. When a message has been received, it has been distorted by noise, and in the high dimensional space the noise distorts the initial transmitted message point in all directions evenly. The final result is that each received message is represented by a sphere. Picking a code corresponds to figuring out how spheres should be placed relative to each other so that they are distinguishable. This situation can be represented by a gumball machine. Shannon's famous work on information theory was frustrating in the sense that he proved that codes exist that can reduce error rates to as low as one may desire, but he did not say how this could be accomplished. Fortunately a large effort by many people established many kinds of communications codes. One of the most famous coding theorists was Hamming. An example of a simple code that protects a message against error is the parity bit. There are many codes in molecular biology besides the genetic code since every molecular machine has its own code.

Further reading

Hamming RW (1986) *Coding and Information Theory*. Prentice-Hall, Inc., Englewood Cliffs NJ.

Shannon CE (1949) Communication in the presence of noise. *Proc. IRE*, 37: 10–21.

See also Molecular Efficiency, Shannon Sphere

CODON

Niall Dillon

Three base unit of the genetic code which specifies the incorporation of a specific amino acid into a protein during mRNA translation.

Within a gene, the DNA coding strand consists of a series of three base units which are read in a 5' to 3' direction in the transcribed RNA and specify the amino acid sequence of the protein encoded by that gene. The genetic code is degenerate with most of the twenty amino acids represented by several codons. Degeneracy is highest in the third base. Initiation and termination of translation are specified by initiator and terminator codons. Codon usage shows significant variation between phyla.

Further reading

Krebs JE (ed.) (2009) *Lewin's Genes X*. Jones and Bartlett Publishers, Inc.

See also Genetic Code

C

CODON USAGE BIAS

Roderic Guigó

Non-randomness in the use of synonymous codons in an organism.

In all genetic codes there are 64 codons encoding about 20 amino acids (sometimes one more or a few less), and a translation termination signal. Often, therefore, more than one codon encodes the same amino acid or the termination signal. Codons that encode the same amino acid are called synonymous codons. Most synonymous codons differ by only one base at their 3' end position. Usage of synonymous codons is non-random. This non-randomness is called codon usage bias. The bias is different in different genomes, which have different 'preferred' synonyms for a given amino acid.

Codon usage bias depends mostly on mutational bias; in other words codon usage reflects mostly the background base composition of the genome. In some prokaryotic organisms, it appears to correlate also with the relative abundance of alternative tRNA isoacceptors (Ikemura, 1985). In general, highly expressed genes tend to show higher codon usage bias.

There are a number of measures of Codon Usage Bias for a given gene. The Codon Adaptation Index, CAI (Sharp and Li, 1987) is one such measure. CAI is a measurement of the relative adaptiveness of the codon usage of a gene toward the codon usage of highly expressed genes. See http://www.molbiol.ox.ac.uk/cu/Indices.html for a list of Codon Usage Indices.

Related websites

Codon Usage Database	http://www.kazusa.or.jp/codon/
Genetic Codes Resource at NCBI	http://www.ncbi.nlm.nih.gov/Taxonomy/taxonomyhome.html/index.cgi?chapter=cgencodes
Analysis of Codon Usage	http://artedi.ebc.uu.se/course/UGSBR/codonusage1.html
Correspondence Analysis of Codon Usage	http://codonw.sourceforge.net/

Further reading

Ikemura T (1985) Codon usage and tRNA content in unicellular and multicellular organisms. *Mol Biol Evol* 2: 13–34.

Sharp PM, et al. (1993) Codon usage: mutational bias, translational selection or both? *Biochem. Soc. Trans.* 21: 835–841.

Sharp PM, Li W-H (1987) The codon adaptation index a measure of directional synonymous codon usage bias, and its potential applications. *Nucleic Acids Res* 15: 1281–1295.

See also Coding Region, Coding Region Prediction, Coding Statistics

COEVOLUTION (MOLECULAR COEVOLUTION)

Laszlo Patthy

Coevolution is a reciprocally induced inherited change in a biological entity (species, organ, cell, organelle, gene, protein, biosynthetic compound) in response to an inherited change in another with which it interacts. Protein–protein coevolution and protein–DNA coevolution are the phenomena describing the coevolution of molecules involved in protein-protein and DNA–protein interactions, respectively. Coevolution of protein residues describes the coevolution of pairs of amino acid residues that interact within the same protein. Coevolution is reflected in cladograms of the interacting entities: the cladograms of coevolving entities are congruent.

Further reading

Dover GA, Flavell RB (1984) Molecular coevolution: DNA divergence and the maintenance of function. *Cell* 38(3), 622–623.

See also Protein–protein Coevolution, Protein–DNA Coevolution, Coevolution of Protein Residues

COEVOLUTION OF PROTEIN RESIDUES

Laszlo Patthy

In general, coevolution is a reciprocally induced inherited change in a biological entity in response to an inherited change in another with which it interacts.

The three-dimensional structure of a protein is maintained by several types of non-covalent interactions (short-range repulsions, electrostatic forces, Van der Waals interactions, hydrogen bonds, hydrophobic interactions) and some covalent interactions (disulfide bonds) between the side-chain or peptide-backbone atoms of the protein. There is a strong tendency for the correlated evolution of interacting residues of proteins. For example, in the case of extracellular proteins, pairs of cysteines forming disulfide bonds are usually replaced or acquired 'simultaneously' (on an evolutionary timescale). Such correlated evolution of pairs of cysteines involved in disulfide-bonds frequently permits the correct prediction of the disulfide-bond pattern of homologous proteins.

Bioinformatic tools have been developed for the identification of coevolving proteins residues in aligned protein sequences. The identification of protein sites undergoing correlated evolution is of major interest for structure prediction since these pairs tend to be adjacent in the three-dimensional structure of proteins.

Further reading

Pollock DD, et al. (1999) Coevolving protein residues: maximum likelihood identification and relationship to structure. *J Mol Biol* 287: 187–198.

Pritchard L, et al. (2001) Evaluation of a novel method for the identification of coevolving protein residues. *Protein Eng* 14: 549–555.

Shindyalov IN, et al. (1994) Can three-dimensional contacts in protein structures be predicted by analysis of correlated mutations? *Protein Eng* 7: 349–358.

Tuff P, Darlu P. (2000) Exploring a phylogenetic approach for the detection of correlated substitutions in proteins. *Mol Biol Evol* 17: 1753–1759.

Yip KY, Patel P, Kim PM, Engelman DM, McDermott D, Gerstein M. (2008) An integrated system for studying residue coevolution in proteins. *Bioinformatics* 24: 290–292.

See also Coevolution

COFACTOR

Jeremy Baum

Cofactors are small organic or inorganic molecules that are bound to enzymes and are active in the catalytic process. Examples include flavin mononucleotide (FMN), flavin adenine dinucleotide (FAD), nicotinamide adenine dinucleotide (NAD), thiamine pyrophosphate (ThP or TPP), coenzyme A, biotin, tetrahydrofolate, vitamin B_{12}, haem, chlorophyll and a variety of iron-sulphur clusters. With the help of cofactors, the enzymes can catalyse chemistry not available with the limited range of amino acid side chains.

COG, *SEE* CLUSTER OF ORTHOLOGOUS GROUPS.

COGNITOR, *SEE* CLUSTER OF ORTHOLOGOUS GROUPS.

COIL (RANDOM COIL)

Roman Laskowski and Tjaart de Beer

Region of irregular, unstructured structure of a protein's backbone, as distinct from the regular regions characterized by specific patterns of main chain hydrogen bonds. Not to be confused with the loop region of a protein.

Further reading

Branden C, Tooze J (1999) *Introduction to Protein Structure.*, Garland Science, New York.

Lesk AM (2000) *Introduction to Protein Architecture.* Oxford University Press, Oxford.

See also Secondary Structure, Main Chain, Loop.

C

COILED-COIL

Patrick Aloy

Coiled-coils are elongated helical structures that contain a highly specific heptad repeat that conforms to a specific sequence pattern. The first and fourth positions within this pattern are hydrophobic while the remaining positions show preferences for polar residues.

Coiled-coils form intimately associated bundles of long alpha-helices and are usually associated with specific functions. Coiled-coil segments should be removed from protein sequences before performing database searches as they can produce misleading results due to their repetitive pattern.

Many programs have been developed to predict coiled-coil segments in protein sequences, *COILS* being one of the most widely used.

The *COILS* Program was developed by Lupas et al. (1991) to predict coiled-coil regions from protein sequences. It assigns a per-residue coiled-coil probability based on the comparison of its flanking sequences with sequences of known coiled-coil proteins.

A more novel method is Multicoil2 by Trigg et al (2011).

Related websites

COILS	http://embnet.vital-it.ch/software/COILS_form.html
Multicoil2	http://groups.csail.mit.edu/cb/multicoil2/cgi-bin/multicoil2.cgi

Further reading

Lupas A, Van Dyke M, Stock J. (1991) Predicting coiled coils from protein sequences. *Science* 252: 1162–1164.

Trigg J, Gutwin K, Keating AE, Berger B (2011) Multicoil2: Predicting coiled coils and their oligomerization states from sequence in the twilight zone. *PLoS ONE* 6(8)

See also Alpha-Helices, Pattern, Hydrophobic, Polar

COINCIDENTAL EVOLUTION, *SEE* CONCERTED EVOLUTION.

COMPARATIVE GENE PREDICTION, *SEE* GENE PREDICTION, COMPARATIVE.

COMPARATIVE GENOMICS

Jean-Michel Claverie

A research area, and the set of bioinformatic approaches, dealing with the global comparison of genome information in term of their structure and gene content.

The information gathered from whole genome comparisons may include:

- the conservation of sequences across species, (e.g. from highly conserved genes to 'ORFans')
- the identification of orthologous and paralogous genes
- the conservation of non-coding regulatory regions
- the conservation of local gene order (co-linearity, synteny)
- the co-evolution patterns of genes (e.g. subset of genes simultaneously retained or lost in genomes)
- the identification of gene fusion/splitting events (e.g. the Rosetta stone method)
- evidence of lateral gene transfer.

Comparative genomic approaches have been used to address a number of very diverse problems such as estimating the number of human genes, identifying and locating human genes, proposing new whole-species phylogenetic classification, predicting functional links between genes from their co-evolution patterns or fusion ('Rosetta Stone') events, recognizing lateral gene transfer, or identifying specific adaptation to exotic life-styles. More information is extracted about each individual sequence as new complete genome sequences become available. Among the most significant discoveries resulting from comparative genomics are i) the large fraction of genes of unknown function, and ii) the fact that every genome, even from small parasitic bacteria, contains a proportion of genes without homologues in any other species.

Related websites

High-Quality Automated and Manual Annotation of Microbial Proteomes (HAMAP)	http://hamap.expasy.org/
Clusters of Orthologous Groups	http://www.ncbi.nlm.nih.gov/COG/

Further reading

Cambillau C, Claverie JM (2000) Structural and genomic correlates of hyperthermostability. *J. Biol. Chem.* 275: 32383–32386.

Marcotte EM et al. (1999) Detecting protein function and protein-protein interactions from genome sequences. *Science* 285 751–753.

Mushegian AR (2007) Foundations of Comparative Genomics. Academic Press, London.

Pellegrini M, et al. (1999) Assigning protein functions by comparative genome analysis: protein phylogenetic profiles. *Proc. Natl. Acad. Sci. USA.* 96: 4285–4288.

Roest Crollius H, et al. (2000) Characterization and repeat analysis of the compact genome of the freshwater pufferfish Tetraodon nigroviridis. *Nat. Genet.* 25: 235–238.

Snel B, et al. (1999) Genome phylogeny based on gene content. *Nat. Genet.* 21: 108–110.

Wilson MD, et al. (2001) Comparative analysis of the gene-dense ACHE/TFR2 region on human chromosome 7q22 with the orthologous region on mouse chromosome 5. *Nucleic Acids Res.* 29,1352–1365.

Worning P, et al. (2000) Structural analysis of DNA sequence: evidence for lateral gene transfer in *Thermotoga maritima*. *Nucleic Acids Res.* 28: 706–709.

COMPARATIVE MODELING (HOMOLOGY MODELING, KNOWLEDGE-BASED MODELING)

Roland Dunbrack

The prediction of a three-dimensional protein structure from a protein's amino acid sequence based on an alignment to a homologous amino acid sequence from a protein with an experimentally solved three-dimensional structure. Such models are relatively accurate and form a rational basis for explaining experimental observations, redesigning proteins to change their performance, and rational drug design.

General steps involved in modelling, after the identification of a known three-dimensional structure of a homolog of the protein that is being modelled, are:

1. Alignment of the target protein to the template protein sequence
2. Transfer of the coordinates of aligned atomic positions.
3. Building of side-chains using rotamer libraries.
4. Building of INDELS (insertion/deletions) or variable (loop) regions.
5. Local and global energy minimization.
6. Assessing the correctness of the fold using programs such as ProCheck or PROSAII.

There are a number of approaches to modeling, including fragment-based modeling, averaged template modeling and single-template modeling. The accuracy of a modeled structure depends on the accuracy of the alignment and the amount and quality of biological knowledge of the target proteins and is influenced by the evolutionary distance separating the sequences.

An alternative strategy is based on deriving distance constraints from the known structure and using these to build a model by simulated annealing or distance geometry methods (see MODELLER).

Related websites

Swiss-Model	http://swissmodel.expasy.org/
ESyPred3D	http://www.fundp.ac.be/sciences/biologie/urbm/bioinfo/esypred/
MODELLER	http://salilab.org/modeller/
YASARA	http://www.yasara.org/

Further reading

Greer J (1990) Comparative modeling methods: application to the family of the mammalian serine proteases. *Proteins* 7: 317–34.

Sali A (1995) Modelling mutations and homologous proteins. *Curr. Opin. Biotech.* 6: 437–451.

Sali A, Blundell TL (1993) Comparative protein modelling by satisfaction of spatial restraints. *J. Mol. Biol.* 234: 779–815.

See also Alignment, Homology, Indel

C

COMPLEMENT

John M. Hancock

As the DNA double helix is made up of two base-paired strands, the sequences of the two strands are not identical but have a strict relationship to one another. For example, a sequence 5'-ACCGTTGACCTC-3' on one strand pairs with the sequence 3'-TGGCAACTGGAG-5'. This second sequence is the known as the complement of the first. Converting a sequence to its complement is a simple operation and may be required, for example, if a sequence has been read in the 3'->5' direction rather than in the conventional direction.

See also DNA Sequence, Reverse Complement

COMPLEX TRAIT, *SEE* MULTIFACTORIAL TRAIT.

COMPLEXITY

Thomas D. Schneider

Like 'specificity', the term 'complexity' appears in many scientific papers, but it is not always well defined. When one comes across a proposed use in the literature one can unveil this difficulty by asking: How would I measure this complexity? What are the units of complexity? An option is to use Shannon's information measure, or explain why Shannon's measure does not cover what you are interested in measuring, then give a precise, practical definition. Some commonly used measures of the complexity of asystem, such as a genetic network, are the numbers of components or interactions in the system. (See also Kolmogorov Complexity: Li & Vitanyi, 1997.)

Further reading

Li M, Vitányi P (1997) *An Introduction to Kolmogorov Complexity and Its Applications.* Springer-Verlag, New York, second edition.

COMPLEXITY REGULARIZATION, *SEE* MODEL SELECTION.

COMPONENTS OF VARIANCE, *SEE* VARIANCE COMPONENTS.

COMPOUND SIMILARITY AND SIMILARITY SEARCHING

Bissan Al-Lazikani

Compound similarity is a measure that indicates how similar two distinct compounds are. The units of the measure are dependent on the method used and can be a percentage similarity, a P-value or a method dependant score. By implication, similar compounds have similar properties and are more likely to bind similar target proteins or have similar binding activities (although there are frequent exceptions to this). Chemical similarity can be computed in many different ways, and can use the 2D or 3D structures of the compounds. Since compound similarity searches are often computed against a large chemical database (akin to a BLAST sequence similarity search), they pose a significant computational time challenge. To overcome this, similarity searching tools often represent compounds as fingerprints. This enables them to use array or matrix comparison methodologies, respectively, to rapidly compute the similarity between any two compounds, e.g. using Tanimoto similarity. All commercial and public chemical data and software suppliers have methods to allow these comparisons.

Other similarity calculations methods can employ graph theory to compare graph representation of the compounds (e.g. RASCAL), electrostatic maps (e.g. EON); shape and volume maps (e.g. ROCS).

Related websites

EON	http://www.eyesopen.com/eon
ROCS	http://www.eyesopen.com/rocs

See also Chemical Hashed Fingerprint, Tanimoto Distance

COMPUTATIONAL BIOLOGY, *SEE* BIOINFORMATICS.

COMPUTATIONAL GENE ANNOTATION, *SEE* GENE PREDICTION.

CONCEPT

Robert Stevens

The International Organization for Standardization defines a concept as a unit of thought.

In practice in computer science we are referring to the encoding of that unit of thought within a system. The word 'concept' is used ambiguously to refer either to categories of things e.g. country, or the individual things themselves e.g. England. On the whole it is better to avoid the use of 'concept' and use *category* and *individual* as defined elsewhere.

See also Category, Individual

C

CONCEPTUAL GRAPH

Robert Stevens

A graphical representation of logic originated by John Sowa based on the 'Existential Grants' graphs proposed by the American philosopher Charles Stanley Pierce.

Subsets of the formalism are often used as an easily understandable notation for representing concepts. Computer implementations for various subsets are available and there is an active user group, but since the full formalism is capable of expressing not just first order but also higher order logics, full proof procedures are computationally intractable. Excellent information is available on line and an interesting web site and there is an active user community.

Related website

Sowa's Conceptual Graphs pages	http://www.jfsowa.com/cg/index.htm

CONCERTED EVOLUTION (COINCIDENTAL EVOLUTION, MOLECULAR DRIVE)

Dov Greenbaum

Concerted evolution explains how it may often seem the case that paralogous genes within one species are more closely related to each other than to members of the same gene family in another species when the gene duplication event that led to the paralogs in multiple genomes preceded the speciation event.

Concerted evolution is a mode of long-term evolution within a multigene family whereby members of the gene family become homogenized over evolutionary time, ie. they undergo genetic exchange, causing their sequence evolution to be concerted over a temporal period.

This results in greater sequence similarity within a species than between species when comparing members of gene families or repetitive sequence families such as satellite DNAs. Concerted evolution is thought to be the result of DNA replication, recombination and repair mechanisms, e.g. gene conversion and unequal crossing over. It requires homogenization – the horizontal transfer of a gene within the genome, and fixation, the spread of the new sequence combination to all (or most) individuals of the species.

Concerted evolution is thought to help spread advantageous mutations throughout a genome. Conversely, though, it erases prior evidence of divergence and the history of gene duplication.

The most striking example of concerted evolution involves ribosomal RNA (rRNA) genes. The rRNA genes constitute a large gene family in most organisms (approximately 400 members in a typical mammalian genome). Yet rRNA genes are typically identical or nearly so within species.

Further reading

Arnheim N, et al. (1980) Molecular evidence for genetic exchanges among ribosomal genes on nonhomologous chromosomes in man and apes. *Proc Natl Acad Sci USA* 77. 7323–7327.

Carson AR, Scherer SW (2009) Identifying concerted evolution and gene conversion in mammalian gene pairs lasting over 100 million years. *BMC Evol Biol* 9: 156.

Dover G (1982) Molecular drive: a cohesive mode of species evolution. *Nature*. 299 111–117.

Liao D (1999) Concerted evolution: molecular mechanism and biological implications. *Am J Hum Genet* 64: 24–30.

See also Birth-and-Death Evolution

CONFIDENCE, *SEE* ASSOCIATION RULE MINING.

CONFORMATION

Roman Laskowski and Tjaart de Beer

A three-dimensional arrangement of atoms making up a particular molecule, or part thereof. Different conformations of the same molecule are those involving re-arrangements of atoms that do not break any covalent bonds or alter the chirality of any of the atoms. Thus, two rotamers of a given protein side chain are different conformations, whereas the D- and L-isomers of an amino acid are not (they correspond to different configurations).

Further reading

Branden C, Tooze J (1998) *Introduction to Protein Structure*. Garland Science New York.

See also Chirality, Rotamers, Side Chain, Structure 3D

CONFORMATIONAL ENERGY

Roland Dunbrack

Potential energy of a flexible molecule dependent on the structure or conformation. In macromolecular modeling programs such as CHARMM and AMBER. this energy is usually calculated as a function of bond lengths, bond angles, dihedral angles, and non-bonded interatomic interactions. For reasonably small molecules, this energy can also be calculated with quantum mechanics packages.

Related website

Alexander MacKerell webpage (CHARMM parameterization)	http://www.pharmacy.umaryland.edu/faculty/amackere/

Further reading

Cornell WD, et al. (1995) A second generation force field for the simulation of proteins, nucleic acids, and organic molecules. *J. Am. Chem. Soc.* 117: 5179–5197.

MacKerell AD, Jr., et al. (1998) All-atom empirical potential for molecular modeling and dynamics studies of proteins. *J. Phys. Chem.* B102: 3586–3616.

C

CONJUGATE GRADIENT MINIMIZATION, *SEE* ENERGY MINIMIZATION.

CONNECTIONIST NETWORKS, *SEE* NEURAL NETWORKS.

CONSENSUS, *SEE* CONSENSUS SEQUENCE, CONSENSUS PATTERN, CONSENSUS TREE.

CONSENSUS CODING SEQUENCE DATABASE, *SEE* CCDS.

CONSENSUS PATTERN (REGULAR EXPRESSION, REGEX)

Teresa K. Attwood

A consensus expression derived from a pattern of residues within a conserved region (motif) of a protein sequence alignment, used as a characteristic signature of family membership. Such patterns retain only the most conserved residue information, discarding the rest. Typically, they are designed to be family-specific, and encode motifs of ~10–20 residues. Usually, patterns take the form of regular expressions (regexs) with the PROSITE syntax. An example is shown here:

R – Y – x – [DT] – W – x – [LIVMF] – [ST] – [TV] – P – [LIVM] – [LIVMNQ] – [LIVM]

Within consensus patterns, the standard IUPAC single-letter code for the amino acids is used; individual letters denote completely conserved amino acid residues at that position in the motif; residues defined within square brackets are allowed at the specified motif position; and the symbol x is a 'wild card' indicating that any residue is allowed at that position. Where wild cards or residues of the same type occur sequentially at a particular location, this is denoted by numbers in parentheses: e.g., x(2) or [FWY]3. Specific residues can also be disqualified from occurring at any position – conventionally, these are grouped in curly brackets: e.g., {PG} indicates that proline and glycine are disallowed.

Patterns like this are simple to derive and use in database searching. However, they suffer from various diagnostic drawbacks. In particular, for a pattern to match a sequence, the correspondence must be exact. This means that any minor change, even if it is conservative (e.g., S in position 4 of the above motif, or N in position 8), will result in a mismatch, even if the rest of the sequence matches the pattern exactly. As a result, closely related sequences may fail to match a pattern if its allowed residue groups are too strict; conversely, random sequences may match if the groups are relaxed to include conservative residue alternatives. Overall, then, a match to a pattern may not necessarily be true, and a mismatch may not necessarily be false; caution (and biological intuition) should therefore be used when interpreting pattern matches.

Consensus patterns are used in the PROSITE database. Their occurrence within user-defined query sequences may be diagnosed using ScanProsite; this tool also allows users to create their own patterns (with PROSITE syntax) to search against UniProtKB, the PDB or their own sequence database (in FastA format).

Related websites

ScanProsite	http://prosite.expasy.org/scanprosite/
Pattern syntax	http://prosite.expasy.org/scanprosite/scanprosite-doc.html
PROSITE Manual	http://prosite.expasy.org/prosuser.html#meth1

Further reading

Attwood TK (2000) The quest to deduce protein function from sequence: the role of pattern databases. *Int. J. Biochem. Cell Biol.* 32(2), 139–155.

Attwood TK, Parry-Smith DJ (1999) *Introduction to Bioinformatics.* Addison Wesley Longman, Harlow, Essex, UK.

Higgs P, Attwood TK (2005) *Bioinformatics and Molecular Evolution.* Blackwell Publishing, Oxford, UK.

See also Multiple Alignment, Amino Acid, Consensus Pattern Rule, Motif, PROSITE

CONSENSUS PATTERN RULE

Teresa K. Attwood

A short consensus expression (typically up to about a dozen residues in length) derived from a generic pattern of residues in a protein sequence alignment, used as a characteristic signature of a non-family-specific trait. Typically, they encode particular functional sites (phosphorylation, glycosylation, hydroxylation, amidation sites, etc.). Usually, rules take the form of regular expressions (regexs) with the PROSITE syntax. Examples are shown here:

Functional site	**Rule**
Protein kinase C phosphorylation site	[ST]-x-[RK]
N-glycosylation site	N-{P}-[ST]-{P}
Asp and Asn hydroxylation site	C-x-[DN]-x(4)-[FY]-x-C-x-C
Amidation site	x-G-[RK]-[RK]

Within such expressions, the standard IUPAC single-letter code for the amino acids is used; amino acid residues defined within square brackets are allowed at the specified position; individual letters denote completely conserved residues at that position; x is a 'wild card' indicating that any residue is allowed at a position (numerals in parentheses show the

C

number of residues or wild cards that may occur sequentially at that location (e.g., x(4)); and curly brackets indicate residues that are disqualified (here, proline is not allowed at positions 2 and 4 of the N-glycosylation site).

The short length of rules means that they do not provide good discrimination – in a typical sequence database, matches to expressions of this type number in the thousands. They cannot therefore be used to show that a particular functional site exists in a protein sequence; rather, they give a guide as to whether such a site *might* exist. Biological knowledge must be used to confirm whether such matches are likely to be meaningful.

Rules are used in the PROSITE database to encode functional sites, such as those mentioned above and many others. The potential occurrence of such functional sites may be highlighted within user-defined query sequences using ScanProsite; this tool provides an option to exclude motifs with a high probability of occurrence, like these, in order that outputs should not be flooded with spurious matches.

Related websites

ScanProsite	http://prosite.expasy.org/scanprosite/
Pattern syntax	http://prosite.expasy.org/scanprosite/scanprosite-doc.html
PROSITE Manual	http://prosite.expasy.org/prosuser.html#meth1

Further reading

Attwood TK (2000) The quest to deduce protein function from sequence: the role of pattern databases. *Int. J. Biochem. Cell Biol.* 32(2), 139–155.

Attwood TK, Parry-Smith DJ (1999) *Introduction to Bioinformatics*. Addison Wesley Longman, Harlow, Essex, UK.

Higgs P, Attwood TK (2005) *Bioinformatics and Molecular Evolution*. Blackwell Publishing, Oxford, UK.

See also Multiple Alignment, Amino Acid, Consensus Pattern, Motif, PROSITE

CONSENSUS SEQUENCE

Dov Greenbaum and Thomas D. Schneider

A consensus sequence is a string of nucleotides or amino acids which best represents a set of multiply aligned sequences.

The consensus sequence is decided by a selection procedure designed to determine which residue or nucleotide is placed at each individual position, usually that which is found at the specific position most often.

Consensus sequences are used to determine and annotate DNA sequences, for example in the case of a TATA sequence in a promoter region, as well as to functionally classify previously unknown peptides such as where a DNA binding consensus sequence defines a sequence as DNA binding and possibly a transcription factor.

The simplest form of a consensus sequence is created by picking the most frequent character at some position in a set of aligned DNA, RNA or protein sequences such as binding sites. Where more than one character type occurs at a position the consensus can be represented by the most frequent character or by a symbol denoting ambiguity (eg IUPAC) depending on the algorithm being employed. For protein sequences the symbol can denote membership of an amino acid family, e.g. the hydrophobics. For DNA sequence assembly projects the algorithm can optionally use the basecall confidence values in conjunction with the frequencies.

A disadvantage of using consensus sequences is that the process of creating a consensus destroys the frequency information and can lead to errors in interpreting sequences. For example, suppose a position in a binding site has 75% A. The consensus would be A. Later, after having forgotten the origin of the consensus while trying to make a prediction, one would be wrong 25% of the time. If this was done over all the positions of a binding site, most predicted sites could be wrong. For example, in Rogan and Schneider (1995) a case is shown where a patient was misdiagnosed because a consensus sequence was used to interpret a sequence change in a splice junction. Figure 2 of Schneider (1997a) shows a Fis binding site that had been missed because it did not fit a consensus model. One can entirely replace this concept with sequence logos and sequence walkers.

Notation:

A[CT]N{A}YR

In this notation, A means that an adenosine is always found in that position; [CT] stands for either cytosine or tyrosine; N stands for any base; and {A} means any base except adenosine. Y represents any pyrimidine, and R indicates any purine.

Related websites

Phform	http://saturn.med.nyu.edu/searching/thc/phform.html
Scanprosite	http://us.expasy.org/tools/scanprosite/
The Consensus Sequence Hall of Fame	http://alum.mit.edu/www/toms/consensus.html
Pitfalls	http://alum.mit.edu/www/toms/pitfalls.html
Colonsplice	http://alum.mit.edu/www/toms/paper/colonsplice/
Walker	http://alum.mit.edu/www/toms/paper/walker/walker.htmlfigr21

Further reading

Aitken A (1999) Protein consensus sequence motifs. *Mol Biotechnol* 12: 241–253.

Bork P, Koonin EV (1996) Protein Sequence Motifs. *Curr Opin Struct Biol* 6: 366–376.

Keith JM, et al. (2002) A simulated annealing algorithm for finding consensus sequences. *Bioinformatics* 18: 1494–1499.

Rogan PK, Schneider TD (1995) Using information content and base frequencies to distinguish mutations from genetic polymorphisms in splice junction recognition sites. *Hum Mut*, 6: 74–76. http://alum.mit.edu/www/toms/paper/colonsplice/

Schneider TD (1997a) Information content of individual genetic sequences. *J. Theor. Biol.*, 189: 427–441. http://alum.mit.edu/www/toms/paper/ri/

Schneider TD (1997b) Sequence walkers: a graphical method to display how binding proteins interact with DNA or RNA sequences. *Nucleic Acids Res* 25: 4408–4415. http://alum.mit.edu/www/toms/paper/walker/, erratum: NAR 26: 1135, 1998.

Schneider TD (2002) Consensus Sequence Zen. *Appl Bioinformatics* 1(3), 111–119.

Staden R (1984) Computer methods to locate signals in nucleic acid sequences. *Nucleic Acid Res* 12: 505–519.

Staden R (1988) Methods to define and locate patterns of motifs in sequences. *Comput Applic Biosci* 4: 53–60.

Staden R (1989) Methods for calculating the probabilities of finding patterns in sequences. *Comput Applic Biosci* 5: 89–96.

See also Box, Complexity, Core Consensus, Score

CONSENSUS TREE (STRICT CONSENSUS, MAJORITY-RULE CONSENSUS, SUPERTREE)

Sudhir Kumar and Alan Filipski

A single tree summarizing phylogenetic relationships from multiple individual phylogenies contained in a tree set.

Several individual phylogenetic trees for the same taxon set may need to be combined into a single tree representing a consensus phylogeny when a tree-building method either generates more than one plausible topology (e.g., Bayesian inference or a set of equally parsimonious trees or a set of pseudoreplicate bootstrap trees needs to be summarized). In general, consensus tree methods intend to extract the number of bipartitions/splits from a tree set that are shared by all trees (strict consensus tree), more than half of the trees (majority rule consensus tree), or the majority of the trees (extended majority rule consensus). Based on those shared bipartitions consensus methods then construct a mostly multifurcating consensus tree.

Also, trees of different size from different genes, with only partially overlapping taxon sets may be combined into a consensus tree, mostly in the context of supertree-methods. When the individual trees have only partial taxon set overlaps, consensus trees are computationally much harder to reconstruct.

Software

Felsenstein J (1993) *PHYLIP: Phylogenetic Inference Package*. Seattle, WA, University of Washington.

Swofford DL (1998) *PAUP*: Phylogenetic Analysis Using Parsimony (and other methods)*. Sunderland, MA., Sinauer Associates.

Aberer A, Pattengale N, Stamatakis A (2010) Parallelized phylogenetic post-analysis on multi-core architectures. *Journal of Computational Science* 1: 107–114.

Further reading

Nei M, Kumar S (2000) *Molecular Evolution and Phylogenetics*. Oxford, Oxford University Press.

Schuh RT (2000) *Biological Systematics: Principles and Applications*. Ithaca, NY, Cornell University Press.

Wilkinson M (1996) Majority-rule reduced consensus trees and their use in bootstrapping. *Mol Biol Evol* 13: 437–444.

See also Phylogenetic Tree

CONSERVATION (PERCENTAGE CONSERVATION)

Dov Greenbaum

The presence of a sequence, or part thereof, in two or more genomes.

When two genes are present in two different genomes, as defined by some threshold of sequence similarity, they are termed conserved. There are different degrees of conservation, the higher the sequence similarity the more conserved the two genes are, indicating the conservation of more elements (nucleotides, amino acids) between the molecular species.

Additionally, while two genes may have different sequences incorporating different amino acids at a specific position, if those amino acids are similar in their chemical properties, ie hydrophobic, and as such do not significantly alter the structure or function of a protein, those two genes are termed conserved. In this case what is conserved is not sequence but (potential) function. Conservation can also be observed at other levels, for example structure, gene organization.

Use of the term conservation implies homology between the species being considered.

The term percentage conservation for two homologous biological sequences represents the degree of identity or similarity of the sequences.

Further reading

Graur D, Li W-H (1999) *Fundamentals of Molecular Evolution*. Sunderland, MA, Sinauer Associates.

CONSTRAINT-BASED MODELING (FLUX BALANCE ANALYSIS)

Neil Swainston

Constraint-based modeling is typically applied to genome-scale metabolic networks and applies linear programming with techniques such as flux balance analysis (FBA) to predict the flow of metabolites through a metabolic network at steady state. FBA relies upon an objective function that the cell is assumed to optimize. In the case of microorganisms, it is assumed that the cell's metabolism is optimized to grow, and hence the objective function is typically the maximization of a hypothetic biomass function, which encapsulates the synthesis of amino acids, nucleotides, lipids, and cofactors that are essential constituents of biomass. FBA can be performed on genome-scale stoichiometric metabolic networks, and in principle relies on little additional experimental data in addition to that underlying the initial genome annotation. Software packages for performing constraint-based modeling include the COBRA Toolbox, the BioMet Toolbox, SurreyFBA, FASIMU and Cycsim.

C

While constraint-based modeling the ability to perform analyses on genome-scale models without the requirement of detailed experimental parameters, it simulates the system at steady state and cannot capture the dynamic behavior of the cell (and components within the cell).

Related websites

COBRA Toolbox	http://opencobra.sourceforge.net/openCOBRA/Welcome.html
BioMet Toolbox	http://129.16.106.142/
SurreyFBA	http://sysbio3.fhms.surrey.ac.uk/
FASIMU	http://www.bioinformatics.org/fasimu/
Cycsim	http://www.genoscope.cns.fr/cycsim/org.nemostudio.web.gwt.App/ App.html

Further reading

Costenoble R, Picotti P, Reiter L (2011) Comprehensive quantitative analysis of central carbon and amino-acid metabolism. *in Saccharomyces cerevisiae* under multiple conditions by targeted proteomics *Mol Syst Biol* 7: 464.

Cvijovic M, Olivares-Hernández R, Agren R, et al. (2010) BioMet Toolbox: genome-wide analysis of metabolism. *Nucleic Acids Res* 38: W144–149.

Feist AM, Palsson BO (2010) The biomass objective function. *Curr Opin Microbiol* 13: 344–349.

Gevorgyan A, Bushell ME, Avignone-Rossa C, Kierzek AM (2011) SurreyFBA: a command line tool and graphics user interface for constraint-based modeling of genome-scale metabolic reaction networks *Bioinformatics* 1,27(3), 433–434.

Ghaemmaghami S, Huh WK, Bower K (2003) Global analysis of protein expression in yeast. *Nature* 425: 737–741.

Hoppe A, Hoffmann S, Gerasch A, Gille C, Holzhütter HG (2011) FASIMU: flexible software for flux-balance computation series in large metabolic networks. *BMC Bioinformatics* 12: 28.

Le Fèvre F, Smidtas S, Combe C, Durot M, d'Alché-Buc F, Schachter V (2009) CycSim – an online tool for exploring and experimenting with genome-scale metabolic models *Bioinformatics* 25(15),1987–1988. http://www.genoscope.cns.fr/cycsim/org.nemostudio.web.gwt.App/App.html

Orth JD, Thiele I, Palsson BØ (2010) What is flux balance analysis? *Nat Biotechnol* 28: 245–248.

Picotti P, Bodenmiller B, Mueller LN, Domon B, Aebersold R (2009) Full dynamic range proteome analysis of *S. cerevisiae* by targeted proteomics *Cell* 138: 795–806.

Quantitative prediction of cellular metabolism with constraint-based models: the Becker SA, Feist AM, Mo ML, Hannum G, Palsson BØ, Herrgard MJ (2007) COBRA Toolbox *Nat Protoc* 2: 727–738.

CONTACT MAP

Andrew Harrison and Christine Orengo

A contact map is a two dimensional matrix which is used to capture information on all residues which are in contact within a protein structure. The axes of the matrix are associated with each residue position in the 3D structure and cells within the matrix are shaded when the associated residues are in contact. Various criteria are used to define a residue contact. For example, the Cα atoms of two residues should be within 10 angstroms distance from each other. Alternatively, any atom from the two residues should be within 8 angstroms from each other. Depending on the fold of the protein, characteristic patterns are seen within the matrix. Contacting residues in secondary structures give rise to a thickening of the central diagonal whilst anti-parallel secondary structures give rise to lines orthogonal to the central diagonal and parallel secondary structures lines which are parallel.

A distance plot is a related 2D representation of a protein, devised by Phillips in the 1970s. In this the matrix cells are shaded depending on the distances between any two residues in the protein. Difference distance plots can be used to show changes occurring in a protein structure, for example on ligand binding.

See also Fold, Secondary Structure

CONTIG

Rodger Staden

An ordered set of overlapping segments of DNA from which a contiguous sequence can be derived. Confusingly, it is presently also used as a shorthand for a single sequence without gaps. Originally (Staden, 1980) a contig was defined to be an ordered set of overlapping sequences from which a contiguous sequence can be derived. Later the meaning was broadened for use in contig mapping to be a set of overlapping clones. The newest usage (as a shorthand for a contiguous sequence) causes great confusion because the objects being defined are so closely related but very different: one is a set, the other a single sequence.

Further reading

Staden R (1980) A new computer method for the storage and manipulation of DNA gel reading data, *Nucleic Acids Res* 8: 3673–3694.

CONTINUOUS TRAIT, *SEE* QUANTITATIVE TRAIT.

CONVERGENCE

Dov Greenbaum

Convergence refers to independent sequences attaining similar traits over evolutionary time.

Functional convergence refers to different protein folds can converge and all attain a similar function.

Structural Convergence refers to differing sequences attaining similar three dimensional structures. A probable example of this are alpha-beta barrels, which frequently re-occur during evolution. This is thought to their stability resulting in the structure being energetically favored.

C

Intuitively, there are six criteria to determine whether sequences have converged: The similarity of DNA sequence, protein sequence, structure, enzyme-substrate interactions catalytic mechanism and same segment in the protein responsible for the protein function.

Conversely divergence refers to proteins that may have had common ancestors but have diverged over time.

Further reading

Holbrook JJ, Ingram VA (1973) Ionic properties of an essential histidine residue in pig heart lactate dehydrogenase. *Biochem J* 131: 729–738.

COORDINATE SYSTEM OF SEQUENCES

Thomas D. Schneider

A coordinate system is the numbering system of a nucleic acid or protein sequence.

Coordinate systems in primary databases such as GenBank and PIR are usually 1-to-*n*, where n is the length of the sequence, so they are not recorded in the database. In the Delila system, one can extract sequence fragments from a larger database. If one does two extractions, then one can go slightly crazy trying to match up sequence coordinates if the numbering of the new sequence is still 1-to-*n*. The Delila system handles all continuous coordinate systems, both linear and circular, as described in LIBDEF, the definition of the Delila database system. For example, on a circular sequence running from 1 to 100, the Delila instruction 'get from 10 to 90 direction -;' will give a coordinate system that runs from 10 down to 1, and then continues from 100 down to 90.

Unfortunately there are many examples in the literature of nucleic-acid coordinate systems without a zero coordinate. A zero base is useful when one is identifying the locations of sequence walkers: the location of the predicted binding site is the zero base of the walker (the vertical rectangle). Without a zero base, it is tricky to determine the positions of bases in a sequence walker. With a zero base it is quite natural. Insertion or deletions will make holes or extra parts of a coordinate system. The Delila system cannot handle these (yet). In the meantime, the sequences are renumbered to create a continuous coordinate system.

Related website

Libdef	http://alum.mit.edu/www/toms/libdef.html

Further reading

Schneider TD, et al. (1982). A design for computer nucleic-acid sequence storage, retrieval and manipulation. *Nucleic Acids Res* 10: 3013–3024.

COPY NUMBER VARIATION

Dov Greenbaum

In some instances, large segments of DNA, ranging in size from 10^3 to 10^6 bases, can vary in the number of times they are represented in an organism's genome. CNVs in some cases may encompass whole genes leading to dosage imbalances with some genes having multiple copies while others may be missing entirely.

Variability in copy numbers is thought to influence varied traits, including disease susceptibility and drug response and outnumbers variability induced by SNPs by 3 to 1. Efforts to understand the mechanisms that result in the formation of CNV may also help better understand human genome evolution

There are many possible sources of CNVs within a genome: they may be inherited or the result of one or a plurality of de novo mutations. Replication mistakes including fork stalling and/or template switching or microhomology-mediated break-induced replication may be some of responsible mechanisms. Other mechanisms may include structural rearrangements e.g., deletions, duplications, inversions, and translocations.

Efforts to discover instances of CNVs within a genome typically use techniques used for discovering large scale variability such as comparative genomic hybridization, array comparative genomic hybridization, and virtual karyotyping with SNP arrays and FISH. While CNVs are viewed as more macro-like structural variations within a genome, genomic sequencing may also be used in their discovery.

Related websites

The Copy Number Variation (CNV) Project	http://www.sanger.ac.uk/research/areas/humangenetics/cnv/
Database of Genomic Variants	http://projects.tcag.ca/variation/
PennCNV: copy number variation detection	http://www.openbioinformatics.org/penncnv/

Further reading

Almal SH, Padh H (2012) Implications of gene copy-number variation in health and diseases. *J Hum Genet* 57(1), 6–13.

Barnes MR, Breen G (2010) A short primer on the functional analysis of copy number variation for biomedical scientists. *Methods Mol Biol* 628: 119–35.

Conrad DF, et al. (2010) Origins and functional impact of copy number variation in the human genome. *Nature* 464, 7289, 704–712.

Daar AS, Scherer SW, Hegele RA (2006) Implications for copy-number variation in the human genome: a time for questions. *Nature Reviews Genetics* 7: 414.

Le Caignec C, Redon R (2009) Copy number variation goes clinical. *Genome Biol* 10(1), 301–303.

Sjödin P, Jakobsson M (2012) Population genetic nature of copy number variation. *Methods Mol Biol* 838: 209–23.

CORE CONSENSUS

Thomas D. Schneider

A core consensus is the strongly conserved portion of a binding site, found by creating a consensus sequence. It is an arbitrary definition as can be seen from the examples in the sequence logo gallery (see Figure B.5 under Binding Site Symmetry). The sequence conservation (measured in bits of information) often follows the cosine waves that represent the twist of B-form DNA. This has been explained by noting that a protein bouncing in and out from DNA must evolve contacts (Schneider 2001). It is easier to evolve DNA contacts that are close to the protein than those that are further around the helix. Because the sequence conservation varies continuously, any cutoff or 'core' is arbitrary.

Related website

oxyr	http://alum.mit.edu/www/toms/papers/oxyr/

Further reading

Schneider TD (2001) Strong minor groove base conservation in sequence logos implies DNA distortion or base flipping during replication and transcription initiation. *Nucleic Acid Res*, 29: 4881–4891. http://alum.mit.edu/www/toms/papers/baseflip/

See also Consensus Sequence

CORRELATION ANALYSIS, *SEE* REGRESSION ANALYSIS, ASSOCIATION RULE MINING, LIFT

COSMIC (CATALOGUE OF SOMATIC MUTATIONS IN CANCER)

Malachi Griffith and Obi L. Griffith

COSMIC operates under the premise that all cancers are a result of DNA sequence abnormalities that confer a growth advantage to cells that harbor them. COSMIC is designed for the storage and visualization of such mutations identified as somatic in human cancers. In particular, COSMIC emphasizes the importance of mutation frequency in interpreting the relative importance of cancer mutations. Mutation data and associated annotations are extracted from both the published literature and systematic screening studies. Mutations may be single nucleotide variants, small insertions or deletions, gene fusions, or larger structural rearrangements. Coordinate and mapping information for all mutations are mapped onto a single version of each gene. Additional information is provided on the types of samples screened, the mutation detection methods employed, and the reliability of the mutation prediction itself as well as the inference that said mutation is somatic and not a germline mutation or polymorphism. As entries are added to COSMIC an attempt is made to standardize and consolidate tissue and histology information to remove redundancies and provide consistency in database queries. The types of samples entered include benign neoplasms as well as invasive tumours, recurrences, metastases and even cancer cell lines. COSMIC is maintained by the Wellcome Trust Sanger Insitute.

Related websites

COSMIC homepage	http://www.sanger.ac.uk/genetics/CGP/cosmic/
COSMIC classic genes	http://www.sanger.ac.uk/genetics/CGP/Classic/
COSMIC systematic screens	http://www.sanger.ac.uk/genetics/CGP/cosmic/papers/
COSMIC FTP	ftp://ftp.sanger.ac.uk/pub/CGP/cosmic

Further reading

Dalgliesh GL, Futreal PA (2007) The continuing search for cancer-causing somatic mutations. *Breast Cancer Res* 9(1), 101.

Forbes S, et al. (2006) Cosmic 2005. *Br J Cancer* 94(2), 318–322.

Forbes SA, et al. (2008) The Catalogue of Somatic Mutations in Cancer (COSMIC). *Curr Protoc Hum Genet* Chapter 10: Unit 10.11.

Forbes SA, et al. (2011) COSMIC: mining complete cancer genomes in the Catalogue of Somatic Mutations in Cancer. *Nucleic Acids Res* 39(Database issue), D945–550.

Stratton M (2008) Genome resequencing and genetic variation. *Nat Biotechnol* 26(1), 65–66.

COVARIATION ANALYSIS

Jamie J. Cannone and Robin R. Gutell

Comparative analysis is used to address a variety of different scientific enquiries. The immediate discussion here addresses the prediction of RNA structure. As noted in 'RNA Structure Prediction' it is known that a large number of RNA (and protein) sequences can form the same higher-order RNA structure. Covariation analysis is a method that identifies positions with similar patterns of variation in an alignment of sequences (see Figure C.2). These positions that 'covary' with one another can potentially form a basepair in a comparative structure model.

Different implementations of the method have been developed. While some methods identify positions with similar patterns of variation regardless of the base pair type and proximity to other base pairs, others only identify G:C, A:U, and G:U base pairs within a regular secondary structure helix. The former method has identified numerous examples of non-canonical base pair exchanges, and the formation of structural elements more complex and interesting than your canonical helix. For example, in addition to the exchange between canonical base pair types A:U <> U:A <> G:C <> C:G, several non-canonical exchanges have been observed. A few are: G:U <> A:C, U:U <> C:C, A:A <> G:G, A:G <> G:A, U:A <> G:G. The majority of the base pairs with a covariation are arranged into a regular secondary structure helix. However, a growing number are part of a tertiary structure. And while the majority of all of the helices are nested within another helix, a small number of helices form a pseudoknot.

As the number of sequences in some of our RNA comparative alignment datasets have grown to more than 100,000, the number of sequences with an exception to a pure covariation has increased. In parallel, our covariation methods have been enhanced, allowing for

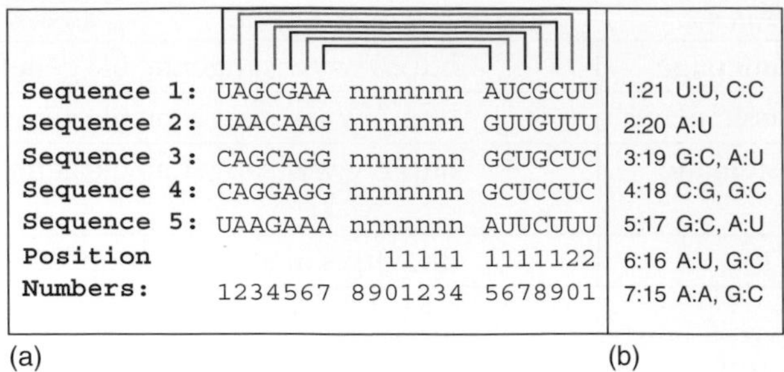

Figure C.2 Examples of Covariation.
Sequence alignment. Five sequences are shown, 5′ at left, 3′ at right. Lines above the alignment connect the columns that are base paired. Nucleotide position numbers are below the alignment. (a) Summary of covariations from the alignment. (b) For each of the seven base pairs, position numbers are followed by the observed base-pair types.

more subtle covariations to be identified. While base pairs with a very strong covariation that contain many covariations (e.g. A:U <> U:A <> G:C <> C:G) at a single base pair, with a minimal number of exceptions (e.g. A:U <> A:A) are almost always present in the high-resolution crystal structures of RNA, some base paired positions in the RNA structure have a higher percentage of exceptions associated with a weaker covariation. In parallel, it has been observed that some sets of positions with a weaker covariation are in proximity in three-dimensional space but do not form base pairs. These 'neighbor effects', indicate that the evolution of nucleotides can influence or be influenced by other nucleotides that are in proximity although they are not base paired.

Our confidence in the prediction of base pairs with covariation analysis, as noted above is proportional to the dependence of the two evolving positions. While more than 97% of the base pairs predicted in some RNA molecules with covariation analysis are present in the high resolution crystal structures, substantiating the covariation methods, the nucleotides that form many tertiary structure base pairs do not have any covariation (see Ribosomal RNA). Either both positions are invariant, or nearly so, or both positions change (or don't change) independently of one another, without any coordination or covariation in the change. Thus other non-covariation analysis methods are required to identify those base pairs and other structural elements that do not have a covariation between the two positions that form a base pair.

One of the methods that gauge the dependence and independence between the evolution of two positions that are proposed to be base-paired is the chi-square statistic. While this method and related methods (e.g. mutual information) measure the frequencies for each base pair type, a more sensitive method determines the changes that have occurred during the evolution of the RNA (see Phylogenetic Event Counting).

Some of the covariation methods evaluate all pairwise comparisons in a sequence alignment for the identification of base pairs composed of positions with similar patterns of variation. This approach does not search for base pairing with any known principles of RNA structure. Thus, the results from this pure covariation analysis were most revealing. The vast majority of all base pairs were composed of the canonical base pair types – G:C, A:U, and G:U. And these base pairs were arranged antiparallel and consecutive with one another to form regular helices. Thus covariation analysis independently identified two of the most fundamental principles in RNA structure – the Watson–Crick base-pairing relationship and the formation of helices from the antiparallel and consecutive arrangement of these base pairs.

Further reading

Chiu DK, Kolodziejczak T (1991) Inferring consensus structure from nucleic acid sequences. *Comput. Appl. Biosci.* 7: 347–352.

Gardner, P, Giegerich R (2004) A comprehensive comparison of comparative RNA structure prediction approaches. *BMC Bioinformatics* 5: 140.

Gutell RR, Power A, Hertz GZ, Putz EJ, Stormo GD (1992) Identifying constraints on the higher-order structure of RNA: continued development and application of comparative sequence analysis methods. *Nucleic Acids Res* 20: 5785–5795.

Gutell RR, Larsen N, Woese CR (1994) Lessons from an Evolving Ribosomal RNA: 16S and 23S rRNA Structure from a Comparative Perspective. *Microbiological Reviews.* 58: 10–26.

Lindgreen S, Gardner PP, Krogh A (2006) Measuring covariation in RNA alignments: physical realism improves information measures. *Bioinformatics* 22: 2988–2995.

See also RNA Structure

CPG ISLAND

Katheleen Gardiner

A 200–2000 bp segment of genomic DNA with base composition >60%GC and CpG dinucleotide frequency 0.6 observed/expected.

These characteristics stand out in the mammalian genome where the average base composition is ~38% GC and the average CpG frequency is 0.2-0.25 observed/expected. CpG islands are found more often in GC-rich R bands, and are most often unmethylated, located at or near the 5' ends of genes and associated with promoters.

Further reading

Antequera F, Bird A (1993) Number of CpG islands and genes in human and mouse. *Proc Natl Acad Sci* 90: 11995–11999.

Ioshikhes IP, Zhang MQ (2000) Large-scale human promoter mapping using CpG islands. *Nat Genet* 26: 61–63.

Takai D, Jones PA (2002) Comprehensive analysis of CpG islands in human chromosomes 21 and 22 *Proc Natl Acad Sci* 99: 3740–3745.

See also Methylation, Promoter

CRM, *SEE* CIS-REGULATORY MODULE.

C

CROSS-REFERENCE (XREF)

Carole Goble and Katy Wolstencroft

A cross-reference is a concept within a document which refers to synonymous information elsewhere. In the life sciences, this is particularly important because biological data is held in heterogeneous data stores at distributed locations. Data about the same biological object is commonly held in multiple and overlapping data resources. Cross-references between these resources allow us to integrate distributed information dynamically, or by collecting all available data together in a data warehouse.

Cross-references in the life sciences are normally provided as mappings of identifiers between different data resources. For example, for the mouse Alpha-actinin-2 protein in Uniprot (ID ACTN2_MOUSE), there are cross references to EMBL (AF248643), KEGG (mmu:11472) and Ensembl (ENSMUSG00000052374), amongst others.

See also Biological Identifiers, Data Integration

CROSS-VALIDATION (K-FOLD CROSS-VALIDATION, LEAVE-ONE-OUT, JACKKNIFE, BOOTSTRAP)

Nello Cristianini

A method for estimating the generalization performance of a machine learning algorithm.

This is done by randomly dividing the data into k mutually exclusive subsets (the 'folds') of approximately equal size. The algorithm is trained and tested k times: each time it is trained on the data set minus a fold and tested on that fold (called also the held-out set). The performance estimate is the average performance for the k folds. When as many folds as data points are used, the method reduces to the procedure known as the leave-one-out, or jacknife method.

Further reading

Mitchell T (1997) *Machine Learning*. McGraw Hill, Maidenhead, UK.

See also Error Measures

C-VALUE, *SEE* GENOME SIZE.

D

DAG, *see* Directed Acyclic Graph.
DAS Services
Data Definition Language, *see* Data Description Language.
Data Description Language (DDL, Data Definition Language)
Data Flow, *see* Stream Mining.
Data Integration
Data Manipulation Language (DML)
Data Mining, *see* Pattern Analysis.
Data Pre-processing
Data Processing
Data Standards
Data Standards in Proteomics
Data Standards in Systems Modeling
Data Stream, *see* Stream Mining.
Data Structure (Array, Associative Array, Binary Tree, Hash, Linked List, Object, Record, Struct, Vector)
Data Warehouse
Database (NoSQL, Quad Store, RDF Database, Relational Database, Triple Store)
Database of Genotypes and Phenotypes, *see* dbGAP.
Database Search Engine (Proteomics) (Peptide Spectrum Match, PSM)
DataMonkey
Dayhoff Amino Acid Substitution Matrix (PAM Matrix, Percent Accepted Mutation Matrix)
dbEST
dbGAP (the Database of Genotypes and Phenotypes)
dbSNP
dbSTS
dbVar (Database of Genomic Structural Variation)
DDBJ (DNA Databank of Japan)
DDL, *see* Data Description Language.
De Novo Assembly in Next Generation Sequencing
Dead-End Elimination Algorithm
Decision Surface
Decision Tree
Decoy Database
Degree of Genetic Determination, *see* Heritability.
Deletion, *see* Indel.
DELILA
DELILA Instructions
DendroPy
Dependent Variable, *see* Label.
Description Logic (DL)
Descriptors, *see* Features.
DGV (Database of Genetic Variants)
Diagnostic Performance, *see* Diagnostic Power.
Diagnostic Power (Diagnostic Performance, Discriminating Power)

Concise Encyclopaedia of Bioinformatics and Computational Biology, Second Edition.
Edited by John M. Hancock and Marketa J. Zvelebil.

Dihedral Angle (Torsion Angle)
Dinucleotide Frequency
DIP
Directed Acyclic Graph (DAG)
Discrete Function Prediction (Function Prediction)
Discriminant Analysis, *see* Classification.
Discriminating Power, *see* Diagnostic Power.
Distance Matrix (Similarity Matrix)
Distances Between Trees (Phylogenetic Trees, Distance)
Distributed Computing
Disulphide Bridge
DL, *see* Description Logic.
DML, *see* Data Manipulation Language.
DNA Array, *see* Microarray.
DNA Databank of Japan, *see* DDBJ.
DNA-Protein Coevolution
DNA Sequence
DNA Sequencing
DnaSP
DOCK
Docking
Domain, *see* Protein Domain.
Domain Alignment, *see* Alignment.
Domain Family
Dot Matrix, *see* Dot Plot.
Dot Plot (Dot Matrix)
Dotter
Downstream
Drug-Like
DrugBank
Druggability
Druggable Genome (Druggable Proteome)
DW, *see* Data Warehouse.
Dynamic Programming

DAG, *SEE* DIRECTED ACYCLIC GRAPH.

DAS SERVICES

Carole Goble and Katy Wolstencroft

The Distributed Annotation System (DAS) is similar to a RESTful Web Service protocol. It provides a communication protocol specifically designed for the exchange and visualization of gene and protein sequences. A DAS client is able to display multiple sources of data annotation on the same view of the gene or protein sequence. Each data annotation source is a DAS server, which stores structured annotations of sequence features and their coordinates in the sequence.

DAS combines data integration and data visualization to provide a uniform view of annotations to its users.

Related website

http://alum.mit.edu/www/toms/delilainstructions.html	http://www.biodas.org

See also RESTful Web Services, Data Integration

DATA DEFINITION LANGUAGE, *SEE* DATA DESCRIPTION LANGUAGE.

DATA DESCRIPTION LANGUAGE (DDL, DATA DEFINITION LANGUAGE)

Matthew He

A Data Description Language or Data Definition Language is a computer language for defining data structures and database schemas. Classically referring to a subset of SQL, it is used to specify the arrangement of data items within a database.

Related website

Wikipedia	http://en.wikipedia.org/wiki/Data_definition_language

See also SQL

DATA FLOW, *SEE* STREAM MINING.

D

DATA INTEGRATION

Carole Goble and Katy Wolstencroft

Data Integration is combining heterogeneous information from multiple sources, to create a composite view of all information available for certain data entities. For example, a protein is expressed in certain tissues and/or cell types and it interacts with other proteins in biological pathways to perform specific molecular functions. Genetically, it is encoded by a particular gene, which is regulated by particular transcription factors and promoters, and it has orthologues and homologues in other organisms. Each piece of information may be contained within one or more different database, which may use different identifiers and names for the same biological entities. Therefore, the challenge is not only gathering the data, but understanding when the same entity is being referred to. Data Integration is a central concern in bioinformatics.

Further reading

Goble C, Stevens R (2008) State of the nation in data integration for bioinformatics. *J Biomed Inform* 41(5): 687–693. Epub 2008 Feb 5. Review.

Sansone S-A, et al. (2012) Toward interoperable bioscience data *Nature Genetics* 44: 121–126. doi:10.1038/ng.1054

See also Identifier, Cross Reference, Database

DATA MANIPULATION LANGUAGE (DML)

Matthew He

Data Manipulation Languages (DML) are a family of computer languages used by computer programs and/or database users to insert, delete and update data in a database (compared to DDLs, which allow users to modify tables). Read-only querying, i.e. `SELECT`, of this data may be considered to be either part of a DML or outside it, depending on the context.

Related website

Wikipedia	http://en.wikipedia.org/wiki/Data_manipulation_language

See also SQL

DATA MINING, *SEE* PATTERN ANALYSIS.

DATA PRE-PROCESSING

Feng Chen and Yi-Ping Phoebe Chen

Data pre-processing is an often neglected but important step in data mining. The data are always inconsistent, incomplete or noisy when they are collected from a variety of data sources. Data pre-processing has to be carried out before a further data analysis. Generally, data pre-processing includes but is not limited to data measurement, missing value processing and noise processing.

Related website

Wikipedia	http://en.wikipedia.org/wiki/Data_pre-processing

See also Missing Value, Noise

DATA PROCESSING

Matthew He

Data processing is defined as the systematic performance of operations upon data such as handling, merging, sorting, and computing. The semantic content of the original data should not be changed, but the semantic content of the processed data may be changed.

Related website

Wikipedia	http://en.wikipedia.org/wiki/Data_processing

DATA STANDARDS

John M. Hancock

A data standard is any specification of a data format that allows error-free parsing and processing by applications. Standards tend to be specific to data classes (e.g. sequence data, microarray data etc.) and can range from simple file formats such as FASTA for DNA and protein sequences to complex XML schemas providing extensive metadata such as minimum information checklists. There has been increasing standardization of data formats by the bioinformatics community in recent years although the process is far from complete.

Related websites

MIBBI	http://mibbi.sourceforge.net/
BioSharing-Standards	http://www.biosharing.org/standards

Further reading

Sansone SA, et al. (2012) Toward interoperable bioscience data. *Nat Genet* 44: 121–126.

Taylor CF, et al. (2008) Promoting coherent minimum reporting guidelines for biological and biomedical investigations: the MIBBI project. *Nat Biotechnol* 26: 889–896.

DATA STANDARDS IN PROTEOMICS

Juan Antonio Vizcaíno

Community data standards in proteomics are developed under the auspices of the HUPO (Human Proteome Organisation) PSI (Proteomics Standards Initiative). Each standard is composed of an output format (typically XML-based), a controlled vocabulary or ontology, and a specific MIAPE (Minimum Information About a Proteomics Experiment) guidelines document. The main proteomics data standards are:

D

- mzML: Standard XML based format for the output of mass spectrometry data. It was built on two previous formats with opposing philosophies (mzXML and mzData), retaining the best technical attributes of the two.
- mzIdentML: Standard XML based format for reporting the search parameters and results of peptide and protein identification data derived from spectrum identification algorithms (mass spectrometry proteomics experiment). It does not support quantitative information, which is provided by a different data standard called mzQuantML.
- PSI MI (PSI Molecular Interaction): Standard format that enables the representation of molecular interactions between different types of molecules, like for example proteins, nucleic acids and chemical compounds. There are two versions of the standard: the first one is PSI-MI XML, which allows the description of highly detailed molecular interaction data and facilitates the data exchange between existing databases. There is also a simpler, tab-delimited format called MITAB, built for those users who require not that detailed information.
- TraML: Standard XML based format for encoding transition lists (and associated metadata), which are used in targeted proteomics approaches such as SRM (Selected Reaction Monitoring).

Related websites

PSI	http://www.psidev.info
mzML	http://www.psidev.info/mzml
mzIdentML	http://www.psidev.info/mzidentml
PSI MI	http://www.psidev.info/node/60
TraML	http://www.psidev.info/traml

Further reading

Deutsch EW, et al. (2012) TraML: a standard format for exchange of selected reaction monitoring transition lists. *Mol Cell Proteomics* 11, R111.015040.

Jones AR, et al. (2012) The mzIdentML data standard for mass spectrometry-based proteomics results. *Mol Cell Proteomics* 11, M111.014381.

Kerrien S, et al. (2007) Broadening the horizon – level 2.5 of the HUPO-PSI format for molecular interactions. *BMC Biol* 5: 44.

Martens L, et al. (2012) mzML – a community standard for mass spectrometry data. *Mol Cell Proteomics* 10, R110.000133.

DATA STANDARDS IN SYSTEMS MODELING

Juan Antonio Vizcaíno

The main data standards in the field are:

- SBML (Systems Biology Mark-up Language): XML-based format for representing biochemical reaction networks. It is suitable for representing mathematical models representing cell signalling or metabolic pathways, biochemical reactions, gene regulation, and many others. SBML is primarily aimed at exchanging information about pathway and reaction models.

- CellML: XML-based format for representing computer-based mathematical models of cellular functions. CellML includes information about the model structure (how the parts of a model are organizationally related to one another), mathematics (equations describing the underlying processes) and metadata describing the model and its parts. It was originally developed for biological data but can represent other types of information such as electrophysiological models. One key difference with SBML is that in CellML, the biological information is entirely stored in metadata rather than the language elements.
- BioPAX (Biological Pathway Exchange): XML-based format that aims to represent biological pathways at the molecular and cellular level, and to facilitate the exchange of pathway related data. It can represent metabolic and signalling pathways, molecular and genetic interactions and gene regulation networks. However, BioPAX does not cover the dynamic and quantitative aspects of biological processes, as SMBL and CellML can do.

Related websites

SBML	http://www.smbl.org
CellML	http://www.cellml.org
BioPAX	http://www.biopax.org

Further reading

Demir E, et al. (2010) The BioPAX community standard for pathway data sharing. *Nat Biotechnol*, 28: 935–942.

Hucka M, et al. (2003) The systems biology markup language (SBML): a medium for representation and exchange of biochemical network models. *Bioinformatics*, 19: 524–531.

Lloyd CM, et al. (2004) CellML: its future, present and past. *Prog Biophys Mol Biol*, 85: 433–450.

DATA STREAM, *SEE* STREAM MINING.

DATA STRUCTURE (ARRAY, ASSOCIATIVE ARRAY, BINARY TREE, HASH, LINKED LIST, OBJECT, RECORD, STRUCT, VECTOR)

David Thorne, Steve Pettifer and James Marsh

A data structure represents in a computer's memory some set of related values from the real world or the problem domain under scrutiny. Data structures are then used by algorithms to store, transfer and analyse this information.

The choice of how to organize and store data in memory can be critical to the success of any data-intensive algorithm. Questions of how often the data might change, how often it is accessed, in what order it will be read, how large or complex it is, and many more besides, inform the developer which data structure(s) to choose for their particular application, or indeed, whether they would be better off implementing their own data structure if their requirements were sufficiently complex. It is frequently impossible to minimize both space and time complexity for any given problem, so the choice often comes down to a trade-off between different factors of the speed, size, and ease of use of data structures.

Although their names may vary from one programming language to another, standard data structures are always available that satisfy the most common use cases. *Records* (or structs) group together related data into self-contained items, and *Objects* extend records to provide code for manipulating them in controlled ways. *Arrays* (also known as vectors) organise their data items in an ordered sequence in a block of contiguous memory that can be accessed very quickly if you know the position of the item you wish to read. Conversely, modifying the order of the items, or introducing new items, tends to be very slow in comparison due to the subsequent reshuffling of memory; searching for arbitrary items without an index can also be very slow, and in the worst case can require the algorithm to query every item in the array to find what it is looking for. *Linked Lists* are superficially similar to arrays in that they provide an ordered sequence of data items, but because they do not store them in contiguous memory locations, direct positional access to items is impossible. Instead, each item is allocated separately in memory, with each holding a link (usually a memory address) to where the next item in the list can be found, meaning inserting new items is a trivial task of modifying a couple of links, and as such is very quick. If a linked list's items also hold links to their predecessors, it is often called a *Doubly Linked List*. Other similar data structures exist: a *Hash* is similar to an array, but its items are stored at positions that relate to the content of those items; a *Binary Tree* stores its items in an ordered hierarchy providing very efficient navigation to items of interest; an *Associative Array* (often called a dictionary) maps keys to related data items using a variety of alternative methods such as the *Red-Black Tree Map* or the *Hash Map*. Depending on the language used, many other structures may be available, each with its own properties that can be advantageous or otherwise in different situations: *Heap*, *Stack*, *Queue*, *Skip List*, *Multi-dimensional Array*, *Graph*, *Multmap*, and *Set* name a few. Choosing the right data structure for the task at hand is an important part of any software solution, and knowing the options can greatly increase a developer's efficiency.

DATA WAREHOUSE

Carole Goble, Katy Wolstencroft and Matthew He

Data warehouses in the life sciences provide a static integration of many different data sources. Individual database schemas or flat file formats are combined into a larger schema. The content of each individual source is then stored in this much larger repository. The BioMart database is an example of a life sciences data warehouse system. BioMart databases are used to integrate all information relating to the genome of a particular species, for example (http://central.biomart.org/martanalysis/#!/Gene%20retrieval/Ensembl)

A data warehouse maintains its functions in three layers: staging, integration, and access. *Staging* is used to store raw data for use by developers. The *integration* layer is used to integrate data and to have a level of abstraction from users. The *access* layer is for getting data out for users. The arrays of heterogeneous biological data stored within a single logical data repository are accessible to different querying and manipulation methods.

The advantage of the warehouse approach is that it provides a 'one-stop-shop' for all data related to a particular biological subject, making it more straightforward to query the data. However, the cost of keeping this information up to date is high. Individual data sources have different update timetables, and small changes in their schemas can have large knock-on effects on updating the data warehouse. Therefore, it is only a powerful technique if it has adequate support from curators.

The data warehouse provides a static and stable view of data integration and is the opposite approach to other methods that dynamically link data at the time of querying, such as Scientific Workflows and Linked Data.

Related website

BioMart	http://www.biomart.org/

See also Scientific Workflow, Linked Data, Data Integration

DATABASE (NOSQL, QUAD STORE, RDF DATABASE, RELATIONAL DATABASE, TRIPLE STORE)

David Thorne, Steve Pettifer and James Marsh

A database is an organised and managed collection of data that provides persistent storage and querying facilities. Databases are frequently optimised for very large numbers of records, and often provide safe simultaneous access to many users. For this reason, almost every major software system relies to some degree on database technology.

The term 'database' covers a multitude of systems providing quite different services over many kinds of data. Relational database management systems (RDBMS) deal with highly structured tables of data where row records adhere to strict schema. They provide guaranteed integrity constraints, safe access to many users simultaneously, fast lookup of indexed fields, and as the name suggests, the ability to enforce relationships between different tables of data. An example of such a relationship would be to have a field in a *proteins* table directly reference a row in a *genes* table, thereby relating proteins to their relevant gene. Every RDBMS provides a structured query language (SQL) for querying the database, which makes use of any available indexes and relationships to join information from multiple tables quickly and seamlessly. On the opposite end of the structural spectrum of databases are systems that do not provide SQL access, often don't guarantee integrity, and sometimes provide no relational control: these are the NoSQL databases. NoSQL systems provide very fast, fault-tolerant, distributed access to often structure-less data, and can range from 'simple' key-value stores, to document and XML storage systems, to graph-based *RDF databases*. This latter class of databases, often described as triple- or quad-stores, allow any kind of data to be stored and related in any way imaginable. This generality of purpose can be a powerful tool, but at the cost of lowering performance in both speed and memory usage. To make full use of the power of RDF databases, the SPARQL protocol and RDF query language (recursively abbreviated as SPARQL) is provided to allow complex querying of arbitrary graphs of data.

As with most technology, the choice of database for any particular use case depends on what kind of data is being stored, and what kind of constraints and services are required. In the best systems, choosing a particular database is also accompanied by a well-designed schema that follows applicable design guidelines (such as minimising redundancy and dependency in relational databases: a process called normalisation).

See also SPARQL, SQL

DATABASE OF GENOTYPES AND PHENOTYPES, *SEE* DBGAP.

DATABASE SEARCH ENGINE (PROTEOMICS) (PEPTIDE SPECTRUM MATCH, PSM)

Simon Hubbard

In proteomic experiments, proteins are characterized from putative identifications of their constituent peptides analysed in the mass spectrometer by comparing experimental peptide spectra to theoretical ones using search engines. These tools take FASTA files of protein amino acid sequences, digest the proteins *in silico* into peptides, and generate the theoretical spectra for comparisons. Most search engines calculate a likelihood score or expectation value to give the user a means to assess the quality of each candidate *Peptide Spectrum Match (PSM)* between theoretical and experimental spectra. The first published algorithm to assign peptide sequences to un-interpreted tandem mass spectra was SEQUEST, whilst other popular tools include Mascot, OMSSA, X!Tandem, Phenyx and SpectrumMill.

Related websites

Matrix Science's Mascot	http://www.matrixscience.com/
OMSSA	http://pubchem.ncbi.nlm.nih.gov/omssa/
X!Tandem	http://www.thegpm.org/tandem/

Further reading

Geer LY, Markey SP, Kowalak JA, Wagner L, Xu M, Maynard DM, Yang X, Shi W, Bryant SH. Open Mass Spectrometry Search Algorithm. *J. Proteome Res.* 2004, 3: 958–964.

Perkins DN, Pappin DJ, Creasy DM, Cottrell JS. Probability-based protein identification by searching sequence databases using mass spectrometry data. *Electrophoresis.* 1999, 20: 3551–67.

Steen H, Mann M. The ABC's (and XYZ's) of peptide sequencing. *Nat Rev Mol Cell Biol.* 2004, 5: 699–711.

DATAMONKEY

Michael P. Cummings

A web-based system for analysing protein coding sequences for detecting signatures of positive and negative selection using numerous statistical models including maximum likelihood and Bayesian approaches.

Related website

DataMonkey home page	http://www.datamonkey.org/

Further reading

Kosakovsky Pond SL, Frost SDW (2005) Not so different after all: a comparison of methods for detecting amino acid sites under selection *Moecular Biology and Evolution* 22: 1208–1222.

See also HyPhy

DAYHOFF AMINO ACID SUBSTITUTION MATRIX (PAM MATRIX, PERCENT ACCEPTED MUTATION MATRIX)

Laszlo Patthy

Amino acid substitution matrices describe the probabilities and patterns displayed by non-synonymous mutations of nucleotide sequences during evolution. Amino acid substitution probabilities in proteins were analyzed by Dayhoff (1978) who tabulated non-synonymous mutations observed in several different groups of global alignments of closely related protein sequences that are at least 85% identical. Mutation data matrices or Percent Accepted Mutation (PAM) matrices, commonly known as Dayhoff matrices, were derived from the observed patterns of non-synonymous substitutions and matrices for greater evolutionary distances were extrapolated from those for lesser ones.

The observed mutational patterns have two distinct aspects: the resistance of an amino acid to change and the pattern observed when it is changed. The data collected on a large number of protein families has shown that the relative mutabilities of the different amino acids show striking differences: on the average, Asn, Ser, Asp, and Glu are most mutable, whereas Trp, Cys, Tyr and Phe are the least mutable. The relative immutability of cysteine can be interpreted as a reflection of the fact that it has several unique, indispensable functions that no other amino acid side-chain can mimic (e.g. it is the only amino acid that can form disulphide bonds). The low mutability of Trp, Tyr and Phe may be explained by the importance of these hydrophobic residues in protein-folding.

The distribution of accepted amino acid replacement mutations suggests that the major cause of preferences in amino acid substitutions is that the new amino acid must function in a way similar to the old one. Accordingly, conservative changes to chemically and physically similar amino acids are more likely to be accepted than radical changes to chemically dissimilar amino acids. Some of the key properties of an amino acid residue that determine its role and replaceability are: size, shape, polarity, electric charge and ability to form salt bridges, hydrophobic bonds, hydrogen bonds, disulphide bonds.

In general, chemically similar amino acids tend to replace one another: within the aliphatic group (Met, Ile, Leu, Val); within the aromatic group (Phe, Tyr, Trp); within the basic group (Arg, Lys); within the acid, acid-amide group (Asn, Asp, Glu, Gln); within the group of hydroxylic amino acids (Ser, Thr) etc. Cysteine practically stands alone. Glycine-alanine interchanges seem to be driven by selection for small side chains, proline-alanine interchanges by selection for small aliphatic side chains etc.

Dayhoff amino acid substitution matrices are used in the alignment of amino acid sequences to calculate alignment scores and in sequence similarity searches.

Further reading

Dayhoff MO (1978) Survey of new data and computer methods of analysis. In: *Atlas of Protein Sequence and Structure*, Natl. Biomed.Res.Found. Wash. DC, Vol 5, suppl. 3.

Dayhoff MO, et al. (1978) A model of evolutionary change in proteins. *Atlas of Protein Sequence and Structure*, Natl. Biomed.Res.Found. Wash. DC, Vol 5, suppl. 3.

Dayhoff MO, et al. (1983) Establishing homologies in protein sequences. *Methods Enzymol* 91: 524–545.

George DG, et al. (1990) Mutation Data Matrix and its uses. *Methods Enzymol* 183: 333–351.

See also Amino Acid Exchange Matrix, BLOSUM Matrix, Non-Synonymous Mutation, Global Alignment, Alignment of Amino Acid Sequences, Alignment Score, Sequence Similarity Search

DBEST

Guenter Stoesser

dbEST is a database containing sequence data and mapping information on Expressed Sequence Tags (ESTs) from various organisms.

ESTs are typically short (about 300–500 bp) single-pass sequence reads from cDNA (mRNA) and are tags of expression for a given cDNA library. ESTs have applications in the discovery of new genes, mapping of genomes, and identification of coding regions in genomic sequences. Sequence data from dbEST are incorporated into the EST division of DDBJ/EMBL/GenBank. dbEST is maintained at the National Center for Biotechnology Information (NCBI), Bethesda (USA).

Related website

dbEST homepage	http://www.ncbi.nlm.nih.gov/dbEST/

Further reading

Adams MD, et al. (1991) Complementary DNA sequencing: expressed sequence tags and human genome project. *Science* 252: 1651–1656.

Boguski MS, et al. (1993) dbEST-database for expressed sequence tags *Nat Genet*. 4: 332–333.

Boguski MS (1995) The turning point in genome research *Trends Biochem Sci*. 20: 295–296.

Schuler GD (1997) Pieces of the puzzle: expressed sequence tags and the catalog of human genes. *J Mol Med*. 75: 694–698.

DBGAP (THE DATABASE OF GENOTYPES AND PHENOTYPES)

Malachi Griffith and Obi L. Griffith

dbGAP is designed to organize results from studies that examine the relationship between genotypes and phenotypes. These studies include genome-wide association studies (GWAS), molecular diagnostics and medical sequencing. All data deposited in dbGAP are designated as either open source or controlled access. In order to obtain controlled access data, the researcher must request permission from a designated data access committee (DAC). Users may submit their own data to dbGAP as long as they meet minimum requirements in describing their study. Data can be browsed or searched by gene symbol, genome position, phenotypic characteristic, or by downloading data via FTP (public access data only). As of this writing, 257 'top-level' studies consisting of 2168 data sets had been entered. Each dbGAP study record includes extensive details on the study design, phenotypic variables considered, analyses conducted and raw data generated. Tools such as an association results browser and phenotype-genotype integrator (PheGenI) are provided to facilitate visualization of the data and to assist in the prioritization of variants and design of GWAS follow-up studies.

Related websites

dbGAP homepage	http://www.ncbi.nlm.nih.gov/gap
dbGAP tutorial	http://www.ncbi.nlm.nih.gov/projects/gap/tutorial/dbGaP_demo_1.htm

Further reading

Wooten EC, Huggins GS (2011) Mind the dbGAP: the application of data mining to identify biological mechanisms. *Mol Interv.* 11(2): 95–102.

Zhang H, et al. (2008) The NEI/NCBI dbGAP database: genotypes and haplotypes that may specifically predispose to risk of neovascular age-related macular degeneration. *BMC Med Genet.* 9:51.

DBSNP

Guenter Stoesser, Obi L. Griffith and Malachi Griffith

dbSNP is a central repository for genetic variations.

The vast majority (>99%) of variations described in dbSNP are single nucleotide polymorphisms (SNPs). Most of the rest are small insertions and deletions. A single nucleotide polymorphism, or SNP (pronounced 'snip') is a small genetic change, or variation that can occur within an organism's DNA sequence by changing one base to another. SNPs occur roughly every 100–300 bp in human chromosome sequences.

dbSNP entries include the sequence information around the polymorphism, the specific experimental conditions necessary to perform an experiment, descriptions of the population containing the variation and frequency information by population or individual genotype. SNP data facilitate large-scale association genetics studies for associating

D

sequence variations with heritable phenotypes. Furthermore, SNPs facilitate research in functional and pharmaco-genomics studies, population genetics, evolutionary biology, positional cloning and physical mapping. As an integrated part of NCBI, the contents of dbSNP are cross-linked to records in GenBank, LocusLink, the human genome sequence and PubMed. The result sets from queries in any of these resources point back to the relevant records in dbSNP. The BLAST algorithm compares a query sequence against all flanking sequence records in dbSNP. dbSNP is collaborating to develop data exchange protocols with other public variation and mutation databases, such as HGBASE (human genic biallelic sequences). dbSNP is maintained at the National Center for Biotechnology Information(NCBI), Bethesda (USA). dbSNP is increasingly used as a reference for the identification of single nucleotide variants in genome or tumor sequencing projects to facilitate the differentiation of somatic mutations from germline variants.

Related website

dbSNP homepage	http://www.ncbi.nlm.nih.gov/SNP

Further reading

Day IN (2010) dbSNP in the detail and copy number complexities. *Hum Mutat*. 31(1): 2–4.

Osier MV, et al. (2001) ALFRED: an allele frequency database for diverse populations and DNA polymorphisms – an update. *Nucleic Acids Res*. 29: 317–319

Saccone SF, et al. (2011) New tools and methods for direct programmatic access to the dbSNP relational database. *Nucleic Acids Res*. *39*(Database issue): D901–7.

Sherry ST, et al. (1999) dbSNP Database for Single Nucleotide Polymorphisms and Other Classes of Minor Genetic Variation *Genome Res*. 9: 677–679.

Sherry ST et al. (2001) dbSNP: the NCBI database of genetic variation *Nucleic Acids Res*. 29: 308–311.

DBSTS

Guenter Stoesser

dbSTS is a database of sequence tagged sites (STS) – short genomic landmark sequences which are used to create high-resolution physical maps of large genomes and to build scaffolds for organizing large-scale genome sequencing.

All STS sequences are incorporated into the STS division of DDBJ/EMBL/GenBank. dbSTS is maintained at the National Center for Biotechnology Information (NCBI), Bethesda (USA).

Related website

dbSTS homepage	http://www.ncbi.nlm.nih.gov/dbSTS/

Further reading

Benson D, et al. (1998) GenBank *Nucleic Acids Res*. 26: 1–7.

Olson M, et al. (1989) A common language for physical mapping of the human genome. *Science* 245: 1434–1435.

DBVAR (DATABASE OF GENOMIC STRUCTURAL VARIATION)

Malachi Griffith and Obi L. Griffith

dbVar is a database of genomic structural variation maintained by the NCBI. A structural variant is defined by dbVar as a region of DNA involved in an inversion, translocation, insertion, or deletion. Although it accepts variants of any size, dbVar recommends that variants smaller than 50 bp be submitted to dbSNP and those larger than 50 bp to dbVar. Both germline and somatic events are stored and unlike the database of genomic variants (DGV), dbVar accepts data for all species, both healthy and diseased individuals, and includes clinical data. dbVar does not restrict variant submissions based on the detection technology employed in the study but the database has been primarily populated by BAC array CGH, representational oligonucleotide microarray analysis (ROMA), fosmid paired-end mapping (PEM), oligo array CGH, the analysis of SNP genotyping data, and PEM using next-generation 454 sequencing. The database comprises three principal data objects: studies, variant regions, and variant instances. A study is a set of variant regions and variant instances that were submitted as a group. A variant region is a region of the genome that the submitter designates as containing variant instances. A variant instance is an actual variant call derived from analysis of raw data. The complete dbVar database is available via FTP on a per study basis and can also be accessed via the dbVar Study Browser.

Related websites

dbVar homepage	http://www.ncbi.nlm.nih.gov/dbvar/
dbVar overview	http://www.ncbi.nlm.nih.gov/dbvar/content/overview/

Further reading

Sayers EW, at al. (2012) Database resources of the National Center for Biotechnology Information. *Nucleic Acids Res.* 40(Database issue): D13–25.

Sneddon TP, Church DM (2012) Online resources for genomic structural variation *Methods Mol Biol.* 838: 273–89.

See also DGV

DDBJ (DNA DATABANK OF JAPAN)

Guenter Stoesser, Obi L. Griffith and Malachi Griffith

The DNA Databank of Japan (DDBJ) is a member of the tripartite International Nucleotide Sequence Database Collaboration, together with NCBI/GenBank (USA) and the EBI/ENA (Europe). Collaboratively, the three organizations collect all publicly available nucleotide sequences and exchange sequence data and biological annotations on a daily basis. DDBJ contains RNA and DNA sequences produced primarily by traditional Sanger sequencing projects but also increasingly includes consensus or assembled contig sequences from next-generation sequencing projects. Major data sources are individual research groups, large-scale genome sequencing centers and the Japanese Patent Office (JPO). Sophisticated submission systems are available (Sakura). Specialized databases provided by DDBJ include

D

the Genome Information Broker (GIB) for complete genome sequence data. The database is maintained at the Center for Information Biology (CIB), a division of the National Institute of Genetics (NIG) in Mishima, Japan. Additional to operating the DNA Databank of Japan, CIB's mission is to carry out research in information biology.

Related websites

DDBJ homepage	http://www.ddbj.nig.ac.jp
DDBJ submission systems and related information	http://www.ddbj.nig.ac.jp/submission-e.html
Search and data analysis	http://www.ddbj.nig.ac.jp/searches-e.html

Further reading

Kaminuma E, et al. (2011) DDBJ progress report. *Nucleic Acids Res.* 39(Database issue): D22–27.

Kaminuma E, et al. (2010) DDBJ launches a new archive database with analytical tools for next-generation sequence data. *Nucleic Acids Res. 38*(Database issue): D33–38.

Kodama Y, et al. (2010) Biological databases at DNA Data Bank of Japan in the era of next-generation sequencing technologies. *Adv Exp Med Biol.* 680: 125–135.

Tateno Y, et al. (2002) DNA Data Bank of Japan (DDBJ) for genome scale research in life science. *Nucleic Acids Res.* 30: 27–30.

Tateno Y, et al. (2002) DNA Data Bank of Japan (DDBJ) in collaboration with mass sequencing teams. *Nucleic Acids Res.* 28: 24–26.

Tateno Y, Gojobori T. (1997) DNA Data Bank of Japan in the age of information biology. *Nucleic Acids Res.* 25: 14–17.

DDL, *SEE* DATA DESCRIPTION LANGUAGE.

DE NOVO ASSEMBLY IN NEXT GENERATION SEQUENCING

Stuart Brown

Next Generation DNA Seqencing (NGS) can be used to study the genomes of new organisms that have never previously been sequenced (*de novo* sequencing). Since NGS starts with randomly sheared genomic DNA, and produces short (25–400) bp reads, building a complete genomic sequence requires assembly of short fragments into a complete genome, which may be composed of one or more chromosomes plus any extra-chromosomal elements such as plasmids, mitochondrial and/or chloroplast DNA, etc.

Traditional fragment assembly methods that were developed for Sanger sequencing rely on finding overlaps between fragments using sequence alignment-based similarity algorithms such as Smith-Waterman. This method makes comparisons between each sequence read and every other read, an all-vs-all strategy that increases the computation with the square of the number of fragments. This is suitable to assemble a few hundred reads for a 5 kb plasmid insertion or a few thousand in a cosmid, but not possible for the hundreds of millions of reads ($\sim 10^{16}$ comparisons) for a NGS run. Methods for *de novo* assembly of genomes from NGS data now make use of De Bruijn graph theory, which creates a map of overlapping short words in the data, then finds overlaps between reads that span adjacent words.

Assembly of billion base pair eukaryotic genomes remains challenging due to the large size and the presence of many types of low complexity and repeated DNA elements including homopolymer runs, simple sequence repeats, telomeres and centromeres, transposons and other insertion sequences, gene duplications, segmental duplications, and partial or complete chromosomal duplications. However, bacterial genomes (of a few million bases pairs) can now be fully sequenced cheaply, and assembled *de novo* with sufficient speed and accuracy to make this a routine tool for microbiology. In fact, even when a reference genome exists for a bacterial species, *de novo* sequencing and assembly methods are often used (rather than mapping reads to the reference genome) so that the full range of sequence diversity in each strain or isolate is captured. Complete genomes for entire collections of bacterial strains can now be sequenced and assembled for a few thousand dollars.

Related website

De Bruijn Graph Theory	http://en.wikipedia.org/wiki/De_Bruijn_graph

See also Next Generation Sequencing (NGS), Reference Genome

DEAD-END ELIMINATION ALGORITHM

Roland Dunbrack

An algorithm for side-chain placement onto protein backbones that eliminates rotamers for some side-chains that cannot be part of the global minimum energy configuration, originally developed by Johan Desmet and colleagues. This method can be used for any search problem that can be expressed as a sum of single-side-chain terms and pairwise interactions. Goldstein's improvement on the original DEE can be expressed as follows. If the total energy for all side chains is expressed as the sum of singlet and pairwise energies,

$$E = \sum_{i=1}^{N} E_{bb}(r_i) + \sum_{i=1}^{N-1} \sum_{j>i}^{N} E_{SC}\left(r_i, r_j\right)$$

then a rotamer r_i can be eliminated from the search if there is another rotamer s_i for the same side chain that satisfied the following equation:

$$E_{bb}(r_i) - E_{bb}(S_i) + \sum_{j=1,\ j\neq i}^{N} \min_{r_j} \left\{ E_{sc}\left(r_i, r_j\right) - E_{sc}\left(s_i, r_j\right) \right\} > 0$$

In words, rotamer r_i of residue i can be eliminated from the search if another rotamer of residue i, s_i, always has a lower interaction energy with all other side chains regardless of which rotamer is chosen for the other side chains. More powerful versions have been

D

developed that eliminate certain pairs of rotamers from the search. DEE-based methods have also proved very useful in protein design, where there is variation of residue type as well as conformation at each position of the protein.

The algorithm can also be applied to other situations where the energy function can be defined in terms of pairwise interactions, such as sequence alignment.

Relevant website

Canonical loop modeling using dead-end	http://antibody.bath.ac.uk/index.html

Further reading

De Maeyer M, et al. (2000) The dead-end elimination theorem: mathematical aspects, implementation, optimizations, evaluation, and performance. *Methods Mol Biol* 143: 265–304.

Desmet J, De Maeyer J, Hazes D, Lasters I. (1992) The dead-end elimination theorem and its use in protein sidechain positioning. *Nature* 356: 539–542.

Goldstein R0. (1994) Efficient rotamer elimination applied to protein side-chains and related spin glasses. *Biophys J* 66: 1335–1340.

Looger LL, Hellinga HW (2001) Generalized dead-end elimination algorithms make large-scale protein side-chain structure prediction tractable: implications for protein design and structural genomics. *J. Mol. Biol.* 307: 429–445.

See also Rotamer

DECISION SURFACE

Concha Bielza and Pedro Larrañaga

A decision surface or decision boundary is a hypersurface that partitions the underlying vector space of predictor variables into different (decision) regions, one for each class label. Data lying on one side of a decision surface are assigned to a different class from those lying on the other.

If the decision surface is a hyperplane, then the classification problem is linear, and the classes are linearly separable.

Further reading

Duda R, Hart PE, Stork DG (2001) *Pattern Classification*. Wiley.

Larrañaga P, Calvo B, Santana R, et al. (2006) Machine learning in bioinformatics. *Briefings in Bioinformatics* 7 (1): 86–112.

See also Machine Learning, Support Vector Machine

DECISION TREE

Feng Chen and Yi-Ping Phoebe Chen

A decision tree is a tree graph, the illustration of which is shown in Figure D.1. The top of the graph is called the root. The end nodes are called leaf points. Other nodes represent

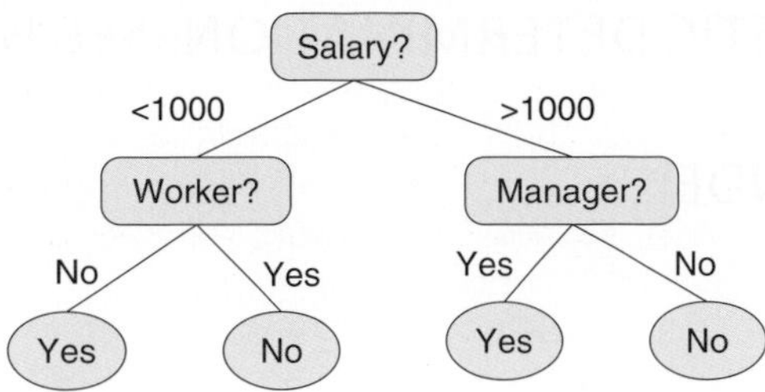

Figure D.1 An Example of Decision Tree.

branches. Every branch indicates a test on one attribute. For example, we use Figure D.1 to answer the question 'What kind of people could buy a computer?'. Then we can find two answers: 1). A person who earns less than 1000 dollars but is not a worker could buy a computer; and 2). A manager whose salary is more than 1000 could buy one.

So, we can use a decision tree to match the attributes of one object to construct a path from root to leaf node when we want to label it. The leaf node represents which class or cluster this object should belong to.

Decision trees are very widely used for classification and prediction for the following reasons: 1). The construction of decision trees is independent of any professional knowledge or parameters; 2). They can deal with high-dimensional data efficiently; 3). The representation of knowledge is very easy to understand; and 4). The key algorithm is simple but effective.

DECOY DATABASE

Simon Hubbard

When performing large numbers of peptide spectrum searches against a FASTA formatted protein database in high throughput proteomics, decoy databases are used to act as a background model of false 'hits' from which False Discovery Rates (FDRs) can be estimated, presuming that all decoy hits above a given threshold score are 'false' by definition. This is a convenient way to control the overall error rate for a large number of database searches, typically limiting the FDR to 1% or 5%. The decoy database is often generated by simply reversing the 'forward' target database and concatenating this onto the end of it. Alternate methods include shuffling the database or generating a random database of equal size. It is desirable to maintain the amino acid composition, number of proteins and number of proteolytic peptides in the decoy database in order to generate meaningful statistics. The forward database is usually referred to as the target database.

Further reading

Elias JE, Haas W, Faherty BK, Gygi SP (2005) Comparative evaluation of mass spectrometry platforms used in large-scale proteomics investigations. *Nat Methods* (9): 667–675.

Käll L, Storey JD, MacCoss MJ, Noble WS (2008) Assigning significance to peptides identified by tandem mass spectrometry using decoy databases. *J. Proteome Res.* 7(1): 29–34.

See also False Discovery Rate (FDR)

D

DEGREE OF GENETIC DETERMINATION, *SEE* HERITABILITY.

DELETION, *SEE* INDEL.

DELILA

Thomas D. Schneider

DELILA stands for DEoxyribonucleic-acid LIbrary LAnguage.

It is a language for extracting DNA fragments from a large collection of sequences, invented around 1980. The idea is that there is a large database containing all the sequences one would like, which we call a 'library'. One would like a particular subset of these sequences, so one writes up some instructions and gives them to the librarian, Delila, which returns a 'book' containing just the sequences one wants for a particular analysis. So 'Delila' also stands for the program that does the extraction. Since it is easier to manipulate Delila instructions than to edit DNA sequences, one makes fewer mistakes in generating one's data set for analysis, and they are trivial to correct. Also, a number of programs create instructions, which provides a powerful means of sequence manipulation. One of Delila's strengths is that it can handle any continuous coordinate system. The 'Delila system' refers to a set of programs that use these sequence subsets for molecular information theory analysis of binding sites and proteins. Delila is capable of making sequence mutations, which can be displayed graphically along with sequence walkers.

A complete definition for the language is available (LIBDEF), although not all of it is implemented. There are also tutorials on building Delila libraries and using Delila instructions. A web-based Delila server is available.

Related websites

NCBI	http://www.ncbi.nlm.nih.gov/
Genbank	http://www.ncbi.nih.gov/Genbank/GenbankOverview.html
Delila	http://alum.mit.edu/www/toms/delila/delila.html
Libdef	http://alum.mit.edu/www/toms/libdef.html
Delila server	http://alum.mit.edu/www/toms/delilaserver.html

Further reading

Schneider TD, et al. (1982) A design for computer nucleic-acid sequence storage, retrieval and manipulation. *Nucleic Acids Res.* 10: 3013–3024.

Schroeder JL, Blattner FR (1982) Formal description of a DNA oriented computer language. *Nucleic Acids Res.* 10: 69–84.

DELILA INSTRUCTIONS

Thomas D. Schneider

Delila instructions are a set of detailed instructions for obtaining specific nucleic-acid sequences from a sequence database. The instructions are written in the computer language Delila. There is a tutorial on using Delila instructions available online.

Related website

DELILA Instructions summary	http://alum.mit.edu/www/toms/delilainstructions.html

Further reading

Schneider TD, et al. (1982) A design for computer nucleic-acid sequence storage, retrieval and manipulation. *Nucleic Acids Res* 10: 3013–3024.

See also DELILA

DENDROPY

Michael P. Cummings

A Python library for phylogenetic-related operations.

It includes classes and functions for the simulation, processing, and manipulation of phylogenetic trees and character matrices, and supports the reading and writing of phylogenetic data in a range of formats.

Related website

DendroPy home page	http://pythonhosted.org/DendroPy/

Further reading

Sukumaran J, Holder MT (2010) DendroPy: A Python library for phylogenetic computing. *Bioinformatics* 26: 1569–1571.

DEPENDENT VARIABLE, *SEE* LABEL.

DESCRIPTION LOGIC (DL)

Robert Stevens

Description logics (DLs) are a form of knowledge representation language describing categories, individuals, and the relationships amongst them in a logical formalism.

They are sub-sets of standard predicate logic. They attempt to give a unified logical basis to the various well known traditions of knowledge representation: frame-based systems, semantic networks. They are derived from work on the KL-ONE family of knowledge representation languages. Different DLs have different levels of expressiveness. All have some operations for defining new concepts from existing concepts and the relations amongst them. The operators available almost always include conjunction and existential quantification, and may include negation, disjunction, universal quantification, and other logical operations on concepts on relations. Due to the logical formalism of the DL, ontologies expressed in a DL can be submitted to a reasoning application. The reasoner can check the logical consistency of the descriptions (concept satisfiability) and infer a subsumption hierarchy (taxonomy) from the description (subsumption reasoning). 'Description logic systems have been used for building a variety of applications including conceptual modeling, information integration, query mechanisms, view maintenance, software management systems, planning systems, configuration systems and natural language understanding. (Description Logic Home Page).'

Related website

Wikipedia	http://en.wikipedia.org/wiki/Description_logic

DESCRIPTORS, *SEE* FEATURES.

DGV (DATABASE OF GENETIC VARIANTS)

Malachi Griffith and Obi L. Griffith

The Database of Genomic Variants (DGV) aims to provide a comprehensive catalogue of human genomic structural variation. DGV considers a structural variant to be DNA rearrangements (CNVs) involving segments greater than 1000 bp and insertions or deletions in the 100–1000 bp range (indels). In contrast to dbVar, DGV is populated by data only from healthy control samples and thus represents useful control data for studies aiming to correlate genomic variation with phenotypic data. Much of the data in DGV was derived by analysis of BAC clone CGH arrays, SNP arrays, and oligonucleotide CGH arrays. As of November 2010, over 100,000 entries had been imported and of these 65% were CNVs, 33% were InDels and the remainder were inversions. The variants in DGV can be visualized using a customized 'gbrowse' genome browser and raw data can be downloaded as tab delimited files. The DGV obtains its data from DGVa, a repository operated by EBI that provides archiving, accessioning and distribution of publicly available genomic structural variants, in all species. DGV provides extra curation and interpretation services and maintains the most comprehensive resource for genomic structural variation in humans. DGV is hosted by the Centre for Applied Genomics in Toronto, Canada.

Related websites

DGV homepage	http://projects.tcag.ca/variation/
DGV overview	http://projects.tcag.ca/variation/project.html
DGV browser	http://projects.tcag.ca/cgi-bin/variation/gbrowse/hg18/

Further reading

Iafrate AJ, et al. (2004) Detection of large-scale variation in the human genome. *Nat Genet.* 36(9): 949–951.

Zhang J, et al. (2006) Development of bioinformatics resources for display and analysis of copy number and other structural variants in the human genome. *Cytogenet Genome Res.* 115(3–4): 205–214.

DIAGNOSTIC PERFORMANCE, *SEE* DIAGNOSTIC POWER.

D

DIAGNOSTIC POWER (DIAGNOSTIC PERFORMANCE, DISCRIMINATING POWER)

Teresa K. Attwood

A measure of the ability of a discriminator to identify true matches and to distinguish them from false matches. In the context of protein sequence pattern recognition, a discriminator is a computational abstraction (a consensus (or regex) pattern, profile, hidden Markov model, fingerprint, etc.) of a conserved motif or set of motifs, or of a domain, used to search either an individual query sequence or a full database for the occurrence of that same, or similar, motif(s) or domain.

For every sequence encountered by such a discriminator, there will be one of four possible outcomes: i) the sequence is a true family member and is correctly diagnosed as such (this result is true-positive); ii) the sequence is not a family member and is not matched by the discriminator (this result is true-negative); iii) the sequence is not a family member, but is matched by the discriminator in error (this result is false-positive); or the sequence is a true family member, but is not diagnosed as such (this result is false-negative).

The goal of pattern recognition methods is to improve diagnostic performance, to maximize the number of true-positive matches captured, while minimizing the number both of false-positive and of false-negative matches. Ideally, there should be perfect separation between true-positive and true-negative results. In practice, however, the populations of each tend to overlap, and scoring thresholds must be used to optimize the balance between the ability of the discriminator to capture the majority of true-positive

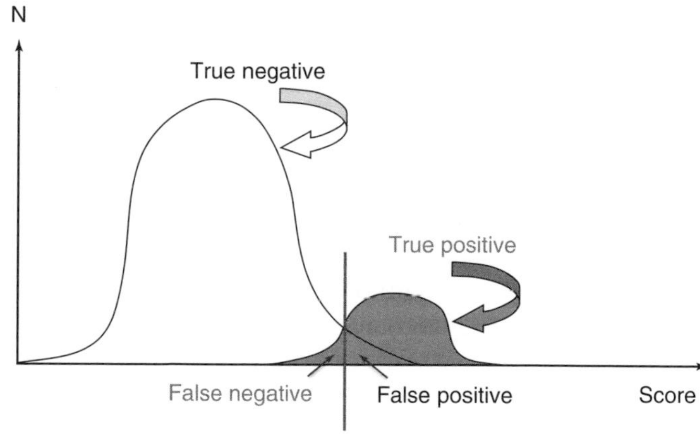

Figure D.2 The diagnostic challenge of separating populations of true-negative and true-positive results, and hence of distinguishing true-positive from false-positive matches. The y axis denotes the number of matches; the x axis gives the match score; the vertical bar denotes a particular scoring threshold, chosen to maximise the number of true-positive matches and minimise the numbers of falsely matched instances (false positives) and of true instances missed in error (false negatives).

instances while keeping the number of false-positive instances to a minimum, as illustrated in Figure D.2.

Given the diagnostic limitations of most sequence analysis methods, it is good practice to use several approaches and to evaluate the reliability of any diagnosis by seeking a consensus.

Related website

Scores and match significance	http://hits.isb-sib.ch/cgi-bin/help?doc=scores.html

Further reading

Attwood TK, Parry-Smith DJ (1999) *Introduction to Bioinformatics*. Addison Wesley Longman, Harlow, Essex, UK.

Higgs P, Attwood TK (2005) *Bioinformatics and Molecular Evolution*. Blackwell Publishing, Oxford, UK.

See also Consensus Pattern, Profile, Hidden Markov Model, Fingerprint, Motif

DIHEDRAL ANGLE (TORSION ANGLE)

Roman Laskowski and Tjaart de Beer

Mathematically, a dihedral angle is the angle between two planes. It is zero if the planes are parallel, and can take values up to 90°.

In chemistry, a dihedral angle can be defined for any chain of four covalently bonded atoms: A–B–C–D. The angle is between the plane containing atoms A, B and C, and that containing atoms B, C and D. It represents the relative rotation of bonds A–B and C–D about the bond B–C.

By convention, dihedral angle values range from −180° to +180°. If, when viewed along the bond B–C, the bond A–B eclipses bond C–D, the value of the dihedral angle is 0°. If the bonds are not eclipsed, the absolute value of the dihedral angle is the angle through which the bond A–B has to be rotated about B–C to achieve eclipse; the sign being positive if the rotation is clockwise, and negative if it is anti-clockwise.

In proteins, dihedral angles are defined for the backbone and side chain atoms. The three backbone dihedral angles, and their defining atoms, are: *phi*, (C_{i-1}–N_i–$C\alpha_i$–C_i), *psi* (N_i–$C\alpha_i$–C_i–N_{i+1}) and *omega* ($C\alpha_i$–C_i–N_{i+1}–$C\alpha_1$. The *phi* and *psi* dihedral angles determine the local secondary structure of the protein chain, and are commonly depicted on a Ramachandran plot. The *omega* dihedral angle is defined about the peptide bond. Because of the partial double bond nature of this bond, it forms a planar, or close to planar, unit. The value of *omega*, therefore, will be close to either *omega* = 180° (for the highly favoured trans conformation) or *omega* = 0° (for the fairly rare cis conformation).

Side chain dihedral angles in proteins are designated by chi_n, where n is the number of the bond counting out from the central Cα atom. Steric hindrance often limits the ranges the values can take, the favoured conformations being known as rotamers.

It is also possible to define a 'virtual' dihedral angle for any set of four atoms, A, B, C and D, where there is no physical bond between atoms B and C. See, for example, the *kappa* and *zeta* virtual dihedral angles.

See also Backbone, Side Chain, Kappa Virtual Dihedral Angle, Zeta Virtual Dihedral Angle, Secondary Structure, Ramachandran Plot, Amide Bond, Rotamer

DINUCLEOTIDE FREQUENCY

Katheleen Gardiner

Frequency at which any two bases are found adjacent on the same strand in a DNA sequence.

The four bases, A, G, C and T, permit 16 dinucleotides (AA, AT, AC, AG, etc). In the mammalian genome, the frequency of the CG dinucleotide is statistically low; it is present at only 1/5–1/4 of the frequency expected from the average genome base composition of ~38%GC. This CpG suppression is related to DNA methylation. CpG dinucleotides are enriched in CpG islands.

Further reading

Nekrutenko A, Li WH (2000) Assessment of compositional heterogeneity within and between eukaryotic genomes *Genome Res* 10: 1986–1995.

Swartz M, Trautner TA, Kornberg A (1962) Enzymatic synthesis of DNA: further studies on nearest neighbor base sequences in DNA *J Biol Chem* 237: 1961–1967.

See also CpG, DNA Methylation, CpG Island

DIP

Patrick Aloy

The Database of Interacting Proteins (DIP) catalogs and classifies experimentally determined protein-protein interactions. It combines information from very different sources like two-hybrid experiments, large-scale studies on protein complexes or literature searches to create a single, consistent and comprehensive set of protein-protein interactions.

To date, DIP contains information about 17838 interactions involving 6812 proteins from 110 different organisms. All the interactions in the DIP are manually curated by both human experts and intelligent knowledge-based computer systems.

The database offers several searching possibilities (e.g. ID, BLAST, motif ...) and graphic tools to display the retrieved interactions.

Relevant website

DIP	http://dip.doe-mbi.ucla.edu

D

Further reading

Xenarios I, Salwinski L, Duan XJ, Higney P, Kim S, Eisenberg D (2002) DIP: The database of interacting proteins. A research tool for studying cellular networks of protein interactions. *Nucleic Acids Res* 30: 303–305.

See also Knowledge-based

DIRECTED ACYCLIC GRAPH (DAG)

Robert Stevens

A DAG is a form of multi-hierarchy of nodes and arcs.

The nodes form the terms or concepts in the ontology and the arcs the relationships. Any one concept can have many parents and many children. More than one kind of relationship can be used. The 'directed' nature of the graph comes from the directional nature of the relationships: A is a parent of B is directed. In other words, the graph has a top and a bottom. The Gene Ontology is referred to as a DAG (see Figure D.3).

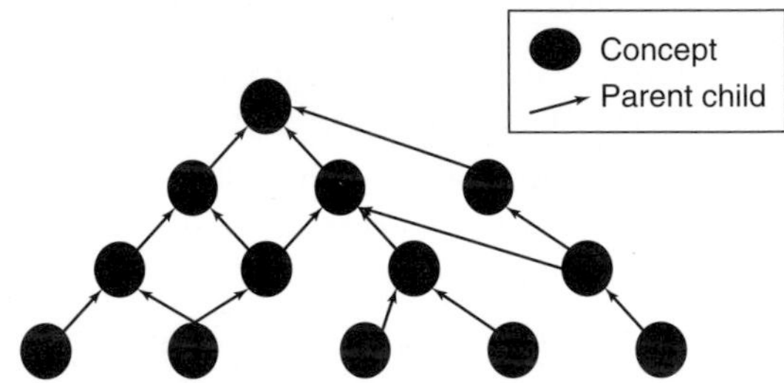

Figure D.3 Organisation of concepts in a directed acyclic graph.

DISCRETE FUNCTION PREDICTION (FUNCTION PREDICTION)

Juan Antonio Garcia Ranea

The act of predicting novel functional aspects of single biological molecules (i.e. proteins and DNA) on the basis of diverse biological data, independent from or in the absence of any direct experimental knowledge. This novel functional knowledge is obtained by using computational methods that exploit different data sources containing information associated with the protein of interest or contextual features of the targeted molecules.

Because the current experimental data is still very limited (e.g. perhaps less than 10% of protein interactions have been experimentally characterised), discrete functional prediction is useful for increasing the efficiency of selecting novel experimental targets, by providing hypotheses about their functions and associations. This is especially useful when characterizing targets that involve experiments that require a large investment in reagents, equipment and data management.

Discrete function prediction methods include mathematical algorithms and computational procedures that mine biological information, searching for statistically significant shared features between the targeted molecules and molecules characterized in biological databases. Detection of a consistent similarity or association enables the transfer of functional knowledge from the characterised molecules to the targeted ones.

There are many approaches for exploiting biological data sources to make inferences about molecular functions and associations, most of which predict protein functions (Lees et al. 2011). Prediction methods can be classified based on the type of data they use:

Sequence homology searches: the first discrete function prediction methods. These make use of sequence information to predict function. *Integration of sequence features*: methods such as machine-learning based on artificial neural networks that integrate amino acid content data, subcellular localization, secondary structure, and other physio-chemical properties to predict function (de Lichtenberg et al., 2003).

Functional association between proteins can be predicted using the following approaches: *Genomic context information*, searching for evolutionary co-dependencies between sequences. These approaches include methods such as: *Comparison of co-occurence (phylogenetic) profiles*, which look for similar presence/absence patterns of genes across species based on the principle that if genes are functionally related, they will tend to be co-inherited (Pellegrini et al., 1999); *Gene fusion methods*, that exploit the evolutionary phenomenum whereby separate but functionally related genes are fused into a single gene in some organisms (Marcotte et al., 1999); *Genomic neighbourhood methods*, which detect genes that are maintained in close proximity, on a chromosome, over long evolutionary time periods due to functional constraints (Ferrer et al., 2010); *Sequence co-evolution methods*, which detect correlated compensatory mutations using multiple alignments of interacting proteins, whereby a mutation of an interaction-mediating residue is compensated by a residue mutation in the binding partner (Pazos et al., 1997); *Commonly occurring domain pairs predictions*, based on domain–domain associations conserved across species with certain domain pairs recurring in multiple protein interactions (Kim et al., 2002); *Inheriting protein interactions from sequence methods*, which search for homologues of proteins with known interactions (interologues), ie interactions that are conserved across species.

Inferring protein interactions from structure, these methods use conserved structural complexes to infer interactions from a known interacting pair to proteins with similar structures (Aloy and Russell, 2003).

Exploiting experimental data: methods which *compare gene expression profiles; text mining* methods which find literature-derived associations; methods which search for *functional semantic similarity* between genes or proteins in structured biological ontologies (Lord et al., 2003); or methods to study the proximity between the nodes of graphs representing protein interaction networks (Ben-Hur & Noble, 2005).

References

Aloy P, Russell RB (2003) InterPreTS: protein interaction prediction through tertiary structure *Bioinformatics* 19 161–162.

Ben-Hur A, Noble WS (2005) Kernel methods for predicting protein–protein interactions *Bioinformatics* 21 (Suppl. 1) i38–46.

D

de Lichtenberg U, Jensen TS, Jensen LJ, Brunak S (2003) Protein feature based identification of cell cycle regulated proteins in yeast. *J Mol Biol* 329: 663–674.

Ferrer L, Dale JM, Karp PD (2010) A systematic study of genome context methods: calibration, normalization and combination *BMC Bioinformatics* 11: 493.

Kim WK, Park J, Suh JK (2002) Large scale statistical prediction of protein–protein interaction by potentially interacting domain (PID) pair *Genome Inform* 13 42–50.

Lees JG, Heriche JK, Morilla I, Ranea JA, Orengo CA (2011) Systematic computational prediction of protein interaction networks. *Phys Biol* 8: 035008.

Lord PW, Stevens RD, Brass A, Goble CA (2003) Investigating semantic similarity measures across the Gene Ontology: the relationship between sequence and annotation *Bioinformatics* 19: 1275–1283.

Marcotte EM, Pellegrini M, Ng HL, Rice DW, Yeates TO, Eisenberg D (1999) Detecting protein function and protein–protein interactions from genome sequences *Science* 285: 751–753.

Pazos F, Helmer-Citterich M, Ausiello G, Valencia A (1997) Correlated mutations contain information about protein–protein interaction J. *Mol. Biol.* 271: 511–523.

Pellegrini M, Marcotte EM, Thompson MJ, Eisenberg D, Yeates TO (1999) Assigning protein functions by comparative genome analysis: protein phylogenetic profiles. *Proc Natl. Acad. Sci. USA* 96 4285–4288.

DISCRIMINANT ANALYSIS, *SEE* CLASSIFICATION.

DISCRIMINATING POWER, *SEE* DIAGNOSTIC POWER.

DISTANCE MATRIX (SIMILARITY MATRIX)

John M. Hancock

In sequence analysis, a matrix containing distance values for all pairs of sequences in a dataset. The distances contained in a distance matrix may be of various types depending on whether the sequences compared are nucleic acids or proteins and the distance measure implemented. The generation of distance matrices is the first step to deriving phylogenetic or gene trees using distance-based methods such as UPGMA or (Bio)NJ as well as being used in some alignment algorithms.

A similarity matrix is the converse of a distance matrix, containing information on sequence similarities (or identities) as opposed to distances.

Related websites

DNADIST	http://cmgm.stanford.edu/phylip/dnadist.html
PROTDIST	http://cmgm.stanford.edu/phylip/protdist.html

See also Evolutionary Distance, Sequence Distance Measures

DISTANCES BETWEEN TREES (PHYLOGENETIC TREES, DISTANCE)

Aidan Budd and Alexandros Stamatakis

Single numerical values that are used for quantifying the topological distance (mostly not taking into account branch lengths) between a pair of trees.

The most commonly used topological distance is the Robinson-Foulds distance that counts the number of non-trivial bipartitions that are present (unique) in one of the two trees but not in both. This non-trivial bipartition count can be normalized to yield a relative dissimilarity value by dividing the unique bipartition count by *2 * (n-3)* where n is the number of taxa. Since an unrooted binary tree has *n-3* inner branches (and hence *n-3* non-trivial bipartitions), a pair of trees that is maximally dissimilar, will not share a single non-trivial bipartition and hence have *2 * (n-3)* unique non-trivial bipartitions.

Another distance measure is the so-called Subtree Pruning Re-grafting distance. This distance counts the minimum number of subtree pruning and re-grafting (subtree re-insertion) operations that are required to transform one tree into the other. Unfortunately, calculating the SPR distance is know to be NP-hard and is thus probably of limited practical interest, since with practical heuristic implementations we do not know how far we are from the absolute minimum SPR distance. The so-called quartet distance represents another topological distance measure between two tree. The quartet distance counts the number of all possible quartets (4 taxon subsets) as induced by the two trees that exhibit a different topology.

Several tree distance measures have been described that take into account both topological and branch length differences between trees, for example the Branch Length Distance or the Minimum Branch Length Distance (K tree score).

Software

RAxML: https://github.com/stamatak/standard-RAxMLMatthews, S., Williams, T. (2010) MrsRF: an efficient MapReduce algorithm for analyzing large collections of evolutionary trees. BMC Bioinformatics 11 (Suppl 1): S15

Stissing, M, Mailund T, Pedersen CNS, Brodal GS, Fagerberg R (2008) Computing the all-pairs quartet distance on a set of evolutionary trees. *Journal of Bioinformatics and Computational Biology* 6(1): 37–50.

Further reading

Brodal GS, Fagerberg R, Nørgaard C, Pedersen S (2003) *Computing the Quartet Distance Between Evolutionary Trees in Time. O(nlogn) Algorithmica, Special issue on ISAAC 2001.*

Bryant D, Tsang J, Kearney PE, Li, M (2000) Computing the quartet distance between evolutionary trees. *Proceedings of the Eleventh Annual ACM-SIAM Symposium on Discrete Algorithms* (N.Y.: ACM Press): 285–286.

Hickey G, Dehne F, Rau-Chaplin A, Blouin C (2008) SPR distance computation for unrooted trees. *Evolutionary Bioinformatics Online* 4: 17.

Kuhner MK, Felsenstein J (1994) A simulation comparison of phylogeny algorithms under equal and unequal evolutionary rates. *Mol Biol Evol* 11: 459–468.

Pattengale ND, Gottlieb EJ, Moret BME (2007) Efficiently computing the Robinson-Foulds metric. *Journal of Computational Biology* 14(6): 724–735.

D

Robinson DF, Foulds LR (1981) Comparison of phylogenetic trees. *Mathematical Biosciences* 53(1–2): 131–147.

Soria-Carrasco V, Talavera G, Igea J, Castresana J. (2007) The K tree score: quantification of differences in the relative branch length and topology of phylogenetic trees. *Bioinformatics* 23: 2954–2956.

See also Bipartition, Split, Consensus Trees

DISTRIBUTED COMPUTING

Steve Pettifer, James Marsh and David Thorne

A distributed system is a computer system in which multiple computational units communicate via a network and cooperate in order to perform a common task or solve a computational problem. There is considerable overlap between the concept of a distributed systems and that of a concurrent or parallel system. Typically, however, parallel/concurrent systems consist of homogeneous components, often connected by bespoke high performance networking, where the computational load of a single algorithm is spread amongst the participating units in order to accelerate its performance. Distributed systems, by contrast, usually comprise more heterogeneous components, often connected by standard wide-area networks where the purpose of distribution is to balance load or to achieve an end that is inherently distributed. The most prevalent and largest distributed system is the World Wide Web, though email, Voice Over IP, distributed multi-player games and peer-to-peer file sharing networks are all examples of large scale distributed systems.

There are two primary architectures for Distributed Systems. In a pure Client/Server architecture, such as the WWW, nodes are either Servers (providing the content of web pages), or Clients (browsing and consuming that content): here communication takes place only between a client and a server, and not between servers nor between clients. In a pure Peer To Peer architecture (such as file-sharing networks), each component potentially interacts with every other component. The majority of real-world distributed systems, hybrid architectures are used (for example Voice Over IP or Instant Messaging), whereby components acting as servers are used to initiate peer-to-peer communication between interacting subgroups.

The term 'cloud computing' has recently been used to describe the provision of computational services to a client via a distributed system. Numerous models of cloud computing exist, typically cloud-based services can be expanded or contracted to suit a client's changing business or research needs. In the 'Infrastructure as a Service' model, raw compute and storage facilities are offered via a distributed mechanism, allowing clients to host data or perform heavy duty computation on a remote service. Examples include Amazon EC2 and Rackspace Cloud). The 'Platform as a Service' model extends this to include specific software stacks (typically pre configured operating systems with programming environments and software servers), enabling clients to offer services without having to host their own hardware. Examples include Google App Engine, Microsoft Azure and Amazon Elastic Beanstalk. Finally the 'Software as a Service' model provides desktop-style software that is hosted remotely, removing the need for clients to install or configure local machines. Examples include Google Apps, Apple iCloud and Salesforce.com.

DISULPHIDE BRIDGE

Jeremy Baum

A disulphide bridge is formed when the side chains of two cysteine residues interact to form a covalent S-S bond. With the exception of bonded cofactors, these are the only covalent bonds occurring in proteins that deviate from the linear polymer primary structure. Such structures typically occur in secreted (i.e. extracellular) proteins.

See also Cysteine, Side Chain, Cofactors

DL, *SEE* DESCRIPTION LOGIC.

DML, *SEE* DATA MANIPULATION LANGUAGE.

DNA ARRAY, *SEE* MICROARRAY.

DNA DATABANK OF JAPAN, *SEE* DDBJ.

DNA-PROTEIN COEVOLUTION

Laszlo Patthy

Coevolution is a reciprocally induced inherited change in a biological entity (species, organ, cell, organelle, gene, protein, biosynthetic compound) in response to an inherited change in another with which it interacts.

Protein-DNA interactions essential for some biological function (e.g. binding of RNA polymerase II and transcription factors to specific sites in the promoter region, binding of a hormone receptor to the appropriate hormone response element) are ensured by specific contacts between a DNA sequence motif and the DNA-binding protein. During evolution gene regulatory networks have expanded through the expansion of the families of nuclear hormone receptors, transcription factors and the coevolution of the DNA-binding proteins with the DNA sequence motifs to which they bind.

For example, the different steroid hormone-receptor systems have evolved through the coevolution of the three interacting partners (the steroid hormone, their protein receptors and their target DNA sites). Phylogenetic analyses of the steroid receptor systems indicate that the full complement of mammalian steroid receptors (androgen receptor, progesterone receptor, glucocorticoid receptor, mineralocorticoid receptor) evolved from an ancestral steroid receptor through a series of gene duplications. Coevolution of the novel hormone receptor (with novel ligand-specificity) and the target DNA-binding sites has led to an increased complexity of gene regulatory networks.

Further reading

Thornton JW (2001) Evolution of vertebrate steroid receptors from an ancestral estrogen receptor by ligand exploitation and serial genome expansions. *Proc Natl Acad Sci U S A* 98: 5671–5676.

Umesono K, Evans RM (1989) Determinants of target gene specificity for steroid/thyroid hormone receptors. *Cell* 57: 1139–1146.

See also Coevolution

DNA SEQUENCE

John M. Hancock and Rodger Staden

The order of the bases, denoted A (adenine), C (cytosine), G (guanine) and T (thymine), along a segment of DNA.

DNA sequences have a directionality resulting from the chemical structure of DNA. By convention a DNA sequence is written in the 5' -> 3' direction. Upper or lower case characters may be used, althought lower case characters are less ambiguous for human reading. Letters other than the four typical ones may be used to represent uncertainty in a sequence.

See also IUPAC-IUB Codes

DNA SEQUENCING

Rodger Staden

The process of discovering the order of the bases in a DNA sequence.

See also Genome Sequencing, Next Generation DNA Sequencing

DNASP

Michael P. Cummings

A program for analysis of nucleotide sequence data that includes a comprehensive set of population genetic analyses.

The program estimates numerous population genetic parameters with associated sampling variance including analyses of polymorphism, divergence, codon usage bias, linkage disequilibrium, gene flow, gene conversion, recombination, and population size. The program also allows for hypothesis tests for departure from neutral theory expectations with and without an outgroup. In addition the program can perform coalescent simulations.

The program accommodates several input and output formats (e.g., NEXUS, FASTA, PHYLIP, NBRF/PIR and MEGA), has a graphical user interface, and is available as a Windows executable file.

Related website

DnaSP home page	http://www.ub.es/dnasp/

Further reading

Librado P, Rozas J (2009) DnaSP v5: A software for comprehensive analysis of DNA polymorphism data. *Bioinformatics* 15: 1451–1452.

D

DOCK

Marketa J. Zvelebil

The DOCK program explores ways in which two molecules, such as a ligand and a protein, can fit together.

DOCK is one of the main efforts in automating the docking process. It is the long term project of Tack Kuntz and colleagues. DOCK uses shape complementarity to search for molecules that can match the shape of the receptor site. In addition it takes into account chemical interactions. The program generates many possible orientations (and more recently, conformations) of a putative ligand within a user-selected region of a receptor structure (protein). These orientations are then, subsequently, scored using several schemes designed to measure steric and/or chemical complementarity of the receptor-ligand complex.

In the first instance, the program DOCK uses molecular coordinates to generate a molecular surface of the protein with the Connolly molecular surface program which describes the site features. Only the surface for the designated active/binding sites is generated. Subsequently, spheres which are defined by the shape of cavities within the surface, are generated and placed into the active site. The centre of the spheres act as the potential location-sites for the ligand atoms. The ligand atoms are then matched with the sphere centres to determine possible ligand orientations. Many orientations are generated for each ligand. Lastly, for each ligand the orientations are scored based on shape, electrostatic potential and force-field potential.

Relevant website

DOCK	http://dock.compbio.ucsf.edu/

Further reading

Kuntz ID (1992) Structure-based strategies for drug design and discovery *Science* 257: 1078–1082.

Kuntz ID., Meng EC., Shoichet BK (1994) Structure-based molecular design. *Acc. Chem. Res* 27(5): 117–123.

Kuntz ID, Blaney JM, Oatley SJ, Langridge R, Ferrin TE (1982) A geometric approach to macromolecule-ligand interactions. *J. Mol. Biol.* 161: 269–288.

Meng EC, Gschwend DA, Blaney JM, Kuntz ID (1993) Orientational sampling and rigid-body minimization in molecular docking. *Proteins.* 17(3): 266–278.

Meng EC, Shoichet BK, Kuntz ID (1992) Automated docking with grid-based energy evaluation. *J. Comp. Chem.* 13: 505–524.

Shoichet BK, Bodian DL, Kuntz ID (1992) Molecular docking using shape descriptors. *J. Comp. Chem.* 13(3): 380–397.

D

DOCKING

Marketa J. Zvelebil

Docking simply attempts to predict the structure of the intermolecular complex formed between the protein and a ligand.

Usually the protein structure is considered to be static or semi-static and the ligand is 'docked' into the binding pocket and its suitability is evaluated by a potential energy function, often semi-empirical in nature. Most docking methods generate a large number of likely ligands, and as a result they have to have a means of scoring each structure to select the most suitable one. Consequently, docking programs deal mainly with the generation and evaluation of possible structures of intermolecular complexes. In conformational flexible docking, where, the ligand is not constrained to be rigid, conformational degrees of freedom have to be taken into account. Monte Carlo methods in combination with stimulated annealing can be used to explore the conformational space of ligands.

Another method deals with building ligand-molecules 'de novo' inside a binding site, introducing fragments that complement the site to optimize intermolecular interactions.

Further reading

Camacho CJ, Vajda S (2002) Protein-protein association kinetics and protein docking. *Curr Opin Struct Biol.* 12(1): 36–40.

Lengauer T, Rarey M (1996) Computational methods for biomolecular docking. *Curr Opin Struct Biol* 6(3): 402–406.

See also DOCK, Virtual Screening, Virtual Library

SwissDock	http://swissdock.vital-it.ch/
GOLD	http://www.ccdc.cam.ac.uk/products/life_sciences/gold/

DOMAIN, *SEE* PROTEIN DOMAIN.

DOMAIN ALIGNMENT, *SEE* ALIGNMENT.

DOMAIN FAMILY

Teresa K. Attwood

A structurally and functionally diverse collection of proteins that share a common region of sequence that mediates a particular structural or functional role. Domain families are fundamentally different from gene families, which are usually structurally and functionally discrete, and, broadly speaking, share similarity along their entire sequence lengths. The distinction between domain families and gene families is not just a semantic one, because different processes underpin their evolution, with different functional consequences.

In this context, domains often refer to protein *modules*. Modules are autonomous folding units believed to have arisen largely as a result of genetic shuffling mechanisms – examples include kringle domains (named after the shape of a

```
            x x x      x x x
          x      C---C       x
         x       x     x       x
        x        Cx  x  x       x
        x        |   x  x       x
        x        Cx  x  x       x
         x       x      x      x
          x      x      x     x
           x      x x x      x
            x               x
              x x       x x
                 C---C
           .....x       x.....
```

Figure D.4 Schematic illustration of part of a protein sequence containing an apple domain: dot characters mark the regions of sequence immediately N- and C-terminal to the domain; x denotes amino acids other than cysteine; the locations of cysteine residues are denoted by the letter C; paired Cs indicate the presence of intra-chain disulphide bridges.

Danish pastry), which are structural units found throughout the blood clotting and fibrinolytic proteins, and believed to play a role in regulation of proteolytic activity; the WW module (named after its two conserved tryptophan residues), which is found in a number of disparate proteins (including dystrophin, the product encoded by the gene responsible for Duchenne muscular dystrophy), and functions by binding to proteins containing particular proline-, phosphoserine- or phosphothreonine-containing motifs; and apple domains (named for their shape in schematic diagrams, as shown in the accompanying Figure D.4), multiple tandem repeats of which are found in the N-termini of the related plasma serine proteases plasma kallikrein and coagulation factor XI, where they mediate binding to factor XIIa, platelets, kininogen, factor IX, heparin and high molecular weight kininogen.

Modules are contiguous in sequence and are often used as building blocks to confer a variety of complex functions on a parent protein, either via multiple combinations of the same module, or by combinations of different modules to form protein mosaics. This accounts for the structural and functional diversity of protein families that share a particular module or domain, and illustrates why knowledge of the function of such a domain does not reveal the overall function of the parent protein.

SMART is an example of a domain family database.

Related website

SMART	http://smart.embl.de/help/smart_about.shtml

Further reading

Apic G, Gough J, Teichmann SA (2001) An insight into domain combinations. *Bioinformatics* 17 Suppl 1: S83–89.

Letunic I, Doerks T, Bork P (2009) SMART 6: recent updates and new developments. *Nucleic Acids Res.* 37(Database issue): D229–232.

See also Gene Family, Protein Domain, Protein Family, Protein Module

DOT MATRIX, *SEE* DOT PLOT.

DOT PLOT (DOT MATRIX)

Jaap Heringa

A way to represent and visualize all possible alignments of two sequences by comparing them in a two-dimensional matrix. One sequence is written out vertically with its amino acids forming the matrix rows, while the other sequence forms the columns. Each intersection of a matrix row with a column represents the comparison of associated amino acids in the two sequences such that all possible local alignments can be found along diagonals paralleling the major matrix diagonal. Overall similarity is discernible by piecing together local sub-diagonals through insertions and deletions. The simplest way of expressing similarity of matched amino acid pairs is to place a dot in a matrix cell whenever they are identical. Such matrices are therefore often referred to as dot matrices and were important early on as the major means to visualize the relationship of two sequences (Fitch, 1966; Gibbs and McIntyre, 1970). More biological insight is typically obtained by using more varying amino acid similarity values than binary identity values. Most dot plot methods therefore refine the crude way of only showing sequence identities by using amino acid exchange odds for each possible exchange. These odds are normally incorporated in a symmetrical 20x20 *amino acids exchange matrix* in which each value approximates the evolutionary likelihood of a mutation from one amino acid into another. To increase the signal-to-noise ratio for dot plots, regions of high similarity can be identified using windows of fixed length which are effectively slid over the two sequences to compare all possible stretches of, for example, five matched residue pairs. Important biological issues include the choice of the amino acid similarity scores to use as well as the adopted length of the windows in the dot matrix analysis. Argos (1987) included physical-chemical amino acid parameters in the calculation of the similarity values. He used these together with windows of different lengths that were tested simultaneously in order to be less dependent on the actual choice of an individual window length. Junier and Pagni (2000) constructed a web-based program Dotlet in which they implemented statistical filtering techniques.

Further reading

Argos P (1987) A sensitive procedure to compare amino acid sequences. *J. Mol. Biol.* 193: 385–396.

Fitch W (1966) An improved method of testing for evolutionary homology. *J. Mol. Biol.* 16: 9–16.

Gibbs AJ, McIntyre GA (1970) The diagram, a method for comparing sequences Its use with amino acid and nucleotide sequences. *Eur. J. Biochem.* 16: 1–11.

Junier T, Pagni M (2000) Dotlet: diagonal plots in a web browser. *Bioinformatics* 16(2): 178–179.

See also Alignment, Global Alignment, Local Alignment, Dynamic Programming

DOTTER

John M. Hancock and M.J. Bishop

A popular implementation of the dot plot method of sequence comparison.

The program accepts input files in FASTA format and can compare DNA or protein sequences or a DNA sequence with a protein sequence. The program can handle sequences of any length and can be run in batch mode for long sequences. Otherwise it can be run interactively. It allows maximum and minimum cutoffs to be set to reduce noise, visualization of alignments giving rise to features in the plot, zooming in, displaying multiple dot plots simultaneously, and colour coding features within the sequence.

Related website

Dotter Web site	http://sonnhammer.sbc.su.se/Dotter.html

Further reading

Sonnhammer ELL and Durbin R (1995) A dot-matrix program with dynamic threshold control suited for genomic DNA and protein sequence analysis. *Gene* 167: GC1–10.

See also Dot Plot, STADEN

DOWNSTREAM

Niall Dillon

Describes a sequence located distal to a specific point in the direction of transcription (i.e. in a 3' direction on the strand being transcribed).

DRUG-LIKE

Bissan Al-Lazikani

An adjective applied to compounds, drug-likeness is a description of a compound indicating that it has physicochemical properties consistent with a therapeutic agent, such as the right lipophilicity, molecular weight, charge distribution etc. It is typically used to describe synthetic, smaller molecular weight compounds (rather than a biotherapeutic). Additionally, it is often used to imply an orally bioavailable compound.

This would typically mean that the compound is around, or smaller than, 500 molecular weight, has approximately Lipinski Rule-of-Five parameters and lacks toxicophores (chemical groups associated with toxicity or adverse reactions) or reactive groups. By implication,

D

the compound also needs to be synthesizable and derivable thus making it amenable for lead optimization.

However, correctly used, the term 'drug-like' is context dependent. A project seeking an intravenously administered drug will define 'drug-like' differently to a project seeking an orally bioactive drug.

Related website

Lipinski's Rule-of-Five	http://en.wikipedia.org/wiki/Lipinski%27s_rule_of_five

Further reading

Brüstle M, Beck B, Schindler T, King W, Mitchell T, Clark T. Descriptors, physical properties, and drug-likeness. *J Med Chem* (2002) 1;45(16):3345–3355. PMID: 12139446.

Ursu O, Rayan A, Goldblum A, Oprea T I (2011) Understanding drug-likeness. *Wiley Interdisciplinary Reviews: Computational Molecular Science* 1: 760–781. doi: 10.1002/wcms.52.

DRUGBANK

Dan Bolser

DrugBank combines detailed drug chemical, pharmacological and pharmaceutical data with drug-target sequence, structure, and pathway information. Data is searchable and browseable via a variety of different tools, and can be downloaded for offline analysis.

Related websites

DrugBank	http://www.drugbank.ca/
Wikipedia	http://en.wikipedia.org/wiki/DrugBank
MetaBase	http://metadatabase.org/wiki/DrugBank

Further reading

Knox C, Law V, Jewison T, et al. (2011) DrugBank 3.0: a comprehensive resource for 'omics' research on drugs. *Nucleic Acids Res.* 39: 1035–1041.

See also ChEMBL

DRUGGABILITY

Bissan Al-Lazikani

An adjective applied to protein targets. Strictly defined, druggability of a protein target is its suitability to be modulated by a drug in a way that causes a therapeutic effect. More commonly, druggablity is used to describe proteins that can be modulated with small synthetic molecules (rather than e.g. a monoclonal antibody). A druggable protein will possess a cavity, or binding pocket, that can accommodate the drug. Thus, this pocket needs to have physicochemical properties compatible with a drug.

Druggability can be determined in multiple ways including:

1. By homology to an existing drug target. This simplest and most commonly used definition of druggability has limitations as it can only identify members of a currently drugged protein family and cannot predict novel families. It also assumes that all members of a family are equally druggable.
2. Structure-based druggability. This uses the 3D structure of a protein to scan for cavities that have suitable physicochemical properties. This method is able to predict novel druggable targets but relies on the existence of a known or well-redicted 3D structure. It is also sensitive to the conformation of the protein in the 3D structure examined.

Further reading

Al-Lazikani B, Gaulton A, Paolini G, et al. (2007) 'The molecular basis of predicting druggability'. In: Lengauer T (ed.) *Bioinformatics: molecular sequences and structures; Vol. 3 The Holy Grail: molecular function.*

Hajduk PJ, Huth JR, Fesik SW (2005) Druggability indices for protein targets derived from NMR-based screening data. *J Med Chem* 7:48(7): 2518–2525.

Hopkins AL, Groom CR (2002) The druggable genome. *Nat Rev Drug Discov* 1(9):727–730.

See also Drug-Like

DRUGGABLE GENOME (DRUGGABLE PROTEOME)

Bissan Al-Lazikani

More correctly, the druggable proteome, is a term that refers to the fraction of the human proteome likely to be druggable using synthetic small molecule.

Further reading

Hopkins AL, Groom CR (2002) The druggable genome. *Nat Rev Drug Discov* 1(9):727–730.

Russ AP, Lampel S (2005) The druggable genome: an update. *Drug Discov Today* 10(23–24): 1607–1610.

See also Druggability

DW, *SEE* DATA WAREHOUSE.

DYNAMIC PROGRAMMING

Jaap Heringa

Optimization technique to calculate the highest scoring or optimal alignment between two protein or nucleotide sequences.

The physicist Richard Bellman first conceived dynamic programming (DP) (Bellman, 1957) and published a number of papers and a book on the topic between 1957 and 1975, followed by Needleman and Wunsch (1970) who introduced the technique to the biological community.

A DP algorithm is guaranteed to yield the maximally scoring alignment of a pair of nucleotide or protein sequences, given an appropriate nucleotide or amino acid exchange matrix and gap penalty values. The algorithm operates in two steps. First a search matrix is set up in the same way as a dot matrix with one sequence displayed horizontally and the other vertically. The matrix is traversed from the upper left to the lower right in a row-wise fashion. Each cell $[i, j]$ in the matrix receives as a score the value composed of the maximum of three values: The score in the preceding diagonal cell $M[i-1, j-1]$ to which the exchange value $s[i, j]$ of the associated matched residue pair of cell $[i, j]$ is added, and respectively the scores in the cell on the left, $M[i, j-1]$; and that in the cell on top, $M[i\text{-}1, j]$, where from the latter two cells a predefined gap penalty value is subtracted. Writing the above in a more explicit form:

$$M[i,j] = \text{Max}\begin{Bmatrix} M[i-j,\ j-1] \\ M[i,\ j-1] - gp \\ M[i-1,\ j] - gp \end{Bmatrix}, \qquad (1)$$

where $M[i,j]$ is the alignment score for the first subsequence from 1 to i and the second subsequence from 1 to j, Max denotes the maximum value of the three arguments between brackets, $s[i,j]$ is the substitution value for the amino acid exchange associated with cell$[i,j]$, and gp is the non-negative penalty value (see Figure D.5).

The choice of proper gap penalty values is closely connected to the residue exchange values used in the analysis. When the search matrix is traversed, the highest scoring matrix cell is selected from the bottom row or the rightmost column and this score is guaranteed to be the optimal alignment score. The second step of a dynamic programming algorithm is usually called the traceback step in which the actual optimal alignment is reconstructed from the matrix cell containing the highest alignment score.

Classical Needleman-Wunsch type dynamic programming algorithms use a two-dimensional search matrix, so that the algorithmic speed and storage requirements are both of the order $N*M$, when two sequences consisting of N and M amino acids in length are matched. The large computer memory requirements of Needleman-Wunsch type algorithms are due to the traceback step, where the matches of the optimal alignment are reconstructed. Furthermore, the amount of computation required makes the dynamic programming technique unfeasible for a query sequence search against a large sequence database on personal computers.

Gotoh (1986, 1987) devised a dynamic programming algorithm that drastically decreased the storage requirements from order N^2 to order N (assuming that two sequences each N amino acids in length are matched) while keeping speed at an order of N^2. Myers and Miller (1988) constructed an even more memory efficient linear space algorithm, based on the Gotoh approach and on a traceback strategy proposed by Hirschberg (1975); which is only slightly slower.

Smith and Waterman (1981) developed the so-called *local alignment* technique in which the most similar regions in two sequences are selected and aligned first. To get from global to local dynamic programming, an important prerequisite is that the amino acids exchange

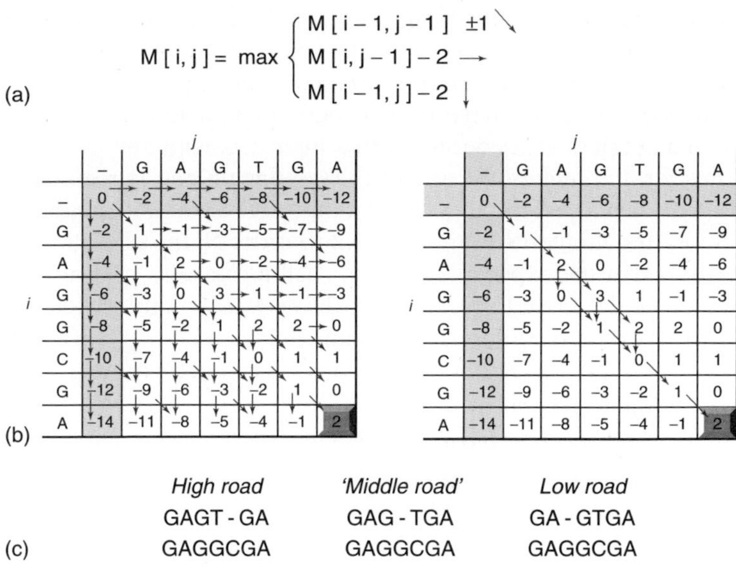

Figure D.5 Dynamic Programming.
(a) The global dynamic programming (DP) formula. Here, a match score of 1 and mismatch score of −1 is taken, and a linear gap penalty value of 2 is used. The direction of matrix traversal is indicated using an arrow for each of the three terms. (b) The search matrix filling stage (left) showing the preceding row and column comprised of proper gap penalty values for gaps preceding either sequence (light-grey row and column), and all search matrix cells filled with proper values after applying the DP formula, and backtracking stage (right) denoting three alternative optimally scoring paths (high road, middle road, and low road) leading to the optimal global alignment score in the bottom rightmost cell. Arrows indicate for each cell from which of three possible cells it derives a maximal value (left). The traceback stage (right) shows the three alternative optimal paths leading to the lower rightmost cell. (c) The corresponding three optimal global alignments, each having an alignment score of 2 (the value in the lower rightmost search matrix cell).

values used must include negative values. Any score in the search matrix lower than zero is to be set to zero. Formula (1) is then changed to:

$$M[i,j] = \mathrm{Max}\begin{Bmatrix} M[i-1,\ j-1] \\ M[i,\ j-1] - gp \\ M[i-1,\ j] - gp \\ 0 \end{Bmatrix}, \tag{2}$$

where Max is now the maximum of four terms. A consequence of this scenario is that the final highest alignment score value does not have to be in the last row or column as in global alignment routines but can be anywhere in the search matrix. The local alignment algorithm relies on dissimilar subsequences producing negative scores which are subsequently discarded by placing zero values in the associated submatrix cells. An arbitrary issue in using the algorithm is deciding the zero cut-off relative to the 20*20 residue exchange weights matrix.

Waterman and Eggert (1987), generalized the local alignment routine by devising an alignment routine which allows the calculation of a user-defined number of top scoring local alignments instead of only the optimal local alignment. The obtained local alignments do not intersect; i.e., they have no matched amino acid pair in common. If during the procedure an alignment is encountered that intersects with any of the top scoring alignments listed thus far, the highest scoring of the conflicting pair is retained in the top list. Huang et al. (1990) developed an implementation (SIM) of the technique in which they reduced the memory requirements from order N2 to order N, thereby allowing very long sequences to be searched at the expense of only a small increase in computational time. Another popular version of the same technique is LALIGN (Huang and Miller, 1991), which is part of the FASTA package (Pearson and Lipman, 1988).

Further reading

Bellman RE (1957) *Dynamic programming*, Princeton University Press, Princeton.

Gotoh O (1986) Alignment of three biological sequences with an efficient traceback procedure. *J. Theor. Biol.* 121: 327–337.

Gotoh O (1987) Pattern matching of biological sequences with limited storage. *Comput Appl Biosci* 3: 17–20.

Hirschberg DS (1975) A linear space algorithm for computing longest common subsequences. *Commun. Assoc. Comput. Mach* 18: 341–343.

Huang X, et al. (1990). A space-efficient algorithm for local similarities. *Comput Appl Biosci* 6: 373–381.

Huang X, Miller W (1991) A time-efficient, linear-space local similarity algorithm. *Adv. Appl. Math.* 12: 337–357.

Myers EW, Miller W (1988) Optimal alignment in linear space. *Comput Appl Biosci* 4: 11–17.

Needleman SB, Wunsch CD (1970) A general method applicable to the search for similarities in the amino acid sequence of two proteins. *J. Mol. Biol.* 48: 443–453.

Pearson WR, Lipman DJ (1988) Improved tools for biological sequence comparison. *Proc. Natl. Acad. Sci. USA* 85: 2444–2448.

Smith TF, Waterman MS (1981) Identification of common molecular subsequences. *J. Mol. Biol.* 147: 195–197.

Waterman MS, Eggert M (1987) A new algorithm for best subsequences alignment with applications to the tRNA-rRNA comparisons. *J. Mol. Biol.* 197: 723–728.

See also Gap Penalty, Multiple Alignment, Pairwise Alignment

E

E

E-M Algorithm, *see* Expectation Maximization Algorithm.

E Value

EBI (EMBL-EBI, European Bioinformatics Institute)

EcoCyc

EDA, *see* Estimation of Distribution Algorithm.

Edge (in a Phylogenetic Tree), *see* Branch (of a Phylogenetic Tree) (Edge, Arc).

EGASP, *see* GASP.

Electron Density Map

Electrostatic Energy

Electrostatic Potential

ELIXIR (Infrastructure for Biological Information in Europe)

Elston-Stewart Algorithm (E-S Algorithm)

EMA, EMAGE, *see* e-Mouse Atlas.

EMBL-Bank, *see* EMBL Nucleotide Sequence Database.

EMBL Database, *see* EMBL Nucleotide Sequence Database.

EMBL-EBI, *see* EBI.

EMBL Nucleotide Sequence Database (EMBL-Bank, EMBL Database)

EMBnet (The Global Bioinformatics Network) (formerly the European Molecular Biology Network)

EMBOSS (The European Molecular Biology Open Software Suite)

e-Mouse Atlas (EMA)

e-Mouse Atlas of Gene Expression (EMAGE), *see* e-Mouse Atlas.

emPAI (Exponentially Modified Protein Abundance Index)

Empirical Pair Potentials

Empirical Potential Energy Function

ENA (European Nucleotide Archive)

ENCODE (Encyclopedia of DNA Elements)

ENCprime/SeqCount

Encyclopedia of DNA Elements, *see* ENCODE.

End Gap

Energy Minimization

Enhancer

Ensembl

Ensembl Genome Browser, *see* Ensembl.

Ensembl Plants

Ensembl Variation, *see* Ensembl.

Ensemble of Classifiers

Entrez

Entrez Gene (NCBI 'Gene')

Concise Encyclopaedia of Bioinformatics and Computational Biology, Second Edition.
Edited by John M. Hancock and Marketa J. Zvelebil.

Entropy

Epaktolog (Epaktologue)

Epistatic Interactions (Epistasis)

Error

Error Measures (Accuracy Measures, Performance Criteria, Predictive Power, Generalization)

E-S Algorithm, *see* Elston-Stewart Algorithm.

EST, *see* Expressed Sequence Tag.

Estimation of Distribution Algorithm (EDA)

Euclidean Distance

Eukaryote Organism and Genome Databases

EUPA (European Proteomics Association)

European Bioinformatics Institute, *see* EBI.

European Molecular Biology Open Software Suite, *see* EMBOSS.

European Nucleotide Archive, *see* ENA.

EuroPhenome

Evolution

Evolution of Biological Information

Evolutionary Distance

Exclusion Mapping

Exome Sequencing

Exon

Exon Shuffling

Expectation Maximization Algorithm (E-M Algorithm)

Exponentially Modified Protein Abundance Index, *see* emPAI.

Expressed Sequence Tag (EST)

Expression Level (of Gene or Protein)

Extended Tracts of Homozygosity

eXtensible Markup Language, *see* XML.

External Branch, *see* Branch of a Phylogenetic Tree.

External Node, *see* Node of a Phylogenetic Tree.

Extrinsic Gene Prediction, *see* Gene Prediction, homology-based.

E-M ALGORITHM, *SEE* EXPECTATION MAXIMIZATION ALGORITHM.

E VALUE

Jaap Heringa

The expectancy value (E value) is a statistical value for estimating the significance of alignments performed in a homology search of a query sequence against a sequence database. It basically estimates for each alignment the number of sequences randomly expected in the sequence database to provide an alignment score, after alignment with the query sequence, at least that of the alignment considered. Such significance estimates are important because sequence alignment methods are essentially pattern search techniques, leading to an alignment with a similarity score even in case of absence of any biological relationship. Although similarity scores of unrelated sequences are essentially random, they can behave like 'real' scores and, for instance, like real scores are correlated with the length of the sequences compared. Particularly in the context of database searching, it is important to know what scores can be expected by chance and how scores that deviate from random expectation should be assessed. If within a rigid statistical framework a sequence similarity is deemed statistically significant, this provides confidence in deducing that the sequences involved are in fact biologically related. As a result of the complexities of protein sequence evolution and distant relationships observed in nature, any statistical scheme will invariably lead to situations where a sequence is assessed as unrelated while it is in fact homologous (false negative), or the inverse, where a sequence is deemed homologous while it is in fact biologically unrelated (false positive). The derivation of a general statistical framework for evaluating the significance of sequence similarity scores has been a major task. However, a rigid framework has not been established for global alignment, and has only partly been completed for local alignment.

Sequence similarity values resulting from global alignments are known to grow linearly with the sequence length, although the growth rate has not been determined. Also, the exact distribution of global similarity scores is yet unknown, and only numerical approximations exist, providing some rough bound on the expected random scores. Since the variance of the global similarity score has not been determined either, most applications derive a sense of the score by using shuffled sequences. Shuffled sequences retain the composition of a given real sequence but have a permuted order of nucleotides or amino acids. The distribution of similarity scores over a large number of such shuffled sequences often approximates the shape of the Gaussian distribution which is therefore taken to represent the underlying random distribution. Using the mean (m) and standard deviation (σ) calculated from such shuffled similarity scores, each real score S can be converted to the z-score using z-score $= (S-m)/\sigma$. The z-score measures how many standard deviations the score is separated from the mean of the random distribution. In many studies, a z-score >6 is taken to indicate a significant similarity.

A rigid statistical framework for local alignments without gaps has been derived for protein sequences following the work by Karlin and Altschul (1990), who showed that the optimal local ungapped alignment score growths linearly with the logarithm of the product of sequence lengths of two considered random sequences: $S \sim \ln(n \bullet m)/\lambda$, where n and m are the lengths of two random sequences, and λ a scaling parameter that depends on the scoring matrix used and the overall distribution of amino acids in the database. Specifically, λ is the unique solution for x in the equation

$$\sum_{ij} p_i p_j e^{s_{i,j} x} = 1$$

where summation is over all amino acid pairs, p_i represents the background probability (frequency) of residue type i, and $s_{i,j}$ is the scoring matrix.

An important contribution for fast sequence database searching has been the realization (Karlin and Altschul, 1990; Dembo and Karlin, 1991; Dembo *et al.*, 1994) that local similarity scores of ungapped alignments follow the extreme value distribution (EVD; Gumbel, 1958). This distribution is unimodal but not symmetric like the normal distribution, because the right-hand tail at high scoring values falls off more gradually than the lower tail, reflecting the fact that a best local alignment is associated with a score that is the maximum out of a great number of independent alignments.

Following the extreme value distribution, the probability of a score S to be larger than a given value x can be calculated as:

$$P(S \geq x) = 1 - \exp(-e^{-\lambda(x-\mu)}),$$

where $\mu = (\ln Kmn)/\lambda$, and K a constant that can be estimated from the background amino acid distribution and scoring matrix (see Altschul and Gish, 1996, for a collection of values for λ and K over a set of widely used scoring matrices). Using the equation for μ, the probability for S becomes

$$P(S \geq x) = 1 - \exp(-Kmne^{-\lambda x}).$$

In practice, the probability $P(S \geq x)$ is estimated using the approximation $1- \exp(-e^{-x}) \approx e^{-x}$, which is valid for large values of x. This leads to a simplification of the equation for $P(S \geq x)$:

$$P(S \geq x) \approx \mathrm{e}^{-\lambda(\mathrm{x}-\mu)} = Kmn\mathrm{e}^{-\lambda x}.$$

The lower the probability for a given threshold value x, the more significant the score S. For example, if the E-value of a matched database sequence segment is 0.01, then the expected number of random hits with score $S \geq x$ is 0.01, which means that this E-value is expected by chance only once in 100 independent searches over the database. However, if the E-value of a hit is 5, then five fortuitous hits with $S \geq x$ are expected within a single database search, which renders the hit not significant. Database searching is commonly performed using an E-value in between 0.1 and 0.001. Low E-values decrease the number of false positives in a database search, but increase the number of false negatives thereby lowering the sensitivity of the search.

Although similarities between sequences can be detected reasonably well using methods that do not allow insertions/deletions in aligned sequences, it is clear that insertion/deletion events play a major role in divergent sequences. This means that accommodating gaps within alignments of distantly related sequences is important for obtaining an accurate measure of similarity. Unfortunately, a rigorous statistical framework as obtained for gapless local alignments has not been conceived for local alignments with gaps. However, although it has not been proven analytically that the distribution of S for gapped alignments can be approximated with the extreme value distribution, there is accumulated evidence that this is the case: For example, for various scoring matrices, gapped alignment similarities have been observed to grow logarithmically with the sequence lengths (Arratia and Waterman, 1994). Other empirical studies have shown it to be likely that the distribution of local gapped similarities follows the extreme value distribution (Smith *et al.*, 1985; Waterman and Vingron, 1994), although an appropriate downward correction for the effective sequence length has been recommended (Altschul and Gish, 1996). The distribution of empirical similarity values can be obtained from unrelated biological sequences (Pearson, 1998). Fitting of the EVD parameters λ and K (see above) can be performed using a linear regression technique (Pearson, 1998), although the technique is not robust against outliers which can have a marked influence. Maximum likelihood estimation (Mott, 1992; Lawless, 1982) has been shown to be superior for EDV parameter fitting and, for example, is the method used to parameterise the gapped BLAST method (Altschul *et al.*, 1997). However, when low gap penalties are used to generate the alignments, the similarity scores can lose their local character and assume more global behaviour, such that the EVD-based probability estimates are not valid anymore (Aratia and Waterman, 1994).

Further reading

Altschul SF, Gish W (1996) Local alignment statistics. In: *Methods in Enzymology*, Doolittle RF (ed.), Vol. 266, pp. 460–480, Academic Press, San Diego.

Altschul SF, Madden TL, Schäffer AA, Zhang J, Zhang Z, Miller W, Lipman DJ (1997) Gapped BLAST and PSI-BLAST: a new generation of protein database search programs. *Nucl. Acids. Res.* 25: 3389–3402.

Arratia R, Waterman MS (1994) A phase transition for the score in matching random sequences allowing depletions. *Ann. Appl. Prob.* 4: 200–225.

Dembo A, Karlin S (1991) Strong limit theorems of empirical functionals for large exceedances of partial sums of i.i.d. variables. *Ann. Prob.* 19: 1737.

Dembo A, Karlin S, Zeitouni O (1994) Limit distributions of maximal non-aligned two-sequence segmental score. *Ann. Prob.* 22: 2022.

Gumbel EJ (1958) *Statistics of Extremes*. Columbia University Press, New York.

Karlin S, Altschul SF (1990) Methods for assessing the statistical significance of molecular sequence features by using general scoring schemes. *Proc. Natl. Acad. Sci. USA* 87: 2264–2268.

Lawless JF (1982) *Statistical Models and Methods for Lifetime Dat*, pp. 141–202. John Wiley & Sons.

Mott R (1992) Maximum-likelihood estimation of the statistical distribution of Smith-Waterman local sequence similarity scores. *Bull. Math. Biol.* 54: 59–75.

Pearson WR (1998) Empirical statistical estimates for sequence similarity searches. *J Mol Biol* 276: 71–84.

E

Smith TF, Waterman MS, Burks C (1985) The statistical distribution of nucleic acid similarities. *Nucleic Acids Res.* 13: 645.

Waterman MS, Vingron M (1994) Rapid and accurate estimates of statistical significance for sequence data base searches. *Proc. Natl. Acad. USA* 91: 4625.

See also BLAST, Homology Search, Sequence Similarity

EBI (EMBL-EBI, EUROPEAN BIOINFORMATICS INSTITUTE)

Guenter Stoesser, Obi L. Griffith and Malachi Griffith

The European Bioinformatics Institute is an international non-profit academic organization that forms part of the European Molecular Biology Laboratory (EMBL).

The EBI covers a wide range of activities in research and services in bioinformatics, including computational and structural genomics, functional genomics as well as building and maintaining comprehensive public biological databases of DNA sequences (European Nucleotide Archive), protein sequences (UniProt), the Interpro resource for protein families, macromolecular structures (formerly MSD, now PDBe), genome annotation (Ensembl), gene expression data (ArrayExpress), biological ontologies such as the Gene Ontology (GO), pathways (Reactome), and scientific literature (CiteXplore).

Data resources are complemented by query and computational analysis servers including the Sequence Retrieval System (SRS) for querying of more than 150 linked biological datasets, data portals such as BioMart, BLAST/FASTA/Smith-Waterman sequence searching, multiple sequence alignment using ClustalW, and structure prediction methods. Most projects at the EBI involve collaborations with major institutes within Europe and worldwide.

The EBI is located on the Wellcome Trust Genome Campus near Cambridge, UK next to the Sanger Institute and the Human Genome Mapping Project Resource Centre (HGMP). Together these institutes constitute one of the world's largest concentrations of expertise in genomics and bioinformatics.

Related websites

ArrayExpress	http://www.ebi.ac.uk/arrayexpress/
BioMart	http://www.biomart.org/
Completed whole genome shotgun sequences	http://www.ebi.ac.uk/genomes/
Database searching, browsing and analysis tools	http://www.ebi.ac.uk/Tools/
EMBL Nucleotide Sequence Database	http://www.ebi.ac.uk/embl/
Ensembl Genome Browser	http://www.ensembl.org/
European Bioinformatics Institute	http://www.ebi.ac.uk
European Nucleotide Archive	http://www.ebi.ac.uk/ena/

Functional Genomics	http://www.ebi.ac.uk/fg/
The Gene Ontology	http://www.geneontology.org/
InterProResource of Protein Domains	http://www.ebi.ac.uk/interpro/
Protein Data Bank Europe (PDBe)	http://www.ebi.ac.uk/pdbe/
Reactome	http://www.reactome.org/
Sequence Retrieval System	http://srs.ebi.ac.uk/
UniProt Knowledgebase	http://www.ebi.ac.uk/uniprot/

Further reading

Brooksbank C, et al. (2010) The European Bioinformatics Institute's data resources. *Nucleic Acids Res.* 38(Database issue): D17–25.

Emmert DB et al (1994) The European Bioinformatics Institute (EBI) databases *Nucl. Acids Res.* 26: 3445–3449.

ECOCYC

Robert Stevens

EcoCyc is an encyclopedia of *E. coli* metabolism, regulation and signal transduction.

It is a knowledge base, whose schema is an ontology of prokaryotic genome, proteins, pathways etc. the ontology is encoded in Ocelot, an OKBC compliant Frame Based knowledge representation language. The individuals conforming to the ontology have been painstakingly curated by hand from the *E. coli* literature. the knowledge base can be used to drive a Web interface for browsing the collected knowledge. Several other prokaryotic organisms have been treated in the same manner and all pathway information has been collected within Metacyc (http://biocyc.org/metacyc/). Given the existence of a Genbank entry for a prokaryotic genome, the annotated genes can be processed with Pathway Tools which compares the new genome with information in Metacyc to automatically generate an organism specific Pathway Genome Database. (NB. EcoCyc is not related to OpenCyc or CycCorp.)

Related Website

EcoCyc	http://ecocyc.org/

Further reading

Karp P, Paley S (1996) Integrated Access to Metabolic and Genomic Data. *J Comput Biol* 3: 191–212,

EDA, *SEE* ESTIMATION OF DISTRIBUTION ALGORITHM.

EDGE (IN A PHYLOGENETIC TREE), *SEE* BRANCH (OF A PHYLOGENETIC TREE) (EDGE, ARC).

EGASP, *SEE* GASP.

ELECTRON DENSITY MAP

Liz Carpenter

X-ray crystallography is a method for solving the structures of macromolecules from a diffraction pattern of X-rays from crystals of the macromolecule. X-rays are scattered by electrons. The diffraction pattern consists of a series of diffraction spots for which an intensity can be measured and a variety of methods can be used to derive the relative phases of the diffraction spots (see Phase Problem). From this information a map of the distribution of electrons within the crystal unit cell can be constructed with the following equation:

$$\rho(x, y, z) = \frac{1}{V}\sum_{h}\sum_{k}\sum_{l}|F(h, k, l)| \cos 2\pi \left[hx + ky + lz - \alpha (h, k, l)\right]$$

where ρ (x,y,z) the density of electrons per unit volume at point x,y,z
V is the volume of the unit cell.
h,k,l are the indices of the diffraction spots.
| F (h,k,l) | is the amplitude of the wave giving rise to the h,k,l diffraction spot.
α (h,k,l) is the phase of the wave giving rise to the diffraction spot h,k,l.

This equation can be used to produce a series of equi-potential contours, the atoms of the macromolecule being found where there is high electron density. Once an initial electron density map has been obtained, the structure of the macromolecule can be built into the map, defining the relative coordinates of most of the atoms in the structure. This structure is then subjected to refinement with programs designed to adjust the positions of the atoms to maximise the agreement between the observed structure factors and those calculated from the model model.

Once a model is available two types of electron density maps can be calculated: the $2F_{obs}-F_{calc}$ map which shows where there is highest density throughout the asymmetric unit and the difference map $mF_{obs}-DF_{calc}$, which shows differences between the observed and calculated electron density. Both maps should be studied on the graphics to identify regions of the structure that could be improved. Several rounds of refinement and model building are usually necessary before the structure converges and no further improvement is possible.

Further reading

Glusker JP, Lewis M, Rossi M (1994) *Crystal Structure Analysis for Chemists and Biologists*, VCH Publishers.

ELECTROSTATIC ENERGY

Roland Dunbrack

The energy of interaction (attraction or repulsion) between charged atoms or molecules. In molecular mechanics functions, this is represented with the Coulomb equation, $E(r_{ij}) = \sum_i \sum_{j>i} kq_iq_j/\varepsilon r_{ij}$, where r_{ij} is the interatomic distance of two charged or partially charged atoms i and j with charges q_i and q_j respectively, ε is the dielectric constant of the medium, and k is the appropriate constant in the units of energy, distance, and charge.

See also Dielectric Constant

ELECTROSTATIC POTENTIAL

Jeremy Baum

When a charge is present on a molecule, chemical group or atom it will have an effect on the surroundings. Two components of this effect are typically discussed - the electrostatic potential and the electric field. The electric field is the gradient of the electrostatic potential, which is a scalar quantity. A charge q_A produces an electrostatic potential of $\varphi = q_A/\varepsilon r$ a distance r away in a medium whose dielectric constant is ε. A second charge q_B that experiences this potential φ will have energy $\varphi q_B = q_A q_B/\varepsilon r$. In enzymes, the electrostatic potential of the surrounding protein structure can substantially alter the protonation equilibrium of a group, and through this control reactivity.

See also Dielectric Constant

ELIXIR (INFRASTRUCTURE FOR BIOLOGICAL INFORMATION IN EUROPE)

Pedro Fernandes

The purpose of ELIXIR is to construct and operate a sustainable infrastructure for biological information in Europe, to support life science research and its translation to medicine and the environment, the bio-industries and society.

ELIXIR unites Europe's leading life science organisations in managing and safeguarding the massive amounts of data being generated every day by publicly funded research. It is a pan-European research infrastructure for biological information. ELIXIR will provide the facilities necessary for life science researchers - from bench biologists to cheminformaticians - to make the most of our rapidly growing store of information about living systems, which is the foundation on which our understanding of life is built (Janet Thornton, ELIXIR Project Coordinator).

Related website

ELIXIR	http://www.elixir-europe.org/

E

ELSTON-STEWART ALGORITHM (E-S ALGORITHM)

Mark McCarthy, Steven Wiltshire and Andrew Collins

A recursive algorithm used in the context of linkage analysis for calculating the likelihood of pedigree data.

The Elston-Stewart algorithm is an arrangement of the likelihood function of a pedigree that expressly treats a simple pedigree as a multigenerational series of sibships, with one parent of each sibship itself being a sib in the preceding sibship in the pedigree as a whole. The likelihood is written in terms of penetrance probabilities, $P(x1 \mid gi)$, and the genotype probabilities, $P(gi \mid \bullet)$, which are the population frequencies of ordered genotypes G_i of individual i, if a founder, or the transmission probabilities of genotypes G_i (as functions of $\boldsymbol{\theta}$, the recombination fractions between the putative disease locus and marker(s)) of individual i, if a non-founder. Writing the pedigree so that parents precede their offspring, the likelihood is given by:

$$L(\theta) = \sum_{g1 \in G1} P(x1 \mid g1)\, P(g1 \mid \bullet) \sum_{g2 \in G2} P(x2 \mid g2)\, P(g2 \mid \bullet) \;\ldots\; \sum_{gn \in G_n} P(xn \mid gn)\, P(gn \mid \bullet)$$

Beginning with the most recent generation in the pedigree (the right-hand side of the equation), the likelihood, over all genotypes, of each sibship is evaluated for all possible genotypes of the linking parent and is then 'peeled' from the pedigree, the conditional likelihoods of the sibships being carried over to the summations for the associated genotypes of the linking parent. This procedure is repeated until the top of the pedigree is reached and the overall likelihood achieved. The Elston-Stewart algorithm makes rapid likelihood evaluation for large pedigrees possible by reducing the number of calculations that would otherwise be necessary if the likelihood were expressed as the multiple sum over all genotypes of the products of penetrance and genotype probabilities. As such, it is central to many computer programs for performing linkage analyses. The basic algorithm has been developed to accommodate complex pedigrees with and without loops, and to allow peeling in all directions. The computational burden of the E-S algorithm increases linearly with pedigree size, but increases exponentially with number of markers, due to the summation terms over all possible genotypes at each step. Despite modifications to facilitate small multipoint calculations, the E-S algorithm remains limited for large multipoint analysis. In this regard, it is complemented by the Lander-Green algorithm, which can evaluate the likelihood of large numbers of markers, but is restricted by pedigree size. Algorithms that combine elements of both the Elston-Stewart and Lander-Green algorithms have been published that allow rapid multipoint likelihood calculation for large pedigrees.

Examples: Craig et al. (1996) used two programs that implement the E-S algorithm – REGRESS and LINKAGE – to perform segregation analysis and genome-wide linkage analysis of a large kindred to identify loci controlling foetal haemoglobin production. Bektas et al. (2001) used VITESSE – which implements a version of the E-S algorithm, improved for speed and multipoint capability – to further map a locus on chromosome 12 influencing type 2 diabetes susceptibility.

Related websites
Programs implementing the E-S algorithm in linkage analyses include:

LINKAGE	ftp://linkage.rockefeller.edu/software/linkage/
VITESSE	http://watson.hgen.pitt.edu/register/docs/vitesse.html

Further reading

Bektas A, et al. (2001) Type 2 diabetes locus on 12q15 Further mapping and mutation screening of two candidate genes. *Diabetes* 50: 204–208.

Craig JE, et al. (1996) Dissecting the loci controlling fetal haemoglobin production on chromosomes 11p and 6q by the regressive approach. *Nat Genet* 12: 58–64.

Elston RC, Stewart J (1971) A general model for the analysis of pedigree data. *Hum Hered* 21: 523–542.

O'Connell JR (2001) Rapid multipoint linkage analysis via inheritance vectors in the Elston-Stewart algorithm. *Hum Hered* 51: 226–240.

O'Connell JR, Weeks DE (1995) The VITESSE algorithm for rapid exact multi-locus linkage analysis via genotype set-recoding and fuzzy inheritance. *Nat Genet* 11: 402–408.

Ott J (1999) *Analysis of Human Genetic Linkage*. Baltimore: Johns Hopkins University Press, pp 182–185.

Stewart J (1992) Genetics and biology: A comment on the significance of the Elston-Stewart algorithm. *Hum Hered* 42: 9–15.

See also Lander-Green Algorithm, Linkage Analysis

EMA, EMAGE, *SEE* E-MOUSE ATLAS.

EMBL-BANK, *SEE* EMBL NUCLEOTIDE SEQUENCE DATABASE.

EMBL DATABASE, *SEE* EMBL NUCLEOTIDE SEQUENCE DATABASE.

EMBL-EBI, *SEE* EBI.

E

EMBL NUCLEOTIDE SEQUENCE DATABASE (EMBL-BANK, EMBL DATABASE)

Guenter Stoesser

The EMBL Nucleotide Sequence Database (EMBL-Bank) is one of the three main databases in the European Nucleotide Archive (ENA) along with the Sequence Read Archive and the Trace Archive. The database is produced in an international collaboration with GenBank (USA) and the DNA Database of Japan (DDBJ). EMBL-Bank contains RNA and DNA sequences produced primarily by traditional Sanger sequencing projects but also increasingly includes consensus or assembled contig sequences from next-generation sequencing projects. EMBL-Bank's major data sources are individual research groups, large-scale genome sequencing centres and the European Patent Office (EPO). Sophisticated submission systems are available for individual scientists (Webin) and large-scale genome sequencing groups. EBI's network services allow free access to the most up-to-date data collection via ftp, email and World Wide Web interfaces. EBI's Sequence Retrieval System (SRS) is an integration system for both data retrieval and applications for data analysis. SRS provides capabilities to search the main nucleotide and protein databases plus many specialized databases by shared attributes and to query across databases. For sequence similarity searching a variety of tools (e.g. Fasta, BLAST) are available which allow external users to compare their own sequences against the latest data in the EMBL Nucleotide Sequence Database and SWISS-PROT. Specialized sequence analysis programs include CLUSTALW for multiple sequence alignment and inference of phylogenies and GeneMark for gene prediction. The database is located and maintained at the European Bioinformatics Institute (EBI) in Cambridge (UK).

Related websites

EMBL Nucleotide Sequence Database Homepage	http://www.ebi.ac.uk/embl/
EMBL Submission systems and related information	http://www.ebi.ac.uk/embl/Submission/
Completed whole genome shotgun sequences	http://www.ebi.ac.uk/genomes/
Database searching, browsing and analysis tools	http://www.ebi.ac.uk/Tools/
European Bioinformatics Institute	http://www.ebi.ac.uk

Further reading

Abola EE, et al. (2000) Quality control in databanks for molecular biology. *BioEssays* 22: 1024–1034.

Hingamp P, et al. (1999) The EMBL nucleotide sequence database: contributing and accessing data. *Mol Biotechnol.* 12: 255–268.

Leinonen R, et al. (2010) Improvements to services at the European Nucleotide Archive. *Nucleic Acids Res.* 38: D39–D45.

Stoesser G, et al. (2002) The EMBL nucleotide sequence database. *Nucleic Acids Res.* 30: 21–26.

Zdobnov, EM, et al. (2002) The EBI SRS server-recent developments. *Bioinformatics.* 18: 368–373.

EMBNET (THE GLOBAL BIOINFORMATICS NETWORK) (FORMERLY THE EUROPEAN MOLECULAR BIOLOGY NETWORK)

Pedro Fernandes

EMBnet is a global network of bioinformatics communities, committed to providing knowledge and services to support and advance research in bioinformatics, life science and biotechnology, registered as the EMBnet Stichting since 1988. It has expanded globally in Europe, Asia, Africa, Australia and the Americas by demand of its members.

EMBnet aims to provide local communities of users and developers with

- education and training
- public domain software
- biotechnology and bioinformatics-related research assistance, bridging the comercial and academic sectors, open access publications
- network infrastructures
- global cooperation via community networking

EMBnet Nodes offer advice, training and workshops to match local capacity and needs.

Its members collaborate to develop good practice and know-how relevant to local communities.

Software and educational tools have originating from EMBnet members: SRS, EMBOSS, CINEMA, UTOPIA, EMBER, etc.

EMBnet open access publications:

EMBnet.journal – *Bioinformatics in Action* (2010–), formerly *EMBnet.news* (1994–2010)
EMBnet Digest
EMBnet also started *Briefings in Bioinformatics* (in 2000)

EMBnet cooperates with ISCB, APBioNet, SoIBio, ASBCB and BTN, and connects to several other collaborative and thematic networks.

Related websites

EMBnet	http://www.embnet.org
EMBnet Digest	http://www.embnet.org/EMBnet.digest
EMBnet Journal	http://journal.embnet.org/index.php/embnetjournal
EMBOSS	http://www.emboss.org/

EMBOSS (THE EUROPEAN MOLECULAR BIOLOGY OPEN SOFTWARE SUITE)

John M. Hancock and M.J. Bishop

A suite of open source programs intended to provide an open source alternative to commercial software.

E

The EMBOSS project developed out of the earlier EGCG project, which aimed to provide extended functionality to the Wisconsin Package making use of its freely available libraries. When the Wisconsin Package moved to commercial distribution, access to its source code was no longer available and the EMBOSS project emerged to provide an open source alternative.

EMBOSS provides approximately 100 programs classified under the following general headings: sequence alignment, rapid database searching with sequence patterns, protein motif identification, including domain analysis, nucleotide sequence pattern analysis, codon usage analysis, rapid identification of sequence patterns in large scale sequence sets, presentation tools for publication. The package is available for a number of Unix platforms.

Related website

EMBOSS Web site	http://emboss.sourceforge.net/

Further reading

Rice P, et al. (2000) EMBOSS: The European Molecular Biology Open Software Suite *Trends Genet.* 16: 276–277.

E-MOUSE ATLAS (EMA)

Dan Bolser

The e-Mouse Atlas is an anatomical atlas of mouse embryo development based on the EMAP ontology of anatomical structure and including the e-Mouse Atlas of Gene Expression, allowing spatial queries to be run over expression data.

Related websites

EMA Title	http://www.emouseatlas.org/
Wikipedia	http://en.wikipedia.org/wiki/EMAGE
MetaBase	http://metadatabase.org/wiki/EMAGE

Further reading

Richardson L, Venkataraman S, Stevenson P, et al. (2010) EMAGE mouse embryo spatial gene expression database: 2010 update *Nucleic Acids Res.* 38: D703–709.

See also MGD, RGD

E-MOUSE ATLAS OF GENE EXPRESSION (EMAGE), *SEE* E-MOUSE ATLAS.

EMPAI (EXPONENTIALLY MODIFIED PROTEIN ABUNDANCE INDEX)

Simon Hubbard

A label-free method to derive relative protein quantitation from observed peptide matches in a database search carried out with collections of mass spectra collected in a high-throughput proteomics experiment. The method uses spectral counting and estimates the protein abundance from the number of observed and observable peptides in a given protein, and is usually only used when there are at least 100 peptide spectra available. The equation is given by:

$$emPAI = 10^{\frac{N_{observed}}{N_{observable}}} - 1$$

where $N_{observed}$ is the number of unique peptide species observed with some minimal score or statistical confidence estimate, and $N_{observable}$ is the number of peptide species that could be observed. The details differ in implementation of how these two numbers are calculated, but usually the observable number is filtered so very small/large peptides and those incompatible with HPLC columns are excluded.

Further reading

Ishihama Y, Oda Y, Tabata T, et al. (2005) Exponentially modified protein abundance index (emPAI) for estimation of absolute protein amount in proteomics by the number of sequenced peptides per protein. *Mol Cell Proteomics* 4: 1265–1272.

See also Quantitative Proteomics

EMPIRICAL PAIR POTENTIALS

Patrick Aloy

Empirical pair potentials (EPP) are mean-force knowledge-based potentials obtained by applying Boltzmann's equation to raw residue-pair statistics derived from sampling databases of known structures.

During the last years, EPPs have been largely applied to many kinds of biological problems (e.g. fold recognition, macromolecular docking, interaction discovery, etc.) and proved to be a very successful scoring system.

In principle, being derived empirically, EPPs incorporate the dominant thermodynamic effects without explicitly having to model each of them. The 'physical' validity of this approach has been largely criticized, as they do not reflect the 'true potentials' when applied to test systems with known potential functions. However, from a pragmatic perspective, as long as the potentials favour the native conformation over others, they are good enough.

Further reading

Reva BA, Finkelstein AV, Skolnik J (2000) Derivation and testing residue-residue mean-force potentials for use in protein structure recognition. *Methods in Molecular Biology* 143: 155–174.

E

EMPIRICAL POTENTIAL ENERGY FUNCTION

Roland Dunbrack

A function that represents conformational energies of molecules and macromolecules, derived from experimental data such as crystal structures and thermodynamic measurements. This is in contrast to *ab initio* potential energy functions, derived purely from quantum mechanical calculations, although some empirical functions may also use *ab initio* calculations to fit parameters in the energy function. The general form of the empirical potential energy function is:

$$U = \sum_{bonds} \frac{1}{2}K_b(b-b_o)^2 + \sum_{\substack{bonds\\angles}} \frac{1}{2}K_\theta(\theta-\theta_o)^2 + \sum_{\substack{dihedral\\angles}} K_\phi\left[1+\cos(n\phi-\delta)\right]$$
$$+ \sum_{\substack{nonbonded\\atompairs\\j>i}} 4\varepsilon_{ij}\left[\left(\frac{\sigma_{ij}}{r}\right)^{12} - \left(\frac{\sigma_{ij}}{r}\right)^{6}\right] + \sum_{\substack{nonbonded\\atompairs\\j>i}} \frac{q_i q_j}{\varepsilon r}$$

The first two terms are harmonic energy terms for all covalent bonds and bond angles, where for each pair and triple of atom types there is a separate value for the force constants K_b and K_θ respectively and the equilibrium values b_o and θ_o. For all sets of four connected atoms (the chain 1-2-3-4 with covalent bonds 1-2, 2-3, and 3-4), dihedral angle energy terms are expressed as the sum of periodic functions, each with its own K_ϕ, periodicity $2\pi/n$, and phase displacement δ. The last two terms are van der Waals energy and Coulombic electrostatic energy. These two terms usually skip all 1-3 and 1-4 terms (those in the same bond angle or dihedral angle). Additional terms may be included in some terms, including 1-3 pseudo-bond harmonic forces, improper dihedral energy terms, and solvation terms.

In molecular mechanics potential energy functions, some hydrogen atoms may not be represented explicitly. Non-polar carbon groups such as methyl and methylene groups are represented as single atoms in the united-atom approximation with van der Waals radius large enough to represent the entire group. In united atom potentials, polar hydrogens such as those on NH and OH groups are represented explicitly, so that polar interactions such as hydrogen bonds can be represented accurately. More modern potentials include all hydrogen atoms explicitly.

Relevant websites

Empirical Force Field Development Resources (CHARMM parameter development)	http://mackerell.umaryland.edu/Empirical_FF_Dev.html
Amber	http://ambermd.org/

Further reading

Cornell WD, et al. (1995) A second generation force field for the simulation of proteins, nucleic acids, and organic molecules. *J. Am. Chem. Soc.* 117: 5179–5197.

MacKerell AD, Jr., et al. (1998) All-atom empirical potential for molecular modeling and dynamics studies of proteins. *J. Phys. Chem.* B102: 3586–3616.

ENA (EUROPEAN NUCLEOTIDE ARCHIVE)

Obi L. Griffith and Malachi Griffith

The European Nucleotide Archive (ENA) constitutes Europe's major nucleotide sequence resource. It is the European member of the tripartite International Nucleotide Sequence Database Collaboration together with DDBJ (Japan) and NCBI/GenBank (USA). Collaboratively, the three organizations collect all publicly available nucleotide sequences and exchange sequence data and biological annotations on a daily basis. The ENA is comprised of three main databases: the Sequence Read Archive (SRA), the Trace Archive and EMBL-Bank. The ENA team develops nucleotide sequence archive data services, as part of the nucleotide section of the Protein and Nucleotide Database (PANDA) group at EBI. They provide data repositories and services to facilitate submission and access to nucleotide sequence, from raw read data, through assembled sequence to functional annotation. Access is provided through a browser, search tools, large-scale file downloads and programmatically through an API. As of October 2010, the ENA contained ~500 billion raw and assembled sequences comprised of ~50 trillion base pairs. By far the fastest growing source of new data, in terms of total base pairs, was the next-generation sequence data in the SRA now accounting for more than 95% of all base pairs. The number of completed genome sequences, perhaps an equally important metric, was over 1,400 for cellular organisms and over 3,000 for viruses and phages.

Related websites

European Nucleotide Archive	http://www.ebi.ac.uk/ena/
Sequence Read Archive	http://www.ncbi.nlm.nih.gov/sra
Trace Archive	http://www.ncbi.nlm.nih.gov/Traces/
EMBL-Bank	http://www.cbi.ac.uk/embl/
PANDA	http://www.ebi.ac.uk/panda/

Further reading

Leinonen R, et al. (2011) The European Nucleotide Archive. *Nucleic Acids Res.* *39*(Database issue): D28–31.

ENCODE (ENCYCLOPEDIA OF DNA ELEMENTS)

Enrique Blanco and Josep F. Abril

The ENCyclopedia Of DNA Elements (ENCODE) Project was launched in 2003 by The National Human Genome Research Institute (NHGRI) in order to fully characterize all functional elements in the human genome sequence and promote new technologies to efficiently generate high-throughput data.

The ENCODE project aims to identify a vast array of elements in the genome: genes, promoters, enhancers, repressors/silencers, exons, origins of replication, sites of replication termination, transcription factor binding sites, methylation sites, deoxyribonuclease I (DNase I) hypersensitive sites, chromatin modifications, and multispecies conserved sequences.

The pilot phase was focused on 30 Mbp that encompass about 1% of the human genome. In total, a set of 44 discrete regions distributed across different chromosomes was selected (14 regions for which abundant biological information was already available and 30 chosen by random-sampling methods, considering homology to mouse genome as another factor). Experimental and computational protocols were implemented and/or assessed in order to scale them up to the whole genome annotation effort, which started in september 2007 with the production phase of the ENCODE project.

In parallel with ENCODE, the modENCODE Project (Model Organism ENCyclopedia Of DNA Elements) was initiated in 2007 to generate a comprehensive annotation of functional elements in the *Caenorhabditis elegans* (worm) and *Drosophila melanogaster* genomes (fruit fly). The major benefit of studying functional elements in model organisms is the ability to effectively validate *in vivo* the features discovered using methods that are often not possible or practical in mammals, paving the way to a more complete understanding of the human genome.

Related website

ENCODE	http://encodeproject.org/ENCODE/

Further reading

The ENCODE Project Consortium (2004) The ENCODE (ENCyclopedia Of DNA Elements) Project. *Science* 306: 636–640.

The ENCODE Project Consortium (2007) Identification and analysis of functional elements in 1% of the human genome by the ENCODE pilot project. *Nature* 447: 799–816.

The ENCODE Project Consortium (2011) A User's Guide to the Encyclopedia of DNA Elements (ENCODE). *PLoS Biol* 9: e1001046.

The modENCODE Consortium (2009) Unlocking the secrets of the genome. *Nature* 459: 927–930.

The modENCODE Consortium (2010). Identification of Functional Elements and Regulatory Circuits by *Drosophila* modENCODE. *Science* 330: 1787–1797.

The modENCODE Consortium (2010). Integrative Analysis of the *Caenorhabditis elegans* Genome by the modENCODE Project. *Science* 330: 1775–1787.

Special Issue on ENCODE from *Genome Research*, June 2007 [http://www.genome.org/content/vol17/issue6/]

See also GENCODE, GASP, Gene Annotation, Gene Annotation, hand-curated, Gene Prediction

ENCPRIME/SEQCOUNT

Michael P. Cummings

A pair of programs to calculate codon bias summary statistics.

SeqCount (sequence count) produces tables of nucleotide and codon counts and frequencies from protein coding sequence data. ENCprime (effective number of codons prime) calculates the codon usage bias statistics effective number of codons $\widehat{N}_c$, effective number of codons prime $\widehat{N}_c'$, scaled χ^2, and $B^*(a)$. $\widehat{N}_c'$ accounts for background nucleotide composition, is independent of gene length, and has a low coefficient of variation.

Input file format is FASTA (SeqCount) or output from SeqCount (ENCprime). The programs are available as ANSI C source code.

Related website

ENCprime home page	http://jnpopgen.org/software/

Further reading

Novembre JA (2002) Accounting for background nucleotide composition when measuring codon usage bias. *Mol. Biol. Evol.* 19: 1390–1394.

ENCYCLOPEDIA OF DNA ELEMENTS, *SEE* ENCODE.

END GAP

Jaap Heringa

End gaps denote gap regions in sequence alignments at either end of a sequence, as opposed to gaps that are inserted in between residues in an alignment. An important issue is whether to penalize end gaps preceding or following a sequence when a global alignment, containing the complete sequences, is constructed. The first implementation of global alignment using the dynamic programming algorithm (Needleman and Wunsch, 1970) and other early versions all applied end gap penalties. However, this is generally not a desirable situation (Abagyan and Batalov, 1997), as many proteins comprise recurring structural domains, often in combination with different complementary domains. For example, if a sequence corresponding to a two-domain protein is globally aligned with a protein sequence associated with one of the two domains only, applying end gaps would be likely to lead to an incorrect alignment in which the single-domain sequence spans the two-domain sequence. Therefore, most modern alignment routines do not penalize end gaps any more, and this mode of alignment is generally referred to as semi-global alignment.

A situation where applying end gaps might be desirable is when aligning tandem repeats, because the various repeat sequences are in fact consecutive parts of a single sequence, so that internal gaps and end gaps cannot be distinguished. End gaps are not an issue in local alignment (Smith and Waterman, 1981) as this technique aligns the most similar regions in a target sequence set only, so that end gaps do not occur.

E

Further reading

Abagyan RA, Batalov S (1997) Do aligned sequences share the same fold? *J. Mol. Biol.* 273: 355–368.

Needleman SB, Wunsch CD (1970) A general method applicable to the search for similarities in the amino acid sequence of two proteins. *J. Mol. Biol.* 48: 443–453.

Smith TF, Waterman MS (1981) Identification of common molecular subsequences. *J. Mol. Biol.* 147: 195–197.

See also Global Alignment, Local Alignment, Dynamic Programming, Gap Penalty

ENERGY MINIMIZATION

Roland Dunbrack

A procedure for lowering the energy of a molecular system by changing the atomic coordinates. Energy minimization requires a defined potential energy function expressed as a function of internal and/or Cartesian coordinates. Because the energy function is complicated and non-linear, energy minimization proceeds in an iterative fashion. The methods differ in how the step direction is chosen and how large the steps are. Most such procedures are not guaranteed to find the global minimum energy of the system, but rather to find a local energy minimum.

Energy minimization methods can be distinguished by whether they use the first and/or second derivatives of the energy function with respect to the coordinates. First-order methods include the steepest descent method and the conjugate gradient method. The steepest descent method changes the atomic coordinates of a molecular system in steps in the opposite direction of the first derivative of the potential energy function. This derivative is the gradient of the energy function with respect to the Cartesian coordinates of all atoms. The step size is varied such that it is increased if the new conformation has a lower energy and decreased if it is higher. Steepest descent minimization will usually not converge but rather sample around a local minimum. The method is fast and is often used to remove close steric contacts or other locally high-energy configurations in protein structures, since it finds the nearest local minimum to the starting position.

The conjugate gradient method is an energy minimization procedure that changes atomic coordinates in steps based on the first derivative of the potential energy and the direction of the previous step. It can be written symbolically as:

$$\delta_i = -\mathbf{g}_i + \delta_{i-1}\,\frac{|\mathbf{g}_i|^2}{|\mathbf{g}_{i-1}|^2}$$

$$\mathbf{r}_{i+1} = \mathbf{r}_i + \alpha\delta_i$$

where $\mathbf{g}_i = \Delta U(\mathbf{r}_i)$, the gradient of the potential energy function U. α determines the step-size, and is often chosen by a line search, that is trying several values, evaluating the energy, and choosing the best value. The purpose of using the previous step is to account for the curvature of the potential energy function without actually calculating the second derivative of this function. Conjugate gradient minimization generally converges faster than steepest descent minimization.

Second order methods that use both the first and second derivatives of the energy function include the Newton-Raphson method and a variation called the adopted-basis Newton-Raphson method. For a one-dimensional system with an approximately quadratic (harmonic) potential, it is possible to move to the minimum in one step:

$$x_{\min} = x_o - \left[\frac{dU}{dx}\right]_{x_o} \Big/ \left[\frac{d^2U}{dx_2}\right]_{x_o}$$

For a molecular system, the change at each step can be expressed as:

$$\delta_i = -\ \mathbf{H}_i^{-1}\mathbf{g}_i$$

where **H** is the Hessian matrix of all second derivatives of the potential energy. The *k,l*-th entry in the Hessian matrix is:

$$\mathbf{H}_{i,kl} = \left\lfloor \frac{\partial^2 U}{\partial x_k \partial x_1} \right\rfloor_{\mathbf{r}_i}$$

where x_k is the x,y, or z coordinate of any atom in the system. The Newton-Raphson method therefore requires $3N$ first derivative terms and $(3N)^2$ second derivative terms, as well as a large matrix inversion. As such it is a very expensive calculation, and may behave erratically far from a local minimum.

One solution to the size problem of the NR method is to perform the calculation on a set of linear combinations of the coordinates, where the number of vectors in the set is very small. These vectors are chosen to be in the direction of the largest changes in recent steps:

$$\Delta\mathbf{r}_i^{(p)} = \mathbf{r}_{i-1} - \mathbf{r}_{i-1-p}$$

where p ranges from 1 to N, the number of vectors in the set, usually between 4 and 10.

The Newton-Raphson equations are then solved for this basis set. This is the adopted-basis Newton Raphson (ABNR) method.

Relevant website

Newton-Raphson Method	http://www.shodor.org/unchem/math/newton/

Further reading

Brooks CL, 3rd, et al. (1988) *Advances in Chemical Physics* 71, New York, Wiley.

ENHANCER

Niall Dillon

Regulatory sequence of a gene responsible for increasing transcription from it.

Operational definition describing a sequence that increases transcription from a promoter that is linked to the enhancer *in cis*. A defining feature of enhancers is that they can exert their effects independently of orientation relative to the promoter and with some flexibility with respect to distance from the promoter. The definition does not specify the nature of the assay used to assess the enhancer effect or the size of the increase in transcription. Analysis of specific enhancers has shown that they contain multiple binding sites for different transcription factors which in turn recruit auxiliary factors. Together, these factors form a complex called the enhanceosome.

Further reading

Carey M (1998) The enhanceosome and transcriptional synergy. *Cell* 92: 5–8.

Panne D (2008) The enhanceosome. *Curr Opin Struct Biol* 18: 236–242.

See also Cis-regulatory Module

ENSEMBL

Guenter Stoesser, Obi L. Griffith and Malachi Griffith

Ensembl is a genome sequence and annotation database including graphical views and web-searchable datasets. Together with the UCSC Genome Browser it represents one of the two primary genome browser systems.

The project creates and applies software systems, which produce and maintain automatic annotation on the genomes of more than 60 species. Manual annotation of select finished genomes is also available through the related Vertebrate Genome Annotation database (VEGA), carried out primarily by the HAVANA group at the Welcome Trust Sanger Institute. As new genomes become available comparative genome views are generated. Ensembl annotates known genes and predicts novel genes based on mRNA and protein evidence from the EMBL Nucleotide Sequence Database, UniProtKB and RefSeq with further annotation of splice variants, regulatory sequences, protein motifs/domains (InterProScan) and with additional links to external resources such as OMIM, EntrezGene, WikiGene, UniGene and more.

Ensembl automatic annotation is available as an interactive web service. Ensembl contigview web pages feature the ability to scroll along entire chromosomes, while viewing all integrated features within a selected region in detail. Users can control which features are displayed and dynamically integrate external data sources such as the Distributed Annotation System (DAS) to easily view and compare annotations from different sources that are distributed across the Internet. The Ensembl resource can also be accessed programmatically through a Perl API and also through BioMart.

Ensembl has been one of the leading sources of human genome sequence annotation and the Ensembl system has been installed around the world in both industrial and academic environments. Ensembl is a joint project between EMBL – EBI and the Sanger Centre.

Related websites

Distributed Annotation System	http://www.biodas.org/
Ensembl BioMart	http://ensembl.org/biomart/martview
Ensembl homepage	http://www.ensembl.org/
Vertebrate Genome Annotation	http://vega.sanger.ac.uk/

Further reading

Ashurst JL, et al. (2005) The Vertebrate Genome Annotation (Vega) database. *Nucleic Acids Res. 33*(Database issue): D459–D465.

Birney E, et al. (2004) An Overview of Ensembl. *Genome Research*. 14(5): 925–928.

Curwen V, et al. (2004) The Ensembl Automatic Gene Annotation System. *Genome Research*. 14(5): 942–950.

Dowell RD, et al. (2001) The Distributed Annotation System. *BMC Bioinformatics*. 2: 7.

Flicek P, et al. (2012) Ensembl 2012. *Nucleic Acids Res. 40*(Database issue): D84–90.

Hammond MP, Birney E (2004) Genome information resources – developments at Ensembl. *Trends Genetics*. 20(6): 268–272.

Hubbard TJP, et al. (2002) The Ensembl genome database project. *Nucleic Acids Res*. 30: 38–41.

Hubbard TJP, Birney E. (2000) Open annotation offers a democratic solution to genome sequencing. *Nature*. 403: 825.

Potter SC, et al. (2004) The Ensembl Analysis Pipeline. *Genome Research*. 14(5): 934–941.

Spudich GM, Fernández-Suárez XM (2010) Touring Ensembl: A practical guide to genome browsing. *BMC Genomics*. 11: 295.

Stalker J, et al. (2004) The Ensembl Web Site: Mechanics of a Genome Browser. *Genome Research*. 14(5): 951–955.

ENSEMBL GENOME BROWSER, *SEE* ENSEMBL.

ENSEMBL PLANTS

Dan Bolser

Ensembl Plants is a division of Ensembl Genomes, a collection of genomes spanning the taxonomic space organized and visualized using the Ensembl technology. The database gathers together the complete genomes of 16 plants in collaboration with Gramene. Automatic analysis applied from the Ensembl framework includes comparative analysis of all proteins and DNA, as well as syntenic analysis of the rice, arabidopsis, and solonaceae genomes. Repeats are identified, and genes are systematically annotated using InterProScan and ncRNA analysis pipelines. Where possible, variation, and functional genomics data are stored and visualized. Advanced querying over all data is permitted via the BioMart infrastructure.

Related website

Ensembl Plants	http://www.ensembl.org/

Further reading

Kersey PJ, Lawson D, Birney E, et al. (2010) Ensembl Genomes: extending Ensembl across the taxonomic space. *Nucleic Acids Res*. 38: 563–569.

Youens-Clark K, Buckler E, Casstevens T, et al. (2010) Gramene database in 2010: updates and extensions. *Nucleic Acids Res* 39: 1085–1094.

See also BioMart, Ensembl, Gramene, MaizeGDB, PlantsDB, SGN

E

ENSEMBL VARIATION, *SEE* ENSEMBL.

ENSEMBLE OF CLASSIFIERS

Concha Bielza and Pedro Larrañaga

An ensemble of classifiers uses multiple models to obtain better predictive performance than could be obtained from any of the constituent models. The idea comes from the *no free-lunch theorem* (Wolpert and MacReady, 1996) in machine learning which states that there is no a learning algorithm that in any domain always induces the most accurate classifier. Therefore, algorithms that make different decisions to complement each other are sought.

Each constituent classifier is accurate on different instances, specializing in problem subdomains. Having strong accurate constituents is not so important. They are chosen for their simplicity leaving the ensemble the task of yielding better results. This is at the expense of working with many models, which perhaps is non-intuitive and harder to interpret. Some examples of ensembles of classifiers are bagging, random forest and boosting.

Further reading

Cornero A, Acquaviva M, Fardin P, et al. (2012) Design of a multi-signature ensemble classifier predicting neuroblastoma patients' outcome. *BMC Bioinformatics* 13(Suppl 4): S13.

Kuncheva L (2004) *Combining Pattern Classifiers*. Wiley.

Shen HB, Chou KC (2006) Ensemble classifier for protein fold pattern recognition. *Bioinformatics* 22(4): 1717–1722.

Wolpert DH, Macready WG (1997) No free lunch theorems for optimization. *IEEE Transactions on Evolutionary Computation* 1(1): 67–82.

See also Random Trees, Bagging, Boosting, Regression Trees

ENTREZ

Dov Greenbaum and John M. Hancock

A large scale data retrieval system, linked to many of the National Center for Biotechnology Information (NCBI) databases.

Entrez allows for interlinking between all of the databases in its system. It provides researchers access to thirty major databases. Entrez also provides access to some other resources including OMIM, PubMed and PubMed Central, online books and MeSH.

Related website

ENTREZ Front Page	http://www.ncbi.nlm.nih.gov/Entrez/

Further reading

Schuler GD, et al. (1996) Entrez: molecular biology database and retrieval system. *Methods Enzymol* 266: 141–162.

ENTREZ GENE (NCBI 'GENE')

Malachi Griffith and Obi L. Griffith

The 'Gene' database is a collection of information that attempts to link gene identifiers to associated map, sequence, expression, structure, function, citation, and homology data. A gene may be entered and assigned a unique identifier based on a defining sequencing, a known physical map position, or inference from phenotypic data. Gene records are not created for whole genome shotgun (WGS) sequenced genomes but a number of other events can trigger the creation of a new gene record including: creation of a new RefSeq record, submission by a genome-specific database, and identification of a gene model by the NCBI Genome Annotation Pipeline. Some components of a gene record are annotated manually and others automatically. Entrez Gene is particularly comprehensive in its attempts to identify the synonyms of each gene and resolve ambiguities in gene naming. Information contained in each Gene record is available in a variety of file formats and available for download by FTP or via NCBI's Entrez and E-Utilities systems. The 'Gene' database is maintained by NCBI.

Related websites

Gene homepage	http://www.ncbi.nlm.nih.gov/gene
Quick start guide	http://www.ncbi.nlm.nih.gov/books/NBK3841/ #EntrezGene.Quick_Start

Further reading

Brown G, et al. (2011) Gene Help: Integrated Access to Genes of Genomes in *the Reference Sequence Collection*. December 15.

Maglott D, et al. (2011) Entrez Gene: gene-centered information at NCBI. *Nucleic Acids Res.* *39*(Database issue): D52–7.

See also Entrez

ENTROPY

Thomas D. Schneider

Entropy is a measure of the state of a system that can roughly be interpreted as its randomness. Since the entropy concept in thermodynamics and chemistry has units of energy per temperature (Joules/Kelvin), while the uncertainty measure from Claude Shannon has units of bits per symbol, it is best to keep these concepts distinct.

See also Molecular Efficiency, Second Law of Thermodynamics, Negentropy, Shannon Entropy

EPAKTOLOG (EPAKTOLOGUE)

Laszlo Patthy

Epaktologs are homologous proteins (genes) related only through the independent acquisition of homologous genetic material. The term is based on the ancient Greek word επακτός, 'imported' since the basis of their homology is the import of homologous genetic material.

Epaktologs show only partial or local homology since their homology is limited to the imported region. Nevertheless the terms 'partial homology' or 'local homology' do not grab the uniqueness of this type of homology since the latter terms may also be valid for cases where two sequences are partially homologous (because one of the homologs lost or gained some region) but at the same time they are orthologous (or paralogous or pseudoparalogous).

In Figure E.1 some epaktologous proteins are shown that are related only through the independent acquisition of Scavenger Receptor Cysteine-Rich (SRCR) domains.

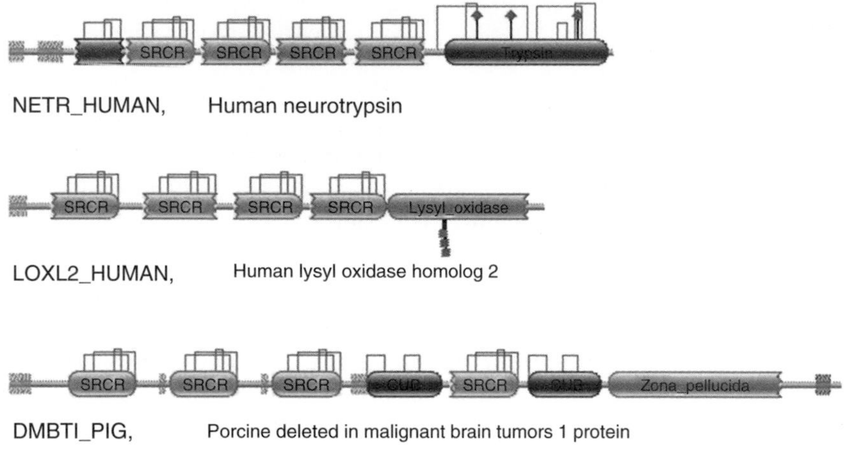

Figure E.1 Epaktologous proteins. (See Colour plate E.1)

Further reading

Nagy A, Bányai L, Patthy L (2011) Reassessing domain architecture evolution of metazoan proteins: major impact of errors caused by confusing paralogs and epaktologs *Genes* 2(3): 516–561.

See also Homologous Genes, Homology, Modular Protein, Module Shuffling, Mosaic Protein, Multidomain Protein, Protein Family, Protein Module

EPISTATIC INTERACTIONS (EPISTASIS)

Dov Greenbaum

Genomic epistatic interactions are thought to influence a number of evolutionary processes. Genes in a genome likely do not all interact with each other in a uniform manner.

Epistatic effects can be classified as: (1) genetic suppression where the phenotype of one gene is suppressed epistatically by a second; (2) genetic enhancement, wherein a phenotype of a gene is enhance or more severe; (3) synthetic lethality wherein the combination of two genes or their mutant alleles fail to compliment each other but do not map to the same loci on the genome; and (4) intragenic complementation wherein two mutations map to the same locus in the genome the two alleles seem to complement in the heteroallelic diploid.

Whereas the majority of genes have few epistatic interactions, a number of genetic hubs have been shown to exhibit an especially large number of epistatic interactions. Those genes that are involved on the giving end of an epistatic interaction are sometimes referred to as modifying genes. Those genes on the receiving end of an epistatic interaction are typically referred to epistatic genes.

Epistasis is somewhat different than simplistic genetic interactions. Epistasis may not necessarily include actual biochemical interactions but may typically refer to statistical probabilities as seen used in population genetics. However, there may remain functional interactions between the genes and the gene products that lead to an epistatic result.

Often tightly linked genetic loci will have epistatic effects from one gene to the other within the supergene family.

Epistatic interactions are thought to be generally trimodal, i.e., pairs of loci show either very strong negative or positive interactions or no interactions at all. Epistatic interactions are generally relatively plastic vis a vis environmental conditions as the interactions are often present in some, but not all, environments.

Related websites

SNPHarvester	http://bioinformatics.ust.hk/SNPHarvester.html
Tools considering Epistatic Effects	http://www.gen2phen.org/book/export/html/5923

Further reading

Ackermann M, Beyer A (2012) Systematic detection of epistatic interactions based on allele pair frequencies. *PLoS Genet* 8(2): e1002463.

Azevedo L, Suriano G, van Asch B, Harding RM, Amorim A. (2006) Epistatic interactions: how strong in disease and evolution? *Trends Genet* 22(11): 581–585.

Snitkin ES, Segrè D (2011) Epistatic interaction maps relative to multiple metabolic phenotypes. *PLoS Genet* 10;7(2): e1001294.

Zhang Y, Jiang B, Zhu J, Liu JS (2011) Bayesian models for detecting epistatic interactions from genetic data. *Ann Hum Genet* 75(1): 183–193.

ERROR

Thomas D. Schneider

In communications, an error is the substitution of one symbol for another in a received message caused by noise. Shannon's channel capacity theorem showed that it is possible to build systems with as low an error as desired, but one cannot avoid errors entirely.

Further reading

Shannon CE (1948) A mathematical theory of communication. *Bell System Tech J* 27: 379–423, 623–656. http://cm.bell-labs.com/cm/ms/what/shannonday/paper.html

Shannon CE (1949) Communication in the presence of noise. *Proc. IRE*, 37: 10–21.

See also Channel Capacity Theorem, Molecular Efficiency

ERROR MEASURES (ACCURACY MEASURES, PERFORMANCE CRITERIA, PREDICTIVE POWER, GENERALIZATION)

Nello Cristianini

In a classifier, we call accuracy (error rate) the expected rate of correct (incorrect) predictions made by the model over a data set drawn from the same distribution that generated the training data. The true accuracy can be only estimated from a finite sample, and different methods can be used. Accuracy is usually estimated by using an independent test set of data that was not used at any time during the learning process. However in the literature more complex accuracy estimation techniques, such as cross validation and the bootstrap, are commonly used, especially with data sets containing a small number of instances. (See Cross-Validation.)

More specific measures of performance can be defined in special cases, for example in the important special case of measuring performance of a 2-class classifier (see Classification). In this case, there are only four possible outcomes of a prediction:

- True positives: (the prediction is positive and the actual class is positive)
- True negatives: (the prediction is negative and the actual class is negative)
- False positives: (the prediction is positive and the actual class is negative)
- False negatives: (the prediction is negative and the actual class is positive).

Let the number of points falling within each of these categories be represented by the variables TP, TN, FP, FN. They can be organized in a so called 'confusion matrix' as follows:

	predicted-negative	predicted-positive
Actual Negative	TN	FP
Actual Positive	FN	TP

Then we can define the following measures of performance of a binary classifier:

Accuracy	A = (TN+TP)/(TN+FP+FN+TP)
True positive rate (Recall, Sensitivity)	TPR = TP/(FN+TP)
True negative rate (Specificity)	TNR = TN/(TN+FP)
Precision	P = TP/(TP+FP)
False positive rate	FPR = FP/(FN+TP)
False negative rate	FNR = FN/(TN+FP)

The 2x2 confusion matrix contains all the information necessary to compute the above measures of performance of a 2-class classifier. More generally it is possible to define a LxL confusion matrix, where L is the number of different label values.

It is also interesting to compare the ranking induced on the data by different functions. Often, classifiers split the data by sorting them, and choosing a cut-off point. The choice of the cut-off (threshold) is the result of a trade-off between sensitivity and specificity in the resulting classifier.

By moving the threshold, one can obtain different levels of sensitivity and specificity. If one plots Sensitivity vs 1-Specificity (or equivalently true positive rate vs false positive rate) for all possible choices of cut-off, then the resulting plot is called a Receiver Operating Characteristic curve (ROC curve), a term derived from applications in radar signal processing). Such a curve shows the trade-off between sensitivity and specificity; the closer it lies to the left and top borders the better is the underlying classifier; the closer it lies to the diagonal the worse (random guessing gives the diagonal ROC curve).

Finally, one can consider the area delimited by the ROC curve as a measure of performance called the ROC ratio (an area of 1 represents a perfect classifier, of 1/2 a random one). This quantity is sometimes called 'discrimination' and is related to the probability that two random points, one taken from the positive and one from the negative class, are both correctly classified by the model.

Further reading

Metz CE (1978) Basic principles of ROC analysis. *Seminars in Nucl Med* 8: 283–298.

Mitchell T *Machine Learning* (1977) McGraw-Hill, Maidenhead, UK.

Provost F, et al. (1998) The case against accuracy estimation for comparing induction algorithms. *Proceedings of the Fifteenth International Conference on Machine Learning*.

E-S ALGORITHM, *SEE* ELSTON-STEWART ALGORITHM.

EST, *SEE* EXPRESSED SEQUENCE TAG.

ESTIMATION OF DISTRIBUTION ALGORITHM (EDA)

Pedro Larrañaga and Concha Bielza

Estimation of distribution algorithms (EDAs) (Larrañaga and Lozano, 2002) are evolutionary algorithms that solve optimization problems by evolving a population of individuals, each of them representing a candidate solution. The scheme of evolution of EDAs is similar to that of genetic algorithms. The main difference is that the EDAs do not need crossover or mutation operators to produce a new population of individuals. Instead, EDAs need to learn the joint probability distribution of selected individuals at each generation, and then by sampling from this probability distribution a new population of individuals is obtained.

EDAs can be categorized according to the complexity of the probabilistic model used to capture the interdependencies between the variables representing the individuals into: univariate, bivariate or multivariate approaches. Univariate EDAs assume that all variables are independent and the joint probability distribution factorizes as a product of univariate marginal probabilities. Bivariate EDAs are able to represent two order dependencies among variables, while multivariate EDAs factorize the joint distribution according to a Bayesian network whose structure represents the conditional independent statements among variables.

Examples on the use of EDAs in genomics include the determination of splice sites (Saeys et al., 2004), and the selection of candidate genes with microarray data (Bielza et al., 2009). In proteomics, EDAs have been applied for example in the prediction of protein structure in simplified models (Santana et al., 2008), and in the protein side chain placement problem.

Further reading

Bielza C, Robles V, Larrañaga P (2009) Estimation of distribution algorithms as logistic regression regularizers of microarray classifiers. *Methods of Information in Medicine* 48(3): 236–241.

Larrañaga P, Lozano JA, (2002) *Estimation of Distribution Algorithms. A New Tool for Evolutionary Computation*. Kluwer Academic Publishers.

Saeys Y, et al. (2004) Fast feature selection using a simple estimation of distribution algorithm: A case study on splice site prediction. *Bioinformatics* 19(Suppl 2): 179–188.

Santana R, Larrañaga P, Lozano JA (2007) Side chain placement using estimation of distribution algorithms. *Artificial Intelligence in Medicine* 39: 49–63.

Santana R, Lozano JA, Larrañaga P (2008) Protein folding in simplified models with estimation of distribution algorithms. *IEEE Transactions on Evolutionary Computation* 12(4): 418–438.

See also Genetic Algorithm

EUCLIDEAN DISTANCE

John M. Hancock

The simple or linear distance between two points in a space. This can be generalized to n-dimensional space as:

$$d(\mathbf{p},\mathbf{q}) = d(\mathbf{q},\mathbf{p}) = \sqrt{\sum_{i=1}^{n}(q_i - p_i)^2}.$$

As such it can be used to estimate a distance between pairs of vectors and is widely used for this.

Related website

Euclidean Distance	http://en.wikipedia.org/wiki/Euclidean_distance

EUKARYOTE ORGANISM AND GENOME DATABASES

Dan Bolser

Several eukaryotic organisms have been studied as models of basic, human and plant biology. One of the simplest free living eukaryotes, bakers yeast (*Saccharomyces cerevisiae*), has been extensively studied as a paradigm of eukaryotic cell biology. There are numerous databases devoted to the data collected for this organism and related yeasts. Mammals, such as mouse and rat, have been studied as models of human metabolism and anatomy, and fly (Drosophila melanogaster) has been used as a model system for studying genetics. In developmental biology, chicken, frog, and mouse have all served as models for understanding human embryonic development.

In plant biology, the model organism *Arabidopsis thaliana* has traditionally been the focus of research. However, with the advent of second generation sequencing technologies, many more agriculturally important plant genomes have been sequenced. Data for these complex organisms are often integrated around the genome sequence, as sequence 'features' or annotations.

An ever increasing number of genome databases from diverse eukaryotic organisms are becoming available. This growing trend has resulted in the formation of several so-called 'clade oriented' genome databases, hosting and annotating collections of genomes from several related organisms.

The associated 'genetic variation' databases have also sprung up, notably the 1000 genomes project in human, and the 1001 wheat genome database, with many more likely to follow.

Related websites

SGD	The Saccharomyces Genome Database.	http://www.yeastgenome.org/
PomBase	Comprehensive database for the fission yeast Schizosaccharomyces pombe.	http://www.pombase.org/
SCPD	The Promoter Database of Saccharomyces cerevisiae.	http://rulai.cshl.edu/SCPD/
YDPM	Yeast Deletion Project.	http://www-deletion.stanford.edu/YDPM/
ALLEN BRAIN ATLAS	Integrated gene expression and neuroanatomical data in brain in mouse, human, and primate.	http://www.brain-map.org/
FLYBRAIN	An atlas and database of the drosophila nervous system.	http://flybrain.neurobio.arizona.edu/
MGD	The Mouse Genome Database, mouse genetic, genomic, and biological information.	http://www.informatics.jax.org/
EMA	The e-Mouse Atlas, a 3-D anatomical atlas of mouse embryo development and anatomical structure.	http://www.emouseatlas.org/

RGD	The Rat Genome Database, combining information on diseases, phenotypes, knockouts, and pathways in rats.	http://rgd.mcw.edu/
DrugBank	Combines detailed drug pharmacological and drug target information.	http://www.drugbank.ca/
FlyBase	A database of drosophila genetics and molecular biology.	http://flybase.org/
GEISHA	A Chicken Embryo Gene Expression Database.	http://geisha.arizona.edu/geisha/
Xenbase	Collection of information on early amphibian development.	http://www.xenbase.org/
TAIR	The Arabidopsis Information Resource.	http://www.arabidopsis.org/
Ensembl Plants	Ensembl genome databases for important plant species.	hup://plants.cnsembl.org/
Gramene	A curated, open-source, data resource for comparative genome analysis in the grasses.	http://www.gramene.org/
PlantsDB	A resource for comparative plant genome research and information for individual plant species.	http://mips.helmholtz-muenchen.de/plant/genomes.jsp
SGN	A clade oriented database containing genomic, genetic, phenotypic and taxonomic information for plant genomes.	http://solgenomics.net/
Ensembl	Genome databases for vertebrates and other eukaryotic species.	http://www.ensembl.org/index.html
Ensembl Plants	A collection of plant genomes using Ensembl technology.	http://plants.ensembl.org/
VEGA	A database of manually annotated vertebrate genomes, including human, mouse and zebrafish.	http://vega.sanger.ac.uk/index.html
OMIM	Online Mendelian Inheritance in Man	http://www.ncbi.nlm.nih.gov/omim
Wikipedia	A popular free encyclopedia.	http://wikipedia.org/wiki/Category:Biological_databases
MetaBase	The database of biological databases	http://metadatabase.org/

EUPA (EUROPEAN PROTEOMICS ASSOCIATION)

Pedro Fernandes

The general objectives of the EuPA are to promote proteomic activities throughout Europe, emphasising the benefits and contribution of proteomics to biological researchers, industry, the general public and politicians. Other areas where the organisation feels activities can help are the promotion of scientific exchange in all areas of biological research, and the consolidation and coordination of duplicate activities by national organizations.

The EuPA's mission is summarized in the following three central points:

- strengthen the national proteomics organizations in their efforts
- coordinate and provide educational programs
- advance the networking of scientists through meetings, workshops and student exchange.

Related website

EUPA	http://www.eupa.org

EUROPEAN BIOINFORMATICS INSTITUTE, *SEE* EBI.

EUROPEAN MOLECULAR BIOLOGY OPEN SOFTWARE SUITE, *SEE* EMBOSS.

EUROPEAN NUCLEOTIDE ARCHIVE, *SEE* ENA.

EUROPHENOME

John M. Hancock

A database of phenotypic information on mouse strains. Established as part of the European projects EUMODIC, which produced phenotype data on 500 mouse lines, EuroPhenome draws data on the phenotypes of mice from a distributed group of phenotyping laboratories, consolidates them, and estimates significant deviations from normality. It provides annotations against individual lines, many of them corresponding to knockout mice generated by the Knockout Mouse Project, using the Mammalian Phenotype Ontology. It provides a number of browsing options, including by gene and by ontology term, and a heatmap-like representation of phenodeviance for all or a selected set of genes. It also provides BioMart access to data and links to OMIM.

As well as providing valuable data, EuroPhenome is an important stepping stone towards the development of the data repository for the International Mouse Phenotyping Consortium (IMPC) which aims to provide systematic phenotype data on 20,000 knockout mouse lines.

E

Related websites

EuroPhenome	http://www.europhenome.org
EMPReSS	http://empress.har.mrc.ac.uk/
EUMODIC	http://www.eumodic.org/
IMPC	http://www.mousephenotype.org/

Further reading

Ayadi A, et al. (2012) Mouse large-scale phenotyping initiatives: overview of the European Mouse Disease Clinic (EUMODIC) and of the Wellcome Trust Sanger Institute Mouse Genetics Project. *Mamm Genome* 23: 600–610.

Brown, SD, Moore MW (2012) Towards an encyclopaedia of mammalian gene function: the International Mouse Phenotyping Consortium. *Dis. Model Mech.* 5: 289–292.

Mallon AM, et al. (2012) Accessing data from the International Mouse Phenotyping Consortium: state of the art and future plans. *Mamm. Genome*, 23: 641–652.

Morgan H, et al. (2010) EuroPhenome: a repository for high-throughput mouse phenotyping data. *Nucleic Acids Res, 38*(Database Issue), D577–D585.

EVOLUTION

A.R. Hoelzel

Evolution is descent with modification leading to the accumulation of change in the characteristics of organisms or populations.

The importance and influence of ideas about evolution changed forever when Charles Darwin proposed the mechanism of natural selection to explain this process.

Evolution is said to be divergent when related species change such that they no longer resemble each other, and convergent when unrelated species change such that they resemble each other more.

Microevolution is thought of as the process of evolutionary change within populations, while macroevolution reflects the pattern of evolution at and above the species level. However there is a continuum from the processes that affect gene frequencies at the population level through to the pattern of lineages in phylogenetic reconstructions. In each case the direction and rate of evolution is influenced by random processes such as genetic drift, and the differential survival of phenotypes through natural selection.

Further reading

Darwin C (1859) *The Origin of Species By Means of Natural Selection, or The Preservation of Favoured Races in The Struggle for Life.*

Futuyma DJ (2009) *Evolution.* Sinauer Associates.

Higgs PG, Attwood TK (2004) *Bioinformatics and Molecular Evolution* Wiley-Blackwell.

See also Allopatric Evolution, Convergence

EVOLUTION OF BIOLOGICAL INFORMATION

Thomas D. Schneider

The information of patterns in nucleic-acid binding sites can be measured as *Rsequence* (the area under a sequence logo). The amount of information needed to find the binding sites, *Rfrequency*, can be predicted from the size of the genome and number of binding sites. *Rfrequency* is fixed by the current physiology of an organism but *Rsequence* can vary. A computer simulation shows that the information in the binding sites (*Rsequence*) does indeed evolve toward the information needed to locate the binding sites (*Rfrequency*) (Schneider 2000).

Related website

EV	http://alum.mit.edu/www/toms/papers/ev/

Further reading

Adami C, et al. (2000) Evolution of biological complexity. *Proc. Natl Acad. Sci. USA*, 97: 4463–4468.

Schneider TD, et al. (1986) Information content of binding sites on nucleotide sequences. *J. Mol. Biol.*, 188: 415–431. http://alum.mit.edu/www/toms/papers/schneider1986/

Schneider TD (2000) Evolution of biological information. *Nucleic Acids Res.*, 28: 2794–2799. http://alum.mit.edu/www/toms/papers/ev/

See also Molecular Efficiency

EVOLUTIONARY DISTANCE

Sudhir Kumar and Alan Filipski

An evolutionary distance measure is a numerical representation of the dissimilarity between two species, sequences, or populations.

Evolutionary distances are often computed by comparing molecular sequences (see Sequence Distance Measures) or by using allele frequency data. A set of pair-wise evolutionary distances may be used to construct a phylogenetic tree (e.g., using neighbor joining methods), and conversely, a phylogenetic tree with branch lengths induces a set of patristic distances on the taxa.

Software

Kumar S, et al. (2001) MEGA2: molecular evolutionary genetics analysis software. *Bioinformatics* 17: 1244–1245.

Further reading

Nei M, Kumar S (2000) *Molecular Evolution and Phylogenetics*. Oxford University Press, Oxford.

Sokal RR, Sneath PHA (1963) *Principles of Numerical Taxonomy*. San Francisco, W.H. Freeman.

See also Sequence Distance Measures

E

EXCLUSION MAPPING

Mark McCarthy, Steven Wiltshire and Andrew Collins

Use of linkage analysis to designate genomic regions for which the evidence against linkage is sufficiently strong as to 'exclude' that region from involvement in disease susceptibility, at least under a specific disease model.

Just as it is possible to use linkage analysis to determine the likely location of a disease-susceptibility gene, it is possible to define genomic locations where such genes are unlikely to be found. Such approaches were originally developed for the analysis of Mendelian diseases, in the context of parametric linkage analyses. In this situation, disease-locus parameters are specified in advance, and it is trivial to exclude regions, on the basis that some values of the recombination fraction (and hence some genomic locations) are incompatible with the experimental data, under the specified disease-locus model. Typically, a LOD score (logarithm of the odds for linkage) of -2 or less is taken as indicating exclusion. In the case of multifactorial traits, where analysis is predominantly based on non-parametric methods and effect size and disease parameters are not specified in advance, exclusion mapping requires designation of a locus effect size (typically parametrized in terms of the locus-specific sibling relative risk, or λ_s). It is then possible to designate those genomic locations which are excluded from containing a susceptibility locus of this magnitude, though such exclusions do not necessarily rule out a locus of more modest effect from that location.

Examples: Hanis, et al. (1995) reported a genome scan for type 2 diabetes conducted in Mexican American families. The strongest evidence for linkage was on chromosome 2q. They were able to exclude 99% of the genome from harbouring a gene with a locus-specific sibling relative risk of 2.8. Equivalent figures for loci of lesser effect were 71% (for $\lambda_s = 1.6$) and 5% (for $\lambda_s = 1.2$). Chung et al. (1995) used homozygosity mapping in consanguineous families to exclude the mineralocorticoid receptor from a role in the aetiology of a form of pseudohypoaldosteronism.

Related websites

Programs for linkage and exclusion mapping

ASPEX	http://aspex.sourceforge.net/
GENEHUNTER	http://www.broadinstitute.org/ftp/distribution/software/genehunter/

Further reading

Chung E, et al. (1995) Exclusion of a locus for autosomal recessive pseudohypoaldosteronism type 1 from the mineralocorticoid receptor gene region on human chromosome 4q by linkage analysis. *J Clin Endocrinol Metab* 80: 3341–3345.

Clerget-Darpoux F, Bonaïti-Pellié C (1993) An exclusion map covering the whole genome: a new challenge for genetic epidemiologists? *Am J Hum Genet* 52: 442–443.

Hanis CL, et al. (1996) A genome-wide search for human non-insulin-dependent (type 2) diabetes genes reveals a major susceptibility locus on chromosome 2. *Nat Genet* 13: 161–171.

Hauser E, et al. (1996) Affected-sib-pair interval mapping and exclusion for complex genetic traits: Sampling considerations. *Genet Epidemiol* 13: 117–137.

See also Linkage Analysis, Genome Scans, Multipoint Linkage Analysis

EXOME SEQUENCING

Andrew Collins

A technique which generates DNA sequence of the protein coding regions of the genome. Protein coding regions constitute about 1% or 30 megabases of the genome contained in about 180,000 exons. Exome sequencing employs 'next generation' sequencing (NGS) which produces 3–4 orders of magnitude more DNA sequence at low cost, than the traditional Sanger sequencing approach. The exome is the best understood component of the genome for relating sequence to function and, similarly, to directly link genetic variants to disease causality. Exome sequencing is particularly powerful for identifying genes underlying Mendelian disease since most such variants are known to alter protein coding sequence. The majority of rare non-synonymous (missense) alleles are considered to be deleterious. Since exome sequencing provides a complete catalogue of such variants across the entire allele frequency spectrum this approach is revolutionizing genome research. Exome sequencing is also applicable to analysis of complex disease in samples of large numbers of cases and controls. For example, it has been shown that rare variants are over-represented in genes already known to contain common variation involved in complex disease.

Examples: Johansen et al. (2010) determined a significantly increased number of rare missense or nonsense mutations in individuals with with hypertriglyceridemia, contrasted with a significantly lower burden in controls, in four genes known to contain common variants for this condition.

Ng, et al. (2009) sequenced the exons of 12 individuals and demonstrated the utility of this technique for identifying genes for Mendelian disorders using only a few unrelated affected individuals.

Further reading

Johansen CT, et al. (2010) Mutation skew in genes identified by genome-wide association study of hypertriglyceridemia. *Nat Genet* 42(8): 684–687.

Kryukov GV, Pennacchio LA, Sunyaev SR (2007) Most rare missense alleles are deleterious in humans: Implications for complex disease and association studies. *Am J Hum Genet* 80(4): 727–739.

Ng SB, et al. (2009) Targeted capture and massively parallel sequencing of twelve human exomes. *Nature* 461(7261): 272–276.

See also Next Generation Sequence Analysis, Next Generation Sequencing, Thousand Genomes Project

EXON

Niall Dillon and Laszlo Patthy

Segment of a eukaryotic gene which, when transcribed, is retained in the final spliced mRNA.

Exons are separated from one another by introns which are removed from the primary transcript by RNA splicing.

E

Eukaryotic genes are split into exons and intervening sequences (introns). The primary RNA transcript contains both exons and introns and the introns are then removed in a complex RNA splicing reaction which is closely coupled to transcription. The coding region of a gene is contained within the exons but exons can also include the 5′ and 3′ untranslated regions of the mRNA. Exons or parts of exons of protein-coding genes that are translated are collectively referred to as protein-coding regions.

Further reading

Krebs JE (ed.) (2009) *Lewin's Genes X*. Jones and Bartlett Publishers, Inc.

See also Intron, CDS

EXON SHUFFLING

Austin L. Hughes and Laszlo Patthy

The process whereby exons of genes are duplicated, deleted, or exons of different genes are joined through recombination in introns.

It has been hypothesized that exon shuffling has played an important role in the evolution of new gene functions, particularly when exon or exons are transferred and encode a domain that constitutes a structural or functional unit. For such cases, the term 'domain shuffling' is frequently used. Shuffling of exons or exon sets encoding complete protein modules leads to module shuffling and to the formation of multidomain proteins with an altered domain organization.

Further reading

Doolittle RF (1995) The multiplicity of domains in proteins. *Annu Rev Biochem* 64: 287–314.

Gilbert W (1978) Why genes in pieces? *Nature* 271: 501.

Patthy L (1995) *Protein Evolution by Exon Shuffling*. Molecular Biology Intelligence Unit, R.G. Landes Company, Springer-Verlag, New York.

Patthy, L. (1996) Exon shuffling and other ways of module exchange. *Matrix Biol.* 15: 301–310.

Patthy L (1999) Genome evolution and the evolution of exon-shuffling – a review *Gene* 238: 103–114.

See also Exon, Intron, Intron Phase, Protein Module, Module Shuffling, Multidomain Protein

EXPECTATION MAXIMIZATION ALGORITHM (E-M ALGORITHM)

Mark McCarthy, Steven Wiltshire and Andrew Collins

An algorithm, much used in various forms of genetic analysis, for the maximum likelihood estimation of parameters in situations where all the information required for the estimation is not available (either because it is missing, or cannot be observed).

In this iterative two-step algorithm, initial values for the parameters whose values are to be estimated are chosen and the function of the parameters, or some part therefore, evaluated – the E step – to obtain expected values for the missing data. These values are then used in the M step to re-estimate the values for the parameters. In turn, these new estimates are used in a second cycle of the algorithm to obtain new expectations of the missing data, and thence, more accurate estimates of the parameters. The cycle is repeated until the algorithm reaches convergence – that is the change in the values of the parameters to be estimated is sufficiently small as to no longer exceed some predetermined threshold – providing maximum likelihood estimates of the parameters of interest.

Examples: The EM algorithm is used by the EH and ARLEQUIN programs to estimate haplotype frequencies (the parameters of interest) from unphased population data. In this case, the unknown or missing information are the haplotypes in individuals who are doubly heterozygous (in whom phase is ambiguous). Initial estimates of the haplotype frequencies are used to obtain expected values for the haplotype distributions in these individuals (the E-step). These expectations are then used to re-estimate the haplotype frequencies (the M step). Dawson et al. (2002), in their construction of a first generation haplotype map of chromosome 19, used the EM algorithm to calculate haplotype frequencies from population and combined population/pedigree data. From these, they were able to determine the pattern of pair-wise linkage disequilibrium across the chromosome.

Related websites

Programs that implement the E-M algorithm for linkage disequilibrium mapping include:

EH-PLUS	http://www.mrc-epid.cam.ac.uk/Personal/jinghua.zhao/software/
ARLEQUIN	http://cmpg.unibe.ch/software/arlequin35/
GOLD	http://www.sph.umich.edu/csg/abecasis/GOLD/

Further reading

Dawson E, et al. (2002) A first-generation linkage disequilibrium map of human chromosome 22. *Nature* 418: 544–548.

Sham P (1998) *Statistics in Human Genetics*. London: Arnold, pp. 151–157.

Weir BS (1996) *Genetic Data Analysis II*. Sunderland: Sinauer Associates, pp. 67–73.

See also Linkage Disequilibrium, Phase, Haplotype

EXPONENTIALLY MODIFIED PROTEIN ABUNDANCE INDEX, *SEE* EMPAI.

EXPRESSED SEQUENCE TAG (EST)

Jean-Michel Claverie

A small (one read) sequence fragment of a cDNA clone insert, selected at random from a cDNA library.

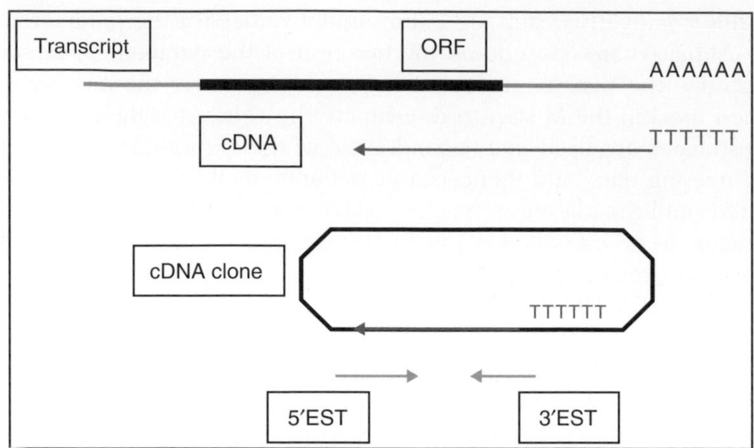

Figure E.2 The EST concept.

ESTs are used according to two different modes: the 'gene discovery' mode and the 'expression profiling' mode. In the 'gene discovery' mode several thousands of inserts of random picked clones from a cDNA library are sequenced, to establish a 'quick and dirty' repertoire of the genes expressed in the organism or tissue under study. The value of this approach is rapidly limited by the redundancy originating from the most abundant transcripts. This problem is partially corrected by the use of 'normalized' cDNA libraries. Prior to the launch of the human genome sequencing projects, large-scale EST sequencing programs from normalized libraries were performed that allowed a tag to be generated for most human genes. These ESTs played an important role in the annotation of the human genome.

In the 'expression profiling' mode the redundancy of ESTs corresponding to the same gene is used as a measurement of expression intensity. ESTs generated from the exact 3′end (3′EST, see Figure E.2) of a given transcript give the most accurate estimate of the number of transcripts corresponding to a given gene. 5′EST (randomly covering other parts of the gene transcripts) are usually preferred for gene discovery, as they usually exhibit a higher fraction of coding regions. It is worth recalling that, in contrast with 3′EST sequences that do correspond to the 3′extremity transcripts (mostly 3′UTR), 5′EST very rarely correspond to the 5′extremity of transcripts.

Library-based approaches are rapidly being replaced by direct sequencing using Next Generation Sequencing approaches.

Related websites

dbEST EST database	http://www.ncbi.nlm.nih.gov/dbEST/
Cancer Genome Anatomy Project	http://www.ncbi.nlm.nih.gov/ncicgap/

Further reading

Adams MD, et al. (1991) Complementary DNA sequencing: expressed sequence tags and human genome project. *Science* 252: 1651–1656.

Adams MD, et al. (1992) Sequence identification of 2,375 human brain genes. *Nature* 355: 632–634.

Bailey LC, et al. (1998) Analysis of EST-driven gene annotation in human genomic sequence. *Genome Res.* 8: 362–376.

Okubo K, et al. (1992) Large scale cDNA sequencing for analysis of quantitative and qualitative aspects of gene expression. *Nat. Genet.* 2: 173–179.

EXPRESSION LEVEL (OF GENE OR PROTEIN)

Matthew He

The expression level of gene or protein is a measure of the presence, amount, and time-course of one or more gene products in a particular cell or tissue. Expression studies are typically performed at the RNA (mRNA) or protein level in order to determine the number, type, and level of genes that may be up-regulated or down-regulated during a cellular process, in response to an external stimulus, or in sickness or disease. Gene Chips, Next Generation Sequencing and proteomics now allow the study of expression profiles of sets of genes or even entire genomes.

See also Microarray, RNA-seq

EXTENDED TRACTS OF HOMOZYGOSITY

Andrew Collins

Multi-megabase long chromosome regions containing only *homozygous* single-nucleotide polymorphisms are frequent in human populations. This is true even of populations that lack appreciable inbreeding. Such regions align with recombinationally suppressed regions of the genome and are therefore likely to reflect the alignment of high frequency ancestral haplotypes that are rarely broken by recombination. The presence of such regions has encouraged efforts to apply homozygosity mapping strategies to outbred populations. Homozygous regions exceeding one megabase in size may harbour rare recessive risk variants contributing to disease such as schizophrenia and late-onset Alzheimer's.

Related website

PLINK program	http://pngu.mgh.harvard.edu/~purcell/plink/ibdibs.shtml

Further reading

Gibson J, Morton NE, Collins, A (2006) Extended tracts of homozygosity in outbred human populations. *Human Molecular Genetics*, 15(5), 789–795.

Lencz T, Lambet C, DeRosse P, et al. (2007) Runs of homozygosity reveal highly penetrant recessive loci in schizophrenia. *Proc Natl Acad Sci USA* 104(50): 19942–19947.

Nalls MA, Guerreiro RJ, Simon-Sanchez J, et al. (2009) Extended tracts of homozygosity identify novel candidate genes associated with late-onset Alzheimer's disease. *Neurogenetics* 10(3): 183–190.

See also Homozygosity Mapping

EXTENSIBLE MARKUP LANGUAGE, *SEE* XML.

EXTERNAL BRANCH, *SEE* BRANCH OF A PHYLOGENETIC TREE.

EXTERNAL NODE, *SEE* NODE OF A PHYLOGENETIC TREE.

EXTRINSIC GENE PREDICTION, *SEE* GENE PREDICTION, HOMOLOGY-BASED.

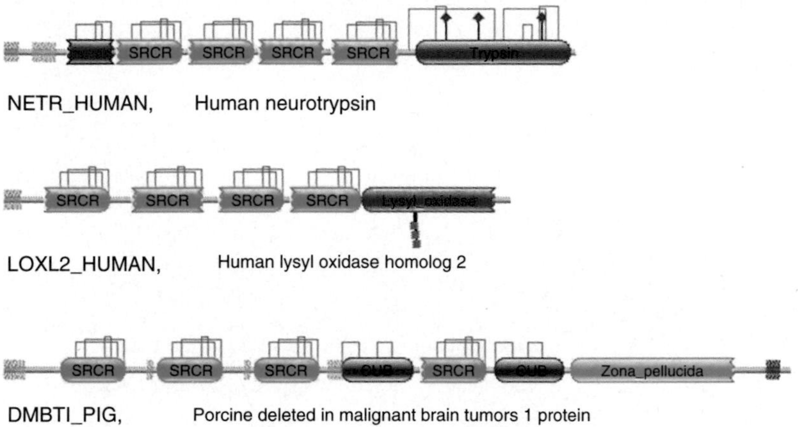

Colour plate E.1 Epaktologous proteins.

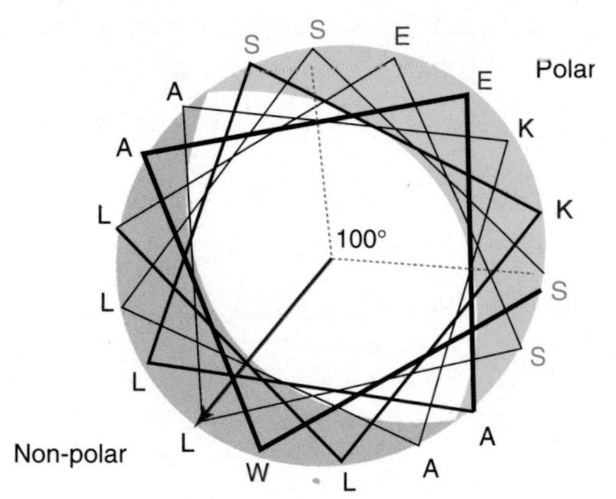

Colour plate H.1 Illustration showing how the amino acids in a particular 18-residue sequence are distributed around the rim of a helical wheel. The angular separation of the amino acid residues is 100° because this wheel depicts an alpha helical conformation. As shown, the polar residues within the given sequence fall on one side of the helix, and the non-polar residues on the other, a distribution characteristic of an amphipathic helix. For the amino acid residue colour scheme, see Visualisation Of Multiple Sequence Alignments.

Zimmerman LY**FP**IV**K**CMH | RA**E**DT**W**SGNQ
Von heinje **F**ILV**W**AMGTS | YQCN**P**H**KE**DR
Efremov I**F**LCMV**W**YHA | GTNRDQS**PEK**
Eisenberg I**F**VL**W**MAGCY | **P**TSH**E**NQD**K**R
Cornette **F**ILVMHYC**W**A | QRTG**E**S**P**N**K**D
Kyte IVL**F**CMAGTS | **W**Y**P**HDN**E**Q**K**R
Rose C**F**IVML**W**HYA | GTSR**P**ND**E**Q**K**
Sweet **F**YILMV**W**CTA | **P**SRHG**K**QN**E**D

Colour plate H.4 Comparison of eight hydropathy scales, ranking the amino acids in terms of their hydrophobic and hydrophilic properties. The scales broadly agree that phenylalanine is strongly hydrophobic and that glutamic acid is strongly hydrophilic. Other amino acids, however, are ambivalent, being sometimes hydrophobic and sometimes hydrophilic. The Zimmerman scale offers a slightly different perspective: e.g., where most scales rank tryptophan as hydrophobic, Zimmerman assigns it greater hydrophilic character; conversely, while most scales rank lysine and proline as hydrophilic, Zimmerman assigns them greater hydrophobic character.

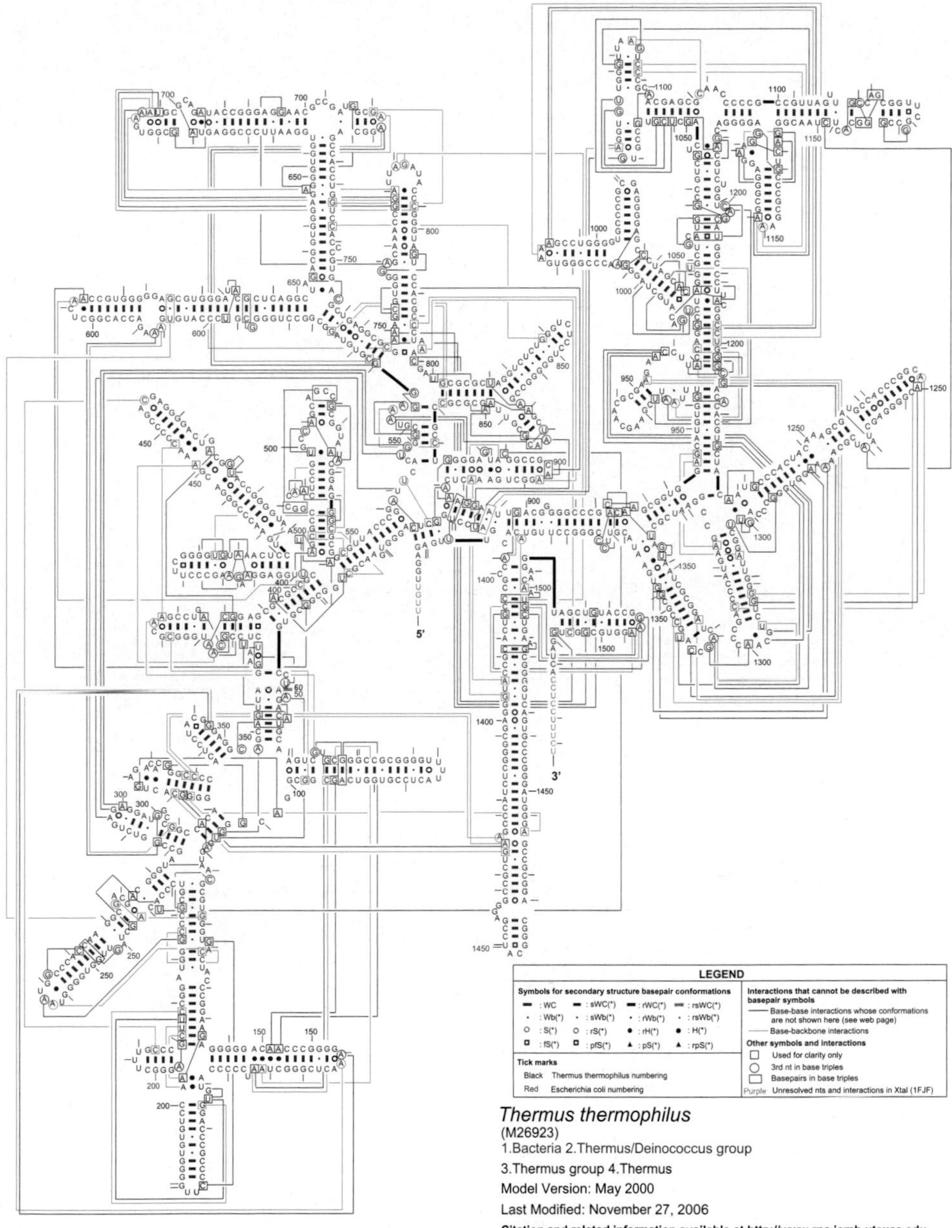

Colour plate R.1 The bacterial Thermus thermophilus 16S rRNA secondary structure, as initially determined with comparative methods and confirmed with the high-resolution crystal structure of the 30S ribosomal subunit. Tertiary structure base-base (blue) and base-backbone (green) interactions are shown with the thin lines connecting nucleotides in the secondary structure. (Source: Gutell RR, Lee JC, Cannone JJ (2002) The Accuracy of Ribosomal RNA Comparative Structure Models.Current Opinion in Structural Biology, 12(3):301–310. Reproduced with permission from Elsevier).

Secondary and tertiary interactions and basepair conformations in 23S rRNA

Haloarcula marismortui rrnB
(AF034620)
1.Archaea 2.Euryarchaeota 3.Halobacteriales
4.Halobacteriaceae 5.Haloarcula
Model Version: May 2000
Last Modified: November 28, 2006
SEE FIGURE LEGENDS ON 3' PART OF THE STRUCTURE
Citation and related information available at http://www.rna.icmb.utexas.edu

Colour plate R.2 The *archaeal Haloarcula marismortui* 23S rRNA secondary structure, as initially determined with comparative methods and confirmed with the high-resolution crystal structure of the 50S ribosomal subunit. Tertiary structure base-base (blue) and base-backbone (green) interactions are shown with the thin lines connecting nucleotides in the secondary structure. (Source: Gutell RR, Lee JC, Cannone JJ (2002) The accuracy of ribosomal RNA comparative structure models. *Current Opinion in Structural Biology*, 12(3): 301–310. Reproduced with permission from Elsevier).

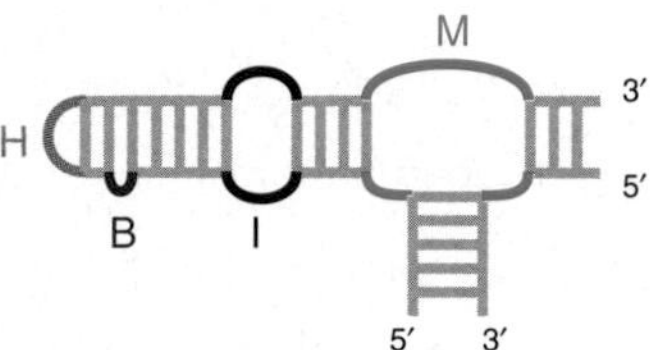

Colour plate R.4 Loop types. This schematic RNA contains one example of each of the four loop types. Colors and labels: B, bulge loop (red); H, hairpin loop (purple), I, internal loop (blue); M, multi-stem loop (green). Stems are shown in grey.

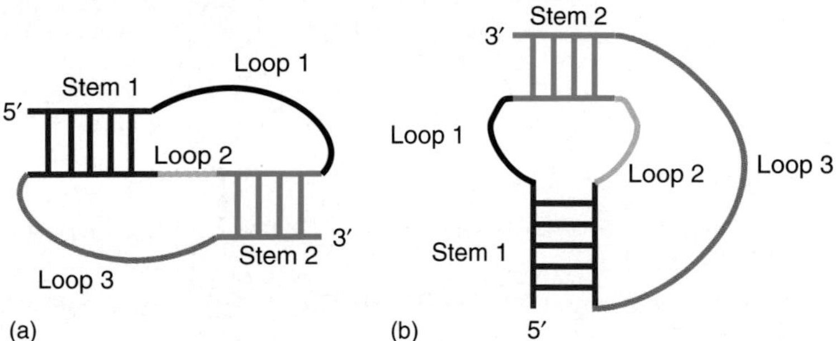

Colour plate R.5 Schematic drawings of a simple pseudoknot. Stems are shown in red and green; loops are shown in blue, orange, and purple. A. Standard format. B. Hairpin loop format (after Hilbers et al. 1998).

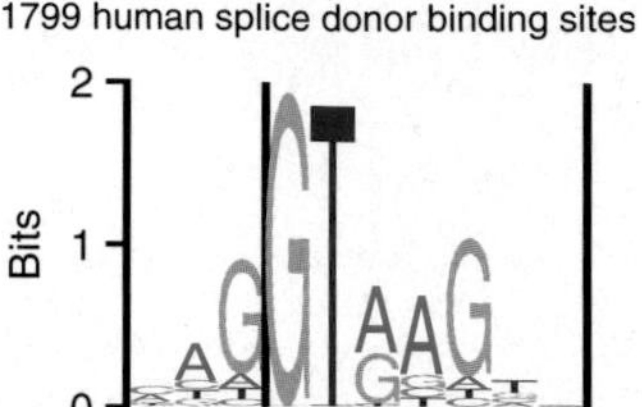

Colour plate S.1 Sequence logo of human splice donor sites (Schneider & Stephens, 1990).

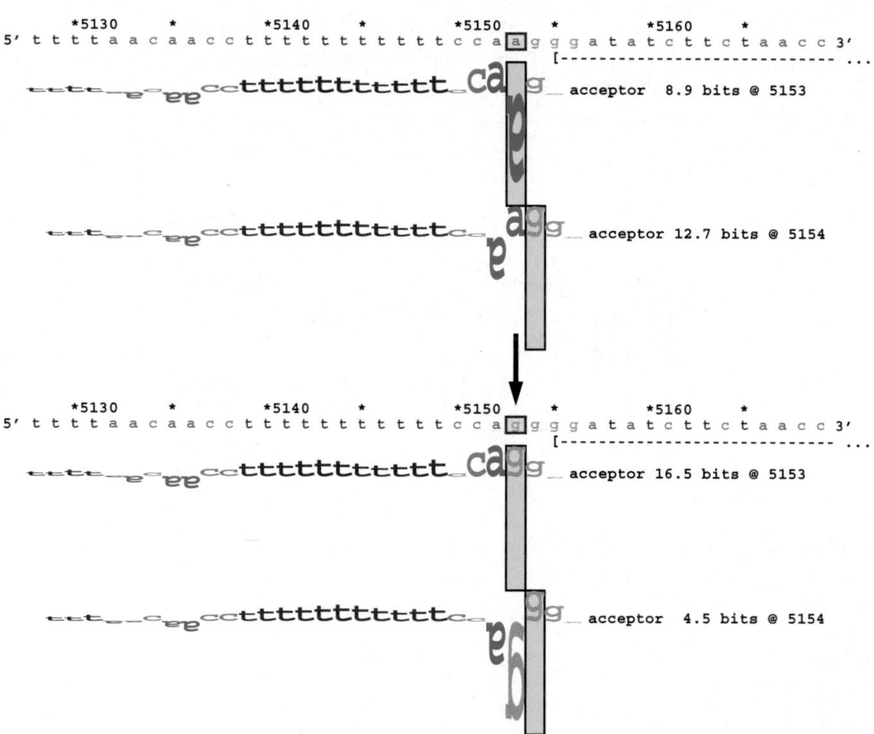

Colour plate S.2 Sequence walkers for human acceptor splice sites at intron 3 of the iduronidase synthetase gene (IDS, L35485).

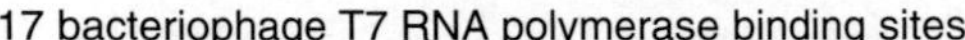

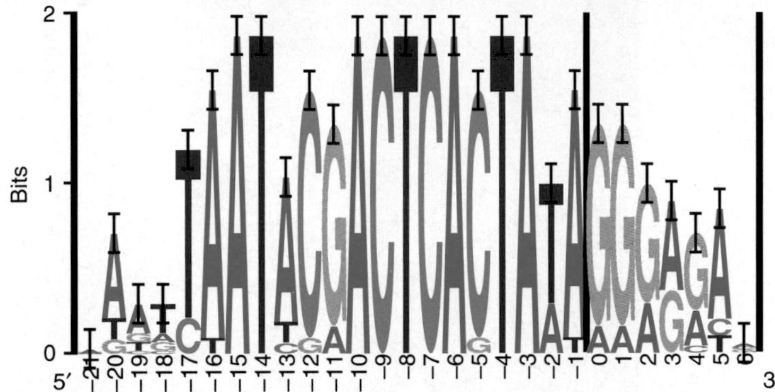

Colour plate S.6 A sequence logo for T7 RNA polymerase binding sites (Schneider & Stormo 1989, Schneider & Stephens, 1990).

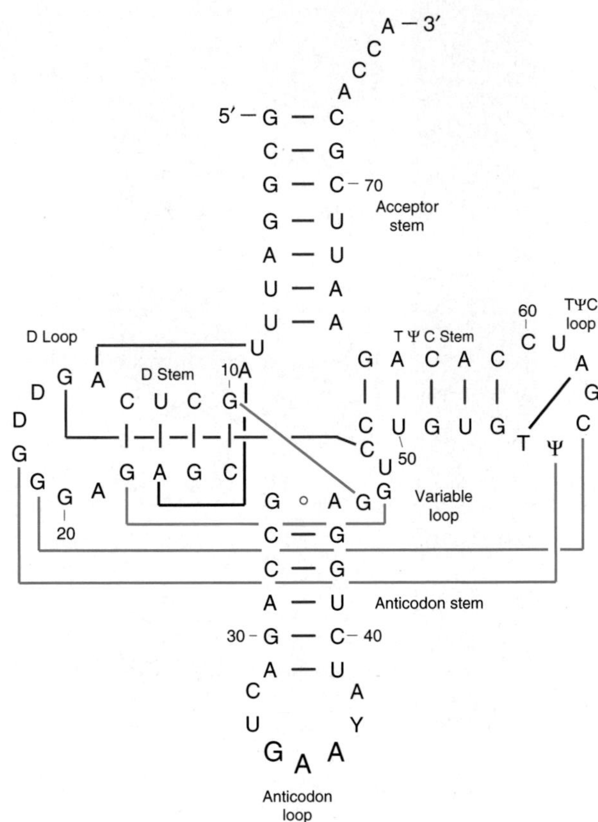

Colour plate T.2 tRNA secondary structure (*Saccharomyces cerevisiae* phenylalanine tRNA). Structural features are labelled.

Colour plate V.1 Part of a multiple sequence alignment represented using different colouring schemes. The top view shows consensus colouring, where alignment columns are shaded according to the number of residues found at each position–the lighter the colour, the more conserved the position. In the middle view, regions are coloured using a hybrid scheme, in which colours may denote both general amino acid property groups and the degree of conservation in a given part of the alignment. In the bottom view, residues are coloured only according to their physicochemical properties. The different colouring approaches offer very different perspectives on the degree of conservation in this portion of the alignment, in some cases suggesting greater degrees of similarity than they really possess.

F

F-Measure
False Discovery Rate Control (False Discovery Rate, FDR)
False Discovery Rate in Proteomics
Family-based Association Analysis
FASTA (FASTP)
FASTP, *see* FASTA.
FDR, *see* False Discovery Rate Control.
Feature (Independent Variable, Predictor Variable, Descriptor, Attribute, Observation)
Feature Subset Selection
FGENES
FigTree
Fingerprint (Chemoinformatics), *see* Chemical Hashed Fingerprint.
Fingerprint of Proteins
Finite Mixture Model
Fisher Discriminant Analysis (Linear Discriminant Analysis)
Flat File Data Formats
FLUCTUATE, *see* LAMARC.
Flux Balance Analysis, *see* Constraint-based Modeling.
Flybase
FLYBRAIN
Fold
Fold Library
Foldback, *see* RNA hairpin.
Folding Free Energy
Founder Effect
Frame-based Language
Free R-Factor
Frequent Itemset, *see* Association Rule Mining.
Frequent Sub-Graph, *see* Graph Mining.
Frequent Sub-Structure, *see* Graph Mining.
Function Prediction, *see* Discrete Function Prediction.
Functional Database
Functional Genomics
Functional Signature
Functome
Fuzzy Logic, *see* Fuzzy Set.
Fuzzy Set (Fuzzy Logic, Possibility Theory)

Concise Encyclopaedia of Bioinformatics and Computational Biology, Second Edition.
Edited by John M. Hancock and Marketa J. Zvelebil.

F-MEASURE

Concha Bielza and Pedro Larrañaga

The F-measure, also called the F-score, is a measure of model performance for 2-class classifiers. It considers both the precision p and the recall r of the model, where p is the number of true positives divided by the number of all real positives (true positives and false negatives) and r is the number of true positives divided by the number of all predicted positives (true and false positives). The F-score can be interpreted as a weighted average of the precision and recall, where it reaches its best value at 1 and its worst value at 0.

The traditional or balanced F_1-measure is the harmonic mean of precision and recall:

$$F_1 = 2 \times \frac{\text{precision} \times \text{recall}}{\text{precision} + \text{recall}}$$

The general formula for $\beta > 0$ is

$$F_\beta = (1+\beta)^2 \times \frac{\text{precision} \times \text{recall}}{(\beta^2 \times \text{precision}) + \text{recall}}$$

Thus, F_2 weights recall higher than precision and $F_{0.5}$ behaves the other way round.

Note that the F-measures do not take the true negative rate into account. Perhaps this is why they are so popular in information retrieval where true negatives are not relevant when searching for important documents.

Further reading

Hakenberg J, Plake C, Leaman R, Schroeder M, Gonzalez G (2008) Inter-species normalization of gene mentions with GNAT. *Bioinformatics* 24: i126–i132.

Kononenko I, Kukar M (2007) *Machine Learning and Data Mining. Introduction to Principles and Algorithms.* Horwood Publishing.

Liu Z, Tan M, Jiang F (2009) Regularized F-measure maximization for feature selection and classification. *Journal of Biomedicine and Biotechnology* 2009: 617946.

See also Error Measures

FALSE DISCOVERY RATE CONTROL (FALSE DISCOVERY RATE, FDR)

John M. Hancock

False Discovery Rate Control is probably the most common statistical procedure used in high-throughput biology to control for multiple testing. It is less conservative than more traditional approaches, such as the Bonferroni correction, which controls Family Wise Errors. Rather than controlling the chance of making even one type I error, FDR Control controls the proportion of type I errors made among all significant results. It therefore results in the loss of fewer true positive results (lower Type II error rate).

F

Algorithms to control FDR include the original Benjamini-Hochberg procedure (Benjamini and Hochberg, 1995), a 'sharpened' variant that adjusts the critical threshold (Benjamini and Hochberg, 2000) and a variant for use when there is dependency between tests (Benjamini and Yekutieli, 2001) but there are a number of other implementations of the general principle.

Related websites

False Discovery Rate	http://en.wikipedia.org/wiki/False_discovery_rate
False Discovery Rate Analysis in R	http://strimmerlab.org/notes/fdr.html

Further reading

Benjamini Y, Hochberg Y (1995) Controlling the false discovery rate: a practical and powerful approach to multiple testing, *Journal of the Royal Statistical Society, Series B (Methodological)*, 57: 289–300.

Benjamini Y, Hochberg Y (2000) On the adaptive control of the false discovery rate in multiple testing with independent statistics J. *Educ. Behav. Statist* 25: 60–83.

Benjamini Y, Yekutieli D (2001) The control of the false discovery rate in multiple testing under dependency, *Annals of Statistics*, 29: 1165–1188.

See also False Discovery Rate in Proteomics

FALSE DISCOVERY RATE IN PROTEOMICS

Simon Hubbard

The rate of false positive matches reported in a ranked list of candidate peptide spectrum matches (PSMs) in a proteomics database search at a given threshold. For example, if a database search yields 1000 PSMs with score greater than a cut-off score T, of which 20 are considered to be false positives (because they are from a decoy database), then the FDR can be estimated as the proportion of false positive (FP) target PSMs divided by the total number of positive target PSMs. In this case, there are 980 target database positive hits, of which an estimated 20 are FPs, and hence the FDR is calculated as $20/(960 + 20) = 0.02$ (or 2%). It is common to set the FDR at a fixed value, rather than the score threshold, to report a set of PSMs at a given fixed FDR (typically 1% or 5%). This is therefore a measure of the global error rate over a whole set of candidate peptide identifications rather than the local error rate of the chances of a single PSM being incorrect. As such, this has become a popular and widely used statistical confidence measure used in proteomics that overcomes the multiple testing problem.

Further reading

Elias JE, Haas W, Faherty BK, Gygi SP. Comparative evaluation of mass spectrometry platforms used in large-scale proteomics investigations. *Nat Methods.* 2005, 2: 667–675.

Käll L, Storey JD, MacCoss MJ, Noble WS. Assigning Significance to Peptides Identified by Tandem Mass Spectrometry Using Decoy Databases. *J. Proteome Res.* 2008, 7: 29–34.

See also False Discovery Rate Control, Decoy Database

FAMILY-BASED ASSOCIATION ANALYSIS

Mark McCarthy, Steven Wiltshire and Andrew Collins

Variant of linkage disequilibrium (association) analysis in which family-based controls are used in place of population-based controls.

The archetypal design for linkage disequilibrium (or association) analysis is the case-control study, in which the aim is to determine whether a particular allele (or haplotype) occurs more frequently in chromosomes from cases (i.e. individuals enriched for disease-susceptibility alleles) than in chromosomes from a sample of control individuals. These controls are typically unrelated individuals, ostensibly from the same population as the cases, either selected at random or on the basis that they do not have the disease phenotype. In principle, however, such a study design is prone to generate spurious associations if the case and control populations are not tightly matched for genetic background, since latent population substructure may lead to case-control frequency differences which reflect differences in ancestry rather than differences in disease status. In other words, associations may be seen between loci which are not in linkage disequilibrium (in the stricter definition of that term). Family-based association methods were devised to counter this possibility: by using appropriately-selected family-based control individuals, it is possible to ensure that positive results are obtained only when the variant genotype and the susceptibility allele are both linked and associated, and thereby protect against population substructure. The transmission disequilibrium test (or TDT), is based on the analysis of trios comprising an affected individual and both parents. Analysis proceeds by comparing, in parents who are heterozygous for the typed variant, those alleles transmitted to affected offspring, with those not so transmitted. Deviation from expectation under the null hypothesis is, under appropriate circumstances, a simultaneous test of both linkage and association. Other family-based association methods have been devised for use in discordant sibling pairs (eg the sibling transmission/disequilibrium test) and in general pedigrees (the pedigree disequilibrium test, or PDT), and for quantitative traits (QTDT). Disadvantages of family-based association methods over the case-control design may include: (a) increased collection costs, especially for late-onset conditions where parents may be rare; (b) increased genotyping costs, since, in the TDT study design, three individuals have to be typed to provide the equivalent of one case-control pair; and (c) lower power, particularly in the case of discordant sibpair methods.

Examples: Huxtable et al. (2000) used parent offspring trios to demonstrate both linkage and allelic association between a regulatory minisatellite upstream of the insulin gene and type 2 diabetes and thereby to consolidate previous findings from case-control studies. An added advantage of the use of trios, apparent in this study, was the ability to examine parent-of-origin effects: susceptibility could be shown to be restricted to alleles transmitted from fathers. Altshuler and colleagues (2000) combined association analysis conducted using a range of case-control and family-based association resources to confirm that the Pro12Ala variant of the PPARG gene is also implicated in type 2 diabetes susceptibility.

F

Related websites

Programs for conducting family-based association tests:

TRANSMIT	www-gene.cimr.cam.ac.uk/clayton/software
PBAT	http://www.biostat.harvard.edu/~clange/default.htm
QTDT	http://www.sph.umich.edu/csg/abecasis/QTDT/
ETDT	http://www.mds.qmw.ac.uk/statgen/dcurtis/software.html

Further reading

Abecasis GR, et al. (2000) A general test of association for quantitative traits in nuclear families. *Am J Hum Genet* 66: 279–292.

Altshuler D, et al. (2000) The common PPARgamma Pro12Ala polymorphism is associated with decreased risk of type 2 diabetes. *Nat Genet* 26: 76–80.

Frayling TM, et al. (1999) Parent-offspring trios: a resource to facilitate the identification of type 2 diabetes genes. *Diabetes* 48: 2475–2479.

Huxtable SJ, et al. (2000) Analysis of parent-offspring trios provides evidence for linkage and association between the insulin gene and type 2 diabetes mediated exclusively through paternally transmitted class III variable number tandem repeat alleles. *Diabetes* 49: 126–130.

Lander ES, Schork NJ (1994) Genetic dissection of complex traits. *Science* 265: 2037–2048.

Lange C, DeMeo D, Silverman EK, Weiss ST, Laird NM (2004) PBAT:Tools for family-based association studies. *Am J Hum Genet* 74(2): 367–369.

Martin ER, et al. (2000) A test for linkage and association in general pedigrees: the pedigree disequilibrium test. *Am J Hum Genet* 67: 146–154.

Morton NE, Collins A (1998) Tests and estimates of allelic association in complex inheritance. *Proc Natl Acad Sci USA* 95: 11389–11393.

Spielman RS, et al. (1993) Transmission test for linkage disequilibrium: the insulin gene region and insulin-dependent diabetes mellitus (IDDM). *Am J Hum Genet* 52: 506–516.

Spielman RS, Ewens WJ (1998) A sibship test for linkage in the presence of association: the sib transmission/disequilibrium test. *Am J Hum Genet* 62: 450–458.

See also Genome-Wide Association Study, Linkage Disequilibrium, Allelic Association, Association Analysis

FASTA (FASTP)

Jaap Heringa

The FASTA program (Pearson and Lipman, 1988) compares a given query sequence with a library of sequences and calculates for each pair the highest scoring local alignment. The speed of the algorithm is obtained by delaying application of the dynamic programming technique to the moment where the most similar segments are already identified by faster and less sensitive techniques. To accomplish this, the FASTA routine operates in four steps. The first step searches for identical words of a user specified length occurring in the query sequence and the target sequence(s). The technique is based on that of Wilbur and Lipman (1983, 1984) and involves searching for identical words (*k*-tuples) of a certain size within a specified bandwidth along search matrix diagonals. For not-too-distant sequences (> 35% residue identity), little sensitivity is lost while speed is greatly increased. The search is performed by a technique known as hash coding or hashing, where a lookup table is constructed for all words in the query sequence, which is then used to compare all encountered words in the target sequence(s). The relative positions of each word in the two sequences are then calculated by subtracting the position in the first sequence from that in the second. Words that have the same offset position reveal a region of alignment between the two sequences. The number of comparisons increases linearly in proportion to average sequence length. In contrast, the time taken in dot matrix and dynamic programming methods increases as the square of the average sequence length. The k-tuple length is user-defined and is usually 1 or 2 for protein sequences (i.e. either the positions of each of the individual 20 amino acids or the positions of each of the 400 possible dipeptides are located). For nucleic acid sequences, the k-tuple is 5–20, and should be longer because short k-tuples are much more common due to the 4 letter alphabet of nucleic acids. The larger the k-tuple chosen, the more rapid but less thorough, a database search.

Generally for proteins, a word length of two residues is sufficient (*ktup*=2). Searching with higher *ktup* values increases the speed but also the risk that similar regions are missed. For each target sequence the 10 regions with the highest density of ungapped common words are determined. In the second step, these 10 regions are rescored using the Dayhoff PAM-250 residue exchange matrix (Dayhoff et al., 1983) and the best scoring region of the 10 is reported under *init1* in the FASTA output. In the third step, regions scoring higher than a threshold value and being sufficiently near each other in the sequence are joined, now allowing gaps. The highest score of these new fragments can be found under *initn* in the FASTA output. The fourth and final step performs a full dynamic programming alignment (Chao *et al.*, 1992) over the final region which is widened by 32 residues at either side, of which the score is written under *opt* in the FASTA output.

The most recent implementation, FASTA3, has the option to perform global alignment in addition to local alignment (Pearson, 2000). It further incorporates statistical estimates of the significance of each pairwise alignment score between a query and database sequence relative to randomly expected scores.

Related website

FASTA	http://www.ebi.ac.uk/Tools/sss/fasta/

Further reading

Chao K-M, Pearson WR, Miller W (1992) Aligning two sequences within a specified diagonal band. *CABIOS* 8: 481–487.

Dayhoff MO, Barker WC, Hunt LT (1983) Establishing homologies in protein sequences. *Methods Enzymol.* 91: 524–545.

Pearson WR (2000) Flexible sequence similarity searching with the FASTA3 program package. *Methods Mol Biol.* 2000;132:185–219.

Pearson WR, Lipman DJ (1988) Improved tools for biological sequence comparison. *Proc. Natl. Acad. Sci. USA* 85: 2444–2448.

Wilbur WJ, Lipman DJ (1983) Rapid similarity searches of nucleic acid and protein data banks. *Proc. Natl. Acad. Sci. USA* 80: 726–730.

Wilbur WJ, Lipman DJ (1984) The context dependent comparison of biological sequences. *SIAM J. Appl. Math.* 44: 557–567.

See also Homology Search, BLAST, Dynamic Programming, Sequence Similarity

FASTP, *SEE* FASTA.

FDR, *SEE* FALSE DISCOVERY RATE CONTROL.

FEATURE (INDEPENDENT VARIABLE, PREDICTOR VARIABLE, DESCRIPTOR, ATTRIBUTE, OBSERVATION)

Nello Cristianini

Attributes describing a data item.

In machine learning, each data item is typically described by a set of attributes, called features (and also descriptors, predictors or independent variables). Often they are organized as vectors of fixed dimension, and the features assume numeric values, although it is possible to have symbolic features (e.g. 'small', 'medium', 'large', or 'green', 'yellow'), or Boolean features (e.g., 'present', 'absent').

When present, the desired response for each input is instead denoted as a 'label' (or response, or dependent variable).

The problem of automatically identifying the subset of features that is most relevant for the task at hand is called 'feature selection' and is of crucial importance in situations where the initial number of attributes is very large, and many of them are irrelevant or redundant, since this may lead to overfitting. This is for example often the case for data generated by DNA microarrays.

Further reading

Duda RO, Hart PE, Stork DG (2001) *Pattern Classification*, John Wiley & Sons, USA.

Mitchell T (1997) *Machine Learning*. McGraw-Hill, Maidenhead, UK.

See also Machine Learning

FEATURE SUBSET SELECTION

Pedro Larrañaga and Concha Bielza

The high dimensional nature of many modeling tasks in Bioinformatics, going from sequence analysis over microarray analysis to spectral analyses and literature mining, has given rise to a wealth of feature selection techniques (Saeys et al., 2007) being presented in the field. The most important objectives of feature selection are: a) to avoid overfitting and improve model performance, b) to provide faster and more cost-effective models, and c) to gain a deeper insight into the underlying processes that generated the data. In the context of classification, feature selection techniques can be organized into three categories, depending on how they combine the feature selection search with the construction of the classification model: filter methods, wrapper methods, and embedded methods.

Filter techniques assess the relevance of features by looking only at the intrinsic properties of the data. In univariate filtering, a feature relevance score is calculated, and low scoring features are removed. Afterwards, the subset of selected features is presented as input to the classification algorithm. A common disadvantage of univariate filter methods is that they ignore the feature dependencies. In order to overcome this problem, a number of multivariate filter techniques have been introduced. Advantages of filter techniques are that they easily scale to very high-dimensional datasets, they are computationally simple and fast, and they are independent of the classification algorithm. As a result, feature selection needs to be performed only once, and then different classifiers can be evaluated.

Wrapper methods embed the model search within the feature subset search. The evaluation of a specific subset of features is obtained by training and testing a specific classification model, so that this approach is tailored to a specific classification algorithm. To search the space of all feature subsets, a search algorithm (deterministic or randomized) is then 'wrapped' around the classification model. Advantages of wrapper approaches include the interaction between feature subset search and model selection, and the ability to take into account feature dependencies. A common drawback of these techniques is that they have a higher risk of overfitting than filter techniques and are very computationally intensive, especially if building the classifier has a high computational cost.

In embedded techniques, the search for an optimal subset of features is built into the classifier construction, and can be seen as a search in the combined space of feature subsets and models. Embedded methods have the advantage that they include the interaction with the classification model, while at the same time they are far less computationally intensive than wrapper methods.

Examples of feature subset selection in bioinformatics include approaches in microarray domains, where univariate filtering (Dudoit et al., 2002), multivariate filtering (Gevaert et al., 2006), wrapper methods (Blanco et al., 2004), and embedded methods (Díaz-Uriarte and Alvarez de Andrés, 2006) have been applied. Other domains that have taken advantage from feature subset selection are mass spectra analysis (Prados et al., 2004), single nucleotide polymorphism analysis (Lee and Shatkay, 2006), and text and literature mining (Han et al., 2006).

Further reading

Blanco R, et al. (2004) Gene selection for cancer classification using wrapper approaches. *International Journal of Pattern Recognition and Artificial Intelligence* 18(8): 1373–1390.

Díaz-Uriarte R., Alvarez de Andrés S (2006) Gene selection and classification of microarray data using random forest. *BMC Bioinformatics* 7:3.

Dudoit J, et al. (2002) Comparison of discriminant methods for the classification of tumors using gene expression data. *Journal of the American Statistical Association* 97(457): 77–87.

Gevaert O, et al. (2006) Predicting the prognosis of breast cancer by integrating clinical and microarray data with Bayesian networks. *Bioinformatics* 22(14): e184–e190.

Han B, et al. (2006) Substring selection for biomedical document classification. *Bioinformatics* 22(17): 2136–2142.

F

Lee PH, Shatkay H (2006) BNTagger: improved tagging SNP selection using Bayesian networks. *Bioinformatics* 22(14): e211–e219.

Prados J, et al. (2004) Mining mass-spectra for diagnosis and biomarker discovery of cerebral accidents. *Proteomics* 4(8): 2320–2332.

Saeys Y, et al. (2007) A review of feature selection techniques in bioinformatics. *Bioinformatics* 23(19): 2507–2517.

FGENES

Enrique Blanco and Josep F. Abril

FGENES is a general-purpose eukaryotic gene prediction program licensed by Softberry Inc.

FGENES produces a list of potential exons with all splicing and translational signals identified on a genomic sequence. Predicted sites and exons are scored using discriminant classifiers that take into account the expected composition of functional elements in a particular genome to filter the most promising ones (Solovyev et al., 1994). Final assembled gene models are obtained using a dynamic programming approach similar to that of Guigó (1998).

The FGENES suite of prediction programs is constituted of several variants of the original approach: (a) FGENESH design is based in other HMM-based gene-finding programs such as GENSCAN in which more prevalence is given to splice sites in contrast to coding potential measures (Salamov and Solovyev, 2000); (b) FGENESH+ can incorporate protein homology information and (c) FGENESH_C allows the use of cDNA/EST data to improve *ab initio* predictions.

The FGENES software has been useful in the annotation of multiple genomes (see below the list of references), including specific versions for bacteria or viruses (FGENESB and FGENESV, respectively). Web servers for most programs are publicly available under certain restrictions.

Related website

FGENES suite	http://www.softberry.com

Further reading

Bowler C, et al. (2008) The Phaeodactylum genome reveals the evolutionary history of diatom genomes. *Nature* 456: 239–244.

Guigó R (1998) Assembling genes from predicted exons in linear time with dynamic programming. *Journal of Computational Biology* 5: 681–702.

International Aphid Genomics Consortium (2010) Genome sequence of the pea aphid Acyrthosiphon pisum. *PLoS Biol.* 8: e1000313.

International Rice Genome Sequencing Project (2005) The map-based sequence of the rice genome. *Nature* 436: 793–800.

Salamov AA and Solovyev VV (2000) *Ab initio* gene finding in *Drosophila* genomic DNA. *Genome Research* 10: 516–522.

Solovyev VV, et al. (1994) Predicting internal exons by oligonucleotide composition and discriminant analysis of spliceable open reading frames. *Nucleic Acids Res.* 22: 5156–5163.

Solovyev V, Kosarev P, Seledsov I, Vorobyev D (2006) Automatic annotation of eukaryotic genes, pseudogenes and promoters. *Genome Biology* 7: S10.

Tribolium Genome Sequencing Consortium (2008) The genome of the model beetle and pest Tribolium castaneum. *Nature* 452: 949–955.

See also Gene Annotation, Gene Prediction, *ab initio*, Gene Prediction, homology-based, Gene Prediction, accuracy

FIGTREE

Michael P. Cummings

A program for viewing and producing publication-ready figures of phylogenetic trees.

The most feature rich program for making phylogenetic tree figures. Written in Java and available for Mac OS, Windows and Linux.

Related website

FigTree home page	http://tree.bio.ed.ac.uk/software/figtree/

FINGERPRINT (CHEMOINFORMATICS), *SEE* CHEMICAL HASHED FINGERPRINT.

FINGERPRINT OF PROTEINS

Teresa K. Attwood

A group of ungapped motifs derived from a set of conserved regions in a protein sequence alignment, used as a characteristic signature of gene- or domain-family membership. The fingerprint concept incorporates the number of motifs, the order in which they occur within an alignment and the relative distances between them, as illustrated in Figure F.1.

Fingerprint motifs are manually excised from alignments and expressed as residue frequency matrices, which are used to trawl UniProtKB in an iterative fashion. At each iteration, information from new sequence matches (beyond those in the initial seed alignment) is added to the motifs, until the scanning process converges, when no further new matches can be found.

Fingerprints offer improved diagnostic reliability over single-motif-based pattern-recognition methods (*e.g.*, such as the consensus patterns embodied in PROSITE) owing to the mutual context provided by motif neighbours: the more motifs matched, the greater the confidence that a diagnosis is correct, provided all the motifs match in the correct order with appropriate distances between them. By contrast with methods that exploit profiles or hidden Markov models, which use more sophisticated scoring methods and are

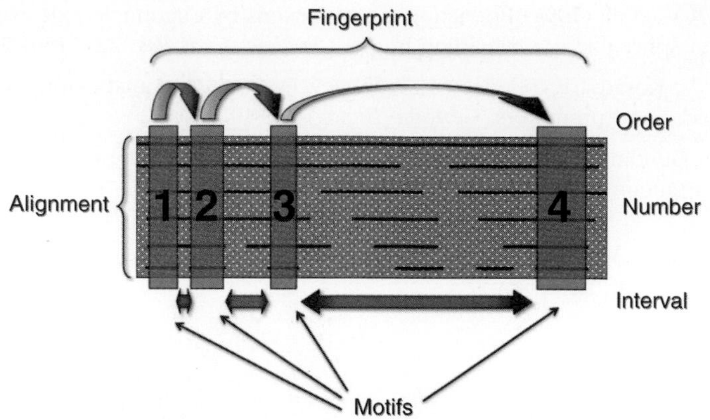

Figure F.1 Schematic illustration of a protein fingerprint: horizontal bars denote the aligned sequences; vertical blocks (numbered 1–4) pinpoint four conserved motifs, which have specific separations, or intervals, and a specific order. The set of motifs, the order in which they appear, and the intervals between them are characteristic of the aligned sequences, and hence together constitute a diagnostic fingerprint.

hence more permissive, fingerprints are selective; they therefore usually perform best in the diagnosis of protein families and subfamilies.

Fingerprints underpin the PRINTS database. They may be searched using the FingerPRINTScan suite, which allows queries of individual sequences against the whole database, or of a single sequence against individual fingerprints.

Related websites

About fingerprints	http://www.bioinf.manchester.ac.uk/dbbrowser/PRINTS/printsman.php
FingerPRINTScan	http://www.bioinf.manchester.ac.uk/dbbrowser/fingerPRINTScan/
PRINTS	http://www.bioinf.manchester.ac.uk/dbbrowser/PRINTS/

Further reading

Attwood TK (2001) A compendium of specific motifs for diagnosing GPCR subtypes. *Trends Pharmacol. Sci.* 22(4): 162–165.

Attwood TK, Bradley P, Flower DR, et al. (2003). PRINTS and its automatic supplement, prePRINTS. *Nucleic Acids Res.* 31(1): 400–402.

Attwood TK, Findlay JBC (1994) Fingerprinting G-protein-coupled receptors. *Protein Eng.* 7(2): 195–203.

Attwood TK, Parry-Smith DJ (1999) *Introduction to Bioinformatics*. Addison Wesley Longman, Harlow, Essex, UK.

Higgs P, Attwood TK (2005) *Bioinformatics and Molecular Evolution*. Blackwell Publishing, Oxford, UK.

Scordis P, Flower DR, Attwood TK (1999) FingerPRINTScan: Intelligent searching of the PRINTS motif database. *Bioinformatics* 15(10): 799–806.

See also Domain Family, Gene Family, Motif, PRINTS, Protein Family

FINITE MIXTURE MODEL

Pedro Larrañaga and Concha Bielza

Finite mixture models (McLachlan and Peel, 2000) provide a natural way of modeling continuous or discrete outcomes that are observed from populations consisting of a finite number of homogeneous subpopulations. The finite mixture model provides a natural representation of the heterogeneity of the population in a finite number of latent classes (or clusters). It concerns modeling a statistical distribution by a mixture (or weighted sum) of other distributions.

In a finite mixture model the p dimensional probability distribution of the population is given by:

$$f(\mathbf{x}; \boldsymbol{\pi}, \boldsymbol{\Theta}) = \Sigma \pi_k f_k(\mathbf{x}; \boldsymbol{\Theta}_k)$$

where each component π_k of the vector $\boldsymbol{\pi} = (\pi_1, \ldots, \pi_K)$ represents the k mixing proportion that observation x belongs to the k subpopulation with distribution fk ($\mathbf{x}$; $\boldsymbol{\Theta}_k$), and verifies $0 \leq \pi_k \leq 1$ and $\Sigma\pi_k = 1$. K denotes the total number of components and the functional form of fk ($\mathbf{x}$; $\boldsymbol{\Theta}$k) is completely known, but for the parameterizing vector $\boldsymbol{\Theta} = (\boldsymbol{\Theta}_1, \ldots, \boldsymbol{\Theta}_K)$. Assuming a random sample of size N extracted from f(x; $\boldsymbol{\pi}$, $\boldsymbol{\Theta}$), the estimation of parameters $\boldsymbol{\pi}$ and $\boldsymbol{\Theta}$ is usually done with the expectation-maximization algorithm (Dempster et al., 1977).

Examples in bioinformatics include the works by Blekas et al. (2003) and Dai et al. (2009).

Further reading

Blekas K, Fotiadis DI, Likas A (2003) Greedy mixture learning for multiple motif discovery in biological sequences. *Bioinformatics* 19(5): 607–617.

Dai X, Erkkilä T, Yli-Harja O, Lähdesmäki H (2009) A joint finite mixture model for clustering genes from independent Gaussian and Beta distributed data. *BMC Bioinformatics* 10:165.

Dempster AP, Laird NM, Rubin DB (1977) Maximum likelihood for incomplete data via the EM algorithm (with discussion). *Journal of the Royal Statistical Society, Series B*, 39: 1–38.

McLachlan G., Peel D (2000) *Finite Mixture Models.* John Wiley and Sons, Inc., New York.

FISHER DISCRIMINANT ANALYSIS (LINEAR DISCRIMINANT ANALYSIS)

Nello Cristianini

The statistical and computational task of assigning data points to one of *2* (or more) predefined classes is called 'classification' or 'discrimination', and the function that maps data into classes is called a 'classifier' or a 'discriminant function'.

Typically the data items are denoted by vectors $x \in \mathcal{R}^n$ and their classes by 'labels' $y \in \{-1, +1\}$. The classifier is then a function $f : \mathcal{R}^n \rightarrow \{-1, +1\}$. When the classifier is based upon a linear function $g_{w,b}(\mathbf{x}) = \langle \mathbf{w}, \mathbf{x} \rangle + b$ parametrized by a vector $\mathbf{x} \in \mathcal{R}^n$ (and

F

possibly by a real number b called 'bias'), this statistical task goes under the name of 'Linear Discriminant Analysis' (LDA), and reduces to the problem of selecting a suitable value of $(\boldsymbol{w},b)$. Many algorithms exist for learning the parameters $(\boldsymbol{w},b)$ from a set of labeled data, called 'training data', among them the Perceptron and the Fisher Discriminant (this giving rise to Fisher Discriminant Analysis, or FDA).

In FDA the parameters are chosen so as to optimize a criterion function that depends on the distance between the means of the two populations being separated, and on their variances. More precisely, consider projecting all the multivariate (training) data onto a generic direction $\boldsymbol{w}$, and then separately observing the mean and the variance of the projections of the two classes. If we denote μ_+ and μ_- the means of the projections of the two classes on the direction $\boldsymbol{w}$, and s_+ and s_- their variances, the cost function associated to the direction $\boldsymbol{w}$ is then defined as $C(\mathbf{w}) = \dfrac{(\mu_+ - \mu_-)^2}{s_+^2 + s_-^2}$. FDA selects the direction $\boldsymbol{w}$ that maximizes this measure of separation.

This mathematical problem can be rewritten in the following way. Define the scatter matrix of the *i-th* class as $S_i = \sum_{x \in \text{class }(i)} (\mathbf{x} - \overline{\mathbf{x}}_i)(\mathbf{x} - \overline{\mathbf{x}}_i)^T$ (where by $\overline{\mathbf{x}}$ we denote the mean of a set of vectors) and define the total within-class scatter matrix as $S_W = S_1 + S_{-1}$. In this way one can write the denominator of the criterion as $s_+^2 + s_-^2 = \mathbf{w}^T S_W \mathbf{w}$ and similarly, the numerator can be written as: $(\mu_+ - \mu_-)^2 = \mathbf{w}^\mathbf{T}(\overline{\mathbf{x}}_1 - \overline{\mathbf{x}}_{-1})(\overline{\mathbf{x}}_1 - \overline{\mathbf{x}}_{-1})^\mathbf{T}\mathbf{w} = \mathbf{w}^T S_B \mathbf{w}$ where S_B is the between-class scatter matrix.

In this way the criterion function becomes $C(w) = \dfrac{\mathbf{w}' S_B \mathbf{w}}{\mathbf{w}' S_W \mathbf{w}}$ (an expression known as a Reyleigh quotient) and its maximization amounts to solving the generalized eigenproblem $S_B \mathbf{w} = \lambda S_w \mathbf{w}$.

This method was introduced by R.A. Fisher in 1936, and can also be used in combination with kernel functions, in this way transforming it into an effective non-linear classifier. It can be proven to be optimal when the two populations are Gaussian and have the same covariance.

Further reading

Duda RO, et al. (2001) Pattern Classification, John Wiley & Sons, New York.

See also Support Vector Machines, Neural Networks, Classification

FLAT FILE DATA FORMATS

Carole Goble and Katy Wolstencroft

A flat file format is a text-based data record, with structured metadata, representing a single object. Often, the flat file represents part of a larger data model (e.g. a single database table).

In the life sciences, flat file databases have been extensively used to describe gene and protein entries in databases such as GenBank, UniProt and EMBL. For example, a Uniprot record describes a protein sequence and a collection of properties defining and describing that protein. Each record has the same set of properties, each of which is denoted by two letter codes at the beginning of a line, followed by a semi-structured property. The first few lines of a Uniprot file have the following structure:

ID ACTN2_MOUSE Reviewed; 894 AA.
AC Q9JI91;
DT 21-FEB-2001, integrated into UniProtKB/Swiss-Prot.
DT 01-OCT-2000, sequence version 1.
DT 20-FEB-2007, entry version 58.
DE Alpha-actinin-2 (Alpha actinin skeletal muscle isoform 2) (F-actin
DE cross-linking protein).
GN Name=Actn2;
OS Mus musculus (Mouse).

'ID' represents the protein identifier in Uniprot format, 'DE' represents the description of the protein, and 'OS' represents the organism/species description. Many of the property fields have predefined formats, making parsing more straightforward.

See also Biological Identifiers, Cross-Reference

FLUCTUATE, *SEE* LAMARC.

FLUX BALANCE ANALYSIS, *SEE* CONSTRAINT-BASED MODELING.

FLYBASE

Guenter Stoesser and Dan Bolser

Flybase is an integrated database for genomic and genetic data on the major genetic model organism *Drosophila melanogaster* and related species. It stores genetic, molecular and descriptive data for *Drosophila*. It integrates standard gene and protein information and provides genome browsing capabilities. The database includes descriptions of mutant phenotypes, expression data, and genetic stock collections. Data are actively maintained and curated using controlled vocabularies.

Following the publication of the *Drosophila* genome sequence, FlyBase has primary responsibility for the continual reannotation of the *D. melanogaster* genome. FlyBase organizes genetic and genomic data on chromosomal sequences and map locations, on the structure and expression patterns of encoded gene products, on mutational and transgenic variants and their phenotypes. The publicly funded stock centres are included, as are numerous cross-links to sequence databases and to homologs in other model organism databases. Images and graphical interfaces within FlyBase include interactive and static regional maps, as well as anatomical drawings and photomicrographs. FlyBase is one of the founding participants in the GO consortium, which provides ontologies for the description of the molecular functions, biological processes and cellular components of gene products. The FlyBase Consortium includes *Drosophila* researchers and computer scientists at Harvard University, University of California, Indiana University, University of Cambridge (UK) and the European Bioinformatics Institute (UK).

Related websites

Flybase homepage	http://flybase.org/
Wikipedia	http://en.wikipedia.org/wiki/FlyBase
MetaBase	http://metadatabase.org/wiki/FlyBase
Berkeley Drosophila Genome Project	http://www.fruitfly.org/
Gene Ontology Consortium	http://www.geneontology.org/

Further reading

Adams MD, et al. (2000) The genome sequence of *Drosophila melanogaster*. *Science*. 287: 2185–2195.

Matthews K, Cook K (2001) Genetic Information Online: A brief introduction to flybase and other organismal databases. In: *Encyclopedia of Genetics* ECR Reeve (ed.) Fitzroy Dearborn Publishers, Chicago.

McQuilton P, et al. The FlyBase Consortium (2011) FlyBase 101 – the basics of navigating FlyBase. *Nucleic Acids Res*. 39:21.

Rubin GM, et al. (2000) Comparative genomics of the eukaryotes. *Science* 287: 2204–2215.

The FlyBase Consortium (2002) The FlyBase database of the Drosophila genome projects and community literature. *Nucleic Acids Res* 30: 106–108.

See also Model Organism Database, GEISHA, Xenbase, modENCODE

FLYBRAIN

Dan Bolser

FLYBRAIN is an atlas of the *Drosophila* nervous system, presenting data for the basic structures in a navigable format with gene expression data where available. Data are presented as schematic diagrams, anatomical descriptions, and various image data integrated into a 3D model in VRML format.

Related websites

FLYBRAIN	http://flybrain.neurobio.arizona.edu
MetaBase	http://metadatabase.org/wiki/FlyBrain

Further reading

Shinomiya K, Matsuda K, Oishi T, Otsuna H, Ito K (2011) Flybrain neuron database: a comprehensive database system of the Drosophila brain neurons. *J Comp Neurol*. 519:807–833.

See also ALLEN BRAIN ATLAS

F

FOLD

Andrew Harrison, Alison Cuff and Christine Orengo

The fold of a protein describes the topology of the 3D domain structure that the polypeptide chain adopts as it folds. The topology refers to both the connectivity of the positions (for example residues or secondary structures) in the domain structure and the arrangements of those positions in 3D coordinate space. Analysis of protein families has revealed that the domain structures of homologous proteins are well conserved during evolution, particularly in the core of the domain, presumably because they are important for the stability and the functions of the proteins. However, since insertions and deletions of residues can occur between related sequences during evolution this can sometimes result in extensive structural embellishments to the core. In this sense the folds of distant structural relatives may not correspond completely over the entire domain. However, it is conventional to describe a group of related domains as having a common fold if their structures are similar over a significant portion. Furthermore, most structural classifications tend to group proteins having similar domain structures, over at least 50–60% of their chains, into the same fold group. Some schools of opinion suggest that it may be not be appropriate to consider discrete fold groups but rather a continuum of structures. In 2012, there are approximately 1280 different fold groups known and some researchers have speculated that there may be a limited number of folds adopted in nature due to constraints on secondary structure packing.

Related websites

DALI server	http://ekhidna.biocenter.helsinki.fi/dali_server/
PDBeFold server	http://www.ebi.ac.uk/msd-srv/ssm/
SCOP server	http://scop.mrc-lmb.cam.ac.uk/scop/
CATH server	http://www.cathdb.info/
PDB server	http://www.rcsb.org/pdb/home/home.do
PDBe server	http://www.ebi.ac.uk/pdbe/

Further reading

Cuff AL, Redfern OC, Greene, L, et al. (2009) The CATH hierarchy revisited - structural divergence in domain superfamilies and the continuity of fold space. *Structure* 17,1051–1062.

Orengo CA (1994) Classification of protein folds. *Current Opinion in Structural Biology* 4: 429–440.

Orengo CA (1999) Protein folds, functions and evolution. J.Mol.Biol. 293: 333–342.

See also Topology, Homology, Indel

FOLD LIBRARY

Andrew Harrison and Christine Orengo

Fold libraries comprise representative 3D structures of each known domain fold. Currently approximately 1,280 unique folds are known. However, there are no commonly agreed quantitative descriptions of protein folds, so that determining whether two proteins share a common fold is a somewhat subjective process and will depend on the methods used for aligning the structures and scoring their similarity. For this reason, different numbers of fold groups are reported in the public structure classification resources. Fold libraries are often used with threading methods which perform a 1D-3D alignment and attempt to 'thread' a query sequence through a 3D target structure to determine whether it is similar to the structure adopted by the query. It is important to use non-redundant data-sets with these methods, both to improve the reliability of the statistics and also to reduce the time required to thread the query against all structures in the library, since threading is a very computationally expensive method.

Related websites

DALI server	http://ekhidna.biocenter.helsinki.fi/dali_server/
PDBeFold server	http://www.ebi.ac.uk/msd-srv/ssm/
SCOP server	http://scop.mrc-lmb.cam.ac.uk/scop/
CATH server	http://www.cathdb.info/
CATHEDRAL server	http://www.cathdb.info/cgi-bin/CathedralServer.pl

Further reading

Thornton, J.M., Orengo, C.A., Todd, A.E. & Pearl, F.M. Protein folds, functions and evolution. *J Mol Biol* 293: 333–42 (1999).

See also Fold, Alignment, Threading.

FOLDBACK, *SEE* RNA HAIRPIN.

FOLDING FREE ENERGY

Antonio Marco and Sam Griffiths-Jones

The Gibbs free energy associated with the folding of an RNA molecule into a given base-paired secondary structure.

A base-paired RNA secondary structure has an associated free energy, often expressed in Kcal/mol. The folding free energy of a structure can be predicted using tools such as mfold (http://mfold.rna.albany.edu/) or the ViennaRNA package (http://www.tbi.univie.ac.at/RNA/). These tools implement a model called the nearest neighbour model, with parameters calculated by extensive experimental determination of the energetic stability of short RNA duplexes. The data show that the dominant stabilizing interaction is base pair stacking. It is usually assumed that the structure adopted by an RNA molecule, for example under physiological conditions, will be the structure with the lowest folding free energy. These tools are therefore widely used to determine the most stable secondary structure of an RNA molecule from its sequence.

The identification of functional RNAs often involves the prediction of their secondary structures. For example, a key step in microRNA gene discovery is the prediction of precursor hairpin structures and the evaluation of their folding free energy; often a stability threshold of -20 Kcal/mol is considered appropriate.

Related websites

Mfold	http://mfold.rna.albany.edu/
ViennaRNA	http://www.tbi.univie.ac.at/RNA/

Further reading

Berezikov E, Cuppen E, Plasterik RHA (2006) Approaches to microRNA discovery. *Nat Genet* 38:S2–S7.

Mathews DH, Sabina J, Zuker M, Turner DH (1999) Expanded Sequence Dependence of Thermodynamic Parameters Improved Prediction of RNA Secondary Structure. *J Mol Biol* 288:911–940.

Washietl S (2010) Sequence and structure analysis of noncoding RNAs. *Methods Mol Biol* 609:285–306.

See also RNA Hairpin, MicroRNA Discovery

FOUNDER EFFECT

A.R. Hoelzel

The founder effect refers to the distortion in allele frequency that can occur when a population is founded by a small sample of individuals from a larger source population.

The high frequency of rare heritable diseases found in some captive breeding populations (e.g. a lethal form of dwarfism called 'chondrodytrophy' in the breeding population of California condors) and in some human populations (e.g. Finland) has been attributed to small founder populations.

The founder effect has also been proposed as a possible mechanism for promoting speciation, especially in the context of founding new populations on islands. For example, this may have been important in the radiation of some of the approximately 800 *Drosophila* species found in the Hawaiian Islands.

Further reading

Avise JC (2004) *Molecular Markers, Natural History and Evolution*. Sinauer Associates.

Hartl DL (2000) *A Primer of Population Genetics*. Sinauer Associates Inc.

See also Population Bottleneck, Evolution

FRAME-BASED LANGUAGE

Robert Stevens

Type of language for representing ontologies.

In a Frame-based Language, a 'frame' represents a concept – category or individual – in the ontology. Properties or relations of a concept are represented by slots in its frame. Slots can be filled by other classes, symbols, strings, numbers, etc. For example, a frame *enzyme*, might have a slot *catalyses* and a slot filler *chemical reaction*. There are usually special slots for the frame name and for the taxanomic 'is-a' relation. Slots are inherited along the taxonomic 'is-a' relation and default slot-fillers can be either inherited or changed in the inheriting frame. Systems such as RiboWeb and EcoCyc are represented in Frame Based Languages. Protégé-2000 is a well-known editor and toolset for building ontologies and applications using a Frame Based Language. OKBC is a standard for exchanging information amongst frame-based languages.

FREE R-FACTOR

Liz Carpenter

The free-R factor is version of the R-factor used in X-ray crystallography to follow the progress of the refinement of a crystallographic structure. The free R-factor is calculated in the same way as an R-factor (see R-factor), using a set of randomly selected reflections, usually 5 or 10 % of the total possible number of reflections. These reflections are excluded from calculations of electron density maps and all refinement steps. The free R-factor is therefore independent of the refinement process and is less biased by the starting model. A drop in the free R-factor is thought to indicate a significant improvement in the agreement between model and data, whereas a decrease in the R-factor could be due to over-refinement.

Generally free R-factors below 30 % are considered to indicate that a model is correct and values above 40 % after refinement suggest that the model has substantial errors. High resolution macromolecular structures can have free R-factors below 20 %.

Further reading

Brünger AT (1992) The Free *R* value: a Novel Statistical Quantity for Assessing the Accuracy of Crystal Structures. *Nature*, 355: 472–474.

Brünger AT (1997) Free *R* Value: cross-validation in crystallography. *Methods in Enzym.* 277: 366–396.

See also R-factor, Refinement, X-ray Crystallography

FREQUENT ITEMSET, *SEE* ASSOCIATION RULE MINING.

FREQUENT SUB-GRAPH, *SEE* GRAPH MINING.

FREQUENT SUB-STRUCTURE, *SEE* GRAPH MINING.

FUNCTION PREDICTION, *SEE* DISCRETE FUNCTION PREDICTION.

F

FUNCTIONAL DATABASE

Dov Greenbaum

Databases that serve to provide a functional annotation of genes and their protein products.

An example of a functional database is the MIPS (Munich Information Center for Protein Sequences) database, which focuses on a few genomes (i.e watercress, yeast, human and neurospora). Swissprot is also a functional database in that it aims to be a 'curated protein sequence database which strives to provide a high level of annotations (such as the description of the function of a protein, its domain structure, post-translational modifications, variants, etc.), a minimal level of redundancy and high level of integration with other databases'.

Functional databases require, most importantly, a clear methodology to describe the exact function of an individual gene. In most cases a gene's function will fit into a hierarchal structure of broad terms and then subsequently narrowing to very well defined nomenclatures. Many databases now use automated processes to define a protein's function, although these databases are still manually curated.

Related websites

SWISSPROT	http://web.expasy.org/docs/swiss-prot_guideline.html
EXProt	http://www.cmbi.kun.nl/EXProt/

Further reading

Kumar A, et al (2000) TRIPLES: a database of gene function in *Saccharomyces cerevisiae*. *Nucleic Acids Res.* 28: 81–84.

Ursing BM, van Enckevort FH, Leunissen JA, Siezen RJ (2002) EXProt: a database for proteins with an experimentally verified function. *Nucleic Acids Res* 30: 50–51.

FUNCTIONAL GENOMICS

Jean-Michel Claverie

A research area, and the set of approaches, dealing with the determination of gene function at a large scale and/or in a high throughput fashion.

The concept of Functional Genomics emerged after the completion of the first whole genome sequences (*H. influenzae*, and *S. cerevisiae*) and the unexpected discovery that a large fraction (25%–50%) of the genes were previously unknown (i.e. were missed by traditional genetics) and had to be functionally characterized. While the goal of classical genetics is to find a sequence for each function, Functional Genomics tries to identify a function for each sequence. Functional Genomics is thus reverse genetics at the genome scale. Functional Genomics is taking advantage of a large (and growing) range of techniques including

F

1. Computational techniques (bioinformatics):
   Sequence similarity searches
   Functional Signature and motif detection
   Phylogenomics
   Chemical library virtual screening (computational docking of small molecule models)
2. Protein-interaction measurements:
   Large-scale chemical library screening
   Two-hybrid system assays
3. Gene expression profiling using:
   Next Generation Sequencing
   Expressed Sequence Tags (ESTs)
   DNA arrays
4. Systematic gene disruption
5. Structural Genomics
6. Phenomics
7. Metabolomics

Each of these approaches only gives a partial view about the functions of the genes, such as 'being a kinase', 'binding calcium', 'being co-expressed with NF kappa-B', 'being involved in liver development', etc. The art of Functional Genomics is thus the integration of multiple functional hints gathered from many different techniques. These various techniques usually address a different level of gene 'function', such as: biochemical (e.g. kinase), cellular (e.g. cell signaling', or pertaining to the whole organism (e.g. liver development).

Related Websites

European Science Foundation Functional Genomics site	http://www.functionalgenomics.org.uk/
Science magazine Functional Genomics site	http://www.sciencemag.org/site/feature/plus/sfg/sfg_archive.zip

Further Reading

Claverie JM (1999) Computational methods for the identification of differential and coordinated gene expression. *Hum. Mol. Genet.* 8: 1821–1832.

Eisenberg D et al. (2000) Protein function in the post-genomic era. *Nature* 405: 823–826.

Legrain P et al. (2001) Protein-protein interaction maps: a lead towards cellular functions. *Trends Genet.* 17: 346–352.

Pevsner, J. (2009) *Bioinformatics and Functional Genomics.* Wiley-Blackwell, p. 984.

FUNCTIONAL SIGNATURE

Jean-Michel Claverie

A sequence pattern that appears to be uniquely (or almost uniquely) associated with a given function in a protein.

The simplest and most ideal form of functional signature is a set of strictly conserved positions within a family of sequences. Allowing a little more variation leads to 'Regular Expression' signatures (such as PROSITE) and to Position Weight Matrices (PWM). Modern Functional Signatures now using more sophisticated implementations such as Hidden Markov Models.

Related websites

NCBI BLAST page	http://www.ncbi.nlm.nih.gov/BLAST/
PROSITE home page	http://www.expasy.ch/prosite/
InterPro home page	http://www.ebi.ac.uk/interpro/
Blocks home page	http://blocks.fhcrc.org/
Pfam home page	http://pfam.sanger.ac.uk/
PRINTS home page	http://bioinf.man.ac.uk/dbbrowser/PRINTS/
ProDom home page	http://prodom.prabi.fr/prodom/current/html/home.php
SMART home page	http://smart.embl-heidelberg.de/

Further reading

Bateman A, et al. (2002) The Pfam protein families database. *Nucleic Acids Res.* 30: 276–280.

Falquet L, et al. (2002) The PROSITE database, its status in 2002. *Nucleic Acids Res.* 30: 235–238.

Gribskov M, Veretnik S (1996) Identification of sequence pattern with profile analysis. *Methods Enzymol.* 266: 198–212.

See also Motif, Gene Family

FUNCTOME

Dov Greenbaum

The functome is described as the list of potential functions encoded by the genome.

The overall purpose of many genomics efforts, including comparative genomics, is to annotate gene sequences, specifically, to do determine the function of unknown genes. The functome is the sum of many other 'omes' as function is described by the conglomeration of other facets of gene annotation including for example subcellular localization, structure, interaction partners, and can be inferred partially by expression analysis through clustering genes with similar expression patterns. Additionally, it incorporates other 'omes' such as the **secretome** or the **metabolome**.

Genomics has many tools to determine the function of previously unknown proteins including: annotation-transfer from known homologues and describing proteins via their function specific **motifs** either in sequence or structure (e.g. transcription factors by DNA binding **motifs**). Extreme caution must be applied when using techniques such as those listed above to describe a protein, specifically not to use weak evidence to functionally annotate a gene and thus possibly corrupt the databases.

There are presently many databases collecting functional data on genes.

Further reading

Greenbaum D, et al. (2001) Interrelating different types of genomic data, from proteome to secretome: 'oming in on function. *Genome Res* 11: 1463–1468.

See also Secretome, Metabolome, Functional Databases

FUZZY LOGIC, *SEE* FUZZY SET.

FUZZY SET (FUZZY LOGIC, POSSIBILITY THEORY)

Feng Chen and Yi-Ping Phoebe Chen

A fuzzy set defines a fuzzy threshold or boundary for every class, the scope of which is between 0 and 1. Such a threshold is called a membership degree. A fuzzy set is often used to describe a situation that does not need an absolute and clear boundary, as shown in Figure F.2 and the following example.

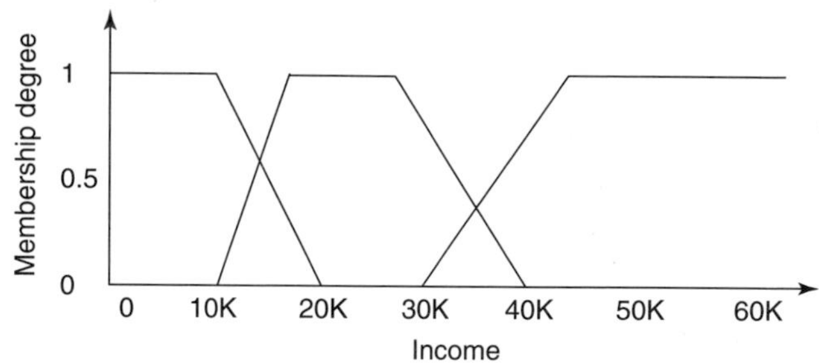

Figure F.2 An example of fuzzy set applied to income.

Suppose the Figure expresses a fuzzy classification of income. Someone whose income is between 0 and 10K belongs definitely to the 'low' class. However whether or not (10K, 20K) is 'low' or 'middle' should be determined by membership degree. Such fuzzy classification is more reasonable for a lot of real cases. In this example, if we define a person who earns 40K as belonging to the 'high' class, then how about one who just gets 39K? He or she will be categorized as 'middle'. This obviously does not conform to our knowledge and habit. So, fuzzy sets can be applied widely in reality, especially in bioinformatics.

Due to the fact that there is still a lot of unknown biological information, our biological dataset is uncertain. In other words, all of our results could be wrong or incomplete. Fuzzy sets can be used to express such uncertainty. For instance, a fuzzy set in ontology can represent the possible similarity between biological terms; a fuzzy set in structural bioinformatics can represent well how one structure transfers to another, for example for protein secondary structures; a fuzzy set in microarray data analysis can tell us what kind of genes are closely relevant to a variety of clusters.

Further reading

Xu D, Keller JM, Popescu M, Bondugula R (2008) *Applications of Fuzzy Logic in Bioinformatics*. Imperial College Press, London.

G

GA, *see* Genetic Algorithm.

Gametic Phase Disequilibrium, *see* Linkage Disequilibrium.

Gap

Gap Penalty

GARLI (Genetic Algorithm for Rapid Likelihood Inference)

Garnier-Osguthorpe-Robson Method, *see* GOR Secondary Structure Prediction Method.

GASP (Genome Annotation Assessment Project, E-GASP)

GBROWSE, *see* Gene Annotation, visualization tools

GC Composition, *see* Base Composition.

GC Richness, *see* Base Composition.

GEISHA

GenBank

GENCODE

Gene Annotation

Gene Annotation, formats

Gene Annotation, hand-curated

Gene Annotation, visualization tools

Gene Cluster

Gene Dispensability

Gene Distribution

Gene Diversity

Gene Duplication

Gene Expression Database (GXD), *see* Mouse Genome Informatics.

Gene Expression Profile

Gene Family

Gene Finding, *see* Gene Prediction.

Gene-finding Format, *see* Gene Annotation, formats.

Gene Flow

Gene Fusion Method

Gene Index

Gene Neighbourhood

Gene Ontology (GO)

Gene Ontology Consortium

Gene Prediction

Gene Prediction, ab initio

Gene Prediction, accuracy

Gene Prediction, alternative splicing

Gene Prediction, comparative

Gene Prediction, homology-based (Extrinsic Gene Prediction, Look-Up Gene Prediction)

Gene Prediction, NGS-based

Gene Prediction, non-canonical

Gene Prediction, pipelines

Gene Size

Gene Symbol

Gene Symbol, human

Concise Encyclopaedia of Bioinformatics and Computational Biology, Second Edition.
Edited by John M. Hancock and Marketa J. Zvelebil.

Gene Transfer Format, *see* Gene Annotation, formats

Genealogical Sorting Index (gsi)

GeneChip, *see* Affymetrix GeneChip™ Oligonucleotide Microarray.

GENEID

General Feature Format, *see* Gene Annotation, Formats.

Generalization, *see* Error Measures.

Genetic Algorithm (GA)

Genetic Code (Universal Genetic Code, Standard Genetic Code)

Genetic Linkage, *see* Linkage

Genetic Network

Genetic Redundancy

Genetic Variation, *see* Variation (Genetic).

GENEWISE

Genome Annotation

Genome Annotation Assessment Project, *see* GASP.

Genome Scans for Linkage (Genome-Wide Scans)

Genome Size (C-Value)

Genome-Wide Association Study (GWAS)

Genome-Wide Scans (linkage), *see* Genome Scans

Genome-Wide Survey

GENOMEGRAPHS, *see* Gene Annotation, visualization tools.

Genomics

Genotype Imputation

GENSCAN

GFF, *see* Gene Annotation, formats.

GFF2PS, *see* Gene Annotation, visualization tools.

GFF3, *see* Gene Annotation, formats.

Gibbs Sampling, *see* Markov Chain Monte Carlo.

Global Alignment

Global Organisation for Bioinformatics Learning, Education & Training (GOBLET), *see* BTN.

Globular

GO, *see* Gene Ontology.

GOBASE (Organelle Genome Database)

GOBLET, *see* BTN.

GOBO, *see* Global Open Biology Ontologies.

GOR Secondary Structure Prediction Method (Garnier-Osguthorpe-Robson Method)

Gradient Descent (Steepest Descent Method)

GRAIL

GRAIL Description Logic

Gramene

Graph Mining (Frequent Sub-Graph, Frequent Sub-Structure)

Graph Representation of Genetic, Molecular and Metabolic Networks

Group I Intron, *see* Intron.

Group II Intron, *see* Intron.

GTF, *see* Gene Annotation, formats.

Guilt by Association Annotation, *see* Annotation Transfer.

Gumball Machine

GWAS, *see* Genome-Wide Association Study.

GWAS Central

GA, *SEE* GENETIC ALGORITHM.

GAMETIC PHASE DISEQUILIBRIUM, *SEE* LINKAGE DISEQUILIBRIUM.

GAP

Jaap Heringa

A meaningful alignment of two or more protein sequences typically contains gaps. The gaps represent sites where the affected sequences are thought to have accumulated an insertion or deletion during divergent evolution. In protein sequence alignments, many gap regions correspond to loop sites in the associated tertiary structures, as protein cores do not easily accommodate insertions or deletions.

In most algorithms for automatic alignment, placing the gaps is penalised through gap penalty parameters, which are set to create a balance between being too strict in creating gaps and being overly permissive. High gap penalties lead to alignments where corresponding motifs might not be in register as this would require too many gaps in the alignment, while low penalties typically yield alignments with many spurious gaps created to align almost any identical residue in the target sequences. The latter alignments often show so-called widow amino acids, which are residues isolated from neighbouring residues by spurious gaps.

See also Gap Penalty, Global Alignment, Indel, Local Alignment

GAP PENALTY

Jaap Heringa and Laszlo Patthy

A penalty used in calculating the score of a sequence alignment.

Since most alignments are scored assuming a Markov process, where the amino acid matches are considered independent events, the product of the probabilities for each match within an alignment is typically taken. Since many of the scoring matrices contain exchange propensities converted to logarithmic values (log-odds), the alignment score can be calculated by summing the log-odd values corresponding to matched residues minus appropriate gap penalties:

$$S_{a,b} = \sum_{i} s(a_l,\ b_l) - \sum_{k} N_k \bullet gp(k)'$$

where the first summation is over the exchange values associated with l matched residues and the second over each group of gaps of length k, with N_k the number of gaps of length k and $gp(k)$ the associated gap penalty.

Needleman and Wunsch (1970) used a fixed penalty value for the inclusion of a gap of any length, while Sellers (1974) added a penalty value for each inserted gap position. Most current alignment routines take an intermediate approach by using the formula $gp(k) = pi + k \cdot pe$, where *pi* and *pe* are the penalties for gap initialisation and extension, respectively. Gap penalties of this latter form are known as affine gap penalties. Many researchers use a value of *pi* 10 to 30 times larger than that of *pe*. Another type of gap penalty is the *concave* gap penalty, which growth logarithmically with the length of the gap: $gp(k) = a + b \cdot \log(k)$. The fact that concave gap penalties increase less with large gap lengths is believed to approximate the evolutionary situation more closely than is the case for the other gap penalty types.

The choice of proper gap penalties is also closely connected to the residue exchange matrix used in the analysis. However, no formal model exists to estimate gap penalty values associated with a particular residue exchange matrix. Terminal gaps at the ends of alignments are usually treated differently to internal gaps, reflecting the fact that the N-terminal and C-terminal regions of proteins are quite tolerant to extensions/deletions.

Further reading

Needleman SB, Wunsch CD (1970) A general method applicable to the search for similarities in the amino acid sequence of two proteins. *J. Mol. Biol.* 48: 443–453.

Sellers PH (1974) On the theory and computation of evolutionary distances. *SIAM J. Appl. Math.* 26: 787–793.

See also Sequence Alignment, Global Alignment, Local Alignment, Amino Acid Exchange Matrix, Dynamic Programming

GARLI (GENETIC ALGORITHM FOR RAPID LIKELIHOOD INFERENCE)

Michael P. Cummings

A program for phylogenetic inference using the maximum likelihood criterion with models of evolution for nucleotide, amino acid, codon and morphology-like data, as well as partitioned models. It uses an evolutionary algorithm largely similar to a genetic algorithm to find the solution maximizing the likelihood score. It is among the most effective and best performing programs available for such analyses.

Related website

GARLI home page	http://code.google.com/p/garli/

Further reading

Zwickl DJ (2006) Genetic algorithm approaches for the phylogenetic analysis of large biological sequence datasets under the maximum likelihood criterion (PhD dissertation). Austin (TX): University of Texas, pp. 1–115.

See also BEAGLE

GARNIER-OSGUTHORPE-ROBSON METHOD, *SEE* GOR SECONDARY STRUCTURE PREDICTION METHOD.

GASP (GENOME ANNOTATION ASSESSMENT PROJECT, E-GASP)

Enrique Blanco and Josep F. Abril

The elaboration of a precise catalog of genes in eukaryote genomes is a difficult task. While the fraction of protein coding sequence along the genome is modest (less that 5% in humans), the transcript maturation leaves introns out and simultaneously produce alternative transcripts that might translate into distinct proteins. Computational gene prediction systems annotate putative coding exons encoded on sequences using statistical and comparative methods previously trained in well-known data sets. However, the lack of biological knowledge on some particular events such as pervasive transcription, alternative splicing, pseudo-genes or selenoproteins makes these predictions partial or incorrect in many cases.

The GASP project was launched in 1999 to evaluate the performance of current computational gene prediction programs on the well-characterized *Adh* gene neighbourhood region of 2.4 Mbp in the *Drosophila melanogaster* genome. This kind of evaluations were inspired by the Critical Assessment of Structure Prediction (CASP, 1995), which became a regular procedure to monitor the state-of-art in methods to model protein structure.

The ENCODE GASP (EGASP) community experiment, within the context of the ENCODE project, developed in 2005 a more exhaustive evaluation procedure on the GENCODE reference set of the human genome spanning 30 Mbp over 44 selected sequences. Later on, in 2008, another international consortium conducted the nematode GASP (nGASP), focusing the interest in the annotation of protein-coding regions from *Caenorhabditis elegans*, four additional *Caenorhabditis* genomes and other nematodes. More recently, the RNA-seq GASP project (RGASP) evaluated the current progress of automatic gene building using massive sequencing expression data to determine alternative transcript isoforms and quantify their abundance.

By and large, the rules of the competition are similar in all GASPs: first, access to a training set of exons is granted to participants in both cases and second, posterior submission of predictions is compared to expert annotations to calculate the accuracy of each system. It was established in all these studies that while a significant fraction of the coding sequence was successfully characterized at nucleotide level, the correct intron/exon structures and alternative splicing events were captured only in about half of genes which suggests substantial room for improvement.

Related websites

GASP	http://www.fruitfly.org/GASP1/
EGASP	http://genome.crg.cat/gencode/workshop2005.html
RGASP	http://www.gencodegenes.org/rgasp/

G

Further reading

Guigó R, Flicek P, Abril JF, et al. (2006) EGASP: the human ENCODE Genome Annotation Assessment Project. *Genome Biology* 7: S2.

Moult J, Pedersen JT, Judson R, Fidelis K (1995) A large-scale experiment to assess protein structure prediction methods. *Proteins* 23: ii–v.

Reese MG, Hartzell G, Harris N, Ohler U, Abril JF, Lewis SE (2000) Genome Annotation Assessment in *Drosophila melanogaster*. *Genome Res.* 10: 483–501.

The nGASP Consortium (2008) nGASP – the nematode genome annotation assessment project. *BMC Bioinformatics* 9: 549.

The RGASP consortium (2012, submitted) RGASP: The RNA-seq genome annotation assessment project.

See also CASP, Gene Prediction accuracy, ENCODE, GENCODE, Gene Prediction, non-canonical

GBROWSE, *SEE* GENE ANNOTATION, VISUALIZATION TOOLS

GC COMPOSITION, *SEE* BASE COMPOSITION.

GC RICHNESS, *SEE* BASE COMPOSITION.

GEISHA

Dan Bolser

GEISHA is the online repository of in-situ hybridization for genes expressed in the chicken embryo during the first six days of development. Data are derived from high throughput screening and literature curation. These data are integrated with other genomics resources.

Related website

GEISHA	http://geisha.arizona.edu/geisha/

Further reading

Darnell DK, Kaur S, Stanislaw S, et al. (2007) GEISHA: an in situ hybridization gene expression resource for the chicken embryo. Cytogenet Genome Res. 117: 30–5.

See also FlyBase, Xenbase, Title2

GENBANK

Guenter Stoesser, Obi L. Griffith and Malachi Griffith

GenBank® is the American member of the tripartite International Nucleotide Sequence Database Collaboration together with DDBJ (Japan) and the European Nucleotide Archive (Europe). Collaboratively, the three organizations collect all publicly available nucleotide sequences and exchange sequence data and biological annotations on a daily basis. GenBank contains RNA and DNA sequences produced primarily by traditional Sanger sequencing projects but also increasingly includes consensus or assembled contig sequences from next-generation sequencing projects. Major data sources are individual research groups, large-scale genome sequencing centres and the US Patent Office (USPTO). Stand-alone (Sequin) and web-based (BankIt) sequence submission systems are available. Entrez is NCBI's search and retrieval system that provides users with integrated access to sequence, mapping, taxonomy and structural data as well as to biomedical literature via PubMed. Sophisticated software tools allow searching and data analysis. NCBI's BLAST program is used for sequence similarity searching and has been instrumental in identifying genes and genetic features. GenBank is the NIH genetic sequence database and is maintained at the National Center for Biotechnology Information (NCBI), Bethesda (USA).

Related websites

National Center for Biotechnology Information	http://www.ncbi.nlm.nih.gov
GenBank homepage	http://www.ncbi.nlm.nih.gov/Genbank/
GenBank Submission Systems	http://www.ncbi.nlm.nih.gov/Sequin/ http://www.ncbi.nlm.nih.gov/BankIt/
Entrez Genome	http://www.ncbi.nlm.nih.gov/genome/
Tools for data mining	http://www.ncbi.nlm.nih.gov/Tools/
Entrez	http://www.ncbi.nlm.nih.gov/Entrez/

Further reading

Benson DA, et al. (2011) GenBank. *Nucleic Acids Res* 39(Database issue): D32–37.

Sayers EW, et al. (2012) Database resources of the National Center for Biotechnology Information. *Nucleic Acids Res.* 40(Database issue): D13–25.

GENCODE

Enrique Blanco and Josep F. Abril

The GENCODE consortium was formed to computationally and experimentally identify and map all protein-coding genes within the ENCODE regions selected for the pilot phase in 2003. The GENCODE reference set represented the standard to which the automated prediction programs were assessed during the ENCODE Genome Annotation Assessment Project (see EGASP).

GENCODE annotations were achieved by a combination of initial manual annotation by the HAVANA team (Sanger Institute, UK), experimental validation and a refinement

G

based on these experimental results. Up to 487 loci were annotated in 44 targeted regions accounting for 1% of the human genome. Only 40% of GENCODE exons were contained in other repositories of genes and 50% of coding loci were experimentally verified.

Further reading

Harrow J, Denoeud F, Frankish A, et al. (2006) GENCODE: producing a reference annotation for ENCODE. *Genome Biology* 7: S4.

See also ENCODE, GASP, Gene Annotation, hand-curated

GENE ANNOTATION

Enrique Blanco and Josep F. Abril

The identification of locations that encode genes along genomic sequences.

Genes can be classified into protein-coding genes and non-coding RNA genes reflecting the biological message encoded on their sequence. Although there are other functional elements in genomes, determining the gene catalog is one of the primary objectives once a genome has been sequenced. While in prokaryotes genes can be distinguished as contiguous open reading frames, gene annotation in eukaryotes is not a straightforward task for several reasons: (1) the fraction of genes is residual in comparison to the total volume of the genome; (2) RNA maturation separates exons (constituting the mRNA) from introns (removed from the spliced transcript); (3) alternative splicing may produce different transcripts of the same gene and (4) non-canonical forms of genes such as pseudogenes or selenoproteins are reported in the genome as well.

Gene annotation of a particular genome is performed employing a mixture of computational approaches that produce an initial set of putative genes and a posterior manual curation by human experts that refine the initial predictions. Experimental work must be incorporated in support of each preliminary annotation on the genome.

Further reading

Ashurst JL, Collins JE (2003) Gene annotation: prediction and testing. *Annual Review of Genomics and Human Genetics* 4: 69–88.

Brent MR (2008) Steady progress and recent breakthroughs in the accuracy of automated genome annotation. *Nature Reviews Genetics* 9: 62–73.

Harrow J, Nagy A, Reymond A, et al. (2009) Identifying protein-coding genes in genomic sequences. *Genome Biology* 10: 201.

See also Gene Prediction, Gene Annotation, hand-curated, Gene Annotation, formats, Gene Annotation, visualization tools

GENE ANNOTATION, FORMATS

Enrique Blanco and Josep F. Abril

The layout in which a file of features describing a particular set of genes must be presented, in order to favor an efficient exchange of information between two bioinformatics applications.

Most annotation formats focus on coordinate pairs that define intervals within nucleotide or protein sequences encoding for a real or predicted feature. Genes are hierarchical structures made up of a variety of components; a gene may encode one or more transcripts, which are in turn may be made up of exons and introns, as well as all the signals defining all boundaries such as splice sites for exons, transcription and translation start/end sites for transcripts, and so on. Gene annotations are usually distributed in plain text files of tabular records in which each line contains the coordinates of a given feature located on a sequence (e.g. gene, transcript, splice site, CDS, etc.). Within each record, its attributes are presented in different columns that are separated by delimiters (i.e. commas, spaces or tabulators). It is fundamental that the value of the same attribute class appears in the same position in distinct records.

The GFF format (General Feature Format or Gene-Finding Format) is a popular gene annotation format that is widely used in the bioinformatics community. In its basic version, GFF records are constituted of nine fields that describe each annotation ('.' is for undefined values):

1. Locus name in which the annotation is present
2. Source of the annotation (program/database)
3. Feature (biological feature such as gene, exon, start codon, ...)
4. Starting position of the feature in the sequence
5. Ending position of the feature in the sequence
6. Score associated to the accuracy of the annotation
7. Strand of the DNA molecule
8. Reading frame (for exons, a number between 0 and 2)
9. Group to link together several records of the same entity (e.g. exons of a gene)

GFF example:

```
chr2L RefSeq gene3 476437 478204 . - . cbt
chr2L RefSeq exon2 476437 478204 . - . cbt
chr2L RefSeq exon1 479406 479682 . - . cbt
```

The GTF format (Gene Transfer Format) is a modification of GFF to hold information about gene structure. Basically, GTF records are GFF records in which the Group attribute is expanded into a list of tag/value pairs delimited by the ';' character. This list must contain at the beginning the identifier for the gene and the transcript to which this feature belongs:

```
chr2L RefSeq gene 476437 478204 . - . gene_id 'cbt';
                                      transcript 'NM_164384'
chr2L RefSeq exon 476437 478204 . - . gene_id 'cbt';
                                      transcript 'NM_164384';
                                      exon_number 2
```

G

```
chr2L RefSeq exon 479406 479682 . - . gene_id 'cbt';
                                      transcript 'NM_164384';
                                      exon_number 1
```

GFF3 introduced the concept of feature parentship via the 'parent' tags, which allows a more hierarchical grouping of the related features (i.e. genes being parent features of sets of transcripts, and those being parents of sets of exons).

Related websites

GFF2 specifications	http://www.sanger.ac.uk/resources/software/gff/spec.html
GTF2 specifications	http://mblab.wustl.edu/GTF22.html
GFF3 specifications	http://www.sequenceontology.org/gff3.shtml

See also Gene Annotation, Gene Prediction, Gene Annotation, visualization tools

GENE ANNOTATION, HAND-CURATED

Enrique Blanco and Josep F. Abril

Manual revision of a set of gene annotations following a protocol based on objective criteria, which is performed by a group of human experts that confirm each feature using experimental data. Validation information is initially retrieved from literature, but the curators can provide candidate sequences for further experimental confirmation.

Hand-curated gene annotation is especially important in non-canonical gene forms that are not appropriately captured by *in silico* approaches, such as selenoproteins, non-canonical splicing variants, pseudogenes, conserved gene families, duplications and non-coding genes. Prior to the acceptance of a particular annotation, it is necessary to find transcriptional or protein homology information confirming its existence.

There are several international efforts to consolidate annotations using manual revision of gene annotations. Among those one should mention the Human and Vertebrate Analysis and Annotation (HAVANA) team, the collaborative consensus coding sequence (CCDS) project and the vertebrate genome annotation (Vega) database.

Further reading

Ashurst JL, Collins JE (2003) Gene annotation: prediction and testing. *Annual Review of Genomics and Human Genetics* 4: 69–88.

Pruitt KD, Harrow J, Harte RA, et al. (2009) The consensus coding sequence (CCDS) project: Identifying a common protein-coding gene set for the human and mouse genomes. *Genome Research* 19: 1316–1323.

Wilming LG, Gilbert JG, Howe K, Trevanion S, Hubbard T, Harrow JL (2008) The vertebrate genome annotation (Vega) database. *Nucleic Acids Research* 36: D753–760.

See also Gene Annotation, Gene Prediction, Gene Prediction, non-canonical, CCDS, HAVANA, VEGA

GENE ANNOTATION, VISUALIZATION TOOLS

Enrique Blanco and Josep F. Abril

Software designed for the graphical representation of multiple biological features annotated along a genomic sequence.

Genome visualization tools generally display the annotations on a particular sequence into a grid of linear tracks. Thus, occurrences for a particular class of elements are represented as a series of boxes in a different track along the sequence (e.g. genes, transcription factor binding sites, etc...). When it is fundamental to graphically arrange relationships between two or more items along the genomes, circular representations are more appropriate. Users and visualization applications exchange information mostly through files generated by other bioinformatics software in standard gene annotation formats such as GFF.

Visualization tools can be classified into (a) applications that produce a high-resolution picture of a particular region for publication (e.g GFF2PS, Circos or GENOMEGRAPHS) and (b) applications that provide options to navigate through the genome annotations into a graphical environment that implements the access to other computational resources and data banks (e.g. APOLLO, IGV, X:MAP and popular genome browsers such as ENSEMBL, UCSC or GBROWSE). A more programmatic approach to annotation visualization can be achieved by using specialized libraries, such as Bio::Graphics.

Related websites

GFF2PS	http://genome.crg.cat/software/gfftools/GFF2PS.html
APOLLO	http://apollo.berkeleybop.org/current/index.html
ARTEMIS	http://www.sanger.ac.uk/resources/software/artemis/
CIRCOS	http://circos.ca/
IGV	http://www.broadinstitute.org/igv/
Gene Structure Display Server (GSDS)	http://gsds.cbi.pku.edu.cn/
FANCYGENE	http://bio.ifom-ieo-campus.it/fancygene
X:MAP	http://annmap.picr.man.ac.uk/

Further reading

Abril JF, Guigó R (2000) gff2ps: visualizing genomic annotations. *Bioinformatics* 16(8): 743–744.

Durinck S, Bullard J, Spellman PT, Dudoit S (2009) GenomeGraphs: integrated genomic data visualization with R. *BMC Bioinformatics* 10: 2.

Krzywinski M, Schein J, Birol I, et al (2009) Circos: an information aesthetic for comparative genomics. *Genome Res.* 19(9): 1639–1645.

Misra S, Harris N (2006) Using Apollo to browse and edit genome annotations. *Curr Protoc Bioinformatics* Chapter 9:Unit 9.5.

Nielsen CB, Cantor M, Dubchak I, Gordon D, Wang T (2010) Visualizing genomes: techniques and challenges. *Nature Methods* 7: S5–S15.

O'Donoghue S, Gavin A, Gehlenborg N, et al. (2010) Visualizing biological data – now and in the future. *Nature Methods* 7: S2–S4.

G

Robinson JT, Thorvaldsdóttir H, Winckler W, et al. (2011) Integrative Genomics Viewer. *Nature Biotechnology* 29: 24–26.

Stein L. (2008) Growing Beautiful Code in BioPerl, In: A. Oram & G. Wilson, editors, *Beautiful Code: Leading Programmers Explain How They Think*, chapter 12, pp: 187–215, O'Reilly Media, Inc.

See also Gene Annotation, Gene Prediction, Gene Annotation, formats

GENE CLUSTER

Katheleen Gardiner

A group of sequence-related genes found in close physical proximity in genomic DNA.

For example, ribosomal RNA genes are clustered on the short arms of the 5 human acrocentric chromosomes. Other well-known gene clusters include the genes for histones and immune system genes such as the MHC.

Further reading

Cooper DN (1999) *Human Gene Evolution.* Bios Scientific Publishers pp 265–285.

Ridley M (1996) *Evolution.* Blackwell Science, Oxford, pp 265–276.

GENE DISPENSABILITY

Dov Greenbaum

Gene dispensability refers to a measure that may be inversely related to the overall importance of a gene to the life of the relevant organism.

The flip side of gene dispensability relates to the essential nature of the gene.

Determining the nature of a gene's dispensability has repercussions for understanding of evolution, cellular molecular physiology and disease.

Typically, this measurement may be approximated by a gene knockout strain of the same or similar organism, the knockout typically under lab conditions.

Recent experiments involving genome-wide mutagenesis suggest that under laboratory conditions, up to about 90% of genes in bacteria and 80% in eukaryotes can be inactivated individually leaving an organism viable, often seemingly unaffected by the loss of the gene.

The consequence of gene dispensability may be related to various attributes of the organism including gene redundancy and the compensatory mechanisms of the organism.

To determine each gene's relevancy to fitness, assays of growth of these organism are conducted under stress conditions, both physical and chemical to tease out the importance of dispensable genes to fitness.

Further reading

Comparitive genomic studies are also used to determine the nature of each gene's dispensiblity vis a vis the organism.

Jewett MC, Forster AC (2010) Update on designing and building minimal cells *Curr Opin Biotechnol* 21: 697–703.

Jordan IK, Rogozin IB, Wolf YI, Koonin, EV (2002) Essential genes are more evolutionarily conserved than are nonessential genes in bacteria. *Genome Res*, 12: 962–968.

Korona R (2011) Gene dispensability. *Curr Opin Biotechnol* 22(4): 547–551.

Makino T, Hokamp K, McLysaght M (2009) The complex relationship of gene duplication and essentiality *Trends Genet* 25: 152–155.

GENE DISTRIBUTION

Katheleen Gardiner

The variable gene density within chromosomes and chromosome bands.

In the human genome, it is strongly non-uniform, with G bands relatively gene-poor, and the most gene-rich regions in T bands. Some regions of R bands on human chromosomes 21 and 22 have been annotated with one gene per 20–40 kb, while in a segment of a large G band on chromosome 21, only two genes were identified within ~7 Mb. The gene distribution varies among chromosomes; >550 genes were annotated within chromosome 22, while in the comparably sized chromosome 21 only 225 genes were found. For their relative sizes, human chromosome 19 appears gene-rich and human chromosome 13 gene-poor. This correlates with the relative density of R bands.

Further reading

Caron H, et al. (2001) The human transcriptome map: clusters of highly expressed genes in chromosome domains *Science* 291: 1289–1292.

Dunham I (1999) The DNA sequence of human chromosome 22 *Nature* 402: 489–495.

Federico C, Andreozzi L, Saccone S, Bernardi G (2000) Gene density in the Giemsa bands of human chromosomes *Chromosome Res* 8: 727–746.

Hattori M, et al. (2000) The DNA sequence of human chromosome 21 *Nature* 405: 311–320.

International Human Genome Sequencing Consortium (2001) Initial sequencing and analysis of the human genome *Nature* 409: 866–921.

Venter C, et al. (2001) The sequence of the human genome *Science* 291: 1304–1351.

GENE DIVERSITY

Mark McCarthy, Steven Wiltshire and Andrew Collins

In population genetics terms, a measure of the extent of genetic variation within populations.

Whereas heterozygosity (see Polymorphism) is often used as a measure of genetic variation at marker loci, gene diversity (sometimes termed average heterozygosity) represents a more general measure of variation. Heterozygosity simply describes the proportion of heterozygous individuals in a sample. Inbreeding, or recent admixture, will however result in a deficiency of heterozygotes and this measure will therefore underestimate the true variation within the population. Gene diversity is formed from the sum of squares of allele frequencies: specifically, where p_i is the frequency of allele i at a locus of interest, the gene diversity at that locus is $1 - \Sigma p^2$. Comparison of these different measures of genetic variation can provide insights into population structure (e.g. evidence for inbreeding).

Related website

ARLEQUIN	http://cmpg.unibe.ch/software/arlequin35/

Further reading

Dutta R, et al. (2002) Patterns of genetic diversity at the nine forensically approved STR loci in the Indian populations. *Hum Biol* 74: 33–49.

Jorgensen TH, et al. (2002) Linkage disequilibrium and demographic history of the isolated population of the Faroe Islands. *Eur J Hum Genet* 10: 381–387.

Weir BS (1996) *Genetic Data Analysis 2: Methods for discrete population genetic data.* Sunderland, MA: Sinauer, pp. 141–159.

See also Polymorphism

GENE DUPLICATION

Dov Greenbaum and Katheleen Gardiner

Gene duplication refers to any duplication of DNA regions encoding genes. As a description of a feature on a chromosome: two closely related gene adjacent in genomic DNA. Alternatively: the process by which duplicated genes (which are not necessarily clustered in the genome) arise.

Typically, gene duplication occurs as an error during a homologous recombination event, a retro-transposition, or in some instances, during the duplication of an entire chromosome.

Evolutionarily, the duplicated gene is often free from the selective evolutionary pressures acting on the original gene; as such, the duplicated gene may be found to accumulate mutations faster than a functional non-duplicated gene given the lack of negative controls on the mutations.

In some examples, without selective pressure, the duplicate genes might pick up deleterious mutations, or in some instances, both the original and the duplicate may pick up complimentary deleterious mutations, such that the combination of the two still allows for a functioning gene product.

Without selective evolutionary pressure acting on the gene, gene duplications may often evolve in many different ways: in some instances, even developing into beneficial novel genes for the organism.

Successive duplications caused by unequal recombination or crossing over will lead to gene clusters and gene families. After duplication, random mutations can lead to sequence divergence and possibly to functional or expression divergence.

Further reading

Cooper DN (1999) *Human Gene Evolution* Bios Scientific Publishers pp. 265–285.

Dickerson JE, Robertson DL (2011) On the origins of Mendelian disease genes in man: the impact of gene duplication *Mol Biol Evol* epub.

Hittinge CT, Carroll SB (2007) Gene duplication and the adaptive evolution of a classic genetic switch, *Nature* 449: 677–681.

Hurles M (2004) Gene duplication: the genomic trade in spare parts *PLoS Biol* e206.

Matthew D, Rasmussen, Manolis Kellis (2012) Unified modeling of gene duplication, loss, and coalescence using a locus tree *Genome Res.* gr.123901.111.

Ridley M (1996) *Evolution.* Blackwell Science, Oxford, pp. 265–276.

Related websites

Gene Duplication (Felix Friedberg, Ed.) (2011)	http://www.intechopen.com/books/show/title/gene-duplication
Yeast Gene Duplications	http://www.gen.tcd.ie/~khwolfe/yeast/

GENE EXPRESSION DATABASE (GXD), *SEE* MOUSE GENOME INFORMATICS.

GENE EXPRESSION PROFILE

Jean-Michel Claverie and John M. Hancock

The measurement of the abundance of transcripts corresponding to the same (usually large) set of genes in a number of different cell types, organs, tissue, physiological or disease conditions.

Gene expression profiling provides information on the variation of gene activity across conditions. At a first level, expression profiling allows the identification of genes that are differentially expressed and the activity of which appear linked to a given condition (e.g. organ or disease). At a second level, the comparison of the activity profiles from different genes allows the discovery of coordinated expression patterns (e.g. co-expression), used to establish functional relationships between genes (e.g. co-expression of genes involved in the same biochemical pathway). Beyond these common principles, the mathematical methods used to analyze expression profiles depend on the technology used to measure expression intensities. Three main classes can be distinguished: sequence tag counting methods (e.g. ESTs and SAGE), hybridization methods (e.g. DNA chips) and sequencing-based (using NGS).

Related websites

GEO database	http://www.ncbi.nlm.nih.gov/geo/
ArrayExpress database	http://www.ebi.ac.uk/arrayexpress/

Further reading

Auer PL, Srivastava S, Doerge RW (2012) Differential expression – the next generation and beyond. *Brief. Funct. Genomics* 11: 57–62.

Brown PO, Botstein D (1999) Exploring the new world of the genome with DNA microarrays. *Nat. Genet.* 21(1 Suppl): 33–37.

Claverie JM (1999) Computational methods for the identification of differential and coordinated gene expression. *Hum. Mol. Genet.* 8: 1821–1832.

Okubo K, et al. (1992) Large scale cDNA sequencing for analysis of quantitative and qualitative aspects of gene expression. *Nat. Genet.* 2: 173–179.

GENE FAMILY

Teresa K. Attwood and Katheleen Gardiner

A group of closely related genes that encode similar products (usually protein but also RNA) and have descended from the same ancestral gene.

Examples of extensive gene families include those that encode:- G protein-coupled receptors, a ubiquitous group of membrane-bound receptors involved in signal transduction; tubulins, structural proteins involved in microtubule formation; and potassium channels, a diverse group of ion channels important in shaping action potentials and in neuronal excitability and plasticity. By contrast with domain families, members of gene or protein families share significant similarity along the entire length of their sequences and consequently exhibit a high degree of structural and functional correspondence.

In order to characterize new members of gene families and their protein products, various bioinformatics tools have become available in recent years. Standard search tools, such as FastA and BLAST, are commonly used to trawl the sequence databases for distant relatives. Other, diagnostically more potent, approaches involve searches of the protein family databases – notably, PROSITE, PRINTS and InterPro. PRINTS, in particular, offers the ability to make fine-grained hierarchical diagnoses, from superfamily, through family, down to individual subfamily levels (consequently, this hierarchy also forms the basis of InterPro).

Members of a gene family may be clustered at one or several chromosome locations or may be widely dispersed. Families may be modest in size and distribution. For example, there are 13 crystallin genes in humans, 4 each clustered on chromosome 2 and chromosome 22, and the rest in individual locations; the HOX genes are located on four chromosomes in groups of 9–10 genes. Larger families include several hundred zinc finger genes, with many clustered on chromosome 19, and more than 1000 olfactory receptor genes. Sequence divergence among members of a gene family may result in functional divergence, expression pattern divergence or pseudogenes.

Relevant websites

PROSITE	http://prosite.expasy.org/
PRINTS	http://www.bioinf.man.ac.uk/dbbrowser/PRINTS/
InterPro	http://www.ebi.ac.uk/interpro/

Further reading

Bockaert J, Pin, JP (1999) Molecular tinkering of G protein-coupled receptors: an evolutionary success. *EMBO J.*, 18(7), 1723–1729.

Cooper DN (1999) Human Gene Evolution. *Bios Scientific Publishers*, pp. 265–285.

Gogarten JP, Olendzenski L (1999) Orthologs, paralogs and genome comparisons. *Curr. Opin. Genet. Dev.*, 9(6), 630–636.

Greene EA, Pietrokovski S, Henikoff S, et al. (1997) Building gene families. *Science*, 278: 615–626.

Henikoff S, Greene EA, Pietrokovski S, Attwood, TK, Bork P, Hood L (1997) Gene families: the taxonomy of protein paralogs and chimeras. *Science*, 278: 609–614.

Ridley M (1996) *Evolution*. Blackwell Science, Oxford, pp. 265–276.

See also FastA, BLAST, PROSITE, PRINTS, InterPro, Domain Family

GENE FINDING, *SEE* GENE PREDICTION.

GENE-FINDING FORMAT, *SEE* GENE ANNOTATION, FORMATS.

GENE FLOW

A.R. Hoelzel

The exchange of genetic factors between populations by migration and interbreeding.

In sexual species, dispersal alone does not necessarily imply gene flow, since successful reproduction is required to transfer alleles. This is also known as 'genetic dispersal'. Gene flow among populations reduces differentiation due to the effects of genetic drift and natural selection. Under the specific conditions set out by Wright's Island model (Wright 1940), the relationship between gene flow and population structure can be defined as follows:

$$Nm = (1 - F_{ST})/(4F_{ST})$$

where N is the effective population size, m is the migration rate (and so Nm is gene flow), and F_{ST} is the inter-population fixation index (the proportion of the total genetic variance accounting for differences among populations). When there is a high degree of structure among populations (F_{ST} is large), gene flow is low and visa versa.

Further reading

Hartl DL (2000) *A Primer of Population Genetics*. Sinauer Associates, Sunderland.

Wright S (1940) Breeding structure of populations in relation to speciation. *Amer. Nat.* 74: 232–248.

See also Evolution

GENE FUSION METHOD

Patrick Aloy

The gene fusion method relies on the principle that there is selective pressure for certain genes to be fused over the course of evolution. It predicts functionally related proteins, or even interacting pairs, by analysing patterns of domain fusion.

If two proteins are found separately in one organism (components) and fused into a single polypeptide chain in another (composite), this implies that they might physically interact or, more likely, have a related function. This approach can be automated and applied large-scale to prokaryotes but not to higher organisms, where the detection of orthologues becomes a difficult problem.

G

Further reading

Enright AJ, Iliopoulos I, Kyrpides N, Ouzounis CA (1999) Protein interaction maps for complete genomes based on gene fusion events. *Nature* 402: 86–90.

Marcotte EM, Pellegrini M, Ng HL, Rice DW, Yeates TO, Eisenberg D (1999) Detecting protein function and protein-protein interactions from genome sequences. *Science* 285: 751–753.

See also Ortholog

GENE INDEX

Matthew He

A gene index is a listing of the number, type, label and sequence of all the genes identified within the genome of a given organism. Gene indices are usually created by assembling overlapping EST (Expressed Sequence Tag) sequences into clusters, and then determining if each cluster corresponds to a unique gene. Methods by which a cluster can be identified as representing a unique gene include identification of long open reading frames (ORFs), comparison to genomic sequence, and detection of SNPs or other features in the cluster that are known to exist in the gene.

See also Expressed Sequence Tag, Open Reading Frame, SNP

GENE NEIGHBOURHOOD

Patrick Aloy

In bacteria and archaea, genes are transcribed in polycistronic mRNAs encoding co-regulated and functionally related genes. Operon structure can be inferred from the conserved physical proximity of genes in phylogenetically distant genomes (i.e. conservation of pairs of genes as neighbours or co-localization of genes in potential operons).

The conservation of gene neighbours in several bacterial genomes can be used to predict functional relationships, or even physical interactions, between the encoded proteins.

As for phylogenetic profiles, this is a *computational genomics* approach that can only be applied to entire genomes (i.e. not individual pairs of proteins). Another limitation of this approach is that cannot be used to predict interactions in eukaryotes.

Further reading

Dandekar T, Snel B, Huynen M, Bork P (1998) Conservation of gene order: a fingerprint of proteins that physically interact. *Trends Biochem Sci* 23: 324–328.

Snel B, Lehmann G, Bork P, Huynen M (2000) STRING: a web-server to retrieve and display the repeatedly occurring neighborhood of a gene. *Nucleic Acids Res* 28: 3442–3444.

See also Operon, Phylogenetics

GENE ONTOLOGY (GO)

Robert Stevens

The Gene Ontology aims to provide a structured, controlled vocabulary.

The **term**s in this ontology are used to annotate gene products or their proxies in the form of a gene, for the attributes of *Molecular function*, *Biological process* and *Cellular component*. Originally developed to provide a shared understanding for the model organisms fly, yeast and mouse, the GO is now used by over 15 model organism databases and non-species specific databases such as SWISS-PROT and INTERPRO.

Related website

Gene Ontology	http://www.geneontology.org

See also Gene Ontology Consortium

GENE ONTOLOGY CONSORTIUM

Robert Stevens

The Gene Ontology Consortium includes representatives of the database groups who participate in the development of the Gene Ontology (GO).

Originally consisting of the model organism databases for fly, yeast and mouse, it has now expanded to more than a dozen groups covering more than 15 organisms. The Consortium manages the content and representation of its ontologies, provides sets of gene product annotation data, and develops software for use with GO. The Consortium also supports the development of additional biological ontologies, such as a chemical ontology and a sequence-feature ontology, through its OBO effort.

Related website

Gene Ontology	http://www.geneontology.org

See also Gene Ontology, Open Biological and Biomedical Ontologies

GENE PREDICTION

Enrique Blanco, Josep F. Abril and Roderic Guigó

Computational procedure to predict the amino acid sequence of the proteins encoded in a query genomic sequence. As the concept of what a gene is has been substantially enriched in the last decade, gene prediction in the post-genomic era can be understood as a computational procedure to locate both coding and non-coding genes on genomic sequences.

Computational gene identification pipelines are built upon the following basic steps:

1. analysis of sequence signals potentially involved in gene specification ('search by signal')

2. analysis of regions showing sequence compositional bias correlated with coding function ('search by content')
3. comparison with known coding sequences ('homology-based gene prediction')
4. comparison with other genomes ('comparative gene prediction')
5. comparison with high-throughput expression data ('NGS-based gene prediction')

Ab initio gene prediction includes approaches 1) and 2), in which genes are predicted without direct reference to sequences external to the query. For that reason, it also often referred to as intrinsic or template gene prediction, as opposed to extrinsic or look up gene prediction. This includes approaches 3), 4) and 5), in which genes are predicted based on information external to the query.

The complexity of gene prediction differs substantially in prokaryotic and eukaryotic genomes. While prokaryotic genes are single continuous ORFs, usually adjacent to each other, eukaryotic genes are separated by long stretches of intergenic DNA and their coding sequences, the exons, are interrupted by large non coding introns. In the gene prediction literature, the term gene is often used to denote only its coding fraction. Similarly, the term exon refers only to the coding fraction of coding exons. See Zhang (2002) for a discussion of the types of exons considered in the bibliography. See Gerstein *et al.* (2007) for a discussion on the current concept of gene after novel biological evidences reported by whole-genome and transcriptome projects, as well as, by ENCODE.

Despite the progress made, computational eukaryotic gene prediction is still an open problem, and the accurate identification of every gene in the higher eukaryotic genomes by computational methods remains a distant goal (see GASP). To perform such evaluation procedures, solid reference gene sets such as GENCODE ensure the fairness of these comparisons.

Only the determination of the mRNA sequence of a gene guarantees the correctness of its predicted exonic structure. Even in this case, a number of important issues remain to be solved: the completeness of the 3′ and 5′ ends of the gene, in particular if only a partial cDNA sequence has been obtained, the possibility of alternative primary transcripts, the determination for each of these of the complete catalog of splice isoforms (of which the sequenced mRNA may correspond to only one among multiple instances), the usage of non-canonical splice sites, the precise site (or the alternative sites) at which translation starts, the possibility of the gene being encoded within an intron, or nested within a gene, or the possibility of the gene not fully conforming to the standard genetic code, such as in the case of selenoproteins. Standard available gene prediction programs do not deal well with any of these issues. The advent of massive sequencing techniques for measuring gene expression, however, promises to overcome these potential pitfalls of current predictive systems.

Further reading

Brent MR (2005) Genome annotation past, present and future: How to define an ORF at each locus. *Genome Res.* 15: 1777–1786.

Brent MR (2008) Steady progress and recent breakthroughs in the accuracy of automated genome annotation. *Nature Reviews Genetics* 9: 62–73.

Burset M, Guigó R (1996) Evaluation of gene structure prediction programs. *Genomics* 34: 353–357.

Gerstein MB, Bruce C, Rozowsky JS, et al. (2007) What is a gene, post-ENCODE? History and updated definition. *Genome Research* 17(6): 669–681.

Guigó R, Flicek P, Abril JF, et al. (2006). EGASP: the human ENCODE Genome Annotation Assessment Project. *Genome Biology* 7: S2.

Harrow J, Denoeud F, Frankish A, et al. (2006) GENCODE: producing a reference annotation for ENCODE. *Genome Biology* 7: S4.

Harrow J, Nagy A, Reymond A, et al. (2009) Identifying protein-coding genes in genomic sequences. *Genome Biology* 10: 201.

Mathe C, Sagot MF, Schiex T, Rouze P (2002) Current methods of gene prediction, their strengths and weaknesses. *Nucl. Acids Res.* 30: 4103–4117.

Zhang MQ (2002) Computational prediction of eukaryotic protein-coding genes. *Nature Reviews Genet* 3: 698–709.

See also Gene Annotation, Gene Prediction, *ab initio*, Gene Prediction, homology-based, Gene Prediction, comparative, Gene Prediction, non-canonical, Gene Prediction, pipelines, Gene Prediction, NGS-based, Gene Prediction, accuracy, GASP, GENCODE

GENE PREDICTION, AB INITIO

Enrique Blanco, Josep F. Abril and Roderic Guigó

Computational procedure to predict the amino acid sequence of the proteins encoded in a query genomic sequence without using information about other sequences external to the query. It is also often referred to as intrinsic or template gene prediction.

Prokaryotic Gene Finding

Ab initio gene prediction in prokaryotic organisms (and, in general, in organisms without splicing) involves the determination of ORFs initiated by suitable translation start sites. The translation start site signal, however, does not carry enough information to discriminate ORFs occurring by chance from those corresponding to protein coding genes. Thus, gene identification requires, in addition, the scoring of putative ORFs according to the species specific protein coding model. For instance, GLIMMER (Salzberg et al., 1998a) and GENEMARK (Borodovsky and McIninch, 1993), two popular programs for prokaryotic gene finding, use non-homogeneous Markov Models as the underlying coding model. In GENEMARK the models are of fixed order five, while GLIMMER uses Interpolated Markov Models, that is, combinations of Markov models from 1st through 8th-order, weighting each model according to its predictive power.

One serious problem facing prokaryotic gene prediction is that species-specific coding models (such as bias in codon usage) may be unavailable prior to the sequencing of a complete genome. To overcome this limitation, gene prediction in complete prokaryotic genomes often proceeds in two steps. First, ORFs corresponding to known genes are identified through sequence similarity database searches. These ORFs are then used to infer the species-specific coding model. In the second step, the inferred coding model is used to score the remaining ORFs. This is the approach used, for instance, in the program ORPHEUS (Frishman et al., 1998).

Eukaryotic Gene Finding

Gene prediction in eukaryotic genomes involves the identification of start and stop translation signals, and splice sites, which are responsible for the exonic structure of genes. These signals, however, do not appear to carry the information required for the recognition of coding exons in genomic sequences (Burge, 1998). Thus, gene prediction in eukaryotes, as in prokaryotes, requires the utilization of some *ad hoc* protein coding model to score potential coding exons.

Typical computational *ab initio* eukaryotic gene prediction involves the following tasks:

- Identification and scoring of suitable splicing sites, start and stop codons along the query sequence
- Prediction of candidate exons defined by these signals
- Scoring of these exons as a function, at least, of the signals defining the exon, and of one or more coding statistics computed on the exon sequence.
- Assembly of a subset of these exon candidates in a predicted gene structure. This operation is produced under a number of constraints, including the maximization of a scoring function which depends, at least, on the scores of the candidate exons.

The particular implementation of these tasks (which often are only implicit) varies considerably between programs. Combining prediction of sequence signals and regions with sequence compositional bias had already been suggested in the mid-eighties as an effective strategy to predict the exonic structure of eukaryotic genes (Nakata et al., 1985). In 1990 Gelfand proved this strategy successfully in nine mammalian genes. The same year, Fields and Soderlund published GeneModeler, a computer program for gene prediction in *Caenorhabditis elegans*. This was the first computer program publicly available for the prediction of the exonic structure of eukaryotic genes. In 1992 Guigó et al. published Geneid, the first generic gene prediction program for vertebrate organisms. A number of accuracy measures were introduced to benchmark Geneid in a large set of genomic sequences. In 1993, Solovyev and colleagues presented a linear discriminant function to combine information about significant triplet frequencies of various functional parts of splice site regions and preferences of oligonucleotides in exons and introns; it was the origin of the Fgenes suite of prediction. The same year, Synder and Stormo published Geneparser, a predecessor of the successful Hidden Markov Models gene predictors. In 1994, Xu and others published Grail2/Gap, built around the popular Grail program for coding region identification. In 1996, Burge and Karlin published GENSCAN. GENSCAN was the first program able to deal with large genomic sequences encoding multiple genes in both strands, and was substantially more accurate that its predecessors. More recently, Snap and mGene have shown modest improvements in the detection of genes in comparison to previous approaches (Korf, 2004; Schweikert et al. 2009).

In any case, the current generation of *ab initio* gene prediction programs is slightly more accurate than Genscan. Their accuracy is remarkable when analysing single gene sequences, but drops significantly when analysing large multi-genic sequences. When these are from human, only about 60–80% of the coding exons are correctly predicted (see GASP for further info). The computational efficiency of the programs, however, has increased substantially during the past decade. While programs in the early nineties were limited to the analysis of sequences only a few kilobases long, programs currently available can analyse chromosome-size sequences in a few minutes, running on standard workstations.

Related websites

Eukaryotic Gene Prediction	
GENEID	http://genome.crg.cat/geneid.html
GENEMARK	http://exon.gatech.edu
GENVIEW	http://www.itb.cnr.it/sun/webgene/
GRAILEXP	http://compbio.ornl.gov/Grail-1.3/
FGENES	http://www.softberry.com/
GENSCAN	http://genes.mit.edu/GENSCAN.html
SNAP	http://korflab.ucdavis.edu/Software/
mGene	http://www.mgene.org
MZEF	http://rulai.cshl.org/tools/genefinder/
MORGAN	http://www.cbcb.umd.edu/~salzberg/morgan.html
HMMGENE	http://www.cbs.dtu.dk/services/HMMgene
EUGENE	http://eugene.toulouse.inra.fr/
AUGUSTUS	http://bioinf.uni-greifswald.de/augustus/
GLIMMERHMM	http://www.cbcb.umd.edu/software/GlimmerHMM/
Prokaryotic Gene Prediction	
GENEMARK	http://exon.gatech.edu
GLIMMER	http://www.cbcb.umd.edu/software/glimmer/
EASYGENE	http://www.cbs.dtu.dk/services/EasyGene/
FRAMED	http://tata.toulouse.inra.fr/apps/FrameD/FD
CRITICA	http://www.ttaxus.com/software.html
PRODIGAL	http://prodigal.ornl.gov/

Further reading

Audic S, Claverie J-M (1998) Self-identification of protein-coding regions in microbial genomes. *Proc Natl Acad Si USA* 95: 10026–10031.

Badger JH, Olsen GJ (1999) Coding region identification tool invoking comparative analysis. *Mol Biol Evol* 16: 512–524.

Borodovsky M, McIninch J (1993) GenMark: Parallel gene recognition for both DNA strands. *Comput Chem* 17: 123–134.

Brent MR (2008) Steady progress and recent breakthroughs in the accuracy of automated genome annotation. *Nature Reviews Genetics* 9: 62–73.

Burge CB (1998) Modeling dependencies in pre-mrna splicing signals. In: Salzberg S, Searls D, Kasif S (eds) *Computational Methods in Molecular Biology*, pp. 127–163, Amsterdam, Elsevier Science.

Burge CB, Karlin S (1997) Prediction of complete gene structures in human genomic DNA. *J Mol Biol* 268: 78–94.

Burge CB, Karlin S (1998) Finding the genes in genomic DNA. *Curr Opin Struct Biol* 8: 346–354.

Claverie JM (1997) Computational methods for the identification of genes in vertebrate genomic sequences. *Hum Mol Genet* 6: 1735–1744.

G

G

Dong S, Searls DB (1994) Gene structure prediction by linguistic methods. *Genomics* 23: 540–551.

Fickett J (1996) Finding genes by computer: The state of the art. *Trends Genet* 12: 316–320.

Fields CA, Soderlund CA (1990) gm: a practical tool for automating DNA sequence analysis. *Comput Appl Biosci* 6: 263–270.

Frishman D, et al. (1998) Combining diverse evidence for gene recognition in completely sequenced bacterial genomes. *Nucleic Acids Res* 26: 2941–2947.

Frishman D, et al. (1999) Starts of bacterial genes: estimating the reliability of computer predictions. *Gene* 234: 257–265.

Gelfand MS (1990) Computer prediction of exon-intron structure of mammalian pre-mrnas. *Nucleic Acids Res* 18: 5865–5869.

Gelfand MS, Roytberg MA (1993) Prediction of the exon-intron structure by a dynamic programming approach. *Biosystems* 30: 173–182.

Guigó R (1997) Computational gene identification: An open problem. *Comput Chem* 21: 215–222.

Guigó R, Knudsen S, Drake N, Smith T (1992) Prediction of gene structure. *J Mol Biol* 226: 141–157.

Henderson J, et al. (1997) Finding genes in DNA with a hidden markov model. *J Comput Biol* 4: 127–141.

Hutchinson GB, Hayden MR (1992) The prediction of exons through an analysis of spliceable open reading frames. *Nucleic Acids Res.* 20: 3453–3462.

Korf I (2004) Gene finding in novel genomes. *BMC Bioinformatics* 5: 59.

Krogh A (1997) Two methods for improving performance of an HMM and their application for gene finding. *ISMB*, 5: 179–186.

Krogh A, et al. (1994) A hidden Markov model that finds genes in *E. coli* DNA. *Nucleic Acids Res.* 22: 4768–4778.

Kulp D, et al. (1996) A generalized hidden markov model for the recognition of human genes in DNA. In: States DJ, Agarwal P, Gaasterland T, Hunter L, Smith R, (eds) *Intelligent Systems for Molecular Biology,* pp 134–142, Menlo Park, California. AAAI Press.

Makarov V (2002) Computer programs for eukaryotic gene prediction. *Briefings Bioinformatics* 3: 195–199.

Milanesi (1999) Genebuilder: interactive in silico prediction of gene structure. *Bioinformatics* 15: 612–621.

Mural RJ (1999) Current status of computational gene finding: a perspective. *Methods Enzymol* 303: 77–83.

Nakata K, et al. (1985) Prediction of splice junctions in mRNA sequences. *Nucleic Acids Res* 13: 5327–5340.

Salzberg SL, et al. (1998a) Microbial gene identification using interpolated Markov models. *Nucleic Acids Res* 26: 544–548.

Salzberg SL, et al. (1998b) A decision tree system for finding genes in DNA *J Comput Biol* 5: 667–680.

Schiex T, et al. (2000) Recherche des gènes et des erreurs de séquençage dans les génomes bactériens GC-riches (et autres...) *Proc. of JOBIM* 321–328. Montpellier. France.

Schweikert G, Zien A, Zeller G, et al. (2009) mGene: Accurate SVM-based gene finding with an application to nematode genomes. *Genome Res.* 19: 2133–2143.

Snyder EE, Stormo GD (1993) Identification of coding regions in genomic DNA sequences: an application of dynamic programming and neural networks. *Nucleic Acids Res* 21: 607–613.

Solovyev VV, et al. (1994) Predicting internal exons by oligonucleotide composition and discriminant analysis of spliceable open reading frames. *Nucleic Acids Res.* 22: 5156–5163.

Thomas A, Skolnick MH (1994) A probabilistic model for detecting coding regions in DNA sequences. *IMA J Math Appl Med Biol.* 11: 149–160.

Xu Y, et al. (1994) Constructing gene models from accurately predicted exons: An application of dynamic programming. *Comput Appl Biosc* 11: 117–124.

Zhang MQ (1997) Identification of protein coding regions in the human genome based on quadratic discriminant analysis. *Proc. Natl. Acad. Sci. USA* 94: 565–568.

Zhang MQ (2002) Computational prediction of eukaryotic protein-coding genes. *Nature Reviews Genet* 3: 698–709.

See also Gene Annotation, Gene Prediction, homology-based, Gene Prediction, comparative, Gene Prediction, non-canonical, Gene Prediction, pipelines, Gene Prediction, NGS-based, Gene Prediction, accuracy, GASP, GENCODE

GENE PREDICTION, ACCURACY

Enrique Blanco, Josep F. Abril and Roderic Guigó

Methods of estimating the accuracy of gene predictions.

To evaluate the accuracy of a gene prediction program on a test sequence, the gene structure predicted by the program is compared with the actual gene structure of the sequence, as, for instance, established with the help of an experimentally validated mRNA. Several metrics have been introduced during the past years to compare the predicted gene structure with the real one.

The accuracy can be evaluated at different levels of resolution. Typically, these are the nucleotide, exon, transcript and gene levels. Those levels offer complementary views of the accuracy of the program. At each level, there are two basic measures: sensitivity and specificity, which essentially measure prediction errors of the first and second kind. Briefly, sensitivity (Sn) is the proportion of real elements (coding nucleotides, exons, transcripts or genes) that have been correctly predicted, while specificity (Sp) is the proportion of predicted elements that are correct. If TP (True Positives) and TN (True Negatives) denote the number of coding and non-coding nucleotides (exons/genes) correctly predicted, FN (False Negatives) the number of actual coding nucleotides (exons/genes) predicted as non-coding, and FP (False Positives), the number of nucleotides (exons/genes) predicted coding that are non-coding, then:

$$Sn = \frac{TP}{TP + FN}$$

G

and

$$Sp = \frac{TN}{TN + FP}$$

Both sensitivity and specificity take values from 0 to 1, with perfect prediction when both measures are equal to 1. Neither of them alone constitutes a good measure of global accuracy, since one can have high sensitivity with little specificity and vice versa. It is desirable to use a single scalar value to summarize both of them. In the gene finding literature, the preferred such measure is the correlation coefficient *CC* at the nucleotide level. *CC* is computed as

$$CC = \frac{(TP \times TN) - (FP \times FN)}{\sqrt{(TP + FN) \times (TN + FP) \times (TP + FP) \times (TN + FN)}}$$

CC ranges from −1 to 1, with 1 corresponding to a perfect prediction, and −1 to a prediction in which each coding nucleotide is predicted as non-coding and vice versa.

To uniformly test and compare different programs, a number of standard sequence data sets have been compiled during the past years to benchmark gene prediction software (see websites below). Thus, distinct international efforts have been developed to standardize the assessment of gene prediction methods (see GASP) and the most accurate computational gene prediction tools exhibit an average accuracy (sensitivity and specificity) at the exon level of about 0.60–0.80 in the human genome.

Related websites

Benchmark data sets	http://genome.crg.cat/datasets/genomics96/
AcE evaluation tool	http://bioinformatics.org/project/?group_id=39
GASP	http://www.fruitfly.org/GASP1/
EGASP	http://genome.crg.cat/gencode/workshop2005.html

Further reading

Bajic V (2000) Comparing the success of different prediction software in sequence analysis: A review. *Briefings Bioinformatics* 1: 214–228.

Baldi P, et al. (2000) Assessing the accuracy of prediction algorithms for classification: an overview. *Bioinformatics* 16: 412–424.

Burset M, Guigó R (1996) Evaluation of gene structure prediction programs. *Genomics* 34: 353–357.

Guigó R, Flicek P, Abril JF, et al. (2006) EGASP: the human ENCODE Genome Annotation Assessment Project. *Genome Biology* 7: S2.

Guigó R, Wiehe T (2003) Gene prediction accuracy in large DNA sequences. In: *Galperin MY*, Koonin EV (eds) Frontiers in Computational Genomic, p. 1–33, Caister Academic Press.

Guigó R, et al. (2000) Gene prediction accuracy in large DNA sequences. *Genome Res* 10: 1631–1642.

Reese MG, et al. (2000) Genome annotation assessment in *Drosophila melanogaster*. *Genome Research*, 10: 483–501.

Rogic S, et al. (2001) Evaluation of gene-finding programs on mammalian sequences. *Genome Res* 11: 817–832.

See also GASP, GENCODE, ENCODE

G

GENE PREDICTION, ALTERNATIVE SPLICING

Enrique Blanco and Josep F. Abril

Computational procedure to predict the set of transcripts and protein products encoded on each putative gene in a query genomic sequence.

Relatively few *ab initio* gene prediction approaches deal with alternative spliced forms in their results (see EuGÈNE in Foissac and Schiex, 2005). Homology-based prediction pipelines are able, instead, to take advantage of expression data to detect putative alternative forms encoded in genomic sequences. Latest developments on gene finding such as COMBINER (Allen et al., 2004) and EVIDENCEMODELER (Haas et al., 2008) can combine information from different sources to build a final prediction (e.g. *ab initio* predictions, protein homology or massive sequencing expression data). Moreover, annotation pipelines such as MAKER (Cantarel et al., 2008), can consider a richer set of evidence. Both combiners and annotation pipelines outperform other approaches in finding untranslated exons and alternative splicing isoforms, although there is still room for improvement on this area as shown by the RGASP gene prediction assessment (see GASP).

Further reading

Arumugam M, Wei C, Brown RH, Brent MR (2006) Pairagon+N-SCAN_EST: a model-based gene annotation pipeline. *Genome Biology* 7: S5.

Cantarel BL, Korf I, Robb SM, et al. (2008) MAKER: an easy-to-use annotation pipeline designed for emerging model organism genomes. *Genome Res.* 18(1): 188–196.

Djebali S, Delaplace F, Roest Crollius H (2006) Exogean: a framework for annotating protein-coding genes in eukaryotic genomic DNA. *Genome Biology* 7: S7.

Foissac S, Schiex T (2005) Integrating alternative splicing detection into gene prediction. *BMC Bioinformatics* 6: 25.

Guigó R, Flicek P, Abril JF, et al. (2006). EGASP: the human ENCODE Genome Annotation Assessment Project. *Genome Biology* 7: S2.

Haas BJ, Salzberg SL, Zhu W, et al. (2008). Automated eukaryotic gene structure annotation using EVidenceModeler and the Program to Assemble Spliced Alignments. *Genome Biol.* 9(1): R7.

Allen JE, Pertea M, Salzberg SL (2004) Computational gene prediction using mutliple sources of evidence. *Genome Research*, 14(1): 142.

Stanke M, Schöffmann O, Morgenstern B, Waack S (2006) Gene prediction in eukaryotes with a generalized hidden Markov model that uses hints from external sources. *BMC Bioinformatics* 7: 62.

See also Alternative Splicing, Gene Annotation, Gene Prediction, Gene Prediction, *ab initio*, Gene Prediction, homology-based, Gene Prediction, non-canonical, GASP

GENE PREDICTION, COMPARATIVE

Enrique Blanco, Josep F. Abril and Roderic Guigó

Computational procedure to predict the amino acid sequence of the proteins encoded in a query sequence of one species using information from the genome of other species.

The rationale behind comparative gene prediction methods is that functional regions, protein coding among them, are more conserved throughout evolution than non-coding ones between genomic sequences from different organisms. This characteristic conservation can be used to identify protein-coding exons in the sequences.

The different methods use quite different approaches:

In a first approach, given two homologous genomic sequences, the problem is to infer the exonic structure in each sequence maximizing the score of the alignment of the resulting amino acid sequences. This problem is usually solved through a complex extension of the classical dynamic programming algorithm for sequence alignment. The program ProGen by Novichkov et al. (2001), and the algorithms by Blayo et al. (2002) and Pedersen and Scharl (2002) are examples of this approach, in which to some extent gene prediction is the result of the sequence alignment.

In a second approach, Pair Hidden Markov Models for sequence alignment, and Generalized HMMs (GHMMs) for gene prediction, are combined into the so-called Generalized Pair HMMs, as in the programs Slam (Pachter et al., 2001) and DOUBLESCAN (Meyer and Durbin, 2002). In this approach, given two homologous genomic sequences, both gene prediction and sequence alignment are obtained simultaneously.

In a third approach, gene prediction is separated from sequence alignment. First, the alignment is obtained between two homologous genomic sequences (also known as syntenic sequences), using some generic sequence alignment program, such as Tblastx, Sim96, Glass or Dialign. Then, gene structures are predicted that are compatible with this alignment, meaning that predicted exons fall in the aligned regions. The programs Rosetta (Batzoglou et al., 2000), CEM (Bafna and Huson, 2000), SGP1 (Wiehe et al., 2001) and AgenDA (Taher et al. 2004) are examples of this approach.

The fourth approach does not require the comparison of two homologous genomic sequences. Rather, a query sequence from a target genome is compared against a collection of sequences from a second (informant, reference) genome (which can be a single homologous sequence to the query sequence, a whole assembled genome, or a collection of shotgun reads), and the results of the comparison are used to modify the scores of the exons produced by underlying *ab initio* gene prediction algorithms. The programs TwinScan (Korf et al., 2001) and SGP2 (Parra et al., 2003) are examples of this approach. When various genomes at the appropriate evolutionary distance are available, both NSCAN (Gross and Brent, 2006) and Contrast (Gross *et al.* 2007) show a significant improvement over pair-wise genome comparisons using multiple genome alignments. Also related to this approach is the program Critica (Badger and Olsen, 1999), for gene prediction in prokaryotic genomes.

Comparative gene prediction methods were used extensively for the first time in the analysis of the human and mouse genomes (Mouse Genome Sequence Consortium, 2002). This analysis underscored the difficulties of using sequence conservation to predict coding regions: depending on the phylogenetic distance separating the genomes, sequence conservation may extend well beyond coding regions. About this issue, the CEGMA (Core Eukaryotic Genes Mapping Approach) project reported a set of highly conserved core proteins that are present in most species to provide reliable exon-intron structures in genomic sequences of eukaryotic genomes (Parra *et al.* 2007).

Related websites

DOUBLESCAN	http://www.sanger.ac.uk/resources/software/doublescan/
AGENDA	http://bibiserv.techfak.uni-bielefeld.de/agenda/
CRITICA	http://www.ttaxus.com/software.html
ROSETTA	http://crossspecies.lcs.mit.edu/
SLAM	http://bio.math.berkeley.edu/slam/mouse/
SGP2	http://genome.crg.cat/software/sgp2/
TWINSCAN	http://mblab.wustl.edu
NSCAN	http://mblab.wustl.edu/nscan
CEGMA	http://korflab.ucdavis.edu/Datasets/cegma/

Further reading

Alexandersson M, et al. (2003) SLAM: cross-species gene finding and alignment with a generalized pair hidden Markov model. *Genome Res.* 13: 496–502.

Badger JH, Olsen GJ (1999) Coding region identification tool invoking comparative analysis. *Mol Biol Evol* 16: 512–524.

Bafna V, Huson DH (2000) The conserved exon method. Proceedings of the eigth intenational conference on Intelligent Systems in Molecular Biology (ISMB), pp. 3–12.

Batzoglou S, et al. (2000) Human and mouse gene structure: Comparative analysis and application to exon prediction. *Genome Res* 10: 950–958.

Blayo P, et al. (2003) Orphan gene finding – an exon assembly approach. *Theor Comp Sci* 290: 1407–1431.

Brent MR (2008) Steady progress and recent breakthroughs in the accuracy of automated genome annotation. *Nature Reviews Genetics* 9: 62–73.

Gross SS, Brent MR (2006) Using multiple alignments to improve gene prediction. *J Comput Biol.* 13: 379–393.

Gross SS, Do CB, Sirota M, Batzoglou S (2007) CONTRAST: a discriminative, phylogeny-free approach to multiple informant de novo gene prediction. *Genome Biology* 8: R269.

Korf I, et al. (2001) Integrating genomic homology into gene structure prediction. *Bioinformatics* 17: S140–148.

Meyer I, Durbin R (2002) Comparative *ab initio* prediction of gene structure using pair HMMs. *Bioinformatics* 18: 1309–1318.

Mouse Genome Sequence Consortium (2002) Initial sequencing and comparative analysis of the mouse genome. *Nature* 420: 520–562.

Novichkov P, et al. (2001) Gene recognition in eukaryotic DNA by comparison of genomic sequences. *Bioinformatics* 17: 1011–1018.

Parra G, Bradnam K, Korf I (2007) CEGMA: a pipeline to accurately annotate core genes in eukaryotic genomes. *Bioinformatics* 23: 1061–1067.

Parra G, et al. (2003) Comparative gene prediction in human and mouse. *Genome Res* 13: 108–117.

Pedersen C, Scharl T (2002) Comparative methods for gene structure prediction in homologous sequences. In: Guigó R, Gusfield D (eds) *Algorithms in Bioinformatics*.

Taher L, Rinner O, Garg S, Sczyrba A, Morgenstern B (2004) AGenDA: gene prediction by cross-species sequence comparison. *Nucl. Acids Res.* 32: W305–W308.

Wiehe T, et al. (2000) Genome sequence comparisons: Hurdles in the fast lane to functional genomics. *Briefings Bioinformatics* 1: 381–388.

Wiehe T, et al. (2001). SGP-1: Prediction and validation of homologous genes based on sequence alignments. *Genome Res* 11: 1574–1583.

See also Gene Annotation, Gene Prediction, Gene Prediction, *ab initio*, Gene Prediction, homology-based, Hidden Markov Model, Alignment, BLAST

GENE PREDICTION, HOMOLOGY-BASED (EXTRINSIC GENE PREDICTION, LOOK-UP GENE PREDICTION)

Enrique Blanco, Josep F. Abril and Roderic Guigó

Computational procedure to predict the amino acid sequence of the proteins encoded in a query genomic sequence using information about other known coding sequences external to the query.

It is also often referred to extrinsic or look-up gene prediction. This includes approaches using homology information, comparative genomics and NGS expression data in which genes are predicted based on external knowledge.

The rationale behind this approach is that those segments in the genomic query similar to known protein or cDNA (EST) sequences are likely to correspond to coding exons. Because the incidence of unspecific EST matches is relatively common, similarity to ESTs is often only considered indicative of coding function when the alignment of the EST on the query sequence occurs across a splice junction.

Translated nucleotide searches, such as those performed by Blastx, constitute one of the simplest homology based gene prediction approaches. These searches are particularly relevant when comparing predicted ORFs in prokaryotic genomes. When dealing with the split nature of the eukaryotic genes, however, Blastx-like searches do not resolve exon splice boundaries well. Thus, a number of programs use the results of translated nucleotide searches against protein sequences (or direct nucleotide searches against ESTs) to modify the exon scoring schema of some underlying *ab initio* gene prediction programs. Genie (Kulp et al., 1997), Grailexp (Xu and Uberbacher, 1997), AAT (Huang et al., 1997), EBEST (Jiang and Jacob, 1998), GIN (Cai and Bork, 1998), GeneBuilder (Milanesi et al., 1999), GRPL (Hooper et al., 2000). HMMGene (Krogh, 2000), Eugene (Schiex et al., 2001), Genomescan (Yeh et al., 2001), Augustus (Stanke *et al.* 2006) and Fgenesh++ (Solovyev *et al.* 2006) are some of the published methods.

The EGASP experiment measured the performance of intrinsic and extrinsic methods concluding that there is still room for the improvement when detecting exon boundaries and alternatively spliced genes. The use of RNA-seq information promises to dramatically improve the quality of homology-based methods in the short-term future.

A more sophisticated approach involves aligning the genomic query against a protein (or cDNA) target (presumably homologous to the protein encoded in the genomic sequence). In these alignments, often referred to as spliced alignments, large gaps corresponding to introns in the query sequence are only allowed at legal splice junctions. SIM4 (Florea et al., 1998), EST_GENOME (Mott, 1997), PROCRUSTES (Gelfand et al., 1996) and GENEWISE (Birney and Durbin, 1997) are examples of this approach (see Spliced Alignment).

Related websites

EUGENE	http://eugene.toulouse.inra.fr/
GENOMETHREADER	http://www.genomethreader.org/
FGENESH++	http://www.softberry.com/
AUGUSTUS+	http://bioinf.uni-greifswald.de/augustus/
GENEBUILDER	http://www.itb.cnr.it/sun/webgene/
GENIE	http://www.fruitfly.org/seq_tools/genie.html
GENOMESCAN	http://genes.mit.edu/genomescan.html
GRAILEXP	http://compbio.ornl.gov/grailexp/
HMMGENE	http://www.cbs.dtu.dk/services/HMMgene/

Further reading

Birney E, Durbin R (1997) Dynamite: a flexible code generating language for dynamic programming methods used in sequence comparison. *ISMB*, 5: 56–64.

Cai Y, Bork P (1998) Homology-based gene prediction using neural nets. *Anal Biochem* 265: 269–274.

Florea L, et al. (1998) A computer program for aligning a cDNA sequence with a genomic DNA sequence. *Genome Res.*, 8: 967–974.

Gelfand MS, et al. (1996) Gene recognition via spliced alignment. *Proc Natl Acad Sci USA* 93: 9061–9066.

Gremme G, Brendel V, Sparks ME, Kurtz S (2005) Engineering a software tool for gene structure prediction in higher organisms. *Information and Software Technology* 47: 965–978.

Guigó R, Flicek P, Abril JF, et al. (2006) EGASP: the human ENCODE Genome Annotation Assessment Project. *Genome Biology* 7: S2.

Guigó R, et al. (2000) Sequence similarity based gene prediction. In: Suhai S (ed.) *Genomics and Proteomics: Functional and Computational Aspects,* pp. 95–105. Kluwer Academic / Plenum Publishing.

Hooper PM, et al. (2000) Prediction of genetic structure in eukariotic DNA using reference point logistic regression and sequence alignment. *Bioinformatics* 16: 425–438.

Huang X, et al. (1997) A tool for analyzing and annotating genomic sequences. *Genomics*, 46: 37–45.

Jiang J, Jacob HJ (1998) Ebest: an automatic tool using expressed sequence tags to delineate gene structure. *Genome Res* 5: 681–702.

Krogh A (2000) Using database matches with hmmgene for automated gene detection in drosophila. *Genome Res* 10: 523–528.

Kulp D, et al. (1996) A generalized hidden markov model for the recognition of human genes in DNA. In: States DJ, Agarwal P, Gaasterland T, Hunter L, Smith R (eds) *Intelligent Systems for Molecular Biology*, pp. 134–142, Menlo Park, California. AAAI press.

Milanesi L, et al. (1999) Genebuilder: interactive in silico prediction of gene structure. *Bioinformatics* 15: 612–621.

Mott R (1997) EST_GENOME: a program to align spliced DNA sequences to unspliced genomic DNA. *Comput Appl Biosci* 13: 477–478.

Schiex T, et al. (2001) *Eugene: an eukaryotic gene finder that combines several sources of evidence*. In: Gascuel O, Sagot M.-F (eds) Lecture Notes in Computer Science, volume 2066 (Computational Biology).

Solovyev V, Kosarev P, Seledsov I, Vorobyev D (2006) Automatic annotation of eukaryotic genes, pseudogenes and promoters. *Genome Biol* S1: S10–12.

Stanke M, Tzvetkova A, Morgenstern B (2006) AUGUSTUS at EGASP: using EST, protein and genomic alignments for improved gene prediction in the human genome. *Genome Biology* S1: S11.

Xu Y, Uberbacher EC (1997) Automated gene identification in large-scale genomic sequences. *J Comput Biol* 4: 325–338.

Yeh R, et al. (2001) Computational inference of homologous gene structures in the human genome. *Genome Res* 11: 803–816.

See also Gene Annotation, Gene Prediction, Gene Prediction, comparative, Gene Prediction, pipelines, GENEWISE, GASP, GENCODE, BLASTX, Dynamic Programming, Spliced Alignment

GENE PREDICTION, NGS-BASED

Enrique Blanco and Josep F. Abril

Computational procedure to predict the location of genes encoded in a query genomic sequence using next-generation sequencing (NGS) information external to the query.

Emerging high-throughput sequencing techniques provide millions of short sequence reads that are extremely useful to study multiple biological problems. RNA-seq, for instance, is of great value for measuring gene expression with great detail in eukaryotic genomes (Mortazavi et al. 2008, Oshlack et al. 2010). However, it is not trivial to estimate the relative abundances of alternative gene transcripts based on how many reads support each one (Trapnell et al. 2012). In this context, gene prediction programs that deal with homology-based information can process such a great volume of massive sequencing data in order to improve current annotations, detect novel genes and provide accurate detection of splice sites (Yassoura et al. 2009). The RNA-seq genome annotation assessment project (RGASP) is currently evaluating the current progress of automatic gene building using RNA-seq data to assembly alternative transcripts and quantify their abundance. Alternatively, ChIP-seq of epigenetic marks involved in chromatin structure demonstrated to be also a good predictive method of gene models as several post-translational modifications of histones are involved in the transcriptional activation of genes (Ernst and Kellis, 2012). In consequence, a mixture of bioinformatics systems processing and integrating both epigenetics information and RNA-seq gene expression data can substantially contribute to refine the annotation of eukaryotic genomes.

Related website

RGASP	http://www.gencodegenes.org/rgasp/

Further reading

Ernst J, Kellis M (2012) ChromHMM: automating chromatin-state discovery and characterization. *Nature Methods* 9: 215–216.

Mortazavi A, Williams BA, McCue K, Schaeffer L, Wold B (2008) Mapping and quantifying mammalian transcriptomes by RNA-seq. *Nat Methods* 5: 621–628.

Oshlack A, Robinson M, Young MD (2010) From RNA-seq reads to differential expression results. *Genome Biology* 2010, 11: 220.

Trapnell C, Roberts A, Goff L, et al. (2012) Differential gene and transcript expression analysis of RNA-seq experiments with TopHat and Cufflinks. *Nature Protocols* 7: 562–578.

Yassoura M, Kaplan T, Fraser H, et al. (2009) *Ab initio* construction of a eukaryotic transcriptome by massively parallel mRNA sequencing. *PNAS* 106: 3264–3269.

See also Gene Annotation, Gene Prediction, GASP

GENE PREDICTION, NON-CANONICAL

Enrique Blanco and Josep F. Abril

Computational identification of genes showing features different from canonical gene models in Eukaryotes.

Since the appearance of pioneering gene prediction programs, these applications have focused mostly on the detection of protein-coding sequences. Therefore, it is reasonable to observe that the 'gene' term is still used in the gene finding literature to denote only its protein-coding fraction. However, emerging massive sequencing technologies are recognizing other forms of genes that should be taken into account when annotating the catalog of genes from one genome, as the definition of what a gene is has substantially evolved since then (Gerstein et al. 2007).

Various non-canonical gene models that need additional processing in the standard gene prediction pipelines have been reported:

1. Selenoproteins are proteins that incorporate the amino acid selenocysteine through a recoding of the UGA codon. Gene predictors able to deal with in-frame TGA codons have shown to be very effective for this purpose (Driscoll and Chavatte, 2004).
2. Non-coding RNAs constitute a vast array of genes that do not necessarily encode proteins (some of the RNA species are reviewed in Kowalczyk et al, 2012). In consequence, while standard gene prediction programs cannot detect them using codon bias sensors, it is possible to identify them with a combination of RNA-seq data and a different DNA composition that includes in-frame stop codons.
3. Pseudogenes are derived from functional genes through retrotransposition or duplication but have lost the original functions of their parental genes. The high degree of evolutionary conservation of pseudogenes and similarity to functional genes are significant enough to confound conventional gene prediction approaches (Zheng et al, 2006).

4. Fusion genes combine part or all of the exons from the transcripts of two collinear gene loci. This class of genes is often the result of a structural variation at chromosomal level, such as a translocation, an interstitial deletion or an inversion. It has been shown that tandem gene pairs may also contribute to increasing the protein catalog in eukaryota (Parra et al, 2006). Recent software developments that predict those cases take advantage of RNA-seq data to improve their performance (Ge et al, 2011; Kim and Salzberg, 2011).
5. Trans-splicing is a transcriptional process where initial exons from one transcript are merged with exons from another transcript, far downstream or even located in another chromosome sequence. For instance, up to 70% of *Caenorhabditis elegans* mRNAs begin with the spliced leader sequence (SL, 22bp), which is not associated with the gene (Hastings, 2005).

Further reading

Brent MR (2008) Steady progress and recent breakthroughs in the accuracy of automated genome annotation. *Nature Reviews Genetics* 9: 62–73.

Driscoll DM, Chavatte L (2004) Finding needles in a haystack. In silico identification of eukaryotic selenoprotein genes. *EMBO reports* 5: 140–141.

Ge H, Liu K, Juan T, Fang F, Newman M, Hoeck W (2011) FusionMap: detecting fusion genes from next-generation sequencing data at base-pair resolution. *Bioinformatics* 27(14): 1922–1928.

Gerstein MB, Bruce C, Rozowsky JS, et al. (2007) What is a gene, post-ENCODE? History and updated definition. *Genome Res.* 17: 669–681.

Gingeras T (2009) Missing lincs in the transcriptome. *Nature Biotechnology* 27: 346–347.

Hastings KEM (2005) SL trans-splicing: easy come or easy go? *Trends Genet.* 21: 240–247.

Kim D, Salzberg SL (2011) TopHat-Fusion: an algorithm for discovery of novel fusion transcripts. *Genome Biol.* 12(8): R72.

Kowalczyk MS, DR Higgs, TR Gingeras (2012) Molecular biology: RNA discrimination. *Nature* 482: 310–311.

Parra G, Reymond A, Dabbouseh N, et al. (2006) Tandem chimerism as a means to increase protein complexity in the human genome. *Genome Res.* 16(1): 37–44.

Zheng D, Frankish A, Baertsch R, et al. (2007) Pseudogenes in the ENCODE regions: consensus annotation, analysis of transcription, and evolution. *Genome Res.* 17: 839–51.

See also Gene Annotation, Gene Prediction, Selenoprotein

GENE PREDICTION, PIPELINES

Enrique Blanco, Josep F. Abril and Roderic Guigó

Automatic set of protocols that annotate the catalog of protein-coding genes into a particular genome.

A number of systems produce and maintain automatic annotation of sequenced genomes. At the core of these systems there is a gene prediction pipeline which integrates information from different gene prediction programs. In the European Bioinformatics Institute (EBI) ENSEMBL analysis pipeline, exon candidates are predicted by GENSCAN, and confirmed through cDNA and protein sequence searches. Then, GENEWISE is used to delineate the exonic structure of the predicted genes. The resulting annotation set is available through the ENSEMBL genes track on their genome browser. Celera used a similar pipeline, OTTO, to annotate the human and mouse genomes. ENSEMBL and OTTO are rather conservative systems, predicting about 25,000–30,000 human genes. The National Center for Biotechnology Information (NCBI) runs a less conservative pipeline based on GENOMESCAN that predicts about 40,000 human genes. The UCSC Genome Browser automatically produces its Known Genes track from processing the catalogs of UniProt proteins, GenBank mRNAs and RefSeq genes. Because of the need for experimental information supporting each prediction, most pipelines produce a very conservative set of annotations.

Given the current status of computational gene prediction, and their rather large ratio of false positives, it is indeed advisable to run a number of different programs on a query sequence and look after consistent predictions. In the past, a few stand-alone systems and public servers offered the option of running a number of gene prediction tools on a sequence query on demand to display the results through a single common interface: GeneMachine, Genotator, Nix, Rummage, Metagene and WebGene were examples of such tools. Currently, due to the abundance of biological information coming from multiple sources (such as gene prediction programs, experimental databases, etc.), bioinformaticians can take advantage of portable genome annotation suites of software instead such as Gaze, Jigsaw, Genomix and Maker that reconstruct the most reliable predicted gene models from a bunch of data.

Related websites

ENSEMBL	http://www.ensembl.org
NCBI annotation pipeline	http://www.ncbi.nlm.nih.gov/genome/guide/build.shtml
UCSC Genome Browser	http://genome.ucsc.edu/
GENEMACHINE	http://genemachine.nhgri.nih.gov/
GAZE	http://www.sanger.ac.uk/resources/software/gaze/
JIGSAW	http://www.cbcb.umd.edu/software/jigsaw/
MAKER	http://www.yandell-lab.org/software/maker.html
WEBGENE	http://www.itb.cnr.it/webgene/

Further reading

Allen JE, Salzberg SL (2005) JIGSAW: integration of multiple sources of evidence for gene prediction. *Bioinformatics* 21: 3596–3603.

Brent MR (2008) Steady progress and recent breakthroughs in the accuracy of automated genome annotation. *Nature Reviews Genetics* 9: 62–73.

Cantarel BL, Korf I, Robb S, et al. (2008) MAKER: An easy-to-use annotation pipeline designed for emerging model organism genomes. *Genome Res.* 18: 188–196.

Coghlan A, Durbin R (2007) Genomix: a method for combining gene-finders' predictions, which uses evolutionary conservation of sequence and intron–exon structure. *Bioinformatics* 23: 1468–1475.

G

Howe KL, Chothia T, Durbin R (2002) GAZE: a generic framework for the integration of gene-prediction data by dynamic programming. *Genome Res.* 12: 1418–1427.

Hsu F, Kent WJ, Clawson H, Kuhn RM, Diekhans M, Haussler D (2006) The UCSC Known Genes. *Bioinformatics* 22: 1036–1046.

Jones J, et al. (2002) A comparative guide to gene prediction tools for the bioinformatics amateur. *Int J Oncology* 20: 697–705.

Makalowska I, Ryan JF, Baxevanis AD (2001) GeneMachine: gene prediction and sequence annotation. *Bioinformatics* 17: 843–844.

See also Gene Annotation, Gene Prediction, Gene Prediction, *ab initio*, Gene Prediction, comparative, Gene Annotation, hand-curated

GENE SIZE

Katheleen Gardiner

The amount of genomic DNA in base pairs containing all exons and introns of a gene.

This includes exons and introns spanned by the 5′ untranslated region, the coding region and the 3′ untranslated region. Gene size in genomic DNA is to be distinguished from that of the mature mRNA or cDNA, which lacks all introns. Gene size is highly variable and is governed by the size and number of introns. Intronless genes can be ~1–5 kb in size (histones), while the largest annotated gene is the 79 intron Duchenne Muscular Dystrophy (DMD) gene that spans 2.4 Mb. Early estimates of gene sizes have probably been low but can now be determined with greater accuracy from the complete genomic sequence. Estimates of average gene size tend to increase as sequence quality of a genome sequence improves and exons initially thought to be from separate genes are merged.

Further reading

Caron H, et al. (2001) The human transcriptome map: clusters of highly expressed genes in chromosome domains *Science* 291: 1289–1292.

Cooper DN (1999) *Human Gene Evolution*. Bios Scientific Publishers.

Dunham I (1999) The DNA sequence of human chromosome 22 *Nature* 402: 489–495.

Hattori M, et al. (2000) The DNA sequence of human chromosome 21 *Nature* 405: 311–320.

International Human Genome Sequencing Consortium (2001) Initial sequencing and analysis of the human genome *Nature* 409: 866–921.

Venter C, et al. (2001) The sequence of the human genome *Science* 291: 1304–1351.

GENE SYMBOL

John M. Hancock

A gene symbol is a standard, shorthand abbreviation for a specific gene in a species that can also act as a unique ID for that gene. These are needed to overcome potential ambiguity in gene naming, allowing researchers including bioinformaticians to be sure they are referring to the same genomic entity. As genes can encode more than one transcript, additional IDs are required to identify individual transcripts. Species or, in some cases, groups of species frequently have official or semi-official groups or committees responsible for accepting proposed new symbols, modifying them where necessary, and promulgating these to their communities and more broadly.

Related website

Wikipedia – Gene Nomenclature	http://en.wikipedia.org/wiki/Gene_nomenclature

See also Gene Symbol, human

GENE SYMBOL, HUMAN

Matthew He

Symbols for human genes are usually designated by scientists who discover the genes. The symbols are created using the Guidelines for Human Gene Nomenclature developed by the Human Genome Organization (HUGO) Gene Nomenclature Committee. Gene symbols usually consist of no more than six upper case letters or a combination of uppercase letters and Arabic numbers. Gene symbols should start with the first letters of the gene name. For example, the gene symbol for insulin is 'INS.' A gene symbol must be submitted to HUGO for approval before it can be considered an official gene symbol.

Related websites

The Department of Energy (DOE) Human Genome Program	http://www.ornl.gov/sci/techresources/Human_Genome/glossary/
National Center for Biotechnology Information (NCBI)	http://www.ncbi.nlm.nih.gov/Education/BLASTinfo/glossary2.html

GENE TRANSFER FORMAT, *SEE* GENE ANNOTATION, FORMATS

GENEALOGICAL SORTING INDEX (GSI)

Michael P. Cummings

A statistic for quantifying the degree of exclusive association among observations in groups that are represented on a phylogenetic tree or gene genealogy.

There are several situations in which it is desirable to quantify the degree of exclusive association on a tree. For example, in the context of evolutionary biology, quantifying the degree of genealogical sorting and allelic histories is important for assessing the degree of differentiation of species and populations or subpopulations, and is essential for understanding conservation status and priority of particular organisms.

G

A measure of genealogical sorting, *gs*, is defined as

$$gs = \frac{n}{\sum_{u=1}^{U}(d_u - 2)}$$

where d is the degree of node u of U total nodes uniting a group (estimated coalescent events) through a most recent common ancestor, and n is the minimum number of nodes (coalescent events) required to unite a group of size $n + 1$ through a most recent common ancestor on a given genealogy.

A normalized statistic to quantify the degree of genealogical sorting along the unit interval, [0, 1], *gsi*, is as follows:

$$gsi = \frac{observed(gs) - \min(gs)}{\max(gs) - \min(gs)}.$$

The genealogical sorting index (gsi) quantifies the historical relationships among groups for any genealogy to enable novel insight into the evolutionary process. The significance of the statistics can be estimated via permutation.

A web-based system is available for performing the calculations.

Related website

genealogical sorting home page	http://www.genealogicalsorting.org/

Further reading

Cummings MP, Neel MC, Shaw KL (2008) A genealogical approach to quantifying lineage divergence. *Evolution* 62: 2411–2422.

GENECHIP, *SEE* AFFYMETRIX GENECHIP™ OLIGONUCLEOTIDE MICROARRAY.

GENEID

Enrique Blanco and Josep F. Abril

Geneid is a general-purpose eukaryotic gene prediction program.

The architecture of Geneid is based on a simple hierarchical design: (1) search splicing signals, start codons, and stop codons, (2) build and score candidate exons, and (3) assemble genes from the exons (Guigó et al., 1992; Blanco et al. 2007).

Geneid uses weight matrices to predict potential splice sites and start codons that are scored as log-likelihood ratios. From the set of predicted sites, the program constructs the initial pool of potential exons. Exons are scored as the sum of the scores of the defining sites, plus the log-likelihood ratio of the Markov model for coding sequences

(Parra et al., 2000). Finally, the gene structure is assembled from the set of predicted exons, maximizing the sum of the scores of the participating exons with a very efficient dynamic programming algorithm (Guigó, 1998). Optionally, GENEID can take into account homology evidence in order to adapt weights given to predicted exons, as defined for the comparative genomics pipeline of SGP2 (Parra et al, 2003).

GENEID was one of the first computational gene identification approaches to predict full exonic structures of vertebrate genes in anonymous DNA sequences: an early version was available as an e-mail server in 1991 (Guigó et al., 1992). In 2000, a new implementation of the system was released, introducing capabilities to include external information that supports genomic reannotation procedures (Blanco et al. 2007). While having accuracy comparable to the most efficient gene-prediction programs, GENEID is very efficient at handling large genomic sequences in terms of memory and speed, offering a vast set of output options for a detailed analysis of gene features in genomic sequences.

GENEID has been useful in several genome annotation projects as a component of the *ab initio* gene prediction pipeline: *Dictyostelium discoideum* (Glökner et al., 2002), *Tetraodon nigroviridis* (Jaillon et al., 2004) and *Paramecium tetraurelia* (Aury et al., 2006), *Solanum lycopersicum* (Mueller et al., 2009). This gene prediction platform contributed to predicting selenoproteins in *Drosophila melanogaster* and *Takifugu rubripes* (Castellano et al., 2001, Castellano et al. 2004).

Related website

Geneid	http://genome.crg.cat/software/geneid/index.html

Further reading

Aury JM, et al. (2006) Global trends of whole-genome duplications revealed by the ciliate *Paramecium tetraurelia. Nature* 444: 171–178.

Blanco E, Parra G, Guigó R (2007) Using geneid to identify genes. *Current Protocols in Bioinformatics* vol. 1, unit 4.3: 1–28. John Wiley & Sons Inc., New York.

Castellano S, Morozova N, Morey M, et al. (2001) In silico identification of novel selenoproteins in the *Drosophila melanogaster* genome. *EMBO reports* 2: 697–702.

Castellano S, Novoselov SV, Kryukov GV, et al. (2004) Reconsidering the evolution of eukaryotic selenoproteins: a novel non-mammalian family with scattered phylogenetic distribution. *EMBO reports* 5: 71–77.

Glöckner, et al. (2002) Sequence and analysis of chromosome 2 of *Dictyostelium discoideum. Nature* 418: 79–85.

Guigó R, Knudsen S, Drake N, Smith TF (1992) Prediction of gene structure. *Journal of Molecular Biology* 226: 141–157.

Guigó R (1998) Assembling genes from predicted exons in linear time with dynamic programming. *Journal of Computational Biology* 5: 681–702.

Jaillon O, et al. (2004) Genome duplication in the teleost fish *Tetraodon nigroviridis* reveals the early vertebrate protokaryotype. *Nature* 431: 916–917.

Mueller L, et al. (2009) A snapshot of the emerging tomato genome sequence. *The Plant Genome* 2: 78–92.

Parra G, Blanco E, Guigó R (2000) Geneid in *Drosophila. Genome Research* 10: 511–515.

Parra G, Agarwal P, Abril JF, Wiehe T, Fickett JW, Guigó R (2003) Comparative gene prediction in human and mouse. *Genome Research* 13: 108–117.

See also Gene Annotation, Gene Prediction, *ab initio*, Gene Prediction, homology-based, Gene Prediction, comparative, Gene Prediction, non-canonical, Gene Prediction, pipelines, Selenoproteins, Gene Prediction, accuracy, GASP, GENCODE

GENERAL FEATURE FORMAT, *SEE* GENE ANNOTATION, FORMATS.

GENERALIZATION, *SEE* ERROR MEASURES.

GENETIC ALGORITHM (GA)

Pedro Larrañaga and Concha Bielza

In genetic algorithms (Holland, 1975; Goldberg, 1989) the search space of a problem can be seen as a collection of individuals. These individuals are represented by character strings, which are often referred to as chromosomes. The purpose of the use of a genetic algorithm is to find the individual from the search space with the best genetic material. The quality of an individual is measured with an evaluation function. The part of the search space to be examined at each generation is called the population.

Roughly, a genetic algorithm works as follows. Firstly, the initial population is chosen, and the quality of each individual inside the population is determined. Next, in every iteration, parents are selected from the population. These parents produce children, which are added to the population. For all newly created individuals of the resulting population a probability near to zero exists that they 'mutate', i.e. that they change their hereditary distinctions. After that, some individuals are removed from the population according to a selection criterion in order to reduce the population to its initial size. One iteration of the algorithm is referred to as a generation. The operators which define the child production process and the mutation process are called the crossover operator and the mutation operator, respectively. Mutation and crossover play different roles in the genetic algorithm. The mutation operator is needed to explore new states and helps the algorithm to avoid local optima. The crossover operator should increase the average quality of the population. By choosing adequate crossover and mutation operators, the probability that the genetic algorithm provides a near-optimal solution in a reasonable number of iterations is enlarged.

Genetic algorithms have been used in several optimization problems in Bioinformatics. Examples include feature selection in microarray data (Jirapech-Umpai and Aitken 2005), mass spectrometry (Petricoin et al., 2002), prediction of protein function from sequence (Chuzhanova et al., 1998) and SNPs (Shah and Kusiak, 2004).

Further reading

Chuzhanova N (1998) Feature selection for genetic sequence classification. *Bioinformatics* 14(2): 139–143.

Goldberg DE (1989) *Genetic Algorithms in Search Optimization and Machine Learning*. Addison-Wesley.

Holland J (1975) *Adaptation in Natural and Artificial Systems*. University of Michigan Press.

Jirapech-Umpai T, Aitken S (2005) Feature selection and classification for microarray data analysis: evolutionary methods for identifying predictive genes. *BMC Bioinformatics* 6(148).

Petricoin E, et al. (2002) Use of proteomics patterns in serum to identify ovarian cancer. *The Lancet* 359(9306): 572–577.

Shah S, Kusiak A (2004) Data mining and genetic algorithm based gene/SNP selection. *Artificial Intelligence in Medicine* 31: 183–196.

GENETIC CODE (UNIVERSAL GENETIC CODE, STANDARD GENETIC CODE)

J.M. Hancock

The "code" by which the molecular machinery of the cell translates protein-coding DNA sequences into amino acid sequences.

Bases in protein-coding regions of genomes are decoded as codons, which consist of a group of three consecutive bases. When processed by the ribosome, each codon can either

Table G.1 The standard genetic code

Codon	AA	Codon	AA	Codon	AA	Codon	AA
AAA	Lys	CAA	Gln	GAA	Glu	UAA	STOP
AAC	Asn	CAC	His	GAC	Asp	UAC	Tyr
AAG	Lys	CAG	Gln	GAG	Glu	UAG	STOP
AAU	Asn	CAU	His	GAU	Asp	UAU	Tyr
ACA	Thr	CCA	Pro	GCA	Ala	UCA	Ser
ACC	Thr	CCC	Pro	GCC	Ala	UCC	Ser
ACG	Thr	CCG	Pro	GCG	Ala	UCG	Ser
ACU	Thr	CCU	Pro	GCU	Ala	UCU	Ser
AGA	Arg	CGA	Arg	GGA	Gly	UGA	STOP
AGC	Ser	CGC	Arg	GGC	Gly	UGC	Cys
AGG	Arg	CGG	Arg	GGG	Gly	UGG	Trp
AGU	Ser	CGU	Arg	GGU	Gly	UGU	Cys
AUA	Ile	CUA	Leu	GUA	Val	UUA	Leu
AUC	Ile	CUC	Leu	GUC	Val	UUC	Phe
AUG	Met	CUG	Leu	GUG	Val	UUG	Leu
AUU	Ile	CUU	Leu	GUU	Val	UUU	Phe

Codons, which are conventionally represented in their RNA form, are arranged alphabetically for ease of reference. Amino acids are represented by their three letter codes. Shaded boxes indicate codons that have been shown to be reassigned in the mitochondrial or nuclear genome of at least one species.

encode an amino acid or signal the termination of a protein sequence. The code by which codons represent amino acids is almost universal in nuclear genomes (see Table G.1) but mitochondria use a different code in which UGA codes for tryptophan, AUA codes for methionine and AGA and AGG are terminators. In addition, a number of variant codes have been discovered in both mitochondrial and nuclear genomes. These are characterized by the reassignment of individual codons, including termination codons, to new amino acids rather than by wholesale reorganization.

Further reading

Barrell BG, et al. (1979) A different genetic code in human mitochondria. *Nature* 282: 189–194.

Knight RD, et al. (2001) Rewiring the keyboard: evolvability of the genetic code. *Nat Rev Genet.* 2: 49–58.

Lewin B (2000) *Genes VII* Oxford University Press, Oxford, Chapters 5–7.

See also Codon

GENETIC LINKAGE, *SEE* LINKAGE

GENETIC NETWORK

Denis Thieffry

One speaks of a genetic interaction occurring between two genes when the perturbation (loss-of-function, over-expression) of the expression of the first gene affects that of the second, something which can be experimentally observed at the phenotypic or at the molecular level. Such interactions can be direct or indirect, i.e. not involving or involving intermediate genes. Turning to the molecular level, one can further distinguish between different types of molecular interactions: protein-DNA interactions (e.g., transcriptional regulation involving a regulatory protein and a DNA cis-regulatory region near the regulated gene), protein-RNA interactions (as in the case of post-transcriptional regulatory interactions), and protein-protein interactions (e.g., phosphorylation of a protein by a kinase, formation of multimeric complexes, etc.).

Molecular interactions between biological macromolecules are often represented as a graph, with genes (or their products) as vertices and interactions between them as edges. Information on gene interactions can also be obtained directly from genetic experiments (mutations, genetic crosses, etc.). Generally, the term genetic network is used to denote a molecular network involving interactions between various types of components (genes, RNA species, proteins). However, this term often refers to an abstract representation of biological networks, with prescribed topological properties but random connections, which is then used to explore connections between network connectivity and dynamical properties.

Regulatory interactions of various types (protein-protein interactions, transcriptional regulation, etc.) form signaling cascades. Signaling molecules are sensed by membrane receptors, leading to (cross-) phoshorylation by kinases (or other types of protein-protein interactions), ultimately leading to the translocation or the change of activity of transcriptional factors, which will in turn affect the transcription of specific subsets of genes in the nucleus. Different regulatory cascades (each forming a graph, typically with a tree topology) can share some elements, thus leading to further regulatory integration, something, which can be represented in terms of acyclic directed graphs. When regulatory cascades ultimately affect the expression of genes encoding for some of their crucial components (e.g., membrane receptors), the system is then better represented by (cyclic) directed graphs.

When facing a genetic network, various biological questions can be addressed in terms of topological comparisons of (sub)graphs, including the identification of functional regulatory motifs such as regulatory circuits. For example, when comparing orthologous or paralogous sets of genes across different organisms, regulatory motifs or metabolic pathways may be conserved beyond the structure or sequence of individual components (genes or molecular species). Formally, such comparative analyses can be expressed in terms of graph theoretic notions such as (sub)graph isomorphism (similar interactive structures) or homeomorphism (similar topologies).

Further reading

Kauffman SA (1993) *The Origins of Order: self organization and selection in evolution*. Oxford University Press, Oxford.

Monod J, Jacob F (1961) Teleonomic mechanisms in cellular metabolism, growth, and differentiation. *Cold Spring Harbor Symposia on Quantitative Biology* 26: 389–401.

Thomas R, D'Ari R (1990) *Biological Feedback*. CRC Press.

GENETIC REDUNDANCY

Katheleen Gardiner

Members of gene family performing the same function, such that deletion of one gene has no effect on organism viability.

For example, the human genome contains hundreds of ribosomal RNA genes in five clusters. The number of rRNA genes in each cluster varies among individuals and entire clusters can be deleted with no obvious negative consequences on the individual. Redundant genes are non-essential. However, for protein coding genes, being non-essential in a controlled laboratory setting may not translate to being redundant in a normal environment.

Further reading

Tautz D (1992) Redundancies, development and the flow of information. *Bioessays.* 14: 263–266.

Tautz D (2000) A genetic uncertainty problem. *Trends Genet.* 16: 475–477.

GENETIC VARIATION, *SEE* VARIATION (GENETIC).

G

GENEWISE

Roderic Guigó

GENEWISE compares a genomic sequence to a protein sequence or to a Hidden Markov Model (HMM) representing a protein domain.

It performs the comparison at the protein translation level, while simultaneously maintaining a reading frame regardless of intervening introns and sequence errors that may cause frame shifts. Thus, GENEWISE does both the gene prediction and the homology comparison together.

GENEWISE is based around the probabilistic formalism of Hidden Markov Models. The underlying dynamic programming algorithms are written using the code generating language DYNAMITE (Birney and Durbin, 1997). GENEWISE is part of the WISE2 package, which includes also ESTWISE, a program to compare EST or cDNA sequences to a protein sequence or a protein domain HMM. GENEWISEDB and ESTWISEDB are the database searching versions of GENEWISE and ESTWISE, respectively. They compare a database of DNA sequences (genomic or cDNA) to a database a protein or protein domains HMMs.

The database search programs in WISE2 are computationally very expensive, and their use may become prohibitive for large collections of sequences. The WISE2 package provides two scripts, BLASTWISE and HALFWISE, that allow users with average computer resources to run database searches with GENEWISE more sensibly. BLASTWISE compares a DNA sequence to a protein database using BLASTX and then calls GENEWISE on a carefully selected set of proteins. Similarly HALFWISE compares a DNA sequence to a database of protein domain HMMs, such as PFAM.

Alternatively, specialized commercial hardware exists from Compugen, Paracel and Time Logic that can approximate a GENEWISE search very closely or even run a full GENEWISE type algorithm. GENEWISE is at the core of the ENSEMBL system and it has been extensively used to annotate the human genome sequence.

Related websites

GENEWISE Home Page	http://www.ebi.ac.uk/Tools/Wise2/
DYNAMITE Home Page	http://www.ebi.ac.uk/Tools/Dynamite/

Further reading

Birney E, Durbin R (1997) Dynamite: a flexible code generating language for dynamic programming methods used in sequence comparison. *ISMB* 5: 56–64.

See also Spliced Alignment, Hidden Markov Models, Gene Prediction, homology-based, BLAST, Alignment, Gene Prediction, pipelines

GENOME ANNOTATION

Jean-Michel Claverie

The process of interpreting a newly determined genome sequence.

Genome annotation is performed in two distinct phases using different software tools. In the first phase, the 'raw' genomic sequence is parsed into elementary components such as protein coding regions, tRNA genes, putative regulatory regions, repeats, etc. The software tools used to parse higher eukaryotic sequences (animal and plants) are more sophisticated (yet still more error-prone) than the programs required for lower eukaryotes, prokaryotes and archea, due to the presence of introns, and more frequent repeats and pseudogenes. Bacterial 'ORFing', for instance, is quite trivial, reaching more than 95% accuracy with the use of low order Markov Models. Higher eukaryotic genome parsing is done by locally fitting a probabilistic multi-feature gene model, supplemented by the detection of local similarity with known protein or RNA sequences, as well as ESTs. High quality gene parsing still requires visual validation and manual curation for higher eukaryotes. In the second phase of the genome annotation process, protein coding genes are translated into putative protein sequences that are exhaustively compared to existing domain and functional motif databases. 'Functional' attributes are mostly inherited from homologous proteins in a few model species (e.g. *E. coli*, yeast, Drosophila, *C. elegans*) that have been previously characterized experimentally. Non-protein coding genes (e.g. tRNA, ribosomal RNA, etc.) are also annotated by homology. The detailed functional annotation of genomes is still a research activity, requiring specialized expertise in cellular physiology, biochemistry and metabolic pathways. Except for the smallest parasitic bacteria, the current genome annotation protocols succeed in predicting a clear function for only half of the genes. Phylogenomics approaches ('guilt-by-association methods') are then used to extend direct functional information through the identification of putative relationships between genes (of unknown or known function).

Related websites

High-Quality Automated and Manual Annotation of Microbial Proteomes (HAMAP)	http://www.expasy.org/sprot/hamap/
Clusters of Orthologous Groups	http://www.ncbi.nlm.nih.gov/COG/
Oak Ridge National Laboratory Genome Analysis page	http://genome.ornl.gov/
NCBI Genomic Sequence Assembly and Annotation Process	http://www.ncbi.nlm.nih.gov/genome/guide/build.html
Ensembl	http://www.ensembl.org/

Further reading

Apweiler R, et al. (2000) InterPro-an integrated documentation resource for protein families, domains and functional sites. *Bioinformatics.* 16: 1145–1150.

Audic S, Claverie JM. (1998) Self-identification of protein-coding regions in microbial genomes. *Proc. Natl. Acad. Sci. USA* 95: 10026–10031.

Burge C, Karlin S. (1997) Prediction of complete gene structures in human genomic DNA. *J. Mol. Biol.* 268: 78–94.

G

Lukashin AV, Borodovsky M. (1998) GeneMark.hmm: new solutions for gene finding. *Nucleic Acids Res.* 26: 1107–1115.

Rogic S, et al. (2001) Evaluation of gene-finding programs on mammalian sequences. *Genome Res.* 11: 817–832.

GENOME ANNOTATION ASSESSMENT PROJECT, *SEE* GASP.

GENOME SCANS FOR LINKAGE (GENOME-WIDE SCANS)

Mark McCarthy, Steven Wiltshire and Andrew Collins

Used, in the context of disease-gene mapping, to describe strategies for susceptibility gene identification which aim to survey the complete genome for linkage and to identify chromosomal regions with a high probability of containing susceptibility variants.

Successful genome scans for linkage have targeted collections of families segregating the disease (or phenotype) of interest and used a set of polymorphic markers arranged at regular intervals across the genome. Disease mapping in this situation can be achieved by typing ~400 highly-polymorphic microsatellite markers, with an average separation of around 10 centimorgans. Since linkage signals typically extend for tens of centimorgans, this strategy has enabled many regions containing major susceptibility genes to be localized. For multifactorial traits, linkage is relatively underpowered and, despite extensive development and application of allele-sharing and related linkage methods, this strategy has now been superseded by association mapping. Numerous genome wide scans that seek evidence of association (or linkage disequilibrium), rather than linkage have been conducted.

Examples: The first example of a genome wide scan for linkage to a multifactorial trait was conducted for type 1 diabetes (Davies et al, 1994). This scan revealed evidence for linkage in several chromosomal regions, though the largest signal by far (as expected) fell in the HLA region of chromosome 6. Since that time around 300 genome linkage scans have been reported for a wide variety of multifactorial (complex) traits.

Further reading

Altmuller J, et al. (2001) Genomewide scans of complex human disease: true linkage is hard to find. *Am J Hum Genet* 69: 936–950.

Concannon P, et al. (1998) A second-generation screen of the human genome for susceptibility to insulin-dependent diabetes mellitus. *Nat Genet* 19: 292–296.

Davies JL, et al. (1994) A genome-wide search for human type 1 diabetes susceptibility genes. *Nature* 371: 130–136.

Risch N, Merikangas K (1996) The future of genetic studies of complex human diseases. *Science* 273: 1516–1517.

Todd JA, Farrall M (1996) Panning for gold: genome-wide scanning for linkage in type 1 diabetes. *Hum Mol Genet* 5: 1443–1448.

Wiltshire S, et al. (2001) A genome-wide scan for loci predisposing to type 2 diabetes in a UK population (The Diabetes (UK) Warren 2 Repository): analysis of 573 pedigrees provides independent replication of a susceptibility locus on chromosome 1q. *Am J Hum Genet* 69: 553–569.

See also Genome-Wide Association, Allele-sharing Methods, Positional Candidate Approach, Linkage, Linkage Analysis, Multipoint Linkage Analysis

GENOME SIZE (C-VALUE)

Katheleen Gardiner

The number of base pairs in a haploid genome (one copy of each pair of all chromosomes) in a given species.

Genome size has also been expressed in picograms (one kb = 10–6 picograms) but this is less informative now that many genomes have been completely sequenced. Mammalian genomes contain ~3 billion base pairs; birds, 1–2 billion.

Further reading

Ridley M (1996) *Evolution* Blackwell Science, Oxford, pp 265–276.

See also C-value Paradox

GENOME-WIDE ASSOCIATION STUDY (GWAS)

Andrew Collins

Genome-wide association mapping has been extensively used to identify numerous common polymorphisms involved in genetically complex (multifactorial) human disease. This has involved the genotyping of hundreds of thousands to millions of single nucleotide polymorphisms (SNPs) across the entire human genome and testing for associations with SNP alleles in very large samples of disease cases and controls. The selection of SNP markers for genotyping arrays to adequately cover the genome with minimum redundancy (because of linkage disequilibrium, LD) has been achieved by detailed understanding of patterns of linkage disequilibrium in the human genome. The Haplotype Map (HapMap) project underpinned this process by delimiting genomic regions with extensive LD which can be 'tagged' by a restricted number of markers, thereby reducing genotyping cost and the number of statistical tests made.

The first significant breakthrough for GWAS was by Klein et al (2005) who identified an important gene (complement factor H, CFH) involved in age-related macular degeneration, a major cause of blindness in the elderly. This genome wide screen comprised 96 disease cases and 50 controls genotyped for 116204 SNPs. However, subsequent studies have employed much larger samples of cases and controls and genotypes since the majority of common variants have been found to contribute much lower disease risks. Because of the huge number of polymorphisms tested in these studies a very substantial statistical correction is required for the number of tests made and small samples of cases and controls are usually under-powered. Bonferroni correction is the most conservative and considers as significant only those SNPs for which $P<0.05/N$, where N is the number of SNPs tested.

G

The false discovery rate (FDR) is a less conservative correction and controls the expected proportion of incorrectly rejected null hypotheses (the number of type 1 errors).

Example: The Wellcome Trust Case Control Consortium (WTCCC) published the first very large GWAS study of seven common diseases in 2007 which identified 24 independent association signals which remained significant after correction.

Further reading

Benjamini Y, Yekutieli D (2001)The control of the false discovery rate in multiple testing under dependency. *Annals of Statistics* 29 (4): 1165–1188.

Klein RJ, Zeiss C, Chew EY, et al. (2005) Complement factor H polymorphism in age-related macular degeneration. *Science* 308(5720): 385–9.

The Wellcome Trust Case Control Consortium (2007) Genome-wide association study of 14,000 cases of seven common diseases and 3,000 shared controls. *Nature* 447: 661–678.

See also Genotype Imputation, Allelic Association, Association Analysis, Linkage Disequilibrium Map, False Discovery Rate Control

GENOME-WIDE SCANS (LINKAGE), *SEE* GENOME SCANS

GENOME-WIDE SURVEY

Dov Greenbaum

Analyses surveying the overall content of genomes, proteomes etc.

Comparative genomic analysis is becoming increasingly more complex given the overwhelming amount of data now being produced for every organism. Methods have been borrowed from other disciplines to help cope with this data deluge.

Similar to censuses taken by governments, genome wide surveys serve to assess the number and nature of distinct parts – finite parts lists – within individual genomes, and between genomes. The power of these surveys lies in the fact that they can condense much of the information unique to a particular genome into key values, thus facilitating genome-genome comparisons.

Surveys assess, in particular, such genomic information as **protein folds, domains, metabolic pathways** and the like; these units generally have a limited repertoire available in nature. Through assessing standardized attributes of each unit, e.g. function, usage and size, these units can be compared with other units both inter- and intra-genomically. This allows science, through investigating shared parts, to globally and comprehensively compare and examine organisms and their genomes.

Furthermore multiple methodologies can be applied to the surveyed information to analyze individual parts, including, for instance, **phylogenetic trees**. Approximately 500 fold types are presently known and the maximum number is thought to be in the few thousand. By calculating the number of these folds within genomes, and subsequently ranking them by their degree of usage by the organism, a researcher can compare diverse genomes, through their specific fold usage, thus determining evolutionary relatedness, or environmental conditions that favour distinct fold types.

Related websites

Partslist	www.partslist.org
GeneCensus	www.genecensus.org

Further reading

Gerstein M, Hegyi H (1998) Comparing genomes in terms of protein structure: surveys of a finite parts list. *FEMS Microbiol Rev* 22: 277–304.

Hegyi H, et al. (2002) Structural genomics analysis: characteristics of atypical, common, and horizontally transferred folds. *Proteins* 47: 126–141.

Qian J, et al. (2001) PartsList: a web-based system for dynamically ranking protein folds based on disparate attributes, including whole-genome expression and interaction information. *Nucleic Acids Res* 29: 1750–1764.

GENOMEGRAPHS, *SEE* GENE ANNOTATION, VISUALIZATION TOOLS.

GENOMICS

Stuart Brown

The application of high-throughput automated technologies to molecular biology and the resulting global approach to the analysis of all genes, all transcripts, or all proteins in an organism.

This is a very broad research area that encompasses many different technologies and types of data. The common theme is the rapid collection of large amounts of data from massively parallel experiments using laboratory robotics and bioinformatics. Genomics technologies include:

- Transcriptomics – the measurement of expression of all genes in an organism by microarray analysis of cDNA
- Proteomics – the measurement of all of the proteins in an organism and the investigation of all protein-protein interactions
- Toxicogenomics – the analysis of the effects of toxic substances on gene expression patterns and the prediction of toxicity based on gene expression studies in model systems and genetic predispositions to toxin sensitivity
- Pharmacogenomics – the study of the interaction between genotype (measured by genome wide SNP profiling) and the response of individuals to drugs.
- Functional Genomics – a high-throughput approach to determining the function of every gene and every protein in an organism and to understand the complex networks of metabolic pathways and regulatory control mechanisms
- Comparative Genomics – comparisons of whole genomes across species to identify similarities in gene sequences, gene structure, and gene expression

G

- Metabolomics – the study of intra- and/or extracellular small molecule content with particular emphasis on end and intermediate products of metabolism.

Further reading

Twyman RM, Primrose SB (2003) *Principles of Genome Analysis*. Blackwell Publishing, Oxford.

GENOTYPE IMPUTATION

Andrew Collins

Genotype imputation enables the testing of disease association with genetic markers that have not been directly genotyped. It is particularly useful for finer-scale mapping, increasing power and for combining evidence across samples (meta-analysis) that may have been genotyped on different platforms. Genome-wide association studies (GWAS) typically genotype only a fraction (~1–10%) of the single nucleotide polymorphisms (SNPs) actually present in an individual sample. Testing in this way using 'tagging' SNPs reduces costs but loses some information that may be recovered by imputation, the inference of 'missing' SNP genotypes by cross-referencing sample haplotypes with more comprehensively haplotyped samples, such as those represented in HapMap and 1000 genomes. Imputation, in the context of GWAS, relies on the identity by descent (IBD) sharing of identical or related short haplotypes between unrelated individuals by comparing sample SNP genotypes with reference panels. Alleles that are missing, because the marker has not actually been genotyped, can then be filled in from matching haplotypes. Where there is ambiguity about haplotype matches imputation programs quantify the ambiguity in probability. Error rates from imputed genotypes have been found to be less than 1% in validation studies. Imputation has been employed in the 1000 genomes project by employing the sequencing of a modest amount of whole-genome shotgun sequence reads (low coverage depth 2–4-fold) with imputation-based methods. In this way polymorphic sites can be identified with high accuracy in a way that is highly cost-effective.

Related websites

IMPUTE	http://mathgen.stats.ox.ac.uk/impute/impute.html
MACH	http://www.sph.umich.edu/csg/abecasis/MACH/tour/imputation.html

Further reading

Burdick JT, Chen WM, Abecasis GR, Cheung VG (2006) In *silico* method for inferring genotypes in pedigrees. *Nat. Genet.* 38: 1002–4.

Li Y, Willer C, Sanna S, Abecasis G (2009) Genotype imputation. *Annu. Rev. Genomics Hum. Genet.* 10: 387–406.

Li Y, Ding J, Abecasis GR (2006) Mach 1.0: rapid haplotype reconstruction and missing genotype inference. *Am. J. Hum. Genet.* 79: S2290.

Marchini J, Howie B, Myers S, McVean G, Donnelly P (2007) A new multipoint method for genome-wide association studies by imputation of genotypes. *Nat. Genet.* 39: 906–13.

Zeggini E, Scott LJ, Saxena R, et al. (2008) Meta-analysis of genome-wide association data and large-scale replication identifies additional susceptibility loci for type 2 diabetes.*Nat. Genet.* 40: 638–45.

See also Identical by Descent, Haplotype, 1000 Genomes Project, HapMap Project, Genome-Wide Association

GENSCAN

Enrique Blanco, Josep F. Abril and Roderic Guigó

GENSCAN is a general-purpose eukaryotic gene prediction program.

For each query sequence, the program determines the most likely gene structure under an underlying Generalized Hidden Markov Model (GHMM). The states in the GENSCAN GHMM correspond to the different functional units of a gene like exons, introns and splice sites. The transitions between the states ensure that the order in which the model visits various states is biologically consistent. In particular, exons are modeled according to a non-homogeneous 5th order Markov Model. Introns are modeled using a homogeneous 5th order Markov Model (see Coding Statistics). To model donor and acceptor splice sites, Maximal Dependence Decomposition is used to obtain a series of Weight Matrices which capture dependencies between positions in these sites.

In addition, the GENSCAN GHMM contains parameters that account for many higher order gene structural properties of genomic sequences, e.g. typical gene density, the typical number of exons per gene, the distribution of exon sizes for different types of exon. Separate sets of model parameters can be used to account for the many substantial differences in gene density and gene characteristics observed in distinct C+G% compositional regions of the human genome and the genomes of other vertebrates. Models exist also for maize and *Arabidopsis* sequences.

The publication of GENSCAN in 1996 was a breakthrough in the field of computational gene prediction. GENSCAN was not only the first program capable of analysing large genomic sequence encoding multiple genes in both strands, but it was also substantially more accurate than the existing programs at that time. GENSCAN is still one of most accurate *ab initio* gene prediction programs and it has been used to annotate many vertebrate and plant genomes. However, although its accuracy when analyzing single gene sequences is quite remarkable, when analysing large genomic sequences in vertebrate genomes the specificity of the program suffers, apparently predicting a large fraction of false positive exons (see Gene Prediction, accuracy).

Other gene prediction programs based on the GENSCAN infrastructure have been proposed in order to introduce additional information into the initial predictive architecture. Thus, GENOMESCAN integrates similarity to known protein sequences into GENSCAN (see Gene Prediction, homology-based), TWINSCAN performs comparative genomics using two genomes and NSCAN implements a similar prediction procedure with multiple informant genomes (see Gene Prediction, comparative).

The GHMM approach has its roots in the program GENEPARSER (Snyder and Stormo, 1993). Other eukaryotic gene prediction programs based on Hidden Markov Models are GENIE (Kulp et al., 1996), HMMGENE (Krogh, 1997), VEIL (Henderson et al., 1997), FGENESH (Salamov and Solovyev, 2000) and GENEMARK.HMM (Lukashin and Borodovsky, 1998).

G

Related websites

GENSCAN Home Page	http://genes.mit.edu/GENSCAN.html
NSCAN Home Page	http://mblab.wustl.edu/nscan

Further reading

Brent MR (2008) Steady progress and recent breakthroughs in the accuracy of automated genome annotation. *Nature Reviews Genetics* 9: 62–73.

Burge CB, Karlin S (1997) Prediction of complete gene structures in human genomic DNA. *J Mol Biol* 268: 78–94.

Gross SS, Brent MR (2006) Using multiple alignments to improve gene prediction. *J Comput Biol*. 13: 379–393.

Henderson J, et al. (1997) Finding genes in DNA with a hidden markov model. *J Comput Biol* 4: 127–141.

Korf I, Flicek P, Duan D, Brent MR (2001) Integrating genomic homology into gene structure prediction. *Bioinformatics*. 2001;17 Suppl 1:S140–148.

Krogh A (1997). Two methods for improving performance of an HMM and their application for gene finding. *ISMB*, 5: 179–186.

Kulp D, et al. (1996) A generalized hidden markov model for the recognition of human genes in DNA. In: States DJ, Agarwal P, Gaasterland T, et al. (eds) *Intelligent Systems for Molecular Biology*, pp. 134–142, *Menlo Park, California*. AAAI Press.

Lukashin A, Borodovsky M (1998) Genemark.hmm: new solutions for gene finding. *Nucleic Acids Res* 26: 1107–1115.

Salamov A, Solovyev V (2000) *Ab initio* gene finding in *Drosophila* genomic DNA. *Nucleic Acid Res* 10: 516–522.

Snyder EE, Stormo GD (1993) Identification of coding regions in genomic DNA sequences: an application of dynamic programming and neural networks. *Nucleic Acids Res* 21: 607–613.

See also Hidden Markov Models, Coding Statistics, Modelling of Sequence Signals, Gene Prediction, *ab initio*, Gene Prediction, homology-based, Gene Prediction, accuracy

GFF, *SEE* GENE ANNOTATION, FORMATS.

GFF2PS, *SEE* GENE ANNOTATION, VISUALIZATION TOOLS.

GFF3, *SEE* GENE ANNOTATION, FORMATS.

GIBBS SAMPLING, *SEE* MARKOV CHAIN MONTE CARLO.

GLOBAL ALIGNMENT

Jaap Heringa

An alignment of two or more sequences over their full lengths, as opposed to local alignment (Smith and Waterman, 1981), which includes a consecutive segment of each sequence. The first automatic method to generate a global alignment for two protein sequences was based upon the Dynamic Programming technique developed by Needleman and Wunsch (1970). Global alignment should be carried out when a sequence set is suspected to be linked through homologous relationships resulting from divergent evolution. If the sequences would have associated tertiary structures with multiple domains, it is important that the domain organisation is co-linear through all sequences and that all domains are present in all sequences.

An alternative to global alignment is *semi-global* alignment, which does not penalize so-called end gaps preceding or following a sequence. If zero end-gap penalty values are used to align the sequences, missing N- or C-terminal domains in one of the two sequences can be tolerated by the automatic alignment methods, in the sense that these domains are unlikely to become matched incorrectly with segments of the other sequence, which is more likely to happen in the case of global alignment, and is sometimes referred to as 'staggered overhang'.

Further reading

Needleman SB, Wunsch CD (1970) A general method applicable to the search for similarities in the amino acid sequence of two proteins. *J. Mol. Biol.* 48: 443–453.

Smith TF, Waterman MS (1981) Identification of common molecular subsequences. *J. Mol. Biol.* 147: 195–197.

See also Dynamic Programming, Optimal Alignment, Local Alignment

GLOBAL ORGANISATION FOR BIOINFORMATICS LEARNING, EDUCATION & TRAINING (GOBLET), *SEE* BTN.

GLOBULAR

Roman Laskowski and Tjaart de Beer

Proteins characterised by a compact 3D conformation of their polypeptide chain are called globular. The compactness is usually achieved by a tight packing of regions of regular secondary structures that places the protein's hydrophobic residues towards the inside of the protein, away from the surrounding solvent, forming a 'hydrophobic core', and the polar residues on, or close to, the surface.

Globular proteins differ markedly from polypeptides with random or simply repetitive conformation. Their overall structure tends to be stable under changes in temperature, pH, or pressure until a point is reached at which they 'denature', losing their globular conformation and consequently biological function.

They form the majority of naturally occurring proteins in solution, and are distinct from structural proteins such as the fibrous proteins.

See also Conformation, Secondary Structure, Residue, Polar, Fibrous

GO, *SEE* GENE ONTOLOGY.

GOBASE (ORGANELLE GENOME DATABASE)

Guenter Stoesser, Obi L. Griffith and Malachi Griffith

The Organelle Genome Database (GOBASE) organizes and integrates data related to organelles.

GOBASE contains all published nucleotide and protein sequences encoded by mitochondrial and chloroplast genomes, selected RNA secondary structures of mitochondria-encoded molecules, genetic maps, taxonomic information for all species whose sequences are present in the database and organismal descriptions of key protistan eukaryotes. Data are integrated and organized in a formal database structure allowing sophisticated biological queries using terms that are inherent in biological concepts. Maintenance of GOBASE ceased at the end of August 2010.

Related websites

GOBASE homepage	http://gobase.bcm.umontreal.ca/
The Organelle Genome Megasequencing Program (OGMP)	http://megasun.bch.umontreal.ca/ogmp/welcome.html

Further reading

O'Brien EA, et al. (2009) GOBASE: an organelle genome database. *Nucleic Acids Res*. 37 (Database issue): D946–950.

GOBLET, *SEE* BTN.

GOBO, *SEE* GLOBAL OPEN BIOLOGY ONTOLOGIES.

GOR SECONDARY STRUCTURE PREDICTION METHOD (GARNIER-OSGUTHORPE-ROBSON METHOD)

Patrick Aloy

The Garnier-Osguthorpe-Robson (GOR) is a method for protein secondary structure prediction from sequence information. It was one of the first automated implementations, a few years after the Chou & Fasman method, and also falls into the categories of 'first generation' structure prediction methods.

The original method (GOR-I) was based on single residue preferences and also took into account the protein type (i.e. relative abundance of helices, beta strands, turns and coils) obtained either by circular dichroism or from a preliminary prediction. All these considerations yielded a three-state-per-residue accuracy of 63%. More recent implementations of the GOR method (GOR-III) include propensities calculated for pairs of residues, increasing the accuracy up to 69.7%. However, more realistic calculations of the performance accuracy have placed the GOR-I and III at levels of 55% and 58% respectively.

Further reading

Garnier J, Osguthorpe DJ, Robson B (1974) Analysis of the accuracy and implications of simple methods for predicting the secondary structure of globular proteins. *J Mol Biol* 120: 97–120.

Gibrat JF, Garnier J, Robson B (1987) Further developments of protein secondary structure prediction using information theory. New parameters and consideration of residue pairs. *J Mol Biol* 198: 425–443.

See also Chou & Fasman Method, Secondary Structure Prediction

GRADIENT DESCENT (STEEPEST DESCENT METHOD)

Nello Cristianini

Simple optimization method for minimizing a function over a large set of parameters whenever such a function is differentiable.

Gradient descent consists of iteratively updating the current estimate, by moving it in the direction of steepest descent, provided by the gradient calculations.

If the function *f(x)* is to be minimized with respect to x, and if the current estimate of the minimum is x_0, a step in the direction of the steepest descent is given by $x \leftarrow x_0 - \alpha \nabla f(x_0)$ where $\nabla f(x_0)$ is the gradient of f at x_0 and α is a parameter that controls the size of such step.

For a suitable choice of the parameter α, this procedure is guaranteed to converge to a local minimum. Its major drawback is that if the function is non-convex, it will not necessarily find the global minimum, and furthermore the solution will be dependent on the initial conditions. Despite this problem, it has been used with success for example forming the basis of neural network training procedures like back propagation, that provides a way to compute the gradient in a feed-forward neural network.

G

Further reading

Bishop C (1996) *Neural Networks for Pattern Recognition*. Oxford University Press, Oxford.

Duda RO, Hart PE, Stork DG (2001) *Pattern Classification*, John Wiley & Sons, New York.

See also Neural Networks

GRAIL

Roderic Guigó

GRAIL is a suite of tools designed to provide analysis and putative annotation of DNA sequences.

The original GRAIL program (GRAIL1) (Uberbacher and Mural, 1991) is a coding region recognition program. It uses a back propagation neural network with two hidden layers to combine the output of seven coding statistics (see Coding Statistics). These statistics (sensors) are computed on a fixed length sliding sequence window. GRAIL can be accessed through a www interface or, more conveniently, through a dedicated X client (XGRAIL) which allows for a high degree of interactivity.

GRAIL2 (Xu et al., 1994a) is a more sophisticated version of GRAIL. It uses variable-length windows tailored to each potential exon candidate on the query sequence. Thus, while GRAIL1 simply predicts coding regions, GRAIL2 is already a coding exon prediction program.

GAP (Xu et al., 1994b) is the GRAIL gene assembly program. It takes exons predicted by GRAIL2 and assembles them into predicted genes.

The GRAIL programs above were extended to the GRAILEXP system (Xu and Uberbacher, 1997). This features a GRAIL-like exon finder, with improved splice site recognition. The gene assembly module has been improved by searching a database of known gene messages and building gene models based on the corresponding alignments. While its primary use is to locate protein-coding genes within DNA sequences, GRAILEXP can also locate EST/mRNA alignments, certain types of promoters, polyadenylation sites, CpG islands, and repetitive elements.

Related website

GRAILEXP	http://compbio.ornl.gov/grailexp/

Further reading

Uberbacher EC, Mural RJ (1991) Locating protein-coding regions in human DNA sequences by a multiple sensor-neural network approach. *Proc Natl Acad Sci USA* 88: 11261–11265.

Xu Y, et al (1994a) Recognizing exons in genomic sequence using GRAIL II. In: Setlow J (ed.) *Genetic Engineering: Principles and Methods*. Plenum Press.

Xu Y, et al (1994b) Constructing gene models from accurately predicted exons: An application of dynamic programming. *Comput Appl Biosci* 11: 117–124.

Xu Y, Uberbacher EC (1997) Automated gene identification in large-scale genomic sequences. *J Comput Biol* 4: 325–338.

See also Coding Region Prediction, Coding Statistics, Gene Prediction, *ab initio*, Gene Prediction, homology-based

GRAIL DESCRIPTION LOGIC

Robert Stevens

GRAIL is a **description logic** with hierarchies for both categories and relations, transitive relations and restricted forms of concept including **axiom**s.

The GALEN Terminology architecture also includes functionality to support lexical operations and multilingual support. The language has primarily been used to produce models of medical terminology (the GALEN Common Reference Model [CRM]). A suite of modelling tools known as the OpenKnoME are available for GRAIL modelling. The GRAIL language was also used to encode the **TAMBIS** ontology. In both GALEN and TAMBIS the **reasoning** facilities of the DL played a vital role inferring the ontology's **taxonomy** from the given descriptions.

Related website

GALEN Common Reference Model	http://www.openGALEN.org

GRAMENE

Dan Bolser

Gramene provides information for grass genomes and tools for identifying and understanding genes, pathways and phenotypes across grass species using manually curated relationships between cereals. Relationships are queried and displayed using controlled vocabularies and ontologies, including the Gene Ontology, Plant Ontology, Trait Ontology, and the Environment Ontology.

Phenotype data are based on QTLs. Sequence-based relationships are displayed using Ensembl technology, and maps are displayed using the comparative map viewer (CMap) from GMOD.

Related websites

GRAMENE	http://www.gramene.org/
MetaBase	http://metadatabase.org/wiki/Gramene

Further reading

Jaiswal P (2011) Gramene database: a hub for comparative plant genomics. *Methods Mol Biol.* 678: 247–275.

See also Ensembl Plants, MaizeGDB, PlantsDB, SGN

G

GRAPH MINING (FREQUENT SUB-GRAPH, FREQUENT SUB-STRUCTURE)

Feng Chen and Yi-Ping Phoebe Chen

Graphs have very wide applications in a variety of fields such as protein structure, biological networks, social networks and so forth. In graph mining, finding frequent sub-graphs, or frequent sub-structure is an important task. A frequent sub-graph is such a sub-graph the support of which is higher than the user-predefined threshold. Figure G.1 is an example of detecting frequent sub-structures in a chemical sample. Sub figures B and C represent the two frequent sub-graphs with supports of 2 and 3 respectively.

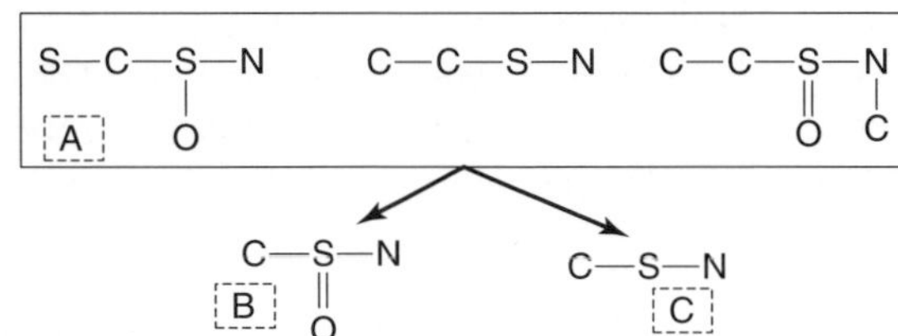

Figure G.1 An example of discovering frequent sub-graphs in a chemical graph database.

Frequent Sub-Graph mining is designed for finding the basic patterns in a graph dataset. In bioinformatics, if we want to mine the structure families in high-dimensional Protein structures, or detect sub-pathways in metabolism networks, then graph mining is a very useful tool.

GRAPH REPRESENTATION OF GENETIC, MOLECULAR AND METABOLIC NETWORKS

Denis Thieffry

Formally, a metabolic or a genetic network can be represented by a **connected graph**. **Vertices** (or **nodes**) then correspond to the different components (genes or metabolic species) of the networks, and the **edges** represent cross-interactions. Interactions can be oriented (one then uses the term **oriented edge** or arc), signed (plus or minus), or labeled in other ways according to functional characteristics (type of molecular mechanism, molecular family, etc.). One then refers to **directed**, **signed**, or **colored connected graphs**, respectively.

In graph theory, one distinguishes between different types of interactive structures or types of **graph topology**, each consisting of a particular type of connected graph: **trees**, **acyclic graphs**, **cyclic graphs**, etc. On another hand, a graph can also be characterized on a local level, by the number of edges involving the same vertex, called the **degree** of a vertex, or, in the case of oriented graphs, the number of incoming edges (**fan-in**) or of out-coming edges (**fan-out**).

When a graph describes regulatory or other types of molecular, metabolic or genetic interactions, one speaks of **graph of interactions**. Vertices then represent the different interacting species (genes, metabolic species, etc.), and the (un-)oriented) edges represent the interactions between these genes or molecular species.

Graph representation can also be used to represent transitions between different genetic or metabolic states. Vertices then represent the state of activity of a metabolic network (e.g. in terms of the different species levels or activities), or of the expression of a genetic network (e.g. in terms of the levels of expression of the different genes), and edges represent mandatory or potential transitions between these different states across time. A specific time-course of the system is thus represented by a pathway in this **state transition graph**. Whenever several edges come out of a single vertex (state), one faces alternative dynamical pathways.

In the context of graph theory, a **connected component** is defined as a maximal subset of edges of a (connected) graph such that the corresponding induced sub-graph cannot be disconnected by deleting any vertex. In the case of a directed graph, one can further define **strongly connected components** as sets of vertices such as each vertex can lead to any other vertex of the set by following the oriented edges. In metabolic or genetic interaction graphs, such connected components may correspond to biologically meaningful functional structures such as protein complexes (graph of protein-protein interactions), genetic regulatory module (genetic regulatory graph), etc.

One further defines a **circuit** of a graph as a subset of the **edges** of this graph forming a closed (circular) chain. From each vertex, a path then leads back to itself through all other vertices of the circuit. When dealing with directed (oriented) graphs, one also uses the term **cycle**. In biology, one distinguishes between two types of regulatory circuits, according to the kind of effect each element has on itself through the other elements of the circuit. On the one hand, each element can have an (indirect) inhibitory effect on itself, amounting to a **negative circuit**. On the other hand, when this effect is activatory, one has a **positive circuit**.

The effect of an element on itself is direct in the case of an **auto-regulation** (corresponding to a **loop** in graph theory), indirect otherwise. To these two classes of regulatory circuit correspond strikingly different dynamic and biological properties. Positive circuits are necessary to generate **multistable** behavior, and are involved in **differentiative** decisions. Negative circuits are needed to generate sustained **oscillatory behavior** and are involved in **homeostatic properties** of molecular and physiological systems. Negative circuits can also contribute to stabilize a steady state.

A **Bayesian graph** (or **Bayesian network**) is a directed acyclic graph in which vertices represent assertions and edges express oriented causal relationships. Such Bayesian networks are used to represent correlation in the expression of genes, which can be computed on the basis of extensive sets of large scale expression measurements (e.g. transcriptomics). Vertices then represent genes and edges expression correlation, and thus potential interactions. This approach has been recently used to reverse-engineer gene cross-regulations on the basis of DNA chip expression data.

Further reading

Diestel R (2000) *Graph Theory* (2nd ed.). Graduate Texts in Mathematics, Springer.

Gross J, Yelen J (1999) *Graph Theory and its Applications*. Discrete Mathematics & Applications Series, Rosen KH (ed.) CRC Press.

Kanehisa M (2000) *Post-Genome Informatics*. Oxford University Press, Oxford.

Thomas R, d'Ari R (1990) *Biological Feedback*. CRC Press.

GROUP I INTRON, *SEE* INTRON.

GROUP II INTRON, *SEE* INTRON.

GTF, *SEE* GENE ANNOTATION, FORMATS.

GUILT BY ASSOCIATION ANNOTATION, *SEE* ANNOTATION TRANSFER.

GUMBALL MACHINE

Thomas D. Schneider

A gumball machine is a model for the packing of Shannon spheres. Each gumball represents one possible message or molecular state (an After Sphere). The radius of the gumball represents the thermal noise. The balls are all enclosed inside a larger sphere (the Before Sphere) whose radius is determined from both the thermal noise and the power dissipated at the receiver (or by the molecule) while it selects that state. The way the spheres are packed relative to each other is the Coding.

Related website

Molecular machines	http://alum.mit.edu/www/toms/molecularmachines.html

Further reading

Schneider TD (1991) Theory of molecular machines. I. Channel capacity of molecular machines. *J. Theor. Biol.*, 148: 83–123. http://alum.mit.edu/www/toms/papers/ccmm/

Schneider TD (1994) Sequence logos, machine/channel capacity, Maxwell's demon, and molecular computers: a review of the theory of molecular machines. *Nanotechnology*, 5: 1–18. http://alum.mit.edu/www/toms/papers/nano2/

See also Channel Capacity, Molecular Efficiency, Molecular Machine Capacity

GWAS, *SEE* GENOME-WIDE ASSOCIATION STUDY.

G

GWAS CENTRAL

Guenter Stoesser, Malachi Griffith and Obi L. Griffith

GWAS Central, formerly HGVBASE (Human Genome Variation Database), includes genomic variation data and summarizes known variations in the human genome as a non-redundant set of records.

The primary purpose of GWAS Central is to facilitate genotype-phenotype association analyses that explore how single nucleotide polymorphisms (SNPs) and other common sequence variations may influence phenotypes such as common disease risk and drug response differences. Variations represented in GWAS Central encompass sequence changes known or suspected to exist in the human genome, including but not limited to SNPs, Indels, and simple tandem repeats and regardless of whether they are known or not known to be functionally neutral or pathogenic. Online search tools facilitate data interrogation by sequence similarity, keyword queries and genome coordinates. All variants have been uniquely mapped to the human draft genome sequence and are referenced to positions in DDBJ/EMBL/GenBank database entries. Additionally, GWAS Central serves as a central depository for mutation collection efforts undertaken by the Human Genome Variation Society (HGVS). GWAS Central is collaborating to develop data exchange protocols with other public variation and mutation databases, such as dbSNP and DGV. Individual genotypes and phenotypes are not provided – only groups of individuals.

Related websites

GWAS Central Homepage	http://www.gwascentral.org/
Human Genome Variation Society	http://www.hgvs.org

Further reading

Thorisson GA (2009) HGVbaseG2P: a central genetic association database. *Nucleic Acids Res.* 37: D797–802.

GWAS, SEE GENOME-WIDE ASSOCIATION STUDY

GWAS CENTRAL

h2, see Heritability.

Hand-curated Gene Annotation, see Gene Annotation, hand-curated.

Haplotype

HapMap Project

Hardy-Weinberg Equilibrium

Haseman-Elston Regression (HE-SD, HE-SS, HE-CP and HE-COM)

Hash, see Data Structure.

HAVANA (Human and Vertebrate Analysis and Annotation)

HE-COM, HE-CP, HE-SD, HE-SS, see Haseman-Elston Regression

Helical Wheel

Heritability (h2, Degree of Genetic Determination)

Heterotachy

HGMD (Human Gene Mutation Database)

HGT, see Horizontal Gene Transfer.

HGVBASE, see GWAS Central.

Hidden Markov Model (HMM, Hidden Semi-Markov Models, Profile Hidden Markov Models, Training of Hidden Markov Models, Dynamic Programming, Pair Hidden Markov Models)

Hierarchy

High-Scoring Segment Pair (HSP)

HIV RT and Protease Sequence Database, see STanford HIV RT and Protease Sequence Database.

HIV Sequence Database

HMM, see Hidden Markov Model.

HMMer

Homologous Genes

Homologous Superfamily

Homology

Homology Modeling, see Comparative Modeling.

Homology Search

Homology-based Gene Prediction, see Gene Prediction, homology-based.

Homozygosity Mapping

Horizontal Gene Transfer (HGT)

HSP, see High-scoring Segment Pair.

HTU, see Hypothetical Taxonomic Unit.

HUGO (The Human Genome Organization)

Human-Curated Gene Annotation, see Gene Annotation, hand-curated.

Human Gene Mutation Database, see HGMD.

Human Genome Variation Database, see HGVBASE.

Concise Encyclopaedia of Bioinformatics and Computational Biology, Second Edition.
Edited by John M. Hancock and Marketa J. Zvelebil.

Human Proteome Organization, see HUPO.

Human Variome Project (HVP)

HUPO (Human Proteome Organization)

HVP, see Human Variome Project.

Hydrogen Bond

Hydropathy

Hydropathy Profile (Hydrophobicity Plot, Hydrophobic Plot)

Hydrophilicity

Hydrophobic Moment

Hydrophobic Scale

Hydrophobicity Plot, see Hydropathy Profile.

HyPhy (Hypothesis Testing Using Phylogenies)

Hypothetical Taxonomic Unit (HTU)

H2, *SEE* HERITABILITY.

HAND-CURATED GENE ANNOTATION, *SEE* GENE ANNOTATION, HAND-CURATED.

HAPLOTYPE

Mark McCarthy, Steven Wiltshire and Andrew Collins

The set of alleles carried on a single chromosome: equivalently, the set of alleles inherited by an individual from a parent.

Haplotype information has been extensively used in the development of single nucleotide (SNP) panels for genome-wide association analysis and in the fine-scale analysis of disease association in specific genomic regions. Analysis at the haplotype level may, on occasion, be more powerful than analyses that examine loci individually. This is particularly so when a gene contains several interacting susceptibility variants, since the combined phenotypic effect may be best understood in terms of haplotypes. Because there is usually linkage disequilibrium between neighbouring loci, certain haplotypes will occur within a population more frequently than expected (on the basis of the individual allele frequencies). It is known that linkage disequilibrium in human populations is arranged into haplotype 'blocks' (regions of marked linkage disequilibrium and limited haplotype diversity separated by intervals that reflect high recombination of limited disequilibrium and extensive diversity). The development of the haplotype map (HapMap) of the genome, charts such blocks and their haplotypic composition. This map has been instrumental in the development of genotype panels for genome-wide association analysis. Since current linkage disequilibrium patterns and haplotype structures reflect the cumulative history, population bottlenecks and migration of the human species, haplotype analyses have been invaluable in helping to reconstruct this history.

Examples: Horikawa et al. (2000) demonstrated association between variants in the calpain 10 gene and type 2 diabetes in Mexican American and European subjects, but the strongest relative risk was observed with particular haplotype combinations: these appeared to explain risk much better than any single marker allele. Rioux et al. (2001) were able to localise a Crohn's disease susceptibility locus to a single haplotype block on chromosome 5q, and to identify the most-strongly associated haplotype: however, almost complete linkage disequilibrium between multiple variants on this haplotype frustrated efforts to determine which variant site was functionally responsible.

Related websites

Programs for haplotype determination:

PHASE	http://www.stats.ox.ac.uk/mathgen/software.html
SNPHAP	http://www-gene.cimr.cam.ac.uk/clayton/software/
HapMap	http://www.ncbi.nlm.nih.gov/SNP/HapMap/index.html
LDMAP	http://cedar.genetics.soton.ac.uk/public_html/helpld.html

H

Further reading

Gabriel SB, et al. (2002) The structure of haplotype blocks in the human genome. *Science* 296: 2225–2229.

Horikawa Y, et al. (2000) Genetic variation in the gene encoding calpain-10 is associated with type 2 diabetes mellitus. *Nat. Genet.* 26: 163–175.

Jakkula E, Rehnström K, Varilo T, Pietiläinen OPH, Paunio T, Pedersen NL, deFaire U, Järvelin M-R, Saharinen J, Freimer N, Ripatti S, Purcell S, Collins A, Daly MJ, Palotie A, Peltonen L (2008) The genome-wide patterns of variation expose significant substructure due to multiple bottlenecks in a founder population. *Am. J. Hum. Genet.*, 83: 787–94.

Johnson GCL, et al. (2001): Haplotype tagging for the identification of common disease genes. *Nat Genet* 29: 233–237.

Patil N, et al. (2001): Blocks of limited hapltoype diversity revealed by high-resolution scanning of human chromosome 21. *Science* 294: 1719–1723.

Rioux JD, et al. (2001) Genetic variation in the 5q31 cytokine gene cluster confers susceptibility to Crohn disease. *Nat. Genet.* 29: 223–228.

Service S, DeYoung J, Karayiorgou M, et al. (2006) Magnitude and distribution of linkage disequilibrium in population isolates and implications for genome-wide association studies, *Nature genetics* 38 (5): 556–560.

See also Phase, Linkage Disequilibrium, Linkage Disequilibrium Map, HapMap project, Association Analysis

HAPMAP PROJECT

Andrew Collins

The International HapMap project has genotyped single nucleotide polymorphisms (SNPs) at high density in DNA samples derived from a range of human populations in order to characterize their genetic structure. Output from the project has been crucial for the development of panels of 'tag' SNPs used to screen case and control samples in genome-wide association studies. Because human populations show limited diversity of haplotypes in many genomic locations, tag SNPs can be selected to uniquely identify these haplotypes and avoid the need to genotype all variants in a region. Thus the genome can be screened effectively with a few hundred thousand tag SNPs rather than the full complement of several million SNPs. The first haplotype map was published in 2005 and genotyped 269 DNA samples from the Yoruba population (Nigeria), Utah (USA), Han Chinese in Beijing (China) and Japanese in Tokyo (Japan). This 'phase I' sample strongly supported the evidence that patterns of LD in human were 'block-like' in regions of low recombination punctuated by recombination hot-spots which locally increase haplotype diversity. Such a pattern was determined for all human populations and LD structure shown to be overwhelmingly determined by recombination.

Related website

International HapMap Project	http://hapmap.ncbi.nlm.nih.gov/

Further reading

Jeffreys AJ, Kauppi L, Neumann R (2001) Intensely punctate meiotic recombination in the class II region of the major histocompatibility complex. *Nat. Genet.* 29(2): 217–22.

Service S, DeYoung J, Karayiorgou M, et al. (2006) Magnitude and distribution of linkage disequilibrium in population isolates and implcations for genome-wide association studies. *Nat Genet* 38(5): 556–560.

Tapper W, Collins A, Gibson J, Maniatis N, Ennis S, Morton NE (2005) A map of the human genome in linkage disequilibrium units. *Proc. Natl. Acad. Sci. USA*, 102(33): 11835–11839.

The International HapMap Consortium (2005) A haplotype map of the human genome. *Nature* 437: 1299–1320.

See also Genotype Imputation, Genome-Wide Association

HARDY-WEINBERG EQUILIBRIUM

A.R. Hoelzel

Description of the population-level equilibrium frequencies of a gene having two or more alleles.

Formulated by an English mathematician, G.H. Hardy and a German physician, W. Weinberg, this equilibrium condition is based on the observation that given certain conditions (random mating and the lack of factors that change allele frequency such as gene flow, mutation, natural selection and genetic drift), genotype frequencies for diploid organisms can be represented by binomial (2 alleles) or multinomial (>2 alleles) functions of allele frequencies. Given two alleles, A_1 and A_2 with frequencies p and q, respectively, the following relationship holds:

Genotypes:	A_1A_1	A_1A_2	A_2A_2	
Genotype Frequencies:	p^2	$+ \; 2pq \; +$	q^2	$= 1$

(expressed in terms of allele frequencies)

This relationship is stable over time, and will revert to the same genotype frequencies after one generation of random mating in a sexual species, or to new genotype frequencies if allele frequencies are changed. Important implications for the study of population genetics include the fact that genotype frequencies can be determined from allele frequencies alone, and that inference about evolutionary processes can be made on the basis of deviation from Hardy-Weinberg equilibrium genotype frequencies.

Further reading

Hartl DL (2000) *A Primer of Population Genetics*. Sinauer Associates, Sunderland.

See also Evolution, Natural Selection

H

HASEMAN-ELSTON REGRESSION (HE-SD, HE-SS, HE-CP AND HE-COM)

Mark McCarthy and Steven Wiltshire

An allele-sharing regression based method for detecting linkage between a marker and a quantitative trait locus.

The original Haseman-Elston regression method (HE-SD) regressed the squared difference of the sibs' quantitative trait values (Y_{ij}) on the estimated proportion of alleles shared identical by descent (IBD) by the sib pair ($\hat{\pi}_{ij}$):

$$Y_{ij} = \alpha + \beta\hat{\pi}_{ij}$$

The gradient of the regression line, β, equals $-2\sigma_g^2(1-2\theta)^2$, where θ is the recombination fraction between the quantitative trait locus (QTL) and the marker, $(1-2\theta)^2$ is the correlation between the proportion of alleles shared IBD at the QTL and the marker locus, and $2\sigma_g^2$ is the proportion of the trait variance attributable to the QTL. The hypothesis $\boldsymbol{\theta} = 0$ (no linkage) can be tested with the ratio of β to its standard error, distributed as Student's t statistic. A significantly negative value of β is evidence for linkage.

Several modifications of the original squared differences Haseman-Elston method exist, differing in the nature of the dependant variable: HE-CP, in which Y_{ij} is the cross product of the mean-centred sibs' trait values; HE-SS, in which Y_{ij} is the square of the sum of the sibs' mean-centred trait values; and HE-COM, in which Y_{ij} is a weighted combination of the squared differences and squared sums. Methods combining squared sums and squared differences have the greatest power. All these methods have been implemented for sib-pairs or small sibships. However, a combined SD-SS implementation that is applicable to general pedigrees has been developed recently.

Examples: Daniels et al. (1996) used the original H-E regression to examine quantitative phenotypes related to asthma and atopy, and found evidence for QTL on chromosome 13 influencing serum levels of IgE and on chromosome 7 influencing bronchial responsiveness to metacholine challenge. Using HE-COM (the combined squared sums/squared differences implementation) Wiltshire et al. (2002) observed evidence for a QTL on chromosome 3 influencing adult stature in a sample of UK pedigrees.

Related websites

Programs for performing original or combined HE regression analyses:

SAGE	http://darwin.cwru.edu/sage/
GENEHUNTER2	http://www.broadinstitute.org/ftp/distribution/software/genehunter/
MERLIN	http://www.sph.umich.edu/csg/abecasis/Merlin/

Further reading

Daniels SE, et al. (1996) A genome-wide search for quantitative trait loci underlying asthma. *Nature* 383: 247–250.

Drigalenko E (1998) How sib pairs reveal linkage. *Am. J. Hum. Genet.* 63: 1242–1245.

Elston RC, et al. (2000) Haseman and Elston revisited. *Genet. Epidemiol.* 19: 1–17.

Haseman JK, Elston RC (1972) The investigation of linkage between a quantitative trait and a marker locus. *Behav. Genet.* 2: 3–9.

Sham PC, Purcell S (2001) Equivalence between Haseman-Elston and variance components linkage analyses for sib pairs. *Am. J. Hum. Genet.* 68: 1527–1532.

Sham PC, et al. (2002) Powerful regression-based quantitative trait linkage analysis of general pedigrees. *Am. J. Hum. Genet.* 71: 238–253.

Visscher PM, Hopper JL (2001) Power of regression and maximum likelihood methods to map QTL from sib-pair and DZ twin data. *Ann. Hum. Genet.* 65: 583–601.

Wiltshire S, et al. (2002) Evidence for linkage of stature to chromosome 3p26 in a large UK family data set ascertained for type 2 diabetes. *Am. J. Hum. Genet.* 70: 543–546.

See also Quantitative Trait, Variance Components, Genome Scan, Identity by Descent, Allele-Sharing Methods

HASH, *SEE* DATA STRUCTURE.

HAVANA (HUMAN AND VERTEBRATE ANALYSIS AND ANNOTATION)

John M. Hancock

The HAVANA group, based at the Wellcome Trust Sanger Institute, provides manual curation of genes in vertebrate genomes. Genomes currently undergoing curation include Human, Mouse and Zebrafish. HAVANA annotations are deposited in the VEGA database and shared with the ENSEMBL and UCSC genome browsers.

Related website

HAVANA	http://www.sanger.ac.uk/research/projects/vertebrategenome/havana/

HE-COM, HE-CP, HE-SD, HE-SS, *SEE* HASEMAN-ELSTON REGRESSION

HELICAL WHEEL

Teresa K. Attwood

Typically, a circular graph containing five turns of α-helix (~20 residues), around which successive residues in a protein sequence are plotted 100 degrees apart (3.6 residues per 360° turn), as shown in Figure H.1. Different parameters can be

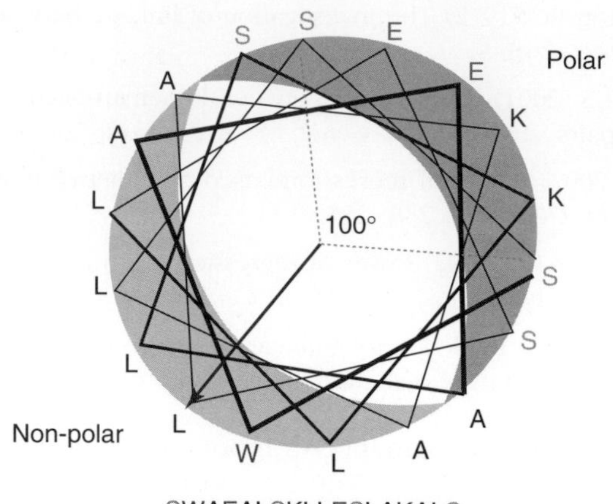

Figure H.1 Illustration showing how the amino acids in a particular 18-residue sequence are distributed around the rim of a helical wheel. The angular separation of the amino acid residues is 100° because this wheel depicts an alpha helical conformation. As shown, the polar residues within the given sequence fall on one side of the helix, and the non-polar residues on the other, a distribution characteristic of an amphipathic helix. For the amino acid residue colour scheme, see Visualisation Of Multiple Sequence Alignments. (See Colour plate H.1)

used to depict helices with different pitch (*e.g.*, 120° for a 3_{10} helix, 160° for β-strands). The graph is plotted such that the viewer looks down the helical axis (see Figure H.1).

The principal use of such graphs is in the depiction of amphipathic helices (*i.e.*, those with both hydrophobic and hydrophilic character). Helical potential is recognised by the clustering of hydrophilic and hydrophobic residues in distinct polar and non-polar arcs. Such arrangements may be seen, for example, in globular proteins, when a helix lies near the surface of the molecule, with one side facing the exterior solvent and the other facing the hydrophobic protein interior. Amphipathic arrangements may also be seen in leucine zipper motifs, where successive leucines, seven residues apart in sequence, form a hydrophobic interface along the helix length, which mediates dimerisation with similar zipper motifs.

Helical wheels may illustrate the presence of amphiphilic character, but are not, in themselves, predictive of helical conformations.

Related website

Helical wheel plotting tool	http://heliquest.ipmc.cnrs.fr/cgi-bin/ComputParamsV2.py

Further reading

Attwood TK, Parry-Smith DJ (1999) *Introduction to bioinformatics.* Addison Wesley Longman, Harlow, Essex, UK.

Mount DM (2004) *Bioinformatics: Sequence and Genome Analysis* (2 ed.). Cold Spring Harbor, NY, Cold Spring Harbor Laboratory Press.

Schiffer M, Edmundson AB (1967) Use of helical wheels to represent the structures of proteins and to identify segments with helical potential. *Biophys. J.* 7: 121–135.

Shultz GE, Schirmer RH (1979) *Principles of Protein Structure*. Springer-Verlag, New York.

See also αHelix, Amphipathic, βStrand, Hydropathy, Hydrophilicity

HERITABILITY (H2, DEGREE OF GENETIC DETERMINATION)

Mark McCarthy, Steven Wiltshire and Andrew Collins

The heritability of a quantitative trait is that proportion of the total variation in a quantitative trait attributable to genetic factors.

If the total variance of the trait is given by σ_p^2, and the total genetic variance (including additive, dominance and epistatic components) is denoted σ_g^2, then broad heritability (or degree of genetic determination – the proportion of the trait variance attributable to all genetic components) is given by σ_g^2/σ_p^2. Narrow heritability (often just called heritability) is the proportion of the trait attributable to the additive genetic variance, σ_a^2, and is given by σ_a^2/σ_p^2. Heritability estimates vary between 0 (no genetic component to the trait) to 1 (no environmental component to the trait). Heritability can be estimated with standard linear regression (of offspring on parent or mid-parent), correlation (between half sibs or full sibs), DeFries-Fulker regression (from twin data), variance components analysis (of pedigree data), and structural equation modelling (of twin or other relative data).

Example: Hirschhorn et al. (2001) calculated heritability estimates for adult stature from samples of Finnish, Swedish and Quebec pedigrees using variance components analysis, to be >95%, 80% and 70%, respectively.

Related websites

Programs that can be used for calculating heritabilities include:

SOLAR	http://txbiomed.org/departments/genetics/genetics-detail?r=37
SAGE	http://darwin.cwru.edu/
MX	http://griffin.vcu.edu/mx/

Further reading

DeFries JC, Fulker DW (1985) Multiple regression analysis of twin data. *Behav. Genet.* 15: 467–473.

DeFries JC, Fulker DW (1988) Multiple regression analysis of twin data: Etiology of deviant scores versus individual differences. *Acta. Genet. Med. Gemellol.* 37: 205–216.

Falconer DS, MacKay TFC (1996) *Introduction to Quantitative Genetics*. Harlow: Prentice Hall, pp 160–183.

Hirschhorn JN, et al. (2001) Genomewide linkage analysis of stature in multiple populations reveals several regions with evidence of linkage to adult height. *Am. J. Hum. Genet.* 69: 106–116.

Plomin R, et al. (2001) *Behavioral genetics*. New York: Worth.

See also Quantitative Trait, Variance Components, Polygenic Inheritance

H

HETEROTACHY

Aidan Budd and Alexandros Stamatakis

A likelihood-based model where different subtrees of a tree evolve under distinct substitution models.

The term heterotachy is orthogonal to among site rate heterogeneity. It denotes the phenomenon that evolutionary rates and substitution models can change along a phylogeny over time. Hence, different subtrees of a phylogeny may evolve according to different models, when climatic conditions on earth were different for instance. The key problems are how to determine the number of models under which a tree evolves and how to assign different models to different regions (subtrees) of a phylogeny. Moreover, if the phylogeny changes during a tree search, the number of models and their assignments to an altered tree topology will need to be re-computed. This is the reason why heterotachy has so far mostly been applied to fixed tree topologies for estimating divergence times. Moreover, because of the high model complexity heterotachy software mainly relies on Bayesian statistics and mostly on reversible-jump MCMC techniques.

DPPDIV software	http://cteg.berkeley.edu/software.html#dppdiv

Software

Pagel M, Meade A (2008) Modelling heterotachy in phylogenetic inference by reversible-jump Markov chain Monte Carlo. *Philosophical Transactions of the Royal Society B: Biological Sciences* 363(1512): 2955–3964.

Further reading

Kolaczkowski B, Thornton J.W. (2008) A mixed branch length model of heterotachy improves phylogenetic accuracy. *Molecular Biology and Evolution* 25(6): 1054–1066.

Lopez P, Casane D, Philippe H (2002) Heterotachy, an important process of protein evolution. *Molecular Biology and Evolution* 19(1): 1–7.

See also Rate Heterogeneity, Mixture Models

HGMD (HUMAN GENE MUTATION DATABASE)

Malachi Griffith and Obi L. Griffith

The Human Gene Mutation Database is an attempt to compile and organize observations of gene mutations responsible for inherited human diseases. Somatic mutations and mutations in the mitochondrial genome are excluded from the database. Each mutation is required to have direct supporting evidence from DNA analysis. Despite use of the word 'mutation' in the database name, HGMD includes records corresponding to disease-associated polymorphisms. At the same time, mutations without a demonstrated phenotypic consequence are usually not included. In cases where the phenotypic consequence of a mutation is not obviously pathological based on the predicted effect on protein structure, additional evidence of a clinical phenotype may be required for inclusion in HGMD in an effort to ensure that the majority of mutations have a causal relationship to a human disease. Mutation types include: missense/nonsense, splicing, regulatory, small insertions and deletions, large insertions and deletions, complex rearrangements, and repeat variations. The HGMD database is manually curated and is available in two forms: a less frequently updated 'public' version maintained at the Institute of Medical Genetics in Cardiff and a second 'professional' version maintained by a commercial partner, BIOBASE. The 'public' version is only freely available to registered users employed by an academic/non-profit organization. At the time of writing the 'public' version contains 85,840 entries and the 'professional version' 120,004 entries.

Related websites

HGMD homepage	http://www.hgmd.org/
HGMD professional	http://www.biobase-international.com/product/human-gene-mutation-database

Further reading

Cooper DN, et al. (2006). The Human Gene Mutation Database (HGMD) and its exploitation in the study of mutational mechanisms. *Curr. Protoc. Bioinformatics*. Chapter 1:Unit 1.13.

George RA, et al. (2008) General mutation databases: analysis and review. *J. Med. Genet.* 45(2): 65–70.

Stenson PD, et al. (2008) The Human Gene Mutation Database: 2008 update. *Genome Med.* 1(1): 13.

HGT, *SEE* HORIZONTAL GENE TRANSFER.

HGVBASE, *SEE* GWAS CENTRAL.

HIDDEN MARKOV MODEL (HMM, HIDDEN SEMI-MARKOV MODELS, PROFILE HIDDEN MARKOV MODELS, TRAINING OF HIDDEN MARKOV MODELS, DYNAMIC PROGRAMMING, PAIR HIDDEN MARKOV MODELS)

Irmtraud Meyer

A Hidden Markov Model (HMM) is a probabilistic method for linearly analyzing sequences.

H

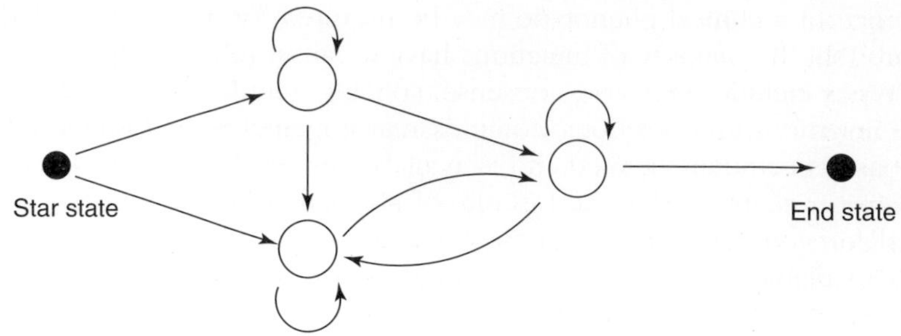

Figure H.2 Figurative representation of a Hidden Markov Model. See following text for details.

A Hidden Markov Model (HMM) M consists of a fixed number of states $Q = \{q_1, \ldots, q_n\}$ which are connected by directed transitions (see Figure H.2). Each state i reads a fixed number of letters Δ_i from an input string S of length L over an alphabet A. States that read no letters are called silent states. Each run of M can be described by a state sequence $\Pi = (\pi_1, \ldots, \pi_m)$ (the state path) which starts in a special silent state (the start state) and proceeds from state to state via transitions between the states until all letters of the input sequence S have been read. The state path ends in another special silent state (the end state). Each state path can be translated into a labeling of the input sequence by assigning to each letter the label of the state that has read the letter. With each transition from a state s to a state t there is associated a transition probability $p_s(t)$ and with every reading of a substring S' by a state s there is associated an emission probability $p_s(S')$. The probabilities of all transitions emerging from each state add up to one and the emission probabilities for reading all possible substrings add up to one for every state. The overall probability of a sequence given a state path is the product of the individual transition and emission probabilities encountered along the state path.

For a given HMM M and an input sequence S, a variety of Dynamic Programming Algorithms can be used to infer different probabilities: The Viterbi Algorithm can be used to determine the most probable state path Π^* (the Viterbi path). This algorithm has a memory and time requirement of the order of the sequence length. Using the Forward and Backward Algorithm, we can calculate the conditional probability that sequence position i is labeled by state k (the posterior probability of state k at position i in the sequence). From these probabilities, the most probable label of each letter in the input sequence can be derived.

When setting up an HMM for a given classification problem, one first defines the states and the transitions between them. The states reflect the different classes which one aims to distinguish and the transitions between them determine all possible linear successions in which the labels may occur in the input sequences. Once the structure of the HMM has been fixed, the transition and emission probabilities have to be specified. This can be done in a variety of ways depending on which information is available on a sufficiently large and representative set of sequences (the training set). If state paths are know which reproduce the correct classification of the training sequences, we can use Maximum Likelihood Estimators to derive the emission and transition probabilities of the HMM. If these state paths are not known, the emission and transition probabilities can be estimated in an iterative way using the Baum-Welch Algorithm. This algorithm iteratively optimizes the parameters such that the training sequences have the maximum likelihood under the model. Unfortunately, the Baum-Welch algorithm can only be shown to converge to a local maximum which may therefore depend on the initial values of the parameters. The Baum-Welch algorithm is a special case of the Expectation Maximization (EM) Algorithm. Instead of optimizing the likelihood of the sequences given the model as is done by the Baum-Welch algorithm, we can alternatively choose to maximize the likelihood of the sequences under the model considering only the Viterbi paths (Viterbi-Training).

HMMS can be expressed as regular grammars. These grammars form the class of transformational grammars within the Chomsky hierarchy of transformational grammars which have the most restrictions. A regular grammar is a set of production rules which analyse a sequence from left to right. The next and more general class of grammars are context-free grammars. They comprise regular grammars as special case, but are also capable of modeling long range correlations within a sequence (this feature is, for example, needed to model the long-range correlations imposed by the secondary structure of an RNA sequence). The computational parsing automatons corresponding to regular grammars are finite state automatons.

Variants of Hidden Markov Models

The above concept of HMMs can be easily extended to analyze several (k) input sequences instead of just one (for k=2 sequences they are called Pair Hidden Markov Models). Each state of a k-HMM then reads substrings of fixed but not necessarily the same length from up to k sequences. The time and memory requirements of the Viterbi algorithm are of the order of the product of the sequence lengths. The memory requirement can be reduced by a factor of one sequence length by using the Hirschberg Algorithm.

Another variant of HMMs are Hidden Semi-Markov Models (HSMMs) which explicitly model the duration of time spent in a state by a probability distribution. The time requirement of the Viterbi algorithm for an HSMM is of the order of the cube of the sequence length in the most general case. The memory requirement is still of the order of the sequence length.

Profile HMMs are another type of HMM. They model a known alignment of several related sequences and are used to test whether or not another sequence belongs to the set of sequences.

Use of HMMs in Bioinformatics

HMMs are used in Bioinformatics for almost any task that can be described as a process which analyses sequences from left to right. Applications of HMMs to biological data include sequence alignment, gene finding, protein secondary structure prediction and the detection of sequence signals such as, for example, translation initiation sites. HMMs are used for aligning sequences in order to detect and test potential evolutionary, functional or other relationships between several sequences (e.g. DNA to DNA alignment, protein to DNA alignment, protein to protein alignment or the alignment of a family of already aligned related proteins to a novel protein sequence). In *ab initio* gene prediction, one input DNA sequence is searched for genes, i.e. classified into protein coding, intronic and intergenic subsequences which form valid splicing patterns. In homology based gene prediction, a known protein sequence is used to predict the location and splicing pattern of the corresponding gene in a DNA sequence. Recently, comparative gene finding methods which search two related DNA sequences simultaneously for pairs of evolutionarily related genes have been developed. HMMs are particularly well suited for complex tasks such as gene finding as they can incorporate probabilistic information derived by dedicated external programs (programs which, for example, predict potential splice sites or translation start signals) into the HMM in order to improve its predictive power.

Further reading

The following two books introduce HMMs in the context of applications to biological data and contain many references which provide good entry points into the original literature:

Durbin R, et al. (1998) *Biological Sequence Analysis*, Cambridge University Press (equally well suited for readers with a biological and mathematical background).

Rabiner L (1989) A tutorial on hidden Markov models and selected applications in speech recognition, *Proc. of the IEEE*, 77: 257–285.

Waterman MS (1995) *Introduction to Computational Biology*, Chapman & Hall (for readers with a mathematical background).

See also, Markov Model

HIERARCHY

Robert Stevens

A grouping of concepts in tree-like structures.

A **tree** structure is a hierarchy in which there is only one parent for each item (also known as a **strict hierarchy, simple hierarchy** or **pure hierarchy**. A hierarchy has an explicit notion of top and bottom and so is often referred to as 'directed'. A tree may be seen in Figure H.3.

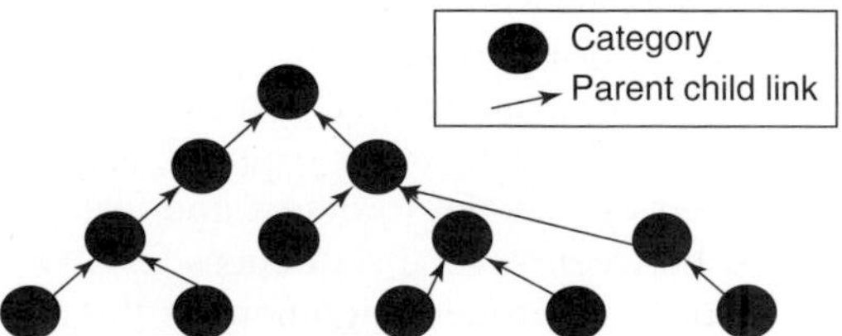

Figure H.3 Organization of concepts in a tree.

HIGH-SCORING SEGMENT PAIR (HSP)

Jaap Heringa

Term introduced by the authors of the homology searching method BLAST (Altschul et al., 1990). High-scoring segments pairs (HSPs) are gapless local pair-wise alignments found by the BLAST method to score at least as high as a given threshold score. An HSP is the fundamental unit of BLAST algorithm output and consists of two sequence fragments of arbitrary but equal length whose alignment is locally maximal and for which the alignment score is beyond a given cut-off score. The alignment score is dependent on a scoring system, which is based on a residue exchange matrix. In the context of the BLAST algorithm, each HSP consists of a segment from the query sequence and one from a database sequence. The sensitivity and speed of the programs can be adjusted using the standard BLAST parameters W, T, and X (Altschul et al., 1990). Selectivity of the programs can be adjusted via the cut-off score.

The approach to similarity searching taken by the BLAST programs is first to look for similar segments (HSPs) between the query sequence and a database sequence, then to evaluate the statistical significance of any matches found, and finally to report only those matches that satisfy a user-selectable threshold of significance. Multiple HSPs involving the query sequence and a single database sequence can be treated statistically in a variety of ways. The BLAST program uses 'Sum' statistics (Karlin and Altschul, 1993), where it is possible that the statistical significance attributed to a set of HSPs is higher than that of any individual member of the set. This corresponds to the fact that occurrence of similarity over a wider region in the alignment of a query and database sequence raises the chance that the latter sequence is a true homologue to the former. A match will only be reported to the user if the significance score is below the user-selectable threshold (E-value).

Further reading

Altschul SF, Gish W, Miller W, Meyers EW, Lipman, DJ (1990) Basic local alignment search tool. *J. Mol. Biol.* 215: 403–410.

Karlin S, Altschul SF (1993) Applications and statistics for multiple high-scoring segments in molecular sequences. *Proc. Natl. Acad. Sci. USA* 90: 5873–5877.

See also Homology Search, BLAST, E-value

HIV RT AND PROTEASE SEQUENCE DATABASE, *SEE* STANFORD HIV RT AND PROTEASE SEQUENCE DATABASE.

HIV SEQUENCE DATABASE

Guenter Stoesser, Malachi Griffith and Obi L. Griffith

The HIV databases contain data on HIV genetic sequences, immunological epitopes, drug resistance-associated mutations, and vaccine trials.

The website gives access to a large number of tools that can be used to analyse these data. The HIV Sequence Database's primary activities are the collection of HIV and SIV sequence data (since 1987), their curation, annotation and computer analysis as well as the production of sequence analysis software. Data and analyses are published in electronic form and also in a yearly printed publication, the HIV Sequence Compendium. A companion database, the HIV Molecular Immunology Database provides a comprehensive annotated listing of defined HIV epitopes. The HIV Sequence Database is maintained at the Los Alamos National Laboratory, New Mexico, USA. In 2011, a next generation sequence archive was added.

Related website

HIV Database homepage	http://www.hiv.lanl.gov/content/index

Further reading

Kuiken C, et al. (eds) (2010) *HIV Sequence Compendium 2010*. Theoretical Biology and Biophysics Group, Los Alamos National Laboratory, NM, LA-UR 10–03684.

HMM, *SEE* HIDDEN MARKOV MODEL.

HMMER

J.M. Hancock and M.J. Bishop

A software package using profile Hidden Markov Models to carry out database searching.

HMMer accepts a multiple alignment as input and uses this to construct a Hidden Markov Model profile of the sequences represented by the alignment. This HMM can then be used to search sequence databases. Version 2.2 of the package (August 2001) consisted of nine main programs: hmmalign, which aligned sequences to a pre-existing model; hmmbuild, which constructed an HMM from a sequence alignment; hmmcalibrate, which calculated E-scores for the database search; hmmconvert, which converted a HMM into a variety of formats; hmmemit, which emitted sequences from a given HMM; hmmfetch, which got an HMM from a database; hmmindex, which indexed an HMM database; hmmpfam, which searched an HMM database for matches to a sequence; and hmmsearch, which searched a sequence database for matches to an HMM.

Version 3 of the package, which was claimed to be as fast as BLAST, was released in 2009.

Related website

HMMER Website	http://hmmer.janelia.org/

Further reading

Eddy SR (1998) Profile hidden Markov models. *Bioinformatics* 14: 755–763.

Eddy SR (2011) Accelerated Profile HMM Searches *PLoS Comput. Biol.* 7(10): e1002195.

See also Hidden Markov Models

HOMOLOGOUS GENES

Austin L. Hughes and Laszlo Patthy

Two genes (or genomic regions) are said to be homologous if they are descended from a common ancestral gene (or genomic region).

In the case of protein-coding genes, the proteins encoded by homologous genes can be said to be homologous. The best evidence of homology of genes is statistically significant sequence similarity that can be explained only by common ancestry. Strictly speaking, homology is an either-or condition. Thus, it is incorrect to speak of about 'percent homology' between of two aligned gene sequences. It should be noted, however, that in the case of genes encoding multidomain proteins it is frequently observed that homology does not hold for the entire length of the related genes: they display only partial homology or local homology.

Homologous genes or proteins of different species that evolved from a common ancestral gene by speciation are said to be orthologs, homologous genes acquired through horizontal transfer of genetic material between different species are called xenologs, homologous genes that arose by duplication within a genome are called paralogs. Partially homologous genes that are related only through the independent acquisition of homologous genetic material are called epaktologs.

See also Homology, Epaktolog, Ortholog, Paralog, Pseudoparalog, Xenolog

HOMOLOGOUS SUPERFAMILY

Andrew Harrison and Christine Orengo

A homologous superfamily consists of a set of proteins related by divergent evolution to a common ancestral protein. However, in contrast to a protein family, the superfamily contains more distant relatives and these may have no detectable sequence similarity. In some remote homologues, the function may also have changed, particularly in paralogous proteins. In order to detect these very remote homologues, sensitive sequence profiles must be used or the structures of proteins must be compared, where they are known. Since structure is much more highly conserved than sequence during evolution, distant homologues can usually be detected from their structural similarity. However, to distinguish homologues from analogues, it is necessary to search for additional evidence to support homology. This may be some unusual sequence pattern associated with ligand binding or some rare structural motif unlikely to have arisen twice by chance within the same analogous structure. Profile or HMM based methods are more sensitive in detecting very remote homologues than pair-wise sequence alignment and can be used to confirm homology suggested by structural similarity. HMM-HMM approaches are currently the most powerful sequence based approaches.

Relevant websites

HOMSTRAD	http://tardis.nibio.go.jp/homstrad/
HHblits	http://toolkit.tuebingen.mpg.de/hhblits

H

Further reading

Chothia C, Gough, J (2009) Genomic and structural aspects of protein evolution. *Biochem. J.* 419(1): 15–28.

Orengo CA, Thornton JM (2005) Protein families and their evolution – a structural perspective (2005) *Annu Rev Biochem.* 74: 867–900.

Thornton JM, Todd AE, Milburn D, et al. From structure to function: Approaches and limitations (2000) *Nature Structural Biology* 7(11): 991–994.

See also Paralog

HOMOLOGY

Jaap Heringa and Laszlo Patthy

Biological entities (systems, organs, genes, proteins) are said to be homologous if they share a common evolutionary ancestor. In molecular biology the term homology in the evolution of molecular sequences implies that the homologous molecular entities (genes, proteins, RNAs) evolved from a common ancestor, and hence assumes that divergent evolution has taken place. Evolutionary events leading to homologous relationships include speciation and gene duplication.

Whereas sequence similarity of genes, RNAs, or proteins is a quantification of evolutionary divergence between sequences expressed using a gradual scale, homology is a binary state: a pair of sequences is either homologous or nonhomologous, so it is not justified to speak about percent homology. However, partial or local homology can occur between sequences, for example in cases of exon shuffling or gene fusion, yielding genes or proteins that share only a certain proportion of homologous sequence. Mutually homologous proteins can be grouped in a homologous protein family. Since protein tertiary structures are more conserved during evolution than their coding sequences, homologous sequences are assumed to share the same protein fold and the same or similar functions. Although it is possible in theory that two proteins evolve different structures and functions from a common ancestor, this situation cannot be traced so that such proteins are seen as unrelated. However, numerous cases exist of homologous protein families where subfamilies with the same fold have evolved distinct molecular functions.

Multi-domain proteins that arose by intragenic duplication of segments encoding a protein domain display internal homology. Mosaic proteins that arose through transfer of genetic material encoding a protein module from one gene to another gene may display a complex network of homologies, with different regions having distinct evolutionary origins. Homologous protein domains derived either through gene duplication, intragenic duplication or intergenic transfer of an ancestral domain constitute a domain family.

The term homology is often used in practice when two sequences have the same structure or function, although in the case of two sequences sharing a common function this ignores the possibility that the sequences are analogues resulting from convergent evolution. Unfortunately, it is not straightforward to infer homology from similarity because enormous differences exist between levels of sequence similarity within homologous families. Many protein families of common descent comprise members sharing pair-wise sequence similarities that are only slightly higher than those observed between unrelated proteins. This region of uncertainty has been characterised to lie in the range of 15 to 25% sequence identity (Doolittle, 1981), and is commonly referred to as the 'twilight zone' (Doolittle, 1987). While sequence identity levels of 30% or higher are indicative of homology (Sander and Schneider, 1991), there are some known examples of homologous proteins with sequence similarities below the randomly expected level given their amino acid composition (Pascarella and Argos, 1992). As a consequence, it is impossible to prove using sequence similarity that two sequences are not homologous.

Further reading

Doolittle RF (1981) Similar amino acid sequences: chance or common ancestry. *Science* 214: 149–159. Doolittle RF (1987) Of URFS and ORFS. *A Primer on How to Analyze Derived Amino Acid Sequences*. University Science Books, Mill Valley, CA.

Fitch WM (2000) Homology: a personal view on some of the problems. *Trends Genet.* 16(5): 227–231.

Pascarella S, Argos P (1992) A data bank merging related protein structures and sequences. *Protein Eng* 5: 121–137.

Reeck GR, de Haen C, Teller DC, et al. (1987) 'Homology' in proteins and nucleic acids: a terminology muddle and a way out of it. *Cell*, 50(5): 667.

See also Domain Family, Mosaic Protein, Multidomain Protein, Epaktolog, Ortholog, Paralog, Protein Domain, Protein Family, Protein Module, Sequence Alignment, Xenolog

HOMOLOGY MODELING, *SEE* COMPARATIVE MODELING.

HOMOLOGY SEARCH

Jaap Heringa

A homology search involves searching with a query sequence against an annotated sequence database. Comparative sequence analysis is a common first step in the analysis of sequence-structure-function relationships in protein and nucleotide sequences. In the quest for knowledge about the role of a certain unknown protein in the cellular molecular network, comparing the query sequence with the many sequences in annotated protein sequence databases often leads to useful suggestions regarding the protein's three-dimensional (3D) structure or molecular function. This extrapolation of the properties of sequences in public databases that are identified as 'neighbours' by sequence analysis techniques has led to the putative characterisation (annotation) of very many sequences. Although progress has been made, the direct prediction of a protein's structure and function is still a major unsolved problem in molecular biology.

Since the advent of the genome sequencing projects, the method of indirect inference by comparative sequence techniques and homology searching has only gained in significance.

A typical application to infer knowledge for a given query sequence is to compare it with all sequences in an annotated sequence database with the aim to find putative homologous sequences. A database search can be performed for a nucleotide or amino acid sequence against an annotated database of nucleotide (*e.g.*, EMBL, GenBank, DDBJ) or protein sequences (*e.g.*, SwissProt, PIR, TrEMBL, GenPept, NR-NCBI, NR-ExPasy). Although the actual pairwise comparison can take place at the nucleotide or peptide level, the most effective way to compare sequences is at the peptide level (Pearson, 1996). This requires that nucleotide sequences must first be translated in all six reading frames followed by comparison with each of these conceptual protein sequences. Although mutation, insertion and deletion events take place at the DNA level, reasons why comparing protein sequences can reveal more distant relationships include: (i) Many mutations within DNA are synonymous, which means that these do not lead to a change of the corresponding amino acids. As a result of the fact that most evolutionary selection pressure is exerted on protein sequences, synonymous mutations can lead to an overestimation of the sequence divergence if compared at the DNA level. (ii) The evolutionary relationships can be more finely expressed using a 20×20 amino acid exchange table than using exchange values among four nucleotides. (iii) DNA sequences contain non-coding regions, which should be avoided in homology searches. Note that the latter is still an issue when using DNA translated into protein sequences through a codon table. However, a complication arises when using translated DNA sequences to search at the protein level because frame shifts can occur, leading to stretches of incorrect amino acids and possibly elongation of sequences due to missed stop codons.

The widely used dynamic programming (DP) technique for sequence alignment is too slow for repeated homology searches over large databases, and may take multiple CPU hours for a single query sequence on a standard workstation. Although some special hardware has been designed to accelerate the DP algorithm, this problem has triggered the development of several heuristic algorithms that represent shortcuts to speed up the basic alignment procedure. The currently most widely used methods to scour sequence databases for homologies include PSI-BLAST (Altschul *et al.*, 1997), an extension of the BLAST technology (Altschul *et al.*, 1990), and FASTA (Pearson and Lipman, 1988). A particular significant feature of these rapid search routines is that they incorporate statistical estimates of the significance of each pairwise alignment score between a query and database sequence relative to random sequence scores (p-values and E-values).

Owing to recent advances in computational performance, procedures for sequence database homology searching have been developed, based on more mathematically rigorous representations of alignments afforded by hidden Markov modelling (HMM). These include SAM-T2K (Karplus *et al.*, 1998), HMMER3 (Finn *et al.*, 2011), and HHsearch (Söding, 2005).

Related websites

BLAST	http://blast.ncbi.nlm.nih.gov/Blast.cgi?CMD=Web&PAGE_TYPE=BlastHome
BLAST FAQ	http://blast.ncbi.nlm.nih.gov/Blast.cgi?CMD=Web&PAGE_TYPE=BlastDocs&DOC_TYPE=FAQ
FASTA	http://www.ebi.ac.uk/Tools/sss/fasta/
HMMER3	http://hmmer.janelia.org/
Hhsearch	http://toolkit.tuebingen.mpg.de/hhpred

Further reading

Altschul SF, Gish W, Miller W, Meyers EW, Lipman DJ (1990) Basic local alignment search tool. *J. Mol. Biol.* 215: 403–410.

Altschul SF, Madden TL, Schäffer AA, et al. (1997) Gapped BLAST and PSI-BLAST: a new generation of protein database search programs. *Nucl. Acids. Res.* 25: 3389–3402.

Eddy SR (1998) Profile hidden Markov models. *Bioinformatics*, 14: 755–763.

Finn RD, Clements J, Eddy SR (2011) HMMER web server: interactive sequence similarity searching. *Nucleic Acids Res. (Web Server Issue)* 39: W29–W37.

Karplus K, Barrett C, Hughey R (1998) Hidden Markov models for detecting remote protein homologies. *Bioinformatics* 14: 846.

Pearson WR, Lipman DJ (1988) Improved tools for biological sequence comparison. *Proc. Natl. Acad. Sci. USA* 85: 2444–2448.

Pearson WR (1996) Effective protein sequence comparison. In: Methods in Enzymology*, Doolittle RF (ed.) Vol. 266*, pp. 227–258, Academic Press, San Diego.

Söding J (2005) Protein homology detection by HMM–HMM comparison. *Bioinformatics* 21: 951–960.

See also BLAST, FASTA, High-scoring Segment Pair, Maximal-scoring Segment Pair

HOMOLOGY-BASED GENE PREDICTION, *SEE* GENE PREDICTION, HOMOLOGY-BASED.

HOMOZYGOSITY MAPPING

Mark McCarthy, Steven Wiltshire and Andrew Collins

A variant of linkage analysis which provides a powerful tool for mapping rare recessive traits within inbred families or populations.

A high proportion of the presentations of rare autosomal recessive traits will occur in consanguineous families (e.g. the offspring of first-cousin marriages), since such pedigree relationships enhance the chance that two copies of the same rare allele will segregate into the same individual. Since disease in this situation results from the relevant part of the same ancestral chromosome being inherited from both parents, the susceptibility gene will lie within a localised region of homozygosity (or, as it sometimes termed, autozygosity). This region can be detectable by undertaking a genome scan for linkage using polymorphic markers. As with any linkage analysis, the strength of the evidence for the localisation of the susceptibility gene is determined using appropriate linkage programs. The HOMOZ program is an example of such a linkage program designed for use in the specialised situation. The particular advantage of homozygosity mapping is that substantial evidence for linkage can be obtained from relatively few affected individuals.

Examples: Novelli et al. (2002) describe their success in mapping mandibuloacral dysplasia, a rare recessive disorder to chromosome 1q21, by homozygosity mapping in five consanguineous Italian families, and the subsequent identification of aetiological mutations in LMNA.

Related websites

HOMOZ (linkage program optimized for homozygosity mapping)	http://www-genome.wi.mit.edu/ftp/distribution/software/
HomozygosityMapper	http://www.homozygositymapper.org/

Further reading

Abney M, et al. (2002) Quantitative-trait homozygosity and association mapping and empirical genomewide significance in large, complex pedigrees: fasting serum-insulin level in the Hutterites. *Am. J. Hum. Genet.* 70: 920–934.

Faivre L, et al. (2002) Homozygosity mapping of a Weill-Marchesani syndrome locus to chromosome 19p13.3-p13.2. *Hum. Genet.* 110: 366–370.

Kruglyak L, et al. (1995) Rapid multipoint linkage analysis of recessive traits in nuclear families, using homozygosity mapping. *Am. J. Hum. Genet.* 56: 519–527.

Lander ES, Botstein D (1986) Mapping complex genetic traits in humans: new methods using a complete RFLP linkage map. *Cold Spring Harbor Symp Quant Biol* 51: 49–62.

Novelli G, et al. (2002) Mandibuloacral Dysplasia is caused by a mutation in LMNA-Encoding Lamin A/C. *Am. J. Hum. Genet.* 71: 426–431.

Seelow D, Schuelke M, Hildebrandt F, Nurnberg P (2009) HomozygosityMapper – an interactive approach to homozygosity mapping. *Nucleic Acids Res.* 37: W593–599.

See also Extended Tracts of Homozygosity, Genome Scan, Linkage Analysis

HORIZONTAL GENE TRANSFER (HGT)

Dov Greenbaum

Horizontal Gene Transfer (HGT) is one of the principal mechanisms for genetic plasticity in many organisms. For example, in some examples it provides increasing antibacterial resistance in bacteria.

HGT can occur via a number of potential mechanisms including, transformation (more typical in prokaryotes than eukaryotes), transduction (the transfer of genetic information via a virus); and some instances, other virus-like agents.

While horizontal gene transfer plays an important role in bacterial evolution and is even common in certain unicellular eukaryotes, it should be noted that remains little evidence that HGT occurs in higher plants and multi-cellular eukaryotes.

One indication that an HGT event has occurred is the determination of the presence of the same gene in organisms that are only very distantly related to each other.

Related website

Nature Focus	http://www.nature.com/nrmicro/focus/genetransfer/index.html

Further reading

Doolittle WF (1998) You are what you eat: a gene transfer ratchet could account for bacterial genes in eukaryotic nuclear genomes. *Trends in Genetics* 14: 307–311.

Gogarten JP, Townsend JP (2005) Horizontal gene transfer, genome innovation and evolution. *Nat. Rev. Microbiol.* 9: 679–687.

Gogarten JP, Townsend JP (2005) Horizontal gene transfer, genome innovation and evolution. *Nat. Rev. Microbiol.* 9: 679–687.

Ge F, et al. (2005) The Cobweb of Life revealed by genome-scale estimates of horizontal gene transfer. *PLoS Biol.* 3: e316.

Jain R, et al. (2003) Horizontal gene transfer accelerates genome innovation and evolution. *Mol. Biol. Evol.* 20: 1598–1602.

Syvanen M, Kado CI (2002) *Horizontal Gene Transfer* 2nd edition, Academic Press.

See also Homology, Pseudoparalog, Xenolog

HSP, *SEE* HIGH-SCORING SEGMENT PAIR.

HTU, *SEE* HYPOTHETICAL TAXONOMIC UNIT.

HUGO (THE HUMAN GENOME ORGANIZATION)

Pedro Fernandes

HUGO was conceived in late April 1988, at the first meeting on genome mapping and sequencing at Cold Spring Harbor. For some time, as the genome initiatives got under way in individual nations, the need for an international coordinating scientific body had been under discussion. The idea of HUGO was particularly Sydney Brenner's. He also suggested the name of the organization and its rather felicitous acronym.

HUGO's missions are:

- to investigate the nature, structure, function and interaction of the genes, genomic elements and genomes of humans and relevant pathogenic and model organisms
- to characterise the nature, distribution and evolution of genetic variation in humans and other relevant organisms
- to study the relationship between genetic variation and the environment in the origins and characteristics of human populations and the causes, diagnoses, treatments and prevention of disease
- to foster the interaction, coordination and dissemination of information and technology between investigators and the global society in genomics, proteomics, bioinformatics, systems biology, and the clinical sciences by promoting quality education, comprehensive communication, and accurate, comprehensive accessible knowledge resources for genes, genomes and disease,
- to sponsor factually grounded dialogues on the social, legal, and ethical issues related to genetic and genomic information and championing the regionally appropriate, ethical utilization of this information for the good of the individual and the society.

Related websites

HUGO	http://www.hugo-international.org/
The HUGO Journal	http://www.thehugojournal.com/

HUMAN-CURATED GENE ANNOTATION, *SEE* GENE ANNOTATION, HAND-CURATED.

HUMAN GENE MUTATION DATABASE, *SEE* HGMD.

HUMAN GENOME VARIATION DATABASE, *SEE* HGVBASE.

HUMAN PROTEOME ORGANIZATION, *SEE* HUPO.

HUMAN VARIOME PROJECT (HVP)

John M. Hancock

The Human Variome Project is a global coordination project whose aim is to support the collection and dissemination of information on disease-causing gene variants. The HVP came into existence because of the incomplete nature of publicly available gene-disease datasets. Some of these are held in dedicated Locus-Specific Databases (LSDBs) but others are published only in scientific papers or appear on individual web sites. The ultimate aim of the project is an open network of LSDBs that can be queried for information on gene-disease relationships. This will be stimulated by supporting the setting up of new LSDBs and the development of standards for the acquisition, description and dissemination of variant data.

Related website

HVP	http://www.humanvariomeproject.org/

Further reading

Byrne M, Fokkema IF, Lancaster O, et al. (2012) VarioML framework for comprehensive variation data representation and exchange. *BMC Bioinformatics*. 13: 254. doi: 10.1186/1471–2105–13–254.

Kohonen-Corish MR, Smith TD, Robinson HM. (2013) Beyond the genomics blueprint: the 4th Human Variome Project Meeting, UNESCO, Paris, 2012. *Genet. Med.* doi: 10.1038/gim.2012.174.

Ring HZ, Kwok PY, Cotton RG. (2006) Human Variome Project: an international collaboration to catalogue human genetic variation. *Pharmacogenomics*. 7: 969–972.

See also, Locus-Specific Database, VarioML, HGMD

HUPO (HUMAN PROTEOME ORGANIZATION)

Pedro Fernandes

HUPO is an international scientific organization representing and promoting proteomics through international cooperation and collaborations by fostering the development of new technologies, techniques and training.

HUPO's activities are aggregated in several initiatives:

- Human Proteome Project (HPP)
- Human Plasma Proteome Project (HPPP)
- Human Liver Proteome Project (HLPP)
- Human Brain Proteome Project (HBPP)
- Human Antibody Initiative (HAI)
- Proteomics Standards Initiative (PSI)
- Human Disease Glycomics/Proteome Initiative (HGPI) – Human Kidney and Urine Proteome Project (HKUPP) Mouse Models of Human Disease
- HUPO CardioVascular Initiative (HCVI)
- Proteome Biology of Stem Cells Initiative

H

- Disease Biomarkers Initiatives
- Initiative on Model Organism Proteomes(iMOP)[end]

HUPO has organized an annual world congress since 2002.

Related website

HUPO	http://www.hupo.org/

HVP, *SEE* HUMAN VARIOME PROJECT.

HYDROGEN BOND

Jeremy Baum

An interaction between two non-hydrogen atoms, one of which (the proton donor) has a covalently bonded hydrogen atom. The other non-hydrogen atom (the proton acceptor) has electrons not involved directly in covalent bonds that can interact favourably with the hydrogen atom. The energy of this interaction is typically up to 10 $kJmol^{-1}$ for uncharged groups, but can be significantly more if charges are present.

This interaction is extremely common in biological systems, as it is a fundamental interaction in water, and within proteins. The oxygen atoms of C=O groups and the nitrogen atoms of N-H groups in protein backbones act as proton acceptors and proton donors respectively. The secondary structures of proteins occur as a result of hydrogen bonding between such backbone groups.

See also Secondary Structure, Backbone

HYDROPATHY

Teresa K. Attwood

Broadly, a term that encapsulates strength of affinity for water: at one end of the spectrum, hydrophilicity denotes a high affinity; at the other, hydrophobicity denotes a low affinity. Depending on the context in which it is being considered, hydropathy may be defined in different ways. For example, it may be estimated from chemical behaviour or amino acid physicochemical properties (solubility in water, chromatographic migration, etc.); or it may be estimated from environmental characteristics of residues in 3D structures, where it carries more complex information than that conferred by a single physical or chemical property (solvent accessibility, propensity to occupy the protein interior, and so on).

There are many different amino acid hydropathic rankings (some examples are given in Figure H.4). At the simplest level, 'internal, external, ambivalent' (where internal=[FILMV],

Figure H.4 Comparison of eight hydropathy scales, ranking the amino acids in terms of their hydrophobic and hydrophilic properties. The scales broadly agree that phenylalanine is strongly hydrophobic and that glutamic acid is strongly hydrophilic. Other amino acids, however, are ambivalent, being sometimes hydrophobic and sometimes hydrophilic. The Zimmerman scale offers a slightly different perspective: e.g., where most scales rank tryptophan as hydrophobic, Zimmerman assigns it greater hydrophilic character; conversely, while most scales rank lysine and proline as hydrophilic, Zimmerman assigns them greater hydrophobic character. (See Colour plate H.4)

external= [DEHKNQR] and ambivalent= [ACGPSTWY]) provides a useful hydrophobicity alphabet. More sensitive scales assign quantitative values to each individual amino acid.

Although the distribution of residues within the available scales is broadly similar, there is no general consensus regarding which residues appear at the most hydrophobic and most hydrophilic extremes. Nevertheless, such scales can be used to plot hydropathy profiles to provide an overview of the hydropathic character of a given protein sequence. There is no 'right' scale but, used together, they can help to yield a consensus of the most significant hydrophobic features within a sequence.

Related website

Hydropathy scale values	http://www.molvis.indiana.edu/C687_S99/hydrophob_scale.html

Further reading

Attwood TK, Parry-Smith DJ (1999) *Introduction to Bioinformatics*. Addison Wesley Longman, Harlow, Essex, UK.

Doolittle RF (ed.) (1990) Molecular evolution: Computer analysis of protein and nucleic acid sequences. In *Methods in Enzymology*, 183, 389. Academic Press, San Diego, CA, USA.

Higgs P, Attwood TK (2005) *Bioinformatics and Molecular Evolution*. Blackwell Publishing, Oxford, UK.

Kyte J, Doolittle RF (1982) A simple method for displaying the hydropathic character of a protein. *J. Mol. Biol.* 157(1): 105–132.

See also Hydropathy Profile, Hydrophilicity, Hydrophobic Scale

H

HYDROPATHY PROFILE (HYDROPHOBICITY PLOT, HYDROPHOBIC PLOT)

Teresa K. Attwood

A graph that provides an overview of the hydropathic character of a protein sequence, where the x-axis represents the sequence and the y-axis the hydrophobicity score. In creating such a graph, a sliding window (whose width may be varied) is scanned across the query sequence and, for each window position, a score is calculated from the hydrophobicity values of the constituent residues according to a particular hydropathic ranking.

Typical profiles show characteristic peaks and troughs, corresponding to the most hydrophobic and most hydrophilic parts of the protein, as shown in Figure H.5. They are therefore especially useful for identifying putative α-helical transmembrane (TM) domains, as denoted by runs of ~20 hydrophobic residues. Because there is variation between hydrophobicity scales, the resulting hydropathy profiles will differ in detail. To achieve the most reliable results, it is advisable to use several different rankings and to seek a consensus between the range of profiles produced – this provides different perspectives

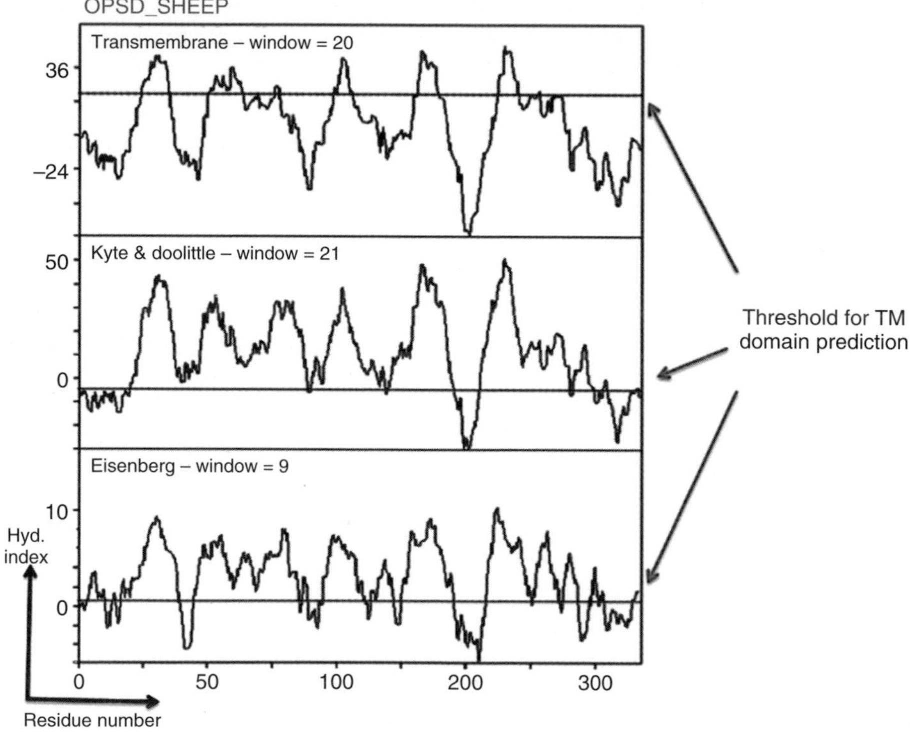

Figure H.5 Comparison of three hydropathy profiles for the protein ovine rhodopsin (OPSD_SHEEP), the sequence of which runs along the x axis. From top to bottom, the hydropathy scales are labelled 'transmembrane' (a scale used particularly to locate transmembrane domains), with a sliding-window size of 20; Kyte and Doolittle, with a sliding-window size of 21; and Eisenberg, with a sliding-window size of 9. The profiles are broadly similar, but the details differ; comparison of plots arising from different scales helps to interpret profiles more reliably. These plots hint at the locations of 6 or 7 transmembrane domains.

on the same sequence, potentially allowing characteristics that are not apparent with one scale to be visualized with one or more of the others (e.g., a relatively hydrophilic TM domain). A combined view is important, as many erroneous TM assignments have been associated, for example, with database sequences based on incautious interpretations of individual profiles.

Related website

TM domain prediction tools	http://web.expasy.org/protscale/

Further reading

Attwood TK, Parry-Smith DJ (1999) *Introduction to Bioinformatics*. Addison Wesley Longman, Harlow, Essex, UK.

Kyte J, Doolittle RF (1982) A simple method for displaying the hydropathic character of a protein. *J. Mol. Biol.* 157(1): 105–132.

See also Hydropathy, Hydrophilicity, Hydrophobic Scale

HYDROPHILICITY

Jeremy Baum

The term hydrophilicity is used in contrast to hydrophobicity, and is applied to polar chemical groups that have favourable energies of solvation in water, and tend to have water/octanol partition coefficients favouring the aqueous phase.

See also Hydrophobicity, Polar

HYDROPHOBIC MOMENT

Jeremy Baum

In an extension of the concept of the hydrophobic scale Eisenberg and co-workers defined a vector quantity for each amino acid residue in a protein structure. This was based on the direction defined from the backbone to the end of the side chain, taken together with a hydrophobic scale. These residue vectors could be summed for pieces of secondary structure such as α-helices, and in experimentally determined structures were found to have characteristic orientations relative to the overall protein fold. Such hydrophobic moments were proposed to be useful in diagnosing incorrect theoretical folds.

See also Hydrophobicity, Back Bone, Side Chain, Secondary Structure

HYDROPHOBIC SCALE

Jeremy Baum

Many workers have deduced numerical scales to define the hydrophobicity of certain chemical groups and of amino acids. Some of these scales have their origins in experimental measurements of for example water/octanol partition coefficients.

See also Hydropathy

HYDROPHOBICITY PLOT, *SEE* HYDROPATHY PROFILE.

HYPHY (HYPOTHESIS TESTING USING PHYLOGENIES)

Michael P. Cummings

Software package for the analysis of sequence data using techniques in phylogenetics, molecular evolution, and machine learning.

Among supported analyses include detection of natural selection (diversifying, purifying, or directional), recombination, co-evolving sites and others.

It can be used via a Graphical User Interface (GUI) or a rich scripting language. Includes support for parallel computing (via MPI) and can be compiled as a shared library and called from other programming environments such as Python or R.

Related website

HyPhy home page	http://www.hyphy.org/w/index.php/Main_Page

See also DataMonkey

HYPOTHETICAL TAXONOMIC UNIT (HTU)

Aidan Budd and Alexandros Stamatakis

A taxonomic unit whose existence is hypothetical or inferred, rather than confirmed directly by observation or measurement. For example, the most recent common ancestor of humans and chimps is a hypothetical taxonomic entity about which we cannot make direct measurements, and is thus an HTU. Internal nodes of phylogenetic trees represent HTUs. Contrast with Operational Taxonomic Units (OTUs).

See also, Operational Taxonomic Unit

I

IBD, *see* Identical by Descent.

IBS, *see* Identical by State.

IC50, *see* Binding Affinity.

ICA, *see* Independent Component Analysis.

Identical by Descent (Identity by Descent, IBD)

Identical by State (Identity by State, IBS)

IGV, *see* Gene Annotation, visualization tools.

IMa2 (Isolation with Migration a2)

IMGT (International Immunogenetics Database)

Imprinting

Imputation, *see* Genotype Imputation.

InChI (International Chemical Identifier)

InChi Key

Indel (Insertion-Deletion Region, Insertion, Deletion, Gap)

Independent Component Analysis (ICA)

Independent Variables, *see* Features.

Individual (Instance)

Individual Information

Information

Information Retrieval, *see* Text Mining.

Information Theory

Initiator Sequence

INPPO (International Plant Proteomics Organization)

Insertion, Insertion-Deletion Region, *see* Indel.

Instance, *see* Individual.

Instance-based Learner, *see* K-Nearest Neighbor Classification.

Integrated Gene Annotation Pipelines, *see* Gene Prediction, Pipelines.

Integrated Gene Prediction Systems, *see* Gene Prediction Systems, Pipelines.

Intelligent Data Analysis, *see* Pattern Analysis.

Interactome

Intergenic Sequence

Interior Branch, Internal Branch, *see* Branch and Phylogenetic Tree.

Interior Node, Internal Node, *see* Node and Phylogenetic Tree.

International Chemical Identifier, *see* InChi.

International Immunogenetics Database, *see* IMGT.

International Plant Proteomics Organization, *see* INPPO.

Concise Encyclopaedia of Bioinformatics and Computational Biology, Second Edition.
Edited by John M. Hancock and Marketa J. Zvelebil.

International Society for Computational Biology (ISCB)

Interolog (Interologue)

InterPro

InterProScan, *see* InterPro.

Interspersed Sequence (Long-Term Interspersion, Long-Period Interspersion, Short-Term Interspersion, Short-Period Interspersion, Locus Repeat)

Intrinsic Gene Prediction, *see* Gene Prediction,
ab initio.

Intron

Intron Phase

IR, *see* Text Mining.

ISCB, *see* International Society for Computational Biology.

Isobaric Tagging

Isochore

IsomiR

Iteration

IUPAC-IUB Codes (Nucleotide Base Codes, Amino Acid Abbreviations)

IBD, *SEE* IDENTICAL BY DESCENT.

IBS, *SEE* IDENTICAL BY STATE.

IC50, *SEE* BINDING AFFINITY.

ICA, *SEE* INDEPENDENT COMPONENT ANALYSIS.

IDENTICAL BY DESCENT (IDENTITY BY DESCENT, IBD)

Mark McCarthy and Steven Wiltshire

A state of nature in which alleles from separate individuals are descended from, and therefore are copies of, the same ancestral allele.

The term IBD is most frequently used in describing the relationship between individuals in terms of allele sharing at particular loci. For any given pair of relatives, there are clear biological constraints to the maximum and minimum numbers of alleles that can be shared IBD at any given locus. For instance, monozygotic twins share both alleles IBD at all loci, whereas full sibs and dizygotic twins may share 2, 1 or 0 alleles IBD at any given locus. A parent and its offspring share one allele IBD at all loci. The parameter $\hat{\pi}_{ij}$ – the estimated proportion of alleles shared IBD by relatives i and j – is the basis of allele-sharing methods of linkage analysis: the correlation coefficient between $\hat{\pi}_{ij}$ at two loci is $(2\Psi - 1)$ where $\Psi = \theta^2 + (1 - \theta)^2$, and $\boldsymbol{\theta}$ is the recombination fraction between the two loci in question. Estimates of the proportion of alleles shared IBD averaged over many markers may also be used to confirm (or refute) familial relationships between purported siblings. Alleles that are identical by descent have the same DNA sequence and are therefore identical by state also.

Related website

Programs for estimating the proportions of alleles shared IBD by purported siblings include:

RELPAIR	http://csg.sph.umich.edu/boehnke/relpair.php

Further reading

Boehnke M, Cox NJ (1997) Accurate inference of relationships in sib-pair linkage studies. *Am J Hum Genet* 61: 423–429.

Sham P (1998) *Statistics in Human Genetics*. London, Arnold, pp 98–106.

See also Identical By State, Recombination, Allele-Sharing Methods, Linkage Analysis, Homology

I

IDENTICAL BY STATE (IDENTITY BY STATE, IBS)

Mark McCarthy and Steven Wiltshire

A state of nature in which alleles from separate individuals are identical in DNA sequence.

Alleles that are identical by state may or may not be descended from the same ancestral copy: if they are copies of the same ancestral allele then they are also identical by descent (IBD). Consider a nuclear family of two sibs with paternal genotype AB and maternal genotype BC at a given locus. If the first sib inherits allele A from the father and B from the mother, and the second sib inherits allele B from the father and C from the mother, then the pair of sibs share one allele–B–IBS (but zero IBD). If the sibs' genotypes had been AB and AC, they would share one allele IBS and one allele–A–IBD; if both sibs had been AB they would share both alleles IBS and both IBD also. It is not always possible to determine the IBD status of a relative pair even though the IBS status is unequivocal: if both parents have genotype AB, and both offspring have genotypes AB, then the two sibs share both alleles IBS, but either no alleles IBD (if each inherited alleles A and B from different parents) or both alleles IBD (if each inherited alleles A and B from the same parent). Early allele-sharing methods for linkage analysis of extended pedigrees utilised IBS sharing, although more powerful, methods are based on allele-sharing IBD.

Further reading

Sham P (1998) *Statistics in Human Genetics*. London, Arnold, pp 98–106.

See also Identical by Descent, Allele-Sharing Methods, Linkage Analysis

IGV, *SEE* GENE ANNOTATION, VISUALIZATION TOOLS.

IMA2 (ISOLATION WITH MIGRATION A2)

Michael P. Cummings

A program for generating posterior probabilities for complex demographic population genetic models using coalescence theory and Markov chain Monte Carlo sampling.

Related website

IMa2 home page	http://astro.temple.edu/~tuf29449/software/software.htm#IMa2

Further reading

Hey J, Nielsen R (2004) Multilocus methods for estimating population sizes, migration rates and divergence time, with applications to the divergence of *Drosophila pseudoobscura* and *D. persimilis*. *Genetics* 167: 747–760.

Hey J, Nielsen R (2007) Integration within the Felsenstein equation for improved Markov chain Monte Carlo methods in population genetics. *P. Natl. Acad. Sci. USA* 104: 2785–2790.

Nielsen R, Wakeley J (2001) Distinguishing migration from isolation. A Markov chain Monte Carlo approach. *Genetics* 158: 885–896.

IMGT (INTERNATIONAL IMMUNOGENETICS DATABASE)

Guenter Stoesser, Malachi Griffith and Obi L. Griffith

The international ImMunoGeneTics Database (IMGT) provides specialized information on Immunoglobulins (Ig), T cell Receptors (TcR) and Major Histocompatibility Complex (MHC) molecules of human and other vertebrate species.

IMGT includes two databases: LIGM-DB, a comprehensive database of nucleic acid sequences for immunoglobulins and T-cell receptors from human and other vertebrates, with translations for fully annotated sequences, and HLA-DB, a database for sequences of the human MHC, referred to as HLA (Human Leucocyte Antigens). IMGT was established in 1989 by the Université Montpellier II and the CNRS (Montpellier, France). IMGT is produced in collaboration with the European Bioinformatics Institute and data is integrated with EMBL nucleotide sequence entries.

Related website

IMGT/HLA Sequence Database	http://www.ebi.ac.uk/imgt/hla/

Further reading

Lefranc MP (2001) IMGT, the international ImMunoGeneTics database. *Nucleic Acids Res.* 29: 207–209.

Robinson J, et al. (2011) The IMGT/HLA Database. *Nucleic Acids Res.* 39(Suppl 1): D1171–176.

IMPRINTING

Mark McCarthy, Steven Wiltshire and Andrew Collins

Process whereby the expression of a gene is determined by its parental origin.

Imprinted genes represent exceptions to the usual rule that both the paternally- and maternally-inherited copies of a gene are equally expressed in the offspring. The imprints inherited from parents are erased in the germ line, and then reset in the gametes by differential methylation of the genes concerned. Since imprinting of a gene or region generally results in reduced function, maternal imprinting generally refers to the situation when only the paternal copy is expressed. Around 50 imprinted loci have been identified in man and other higher mammals. The evolutionary basis for the development of imprinting remains uncertain, but the observation that many imprinted genes are important regulators of early growth suggests that it may have evolved to mediate the conflict between the differing costs and benefits of excessive fetal growth for father and mother. Breakdown of normal imprinting mechanisms (which may arise via defects in methylation; duplication; translocation; or defects in meiosis that lead to uniparental disomy) such that individuals inherit either zero or two expressed copies of a usually imprinted gene, is responsible for a number of diseases.

Examples: Classical examples of imprinting include the chromosome 15q region whereby deletion of the paternally (expressed) copy of the imprinted SNRPN gene (and of other contiguous genes) leads to Prader-Willi syndrome. In contrast, deletion of the same region on the maternal chromosome leads (due to loss of maternal UBE3A brain-specific expression) to Angelman syndrome. Loss of IGF2 imprinting in somatic cells is associated with development of a renal malignancy of childhood (Wilms' tumour). Finally, expression of two paternal copies of the imprinted ZAC gene on chromosome 6q24 (through duplication or uniparental disomy) leads to the condition of transient neonatal diabetes mellitus.

Related website

Catalogue of imprinted genes	http://igc.otago.ac.nz/home.html

Further reading

Gardner RJ, et al. (2000) An imprinted locus associated with transient neonatal diabetes mellitus. *Hum Mol Genet* 9: 589–596.

Morrison IM, et al. (2001) The imprinted gene and parent-of-origin effect database. *Nucleic Acids Res* 29: 275–276.

Nicholls RD, Knepper JL (2001) Genome organization, function, and imprinting in Prader-Willi and Angelman syndromes. *Annu Rev Genomics Hum Genet* 2: 153–175.

IMPUTATION, *SEE* GENOTYPE IMPUTATION.

INCHI (INTERNATIONAL CHEMICAL IDENTIFIER)

Bissan Al-Lazikani

InChI (International Chemical Identifier) is a non-proprietary identifier for chemical substances that represents the chemical structure of a covalently bonded compound. It is used to describe, store and cross-reference compounds between different databases. It was created by IUPAC in order to enable printing and electronic data distribution and linking. IUPAC also provides software to generate InChIs from fully enumerated chemical structures.

One of the advantages of InChIs is the annotated 'layers' of which an InChI is constructed. 'Layers' are separated by a '/' character. Firstly the version of the InChi is stated (e.g. 1 or 1S for 'Standard InChi'). There are four major layers that may contain sublayers. Each layer or sublayer is separated by a '/' and usually most start with a lowercase letter indicating the layer in question (e.g. /c is the connectivity layer and /h is the hydrogen layer). The layers are as follows: The main layer containing chemical formula and connections; the charge layer describing net charges; the stereochemical layer describing relative stereochemistry; the isotopic layer and fixed Hydrogen layer.

Examples:

Benzene: InChI = 1S/C6H6/c1-2-4-6-5-3-1/h1-6H
L-Proline: InChI = 1S/C5H9NO2/c7-5(8)4-2-1-3-6-4/h4,6H,1-3H2,(H,7,8)/t4-/m0/s1

See also InChIKey, SMILES

INCHI KEY

Bissan Al-Lazikani

A computer hashed representation of the InChI. This enables faster comparison to identify identical InChIs. It is composed of a 14-character string encoding the molecular skeleton of a compound; followed by a '-' then an eight-character string encoding sterochemical and isotopic information; another '-' followed by a letter indicating whether this is derived from a Standard InChI or a non-standard InChI, followed by the version number.

Examples:

Benzene: InChIKey = UHOVQNZJYSORNB-UHFFFAOYSA-N
Proline: InChIKey = ONIBWKKTOPOVIA-BYPYZUCNSA-N

While an InChI can usually be used to reproduce an accurate image of the structure, an InChIkey cannot.

See also, InChI

INDEL (INSERTION-DELETION REGION, INSERTION, DELETION, GAP)

Teresa K. Attwood

An acronym denoting evolutionary insertion or deletion events, where the mode of selection is unknown. DNA sequences evolve not only as a result of point mutations, but also as a consequence of sequence expansions and/or contractions. Frequently, therefore, evolutionarily related sequences have different lengths. In some cases, it may be impossible to infer whether a particular sequence lost one or more nucleotides relative to another, or whether the other gained them. For example, for sequences including the nucleotides TATA and TATAT, did the first lose a T via a deletion event, or did the second gain a T via an insertion event? Only knowledge of the ancestral sequence can help to determine whether the length difference results from an insertion or a deletion. This ambiguity was originally captured in the term indel.

Some of this subtlety is lost in common usage, where it often implies an insertion-deletion region within a multiple sequence alignment, as denoted by a 'gap' character. During the process of creating multiple alignments, sequences with different lengths must be brought into the correct register by inserting gaps into shorter sequences relative to longer ones. The more divergent the sequences, the greater the number of gaps that are likely to be needed to achieve the optimal alignment. This presents particular challenges to alignment algorithms, which assign different values for opening gaps and for extending them, and makes reliable alignment of divergent sequences particularly difficult. However, given the evolutionary complexities, the notion of what constitutes a 'good alignment' is still controversial.

Notwithstanding the ambiguities, the consequence of the presence of indels within a sequence alignment is clear: islands of conservation (motifs) become visible against a backdrop of mutational change. Thus indels tend to mark the boundaries of the core structural and functional motifs within alignments (see Figure I.1). They may therefore be used to

I

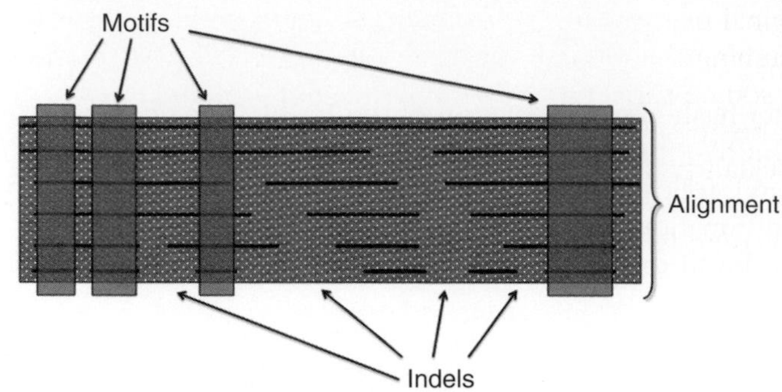

Figure I.1 Schematic diagram showing a portion of a sequence alignment containing a number of indel regions: horizontal bars denote the aligned sequences; vertical blocks pinpoint islands of conservation — motifs — that emerge against the backdrop of insertion and deletion events.

guide the design of appropriate diagnostic signatures for particular aligned families by highlighting critically conserved motifs and domains. Often, indel regions correspond to loop sites in the tertiary structures of aligned sequences, as protein cores do not easily accommodate insertions or deletions.

Further reading

Doolittle RF (ed.) (1990) Molecular evolution: Computer analysis of protein and nucleic acid sequences. In: *Methods in Enzymology*, Vol.183, pp. 133–134. Academic Press, San Diego, CA, USA.

Higgs P, Attwood TK (2005) *Bioinformatics and Molecular Evolution*. Blackwell Publishing, Oxford, UK.

Lawrence CB, Goldman DA (1988) Definition and identification of homology domains. *Comput. Appl. Biosci.* 4(1): 25–33.

See also Multiple Alignment, Fingerprint, Motif

INDEPENDENT COMPONENT ANALYSIS (ICA)

Concha Bielza and Pedro Larrañaga

This is a computational method that looks for linear projections of the data such that they should be as nearly statistically independent as possible. In linear ICA, the data matrix is assumed to be a linear combination of non-Gaussian independent components. Our data can be viewed as indirect measurements arising from an underlying source, which typically cannot be directly measured. The measured 'signals' in the data tend to be 'more Gaussian' than the original source components due to the Central Limit Theorem. ICA algorithms find the independent components or sources by maximizing the statistical independence of the estimated components. This is shown to be equivalent to maximizing their departures from Gaussianity. The projected data will look as far from Gaussian as possible.

Unlike principal component analysis, the goal of ICA is not necessarily dimensionality reduction.

Its original use was for separating a multivariate mixed signal into additive subcomponents assuming the mutual statistical independence of the non-Gaussian source signals, i.e. blind source separation. Typical examples are mixtures of simultaneous speech signals that have been picked up by several microphones from people talking simultaneously in a room (cocktail party problem), brain waves recorded by multiple sensors, or parallel time series obtained from some industrial process. Other ICA applications are in exploratory data analysis and in feature extraction. Popular ICA algorithms are infomax, FastICA and JADE.

Further reading

Hyvärinen A, Karhunen J, Oja E (2001) *Independent Component Analysis*. Wiley.

Mantini D, Petrucci F, Del Boccio P, et al. (2008) Independent component analysis for the extraction of reliable protein signal profiles from MALDI-TOF mass spectra. *Bioinformatics* 24(1): 63–70.

Yao F, Coquery F, Lê Cao K-A (2012) Independent principal component analysis for biologically meaningful dimension reduction of large biological data sets. *BMC Bioinformatics* 13: 24.

See also Principal Component Analysis

INDEPENDENT VARIABLES, *SEE* FEATURES.

INDIVIDUAL (INSTANCE)

Robert Stevens

A particular member of a **category** either because it possesses the properties of that category or is part of the enumeration of category members.

'England' is an individual of the category *Country*; 'Professor Michael Ashburner' is an individual of the category *Person*; a single *E. coli* cell is an individual of the category *Bacterium*. The distinction between **category** and **individual** is often difficult to determine. Some systems refer to individuals as 'instances'. For most purposes the two words are interchangeable.

INDIVIDUAL INFORMATION

Thomas D. Schneider

Individual information is the information that a single binding site contributes to the sequence conservation of a set of binding sites. This can be graphically displayed by a sequence walker. It is computed as the decrease in surprisal between the Before State and the After State. The technical name is R_i.

Further reading

Schneider TD (1997) Information content of individual genetic sequences. *J. Theor. Biol.*, 189: 427–441. http://alum.mit.edu/www/toms/papers/ri/

Schneider TD (1999) Measuring molecular information. *J. Theor. Biol.*, 201: 87–92. http://alum.mit.edu/www/toms/papers/ridebate/

I

INFORMATION

Thomas D. Schneider

Information is measured as the decrease in uncertainty of a receiver or molecular machine in going from the before state to the after state. It is usually measured in bits per second or bits per molecular machine operation.

Related website

Information Is Not Entropy, Information Is Not Uncertainty!	http://alum.mit.edu/www/toms/information.is.not.uncertainty.html

See also Information Theory, Evolution of Biological Information, Molecular Efficiency

INFORMATION RETRIEVAL, *SEE* TEXT MINING.

INFORMATION THEORY

Thomas D. Schneider

Information theory is a branch of mathematics founded by Claude Shannon in the 1940s. The theory addresses two aspects of communication: 'How can we define and measure information?' and 'What is the maximum information that can be sent through a communications channel?' (channel capacity). John R. Pierce, an engineer at Bell Labs in the 1940s, wrote an excellent introductory book about information theory.

Related websites

Information Theory References	http://alum.mit.edu/www/toms/bionet.info-theory.faq.html
Primer on Information Theory	http://alum.mit.edu/www/toms/papers/primer/
Information theory resources	http://alum.mit.edu/www/toms/itresources.html

Further reading

Cover TM, Thomas JA (1991) *Elements of Information Theory*. John Wiley & Sons, Inc., New York.

Pierce JR (1980) *An Introduction to Information Theory: Symbols, Signals and Noise*. Dover Publications, Inc., New York, second edition.

See also Molecular Efficiency, Molecular Information Theory

INITIATOR SEQUENCE

Niall Dillon

Short consensus sequence located around the transcription initiation site of RNA polymerase II transcribed genes that lack a TATA box and involved in specifying the site of initiation of transcription.

Two different elements have been identified that are involved in specifying the transcriptional initiation site of polymerase II transcribed genes in eukaryotes. These are the TATA box (located 30–35 bp upstream from the transcription start site) and the intitator. Promoters can contain one or both of these elements and a small number have neither. The intitator and the TATA box bind the multi-protein complex TFIID, which is involved in specifying the initiation site.

Further reading

Groschedl R, Birnstiel M (1980) Identification of regulatory sequences in the prelude sequences of an H2A histone gene by the study of specific deletion mutants in vivo. *Proc. Natl. Acad Sci. U S A.* 77: 1432–1436.

Smale S, Baltimore D (1989) The 'initiator' as a transcription control element. *Cell* 57: 103–113.

Smale S, et al. (1998) The initiator element: a paradigm for core promoter heterogeneity within metazoan protein-coding genes. *Cold Spring Harb. Symp. Quant. Biol.* 63: 21–31.

See also TATA Box

INPPO (INTERNATIONAL PLANT PROTEOMICS ORGANIZATION)

Pedro Fernandes

INPPO is a global plant proteomics organization to properly organize, preserve and disseminate collected information on plant proteomics. It was established in 2011 to assist in the essential role of proteomics in understanding the biology of plants.

The major INPPO initiatives are:

- to further intensify successful ongoing cooperation in the field of plant proteomics for both model and crop plants.
- to promote the establishment of national plant proteomics organizations.
- to develop an open partnership around the globe.
- to bridge the gap between academy and industry.
- to establish centralized databases at several locations (Americas, Europe, Asia-Pacific and Australia) with their real time integration.
- to organize workshops at national and international levels to train manpower and exchange information.
- to integrate proteomics-related activities and disseminate them to partners through the INPPO website.
- to bring proteomics to every laboratory working on plants around the globe.

I

- to outreach to the younger generation students at the school, college and university levels.
- to help translate proteomics knowledge into biology and vice versa.

Related website

INPPO	http://www.inppo.com

INSERTION, INSERTION-DELETION REGION, *SEE* INDEL.

INSTANCE, *SEE* INDIVIDUAL.

INSTANCE-BASED LEARNER, *SEE* K-NEAREST NEIGHBOR CLASSIFICATION.

INTEGRATED GENE ANNOTATION PIPELINES, *SEE* GENE PREDICTION, PIPELINES.

INTEGRATED GENE PREDICTION SYSTEMS, *SEE* GENE PREDICTION SYSTEMS, PIPELINES.

INTELLIGENT DATA ANALYSIS, *SEE* PATTERN ANALYSIS.

INTERACTOME

Dov Greenbaum

The list of interactions between all macromolecules of the cell.

This includes, but is not limited to, protein-protein, protein-DNA, and protein complex interactions. As many relatively simple organisms seem to have genes and protein counts on par with more complex organisms, it is thought that much of the complexity results, not from the number of proteins but rather from the degree and complexity of their interactions.

Given that much of a protein's function can sometimes be deduced from its interactions with other macromolecules in the cell, the elucidation of the interactome is important for the further annotation of the **genome**. Additionally, the sum total of interactions in an organism can be used as a tool to compare distinct organisims, defining their specific networks of protein interactions, and distinguishing genomes from each other on the basis of differing interactions.

There are presently many high throughput techniques for discovering protein protein interactions, e.g. **yeast two hybrid** or TAP tagging. Due to limitations in the experiments many high throughput experiments are prone to high levels of both false positives and false negatives. It is imperative when studying interactomes that protein interactions be verified through multiple methods to minimize the risk of incorporating false and misleading information into the databases.

Presently there are many databases which serve to collect, catalogue and curate the known protein-protein interactions. These include DIP, MIPS, and BIND.

Related websites

Database of Interacting Proteins	http://dip.doe-mbi.ucla.edu/dip/Main.cgi
BIND–The Biomolecular Interaction Network Database	http://bond.unleashedinformatics.com/Action?
MIPS Interaction Tables	http://mips.helmholtz-muenchen.de/genre/proj/mpact/
Wikipedia gives a longer list of databases	http://en.wikipedia.org/wiki/Interactome

Further reading

Gerstein M, et al. (2002) Proteomics. Integrating interactomes. *Science* 295: 284–287.

Ito T, et al. (2000) A comprehensive two-hybrid analysis to explore the yeast protein interactome. *Proc Natl Acad Sci U S A* 98: 4569–4574.

INTERGENIC SEQUENCE

Katheleen Gardiner

Regions in genomic DNA between genes.

The region between the promoter of one gene and the 3' end of the adjacent gene when two genes are arranged head to tail, or the distance between promoters when tow genes are arranged head to head. Intergenic distances are highly variable in size and related to base composition. In GC-rich regions, they tend to be short (a few kb) or non-existent (where adjacent genes overlap). In AT-rich regions, intergenic distances can be very large (tens to hundreds of kb). Intergenic regions may contain regulatory sequences or may be functionless.

Further reading

Dunham I (1999) The DNA sequence of human chromosome 22 *Nature* 402: 489–495.

Hattori M, et al. (2000) The DNA sequence of human chromosome 21 *Nature* 405: 311–320.

INTERIOR BRANCH, INTERNAL BRANCH, *SEE* BRANCH AND PHYLOGENETIC TREE.

INTERIOR NODE, INTERNAL NODE, *SEE* NODE AND PHYLOGENETIC TREE.

INTERNATIONAL CHEMICAL IDENTIFIER, *SEE* INCHI.

INTERNATIONAL IMMUNOGENETICS DATABASE, *SEE* IMGT.

INTERNATIONAL PLANT PROTEOMICS ORGANIZATION, *SEE* INPPO.

INTERNATIONAL SOCIETY FOR COMPUTATIONAL BIOLOGY (ISCB)

Pedro Fernandes

The ISCB is a scholarly society for researchers in computational biology and bioinformatics. Founded in 1997, the society's core mission is to contribute to the scientific understanding of living systems through computation.

ISCB seeks to communicate the significance of computational biology to the larger scientific community, to governmental organizations, and to the general public; the society serves its members locally, nationally, and internationally; it provides guidance for scientific policies, publications, meetings, and distributes information through multiple platforms. ISCB organizes the Intelligent Systems for Molecular Biology (ISMB) conference every year, a growing number of smaller, more regionally or topically focused annual and bi-annual conferences, and has two official journals: *PLoS Computational Biology* and *Bioinformatics*. The society awards two prizes each year, the Overton Prize and the Accomplishment by a Senior Scientist Award, and it inducts Fellows, to honor members that have distinguished themselves through outstanding contributions to the fields of computational biology and bioinformatics.

Related websites

ISCB	http://www.iscb.org/
ISMB Conference	http://www.iscb.org/about-ismb
ISCB Meetings	http://www.iscb.org/high-quality-meetings
ISCB Annual Awards	http://www.iscb.org/iscb-awards
ISCB Fellows	http://www.iscb.org/iscb-fellows
ISCB Student Council	http://www.iscbsc.org/
PLoS Computational Biology	http://www.ploscompbiol.org/
OUP Bioinformatics	http://bioinformatics.oxfordjournals.org/

INTEROLOG (INTEROLOGUE)

Dov Greenbaum

Homolog sets of proteins that are know to interact within a different organism.

It is thought that much of the diversity between different organisms can be traced to their protein-proteins interactions. Although there are many high throughput techniques designed to study protein interactions, there is also an effort to transfer annotation from one set of interactions to similar protein sets.

Logically, if two proteins interact in one organism, two functionally and structurally similar proteins can be thought to interact in another, thus hypothesizing a novel interactions without strict experimental information–similar to transferring structural annotation based on sequence homology. With the knowledge of interologues it may be possible to predict the function of a gene.

Further reading

Matthews LR, et al. (2001) Identification of potential interaction networks using sequence-based searches for conserved protein-protein interactions or 'interologs'. *Genome Res* 11: 2120–2126.

INTERPRO

Teresa K. Attwood

An integrated documentation and diagnostic resource for gene and domain families, and protein functional sites, developed initially as a means of rationalising the complementary work of the PROSITE, PRINTS, Pfam and ProDom databases (see Figure I.2). During the last decade, the database has expanded significantly, and now includes around a dozen partners, including those that provide protein structural information (e.g., Gene3D). InterPro thus combines databases with different underpinning methodologies and a varying degree of biological information – the annotation included in its entries draws heavily on that provided by PROSITE and PRINTS.

The methodologies used by InterPro's member databases include consensus (regex) patterns, fingerprints, profiles and hidden Markov models (HMMs). Diagnostically, these techniques have different areas of optimum application: some focus on conserved functional

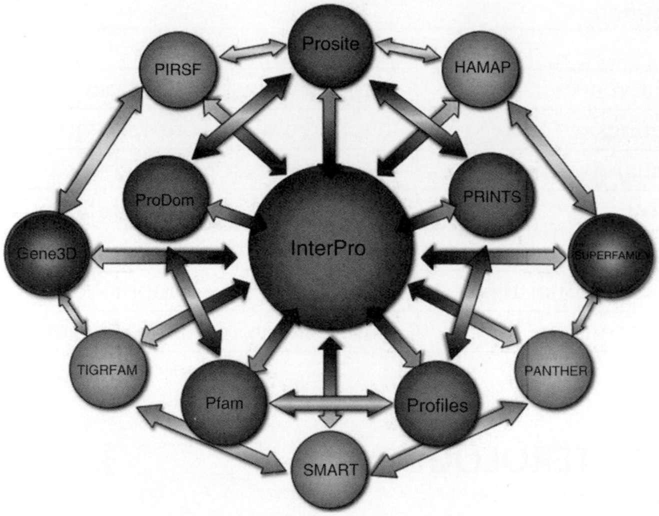

Figure I.2 Illustration showing the member databases of InterPro. The five founding partners are highlighted in the inner ring; more recent partner databases are shown in the outer ring; the databases at the horizontal extremes provide structure-based information to the resource. The arrows denote the flow of information both between the central hub and the member resources, and between member databases. The complex interplay between InterPro and its partners creates a diagnostic tool that is much greater than the sum of its parts.

sites (e.g., PROSITE); some focus on divergent domains (e.g., Pfam); and others focus on families, specialising in hierarchical definitions from superfamily down to subfamily levels in order to pin-point specific functions (e.g., PRINTS) – this hierarchy forms the backbone of InterPro's gene family organisation. ProDom exploits a different approach, instead using an automatic algorithm to cluster similar sequences in UniProtKB – this allows the resource to be relatively comprehensive, because it does not depend on manual crafting, validation and annotation of family discriminators.

InterPro is hosted at the European Bioinformatics Institute, UK. Its entries have unique accession numbers and include functional descriptions, literature references, links back to the relevant member database(s), and match-lists against UniProtKB, with links to graphical views of the results. The database may be interrogated with simple keywords, or searched with query sequences using InterProScan. By uniting complementary databases, InterPro capitalises on their individual strengths, producing a powerful, integrated diagnostic tool that is greater than the sum of its parts.

Related websites

InterPro	http://www.ebi.ac.uk/interpro/
InterPro text search	http://www.ebi.ac.uk/interpro/search.html
InterProScan	http://www.ebi.ac.uk/Tools/pfa/iprscan/

Further reading

Greene EA, Pietrokovski S, Henikoff S, et al. (1997) Building gene families. *Science* 278: 615–626.

Henikoff S, Greene EA, Pietrokovski S, Attwood TK, Bork P, Hood L (1997) Gene families: the taxonomy of protein paralogs and chimeras. *Science* 278: 609–614.

Hunter S, Apweiler R, Attwood TK, et al. (2009) InterPro: the integrative protein signature database. *Nucleic Acids Res*. 37: D211–215.

Tatusov RL (1997) A genomic perspective on protein families. *Science* 278: 631–637.

Zdobnov EM, Apweiler R (2001) InterProScan–an integration platform for the sigature-recognition methods in InterPro. *Bioinformatics* 17(9): 847–848.

See also Domain Family, Gene Family, Pfam, ProDom, PROSITE, PRINTS, Protein Family.

INTERPROSCAN, *SEE* INTERPRO.

INTERSPERSED SEQUENCE (LONG-TERM INTERSPERSION, LONG-PERIOD INTERSPERSION, SHORT-TERM INTERSPERSION, SHORT-PERIOD INTERSPERSION, LOCUS REPEAT)

Dov Greenbaum

Sequences comprising repeated sequences interspersed with unique sequences.

The coding regions of the genomes of many insects and plants have either short or long period interspersion patterns. Typically, smaller genomes contain long-period interspersions and larger genomes have short-period interspersions.

A long-period interspersion is defined as repetitious DNA sequences greater than 5.6 kb that alternate with longer stretches of non-repetitious DNA. Short-period interspersions are composed of unique sequences of DNA that alternate with shorter repetitious sequences.

Further reading

Britten RJ, Davidson EH (1976) DNA sequence arrangement and preliminary evidence on its evolution. *Fed Proc* 35: 2151–2157.

INTRINSIC GENE PREDICTION, *SEE* GENE PREDICTION, AB INITIO.

INTRON

Niall Dillon and John M. Hancock

Sequence within a gene that separates two exons.

Eukaryotic genes are split into exons and intervening sequences (introns). The primary RNA transcript contains both exons and introns and the introns are then spliced out in a complex splicing reaction which is closely coupled to transcription. In the majority of genes, the sequence of the intron begins with GT and ends with AG (the GT-AG rule).

Three classes of intron are recognised: nuclear introns, Group I introns and Group II introns.

Nuclear introns are the common type of introns in eukaryotic nuclear genes and require the spliceosome (a macromolecular complex of proteins and small nuclear RNAs (snRNAs)) to be spliced.

Group I introns are self-splicing introns that have a conserved secondary structure. They are found in RNA transcripts of protozoa, fungal mitochondria, bacteriophage T4 and bacteria.

Group II introns are found in fungal mitochondria, higher plant mitochondria and plastids. They have a conserved secondary structure and may or may not require the participation of proteins in the splicing reaction.

Further reading

Krebs JE (ed.) (2009) *Lewin's Genes X*. Jones and Bartlett Publishers, Inc.

See also Exon.

INTRON PHASE

Austin L. Hughes

The phase of an intron is a way of describing the position of the intron relative to the reading frame of the exons.

There are three types of intron phase depending on where the intron is located. In phase 0, the intron lies between codons. In phase 1, the intron occurs between the first and second bases of a codon. In phase 2, the intron occurs between the second and third bases of a codon. An exon in which both upstream and downstream introns have the same phase is called a symmetrical exon. If an exon from one gene is inserted into an intron of another gene (as has been hypothesized to occur in cases of exon shuffling), the exon must be symmetrical if the new exon is to be functional in its new location.

Further reading

Long M, et al. (1995) Intron phase correlations and the evolution of the intron/exon structure of genes. *Proc Natl Acad Sci USA* 92: 12495–12499.

Patthy L (1991) Modular exchange principles in proteins. *Curr Opin Struc Biol* 1: 351–362.

See also Exon shuffling, Intron

I

IR, *SEE* TEXT MINING.

ISCB, *SEE* INTERNATIONAL SOCIETY FOR COMPUTATIONAL BIOLOGY.

ISOBARIC TAGGING

Simon Hubbard

A quantitative proteomics method for labelling proteome peptides with a series of tags that share the same mass, constituted by reporter and balancer groups whose individual masses vary but sum to the same value. In the tandem mass spectrometry experiment, the labile reporter and balancer tags are lost from the peptide at the fragmentation stage to yield the usual ions series from which Peptide Spectrum Matches (PSMs) are obtained, along with reporter ion signal at the low mass end of the tandem MS spectrum. The intensity of these reporter ion peaks is used to quantify the level of the parent peptide from which these ions were derived, and by inference, their parent proteins. Examples of this technique are iTRAQ™ and Tandem Mass Tags (TMT).

See also Quantitative Proteomics

ISOCHORE

Katheleen Gardiner

Within a genome, a large DNA segment (>300 kb) that is homogeneous in base composition.

Genomes of warm blooded vertebrates are mosaics of isochores belonging to four classes: the L isochores are AT-rich (~35%–42% GC), and the H1, H2 and H3 isochores that are increasingly GC-rich (42%–46%, 47%–52% and >52%, respectively). Isochores may be reflected in chromosome band patterns. G bands appear to be uniformly composed of L isochores, while R bands are heterogeneously contain all classes. The brightest R bands, T bands, contain H2 and H3 isochores.

Further reading

Bernardi G (2000) Isochores and the evolutionary genomics of vertebrates *Gene* 241: 3–17.

Eyre-Walker A, Hurst LD (2001) The evolution of Isochores *Nat Rev Genet* 2: 549–555.

Oliver JL, Bernaola-Galvan P, Carpena P, Roman-Roldan R (2001) Isochore chromosome maps of eukaryotic genomes *Gene* 276: 47–56.

Pavlicek A, Paces J, Clay O, Bernardi G. (2002) A compact view of isochores in the draft human genome sequence *FEBS Lett* 511: 165–169.

I

ISOMIR

Ana Kozomara and Sam Griffiths-Jones

Different processed microRNAs generated from the same precursor hairpin.

Many different reads from deep sequencing experiments map to each microRNA precursor sequence. These reads exhibit significant heterogeneity, with different 5' and 3' ends. These different forms are referred to as isomiRs. The functional significance of isomiR sequences is unknown.

Further reading

Fernandez-Valverde SL, Taft RJ, Mattick JS (2010) Dynamic isomiR regulation in Drosophila development. *RNA* 16: 1881–1888.

Kozomara A, Griffiths-Jones S (2011) miRBase: integrating microRNA annotation and deep-sequencing data. *Nucleic Acids Res* 39: D152–D157.

See also MicroRNA, MicroRNA duplex, Mature microRNA

ITERATION

Matthew He

Iteration is a series of steps in an algorithm whereby the processing of data is performed repetitively until the result exceeds a particular threshold. Iteration is often used in multiple sequence alignments whereby each set of pairwise alignments are compared with every other, starting with the most similar pairs and progressing to the least similar, until there are no longer any sequence-pairs remaining to be aligned.

IUPAC-IUB CODES (NUCLEOTIDE BASE CODES, AMINO ACID ABBREVIATIONS)

John M. Hancock

Codes used to represent nucleotide bases and amino acid residues.

The process of sequencing DNA or protein molecules does not always produce unambiguous results. Similarly, the representation of consensus sequences of either type necessitates a simple way of representing sequence redundancy at a position. To produce a standard nomenclature that could accommodate these requirements, the IUPAC (International Union of Pure and Applied Chemistry) and IUB (International Union of Biochemistry) Commission on Biochemical Nomenclature agreed standard nomenclatures for nucleic acids and amino acids. For nucleic acids they agreed a single nomenclature, while for amino acids they agreed two parallel nomenclatures, a single letter and a three letter code, the three letter code being more descriptive. See Tables I.1 and I.2.

Table I.1 Commonly used IUPAC-IUB codes for nucleic acid bases

Code	Base
a	adenine
c	cytosine
g	guanine
t	thymine in DNA; uracil in RNA
m	a or c
r	a or g
w	a or t
s	c or g
y	c or t
k	g or t
v	a or c or g; not t
h	a or c or t; not g
d	a or g or t; not c
b	c or g or t; not a
n	a or c or g or t

Table I.2 Commonly used IUPAC-IUB codes for amino acids

Code 3 Letter	1 Letter	Amino Acid
Ala	A	Alanine
Arg	R	Arginine
Asn	N	Asparagine
Asp	D	Aspartic acid (Aspartate)
Cys	C	Cysteine
Gln	Q	Glutamine
Glu	E	Glutamic acid (Glutamate)
Gly	G	Glycine
His	H	Histidine
Ile	I	Isoleucine
Leu	L	Leucine
Lys	K	Lysine
Met	M	Methionine
Phe	F	Phenylalanine
Pro	P	Proline
Ser	S	Serine
Thr	T	Threonine
Trp	W	Tryptophan
Tyr	Y	Tyrosine
Val	V	Valine
Asx	B	Aspartic acid or Asparagine
Glx	Z	Glutamine or Glutamic acid.
Xaa	X	Any amino acid.
TERM		termination codon

Related websites

Amino Acid nomenclature	http://www.chem.qmul.ac.uk/iupac/AminoAcid/
Nucleic Acid nomenclature	http://www.chem.qmul.ac.uk/iupac/misc/naabb.html

Further reading

Cornish-Bowden A (1985) Nomenclature for incompletely specified bases in nucleic acid sequences: recommendations 1984 *Nucleic Acids Res* 13: 3021–3030.

IUPAC-IUB Joint Commission on Biochemical Nomenclature (JCBN). (1984) Nomenclature and symbolism for amino acids and peptides. Recommendations 1983. *Eur J Biochem.* 138: 9–37.

See also Base Pair, Amino Acid

J

Jaccard Distance (Jaccard Index, Jaccard Similarity Coefficient)
Jackknife
JASPAR
JELLYFISH
jModelTest
Jpred, *see* Web-based Secondary Structure Prediction Programs.
Jumping Gene, *see* Transposable Element.
Junk DNA

Concise Encyclopaedia of Bioinformatics and Computational Biology, Second Edition.
Edited by John M. Hancock and Marketa J. Zvelebil.

J

JACCARD DISTANCE (JACCARD INDEX, JACCARD SIMILARITY COEFFICIENT)

John M. Hancock

The Jaccard Distance is based on the Jaccard Index which measures the ratio of the size of the intersection of the two data sets divided by their union:

$$J(A, B) = \frac{|A \cap B|}{|A \cup B|}$$

The Jaccard Distance is:

$$J_\delta(A, B) = 1 - J(A, B) = \frac{|A \cup B| - |A \cap B|}{|A \cup B|}$$

Related website

Wikipedia	http://en.wikipedia.org/wiki/Jaccard_index

See also Tanimoto Distance

JACKKNIFE

Dov Greenbaum

A method used when attempting to determine a confidence level of a proposed relationship within a phylogeny.

Jackknifing is similar to bootstrapping, but differs in that it only removes (iteratively and randomly) one value and uses the new smaller data set to calculate the phylogenetic relationship. After multiple repetitions a consensus tree can be built. The degree of variance of the sample can then be determined by comparing each branch in the original tree to the number of times it was found via the jackknifing algorithm.

See also Bootstrap, Cross-Validation

JASPAR

Obi L. Griffith and Malachi Griffith

The JASPAR CORE database contains a curated, non-redundant set of profiles, derived from published collections of experimentally defined transcription factor binding sites for eukaryotes. Whereas databases like Oreganno and PAZAR contain individual binding sites for transcription factors, JASPAR contains transcription factor DNA-binding preferences, modeled as matrices. These can be converted into Position Weight Matrices (PWMs or PSSMs), used for scanning genomic sequences. Binding sites were historically determined by experiments designed specifically for that purpose (e.g., SELEX). New high-throughput techniques, like ChIP-seq, are increasingly being used. In addition to the CORE database, which focuses on individual transcription factors, JASPAR also includes several other modules with models for transcription factor families, phylogenetically conserved gene upstream elements, RNA polymerase II core promoter elements, highly conserved non-coding elements, splice sites, and others. JASPAR was conceived as an alternative to other closed and commercial models (e.g., Transfac). For related projects, see Oreganno and PAZAR.

Related website

JASPAR	http://jaspar.cgb.ki.se/

Further reading

Portales-Casamar E, et al. (2010) JASPAR 2010: the greatly expanded open-access database of transcription factor binding profiles. *Nucleic Acids Res.* 38(Database issue):D105–110.

JELLYFISH

Michael P. Cummings

A program for counting substrings (*k*-mers) in DNA sequence data.

It makes use of compare-and-swap CPU instructions for parallel execution and efficient encoding resulting in less memory usage and faster execution compared to other algorithms.

The program is written in C++, runs on 64-bit Intel-compatible processors under Linux, FreeBSD or Mac OS, and requires GNU GCC to compile.

Related website

JELLYFISH home page	http://www.cbcb.umd.edu/software/jellyfish/

Further reading

Marcais G, Kingsford C (2011) A fast, lock-free approach for efficient parallel counting of occurrences of k-mers. *Bioinformatics* 27: 764–770.

JMODELTEST

Michael P. Cummings

A program for selection of nucleotide substitution models.

Models are evaluated under different criteria, including hierarchical and dynamical likelihood ratio tests (hLRT and dLRT), Akaike and Bayesian information criteria (AIC and BIC), and a decision theory method (DT). The program provides numerous statistics for the evaluated models and model parameters.

Executables are available for Mac OS, Windows and Linux. The program subsumes and extends the capabilities of ModelTest.

Related website

jModelTest home page	http://code.google.com/p/jmodeltest2/

Further reading

Darriba D, Taboada GL, Doallo R, Posada D (2012) jModelTest 2: more models, new heuristics and parallel computing. *Nat Method*. 9: 772.

JPRED, *SEE* WEB-BASED SECONDARY STRUCTURE PREDICTION PROGRAMS.

JUMPING GENE, *SEE* TRANSPOSABLE ELEMENT.

JUNK DNA

Dov Greenbaum

Non-coding DNA includes introns transposable elements, pseudogenes, repeat elements, satellites, UTRs hnRNAs LINEs SINEs, as well as unidentified junk and makes up approximately 97% of the human genome. Junk DNA is ubiquitous and extends to all forms of life, making it an exciting evolutionary phenomenon.

The most functionless theory of Junk DNA claims that it is just that, a generous juxtaposition of non-functional junk. This useless DNA grows in the genome until the costs of replicating it become too great to maintain. Thus organisms that develop at a slower rate tolerate more junk and use it to their advantage to slow down the rate of development via increased cell cycle length.

Some scientists have posited that junk has only a passive purpose. The total genome size is related to a number of organismal and cellular level traits, thereby suggesting that there is a selective advantage in larger genomes, including those that result from junk DNA filler. One theory claims that Junk absorbs harmful chemicals that could affect genuine genes. This has been refuted; in fact, larger genomes are subjected to more physical and chemical damage, outweighing the bodyguard function. Moreover, it has been shown that most mutations that reduce viability occur in non-coding DNA, possibly indicating that Junk plays an active role in the genome.

Another postulated purpose hypothesizes Junk as a sink for DNA-tropic proteins, thereby buffering the effect of intracellular solute concentrations on nuclear machinery. This energy independent function of Junk could allow for a reduced basal metabolic rate and therefore an evolutionary niche.

Researchers studying Junk tend to favor searching through repeat elements for function. Repeat elements are thought to be involved in chromosomal integrity. Many are also found in the heterochromatin and may be involved in centromeric activity and chromosome pairing, both have relevance to evolutionary divergence and speciation through manipulation of chromosomes.

Additionally, non-repetitive Junk may also have functions that can be elucidated. It was determined, that non-coding DNA has a high GC concentration. Theoretically ORFs within these regions would have higher adaptive plasticity, as they could use GC rich DNA to vary their final products for selective advantage via alternative splicing.

Junk sequences may also function as cis-acting transcriptional regulators. One theory posits that base pair distribution in non-coding DNA has an effect on gene transcription through a thermodynamic process. Moreover, the movement of transposable Junk results in a dynamic system of gene activation, which allows for the organism to adapt to its environment without redesigning its hardwired system of gene activators.

In 1994 it was proposed that Junk might be similar to natural languages since it follows, among other things, Zipf's law. The researchers cited this as a possible proof that there exists one or more structured language in our Junk. This idea was refuted in many letters that claimed, among other things, that Junk does not fit Zipf's law any better than coding DNA.

Similarly, researchers, using a variety of statistical techniques have deduced long-range correlations in Junk DNA. This may be attributable to the new found functions of many intergenic non-coding regions, which include replication, chromosome segregation, recombination, chromosome stability and interaction with the nuclear matrix. All of these require the 'high redundancy low information' sequence inherent in Junk.

Further reading

Veitia RA, Bottani S. (2009) Whole genome duplications and a 'function' for junk DNA? Facts and hypotheses. *PLoS One*. Dec 14;4(12):e8201.

Vickers KC, Palmisano BT, Remaley AT. (2010) The role of noncoding "junk DNA" in cardiovascular disease. *Clin Chem*. 6(10):1518–1520. Epub 2010.

Khajavinia A, Makalowski W (2007) What is "junk" DNA, and what is it worth?. *Sci Am*. 296(5):104.

K

K-Fold Cross-Validation, *see* Cross-Validation.

K-Means Clustering, *see* Clustering.

K-Medoids

K-Nearest Neighbor Classification (Lazy Learner, KNN, Instance-based Learner)

Kappa Virtual Dihedral Angle

Karyotype

K_d, *see* Binding Affinity.

Kernel-based Learning Method, *see* Kernel Method.

Kernel Function

Kernel Machine, *see* Kernel Method.

Kernel Method (Kernel Machine, Kernel-based Learning Method)

K_i, *see* Binding Affinity.

KIF, *see* Knowledge Interchange Format.

Kin Selection

Kinetic Modeling

Kinetochore

Kingdom

KNN, *see* K-Nearest Neighbor Classification, Lazy Learner, Instance-based Learner.

Knowledge

Knowledge Base

Knowledge-based Modeling, *see* Homology Modeling.

Knowledge Interchange Format (KIF)

Knowledge Map, *see* Bayesian Network.

Knowledge Representation Language (KRL)

Kozak Sequence

KRL, *see* Knowledge Representation Language.

Concise Encyclopaedia of Bioinformatics and Computational Biology, Second Edition.
Edited by John M. Hancock and Marketa J. Zvelebil.

K-FOLD CROSS-VALIDATION, *SEE* CROSS-VALIDATION.

K-MEANS CLUSTERING, *SEE* CLUSTERING.

K-MEDOIDS

Concha Bielza and Pedro Larrañaga

This is a clustering partitional algorithm similar to k-means that groups the data set of objects into a predefined number of k clusters. Both k-means and k-medoids attempt to minimize the distance (or dissimilarity) between points labeled to be in a cluster and a point designated as the center of that cluster. However, the k-medoids method is more robust to noise and outliers than k-means because it minimizes a sum of pairwise dissimilarities instead of a sum of squared Euclidean distances. Also, the medoid is the center of each cluster and its average dissimilarity to all the objects in the cluster is minimal. Medoids are always members of the data set and it makes sense when a mean or centroid cannot be defined such as in the gene expression context.

k-medoids basically works iteratively. First, a set of medoids (data points) is chosen at random. Second, each object is associated to the closest medoid (using any distance metric or similarity measure). Third, for each non-medoid a possible swapping of the non-medoid and the medoid is evaluated. The process is repeated until there is no change in the medoid. The most representative algorithm is the Partitioning Around Medoids (PAM) algorithm.

Further reading

Huang D, Pan W (2006) Incorporating biological knowledge into distance-based clustering analysis of microarray gene expression data. *Bioinformatics* 22(10): 1259–1268.

Kohlhoff KJ, Sosnick MH, Hsu WT, Pande VS, Altman RB (2011) CAMPAIGN: An open-source library of GPU-accelerated data clustering algorithms. *Bioinformatics* 27(16): 2322–2323.

Van Der Lann MJ, Pollard KS, Bryan JE (2003) A new partitioning around medoids algorithm. *Journal of Statistical Computation and Simulation* 73(8): 575–584.

See also Clustering

K-NEAREST NEIGHBOR CLASSIFICATION (LAZY LEARNER, KNN, INSTANCE-BASED LEARNER)

Feng Chen and Yi-Ping Phoebe Chen

K-nearest neighbor classification is for classifying objects based on closest training examples. The training examples are vectors in a multidimensional feature space, each with a class label. The training phase of the algorithm consists only of storing the

K

feature vectors and class labels of the training samples. In the classification phase, k is a pre-defined parameter, and an unlabeled vector is categorized by assigning the label which is most frequent among the k training samples nearest to that query point. During this process, Euclidean distance is always used to indicate how close two vectors are.

The best choice of k depends upon the data. Larger values of k often reduce the effect of noise on the classification although they also make boundaries between classes less distinct. A satisfying k can be chosen by various heuristic techniques such as cross-validation.

The main drawback of kNN is that the classes with the more frequent examples tend to dominate the prediction of the new vector because they are very likely to come up in the k nearest neighborus when the neighbours are computed.

Further reading

Shakhnarovish G, Darrell P, Indyk T (eds) (2005) *Nearest-Neighbor Methods in Learning and Vision*. MIR Press 2005.

See also Cross-Validation, K-Fold Cross-Validation

KAPPA VIRTUAL DIHEDRAL ANGLE

Roman Laskowski and Tjaart de Beer

Defined by four successive Cα atoms along a protein chain: $C\alpha_{i-1} - C\alpha_i - C\alpha_{i+1} - C\alpha_{i+2}$.

KARYOTYPE

Katheleen Gardiner

The number and structure of the complement of chromosomes from a cell, individual or species.

It is determined by examining metaphase chromosomes under the light microscope after they have been stained to produce chromosome banding patterns. Structural features include the relative sizes of each chromosome and the locations of each centromere, plus the identity of the sex chromosomes. Each species has a unique normal karyotype that is invariant among individuals of the same sex. For example, the human karyotype is composed of 23 pairs of chromosomes, 22 autosomes and two sex chromosomes. Members of an autosome pair appear identical, contain the same set of genes and are referred to as homologous chromosomes. Sex chromosomes are usually significantly different in size and gene content.

Further reading

Miller OJ, Therman E (2000) *Human Chromosomes* (Springer).

Scriver CR, Beaudet AL, Sly WS, Valle D (eds) (1995) *The Metabolic and Molecular Basis of Inherited Disease*. Springer.

Standing Committee on Human Cytogenetic Nomenclature (1985) *An International System for Human Cytogenetic Nomenclature*. Karger.

Wagner RP, Maguire MP, Stallings RL (1993) *Chromosomes – a synthesis* (Wiley-Liss)

K_D, *SEE* BINDING AFFINITY.

KERNEL-BASED LEARNING METHOD, *SEE* KERNEL METHOD.

KERNEL FUNCTION

Nello Cristianini

Functions of two arguments that return the inner product between the images of such arguments in a vector space.

If the map embedding the two arguments is called ϕ, we can write a kernel function as $k(x,z) = \langle\phi(x), \phi(z)\rangle$. The arguments can be vectors, strings, or other data structures. These functions are used in Machine Learning and Pattern Analysis, as the essential building block of algorithms in the class of Kernel Methods.

A simple example of a kernel function is given by the following map from a 2-dimensional space to a 3-dimensional one: $\phi(x_1, x_2) = (x_1^2, x_2^2, \sqrt{2}x_1x_2)$. The inner product in such a space can easily be computed by a kernel function without explicitly rewriting the data in the new representation.

Consider two points:

$$x = (x_1, x_2)$$

$$y = (y_1, y_2)$$

and consider the kernel function obtained by squaring their inner product:

$$\begin{aligned}\langle x, z\rangle^2 &= (x_1z_1 + x_2z_2)^2 \\ &= x_1{z_1}^2 + x_2{z_2}^2 + 2x_1z_1x_2z_2 \\ &= \langle({x_1}^2, {x_2}^2, \sqrt{2}x_1x_2), ({z_1}^2,\ {z_2}^2,\ \sqrt{2}z_1z_2)\rangle \\ &= \langle\phi(x), \phi(z)\rangle\end{aligned}$$

This corresponds to the inner product between two 3-dimensional vectors, and is computed without explicitly writing their coordinates. By using a higher exponent, it is possible to virtually embed those two vectors in a much higher dimensional space, at a very low computational cost.

More generally, one can prove that every symmetric and positive definite function $k(x,z)$ is a valid kernel, that is a map $\phi : \mathrm{X} \rightarrow \mathrm{R}^n$ exists such that $k(x,z) = \langle\phi(x), \phi(z)\rangle$, where x $\in$X and z$\in$X. Examples of kernels include the Gaussian $k(x, z) = e^{-\frac{\|x-y\|^2}{2\sigma^2}}$, the general polynomial kernel $k(x,z) = (\langle x,z\rangle + 1)^d$ and many others, including kernels defined over sets of sequences that have been used for bioinformatics applications such as remote protein homology detection.

K

Related website

Kernel Machines web site	www.kernel-machines.org

Further reading

Cristianini N, Shawe-Taylor J (2000) *An Introduction to Support Vector Machines.* Cambridge University Press.

See also Kernel Method, Support Vector Machine

KERNEL MACHINE, *SEE* KERNEL METHOD.

KERNEL METHOD (KERNEL MACHINE, KERNEL-BASED LEARNING METHOD)

Nello Cristianini

A class of algorithms for machine learning and pattern analysis based on the notion of the kernel function.

The main idea underlying the use of kernel methods is to embed the data into a vector space where an inner product is defined. Linear relations among the images of the data in such a space are then sought. The mapping (or embedding) is performed implicitly, by defining a kernel function $K(x,z) = \langle \phi(x), \phi(z) \rangle$.

By mapping the data into a suitable space, it is possible to transform non-linear relations within the data into linear ones. Furthermore the input domain does not need to be a vector space, so that statistical methods developed for data in the form of inputs can be applied to data structures such as strings by using kernels.

Kernel-based algorithms only require information about the inner-products between data points. Such information can often be obtained at a computational cost that is independent of the dimensionality of that space. The kind of relations detected by kernel methods include classifications, regressions, clustering, principal components, canonical correlations and many others.

Figure K.1 shows the basic idea of a kernel induced embedding into a feature space: that by using a non-linear map some problems can be simplified.

In the kernel approach, the linear functions used to describe the relations found in the data are written in the form $f(x) = \sum \alpha_i K(x_i, x)$ called the dual form, where K is the chosen kernel, α_i are the dual coordinates, and x_i the training points.

Most kernel-based learning algorithms, such as Support Vector Machines, reduce their training phase to optimizing a convex cost function, hence avoiding one of the main computational pitfalls of Neural Networks. Since they often make use of very high dimensional spaces, Kernel Methods run the risk of overfitting. For this reason, their design needs to incorporate principles of statistical learning theory, that help identify the crucial parameter that need to be controlled in order to avoid this risk.

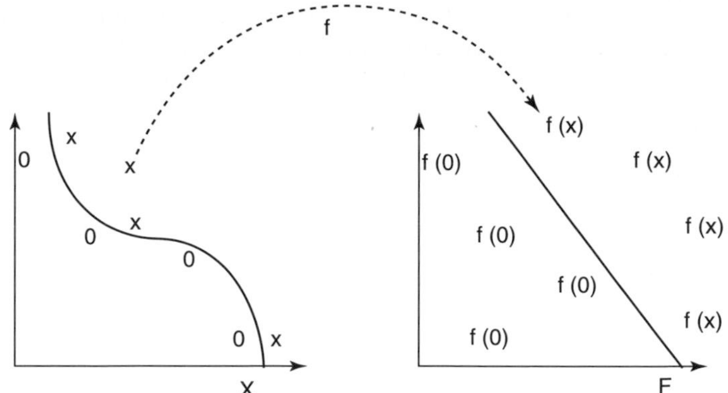

Figure K.1 The embedding of data into a feature space may simplify the classification problem.

Related website

Kernel Machines web site	www.kernel-machines.org

Further reading

Cristianini N, Shawe-Taylor J. (2000) *An Introduction to Support Vector Machines*. Cambridge University Press.

K_I, *SEE* BINDING AFFINITY.

KIF, *SEE* KNOWLEDGE INTERCHANGE FORMAT.

KIN SELECTION

A.R. Hoelzel

A mechanism for the adaptive evolution of genes through the survival of relatives.

The central idea is said to have been worked out by the British geneticist J.B.S. Haldane in a pub one evening, where he announced that he would lay down his life for two brothers, eight cousins, etc. Kin selection provides a model for the evolution of altruism, whereby an altruist provides a fitness benefit to a recipient, at a fitness cost to the altruist. This can evolve by kin selection if the ratio of the recipient's gains to the altruist's cost is greater than the reciprocal of the coefficient of relatedness between the two, and this is known as Hamilton's rule (Hamilton 1964):

$$B/C > 1/r$$

K

It is based on the supposition that evolution should make no distinction between genes in direct compared to indirect descendent kin. Provided that the representation of the gene increases in the next generation (through inclusive fitness), traits that reduce individual fitness can evolve.

Further reading

Hamilton WD (1964) The genetical theory of social behaviour. *J. Theor. Biol.* 7: 1–52.

See also Evolution, Natural Selection

KINETIC MODELING

Neil Swainston

Kinetic modeling is a systems biology paradigm used to model the dynamic behavior of a system, such as the variation in metabolite concentrations over time following a perturbation. Due to the increased complexity of kinetic modeling, the systems considered are typically smaller than those in constraint-based modeling, often focusing on a single pathway. One approach to kinetic modeling is to represent the system as a collection of individual reactions, but with each reaction being defined by a non-linear kinetic rate law, with the rate of change of the reaction participant concentrations over time being defined by ordinary differential equations (ODEs). By solving the collection of ODEs, the dynamic behavior of the reaction participants can be simulated.

While many kinetic parameters have already been measured and are publicly available in resources such as BRENDA and SABIO-RK, the range of data available is currently insufficient for genome-scale kinetic modeling to be realized. Additionally, the kinetic laws represented by ODEs have a dependency upon enzyme concentration. Although some enzyme concentrations have been measured and are publicly available, the coverage of such data sets is even smaller than that of kinetic parameters. Consequently, experimental determination of both kinetic parameters and enzyme concentrations remain necessary steps in the development of accurate kinetic models. Kinetic modeling can be performed by software tools such as COPASI and the web application JWS Online (http://www.jjj.bio.vu.nl/). The database BioModels (www.biomodels.org) collects molecular models by using a standardized format (Li, Donizelli et al. 2010).

Related websites

JWS Online	http://www.jjj.bio.vu.nl/
BioModels	www.biomodels.org

Further reading

Hoops S, Sahle S, Gauges R, et al. (2006) COPASI – a COmplex PAthway SImulator. *Bioinformatics* 22: 3067–3074.

Klipp, E. (2007) Modelling dynamic processes in yeast. *Yeast* 24(11): 943–959.

Mendes P, Messiha H, Malys N, Hoops S. (2009) Enzyme kinetics and computational modeling for systems biology. *Methods Enzymol.*, 467: 583–599.

Olivier BG, Snoep JL. (2004) Web-based kinetic modelling using JWS Online. *Bioinformatics* 20: 2143–2144.

Rojas I, Golebiewski M, Kania R, et al. (2007) Storing and annotating of kinetic data. *Silico Biol.* 7: S37–44.

Schomburg I, Chang A, Schomburg D. (2002) BRENDA, enzyme data and metabolic information. *Nucleic Acids Res.* 30: 47–49.

Teusink B, Passarge J, Reijenga CA, et al. (2000) Can yeast glycolysis be understood in terms of in vitro kinetics of the constituent enzymes? Testing biochemistry. *Eur J Biochem.* 267: 5313–5329.

van Riel NA (2006) Dynamic modelling and analysis of biochemical networks: mechanism-based models and model-based experiments. *Brief Bioinform.* 7(4): 364–74.

KINETOCHORE

Katheleen Gardiner

Region of the centromere to which spindle fibers attach during mitosis and meiosis and required for segregation of chromatids to daughter cells.

The kinetochore is composed of chromatin (DNA and proteins) complexed with additional specialized proteins.

Further reading

Miller OJ, Therman E (2000) *Human Chromosomes* (Springer).

Schueler MG, Higgens AW, Rudd MK, Gustashaw K, Willard HF (2001) Genomic and genetic definition of a functional human centromere Science 294: 109–115.

Scriver CR, Beaudet AL, Sly WS, Valle D (eds) (1995) *The Metabolic and Molecular Basis of Inherited Disease.*

Sullivan BA, Schwartz S, Willard HF (1996) Centromeres of human chromosomes Environ Mol Mutagen 28: 182–191.

Wagner RP, Maguire MP, Stallings RL (1993) *Chromosomes – a synthesis* (Wiley-Liss).

KINGDOM

Dov Greenbaum

Traditionally the highest level of taxonomic classification.

Until the twentieth century biology divided life into two classes, animal and plant. In 1969 Robert Whittaker proposed dividing the kingdoms into 5 separate groups. An alternative approach, by Carl Woese, was to divide all organisms into three domains or kingdoms. Woese, stating that the 'five-kingdom scheme is essentially a mixture of taxonomic apples and oranges', sought to create a theory that allowed useful predictions to be made from its organization – 'The incredible diversity of life on this planet, most of which is microbial, can only be understood in an evolutionary framework.' Although each of the three kingdoms is composed of distinct and diverse organisms, they have unique and unifying characteristics.

Eukaryotes (including the Whittaker kingdoms of Protista, Plantae, Animalia and Fungi) are defined by those organisms with nuclei, DNA bound by histones and forming chromosomes, cytoskeletons and internal membranes (i.e. organelles) Prokaryotes and Archea, both lacking the above components, were once considered a single group until Woese and colleagues found, using the 16S ribosomal RNA, that Archea, which lived at high temperatures or produced methane could be defined as a separate group from the other bacteria. Archea are divided into extreme thermophiles, extreme halophiles and methanogens. Further analysis has shown that Archea differ dramatically from Bac teria and are often a mosaic of bacterial and eukaryotic features, thus genes and structures are similar either to Prokayotes, Eukayotes or neither.

Although subsequent systems have been proposed, none have usurped this present system. The classical phylogenetic tree based on the ribosome (a cellular component common to all kingdoms whose RNA changes slowly, thus the greater variation between the genetic sequences, the greater the evolutionary distance) shows how each of these kingdoms cluster.

Related website

Biodiversity: The Three Domains of Life	http://www.biology.iupui.edu/biocourses/N100/2k23domain.html

Further reading

Olsen GJ, Woese CR (1993) Ribosomal RNA: a key to phylogeny FASEB J 7: 113–123.

Woese CR, Kandler O, Wheelis ML (1990) Towards a natural system of organisms: Proposal for the domains Archaea, Bacteria, and Eucarya. *Proc. Natl. Acad. Sci. USA*. 87: 4576–4579.

KNN, *SEE* K-NEAREST NEIGHBOR CLASSIFICATION, LAZY LEARNER, INSTANCE-BASED LEARNER.

KNOWLEDGE

Robert Stevens

What we understand about a domain of interest.

Normally, we look at three levels of interpretation: data, information and knowledge. Data can be thought of as the stimuli that reach our senses – the patterns of light that reach our eyes or sound waves that reach our ears. Similarly, data are the patterns of bits held in a computer or the raw results gathered from an experiment. The notion of information adds interpretation to these data often in the form of classification. These patterns of light are a person; these sounds are words (nouns, verbs, etc.); these bits are an integer or a 'person' object. Knowledge is what we understand about a piece of information. The information entity *person* is called 'Robert Stevens'; he is a middle-aged man; he has a bachelors degree in biochemistry and a doctorate in computer science; he is a lecturer in bioinformatics at the University of Manchester. It is this knowledge that we wish to capture in an ontology.

KNOWLEDGE BASE

Robert Stevens

Classically, a knowledge base is an ontology together with facts about the categories in it and their instances.

The ontology provides a model or schema, and the instances collected must conform to that schema. Systems such as Protégé-2000 provide an ontology modelling environment and the ability to create forms for collecting instances conforming to the model. It is the use of reasoning to make inferences over the facts that distinguishes a knowledge base from a database.

KNOWLEDGE-BASED MODELING, *SEE* HOMOLOGY MODELING.

KNOWLEDGE INTERCHANGE FORMAT (KIF)

Robert Stevens

Knowledge Interchange Format (KIF) provides a declarative language for describing knowledge and for the interchange of knowledge among disparate programs. KIF has a declarative semantics (i.e. the meaning of expressions in the representation can be understood without appeal to an interpreter for manipulating those expressions). It is logically comprehensive (i.e. it provides for the expression of arbitrary sentences in first-order predicate calculus), and it provides for the representation of knowledge about knowledge. KIF is not intended as a primary language for interaction with human users (though it can be used for this purpose). Different programs can interact with their users in whatever forms are most appropriate to their applications (for example frames, graphs, charts, tables, diagrams, natural language, and so forth) [bec02].

Related website

Knowledge Interchange Format (KIF)	http://logic.stanford.edu/kif/kif.html

Further reading

Bechhofer S (2002) Ontology Language Standardisation Efforts; OntoWeb Deliverable D4.0; http://ontoweb.aifb.uni-karlsruhe.de/About/Deliverables/d4.0.pdf.

KNOWLEDGE MAP, *SEE* BAYESIAN NETWORK.

KNOWLEDGE REPRESENTATION LANGUAGE (KRL)

Robert Stevens

An ontology is couched in terms of concepts and relationships. This conceptualization of a domain's knowledge needs to be encoded so that it can be stored, transmitted and used by humans, computers or both. In biological databases, the usual form for representing knowledge is English. This form of representation is not very amenable to computational processing, as well as having the disadvantage of ambiguity in interpretation by humans. Therefore, a knowledge representation language with defined semantics, amenable to computational processing, should be used to encode knowledge. There are many kinds of knowledge representation language and they have different capabilities and expressiveness. We can divide Knowledge Representation Languages into three broad categories:

1. An informal Knowledge Representation Language may be specified by a catalogue of terms that are either undefined or defined only by statements in a natural language. The Gene Ontology would fall into this category.
2. A formal ontology is specified by a collection of names for concept and relation types organized in a partial ordering by the type-subtype relation. RiboWeb and Ecocyc are examples of this kind of ontology.
3. Formal ontologies are further distinguished by the way the relationship between subtype and supertypes is established. In an asserted formal ontology all subtype-supertype relations are asserted by the ontology author. In an axiomatized ontology additional subtype-supertype relationships can be inferred by a reasoner based on axioms and definitions stated in a formal language, such as a description logic. The TAMBIS ontology is an example of this type.

Formal ontology languages have a formal semantics, which lend a precise meaning to statements in the language. These statements can be interpreted and manipulated by a machine without ambiguity.

There is no agreed vocabulary for talking about Knowledge Representation Languages. Different languages use different terms for the same notion or the same term for different notions. Table K.1 summarizes some of these variations in usage across Knowledge Representation Languages.

Further reading

Ringland GA, Duce DA (1988) *Approaches to Knowledge Representation: An Introduction*. John Wiley, Chichester.

Table K.1 Terminology used in Knowledge Representation Languages

Meaning	*DLs*	*OWL*	*Grail*	*Frames*	*UMLclass diagrams*
Category	Concept	Class	Category	Class	Class
Individual	individual	individual	(individual)	instance	instance
Relationship type	Role	property	Attribute	slot	relation
Relationship	existential restriction	same	criterion	filled-slot	relation
Range constraint	universal restriction	same	(sanction[1])	various facets depending on value type. Typically 'allowed classes'	(implied)
Sanctioned	—	—	sanction	(implied)	(implied)
some	some	has-class some Values From	topic Necessarily which	filled slot[2]	1..* mandatory
only	all	to-class all Values From	(sensibly) 'sanctions'	range facet	-
at-least n	at-least n	at-least n	—	min cardinality facet = n	n..*
at-most	at-most n	at-most n	—	max cardinality facet = n	x..n
default	—	—	(extrinsics)	(implied)	—
any	—	—	(extrinsics)	(ambiguous)	(ambiguous)
and	and	and	(implied)	(implied)	(implied)
or	or	or	—	—	—
not	not	complement —¬	—	—	—

[1] Sanctions in GRAIL provide a check on domain and range constraints but are not used in classification.
[2] The semantics of filling a slot with a class are ambiguous in most frame systems. Usually the most satisfactory translation into DLs or OWL is as existential restrictions.

K

KOZAK SEQUENCE

Niall Dillon

Consensus sequence located around the site of initiation of eukaryotic mRNA translation.

The Kozac sequence is involved in recognition of translation start site by the ribosome. Using comparisons of large numbers of genes and site-directed mutagenesis, the optimal Kozak consensus sequence has been identified as CCA/GCCAUGG. The small 40S ribosomal subunit, carrying the methionine tRNA and translation initiation factors is thought to engage the mRNA at the capped 5′ end and then scan linearly until it encounters the first AUG (initiator) codon. The Kozak sequence modulates the ability of the AUG codon to halt the scanning 40S subunit.

Further reading

Kozak M (1987) An analysis of 5′-noncoding sequences from 699 vertebrate messenger RNAs. *Nucleic Acids Res.* 15: 8125–8148.

Kozak M (1997) Recognition of AUG and alternative initiator codons is augmented by G in position +4 but is not generally affected by the nucleotides in positions +5 and +6. *EMBO J.* 16: 2482–2492.

KRL, *SEE* KNOWLEDGE REPRESENTATION LANGUAGE.

L

L-G Algorithm, *see* Lander-Green Algorithm.
Label (Labeled Data, Response, Dependent Variable)
Labeled Data, *see* Label.
Labeled Tree
Laboratory Information Management System (LIMS)
LAMARC
Lander-Green Algorithm (L-G Algorithm)
Lattice
Lazy Learner, *see* K-Nearest Neighbor Classification.
LD, *see* Linkage Disequilibrium.
Lead Optimization
Leaf, *see* Node and Phylogenetic Tree.
Leave-One-Out Validation, *see* Cross-Validation.
Leiden Open Variation Database, *see* Locus-Specific Database.
Lexicon
Ligand Efficiency
LIMS, *see* Laboratory Information Management System.
LINE (Long Interspersed Nuclear Element)
Linear Discriminant Analysis, *see* Fisher Discriminant Analysis.
Linear Regression and Non-Linear Regression, *see* Regression Analysis.
Linkage (Genetic Linkage)
Linkage Analysis
Linkage Disequilibrium (LD, Gametic Phase Disequilibrium, Allelic Association)
Linkage Disequilibrium Analysis, *see* Association Analysis.
Linkage Disequilibrium Map
Linked Data
Linked List, *see* Data Structure.
Lipinski Rule of Five, *see* Rule of Five.
Local Alignment (Local Similarity)
Local Similarity, *see* Local Alignment.
Locus Repeat, *see* Interspersed Sequence.
Locus-Specific Database (Locus-Specific Mutation Database, LSDB)
LOD Score (Logarithm of Odds Score)
Log Odds Score, *see* Amino Acid Exchange Matrix, LOD Score.
LogDet, *see* Paralinear Distance.
Logical Modeling of Genetic Networks
Logo, *see* Sequence Logo.
LogP (ALogP, CLogP)

Concise Encyclopaedia of Bioinformatics and Computational Biology, Second Edition.
Edited by John M. Hancock and Marketa J. Zvelebil.

Long-Period Interspersion, *see* Interspersed Sequence.

Long-Term Interspersion, *see* Interspersed Sequence.

Look-Up Gene Prediction, *see* Gene Prediction, Homology-based.

Loop

Loop Prediction/Modeling

LOVD, *see* Locus-Specific Database.

Low Complexity Region

LSDB, *see* Locus-Specific Database.

L-G ALGORITHM, *SEE* LANDER-GREEN ALGORITHM.

LABEL (LABELED DATA, RESPONSE, DEPENDENT VARIABLE)

Nello Cristianini

Term used in machine learning to designate an attribute of data that requires to be predicted.

In machine learning and statistical data analysis, a common task is to learn a function y = f(x) from example pairs (x,y). The second part of this pair is called the response, or the label of the data point x. Datasets of examples of the type (x,y) are called 'labeled'. In statistics the labels are often called 'dependent variables'.

This form of machine learning goes under the name of 'supervised learning', since the desired response for each training point is given, as a form of 'supervision' to the learning process.

It is opposite to 'unsupervised' learning, where only the observations {x} are given, and the algorithm is requested to discover generic relations within them, for example clusters.

Further reading

Mitchell T *Machine Learning*, McGraw Hill, 1997.

See also Machine Learning, Pattern Analysis, Feature

LABELED DATA, *SEE* LABEL.

LABELED TREE

Aidan Budd and Alexandros Stamatakis

A tree in which some or all taxonomic units (i.e., nodes of the tree) are named, that is, they are associated with a label. An unlabeled tree has no labels associated with any of its nodes (Figure L.1).

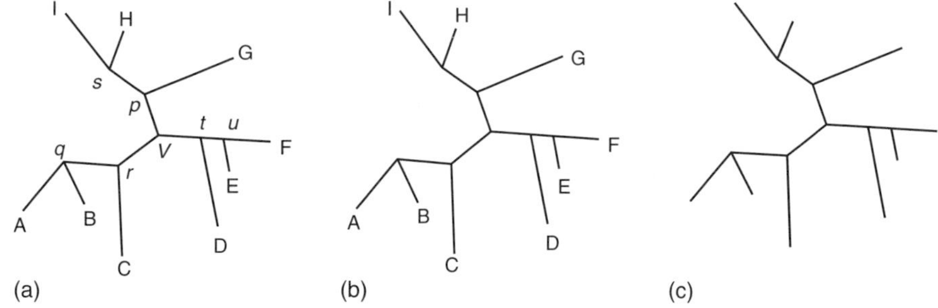

Figure L.1 Figure shows three trees, (a–c). All trees have the same topology. Trees (a) and (b) are labelled; on both trees, all operational taxonomic units/terminal nodes are labelled (A–I), and on tree (b) the hypothetical taxonomic units/internal nodes (q–v) are also labelled. Tree (c) is unlabelled, as no labels are associated with any of its nodes.

See also Phylogenetic Tree

L

LABORATORY INFORMATION MANAGEMENT SYSTEM (LIMS)

John M. Hancock

A type of application and database designed to capture information about standard processes taking place in a laboratory.

Historically, LIMS were employed in analytical laboratories, process testing labs, quality assurance labs etc. where the accurate tracking of experiments was essential. Functions of a LIMS can include direct acquisition of data from instruments, recording the details of experiments carried out and producing reports. By keeping a permanent record LIMS ensure that critical information is not lost or overwritten. These sorts of functionality have not traditionally been needed in biological research laboratories, but the advent of high-throughput functional genomics and other facilities has resulted in an increase in LIMS use in these settings. Many LIMS are commercial products, and some are available from suppliers of high-throughput equipment, although these are often not appropriate for the full range of experiments that may be carried out in a functional genomics laboratory. Alternatively LIMS may be developed in house for large specialized facilities, although this is labour-intensive and expensive for fully functional systems.

Related website

Limsfinder	http://www.limsfinder.com/

Further Reading

Goodman N, et al. (1998) The LabBase system for data management in large scale biology research laboratories. *Bioinformatics*. 14: 562–574.

Strivens MA, et al. (2000) Informatics for mutagenesis: the design of mutabase – a distributed data recording system for animal husbandry, mutagenesis, and phenotypic analysis. *Mamm Genome*. 11: 577–583.

Taylor S, et al. (1998) Automated management of gene discovery projects. *Bioinformatics*. 14: 217–218.

LAMARC

Michael P. Cummings

A program for estimating population-genetic parameters such as population size, population growth rate, recombination rate, and migration rates.

Based on coalescent theory, the analyses use maximum likelihood or Bayesian inference. Different data types are accommodated, including DNA sequence data, SNPs (single nucleotide polymorphisms), microsatellite data, electrophoretic data.

Available as C++ source code and executables for Mac OS, Windows and Linux. Some of the output is compatible with Tracer.

Related website

LAMARC web site	http://evolution.genetics.washington.edu/lamarc/index.html

Further reading

Kuhner MK (2006) LAMARC 2.0: maximum likelihood and Bayesian estimation of population parameters. *Bioinformatics* 22: 768–770.

Kuhner MK, et al. (1995) Estimating effective population size and mutation rate from sequence data using Metropolis-Hastings sampling. *Genetics* 140: 1421–1430.

Kuhner MK, et al. (1998) Maximum likelihood estimation of population growth rates based on the coalescent. *Genetics* 149: 429–434.

Kuhner MK, et al. (2001) Maximum likelihood estimation of recombination rates from population data. *Genetics* 156: 1393–1401.

See also Tracer

LANDER-GREEN ALGORITHM (L-G ALGORITHM)

Mark McCarthy, Steven Wiltshire and Andrew Collins

Used in the context of linkage analysis, an algorithm for calculating the likelihood of a pedigree based on a hidden Markov chain of inheritance distributions for a map of ordered marker loci.

For each position in a given marker map, the Lander-Green algorithm defines the distribution of binary inheritance vectors describing, for the entire pedigree at once, all possible inheritance patterns of alleles at that position in terms of their parental origin. In the absence of any genotype data, all inheritance vectors compatible with Mendelian segregation have equal prior probabilities. However, in the presence of genotype data, the posterior distribution of inheritance vectors, conditional on the observed genotypes, will have non-equal vector probabilities. In ideal circumstances – fully informative markers, no missing data, no missing individuals – it may be possible to identify unambiguously the one true inheritance vector for any given position: in real-life data, this is unusual. The Lander-Green algorithm models these inheritance distributions at positions across the marker map, in terms of a Hidden Markov Model, with the transition probabilities between states (the inheritance distributions) being functions of the recombination fractions between the loci. The likelihood of the inheritance distribution at any locus, conditional on the whole marker set, can be calculated from this model.

This likelihood function forms the basis of several software implementations of parametric and non-parametric linkage analyses. The computational burden of the L-G algorithm increases linearly with marker number, but increases exponentially with pedigree size, and is therefore ideally suited to multipoint linkage analyses in medium-sized pedigrees. In this regard, it complements the Elston-Stewart algorithm, which can calculate likelihoods rapidly for large pedigrees, but can cope with relatively few markers at a time.

Examples: Hanna et al. (2002) used GENEHUNTER PLUS – implements the L-G algorithm – to perform genome-wide parametric and non-parametric linkage analyses of seven extended pedigrees for loci influencing susceptibility to obsessive-compulsive disorder. Wiltshire et al (2002) used ALLEGRO – a related program – to study a large collection of UK sibpair families for evidence of linkage to type 2 diabetes.

L

Related websites

Software that implement the Lander-Green algorithm in multipoint linkage analyses include:

Genehunter 2	www-genome.wi.mit.edu/ftp/distribution/software/genehunter/
Genehunter Plus	http://galton.uchicago.edu/genehunterplus/
Merlin	http://www.sph.umich.edu/csg/abecasis/Merlin/index.html

Further reading

Abecasis GR, et al. (2002) Merlin – rapid analysis of dense genetic maps using sparse gene flow trees. *Nat Genet* 30: 97–101.

Hanna GL, et al. (2002) Genome-wide linkage analysis of families with obsessive-compulsive disorder ascertained through pediatric probands. *Am J Med Genet* 114: 541–552.

Kruglyak L, et al. (1995) Rapid multipoint linkage analysis of recessive traits in nuclear families, including homozygosity mapping. *Am J Hum Genet* 56: 519–527.

Kruglyak L, et al. (1996) Parametric and nonparametric linkage analysis: a unified multipoint approach. *Am J Hum Genet* 58: 1347–1363.

Lander ES, Green P (1987) Construction of multilocus genetic maps in humans. *Proc Natl Acad Sci USA* 84: 2363–2367.

Wiltshire S, et al. (2001) A genome-wide scan for loci predisposing to type 2 diabetes in a UK population (The Diabetes (UK) Warren 2 Repository): analysis of 573 pedigrees provides independent replication of a susceptibility locus on chromosome 1q. *Am J Hum Genet* 69: 553–569.

See also Elston-Stewart Algorithm, Linkage Analysis, Allele-Sharing Methods, Genome-Scan

LATTICE

Robert Stevens

1. In mathematics, a **directed acyclic graph** which is 'closed' at both top and bottom. (formally, a directed acyclic graph in which every node has a unique least common subsumer and greatest common subsumee.)

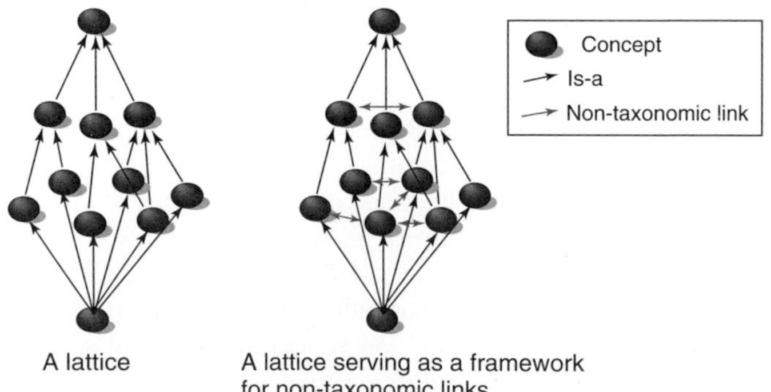

Figure L.2 Organization of concepts into a lattice.

2. Informally in knowledge representation a multiple hierarchy, where the hierarchy is based upon one 'taxonomic' relationship (usually subtype), and forms a framework for other 'non-taxonomic' relationships. (For knowledge representation the distinction between a 'lattice' and an 'acyclic directed graph' makes little difference, since any directed acyclic graph can be trivially converted to a lattice by providing a 'bottom' node linked to each leaf node. However, software often depends on the extensive body of mathematical theory around lattices which depends on their being closed at the bottom and at the top. See Figure L.2.

LAZY LEARNER, *SEE* K-NEAREST NEIGHBOR CLASSIFICATION.

LD, *SEE* LINKAGE DISEQUILIBRIUM.

LEAD OPTIMIZATION

Matthew He

Lead optimization is a process of converting a putative lead compound ('hit') into a therapeutic drug with maximal activity and minimal side effects, typically using a combination of computer-based drug design, medicinal chemistry and pharmacology.

LEAF, *SEE* NODE AND PHYLOGENETIC TREE.

LEAVE-ONE-OUT VALIDATION, *SEE* CROSS-VALIDATION.

LEIDEN OPEN VARIATION DATABASE, *SEE* LOCUS-SPECIFIC DATABASE.

LEXICON

Robert Stevens

The collection of terms used to refer to the concepts of an ontology. This can also be thought of as the vocabulary delivered by an ontology. It is often important to make concepts independent of any one language; for example the concept of 'Leg' would be linked to 'leg' in an English lexicon and 'jambe' in a French lexicon. The lexicon is also where information about synonyms and other linguistic information is held. (Not all ontologies distinguish clearly between the lexicon and the ontology proper. The practice is, however, to be encouraged.)

L

LIGAND EFFICIENCY

Bissan Al-Lazikani

A calculated measure reflecting the binding affinity achieved by a small molecule to its receptor per size unit. Most directly, this can be calculated by dividing the experimentally derived Gibbs free energy (ΔG) by the number of heavy (non-hydrogen) atoms of the compound (N):

$$LE = \frac{\Delta G}{N}$$

More crudely, it can be calculated by simply dividing the binding affinity (e.g. pIC_{50}) by the number of heavy weight atoms. Thus an efficient molecule will achieve better binding affinity per heavy atom or molecular weight unit.

Ligand efficiency is an important measure during drug design and lead optimization because, an efficient starting compound allows medicinal chemists to chemically modify it in order to achieve better properties in the final drug (e.g. target selectivity, oral bioavailability etc.) without losing affinity to the original target.

Further reading

Abad-Zapatero C, Perišić O, Wass J, Bento AP, Overington J, Al-Lazikani B, Johnson ME (2010) Ligand efficiency indices for an effective mapping of chemico-biological space: the concept of an atlas-like representation. *Drug Discov Today*. 15(19–20): 804–11. PubMed PMID: 20727982.

Kuntz ID, Chen K, Sharp KA, Kollman PA. (1999) The maximal affinity of ligands. *Proc Natl Acad Sci U S A*. 96(18): 9997–10002. PMID: 10468550;

LIMS, *SEE* LABORATORY INFORMATION MANAGEMENT SYSTEM.

LINE (LONG INTERSPERSED NUCLEAR ELEMENT)

Dov Greenbaum

A class of large transposable elements common in the human genome.

Long interspersed elements differ from other retrotransposons in that they do not have long terminal direct repeats (LTRs). In humans, LINEs are one of the major sources of insertional mutagenesis in both somatic and germ cells.

There are two major LINE transposable elements in the human genome, L1 and L2. The approximately half a million repeats take up more than eighteen percent of the human genome. Because of their high frequency in the genome, LINE methylation serves as a useful surrogate marker of global methylation; loss of methylation is consistently observed in human cancers. Thus LINEs have also been show to have clinical significance, particuarlu in cancer prognosis.

While other transposons have enhancers and promoters within their LTRs, LINEs differ in that they only have internal promoters.

LINES come in two flavours: there are full-length copies of these elements and derivatives that have lost some of their sequence. The full length copies (approximately 6 kb) contain two open reading frames: one that is similar to the viral gag gene and another with sequence similarity to reverse transcriptase. These genes allow LINEs to retrotransposase themselves and, to a much smaller degree, other DNA sequences, resulting in processed pseudogenes

Recent research has shown that specific LINE-1 retroposons in the human genome are actively transcribed; its suggests that retrotransposable elements may serve as a critical epigenetic determinant in the chromatin remodeling events leading to neocentromere formation.

Further reading

Chueh AC, Northrop EL, Brettingham-Moore K, et al. (2009) LINE retrotransposon RNA is an essential structural and functional epigenetic component of a core neocentromeric chromatin. *PLoS Genetics* 5 (1): e1000354.

Cooper DN (1999) *Human Gene Evolution*. Bios Scientific Publishers, pp. 265–285.

Kazazian HHJ, Moran JV (1998) The impact of L1 retrotransposons on the human genome. *Nat. Genet.* 19: 19–24.

Moran JV, et al. (1996) High frequency retroposition in cultured mammalian cells. *Cell* 87: 917–927.

Ridley M (1996) Evolution Blackwell Science Inc., pp. 265–276.

Rowold DJ, Herrera RJ (2000) Alu elements and the human genome. *Genetics* 108: 57–72.

Singer MF (1982) SINEs and LINEs: highly repeated short and long interspersed sequences in mammalian genomes. *Cell* 28 (3): 433–434.

LINEAR DISCRIMINANT ANALYSIS, *SEE* FISHER DISCRIMINANT ANALYSIS.

LINEAR REGRESSION AND NON-LINEAR REGRESSION, *SEE* REGRESSION ANALYSIS.

LINKAGE (GENETIC LINKAGE)

Mark McCarthy and Steven Wiltshire

Genetic linkage describes the phenomenon whereby loci lying on the same chromosome do not show independent segregation during meiosis.

Assume that an individual inherits, at two loci, A and B, alleles A_p and B_p from its father and A_m and B_m from its mother. If the individual produces gametes which are non-recombinant (containing either A_p and B_p or A_m and B_m) and recombinant (containing either A_p and B_m or A_m and B_p) in equal proportions, loci A and B are being inherited independently of one another, and there is no linkage between them. In other words, they are unlinked and the probability of recombination between them, designated θ, is 0.5. We may conclude that these two loci are on separate chromosomes (or possibly far apart on the same chromosome). If non-recombinant gametes are produced in greater proportion to recombinant gametes, the segregation of loci A and B is not independent, and there is genetic linkage between them. The recombination fraction between them, θ, is < 0.5, and we conclude that the two loci are on the same chromosome. When $\theta = 0$, the loci concerned are in complete genetic linkage: there is no recombination between them, and only non-recombinant gametes are produced. The proportion of recombinant gametes increases with the 'genetic' distance between the loci, reflecting the increased probability that a crossover event will have intervened during meiosis. The evidence for linkage between loci can be quantified with a LOD (logarithm of odds) score. Detecting and quantifying the evidence for linkage between loci forms the basis of linkage analysis, but requires that it be possible to distinguish between the alleles at each locus.

Example: In a genome-wide parametric linkage analysis of a large Australian Aboriginal pedigree for loci influencing susceptibility to type 2 diabetes, Busfield et al. (2002) detected linkage to marker D2S2345 with a maximum two-point LOD score of 2.97 at a recombination fraction of 0.01. For other examples, see under linkage analysis.

Further reading

Busfield F, et al. (2002) A genomewide search for type 2 diabetes-susceptibility genes in indigenous Australians. *Am J Hum Genet* 70: 349–357.

Lander ES, Schork NJ (1994) Genetic dissection of complex traits. *Science* 265: 2037–2048.

Ott J (1999) *Analysis of human genetic linkage*. Baltimore, The Johns Hopkins University Press, pp 1–23.

See also Recombination, LOD Score, Linkage Analysis.

LINKAGE ANALYSIS

Mark McCarthy, Steven Wiltshire and Andrew Collins

Method for identifying the genomic position of genetic loci which relies on the fact that genes which lie close to each other (i.e. are in linkage) are only rarely separated by meiotic recombination and will therefore co-segregate within pedigrees.

By charting the segregation of variant sites within families, it becomes possible to identify loci which are in linkage and determine their relative genomic location. One use of such methods has been to build genetic maps of polymorphic markers. Given such marker maps, it becomes possible, to define the chromosomal location of loci of unknown position, such as disease-susceptibility genes, by comparing the pattern of segregation of disease (and by inference of disease-susceptibility genes) with that of markers of known position. Broadly, there are two main flavours of linkage analysis. Parametric linkage analysis, typically used in the analysis of Mendelian diseases, requires specification of a credible disease locus model (in which parameters such as dominance, allele frequency and penetrance are prescribed), and the evidence for linkage (typically expressed as a LOD – or logarithm of the odds – score) is maximized with respect to the recombination distance between the loci. Software programs such as LINKAGE and GENEHUNTER are available for this. Non-parametric linkage analysis applied to multifactorial traits, requires no prior specification of a disease-locus model, and seeks to detect genomic regions at which phenotypically similar relatives show more genetic similarity than expected by chance (see allele-sharing methods).

Examples: In their classical paper, Gusella and colleagues demonstrated linkage between Huntington's disease and a marker on chromosome 4q. A decade later, positional cloning efforts finally identified the aetiological gene, huntingtin. Hanis and colleagues (1996) reported their genome scan for type 2 diabetes in Mexican American families that identified a significant linkage to diabetes on chromosome 2q. Subsequent work has indicated that variation in the calpain-10 gene is the strongest candidate for explaining this linkage.

Related websites

Programs for linkage analysis:

Genehunter	http://linkage.rockefeller.edu/soft/gh/
Merlin	http://www.sph.umich.edu/csg/abecasis/Merlin/index.html
Jurg Ott's Linkage page	http://lab.rockefeller.edu/ott/

Further reading

Gusella JF, et al. (1983) A polymorphic DNA marker genetically linked to Huntington's disease. *Nature* 306: 234–238.

Hanis CL, et al. (1996) A genome wide search for human non-insulin-dependent (type 2) diabetes genes reveals a major susceptibility locus on chromosome 2. *Nat Genet* 13: 161–171.

Kong A, et al. (2002) A high-resolution recombination map of the human genome. *Nat Genet* 31: 241–247.

Kruglyak L, et al. (1996) Parametric and non-parametric linkage analysis: a unified multipoint approach. *Am J Hum Genet* 58: 1347–1363.

Lander ES, Schork NJ (1994) Genetic dissection of complex traits. *Science* 265: 2037–2048.

Morton NE (1956) The detection and estimation of linkage between the genes for elliptocytosis and the Rh blood type. *Am J Hum Genet* 8: 80–96.

Ott J (1999) *Analysis of Human Genetic Linkage* (Third Edition). Baltimore: Johns Hopkins University Press.

Sham P (1998) *Statistics in Human Genetics*. London, Arnold pp 51–144.

See also Genome Scans, Allele-Sharing Methods, Multipoint Linkage Analysis, Linkage

LINKAGE DISEQUILIBRIUM (LD, GAMETIC PHASE DISEQUILIBRIUM, ALLELIC ASSOCIATION)

Mark McCarthy, Steven Wiltshire and Andrew Collins

The statistical association of alleles at two (or more) polymorphic loci.

Consider two biallelic loci, with alleles A,a (frequencies P_A, P_a) and B,b (frequencies P_B, P_b). If, within a population of individuals (or chromosomes), alleles at these two loci are seen to be independent of one other, then the frequency of the haplotype AB (that is, P_{AB}) will be the product of the respective allele frequencies, P_AP_B. However, if the alleles are not independent of one another, the frequency of the haplotype AB will differ from P_AP_B by a non-zero amount, D_{AB}:

$$D_{AB} = P_{AB} - P_AP_B$$

If $D_{AB} \neq 0$ then the alleles at loci A and B are in linkage, or gametic phase, disequilibrium; if $D_{AB} = 0$, then the alleles are said to be in linkage equilibrium. For the example of two biallelic markers, $D_{ab} = D_{AB}$, and $D_{aB} = D_{Ab} = -(D_{AB})$. Additional linkage disequilibrium parameters need to be calculated for markers with more than two alleles, and for considerations of more than two markers.

A host of different measures of LD exist, with differing properties. Often, the extent of LD is expressed as a proportion, D′, of its maximum or minimum possible values given the frequencies of the alleles concerned:

$$D'_{AB} = D_{AB} / \max(-P_A P_B, -P_aP_b)$$

if $D_{AB} < 0$, or

$$D_{AB} = D_{AB} / \min(P_a P_B, P_AP_b)$$

if $D_{AB} > 0$.

A second measure of LD is r^2, calculated from D, thus:

$$r^2 = D^2 / P_A P_a P_B P_b$$

r^2N is distributed as chi-square, where N is the total number of chromosomes.

LD is a feature of a particular population of individuals (or more strictly, their chromosomes) and arises in the first place through mutation (which creates a new haplotype), and is then sustained and bolstered by events and processes modifying the genetic composition of a population during its history – these include periods of small population size ('bottlenecks'), genetic admixture (due to interbreeding with a distinct population) and stochastic effects ('genetic drift'). At the same time, any LD established is gradually dissipated by the actions of recombination and, much less strongly, by further mutation. In most current human populations, LD extends over a typically short range (tens of kilobases).

Note that, although the term linkage disequilibrium is often reserved to describe associations between linked loci, it is sometimes used more broadly, in which case it equates to allelic association. Association (or linkage disequilibrium) mapping exploits linkage disequilibrium (in the stricter sense) to map disease-susceptibility genes.

Examples: Abecasis et al. (2001) calculated pair-wise between markers in three genomic regions and observed patterns of measurable LD extending for distances exceeding 50kb. Gabriel et al. (2002) and Dawson et al. (2002) have used pair-wise measure of LD to investigate fine scale LD structure of human chromosomes.

Related websites

EH+	http://www.mrc-epid.cam.ac.uk/~jinghua.zhao/software.htm
ARLEQUIN	http://cmpg.unibe.ch/software/arlequin35/
GOLD	http://www.sph.umich.edu/csg/abecasis/GOLD/

Further reading

Abecasis GR, et al. (2001) Extent and distribution of linkage disequilibrium in three genomic regions. *Am J Hum Genet* 68: 191–197.

Dawson E, et al. (2002) A first-generation linkage disequilibrium map of human chromosome 22. *Nature* 418: 544–548.

Devlin B, Risch N (1995) A comparison of linkage disequilibrium measures for fine-scale mapping. *Genomics* 29: 311–322.

Gabriel SB, et al. (2002) The structure of haplotype blocks in the human genome. *Science* 296: 2225–2229.

Ott J (1999) *Analysis of human genetic linkage*. Baltimore: The Johns Hopkins University Press pp 280–291.

Weir BS (1996) *Genetic Data Analysis II*. Sunderland: Sinauer Associates pp 112–133.

See also Genome-Wide Association, Allelic Association, Haplotype, Phase, Recombination, Linkage Disequilibrium Map

LINKAGE DISEQUILIBRIUM ANALYSIS, *SEE* ASSOCIATION ANALYSIS.

LINKAGE DISEQUILIBRIUM MAP

Andrew Collins

Linkage disequilibrium maps describe patterns of LD at high resolution in a form which is analogous to the linkage map. The decline of association in an interval between single nucleotide polymorphism (SNP) markers is represented using an adaptation of the formula first used by Malecot (1948) to represent isolation by distance. Association is modeled as:

$$\widehat{\rho} = (1 - L)\, Me^{-sd} + L$$

where $\widehat{\rho}$ is association between a pair of SNPs, L is the asymptote (background association not due to linkage), M is the intercept, ϵ is the decline of association with distance and d is the distance between a pair of SNPs in kilobases. A distance for the interval in linkage disequilibrium units (LDU) is the product $\epsilon\ d$. Genetic distances in LDU maps are additive at successive SNP intervals. For every interval between two adjacent SNPs an LDU distance is calculated, which in part reflects the recombination frequency in genomic region within that interval. One LDU is the distance over which LD declines to background levels. LDU maps, when constructed for whole chromosomes show contours which closely correspond to those of the genetic linkage map in centimorgans reflecting the dominant role of recombination in defining the haplotype block structure. Increased power for association mapping of disease genes has been demonstrated when using an underlying LDU map.

Further reading

Collins A, Lau W, De La Vega FM (2004) Mapping genes for common diseases: the case for genetic (LD) maps. *Human Heredity* 58(1): 2–9.

De La Vega FM, Isaac H, Collins A, et al. (2005) The linkage disequilibrium maps of three human chromosomes across four populations reflect their demographic history and a common underlying recombination pattern. *Genome Research* 15(4): 454–456.

Lonjou C, Zhang W, Collins A, et al. (2003) Linkage disequilibrium in human populations. *Proceedings of the National Academy of Sciences of the United States of America*100(10): 6069–60674.

Malecot G (1948) *Les mathématiques de l'hérédité*. Masson, Paris.

Maniatis N, Collins A, Gibson J, Zhang W, Tapper W, Morton NE. (2004) Positional cloning by linkage disequilibrium. *Am J Hum Genet* 74(5): 846–855.

Zhang W, Collins A, Maniatis N, Tapper W, Morton NE (2002) Properties of linkage disequilibrium (LD) maps. *Proceedings of the National Academy of Sciences of the United States of America* 99(26): 17004–07.

See also Genome-Wide Association, Linkage Disequilibrium

LINKED DATA

Carole Goble and Katy Wolstencroft

Linked Data is a method for dynamically integrating data on the web using URIs (Universal Resource Identifiers) and RDF (Resource Description Framework). Rather than gathering related data in a data warehouse, the Linked Data approach provides recommendations for exposing, sharing, and connecting pieces of data using the same technologies and formats. If two or more resources have exposed their data in this structured format, they can be linked and queried over dynamically.

Many Life Science datasets have been published as Linked Data and biological data makes up more than a quarter of the Linked Data currently available on the Web. The most widely known biology Linked Data sets are: Bio2RDF, Chem2Bio2RDF and LinkedLifeData.

Related websites

Linked Data	http://linkeddata.org
LinkedLifeData	http://linkedlifedata.com/
Bio2RDF	http://bio2rdf.org/
WikiSpaces	http://chem2bio2rdf.wikispaces.com/

Further reading

Heath T, Bizer C (2011) Linked Data: Evolving the Web into a Global Data Space (1st edition). Synthesis Lectures on the Semantic Web: Theory and Technology, 1:1: 1–136. Morgan & Claypool. http://linkeddatabook.com/editions/1.0/

See also Data Integration, Data Warehouse, Scientific Workflow, RDF, Identifer

LINKED LIST, *SEE* DATA STRUCTURE.

LIPINSKI RULE OF FIVE, *SEE* RULE OF FIVE.

LOCAL ALIGNMENT (LOCAL SIMILARITY)

Jaap Heringa

An alignment of the most similar consecutive segments of two or more sequences, as opposed to global alignment, which is an alignment of sequences over their entire lengths. A local alignment should be attempted when the target sequences have different lengths, so that different domain organisations or different repeat copy numbers could be present. Local alignment is also the method of choice if permuted domains are suspected. It is further appropriate when sequences are extremely divergent, such that evolutionary memory might be retained in some local fragments only.

The first automatic method to assess local similarity and generate a local alignment for two protein sequences was devised by Smith and Waterman (1981). The technique is an adaptation of the Dynamic Programming technique developed by Needleman and Wunsch (1970) for global alignment. The Smith-Waterman technique calculates a single optimal local alignment containing one subsequence from each aligned sequence. The Smith-Waterman algorithm has been extended in various techniques to compute a list of top-scoring pair-wise local alignments (Waterman & Eggert, 1987; Huang *et al.*, 1990; Huang & Miller, 1991). Alignments produced by the latter techniques are non-intersecting; i.e., they have no matched pair of amino acids in common. Fast sequence database search methods, such as BLAST and FASTA, are heuristic approximations of the Smith-Waterman local alignment technique.

Further reading

Huang X, Hardison RC, Miller W (1990). A space-efficient algorithm for local similarities. *CABIOS* 6: 373–381.

Huang X, Miller W (1991) A time-efficient, linear-space local similarity algorithm. *Adv. Appl. Math.* 12: 337–357.

Needleman SB, Wunsch CD (1970) A general method applicable to the search for similarities in the amino acid sequence of two proteins. *J. Mol. Biol.* 48: 443–453.

Smith TF, Waterman MS (1981) Identification of common molecular subsequences. *J. Mol. Biol.* 147: 195–197.

Waterman MS, Eggert M (1987) A new algorithm for best subsequences alignment with applications to the tRNA-rRNA comparisons. *J. Mol. Biol.* 197: 723–728.

See also Alignment, Dynamic Programming, Homology Search

L

LOCAL SIMILARITY, *SEE* LOCAL ALIGNMENT.

LOCUS REPEAT, *SEE* INTERSPERSED SEQUENCE.

LOCUS-SPECIFIC DATABASE (LOCUS-SPECIFIC MUTATION DATABASE, LSDB)

John M. Hancock

Locus-Specific Databases collect human mutations and their clinical or phenotypic effects at specific loci. Although described as locus-specific they also tend to concentrate on particular clinical phenotypes, so that there are at least five databases dealing with mutations in the TP53 gene, for example. In total over 6,000 LSDBs are listed by the Leiden University Medical Center.

Many, but not all, LSDBs make use of the LOVD (Leiden Open Variation Database) infrastructure which is free to use and provides a number of built-in utilities such as graphical displays and utilities, sequence variant tables, search functions and the ability to link to other resources. Many LOVD-based LSDBs are hosted at the central LOVD server.

Related websites

Leiden list of LSDBs	http://grenada.lumc.nl/LSDB_list/lsdbs
LOVD home page	http://www.lovd.nl/3.0/home

Further reading

Fokkema IF, Taschner PE, Schaafsma GC, Celli J, Laros JF, den Dunnen JT (2011). LOVD v.2.0: the next generation in gene variant databases. *Hum Mutat*. 32: 557–563.

LOD SCORE (LOGARITHM OF ODDS SCORE)

Mark McCarthy, Steven Wiltshire and Andrew Collins

Logarithm of odds for linkage between two loci against no linkage.

LOD score is a general term for the decimal logarithm of a likelihood-ratio-based linkage statistic and is a measure of the evidence for linkage. The precise nature of the LOD score depends on the nature of the analysis, namely parametric or non-parametric linkage analysis.

In parametric linkage analysis, typically used in the study of Mendelian diseases, the classical LOD score, $Z(\theta)$, is given by:

$$Z(\theta) = \log 10[L(\theta < 0.5)/L(\theta = 0.5)]$$

where $L(\theta)$ is the likelihood of the data given the value of θ, the recombination fraction between the two loci. The LOD score reaches its maximum value, Z_{max} at the maximum likelihood estimate of θ. Several variants of the parametric LOD score exist including the HLOD (heterogeneity LOD), in which a proportion of the pedigrees analysed are modelled as having no linkage between the two loci; and ELOD (expected LOD), which yields a measure of the informativeness of the pedigree for linkage. Generally, a parametric LOD score of 3 or more is taken a significant evidence for linkage.

There are two common LOD score statistics in use in non-parametric, allele-sharing methods of qualitative trait linkage analysis: the 'maximum LOD score' (or MLS) and the 'allele-sharing LOD score' (or LOD*). The MLS is based on the likelihood ratio of observed allele sharing for 0, 1 and 2 alleles IBD to that expected under the null hypothesis of no linkage:

$$MLS = \log 10\,[(z_0^{n0}\, z_1{}^{n1}\, z_2^{n2}) \,/\, (0.25\,^{n0} 0.50\,^{n1}\, 0.25^{n2})]$$

where z_0, z_1 and z_2 are the allele sharing proportions for 0,1 and 2 alleles IBD under the alternative hypothesis, and n0, n1 and n2 are the number of sib pairs falling into each of the three IBD categories. The MLS can be maximised over one (z_0) or two (z_0 and z_1) parameters depending on whether dominance at the trait locus is being modelled. The allele-sharing LOD score is a re-parameterisation of the non-parametric linkage score calculated by the program GENEHUNTER in terms of an allele-sharing parameter, $\widehat{\delta}$. Two versions of this re-parameterization exist – linear and exponential.

One of the two variants of linkage analysis of quantitative traits, namely variance components analysis, also uses LOD scores to measure the evidence for linkage of a quantitative trait locus to marker loci.

Examples: Busfield et al. (2002) used parametric linkage analysis in a large aboriginal Australian population to detect evidence for linkage to type 2 diabetes susceptibility loci, quoting classical LOD scores. Pajukanta et al. (2000) used allele-sharing methods to perform a genome scan for loci influencing premature coronary heart disease in Finnish populations, quoting MLS scores. Hanna et al. (2000) used both parametric and non-parametric linkage analyses to detect loci in a genome scan for susceptibility loci influencing obsessive-compulsive disorder, quoting both classical parametric LOD score and non-parametric allele-sharing LOD scores.

Further reading

Busfield F, et al. (2002) A genomewide search for type 2 diabetes-susceptibility genes in indigenous Australians. *Am J Hum Genet* 70: 349–357.

Hanna GL, et al. (2002) Genome-wide linkage analysis of families with obsessive-compulsive disorder ascertained through pediatric probands. *Am J Med Genet* 114: 541–552.

Kong A, Cox NJ (1997) Allele-Sharing Models: LOD Scores and Accurate Linkage Tests. *Am J Hum Genet* 61: 1179–1188.

Morton NE (1955) Sequential tests for the detection of linkage. *Am. J. Hum. Genet.* 7: 277–318.

Nyholt DR (2000) All LODs are not created equal. *Am J Hum Genet* 67: 282–288.

Ott J (1999) *Analysis of Human Genetic Linkage*. Baltimore: The Johns Hopkins University Press.

L

Pajukanta P, et al. (2000) Two loci on chromosomes 2 and X for premature coronary heart disease identified in early- and late-settlement populations of Finland. *Am J Hum Genet* 67: 1481–1493.

See also Linkage, Recombination, Linkage Analysis

LOG ODDS SCORE, *SEE* AMINO ACID EXCHANGE MATRIX, LOD SCORE.

LOGDET, *SEE* PARALINEAR DISTANCE.

LOGICAL MODELING OF GENETIC NETWORKS

Denis Thieffry

Mathematical equations can be used to model the evolution of biological components (variables) interacting across time and possibly space. The most common approach relies on differential equations.

Logical modeling is particularly appropriate when the information at hand is mainly qualitative or if one focuses on qualitative aspects of the network behavior. In the simplest and most common framework, each component is represented by a Boolean variable, which can take only two different values: the value 0 denotes a component absent or inactive, whereas the value 1 denotes a component present or active. The effect of different combinations of regulatory interactions on the evolution of a given component can then described in terms of a logical equation defining the target value of the corresponding variable depending on the presence or absence of its regulators. This can formulated as a logical function using the Boolean operators AND, OR, and NOT. Alternatively, for small systems, one can use a truth table listing the target values for each combination of the values of the variables.

Regarding the simulation of such Boolean systems, two main updating approaches used:

1. The synchronous approach considers all equations at once and updates all variables simultaneously, according to the values calculated on the basis of the formula $[x_{t+1} = B(x_t)]$, where x denotes a vector encompassing all variables, and B a Boolean function). Starting from an initial state, the behaviour is thus completely deterministic.
2. The asynchronous approach considers only single variable updates whenever the Boolean function and the state of the system differ. Multiple pathways might thus be followed from some initial states, potentially leading to different asymptotic behaviours (non-deterministic trajectories).

Table L.1 Logical modeling of genetic networks – selection of software tools

Name	Characteristics	URL
Biocham	Modeling environment encompassing a rule-based language, several simulators, and model checking tools (multiple semantics, including Boolean)	http://contraintes.inria.fr/BIOCHAM/
CellNetAnalyser	MatLab toolbox enabling the analysis of information flows in Boolean signaling networks	http://www.mpi-magdeburg.mpg.de/projects/cna/cna.html
ChemChain	Boolean network simulations and analysis	http://www.bioinformatics.org/chemchains/
DDLab	Synchronous simulations of logical models	http://www.ddlab.com/
GINsim	Java software enabling the definition, simulation and various kinds of analyses of multi-level logical models	http://www.ginsim.org/

Whatever the updating assumption, the system will present the same stable states, i.e. the states for which the logical variables and the corresponding logical functions are equal. However, different dynamics in terms of transient or potential cyclic behaviors may result depending on the updating assumption used. Furthermore, various refined updating schemes have been proposed over the last decades, making use of time delays, continuous clocks, or transition priority classes.

In general, logical simulations are represented in terms of state transition graphs, where each vertex represents a logical state (vector encompassing a value for each variable), whereas arcs between states denote enabled transitions. Stable states then correspond to single terminal states, whereas logical attractors (including stable cycles) correspond to (more complex) terminal, strongly connected components.

The logical approach has been generalized in order to encompass multi-level variables, (i.e. variables taking more than two integer values: 0, 1, 2, . . .), as well as the explicit consideration of threshold values (at the interface between logical values) or unknown values (fuzzy logics). Finally, logical parameters have been introduced to cover families of logical functions. Various software programs are currently available to define, simulate and analyse logical models (see Table L.1 for a selection of public tools).

Further reading

Kauffman SA (1993) *The Origins of Order: Self-Organization and Selection in Evolution*. Oxford University Press.

Thomas R, d'Ari, R (1990) *Biological Feedback*. CRC Press.

LOGO, *SEE* SEQUENCE LOGO.

L

LOGP (ALOGP, CLOGP)

Bissan Al-Lazikani

The partition coefficient is an experimentally derived property of an un-ionized compound, reflecting how hydrophobic the compound is. It is calculated as the logarithm of the ratio of measured concentration in the two phases of a mixture of two immiscible solvents, typically water and octanol:

$$LogP_{oct/wat} = Log\left(\frac{[compound]_{oct}}{[coupound]_{wat}}\right)$$

This is an important measure in drug discovery because it dictates how much of an orally administered compound is likely to be absorbed and/or metabolized reach its intended molecular target. A very low LogP means that a compound is unlikely to pass the lipid bilayers in the gut. If the LogP were too high, the compound will be highly metabolized and not retain sufficient quantities to modulate its intended target.

In chemoinformatics, the LogP of a compound is estimable from its chemical structure. There are different methods of performing calculations, the most popular of which are AlogP and ClogP. AlogP is an atom-based calculation that uses the contribution of individual atoms in the structure to calculate an overall LogP. CLogP is calculated using the individual contribution of non-overlapping fragments in the structure.

LONG-PERIOD INTERSPERSION, *SEE* INTERSPERSED SEQUENCE.

LONG-TERM INTERSPERSION, *SEE* INTERSPERSED SEQUENCE.

LOOK-UP GENE PREDICTION, *SEE* GENE PREDICTION, HOMOLOGY-BASED.

LOOP

Marketa J. Zvelebil

Secondary structures such as α-helices and β-strands or sheets are connected to each other by structural segments called loops. Although sometimes referred to as coils, loops often have at least some definite structure while coil regions are unstructured. Insertions and deletions (INDELS) often occur in loop (or coil) regions. Loops are often found on the surface of the protein (but not always) and are therefore rich in polar or charged residues. Loops often are involved in ligand binding or form part of an active site of the protein.

Relevant website

Loop Database	http://mdl.ipc.pku.edu.cn/moldes/oldmem/liwz/home/loop.html

Further reading

Berezovsky IN, Grosberg AY, Trifonov EN (2000) Closed loops of nearly standard size: common basic element of protein structure. *FEBS Letters* 466: 283–286.

Ring CS, Cohen FE (1994) Conformational sampling of loop structures using genetic algorithms. *Israel J Chem* 34: 245–252.

Ring CS, Kneller DG, Langridge R, Cohen FE (1992) Taxonomy and conformational analysis of loops in proteins. *J Mol Biol*, 224: 685–699.

See also Coil, Indel, Loop Prediction or Modeling

LOOP PREDICTION/MODELING

Roland Dunbrack

A step in comparative modeling of protein structures in which loop or coil segments between secondary structures are built. Alignment of a target sequence to be modeled and a structure to be used as a template or parent will produce insertions and deletions most commonly in coil segments of the template protein. Since the target and template loops are of different lengths, loops of the correct length and sequence must be constructed and fitted onto the template. This is performed first by identifying anchor points, or residues that will remain fixed at or near their Cartesian coordinate positions in the template structure. The loop is then constructed to connect the anchor points with the appropriate sequence.

Most loop prediction methods can be classified either as database methods or construction methods. In database methods, the Protein Data Bank is searched for loops of the correct length that will span the anchor points within some tolerance. The loop is then positioned and adjusted to fit the actual anchors in the template structure. In construction methods, a segment is built with no reference to a loop in a known structure, but rather by building a segment that will fit the anchor points with reasonable stereochemical quality. This can be performed with Monte Carlo simulations, molecular dynamics, or energy minimization techniques.

Further reading

Fiser A, et al. (2000) Modeling of loops in protein structures. *Prot. Science* 9: 1753–1773.

Moult J, James MNG (1986) An algorithm for determining the conformation of polypeptide segments in proteins by systematic search. *Proteins* 1: 146–163.

Vlijmen HWT, Karplus M (1997) PDB-based protein loop prediction: Parameters for selection and methods for optimization. *J. Mol. Biol.* 267: 975–1001.

Xiang Z, et al. (2002). Extending the accuracy limits of prediction for side-chain conformations. *Proc. Natl. Acad. Sci. USA* 99: 7432–7437.

See also Comparative Modeling, Target, Parent, Anchor Points, Loop

L

LOVD, *SEE* LOCUS-SPECIFIC DATABASE.

LOW COMPLEXITY REGION

Jaap Heringa and Patrick Aloy

The major source of deceptive alignments is the presence within proteins of regions with highly biased amino acid composition (Altschul *et al*., 1994)). If such a region within a query sequence is included during a homology search of that query sequence against a sequence database, biologically unrelated database sequences containing similarly biased regions are likely to score spuriously high, often rendering the search meaningless. For this reason, the PSI-BLAST program filters out biased regions of query sequences by default, using the SEG program (Wootton and Federhen., 1993, 1996). Because the SEG parameters have been set conservatively to avoid masking potentially important regions, some bias may remain so that compositionally biased spurious hits can still occur, for example with protein sequences that have a known bias, such as myosins or collagens.

The SEG filtering method (ftp://ftp.ncbi.nih.gov/pub/seg/seg/) can be used with parameters that eliminate nearly all biased regions, and the user can apply locally other filtering procedures, such as COILS (Lupas, 1996), which delineates putative coiled-coil regions, before submitting the appropriately masked sequence to PSI-BLAST. Low complexity sequences found by filtering are substituted by PSI-BLAST using the letter 'N' in nucleotide sequence (e.g., 'NNNNNNNNNNNNN') and the letter 'X' in protein sequences (e.g., 'XXXXXXXXX').

A well-known low complexity region is the polyglutamine repeat which forms insoluble fibers of beta-sheets that can cause neurodegenerative diseases.

Related website

seg	ftp://ftp.ncbi.nih.gov/pub/seg/seg/

Further reading

Altschul SF, et al. (1994) Issues in searching molecular sequence databases. *Nat. Genet.* 6: 119–129.

Lupas A (1996) Prediction and analysis of coiled-coil structures. *Methods Enzymol.* 266: 513–525.

Perutz MF (1999) Glutamine repeats and neurodegenerative diseases: molecular aspects. *Trends Biochem. Sci.* 24: 58–63.

Wootton JC, Federhen S (1993) Statistics of local complexity in amino acid sequences and sequence databases. *Comput. Chem.* 17: 149–163.

Wootton JC, Federhen S (1996) Analysis of compositionally biased regions in sequence databases. *Methods Enzymol.* 266: 554–571.

See also BLAST, Homology Search, RepeatMasker, Sequence Simplicity, Indel

LSDB, *SEE* LOCUS-SPECIFIC DATABASE.

M

M

MacClade
Machine Learning
Majority-Rule Consensus Tree, *see* Consensus Tree.
Mammalian Gene Collection, *see* MGC.
Mammalian Promoter Database, *see* MpromDB.
Manual Gene Annotation, *see* Gene Annotation (hand-curated).
Map Function
Mapping by Admixture Linkage Disequilibrium, *see* Admixture Mapping.
Mark-up Language
Marker
Markov Chain
Markov Chain Monte Carlo (MCMC, Metropolis-Hastings, Gibbs Sampling)
Markov Model, *see* Hidden Markov Model, Markov Chain.
Mathematical Modeling (of Molecular/ Metabolic/Genetic Networks)
Mature microRNA
Maximal Margin Classifier, *see* Support Vector Machine.
Maximum Likelihood Phylogeny Reconstruction
Maximum Parsimony Principle (Parsimony, Occam's Razor)
MaxQuant
MCMC, *see* Markov Chain Monte Carlo.
MEGA (Molecular Evolutionary Genetics Analysis)
Mendelian Disease
MEROPS
Mesquite
Message
Metabolic Modeling
Metabolic Network
Metabolic Pathway
Metabolome (Metabonome)
Metabolomics Databases
Metabolomics Software
Metabonome, *see* Metabolome.
Metadata
Metropolis-Hastings, *see* Markov Chain Monte Carlo.
MGC (Mammalian Gene Collection)
MGD (Mouse Genome Database)
MGED Ontology
Microarray
Microarray Image Analysis
Microarray Normalization
Microfunctionalization

Concise Encyclopaedia of Bioinformatics and Computational Biology, Second Edition.
Edited by John M. Hancock and Marketa J. Zvelebil.

MicroRNA

MicroRNA Discovery

MicroRNA Family

MicroRNA Prediction, *see* MicroRNA Discovery.

MicroRNA Seed

MicroRNA Seed Family, *see* MicroRNA Family.

MicroRNA Target

MicroRNA Target Prediction

Microsatellite

Midnight Zone

MIGRATE-N

MIME Types

Minimum Evolution Principle

Minimum Information Models

Minisatellite

miRBase

Mirtron

Missing Data, *see* Missing Value.

Missing Value (Missing Data)

Mitelman Database (Chromosome Aberrations and Gene Fusions in Cancer)

Mixture Models

MM, *see* Markov Chain.

MOD, *see* Model Organism Database.

Model Order Selection, *see* Model Selection.

Model Organism Database (MOD)

Model Selection (Model Order Selection, Complexity Regularization)

Modeling, Macromolecular

Models, Molecular

Modeltest

ModENCODE, *see* ENCODE.

Modular Protein

Module Shuffling

Mol Chemical Representation Format

Molecular Clock (Evolutionary Clock, Rate of Evolution)

Molecular Coevolution, *see* Coevolution.

Molecular Drive, *see* Concerted Evolution.

Molecular Dynamics Simulation

Molecular Efficiency

Molecular Evolutionary Mechanisms

Molecular Information Theory

Molecular Machine

Molecular Machine Capacity

Molecular Machine Operation

Molecular Mechanics

MOLECULAR NETWORK, *see* Network.

Molecular Replacement

Monophyletic Group, *see* Clade.

Monte Carlo Simulation

Motif

Motif Discovery

Motif Enrichment Analysis

Motif Search

Mouse Genome Database, *see* Mouse Genome Informatics.

Mouse Genome Informatics (MGI, Mouse Genome Database, MGD)

Mouse Tumor Biology (MTB) Database, *see* Mouse Genome Informatics.

MouseCyc, *see* Mouse Genome Informatics.

MPromDB (Mammalian Promoter Database)

MrBayes

Multidomain Protein

Multifactorial Trait (Complex Trait)

Multifurcation (Polytomy)

Multilabel Classification

Multilayer Perceptron, *see* Neural Network.
Multiple Alignment
MULTIPLE HIERARCHY (Polyhierarchy)
Multiplex Sequencing
Multipoint Linkage Analysis
Murcko Framework, *see* Bemis and Murcko Framework.
Mutation Matrix, *see* Amino Acid Exchange Matrix.

MACCLADE
Michael P. Cummings

A program for the study of phylogenetic trees and character evolution.

The program has a rich set of features for graphical-based tree manipulation and exploration (e.g., move, clip, collapse branches), and examination of character patterns by mapping characters on trees or through charts and diagrams. The program accommodates multiple input and output formats for nucleotide and amino acid sequence data, as well as general data types, and has broad data editing capabilities. Many options are available for tree formatting and printing.

The program features an extensive graphical user interface, and is available as an executable file for Mac OS up to version 10.6. There is a lengthy book providing extensive details about program features, examples of their use, and background information.

Related website

MacClade web site	http://macclade.org/macclade.html

Further reading

Maddison DR, Maddison WP (2001) *MacClade 4: Analysis of Phylogeny and Character Evolution* Sinauer Associates.

MACHINE LEARNING
Nello Cristianini

Branch of artificial intelligence concerned with developing computer programs that can learn and generalize from examples.

By this it is meant the acquisition of domain-specific knowledge, resulting in increased predictive power. Limited to the setting when the examples are all given together at the start, it is a valuable tool for Data Analysis. Many algorithms have been proposed, all aimed at detecting relations (patterns) in the training data and exploiting them to make reliable predictions about new, unseen data.

Two stages can be distinguished in the use of learning algorithms for data analysis. First a training set of data is provided to the algorithm, and used for selecting a 'hypothesis'. Then such a hypothesis is used to make predictions on unseen data, or tested on a set of known data to measure its predictive power.

A major problem in this setting is that of overfitting or overtraining, when the hypothesis selected reflects specific features of the particular training set due to chance and not to the underlying source generating it. This happens mostly with small training samples, and leads to reduced predictive power.

Motivated by the need to understand overfitting and generalization, in the last few years significant advances in the mathematical theory of learning algorithms have brought this field very close to certain parts of statistics, and modern machine learning methods tend to be less motivated by heuristics or analogies with biology (as was the case – at least originally – for neural networks or genetic algorithms) and more by theoretical considerations (as is the case for support vector machines and graphical models).

The output of a machine learning algorithm is called a hypothesis, or sometimes a model. A common type of hypothesis is a classifier, that is a function that assigns inputs to one of a finite number of classes.

Further reading

Cristianini N, Shawe-Taylor J (2000) *An Introduction to Support Vector Machines* Cambridge University Press.

Duda RO, Hart PE, Stork DG (2001) *Pattern Classification*, John Wiley & Sons, USA.

Mitchell T (1997) *Machine Learning*, McGraw Hill.

MAJORITY-RULE CONSENSUS TREE, *SEE* CONSENSUS TREE.

MAMMALIAN GENE COLLECTION, *SEE* MGC.

MAMMALIAN PROMOTER DATABASE, *SEE* MPROMDB.

MANUAL GENE ANNOTATION, *SEE* GENE ANNOTATION (HAND-CURATED).

MAP FUNCTION

Mark McCarthy and Steven Wiltshire

A mathematical formula for converting between a recombination fraction and a genetic – or map – distance measured in Morgans (M) or centimorgans (cM).

A map function describes the mathematical relationship, for two loci under study, between the non-additive recombination fraction, θ, and the additive map distance, x. The map distance is the mean number of crossovers occurring between the loci on a single chromatid per meiosis. The map function is required because θ has a maximum of 0.5, whilst, on larger chromosomes at least, several crossovers are possible. The most commonly used map functions differ in the extent to which they allow for the phenomenon of interference (that is, the tendency for crossover events rarely to occur in close proximity to each other). Haldane's map function assumes no interference (i.e. all crossovers are independent), and is given by:

$$x_H = -\frac{1}{2}\ \ln(1 - 2\theta)$$

for 0 ≤ < 0.5, ∞ otherwise, with the inverse being:

$$\theta = \frac{1}{2}\ [1 - \exp(-2|x_H|)]$$

Kosambi's map function assumes a variable level of interference, expressed as $2q$:

$$xk = \frac{1}{2}\ \ln\left(\frac{1+2\theta}{1-2\theta}\right)$$

for 0 ≤ < 0.5, ∞ otherwise, with the inverse being:

$$\theta \ = \frac{1}{2}\frac{(\exp(4xk) - 1)}{(\exp(4xk) + 1)}$$

Other map functions exist, differing in their modelling of interference (Carter-Falconer, Felsenstein) or the assumptions regarding the distribution of chiasmata between loci (Sturt). However, Kosambi's and Haldane's map functions, described above, are the most widely used, with Kosambi's function tending to give more realistic distances, although at small values of θ, there is little practical difference between the two.

Examples: Consider two loci with $\theta = 0.3$. Haldane's map function gives $x_H = 45.8$ cM, whereas Kosambi's function gives $x_K = 34.7$ cM. If the recombination fraction were 0.03, the distances would be $x_H = 3.1$ cM and $x_K = 3.0$ cM.

Kosambi's map function has been used in the construction of Généthon (Dib et al. 1996), Marshfield (Broman et al. 1998) and deCODE (Kong et al. 2002) marker maps. Haldane's map function is used in programs such as GENEHUNTER, ALLEGRO and MERLIN during multipoint IBD estimation between markers since the Lander-Green algorithm, used by these programs to calculate pedigree likelihoods, implicitly considers recombination events in adjacent intervals to be independent (i.e. no interference).

Further reading

Broman KW, et al. (1998) Comprehensive human genetic maps: individual and sex-specific variation in recombination. *Am. J. Hum. Genet.* 63: 861–869

Dib C, et al. (1996) A comprehensive genetic map of the human genome based on 5,264 microsatellites. *Nature* 380: 152–154.

Kong A, et al. (2002) A high-resolution recombination map of the human genome. *Nat. Genet.* 31: 241–247.

Ott J (1999) *Analysis of human genetic linkage.* Baltimore: The Johns Hopkins University Press pp17–21.

Sham P (1998) *Statistics in Human Genetics.* London: Arnold pp 54–58.

See also Recombination, Linkage Analysis, Marker

M

MAPPING BY ADMIXTURE LINKAGE DISEQUILIBRIUM, *SEE* ADMIXTURE MAPPING.

MARK-UP LANGUAGE

Carole Goble and Katy Wolstencroft

Minimum Information Models specify what metadata are important in describing a particular type of data and/or experiment, but they do not specify the format of those descriptions. Mark-up languages, such as MAGE-ML (MicroArray and Gene Expression Markup Language) and SBML (Systems Biology Markup Language), provide a structured syntax for metadata.

Mark-up languages are typically written in XML (hence the names of many end in -ML), and they are designed to facilitate the exchange, pooling and querying of related data. For example, SBML provides an XML format to describe metabolic reactions and species in those reactions. This means scientists can easily compare SBML files for the same species, reactions and other parameters. More importantly, the common format of SBML means that files can be uploaded and used in a large collection of analysis tools. If an SBML file is annotated to MIRIAM standard (Minimum Information Required in the Annotation of Models), SBML files can be cross-linked to public databases and annotation/visualization tools to display the resulting model structures.

More recently, some -ML languages have been replaced with tabular formats (for example MAGE-TAB). This provides a more accessible structure for annotating data, since spreadsheet data management is commonplace.

Further reading

Ball CA, Brazma A (2006) MGED standards: work in progress. *OMICS* 10(2):138–44. Review.

Hucka, et al. (2003) The systems biology markup language (SBML) a medium for representation and exchange of biochemical network models. *Bioinformatics* 19(4):524–31.

Related websites

MAGE	http://www.mged.org/Workgroups/MAGE/mage.html
SBML	http://sbml.org
XML schemas at MIBBI	http://mibbi.sourceforge.net/portal.shtml

See also Minimum Information Models

MARKER

Mark McCarthy, Steven Wiltshire and Andrew Collins

Any polymorphism of known genomic location, such that it can be used for disease-gene mapping.

Polymorphic sites, whether or not they represent sites of functional variation, are essential tools for gene mapping efforts as they can be used to follow chromosomal segregation and infer the likely position of recombinant events within pedigrees and populations. To be useful, markers must: (a) have a known, and unique, chromosomal location; (b) show at least a moderate degree of polymorphism; (c) have a low mutation rate; (d) be easily typed (most usually by PCR-based methods). The types of markers most frequently used in linkage mapping analyses are microsatellites (usually di-, tri- or tetra-nucleotide repeats), which are often highly polymorphic and well suited to genomewide linkage analysis. Single nucleotide polymorphisms (SNPs), are less polymorphic, but much more frequent, representing the majority of functional variation, and have been extensively used in association analyses. The HapMap project has characterized millions of SNPs which have underpinned the development of arrays for genome-wide association studies.

Examples: Researchers in Iceland (Kong et al., 2002) published a high-density genetic map of over 5000 microsatellites, providing a dense framework of robust and polymorphic markers suitable for linkage studies.

Related websites

HapMap	http://hapmap.ncbi.nlm.nih.gov/
DbSNP (database of single nucleotide polymorphisms)	http://www.ncbi.nlm.nih.gov/projects/SNP/

Further reading

Dib C, et al. (1996) A comprehensive genetic map of the human genome based on 5,264 microsatellites. *Nature* 380: 152–154.

Kong A, et al. (2002) A high-resolution recombination map of the human genome. *Nat. Genet.* 31: 241–247.

Kruglyak L (1997) The use of a genetic map of biallelic markers in linkage studies. *Nat. Genet.* 17: 21–24.

The International HapMap Consortium (2010) Integrating common and rare genetic variation in diverse human populations. *Nature* 467: 52–58.

The Utah Marker Development Group (1995) A collection of ordered tetranucleotide-repeat markers from the human genome. *Am. J. Hum. Genet.* 57: 619–628.

See also The HapMap Project, Genome-Wide Association, Genome Scans, Polymorphism

MARKOV CHAIN

John M. Hancock

A Markov chain is a series of observations in which the probability of an observation occurring is a function of the previous observation or observations.

A DNA sequence can be considered to be an example of a Markov chain as the likelihood of observing a base at a particular position depends strongly on the nature of the preceding base. This is the basis of the non-random dinucleotide frequency distribution of most DNA sequences.

Strictly, the above definition defines a first order Markov chain (Markov chain of order 1). Higher-order Markov chains can also be defined, in which the probability of an observation depends on a larger number of preceding observations. Again, DNA sequences can be taken as an example as base frequencies are strongly influenced by the preceding five bases, making DNA sequences equivalent to Markov chains of order five.

The process of generating a Markov chain is known as a Markov process.

Related website

Wikipedia	http://en.wikipedia.org/wiki/Markov_chain

See also Dinucleotide Frequency

MARKOV CHAIN MONTE CARLO (MCMC, METROPOLIS-HASTINGS, GIBBS SAMPLING)

John M. Hancock

Markov Chain Monte Carlo algorithms are a class of algorithms for estimating parameters of a model. They operate by generating a series of estimates of the underlying parameter(s) which are then tested for the goodness of fit of the outcomes they produce to the expected output distribution. The estimates are not generated independently but result from 'moves' from a preceding state, the series of parameter estimates therefore corresponding to Markov Chains. If an MCMC simulation is run for long enough, the distribution it will produce (stationary distribution) is a good approximation of the likelihood distribution of the parameter(s) being estimated. MCMC is widely used for estimating parameters in Bayesian analysis.

Two commonly applied classes of MCMC are the Metropolis-Hastings and Gibbs Sampling approaches. In Metropolis-Hastings, moves are, broadly, accepted if they improve the goodness of fit to the model, although moves that do not improve goodness of fit are accepted with reduced probability. In Gibbs sampling, rather than roaming across the parameter space, individual parameters are changed in turn based on their known (or assumed) conditional distributions.

Relevant websites

Markov chain Monte Carlo	http://en.wikipedia.org/wiki/Markov_chain_Monte_Carlo
Metropolis-Hastings	http://en.wikipedia.org/wiki/Metropolis-Hastings_algorithm
Gibbs Sampling	http://en.wikipedia.org/wiki/Gibbs_sampling

MARKOV MODEL, *SEE* HIDDEN MARKOV MODEL, MARKOV CHAIN.

MATHEMATICAL MODELING (OF MOLECULAR/ METABOLIC/GENETIC NETWORKS)

Denis Thieffry

To integrate data on molecular, metabolic or genetic interactions between individual components, one can write mathematical equations modeling the evolution of each molecular species (variable) across time and possibly space. This allows study, simulation and even predictions about the temporal behavior of the system in response to various types of perturbations. The mathematical formalisms used can be grouped into several classes according to levels of detail (qualitative or quantitative) and basic assumptions (deterministic or stochastic).

In the case of quantitative mathematical modeling, Ordinary Differential Equations (ODE) are most often used, in particular when dealing with (bio)chemical reactions; these express the evolution (time derivative) of each variable of the system (e.g. the concentration of a molecular species) as a function of other variables of the system. When dealing with regulatory systems, these functions are usually non-linear, involving products or power of variables to model cooperative behavior. Non-linearity complicates the analysis of such sets of equations, and the modeler has to rely on the use of numerical simulation techniques. Parameters (supposedly constant) are included in the equations to modulate the weight of the different terms involved in each equation. On the basis of this formal description, one can select parameter values and an initial state (a set of values for the different variables) to perform a simulation, with the help of publicly available software (see Table M.1). Simulations can be shown in the form of the evolution of the system as a function of time, or in the form of a trajectory in the space defined by the ranges of values for the variables of the system (variable space or phase portrait). Steady states can be located thanks to numerical iteration methods used by physicists.

To represent spatial effects such as diffusion or transport one can use Partial Differential Equations (PDE) expressing the variation of the different molecular species in different directions (usually the three axes of the Euclidean space, or referring to polar coordinates). As in the case of ODE systems, it is possible to simulate PDE, but one then needs to specify boundary conditions in addition to the values of the different parameters of the system. One famous, historical example of of PDE application to (bio)chemical problem can be found in the work of Alan Turing (1952) on pattern formation defined by reaction-difusion systems.

Table M.1 Examples of Simulation Software

Name	Type of simulation/analysis	URL
Gepasi	Continuous biochemical simulations	http://www.gepasi.org/
StochSim	Stochastic simulation of molecular networks	http://www.pdn.cam.ac.uk/groups/comp-cell/StochSim.html
XPPAUTO	Differential equations + bifurcation analyses	http://www.math.pitt.edu/~bard/xpp/xpp.html
GRIND	Integration of differential equations	http://www-binf.bio.uu.nl/rdb/grind.html
DDLab	Boolean networks simulation	http://www.ddlab.com/
GINsim	Logical simulations + analysis	http://gin.univ-mrs.fr/

If the modeler wishes to focus on qualitative aspects of the behavior of a regulatory system, he can use Boolean equations instead of quantitative differential equations. Each component is represented by a Boolean variable which can take only two different values: 0 if the component is absent or inactive, 1 otherwise. The effect of other components on the evolution of a given variable is then described in terms of a logical function, built with operators such as AND, OR, etc. When studying the evolution of these logical systems, two opposite treatments of time are used: (1) a synchronous approach considers all equations at once and update all variables simultaneously, according to the values calculated on the basis of the functions (left-hand terms in the logical equations). (2) an asynchronous approach consists in selecting only one transition each time the system is in a state in which several variable changes are possible. In the context of these different updating assumptions, the system will present the same stable states, defined as the states where there is equality between the value of the logical variable and the corresponding logical function. To these different updating assumptions, however, correspond different dynamics in terms of transient or potential cyclic behavior. The logical approach has been generalized to encompass multi-level variables, (i.e. variables taking more than two integer values: 0, 1, 2, ...), as well as the explicit consideration of threshold values (at the interface between logical values). Finally, logical parameters can be introduced in order to cover families of logical functions.

When studying interactions between components present at very low numbers, it is not possible anymore to assume a continuous range of concentration. It is then necessary to take into account explicitly the probability of encounters between two molecules to describe their interaction. This is done by using stochastic equations. In this context, the system is described as a list of molecular states, plus the distributions of probabilities associated to each state transition. As in the case of non-linear differential equations, stochastic systems can generally not be treated analytically and the modeler has to rely on numerical simulations. Relatively heavy from a computational point of view, such simulations has been performed only for a limited number of experimentally well-defined networks, including the genetic network controlling the lysis-lysogeny decision in bacteriophage lambda, and the phosphorylation cascade involved in bacterial chemotaxis.

In the context of dynamical systems, one says that a state is steady when the time derivative of all variables are nil (in the case of ODE or PDE), or when the values of all the logical variables equal those of the corresponding functions. In the differential context (ODE), it is possible to further evaluate the stability of a steady state by analyzing the effects of perturbations around this steady state (usually on the basis of a linear approximation of the original equations). Mathematically, one can delineate these dynamical properties by analyzing the Jacobian matrix of the system, which gives the partial derivative of each equation (row) according to each variable (column) (an approach called stability linear analysis). On the basis of this matrix, one can compute the characteristic equation of the system. The types of roots of this equation at steady state, called the Eigen values, are characteristic of specific types of steady state (see Table M.2 in the case of two-dimentional systems).

Once the steady states have been located, other numerical tools enable the modeler to progressively follow their displacement, change of nature, or disappearance as a parameter of the system is modified, something which is represented in the form of a bifurcation diagram.

Table M.2 Main Types of Steady States in the Case of Two-dimensional ODE Systems

Type of steady states (2 dimensions)	Local dynamical properties	Roots of the characteristic equations (Eigen values)
Stable node	Attractive along all directions	2 real negative roots
Stable focus	Attractive in a periodic way	2 complex roots with negative real parts
Unstable focus	Repulsive in a periodic way	2 complex roots with positive real parts
Saddle point	Attractive along one direction,repulsive along the orthogonal direction	1 positive and 1 negative real root
Peak	Repulsive along all directions	2 positive real roots

Related to the notion of steady state is that of an attractor, defined as a set of states (points in the phase space), invariant under the dynamics, towards which neighboring states asymptotically approach in the course of dynamic evolution. An attractor is defined as the smallest unit which cannot be decomposed into two or more attractors. The (inclusive) set of states leading to a given attractor is called its basin of attraction. In the simplest case, a system can have a unique, stable state, which then constitutes the unique attractor of the system, the rest of the phase space defining its basin of attraction.

The list of steady states, their properties, and the extension of the corresponding basins of attractions define the main qualitative dynamical features of a dynamical system. These features depend on the structure of the equations (presence of feedback circuits, of non-linearities, etc.), but also on the values selected for the different parameters of the system. When a system has several alternative steady states for given parameter values, one speaks of multistationarity, a property often advanced to explain differentiation and development. On another hand, cyclic attractors are associated with molecular clocks such as those controlling the cell cycle or circardian rythms. When the most important dynamical features are only marginally affected by ample changes of parameter values, one says that the system is robust. Robustness appears to be a general property of many biological networks.

More recently, metabolic networks have also been described in terms of Petri Nets, which can be defined as digraphs (i.e. graphs involving two different types of vertices, each connecting to the other, exclusively). The first type of vertex, called places, corresponds to molecular species, whereas the second type of vertices, called transitions, represents reactions. The places contain resources (molecules), which can circulate along the edges connecting places to transitions, according to rules associated to the transition vertices. Starting from an initial state, transition rules are used to perform numerical simulations, under deterministic or stochastic assumptions. Petri Nets have recently been applied to metabolic graphs comprising up to several thousand nodes.

Applied exclusively to metabolic networks, the Metabolic Control Analysis (MCA) is a phenomenological quantitative sensitivity analysis of fluxes and metabolite concentrations. In MCA, one evaluates the relative control exerted by each step (enzyme) on the system's variables (fluxes and metabolite concentrations). This control is measured by applying a perturbation to the step being studied and measuring the effect on the variable of interest after the system has settled to a new steady state. Instead of assuming the existence of a unique rate-limiting step, it assumes that there is a definite amount of flux control and that this is spread quantitatively among the component enzymes.

M

Further reading

Heinrich R, Schuster S (1996) *The Regulation of Cellular Systems*. Chapman & Hall.

Kaplan D, Glass L (1995) *Understanding Nonlinear Dynamics*. Springer-Verlag.

Kauffman SA (1993) *The Origins of Order: Self-Organization and Selection in Evolution*. Oxford University Press.

MATURE MICRORNA

Sam Griffiths-Jones

The mature microRNA is a ~22nt single-stranded RNA molecule that interacts with complementary sites in target mRNAs to guide their repression or cleavage by the RNA induced silencing complex (RISC). Mature microRNAs are produced by cleavage of precursor microRNAs by the Dicer enzyme in the cytoplasm.

See also Guide Strand, MicroRNA, Precursor MicroRNA, Dicer, RISC

MAXIMAL MARGIN CLASSIFIER, *SEE* SUPPORT VECTOR MACHINE.

MAXIMUM LIKELIHOOD PHYLOGENY RECONSTRUCTION

Sudhir Kumar and Alan Filipski

A method for selecting a phylogenetic tree relating a set of sequence data based on an assumed probabilistic model of evolutionary change and the maximum likelihood principle of statistical inference.

Maximum likelihood is a standard statistical method for parameter estimation using a model and a set of empirical data. The parameter values are chosen so as to maximize the probability of observing the data given the hypothesis. In phylogenetics, this method has been applied to finding the phylogeny (tree topology) with the highest likelihood and estimating branch lengths as well as other parameters (e.g., the alpha shape parameter of the GAMMA distribution of rate heterogeneity) of the evolutionary process. Because of the large number of possible alternative tree topologies the problem of finding the Maximum Likelihood tree is known to be NP-hard.

Software

Guindon S, Dufayard JF, Lefort V, Anisimova M, Hordijk W, Gascuel O (2010) New algorithms and methods to estimate maximum-likelihood phylogenies: assessing the performance of PhyML 3.0. *Systematic Biology* 59(3): 307–321.

Jobb G, Von Haeseler A, Strimmer K (2004) TREEFINDER: a powerful graphical analysis environment for molecular phylogenetics. *BMC Evolutionary Biology* 4(1): 18.

Price MN, Dehal PS, Arkin AP (2010) FastTree 2-approximately maximum-likelihood trees for large alignments. *PloS One* 5(3): e9490.

Stamatakis A (2006) RAxML-VI-HPC: Maximum likelihood-based phylogenetic analyses with thousands of taxa and mixed models. *Bioinformatics* 22(21): 2688–2690.

Vinh LS, Haeseler A (2004) IQPNNI: moving fast through tree space and stopping in time. *Molecular Biology and Evolution* 21(8): 1565–1571.

Zwickl D (2006) Genetic algorithm approaches for the phylogenetic analysis of large biological sequence datasets under the maximum likelihood criterion. PhD thesis, the University of Texas at Austin.

Further reading

Felsenstein J (1981) Evolutionary trees from DNA sequences: a maximum likelihood approach. *Journal of Molecular Evolution* 17(6): 368–376.

Felsenstein J (2004) *Inferring Phylogenies*. Sinauer Associates.

Roch S (2006) A short proof that phylogenetic tree reconstruction by maximum likelihood is hard. *IEEE/ACM Transactions on Computational Biology and Bioinformatics* 3(1): 92–94.

Yang Z (1996) Maximum-likelihood models for combined analyses of multiple sequence data. *J Mol Evol* 42: 587–596.

See also Phylogenetic Tree

MAXIMUM PARSIMONY PRINCIPLE (PARSIMONY, OCCAM'S RAZOR)

Sudhir Kumar, Alan Filipski and Dov Greenbaum

A method of inferring a phylogenetic tree topology over a set of sequences in which the tree that requires the fewest total number of evolutionary changes to explain the observed sequences is chosen.

Parsimony is the law of Occam's razor, whereby the simplest explanation/model is preferred to more complex explanation. Thus in phylogeny reconstruction a tree that explains the data by the lowest number of mutations is preferred over more complex trees.

An efficient algorithm for estimating the number of nucleotide or amino acid changes in a given tree is available. Finding the best topology (with the least mutations, lowest parsimony score) under parsimony requires enumeration and computation of the parsimony score on all alternative possible unrooted trees. This is the reason why the parsimony-based tree inference problem is known to be NP-hard.

For computing the parsimon score on a tree, only parsimony informative sites need to be considered; a site is called informative (or parsimony-informative) if at least two different states occur at least twice each at that site. For non-informative sites, the parsimony score of a site will be constant for all possible trees and does hence not need to be considered.

Parsimony methods do not perform well in the presence of too much homoplasy, that is, identity of character states that is not attributed to common ancestry.

This can arise in molecular sequences by multiple substitutions or parallel and convergent changes.

Another shortcoming of parsimony is the phenomenon known as long-branch attraction. Non-sister taxa may erroneously cluster together in the phylogenetic tree because of similarly longer branches when compared to their true sister taxa. This effect was first demonstrated for the maximum parsimony method, but is known to also occur in statistically rigorous methods.

Software

Maddison WP, Maddison DR (1992) *MacClade: analysis of phylogeny and character evolution*. Sunderland, Mass., Sinauer Associates.

Parsimonator open source code: https://github.com/stamatak/Parsimonator-1.0.2 Goloboff PA, Farris JS, Nixon KC (2008) TNT, a free program for phylogenetic analysis. *Cladistics* 24(5): 774–786.

Swofford DL (1998) *PAUP*: Phylogenetic Analysis Using Parsimony (and other methods)*. Sunderland, MA., Sinauer Associates.

Further reading

Day W, Jonhson D, Sankoff D (1986) The computational complexity of inferring rooted phylogenies by parsimony. *Mathematical Biosciences* 81: 33–42.

Densmore LD 3rd (2001) Phylogenetic inference and parsimony analysis. *Methods Mol Biol* 176: 23–36.

Felsenstein J (1978) Cases in which parsimony or compatibility methods will be positively misleading. *Systematic Biology* 27(4): 401–410.

Fitch WM (1971) Toward defining course of evolution – minimum change for a specific tree topology. *Syst Zool* 20: 406–416.

Hendy MD, Penny D (1989) A framework for the quantitative study of evolutionary trees. *Syst Zool* 38: 297–309.

Kluge AG, Farris JS (1969) Quantitative Phyletics and Evolution of Anurans. *Syst Zool* 18: 1–32.

Nei M, Kumar S (2000) *Molecular Evolution and Phylogenetics*. Oxford University Press.

Stewart CB (1993) The powers and pitfalls of parsimony. *Nature* 361: 603–607.

See also Phylogenetic Tree

MAXQUANT

Simon Hubbard

Software tool for processing mass spectrometry data to generate quantitation data for proteome proteins. Maxquant is designed for handling large, complex datasets, is freely available, and can handle both label-mediated and label-free data. It is developed by Jürgen Cox at the Max Planck Institute, Martinsreid, Germany, and is targeted at high-resolution mass spectrometry data.

Related website

MaxQuant	http://maxquant.org

Further reading

Cox J, Mann M. MaxQuant enables high peptide identification rates, individualized p.p.b.-range mass accuracies and proteome-wide protein quantification. *Nat Biotechnol.* 2008, 26: 1367–1372.

See also Quantitative Proteomics

MCMC, *SEE* MARKOV CHAIN MONTE CARLO.

MEGA (MOLECULAR EVOLUTIONARY GENETICS ANALYSIS)

Michael P. Cummings

A program for conducting comparative analysis of DNA and protein sequence data. Version 5 of the software was released in 2011.

The program features phylogenetic analysis using parsimony, or a choice of a variety of distance models and methods (neighbor-joining, minimum evolution, UPGMA (Unweighted Pair Group Method with Arithmetic Mean). There are capabilities for summarizing sequence patterns (e.g., nucleotide substitutions of various types, insertions and deletions, codon usage). The program also allows testing of the molecular clock hypothesis and allows some tree manipulation.

The program has a graphical user interface and is available for various platforms.

Related website

MEGA web site	http://www.megasoftware.net/

Further reading

Tamura K, Peterson D, Peterson N, Stecher G, Nei M, Kumar S (2011) MEGA5: Molecular Evolutionary Genetics Analysis using Maximum Likelihood, Evolutionary Distance, and Maximum Parsimony Methods. *Molecular Biology and Evolution* 28: 2731–2739.

MENDELIAN DISEASE

Mark McCarthy, Steven Wiltshire and Andrew Collins

Diseases (or traits) that display classical Mendelian segregation patterns within families, usually because susceptibility in any given pedigree is determined by variation at a single locus.

Mendelian traits are usually the result of mutations causing dramatic changes in the function of genes or their products, with substantial consequences for phenotype. As a result, these conditions tend to be rare, though such deleterious alleles may be maintained through a variety of mechanisms including high mutation rates or (in the case of some recessive disorders) evolutionary advantages associated with the heterozygous state. Depending on their chromosomal location, and whether or not disease is associated with loss of one or two copies of the normal gene, Mendelian traits are characterised as autosomal or sex-linked, and dominant or recessive. The genes responsible for over 1300 Mendelian traits have now been identified through a combination of positional cloning and candidate gene methods: these are catalogued in databases such as OMIM. The characteristic feature of Mendelian traits is a close correlation between genotype and phenotype (in contrast with multifactorial traits), although genetic heterogeneity (where a Mendelian trait is caused by defects in more than one gene) and the effects of modifier loci and environment, mean that the distinction between Mendelian and multifactorial traits can be somewhat arbitrary.

Examples: The identification of the CFTR gene as the basis for the development of cystic fibrosis – one of the most common, severe Mendelian traits – is one of the classical stories in modern molecular biology (Kerem et al, 1989). Retinitis pigmentosa and maturity onset diabetes of the young (Owen and Hattersley, 2001) are examples of Mendelian traits featuring substantial locus heterogeneity. Finally, thalassaemia (Weatherall, 2000) represents a group of conditions which combine features of both Mendelian and multifactorial traits.

Related websites

Catalogues of human genes and associated phenotypes:

Online Mendelian Inheritance in Man	http://www.ncbi.nlm.nih.gov/omim/
GeneCards	http://www.genecards.org/

Further reading

Kerem B, et al. (1989) Identification of the cystic fibrosis gene: genetic analysis. *Science* 245: 1073–1080.

Owen K, Hattersley AT (2001) Maturity-onset diabetes of the young: from clinical description to molecular genetic characterization. *Best Pract. Res. Clin. Endocrinol Metab.* 15: 309–323.

Peltonen L, McKusick V (2001) Genomics and medicine. Dissecting human disease in the postgenomic era. *Science* 291: 1224–1229.

Weatherall DJ (2000) Single gene disorders or complex traits: lessons from the thalassaemias and other monogenic diseases. *Brit. Med. J.* 321: 1117–1120.

See also Multifactorial Trait, Linkage, Candidate Gene, Linkage Analysis

MEROPS

Rolf Apweiler

The MEROPS database is a specialised protein sequence database covering peptidases.

The MEROPS database provides a catalogue and structure-based classification of peptidases. An index by name or synonym gives access to a set of files, each providing information on a single peptidase, including classification and nomenclature, and hypertext links to the relevant entries in other databases. The peptidases are classified into families based on statistically significant similarities between the protein sequences in the 'peptidase unit', the part most directly responsible for activity. Families that are thought to have common evolutionary origins and are known or expected to have similar tertiary folds are grouped into clans.

Relevant website

Merops	http://merops.sanger.ac.uk/

Further reading

Rawlings ND, Barrett AJ, Bateman A. (2012) MEROPS: the database of proteolytic enzymes, their substrates and inhibitors. *Nucleic Acids Res* 40: D343–D350.

See also Protein Family, Sequence (proteins), Folds

MESQUITE

Michael P. Cummings

A modular software system for evolutionary biology, with an emphasis on phylogenetic analysis and population genetics.

Analyses include: reconstruction of ancestral states (parsimony, likelihood); tests of process of character evolution, including correlation; analysis of speciation and extinction rates; simulation of character evolution; multivariate analyses of morphometric data; coalescence simulations and other calculations; and tree comparisons and simulations.

Written in Java and platform independent via Java virtual machine.

Related website

Mesquite home page	http://mesquiteproject.org/mesquite/mesquite.html

MESSAGE

Thomas D. Schneider

In communications theory, a message is a series of symbols chosen from a predefined alphabet.

In molecular biology the term 'message' usually refers to a messenger RNA. In molecular information theory, a message corresponds to an After State of a molecular machine. In information theory, Shannon proposed to represent a message as a point in a high dimensional space (see Shannon, 1949). For example, if we send three independent voltage pulses, their heights correspond to a point in three-dimensional space. A message consisting of 100 pulses corresponds to a point in 100 dimensional space. Starting from this concept, Shannon derived the channel capacity.

Further reading

Shannon CE (1949) Communication in the presence of noise. *Proc. IRE*, 37: 10–21.

See also Molecular Efficiency, Molecular Machine Capacity, Shannon Sphere, Noise

M

METABOLIC MODELING

Neil Swainston

The quantitative description and analysis of a metabolic system. In a biological context, a mathematical metabolic model can be used to predict the effect of a gene knockout or pharmaceutical intervention, or could be analysed to determine which aspects of the model (reactions or pathways) contribute most to increasing the yield of an industrially relevant product. Metabolic modeling in systems biology can take a number of forms, including constraint-based modeling and →kinetic modeling (ODE-based).

See also Constraint-based Modelling, Kinetic Modelling

METABOLIC NETWORK

Neil Swainston

The complete interconnected collection of metabolic reactions that define the metabolic capabilities of a given organism. Unlike for metabolic pathways, there is no artificial restriction of the scope of a metabolic network. A metabolic network can therefore be constructed by aggregating the set of all metabolic pathways of a given organism.

Metabolic networks can be inferred from an organism's genome, giving rise to genome-scale metabolic reconstructions that can be analyzed through constraint-based metabolic modeling approaches. Such reconstructions often go beyond the collection of metabolic reactions catalogued in data resources such as KEGG and MetaCyc, and their development often relies on additional manual literature searches and curation to uncover metabolic capabilities that are specific to an individual species.

Like metabolic pathways, individual reactions in metabolic networks are commonly associated with one or more enzymes that catalyze the reaction. Furthermore, metabolic networks of eukaryotes often consider intracellular compartmentalization, and also contain definitions of intracellular transport reactions between organelles, and – if applicable – the transport proteins that mediate such reactions.

Further reading

Metabolic modelling of microbes: the flux-balance approach. (2002) Edwards JS, Covert M, Palsson B. *Environ Microbiol.* 4, 133–140.

Kanehisa M, Goto S. (2000) KEGG: Kyoto Encyclopedia of genes and genomes. *Nucleic Acids Res.* 28, 27–30.

Krieger CJ, Zhang P, Mueller LA (2004) MetaCyc: a multiorganism database of metabolic pathways and enzymes. *Nucleic Acids Res.* 32, D438–442.

Reconstruction of biochemical networks in microorganisms. Feist AM, Herrgård MJ, Thiele I, Reed JL, Palsson BØ. (2009) *Nat. Rev. Microbiol.* 7, 129–143.

Thiele I, Palsson BØ (2010) Reconstruction annotation jamborees: a community approach to systems biology. *Mol Syst Biol.* 6, 361.

See also Metabolic Modeling

METABOLIC PATHWAY

Neil Swainston

An interconnected collection of metabolic reactions that take place in the cell. Individual reactions are defined by their collection of substrates and products, and are often associated with one or more metabolic enzymes that catalyse the reaction.

A metabolic pathway is largely conceptual and their limits within the larger metabolic network are arbitrary and based on historical contingency. Often pathways are roughly linear sequences of reactions, considered to describe flow of energy and mass in a single direction, e.g. towards a specific end product; but many reactions are thermodynamically reversible, and the pathways may also include regulatory feedback loops, such that a given product can act as a substrate or an enzymatic inhibitor or activator at considerable distance in the pathway.

The major metabolic pathways are well studied and in many cases evolutionarily conserved across species. A number of computational resources exist that classify metabolic pathways, amongst which KEGG and MetaCyc are the most comprehensive and widely used.

Further reading

Kanehisa M, Goto S (2000) KEGG: Kyoto Encyclopedia of genes and genomes. *Nucleic Acids Res.* 28, 27–30.

Krieger CJ, Zhang P, Mueller LA (2004) MetaCyc: a multiorganism database of metabolic pathways and enzymes. *Nucleic Acids Res.* 32, D438–442.

See also Metabolic Network

METABOLOME (METABONOME)

Dov Greenbaum

The quantitative complement of all small molecules present in a cell in a specific physiological state.

It has been claimed that knowledge of the translatome and transcriptome alone are not enough to fully describe the cell in any given state, and that this requires the further analysis of the metabolome as well. While those studying the metabolome have the advantage that the overall size of the population is generally less than either that of the translatome or transcriptome, there is no direct link between the genome and the metabolome. Still, researchers have looked at the effect that deleted genes have on the metabolome.

The metabolome is generally measured experimentally using biochemical analyses including, but not limited to, mass spectrometry, NMR and fourier transform infrared spectrometry.

Metabolic Flux Analysis attempts to quantify the intracellular metabolic fluxes by measuring extracellular metabolite concentrations in combination with intracellular reaction stoichiometry (assuming that the system is in a steady state).

The analysis of the metabolome of each individual organism will give insight into the metabolic processes of the cells and allow for comparisons of multiple organisms by way of their common metabolites and small molecules.

M

Further reading

Fiehn O (2001) Combining genomics, metabolome analysis, and biochemical modelling to understand metabolic networks. *Comp. Func. Genomics* 2: 155–168.

Mendes P (2002) Emerging bioinformatics for the metabolome. *Brief Bioinform* 3: 134–145.

Tweeddale H (1998) Effect of slow growth on metabolism of *Escherichia coli*, as revealed by global metabolite pool ('metabolome') analysis. *J Bacteriol* 180: 5109–5116.

See also Constraint-based Modeling

METABOLOMICS DATABASES

Darren Creek

There is no single definitive metabolomics database, however there are numerous useful metabolite databases, each with a different focus. Organism-specific metabolite databases are generated by curation of the biochemical literature and/or reconstructions of metabolic pathways from genome annotations (e.g. KEGG, BioCyc). Metabolomics-specific databases have been developed with experimental information to assist with metabolite identification, whilst often including biochemical data to assist with interpretation. Metabolite libraries are particularly important for metabolite identification, and a number of metabolite-specific spectral databases have been developed for NMR, GC-MS and MS/MS (and MSn) data. It is likely that many naturally occurring metabolites are unknown, and hence not present in existing metabolite databases, hence non-specific chemical databases are often required.

See Table M.3.

Table M.3 Useful Compound Databases for Metabolomics

Database	Website	Focus
KEGG	www.genome.jp/kegg/	Metabolic pathways
BioCyc (inc. HumanCyc, MetaCyc, etc)	biocyc.org/	Metabolic pathways
Reactome	www.reactome.org	Metabolic pathways
HMDB	www.hmdb.ca	Human metabolites
BiGG	bigg.ucsd.edu	Human pathways
KnapSack	kanaya.aist-nara.ac.jp/KNApSAcK/	Species-metabolite relationships
Lipidmaps	www.lipidmaps.org	Lipids
ChEBI	www.ebi.ac.uk/chebi/	Chemicals of biological interest
Chemspider	www.chemspider.com/	All chemicals
Pubchem	pubchem.ncbi.nlm.nih.gov/	All chemicals
Metlin	metlin.scripps.edu/	MS/MS
Massbank	www.massbank.jp	MS/MS
Golm	gmd.mpimp-golm.mpg.de	GC-MS
NIST	www.nist.gov/srd/nist1a.cfm	GC-MS
MMCD (Madison)	mmcd.nmrfam.wisc.edu/	NMR, MS
BioMagResBank (BMRB)	www.bmrb.wisc.edu	NMR
BML-NMR (Birmingham)	www.bml-nmr.org	NMR

Related website

Metabolomics Society	www.metabolomicssociety.org/database

Further reading

Go E (2010) Database resources in metabolomics: an overview. *Journal of Neuroimmune Pharmacology* 5: 18–30.

Halket JM, Waterman D, Przyborowska AM, Patel RK, Fraser PD, Bramley PM (2005) Chemical derivatization and mass spectral libraries in metabolic profiling by GC/MS and LC/MS/MS. *J Exp Bot.* 56: 219–243.

Tohge T. Fernie AR (2009) Web-based resources for mass-spectrometry-based metabolomics: A user's guide. *Phytochemistry* 70: 450–456.

METABOLOMICS SOFTWARE

Richard Scheltema

The data collected by mass spectrometry is extremely complex, often consists of many individual measurements and requires automated processing to interpret. For NMR and GC-MS specialized software is usually available with the equipment, but for LC-MS metabolomics, a very active bioinformatics community is producing open-source solutions for many analysis steps. Metabolomics software contributes to the processing of LC-MS metabolomics data in multiple steps:

1. Mass peak detection (centroiding): An individual scan (x-axis mass-over-charge; y-axis intensity) of a mass spectrometer consists of peaks spread out in the mass-over-charge range. The narrower these peaks, the higher the resolution of the mass spectrometer and the better individual substances can be distinguished. The first task of data analysis software is to detect these peaks and calculate a single mass-over-charge value and intensity for each of these peaks. Especially for low-resolution data, multiple substances can make up a single peak, which need to be separated using, e.g., statistical mixture models.
2. Chromatographic peak detection: Compounds elute over a certain time window, and software is available to detect and quantify the signal along the chromatographic dimension, at the same time often separating compounds that elute in overlapping windows.
3. Mass and retention time alignment: Across multiple measurements mass accuracy and retention time tend to drift. Using internal standards and/or background ions present over the complete chromatogram systematic trends in mass accuracy can be corrected. As the retention time drift typically does not follow a linear pattern, more elaborate approaches are usually required to correct for it.
4. Isotope/derivative peak detection: A single substance can give rise to many peaks in a single measurement, consisting of isotopes, fragments and adducts. In order to prevent false positive identification of these peaks as (novel) metabolites, these peaks need to be identified and removed from or used during the identification process. Mass spectrometry combined with chromatography provides information, in the form of retention time and peak shape information, which can be used to detect such dependent peaks.

M

Table M.4 Commonly used Metabolomics Software

Name	License	Description
XCMS	GPL	Supports the full analysis of RAW data in either R or through a web platform. The R platform is extensively used for visualization of the results and requires knowledge in this environment. This is alleviated with the availability of the web service. Supports: NetCDF, mzXML, mzML. http://metlin.scripps.edu/xcms/ http://masspec.scripps.edu/xcms/download.php
mzMatch	GPL	A modular collection of tools for the full analysis of high resolution mass spectrometry data. The project provides a novel file format allowing for the exchange of intermediate data between tools and researchers; the environment for example has a complete coupling with the XCMS toolbox. A full user interface environment is provided and an open source Java library supports the development of novel processing and interpretation tools. Visualization tools are also available through the user interface in R or Excel (Ideom). Supports: mzML, mzXML. http://mzmatch.sourceforge.net/
MZmine	GPL	A user interface environment for the full analysis of mass spectrometry data developed in parallel with the mzXML file format. Offers some capabilities with regards to visualization. Supports: mzML, mzXML http://mzmine.sourceforge.net/
OpenMS	LGPL	An open source C++ library with limited user interface support for the full analysis of mass spectrometry data. Supports: mzML, mzXML, NetCDF. http://open-ms.sourceforge.net/
Sirius	Unknown	A Java user interface environment for the *de novo* identification of metabolites from mass spectrometry scan data. It provides de-isotoping and fragment analysis to accomplish this task. Supports: peak lists. http://bio.informatik.uni-jena.de/sirius

5. Normalization: To correct for changes in machine sensitivity over time, especially for large batches of measurements, normalization is often necessary, ideally based on internal standards. An alternative approach is to intersperse a standard sample throughout the measurements, which can be used by the software to detect drifts in intensity.
6. Identification: Strategies like metabolic fingerprinting and metabolic footprinting often do not require the unambiguous identification of metabolites. However, it is essential for biological interpretation and the use of data in metabolic modeling. For high-resolution mass spectrometry platforms tentative identification is often possible based on accurate mass and retention time. In addition, isotope profiles and matches to metabolite libraries provide helpful information.

Currently a number of software solutions exist that help with the automatic analysis of metabolomic data; a number of important freely available efforts are listed in Table M.4.

Further reading

Benton HP, Wong DM, Siuzdak G (2008) XCMS2: Processing tandem mass spectrometry data for metabolite identification and structural characterization *Analytical Chemistry*, 80(16), 6382–6389.

Böcker C, Letzel MC, Lipták Z, Pervukhin A (2009) SIRIUS: decomposing isotope patterns for metabolite identification. *Bioinformatics*, 25(2) 218–224.

Pluskal T, Castillo S, Villar-Briones A, Orešič M (2010) MZmine 2: Modular framework for processing, visualizing, and analyzing mass spectrometry-based molecular profile data. *BMC Bioinformatics* 11: 395.

Scheltema RA, Jankevics A, Jansen RC, Swertz MA, Breitling R (2011) PeakML/mzMatch: A File Format, Java Library, R Library, and Tool-Chain for Mass Spectrometry Data Analysis. *Analytical Chemistry* 83 (7), 2786–2793.

Scheltema RA, Decuypere A, DujardinJC, Watson D, Jansen RC, Breitling R (2009) Simple data-reduction method for high-resolution LC–MS data in metabolomics; *Bioanalysis* 1(9), 1551–1557.

Sturm M, Bertsch M, Gröpl C, et al. (2008) OpenMS – An open-source software framework for mass spectrometry; *BMC Bioinformatics*, 9: 1–11.

METABONOME, *SEE* METABOLOME.

METADATA

Robert Stevens

Metadata is literally 'data about data'. An ontology may be thought about as providing metadata about the data about individuals held in a knowledge base. A database schema forms part of the metadata for a database. The ontology or schema itself may have further metadata, such as version number and source. An important class of metadata concerns the authorship and editorial status of works. The Dublin Core is a standard for editorial metadata used to describe library material such as books and journal articles. It includes such information as author, date, publisher, year, subject and a brief description.

M

Related website

Dublin Core	http://dublincore.org

METROPOLIS-HASTINGS, *SEE* MARKOV CHAIN MONTE CARLO.

MGC (MAMMALIAN GENE COLLECTION)

Malachi Griffith and Obi L. Griffith

The goal of the Mammalian Gene Collection is to create a physical collection of full-length protein-coding cDNA clones representing each human, mouse and rat gene. The project targeted at least one clone for every RefSeq sequence. When multiple isoforms were known for a single gene locus, the longest representative RefSeq was generally selected. Clones were obtained by screening of cDNA libraries, by transcript-specific RT-PCR cloning, or DNA synthesis of cDNA inserts. Each resulting clone was sequenced to high quality standards and the resulting sequences were deposited in GenBank. The MGC project is lead by the National Institutes of Health (NIH) and the physical collection of clones is available for purchase as a collection or individually by several suppliers (see link below).

Related websites

MGC homepage	http://mgc.nci.nih.gov/
MGC suppliers	http://mgc.nci.nih.gov/Info/Buy
MGC project summary	http://mgc.nci.nih.gov/Info/Summary

Further reading

Baross A, et al. (2004) Systematic recovery and analysis of full-ORF human cDNA clones. *Genome Res.* 14:2083–2092.

MGC Project Team (2004) The status, quality, and expansion of the NIH full-length cDNA project: the Mammalian Gene Collection (MGC). *Genome Res.* 14:2121–2127.

MGD (MOUSE GENOME DATABASE)

Guenter Stoesser

The Mouse Genome Database (MGD) represents the genetics and genomics of the laboratory mouse, a key model organism.

MGD provides integrated information of genotype (nucleotide sequences) to phenotype including curation of genes and gene products as well as relationships between genes, sequences and phenotypes. MGD maintains the official mouse gene nomenclature and collaborates with human and rat genome groups to curate relationships between these genomes and to standardize the representation of genes and gene families. MGD includes information on genetic markers, molecular segments (probes, primers, YACs and STSs), description of phenotypic effects of mutations from the Mouse Locus Catalog (MLC), comparative mapping data, graphical displays of linkage, cytogenetic and physical maps, and experimental mapping data. Collaborations include the Gene Expression Database (GXD), the Mouse Genome Sequencing (MGS) project and the Mouse Tumor Biology (MTB) database as well as SWISS-PROT (EBI) and LocusLink (NCBI) to provide an integrated information resource for the laboratory mouse. MGD is part of the Mouse Genome Informatics (MGI) project based at the Jackson Laboratory in Maine, USA.

Related websites

MGD homepage	http://www.informatics.jax.org
The Mouse Gene Expression Database (GXD)	http://www.informatics.jax.org/expression.shtml
MGI Mouse Genome Browser on build 37	http://gbrowse.informatics.jax.org/cgi-bin/gbrowse/mouse_current/

Further reading

Eppig JT, et al.; the Mouse Genome Database Group (2012) The Mouse Genome Database (MGD) comprehensive resource for genetics and genomics of the laboratory mouse. *Nucleic Acids Res.* 40(D1): D881–D886.

See also Model Organism Database

MGED ONTOLOGY

Robert Stevens

The primary purpose of the MGED Ontology is to provide standard terms for the annotation of microarray experiments. The terms taken from the MGED ontology is intended to be compliant with the MAGE standard for the representation of micro-array experiments. These terms enable structured queries of elements of the experiments. Furthermore, the terms also enable unambiguous descriptions of how the experiment was performed. The terms are provided in the form of an ontology, where terms are organized into classes with properties and are defined. For descriptions of biological material (biomaterial) and certain treatments used in the experiment, terms may come from external ontologies. Software programs utilizing the ontology are expected to generate forms for annotation, populate databases directly, or generate files in the established MAGE-ML format. Thus, the ontology is to be used directly by investigators annotating their microarray experiments as well as by software and database developers and therefore will be developed with these very practical applications in mind.

Related websites

MGED Ontology	http://mged.sourceforge.net/ontologies/MGEDontology.php
MAGE Standards	http://www.fged.org/

M

MICROARRAY

Stuart Brown

A collection of tiny DNA spots attached to a solid surface that is used to measure gene expression by hybridization with a labeled cDNA mixture. Global measurement of differential gene expression across treatments, developmental stages, disease states, etc, is relevant to a very wide range of biological and clinical disciplines.

In general, a DNA *array* is an ordered arrangement of known DNA sequences chemically bound to a solid substrate. A *microarray* has *features* (spots with many copies of a specific DNA sequence) that are typically less than 200 microns in diameter, so that a single array may have tens of thousands of spots on a square centimeter of a glass slide. The most common use of this technology is to measure gene expression, in which case the bound DNA sequences are intended to represent individual genes. The array is then hybridized with labeled cDNA that has been reverse transcribed from a cellular RNA extract. The amount of labeled cDNA bound to each spot on the array is measured, generally by fluorescent scanning, and comparisons are made as to the relative amount of fluorescence for each gene in cDNA samples from various experimental treatments.

The DNA sequences bound on the array are referred to as *probes* because they are of known sequence while the experimental sample of labeled cDNA is called the *target* as it is uncharacterized. The bound probe DNA may be entire cloned cDNA sequences or short oligonucleotides designed to represent a portion of a cDNA. If oligonucleotides are used, they may be synthesized directly on the array, or synthesized in individual batches and applied to the array by a robotic spotter.

Bioinformatics challenges for the analysis of microarray data include image analysis to quantify the fluorescent intensity of each spot and subtract local background, normalization between two colours on one array or across multiple arrays, calculation of fold change and statistical significance for each gene across experimental treatments, clustering of gene expression patterns across treatments, and functional annotation of co-regulated gene clusters. An additional challenge is the development of data annotation standards, file formats, and databases to allow for the comparison of experiments performed by different investigators using different biological samples, different treatments, and different microarray technologies.

Related websites

Leming Shi's 'Genome Chip' resources	http://www.gene-chips.com/
GRID IT – Introduction to microarrays:	http://www.bsi.vt.edu/ralscher/gridit/intro_ma.htm

Further reading

Bowtell D, Sambrook J (2003) *DNA Microarrays: A Molecular Cloning Manual*. Cold Spring Harbor Laboratory Press, pp 712.

Schena M (2002) *Microarray Analysis*, John Wiley & Sons, pp 448.

See also Gene Expression Profiling, Spotted cDNA Microarray, Affymetrix GeneChip™ Oligonucleotide Microarray, Microarray Image Analysis, Microarray Normalization

MICROARRAY IMAGE ANALYSIS

Stuart Brown

The application of image analysis software to quantitate the signals due to hybridization of a complex mixture of fluorescently tagged target cDNAs to each of the spots of a microarray.

Once a microarray chip has been hybridized with a fluorescently labeled cDNA and scanned, a digital image file is produced. The initial step in data analysis requires that the signal intensity of each spot be quantitated using some form of densitometry. The image analysis software must achieve several tasks: identify the centers and boundaries of each spot on the array, integrate the signal within the boundaries of each spot, identify and subtract background values, and flag abnormal spots that may contain image artifacts.

Many different algorithms for edge detection and signal processing have been adapted to microarray analysis (e.g. adaptive threshold segmentation, intensity histograms, spot shape selection, and pixel selection), but no rigorous comprehensive analysis of the various methods has been conducted. In general, investigators require a highly automated image processing solution that applies a common template and a uniform analysis procedure to every hybridization image in a microarray experiment in order to minimize error and variability being added at the image analysis stage of the protocol. Spotted arrays are subject to considerably more variability than *in situ* synthesized oligonucleotide arrays since there is the potential for each spot to take on an irregular shape and for signal from adjacent spots to overlap. Many of the manufacturers of fluorescent scanners used for microarray analysis have created image analysis software optimized for the characteristics of their scanners. A variety of free software for microarray image analysis has also been produced by academic groups including Scanalyze by Mike Eisen and Spot by Jean Yang.

Further reading

Kamberova G, Shah S. (2002) *DNA Array Image Analysis: Nuts & Bolts*. DNA Press, pp 280.

Yang YH, et al. (2001) Analysis of cDNA microarray images. *Brief Bioinform*. 2: 341–349.

MICROARRAY NORMALIZATION

Stuart Brown

A mathematical transformation of microarray data to allow for accurate comparisons between scanned images of arrays that have different overall brightness.

A microarray experiment can measure the differential abundance of transcripts from many different genes within the total population of mRNA across some set of experimental conditions. However, this requires making comparisons of the images created by scanning fluorescent intensities of labeled target mRNA bound by a particular probe in several different array hybridizations. Spotted arrays that are hybridized with a mixture of two cDNA extractions labeled with two different fluorescent dyes must be scanned in two colours. Since the two dyes were added to the cDNAs in separate labeling reactions and the two dyes have different fluorescent properties, the relative signal strengths of the two colors on each spot can only be compared after the two images are normalized to the same average brightness.

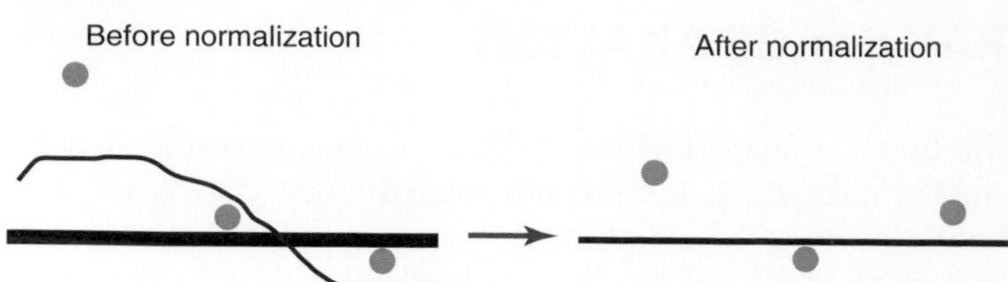

Figure M.1 Representation of microarray intensity levels before (left) and after (right) normalization. The wavy line represents the background intensity level before normalization which is corrected on the right hand side.

There are many experimental variables (systematic errors) that can effect the intensity of fluorescent signal detected in a microarray hybridization experiment including the chemical activity of the fluorescent label, labeling efficiency, hybridization and washing efficiency, and the performance of the scanner. Many of these variables have a non-specific effect on all measurements in a particular array, so that they can be controlled by scaling the intensity values across a set of array images to the same average brightness. However, the variability in fluorescent measurements between arrays often have non-linear and/or intensity-dependent relationships.

The most common method of microarray normalization is to divide all measurements in each array by the mean or median of that array and then multiply by a common scaling factor. The Affymetrix Microarray Analysis Suite software scales probe set signals from GeneChip arrays to a median value of 2500 (MAS 4) or 500 (MAS 5) by default. A more accurate normalization that corrects for intensity-dependent variation can be calculated by fitting the data for two arrays to a smooth curve on a graph of the intensity of each probe on array A vs. array B using a local regression model (loess). The differences, A-B are then normalized by subtracting the fitted curve (quantile normalization). This method can be further refined by fitting the curve to the graph of A-B (fold change) vs. A+B (intensity) for each probe. In the dChip software, Li and Wong normalize a set of arrays to a smooth curve fit to the expression levels of an 'invariant set' of probes with small rank differences across the arrays. The method of normalization applied to a set of array data can dramatically affect the fold change and statistical significance observed for genes with low levels of expression Figure M.1.

Related websites

Standardization and Normalization of MicroArray Data	http://pevsnerlab.kennedykrieger.org/snomadinput.html
Bioconductor	http://www.bioconductor.org

Further reading

Bolstad BM, et al. (2003) A comparison of normalization methods for high density oligonucleotide array data based on variance and bias. *Bioinformatics* 19: 185–193.

Colantuoni C, et al. (2002) SNOMAD (Standardization and NOrmalization of MicroArray Data) Web-accessible gene expression data analysis. *Bioinformatics.* 18: 1540–1541.

Gautier L, et al. (2002) Textual description of Affy. www.bioconductor.org/repository/devel/vignette/affy.pdf

Schuchhardt J, et al. (2000) Normalization strategies for cDNA microarrays. *Nucleic Acids Res*, 28, E47.

MICROFUNCTIONALIZATION

John M. Hancock

After gene duplication the resulting paralogues undergo functional diversification due to a mixture of genetic drift and natural selection. Classically two outcomes have been described: neofunctionalization, whereby one gene evolves to adopt a new function, and subfunctionalization, whereby the function of the ancestral gene becomes split between the duplicates.

This model ignores an important class of duplicate genes in which all paralogues contribute to essentially the same organismal function. A good example is the olfactory receptor gene family, of which the human genome contains 400–1000 copies. These genes all encode similar proteins with different specificites which contribute collectively to odour detection. The term microfunctionalization was proposed to describe the process of developing such subtle functional differences within a gene family with essentially homogeneous functions.

Further reading

Hancock JM (2005) Gene factories, microfunctionalization and the evolution of gene families. *Trends Genet.* 21: 591–595.

See also Gene Duplication, Paralog, Neofunctioalization, Subfunctionalization

MICRORNA

Sam Griffiths-Jones

MicroRNAs are short, approximately 22nt single-stranded RNA molecules, found in the genomes of animals, plants and some viruses. MicroRNAs are produced by a complex and well-studied biogenesis pathway (see MicroRNA biogenesis). Mature microRNAs associate with the RNA induced silencing complex (see RISC) and guide its function by base-pairing to complementary regions in protein-coding mRNAs (see MicroRNA targets). RISC acts to repress the translation or cause degradation of target transcripts. MicroRNAs therefore provide a mechanism for the post-transcriptional regulation of many genes.

MicroRNAs as a class of genes were discovered in 2001. By 2011, the miRBase database (see miRBase) annotated over 1500 microRNA loci in the human genome, around 200 in fly and worm, and up to 600 in some plants (see MicroRNA discovery). It is predicted that over one third of all human protein-coding genes are regulated by microRNAs. MicroRNAs have been implicated in a vast range of biological processes, including development and disease.

Related website

miRBase	http://www.mirbase.org/

See also MicroRNA target, miRBase, MicroRNA Discovery

MICRORNA DISCOVERY

Antonio Marco and Sam Griffiths-Jones

Detection of genomic loci that potentially encode microRNAs.

MicroRNA genes can be predicted and annotated using a variety of methods, including *de novo* prediction, expression analysis, cloning and sequencing, and deep sequencing methods. Since 2007, the overwhelming majority of novel microRNAs have been annotated using small RNA deep sequencing data. A number of methods have been developed to search for microRNA signatures in the patterns of reads mapped to the genome. These patterns are then usually combined with predicted RNA hairpin structures to annotate microRNA candidate loci. For example, a hairpin prediction with multiple ~22nt reads mapped to each arm of the hairpin (representing the mature microRNA duplex), with the 2 nt 3' overhangs in the inferred secondary structure predicted by Drosha and Dicer processing, is usually taken as good evidence of a microRNA. However, different programs and different groups use different evaluation benchmarks and different heuristics. Consequently, the collection of putative microRNAs may vary among programs and settings. The following table lists microRNA prediction programs that make use of expression information from deep sequencing experiments:

miRDeep	http://www.mdc-berlin.de/rajewsky/miRDeep
miRCat	http://srna-tools.cmp.uea.ac.uk/animal/cgi-bin/srna-tools.cgi
MicroInspector	http://bioinfo.uni-plovdiv.bg/microinspector/
miRAnalyzer	http://web.bioinformatics.cicbiogune.es/microRNA/
mirExplorer	http://rnaqueen.sysu.edu.cn/mirExplorer/mirExplorer.php
miRTRAP	http://flybuzz.berkeley.edu/miRTRAP.html
MIReNA	http://www.ihes.fr/~carbone/data8/

Prior to the explosion of deep sequencing technologies and data, most microRNAs were identified by small RNA cloning and sequencing. *De novo* computational prediction of microRNA loci is not generally considered sufficient for a confident microRNA annotation. However, such approaches may be useful for species for which expression information is not yet available, or as a starting point for subsequent validation by Northern blot, microarray, or RT-PCR, for example. The table below lists some of the available tools for microRNA prediction based exclusively on computation:

MiRscan	http://genes.mit.edu/mirscan/
MiPred	http://www.bioinf.seu.edu.cn/miRNA/
miRFinder	http://www.bioinformatics.org/mirfinder/
mirExplorer	http://rnaqueen.sysu.edu.cn/mirExplorer/mirExplorer.php
MiRAlign	http://bioinfo.au.tsinghua.edu.cn/miralign/

MicroRNAs may also be predicted based on homology. Some programs to predict homologous microRNAs are listed in the following table:

MapMi	http://www.ebi.ac.uk/enright-srv/MapMi/
miROrtho	http://cegg.unige.ch/mirortho
miRNAminer	http://groups.csail.mit.edu/pag/mirnaminer/
CoGemiR	http://cogemir.tigem.it/

The miRBase database accepts microRNA annotations based on consensus criteria agreed by the microRNA community. These criteria are under constant review, particularly in the light of the availability of ever-increasing depth of sequencing.

Further reading

Ambros V, Bartel B, Bartel DP, et al. (2003) A uniform system for microRNA annotation. *RNA* 9: 277–279.

Berezikov E, Cuppen E, Plasterik RHA (2006) Approaches to microRNA discovery. *Nat Genet* 38: S2–S7.

Creighton CJ, Reid JG, Gunaratne PH (2009) Expression profiling of microRNAs by deep sequencing. *Brief Bioinform* 10: 490–497.

Meyers BC, Axtell MJ, Bartel B, et al. (2008) Criteria for annotation of plant MicroRNAs. *Plant Cell* 20: 3186–3190.

See also MicroRNA, MicroRNA Biogenesis, miRBase

MICRORNA FAMILY

Antonio Marco and Sam Griffiths-Jones

A microRNA family is a group of related microRNA sequences. The term is ambiguous, and may refer to evolutionary or functional levels of relationship.

As for protein-coding genes, novel microRNAs often originate by gene duplication. Therefore, many different microRNAs come from the same ancestral microRNA and, hence, they belong to the same family (i.e. they are homologous). Some microRNA prediction algorithms are based on the detection of homologous microRNAs within or between species.

The function of a microRNA is largely specified by only 6–8 nts of the mature sequence (often called the seed region). Non-homologous microRNA sequences (i.e. not derived from a common ancestor) may share identical or nearly identical seed regions, and this

M

provides an alternative definition of microRNA families. Members of so-called seed families are predicted to share targets and therefore a common function.

It is recommended that the term 'microRNA family' is reserved for homologous microRNA sequences, while the term 'microRNA seed family' is used to denote mature microRNA sequences that share seed sequence similarity.

Further reading

Berezikov E (2011) Evolution of microRNA diversity and regulation in animals. *Nat Rev Genet* 12: 846–860.

Hertel J, Lindemeyer M, Missal K, et al. (2006). The expansion of the metazoan microRNA repertoire. *BMC Genomics* 7: 25.

Kozomara A, Griffiths-Jones S (2011) miRBase: integrating microRNA annotation and deep-sequencing data. *Nucleic Acids Res* 39: D152–D157.

See also MicroRNA, MicroRNA discovery, MicroRNA seed

MICRORNA PREDICTION, *SEE* MICRORNA DISCOVERY.

MICRORNA SEED

Ana Kozomara and Sam Griffiths-Jones

The 5′ region of a mature microRNA, usually defined as bases 2 to 7. The microRNA seed region has been shown to be critical for microRNA target recognition. The seed sequence is highly conserved amongst microRNAs from the same family and almost perfectly complementary to the targeted regions in mRNA. Some microRNA target prediction algorithms consider only the complementarity of the seed to the target site.

See also MicroRNA Target, MicroRNA Target Prediction, Mature microRNA, MicroRNA Family

MICRORNA SEED FAMILY, *SEE* MICRORNA FAMILY.

MICRORNA TARGET

Antonio Marco and Sam Griffiths-Jones

A transcript regulated by a microRNA. The term is also used to refer to the specific sequence region in the transcript to which the microRNA binds – the latter is more specifically termed the microRNA target site.

MicroRNAs mostly target protein-coding transcripts. A microRNA binds by sequence complementarity to the mRNA, typically in the 3′ untranslated region (3′ UTR) in animals, and in the coding region or UTRs in plants. Plant microRNAs usually bind their targets with near perfect complementarity, whereas animal microRNAs exhibit imperfect base-pairing with their target sites. In animals, the 5′ region of the microRNA, known as the microRNA seed and encompassing bases 2 to 7 of the microRNA, has been shown to be the primary determinant of target site specificity. The data suggest that mismatches or G:U base pairs between the seed region and target site compromise the ability to direct RISC-mediated repression or degradation. The 3′ region of the microRNA usually has only partial complementarity to the target site. Target sites with these properties are often called canonical seed sites. Other non-canonical sites have been described (see references below for a comprehensive review). Perfect complementarity with a target site is suggested to lead to transcript degradation (as in the siRNA pathway), whereas imperfect binding is thought to lead to translational repression.

Experimental characterization of microRNA targets is time consuming and not easy to parallelize. Only a few hundred target sites have been experimentally validated. Consequently, computational prediction of targets is of paramount importance for the study of microRNA function.

Further reading

Bartel DP (2009) MicroRNAs: target recognition and regulatory functions. *Cell* 136: 215–233.

Mallory AC, Bouché N (2009) MicroRNA-directed regulation: to cleave or not to cleave. *Trends Plant Sci* 13: 359–367.

See also MicroRNA target prediction, MicroRNA, Mature microRNA

MICRORNA TARGET PREDICTION

Antonio Marco and Sam Griffiths-Jones

Prediction of putatively targeted transcripts based on rules of microRNA/target interaction.

The experimental identification of a small number of microRNA target sites led to the characterization of common properties among targets. MicroRNAs are complementary to their target sites. In plants this complementarity is almost perfect across the whole site, whereas in animals the complementarity is imperfect. Indeed, the most important region for microRNA targeting has been shown to be the so-called 'seed region', a six-nucleotide region encompassing bases 2 to 7 of the mature microRNA. The remainder of the microRNA may have only partial complementarity to its target.

Different microRNA target prediction algorithms are based on different rules and assumptions; for example, whether to allow mismatches or G:U pairs in the seed region, and whether the microRNA and its target sites are conserved in related species. It is often therefore useful to obtain prediction results from multiple methods. However, the combination of algorithms (i.e. only considering targets detected by several methods) has been shown to limit the usefulness of the prediction set, and is therefore not recommended. The following table lists the most popular algorithms and tools for microRNA target prediction. Because plants and animals have different principles of microRNA targeting, we must use different types of prediction programs. Since plant microRNA have almost perfect complementarity with their targets, common sequence similarity tools such as BLAST and FASTA may provide useful results.

M

Relevant websites

Animals	DIANA-microT	http://diana.cslab.ece.ntua.gr/microT/
	miRanda	http://www.microrna.org/
	Pictar	http://pictar.mdc-berlin.de/
	RNA22	http://cbcsrv.watson.ibm.com/rna22.html
	RNAhybrid	http://bibiserv.techfak.uni-bielefeld.de/rnahybrid/
	PITA	http://genie.weizmann.ac.il/pubs/mir07/mir07_data.html
	TargetScan	http://www.targetscan.org/
Plants	PatScan	http://blog.theseed.org/servers/2010/07/scan-for-matches.html
	psRNATarget	http://plantgrn.noble.org/psRNATarget/

Further reading

Alexiou P, Maragkakis M, Papadopoulos GL, Reczko M, Hatzigeorgiou AG (2009) Lost in translation: an assessment and perspective for computational microRNA target identification. *Bioinformatics* 25: 3049–3055.

Bartel DP (2009) MicroRNAs: target recognition and regulatory functions. *Cell* 136: 215–233.

Dai X, Zhuang Z, Zhao PX (2011) Computational analysis of microRNA targets in plants: current status and challenges. *Brief Bioinform* 12: 115–121.

See also MicroRNA Discovery, MicroRNA target

MICROSATELLITE

Katheleen Gardiner

DNA sequences that are internally repetitive.

Microsatellites consist of tandem (end-to-end) repeats of short sequence motifs, such as CA. Because of their hypervariability and their wide distribution, many are useful in genetic linkage studies and as genetic disease markers. The frequencies of the different repeated motifs, mutation rates, and microsatellite length distributions differ between species.

Concentrations and types of repetitive sequences found within genomes vary between species. Concentrations depend to a limited extent on genome size but differences in the molecular mechanisms generating them, and in some cases selective pressures, may also be important.

It should be noted that repetitive sequences are to be distinguished from repeated sequences, which are sequences that occur large numbers of times within an individual genome, although many repeated sequences are also internally repetitive. Cryptically simple regions are also detected, corresponding to local regions rich in a small number of short motifs that are not arranged in a tandem manner.

Repetitive (and repeated) sequences can be detected by matching to a pre-existing database or by searching for tandemly arranged patterns. A further development is to identify cryptically simple regions by identifying regions within sequences that contain statistically significant concentrations of sequence motifs. A more abstract feature of a sequence, its repetitiveness, can also be measured, for example by the Relative Simplicity Factor of the SIMPLE program.

Relevant websites

RepeatMasker	http://www.repeatmasker.org/cgi-bin/WEBRepeatMasker
CENSOR	http://www.girinst.org/censor/help.html
Tandem Repeats Finder	http://tandem.bu.edu/trf/trf.html
SIMPLE	http://www.biochem.ucl.ac.uk/bsm/SIMPLE/index.html

Further reading

Charlesworth B, Sniegowski P, Stephan W (1994) The evolutionary dynamics of repetitive DNA in eukaryotes *Nature* 371: 215–220.

Dib C, et al. (1996) A comprehensive genetic map of the human genome based on 5264 microsatellites *Nature* 380: 152–154.

Goldstein DB, Schlötterer C (1999) *Microsatellites: Evolution and Applications*. Oxford University Press.

Richard G-F, Paques F (2000) Mini- and microsatellite expansion: the recombination connection *EMBO Rep* 1: 122–126.

See also Genome Sequence, Minisatellite, SIMPLE

MIDNIGHT ZONE

Teresa K. Attwood

The region of protein sequence identity (typically below 10%) where sequence comparison methods fail completely to detect structural similarity. In many cases, protein sequences have diverged to such an extent that their evolutionary relationships are apparent only at the level of shared structural features. Such characteristics cannot be detected even using the most sensitive sequence comparison methods. Consequently, the Midnight Zone denotes the theoretical limit to the effectiveness of sequence analysis techniques. Within this region, fold-recognition algorithms may be employed to ascertain whether sequences are likely to be homologous, essentially by determining their compatibility with particular folds.

Because the Midnight Zone is populated by protein structures between which there is no detectable sequence similarity, it is sometimes unclear whether structural relationships are the result of divergent or convergent evolution. Within a range of structurally similar but sequentially dissimilar proteins, patterns of conservation of residues at aligned positions have been studied. This has shown that, while moderately conserved positions are likely to have arisen by chance, those that are highly conserved reveal distinct features associated with structure and function: a relatively high fraction of structurally aligned positions are buried within the protein core; and glycine, cysteine, histidine and tryptophan are significantly over-represented, suggesting residue-specific structural and functional roles.

Further reading

Doolittle, RF (1994) Convergent evolution: the need to be explicit. *TIBS*. 19: 15–18.

Friedberg I, Kaplan T, Margalit H (2000) Glimmers in the midnight zone: characterization of aligned identical residues in sequence-dissimilar proteins sharing a common fold. *Proc. Int. Conf. Intell. Syst. Mol. Biol.*, 8: 162–170.

Friedberg, I, Margalit H (2002) Persistently conserved positions in structurally similar, sequence dissimilar proteins: roles in preserving protein fold and function. *Protein Sci.*, 11(2) 350–360.

Rost B (1997) Protein structures sustain evolutionary drift. *Fold Des.*, 2: 519–524.

Rost B (1998) Marrying structure and genomics. *Structure*, 6(3) 259–263.

Rost B (1999) Twilight zone of protein sequence alignments. *Protein Eng.*, 12(2) 85–94.

See also Fold Recognition, Threading, Twilight Zone

MIGRATE-N

Michael P. Cummings

A program for estimating effective population size and migration rates among populations.

Based on coalescent theory and use maximum likelihood or Bayesian inference. The general problems analyzed take the general form of migration matrix model with asymmetric migration rates and different subpopulation sizes. Different data types are accommodated, including DNA sequence data, SNPs (single nucleotide polymorphisms), and microsatellite data. Additional features include statistics for hypothesis testing, various summary statistics, graphics, and output compatible with Tracer and Figtree.

Related website

MIGRATE-N home page	http://popgen.sc.fsu.edu/Migrate/Migrate-n.html

Further reading

Beerli P (2006) Comparison of Bayesian and maximum-likelihood inference of population genetic parameters. *Bioinfomratics* 22: 341–345.

Beerli P, Palczewski M (2010) Unified framework to evaluate panmixia and migration direction among multiple sampling locations. *Genetics* 185: 313–326.

See also FigTree, Tracer

MIME TYPES

Eric Martz

Providers of information sent through the Internet label their information with MIME types in order to inform the recipients how to process that information. MIME types are attached as prefixes to transmitted information in order to identify the format in which it is represented, and therefore to some extent, the type of information contained.

'MIME' stands for Multipart Internet Mail Extensions, which were originally developed to enable content beyond plain text (US-ASCII) to be included in electronic mail. The ambiguities and limitations of MIME types have encouraged development of a better information exchange standard. As the web evolves, XML (eXtensible markup language) is replacing MIME types.

When information is requested through the Internet, the user or receiving program needs to know how to process and interpret the information received. When a server computer sends a data file requested by a client computer or web browser, the server typically prefixes the data with a tag that specifies the type of information, or MIME type, contained in the file. Typically, the server makes this determination based on name of the data file requested, specifically, the filename extension (last 3 characters, following a period), using a list maintained on the server. Each server must be configured with a suitable list matching the files it serves.

If the client or browser does not handle files properly, one possible reason is that the server is serving the file with an incorrect MIME type. Typically, the server administrator needs to add some new MIME types to the server's list. The MIME type specified by the server can be seen in Netscape (View, Page Info). Another possible reason is that the browser lacks the ability to process that type of information correctly – for example, perhaps a plugin such as Chime or a helper application such as RasMol [see visualization] needs to be installed. The client computer and its web browsers also contain lists of MIME

Table M.5 Type of information in file

MIME type	Customary filename extension	Type of information in file
text/plain	.txt	Text with no special formatting
text/html	.htm, .html	Text containing hypertext markup language (HTML) tags that specify formatting in a web browser
image/gif	.gif	Image in graphics interchange format (GIF)
chemical/x-mmcif	.mcif	Macromolecular crystallographic interchange format for annotated atomic coordinates
chemical/x-pdb	.pdb	Atomic coordinates in Protein Data Bank format (most covalent bonds not explicit)
chemical/x-mdl-molfile	.mol	Atomic coordinates with explicit bond information, typically for small molecules
application/x-spt	.spt	Script commands for the RasMol family of molecular visualization programs
application/x-javascript	.js	Javascript (commands in the browser's programming language)

M

types from which they determine the appropriate function or program to invoke when a particular MIME type of information is received. Some common MIME types are (see Table M.5):

Relevant websites

Rzepa HS, Chemical MIME Home Page	http://www.ch.ic.ac.uk/chemime/
Internet Assigned Numbers Authority, MIME Media Types	http://www.iana.org/assignments/media-types/

Further reading

Rzepa HS, Murray-Rust P, Whitaker BJ (1998) The Application of Chemical Multipurpose Internet Mail Extensions (Chemical MIME) Internet Standards to Electronic Mail and World-Wide Web information exchange. *J. Chem. Inf. Comp. Sci.* 38: 976–982.

See also Visualization, molecular, XML (eXtensible Markup Language)

Eric Martz is grateful for help from Eric Francoeur, Peter Murray-Rust, Byron Rubin, and Henry Rzepa.

MINIMUM EVOLUTION PRINCIPLE

Sudhir Kumar and Alan Filipski

A phylogenetic reconstruction principle based on minimizing the sum of branch lengths of a tree.

Of all possible tree topologies over a given set of taxa, the Minimum Evolution (ME) topology is the one that requires the smallest sum of branch lengths, as estimated by mean-squared error or similar methods. Mathematically, it can be shown that, when unbiased estimates of evolutionary distances are used, the expected value of the sum of branch lengths is smaller for the true topology than that for a any given incorrect topology. In general, phylogeny reconstruction by this method requires the enumeration of all candidate trees (it is known to be an NP-hard problem) and can require unacceptable computational effort for large problems. Typically, heuristics are used to reduce the size of the search space and obtain an approximate (mostly suboptimal) solution. A number of heuristic variants of this approach are available that limit the space of phylogenetic trees being searched, for example, the neighbor-joining method.

Software

Kumar S, et al. (2001) MEGA2: molecular evolutionary genetics analysis software. *Bioinformatics* 17: 1244–1245.

Swofford DL (1998) *PAUP*: Phylogenetic Analysis Using Parsimony (and other methods)*. Sunderland, MA., Sinauer Associates.

Further reading

Nei M. Kumar S (2000) *Molecular Evolution and Phylogenetics*. Oxford University Press.

Rzhetsky A, Nei M (1993) *METREE: Program package for inferring and testing minimum evolution trees* Version 1.2. University Park, PA, The Pennsylvania State University.

See also Phylogenetic Tree, Branch Length Estimation, Neighbor Joining Method, Maximum Likelihood

MINIMUM INFORMATION MODELS

Carole Goble and Katy Wolstencroft

Scientific data is only properly interpretable and re-usable if it can be considered in the context of the original experiments that produced it. As data becomes more complex and higher throughput (particularly in 'omics) it must be described with sufficient metadata to understand the experimental context. Therefore, for different types of experiment, different metadata is important. A Minimum Information Model is a pragmatic approach to ensuring data is adequately annotated before submission to public repositories or journals. Instead of attempting to specify every type of metadata for an experiment, a minimum model specifies only what is required for interpretation. This reduces the annotation overheads for submitting scientists and ensures that all data collected has a useful minimum standard of annotation.

MIAME (the Minimum Information About a Microarray Experiment) was the first Minimum Information Model to be developed. Scientists publishing transcriptomics papers, or submitting microarray data to Array Express (http://www.ebi.ac.uk/arrayexpress/) or GEO (http://www.ncbi.nlm.nih.gov/geo/) must conform to MIAME, so it has widespread adoption in the community. Since MIAME, more than 30 other minimum models have been produced. These are all managed under the umbrella organization of MIBBI (Minimum Information for Biological and Biomedical Investigations).

Related websites

MIBBI	http://www.mibbi.org
BioSharing	http://www.biosharing.org

Further reading

Brazma A (2009) Minimum information about a microarray experiment (MIAME) –successes, failures, challenges. *ScientificWorldJournal*. 29: 420–423.

Sansone, S.-A., et al. (2012) Toward interoperable bioscience data, *Nature Genetics*, 44: 121–126.

Taylor CF, et al. (2008) Promoting coherent minimum reporting guidelines for biological and biomedical investigations: the MIBBI project. *Nat Biotechnol*. 26: 889–896.

M

MINISATELLITE

Katheleen Gardiner

Tandem repeats of intermediate length motifs approximately 10–100 bp in length.

Similar to microsatellites in that they are dispersed throughout the genome (although most common near telomeres) and variable among individuals. For these reasons they have found application in genetic linkage and disease mapping as well as in forensics. Minisatellites may show much higher levels of length polymorphism than microsatellites or may show none at all. Their level of polymorphism appears to relate to the local frequency of recombination, especially gene conversion.

Further reading

Buard J, Jeffreys A (1997) Big, bad minisatellites *Nat. Genet.* 15: 327–328.

Charlesworth B, Sniegowski P, Stephan W (1994) The evolutionary dynamics of repetitive DNA in eukaryotes *Nature* 371: 215–220.

Jeffreys A, Wilson V, Thein SL (1985) Hypervariable minisatellite regions in human DNA. *Nature* 314: 67–73.

Richard G-F, Paques F (2000) Mini- and microsatellite expansion: the recombination connection *EMBO Rep* 1: 122–126.

MIRBASE

Ana Kozomara and Sam Griffiths-Jones

miRBase is the reference database for microRNA sequences and annotation, and the central repository of all published microRNA sequences. miRBase provides a gene name service – novel microRNAs are submitted prior to publication, such that official and consistent names can be assigned for use in manuscripts and databases. miRBase stores and distributes mature and hairpin precursor sequences, genomic coordinates, literature references regarding their discovery, and links to other related databases. miRBase provides an easy-to-use web interface, tools to query the data by sequence similarity or text search, and bulk downloads of all data. miRBase also aggregates data from deep sequencing experiments that validate the expression of mature microRNAs, and allows relative expression levels to be queried. Release 18 of miRBase (November 2011) contains over 18,000 microRNA hairpin precursor sequences and over 21,000 mature microRNAs from over 160 species.

Related website

miRBase	http://www.mirbase.org/

Further reading

Griffiths-Jones S (2004) The microRNA Registry. *Nucleic Acids Res* 32:D109–D111.

Griffiths-Jones S, Grocock RJ, van Dongen S, Bateman A, Enright AJ (2006) miRBase: microRNA sequences, targets and gene nomenclature. *Nucleic. Acids. Res.* 34:D140–D144.

Griffiths-Jones S, Saini HK, van Dongen S, Enright AJ (2008) miRBase: tools for microRNA genomics. *Nucleic. Acids. Res.* 36:D154–D158.

Kozomara A, Griffiths-Jones S (2011) miRBase: integrating microRNA annotation and deep-sequencing data. *Nucleic. Acids. Res.* 39:D152–D157.

MIRTRON

Antonio Marco and Sam Griffiths-Jones

A precursor microRNA, located in an intron of a protein-coding gene, that is excised from the primary transcript by the spliceosomal machinery.

Primary microRNAs are generally processed by the Drosha RNase III enzyme in the nucleus to produce the precursor microRNA hairpin. Mirtrons are intronic microRNA sequences that are recognized and processed by the spliceosome, bypassing Drosha processing. Splicing defines the ends of the precursor microRNA, and thus one end of each of the mature miRNA duplex sequences. Mirtrons have been found in different animal lineages including insects and mammals, and are likely to have emerged independently several times during evolution. Only one instance of a plant mirtron has been observed. Variants of the canonical mirtron pathway have been observed: tailed mirtrons are intronic microRNA sequences where one end of the precursor is defined by splicing, and the other by subsequent processing. For example, mir-1017 in fruit flies is a $3'$ tailed mirtron.

Mirtrons are computationally predicted by searching for hairpins in short intron sequences. The splice site predictions provide additional signal, such that the *de novo* prediction of mirtrons is more accurate than the prediction of canonical microRNAs. However, evidence of expression of mature microRNAs from predicted microRNA loci (for example, from small RNA deep sequencing experiments) is always preferred. In the case of mirtrons, small RNA reads are expected to directly abut intron-exon boundaries.

Further reading

Berezikov E, Chung W-J, Willis J, Cuppen E, Lai EC (2007) Mammalian mirtron genes. *Mol Cell* 28: 328–336.

Ruby JG, Jan CH, Bartel DP (2007) Intronic microRNA precursors that bypass Drosha processing. *Nature* 448: 83–86.

Westholm JO, Lai EC (2011) Mirtrons: microRNA biogenesis via splicing. *Biochimie* 93: 1897–1904.

See also MicroRNA Prediction

MISSING DATA, *SEE* MISSING VALUE.

MISSING VALUE (MISSING DATA)

Feng Chen and Yi-Ping Phoebe Chen

Missing values occur when no data value is stored for the variable in the current observation, which is a common occurrence and can have a significant effect on the conclusions that can be drawn from the data.

Missing values may occur because of the following reasons: 1). No information is provided; 2). Some mistake is made during data collection; 3). The data are not updated in time; or other particular occasions.

Different reasons lead to different techniques to deal with missing data. First, we may delete the whole item if one of the attributes in it is missing. Second, we can use the mean of the attribute as the missing value. For instance, the managers in Wal-Mart might treat 50,000 dollars as the missing salary value of one customer when they find the average salary of all the known customers is 50,000 dollars. Third, we can employ diverse statistical methods, such as regression (see Regression) or Bayes (see Bayes' Theorem), for a further understanding of the data so that we can predict the missing values much more efficiently and effectively.

However, not all missing values have to be fixed because some can represent the novel knowledge. When you are applying for a credit card the form of which requires your registration number, you have to leave it blank if you do not have one. Such missing values are meaningful as they express useful information.

In bioinformatics, missing values occur as inevitable statistical errors or incomplete original data. For a comprehensive and correct analysis, the general strategy is to use statistical theory, such as Bayesian (see Bayes' Theorem, Bayesian Classifier) or SVM (see Support Vector Machine), to estimate missing values.

Further reading

Oba S, Sato M, Takemasa I, et al. (2003) A Bayesian missing value estimation method for gene expression profile data. *Bioinformatics*, 19(16): 2088–2096.

Rubin DB, Little RJ A (2002) *Statistical Analysis with Missing Data* (2nd ed.). New York: Wiley, 2002.

Troyanskaya O, Cantor M, Sherlock G, et al. (2001 *Missing Value Estimation Methods for DNA Microarrays. Bioinformatics*, 17(6): 520–525.

Zhang S, Qin Z, Ling C, Sheng S (2005) *'Missing is Useful': Missing Values in Cost-sensitive Decision Trees. IEEE Transactions on Knowledge and Data Engineering*, 17(12): 1689–1693.

See also Regression, Bayes' Theorem, Bayesian Classifier, Support Vector Machine

MITELMAN DATABASE (CHROMOSOME ABERRATIONS AND GENE FUSIONS IN CANCER)

Malachi Griffith and Obi L. Griffith

The Mitelman Database consists of data manually obtained from the literature that relates chromosomal aberrations and gene fusions to tumor characteristics. More than 60,000 cases and 706 gene fusions have been entered in the database. In collaboration with the Cancer Genome Anatomy Project (CGAP), six web search tools have been created to assist in the mining of these data. These tools allow the user to search by cancer case, molecular biology associations (e.g. particular gene rearrangements), clinical associations (e.g. prognosis, tumor grade, etc.), recurrence, and literature references.

Related websites

Mitelman database	http://cgap.nci.nih.gov/Chromosomes/Mitelman
Mitelman database intro	http://cgap.nci.nih.gov/Chromosomes/AllAboutMitelman
CGAP homepage	http://cgap.nci.nih.gov/cgap.html

Further reading

Mitelman F, et al. (eds) (2012) Mitelman Database of Chromosome Aberrations and Gene Fusions in Cancer. http://cgap.nci.nih.gov/Chromosomes/Mitelman

MIXTURE MODELS

Aidan Budd and Alexandros Stamatakis

A statistical model that mixes the signals of various plausible simpler models, mostly used for computing the likelihood at a site.

This term mainly refers to mixing/combing the likelihood scores of several protein substitution models and/or pre-computed amino acid frequencies for calculating the likelihood score at each site of a protein sequence alignment. Because of the high model complexity, protein mixture models are mostly implemented in Bayesian MCMC frameworks, albeit Maximum Likelihood approaches have also been proposed. The main argument in favour of those more complex models is that a single protein substitution matrix may not be sufficient for capturing the complexity of the evolutionary processes. This approach exhibits some computational analogies, in particular with respect to increased computational and memory requirements, with the GAMMA model of rate heterogeneity. Bayesian methods can also be used to integrate over single assignments of different protein substitution matrices or site frequencies to sites of an alignment. The MCMC framework can also be used to determine an adequate number of such model categories via reversible-jump MCMC procedures.

Software

MrBayes, PhyloBayes, PHYML

Further reading

Lartillot N, Lepage T, Blanquart S (2009) PhyloBayes 3: a Bayesian software package for phylogenetic reconstruction and molecular dating. *Bioinformatics*, 25(17): 2286–2288.

Lartillot N, Philippe H (2004) A Bayesian mixture model for across-site heterogeneities in the amino-acid replacement process. *Molecular Biology and Evolution*, 21(6): 1095–1109.

Le SQ, Lartillot N, Gascuel O (2008) Phylogenetic mixture models for proteins. *Philosophical Transactions of the Royal Society B: Biological Sciences*, 363(1512): 3965–3976.

Wang HC, Li K, Susko E, Roger A (2008) A class frequency mixture model that adjusts for site-specific amino acid frequencies and improves inference of protein phylogeny. *BMC Evolutionary Biology*, 8(1): 331.

See also Rate Heterogeneity, Heterotachy

MM, *SEE* MARKOV CHAIN.

MOD, *SEE* MODEL ORGANISM DATABASE.

M

MODEL ORDER SELECTION, *SEE* MODEL SELECTION.

MODEL ORGANISM DATABASE (MOD)

Obi L. Griffith and Malachi Griffith

Model Organism Databases typically center around the genome, transcriptome, or proteome sequence together with associated annotations (genes, proteins, regulatory elements, etc.) for a particular species or related group of species that may or may not be considered model organisms. Large projects such as Ensembl or the UCSC Genome Browser can be considered collections of MODs with a common set of visualizations and analysis tools. However, in addition to these collections, many individual MODs exist each targeted towards a specific species and its associated research community. Examples for 'classic' model organisms include FlyBase for Drosophila species, SGD for *Saccharomyces cerevisiae*, MGD for mouse, RGD for rat, TAIR for Arabidopsis and WormBase for *Caenorhabditis elegans* and related nematodes. More recent additions include BeeBase, BeetleBase, BGD for bovine species, CGD for Candida species, PossumBase, Soybase, and Xenbase for Xenopus species to name only a few. The Generic Model Organism Database (GMOD) project provides a set of common, reusable software tools and database schemas for the development of MODs to help manage data and allow users to query and access data. A component of GMOD is the Generic Genome Browser (GBrowse) that provides a common set of software for visualizing DNA, protein or other sequence features within the context of a reference sequence.

Related websites

Generic Model Organism Database Project	http://gmod.org/
Generic Genome Browser	http://gbrowse.org/

Further reading

Donlin MJ (2009) Using the Generic Genome Browser (GBrowse). *Curr Protoc Bioinformatics.* Chapter 9: Unit 9.9.

O'Connor BD, et al. (2008) GMODWeb: a web framework for the Generic Model Organism Database. *Genome Biol.*, 9(6): R102.

MODEL SELECTION (MODEL ORDER SELECTION, COMPLEXITY REGULARIZATION)

Nello Cristianini

In machine learning, the problem of finding a class of hypothesis whose complexity matches the features (size, noise level, complexity) of the dataset at hand.

Too rich a model class will result in overfitting, too poor a class will result in underfitting. Often the problem is addressed by using a parametrized class of hypotheses, and then tuning the parameter using a validation set.

In neural networks this coincides with the problem of choosing the network's architecture, in kernel methods with the problem of selecting the kernel function. Prior knowledge about the problem can often guide in the selection of an appropriate model.

Further reading

Duda RO, Hart PE, Stork DG (2001) *Pattern Classification*, John Wiley & Sons, USA.

Mitchell T (1997) *Machine Learning*, McGraw Hill.

MODELING, MACROMOLECULAR

Eric Martz

Generating a three-dimensional structural model for an amino acid or nucleotide sequence from empirical structure data or from theoretical considerations.

The most reliable and accurate sources for macromolecular models are empirical, most commonly X-ray crystallography or nuclear magnetic resonance (NMR). Even these involve modeling to fit experimental results, and vary in reliability (e.g. resolution, disorder) and occasionally err in interpretation. Furthermore, crystallography is often precluded by the difficulty of obtaining high-quality singular crystals, while NMR is generally limited to molecules no greater than 30 kiloDaltons that remain soluble without aggregation at high concentrations. Therefore, for the vast majority of proteins, the only way to obtain a model is from theory.

Theoretical modeling can bc based upon an empirical structural for a template molecule having sufficient sequence similarity to the unknown target, if one is available. This is called comparative modeling (or homology modeling/ knowledge based modeling) and is usually reliable at predicting the fold of the backbone, thereby identifying residues that are buried, or on the surface, or clustered together. It is unreliable at predicting details of side chain positions.

In the absence of a sequence-similar empirical template (by far the most common case), the only option for obtaining a three-dimensional model is *ab initio* theoretical modeling. Typically, it is based in part upon generalizations and patterns derived from known empirical structures. Theoretical modeling has fair success at predicting protein secondary structure, but rather limited success at predicting tertiary or quaternary structure of proteins. A series of international collaborations performs Critical Assessment of techniques for protein Structure Prediction (CASP)

Further reading

Baker D, Sali A. (2001) Protein structure prediction and structural genomics. *Science* 294: 93–96.

Sali, A (1995) Modelling mutations and homologous proteins. *Curr. Opin. Struct. Biol.* 6: 437–451.

See also Comparative Modeling, Docking, Models, molecular, Secondary Structure Prediction, Atomic Coordinate File

Eric Martz is grateful for help from Eric Francoeur, Peter Murray-Rust, Byron Rubin and Henry Rzepa.

M

MODELS, MOLECULAR

Eric Martz

Three-dimensional depictions of molecular structures.

There are many styles for rendering molecular models, including those with atomic-level detail such as space-filling models, ball and stick models, and wire-frame models (also called 'stick models' or 'skeletal models'), as well as surface models (see Figure M.2). Effectively perceiving the polymer chain folds of macromolecules requires simplified models, such as backbone models and schematic ('cartoon') models. Such models can be made of solid physical materials, or represented on a computer screen. Physical models allow both visual and tactile perception of molecular properties, while computer models can be rendered in different styles and color schemes at will. Molecular models are usually based upon an atomic coordinate file.

Color schemes are typically employed in molecular models in order to convey additional information. Elements can be identified in models with atomic-level detail. The most common element colour scheme originated with space-filling physical models, termed Corey-Pauling-Koltun (CPK) models. In this scheme, carbon is black or gray, nitrogen is blue, oxygen is red, sulphr and disulfide bonds are yellow, phosphorus is orange, and hydrogen is white. This scheme has been adopted in most visualization software. Since most negative charge in proteins is on oxygen atoms, and most positive charge on protonated amino radicals, red and blue are common colours for anionic and cationic charges, respectively, used for example on surface models. Since red and blue, when mixed, make magenta, magenta is logical when a single colour is desired to distinguish polar or charged regions from apolar regions. Gray (signifying the preponderance of carbon) is the customary colour for apolar regions. In RasMol and its derivatives (visualization), alpha helices are coloured red, and beta strands yellow. These and other colour schemes have been unified in a proposed set of open-source, non-proprietary standard colour schemes for molecular models, designated the DRuMS Color Schemes (see Websites).

→

Figure M.2 Six types of molecular models (q.v.). On the left is a three amino acid peptide (Tyr Lys Leu, residues 3–5 of the protein on the right). On the right is a small protein, the immunoglobulin G-binding domain of streptococcal protein G (PDB identification code 1PGB). Notice how the schematic and backbone models enable the main chain of the protein to be discerned, in contrast to the overly detailed atomic-resolution models (wire-frame, ball and stick, spacefill) that are more suitable for small molecules such as the tripeptide at left, or details of functional sites on proteins (not shown). Atoms and bonds are coloured with the CPK scheme (carbon gray, alpha carbons dark gray, nitrogen blue, oxygen red. In the schematic model (and protein backbone), beta strands are yellow and the alpha helix is red. Images prepared with Protein Explorer (http://proteinexplorer.org).
All models are in divergent stereo. To see in stereo: nearly touch the paper with your nose between the images of a pair, relaxing your eyes while gazing into the distance (so the images are blurry). Slowly move the paper away from your eyes. You will see three images of the model; attend to the center of the three images. As the paper is moved just far enough away to focus clearly, the centre image will appear three-dimensional. Seeing in this way is very difficult for about one in three people; it is easier for the majority, but takes some practice. Special stereo-viewing lenses, if available, may also be used.

M

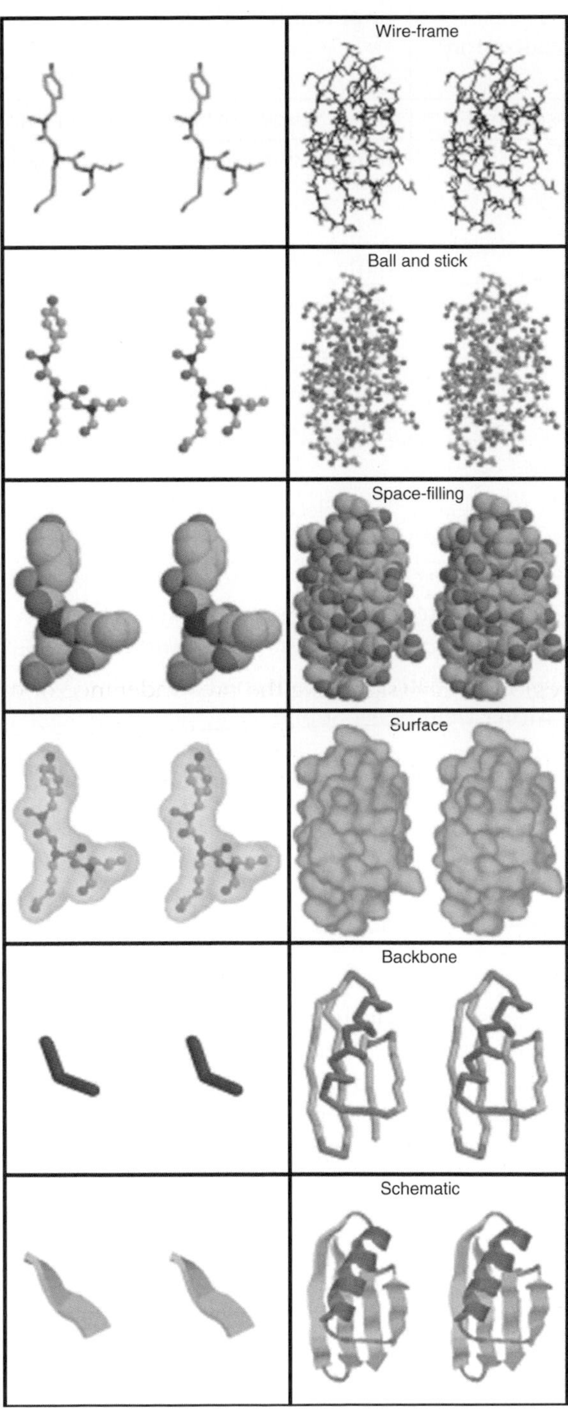
Wire-frame
Ball and stick
Space-filling
Surface
Backbone
Schematic

M

Related websites

World Index of Molecular Visualization Resources	http://molvis.sdsc.edu/visres/
DruMS Color schemes	http://www.umass.edu/molvis/drums
History of Macromolecular Visualization	http://www.umass.edu/microbio/rasmol/history.htm
ModView	http://salilab.org/modview/
Physical Molecular Models	http://www.netsci.org/Science/Compchem/feature14b.html

Further reading

Petersen QR (1970) Some reflections on the use and abuse of molecular models. *J. Chem. Education* 47: 24–29.

Smith DK (1960) Bibliography on Molecular and Crystal Structure Models. Vol. 14, *National Bureau of Standards Monograph*. Washington, DC.

Walton A (1978) *Molecular Crystal Structure Models*. Ellis Horwood, Chichester, UK.

Ilyin VA, Pieper U, Stuart AC, Martí-Renom MA, McMahan L, Sali A (2002) ModView, visualization of multiple protein sequences and structures. *Bioinformatics* 19: 165–166.

See also Atomic Coordinate File, Modeling, molecular, Visualization, molecular, Space-Filling, Alpha Helix, Beta Strand.

Eric Martz is grateful for help from Eric Francoeur, Peter Murray-Rust, Byron Rubin and Henry Rzepa.

MODELTEST

Michael P. Cummings

A program that assists in evaluating the fit of a range of nucleotide substitution models to DNA sequence data through a hierarchical series of hypothesis tests.

Two test statistics, likelihood ratio and Akaike information criterion (AIC), are provided to compare model pairs that differ in complexity. The program is used together with PAUP*, typically as part of data exploration prior to more extensive phylogenetic analysis.

The program is available as executable files for several platforms. The functionally of the program has been subsumed and extended by the program jModelTest.

Further reading

Akaike H (1974) A new look at the statistical model identification. *IEEE Trans. Autom. Contr.* 19: 716–723.

Posada D, Crandall KA (1998) MODELTEST: testing the model of DNA substitution. *Bioinformatics* 14: 817–818.

See also jModelTest

MODENCODE, *SEE* ENCODE.

MODULAR PROTEIN

Laszlo Patthy

Multidomain proteins containing multiple copies and/or multiple types of protein modules.

See also Multidomain Protein, Protein Module

MODULE SHUFFLING

Laszlo Patthy

During evolution protein modules may be shuffled either through exon shuffling or exonic recombination to create multidomain proteins with various domain combinations.

Related website

SMART	http://smart.embl-heidelberg.de

Further reading

Patthy L (1991) Modular exchange principles in proteins. *Curr. Opin. Struct. Biol.* 1: 351–361.

Patthy L (1996) Exon shuffling and other ways of module exchange. *Matrix Biol.* 15: 301–310.

Schultz J, et al. (2000) SMART: A Web-based tool for the study of genetically mobile domains *Nucleic Acids Res*, 28: 231–234

See also Protein Module, Exon Shuffling, Multidomain Protein

MOL CHEMICAL REPRESENTATION FORMAT

Bissan Al-Lazikani

A proprietary chemical representation of chemical structures that is fully representative of all properties of the compound including atom coordinates, charges, stereochemistry and tautomers. Originally developed by MDL (now Accelrys) it dominated the chemical representation field until the recent development of the non-proprietary InChIs. The format consists of separate blocks: the first three lines are reserved for names and identifiers, followed by one line defining atom counts and file version number, atom 3D coordinates and properties and bond connections. As seen in the example in Figure M.3, the Z-coordinates and atom properties can be left blank (set to 0).

See also SDF, SMILES, InChI, Virtual Library

M

```
Mrv0541 04051115152D

6 6 0 0 0 0            999 V2000
 0.7145   0.4125   0.0000 C  0 0 0 0 0 0 0 0 0 0 0 0
 0.7145  -0.4125   0.0000 C  0 0 0 0 0 0 0 0 0 0 0 0
-0.0000  -0.8250   0.0000 C  0 0 0 0 0 0 0 0 0 0 0 0
-0.7145  -0.4125   0.0000 C  0 0 0 0 0 0 0 0 0 0 0 0
-0.7145   0.4125   0.0000 C  0 0 0 0 0 0 0 0 0 0 0 0
 0.0000   0.8250   0.0000 C  0 0 0 0 0 0 0 0 0 0 0 0
1 2 2 0 0 0 0
1 6 1 0 0 0 0
2 3 1 0 0 0 0
3 4 2 0 0 0 0
4 5 1 0 0 0 0
5 6 2 0 0 0 0
M  END
```

Figure M.3 The mol file for benzene.

MOLECULAR CLOCK (EVOLUTIONARY CLOCK, RATE OF EVOLUTION)

Sudhir Kumar and Alan Filipski

The approximate linear relationship between the evolutionary distance and time in a rooted phylogeny.

Evolutionary rate is often measured in terms of the number of substitutions per site per abstract unit of time. It is obtained by dividing the evolutionary distance by twice the divergence time between taxa (a factor of two appears because evolutionary distance captures changes in both evolutionary lineages). In protein-coding DNA sequences, rates may be determined for both synonymous and non-synonymous change by using the appropriate distance measure.

The molecular clock hypothesis is consistent with the neutral theory of evolutionary change in the presence of relatively constant functional constraints. A rooted phylogenetic tree satisfies the molecular clock assumption if all leaf nodes are equidistant from the root.

Tests are available to determine if two lineages are evolving at the same rate. Muse and Weir (1992) developed a test to measure the significance of observed rate differences between two taxa, based on likelihood ratios. Tajima (1993) presents a simple non-parametric test for rate equality. More complex tests, that deploy relaxed molecular clocks, MCMC frameworks, or more elaborate likelihood models (that don not assume a molecular clock) have recently been developed.

Software

Drummond A, Rambaut A (2007) BEAST: Bayesian evolutionary analysis by sampling trees. *BMC Evolutionary Biology* 7(1): 214.

Sanderson MJ (2003) r8s: inferring absolute rates of molecular evolution and divergence times in the absence of a molecular clock. *Bioinformatics* 19(2): 301–302.

Further reading

Kumar S, Subramanian S (2002) Mutation rates in mammalian genomes. *Proc. Natl. Acad. Sci. USA.* 99: 803–808.

Li W-H (1997) *Molecular Evolution.* Sunderland, Mass., Sinauer Associates.

Muse SV, Weir BS (1992) Testing for equality of evolutionary rates. *Genetics* 132: 269–276.

Nei M, Kumar S (2000) *Molecular Evolution and Phylogenetics.* Oxford University Press.

Tajima F (1993) Simple methods for testing the molecular evolutionary clock hypothesis. *Genetics* 135: 599–607.

Takezaki N, et al. (1995) Phylogenetic test of the molecular clock and linearized trees. *Mol Biol Evol* 12: 823–833.

Zuckerkandl E, Pauling L (1965) Evolutionary divergence and convergence in proteins. In *Evolving genes and proteins* (eds. V. Bryson and H. Vogel), pp. 97–166. Academic Press.

See also UPGMA, Heterotachy

MOLECULAR COEVOLUTION, *SEE* COEVOLUTION.

MOLECULAR DRIVE, *SEE* CONCERTED EVOLUTION.

MOLECULAR DYNAMICS SIMULATION

Roland Dunbrack

Simulation of molecular motion, performed by solving Newton's equation's of motion for a system of molecules. The simulation requires a potential energy function that expresses the energy of the system as a function of its coordinates. Simulations usually start with all atoms given a random velocity and direction of motion drawn from a distribution at the temperature of the simulation.

To simulate an infinite system, periodic boundary conditions are used by enclosing a molecular system in a box, and virtually reproducing the system infinitely in each Cartesian direction. For instance, molecules on the right face of the box feel the presence (in terms of the energy function) of atoms on the left face of the box from a virtual copy of the box to the right of the system.

Molecular dynamics simulations can be performed at constant pressure rather than constant volume. The volume, V, of the unit cell is allowed to fluctuate under a piston of mass M, kinetic energy, $K = M(dV/dT)^2/2$ and potential energy $E(V) = PV$.

Further reading

Karplus M, McCammon JA (2002) Molecular dynamics simulations of biomolecules. *Nat. Struct. Biol.* 9: 646–652.

Nose S, Klein ML (1986) Constant-temperature-constant-pressure molecular-dynamics calculations for molecular solids: Application to solid nitrogen at high pressure. *Physical Review B. Condensed Matter* 33: 339–342.

See also Potential Energy Function

M

MOLECULAR EFFICIENCY

Thomas D. Schneider

The Carnot efficiency is determined from a temperature difference, so it is not appropriate to use for molecular machines since they function at one temperature. Instead, an isothermal efficiency first used to understand satellite communications can be applied to molecular states. The formula for the theoretical upper bound of the isothermal efficiency is

$$\varepsilon_t = \ln(P_y/N_y + 1)/(P_y/N_y)$$

where P_y is the energy dissipated by the molecule as it makes selections between states and N_y is the thermal energy flowing through the molecule that interferes with the molecular machine during the selection process. Thomas Schneider discovered that when molecules select between two or more states, their isothermal efficiency is around 70%. From the above formula, a limitation on the evolutionary maximization of the efficiency at 70% corresponds to a lower bound on the P_y/N_y ratio. That is, the experimental observations imply that $P_y > N_y$. Schneider recognized that if the power exceeds the noise, a molecule can make state choices distinctly. The key concept is that biological systems require distinct molecular states and this leads to a maximum 70% efficiency.

For example rhodopsin, the light sensitive molecule in the retina of the eye, has two states: 'in the dark' and 'having seen a photon'. If these two states were not distinct, one would see imaginary flashes of light when in the dark. So these states should evolve to be distinct and the isothermal efficiency of rhodopsin cannot exceed ln2=0.69. Indeed, Dartnall's measurements from 12 species gave efficiencies of 66±2%, which is close to the theoretical maximum. Surprisingly this means that for every 100 photons absorbed by rhodopsin, only 66 are used to switch states.

A second example is the selective binding of the restriction enzyme EcoRI to the DNA sequence 5' GAATTC 3'. To select the first base G from the four nucleotides takes 2 bits. There are 6 such selections so EcoRI makes 12 bits of decisions.

When a coin is placed on a table as either heads or tails, one bit of information can be recorded. For the coin to stay in place, it must lose both kinetic and potential energy. From the second law of thermodynamics, the minimum energy that must be dissipated is $Emin = k_B T \ln 2$ joules per bit where k_B is Boltzmann's constant, T is the absolute temperature and ln2 sets the units of bits. This can be used as an ideal conversion factor between energy and information.

The binding energy dissipated by EcoRI in transition from non-specific to specific DNA binding has been measured and it corresponds to 17.3 bits. Thus EcoRI could make 17.3 yes-no decisions for the energy it uses, but it only makes 12 bits and so the efficiency is 12/17.3 = 69%. Like rhodopsin, EcoRI does not exceed the theoretical upper limit for selection between states.

The observation of 70% efficiency implies that the states of rhodopsin and EcoRI are distinct and it explains this widely known fact of molecular biology. From the derivation of this result (packing of spheres in a high dimensional space), we know that each molecular interaction must use a code similar to those used in communications systems. For a review see Schneider (2010b).

Further reading

Dartnall HJA (1968) The photosensitivities of visual pigments in the presence of hydroxylamine. *Vision Res*, 8: 339–358.

Pierce JR, Cutler CC (1959) Interplanetary communications. In: FI Ordway, III, editor, *Advances in Space Science, Vol. 1,* pages 55–109, New York, Academic Press.

Schneider, TD (1991) Theory of molecular machines. I. Channel capacity of molecular machines. *J. Theor. Biol.*, 148: 83–123.

Schneider TD (1991) Theory of molecular machines. II. Energy dissipation from molecular machines. *J. Theor. Biol.* 148: 125–137.

Schneider TD (2010a) 70% efficiency of bistate molecular machines explained by information theory, high dimensional geometry and evolutionary convergence. *Nucleic Acids Res.*, 38:5995–6006.

Schneider TD (2010b) A brief review of molecular information theory. *Nano Communication Networks*, 1: 173–180.

Schneider TD, Stormo GD, Gold L, Ehrenfeucht A. (1986) Information content of binding sites on nucleotide sequences. *J. Mol. Biol.*, 188: 415–431.

See also: Binding Site, Bit, Channel Capacity (Channel Capacity Theorem), Code, Coding, see Coding Theory, Coding Theory (Code, Coding), Entropy, Error, Evolution of Biological Information, Gumball Machine, Information, Information Theory, Message, Molecular Information Theory, Molecular Machine, Molecular Machine Capacity, Molecular Machine Operation, Noise, Rsequence, Second Law of Thermodynamics, Sequence Logo, Shannon Entropy (Shannon Uncertainty), Shannon Sphere, Shannon Uncertainty, see Shannon Entropy, Signal-To-Noise Ratio, Thermal Noise, Uncertainty

MOLECULAR EVOLUTIONARY MECHANISMS

Dov Greenbaum

The processes giving rise to changes in the sequences of DNA, RNA and protein molecules during evolution.

There are many mechanisms through which evolutionary changes can take place. These mechanisms can result in either a change in the gene itself – which may or may not have a phenotypic effect, affect the expression of the gene, or on a more global scale affect the frequency of a gene within a population.

These include:

- Selective pressure – influential extrinsic factors such as environmental factors. Selective pressure has many results:
  - i. Directional
  - ii. Stabilizing
  - iii. Disruptive
- Recombination/duplication – the combination of different gene segments or the duplication of a gene or a gene segment.
- Replication
- Deletion
- Mutation

iv. Neutral – a neutral mutation does not effect the function of a gene
v. Null – the wild type function is lost due to this mutation (also called a loss of function mutation)
vi. Spontaneous – spontaneous mutations arise randomly with a very minimal frequency (1/106) genomic sequences.
vii. Point – sometimes natural, but sometimes experimentally induced, a point mutation causes a mutation in a precise area of the sequence and to only one nucleotide. A transition point mutation is when a purine is substituted for another purine or a pyrimidine for another pyrimidine. A transversion occurs when a purine is substituted for a pyrimidine or vice versa.
viii. Frameshift – where there is a shift in the coding sequence (i.e. the insertion or deletion of a nucleotide). Given that each set of three nucleotides codes for an amino acid, the resulting shift causes the sequence to code for totally different amino acids after the insertion or deletion event; this results in a new reading frame
ix. Constitutive – that is the gene in question is either turned on or off independent of upstream regulatory factors

- Domain-shuffling – a new protein is the outcome of bringing different independent domains together.
- Rearrangement at the chromosome level. This refers to changes that can be seen cytogenetically. They include:
  x. Deletion – loss of sequence – these can be large losses stretching from single bases to genes to millions of bases
  xi. Duplication copying of a sequence when this occurs to the whole gene, a paralog is formed
  xii. Insertion inserting a new sequence into the middle of gene sequence this can have minimal or major effects
  xiii. Inversion flipping the sequence so that it is read backwards
  xiv. Translocation moving the sequence to another area in the genome
  xv. Transposition occuring when a segment of a chromosome is broken off and moved to another place within that chromosome
- Resistance transfer – the transfer of drug resistance through methods such as horizontal gene transfer can significantly change the lifestyle of microorganisms.
- Horizontal transfer (lateral transfer) – the process of transferring a gene between two independent organisms. This is in contradistinction to the transfer of genes from a parent to offspring – vertical gene transfer
- Silencing – the inactivation of a gene

Studying these natural events, along with an analysis population genetics allows one to study the evolution of an organism.

Further reading

Graur D, Li W-H (1999) *Fundamentals of Molecular Evolution*. Sinauer Associates.

MOLECULAR INFORMATION THEORY

Thomas D. Schneider

Molecular information theory is information theory applied to molecular patterns and states. For a review see Schneider (1994).

Related website

Tom Schneider's site	http://alum.mit.edu/www/toms/

Further reading

Schneider TD (1994) Sequence logos, machine/channel capacity, Maxwell's demon, and molecular computers: a review of the theory of molecular machines. *Nanotechnology* 5: 1–18.

See also Molecular Efficiency

MOLECULAR MACHINE

Thomas D. Schneider

A molecular machine can be defined precisely by six criteria:

1. A molecular machine is a single macromolecule or macromolecular complex.
2. A molecular machine performs a specific function for a living system.
3. A molecular machine is usually primed by an energy source.
4. A molecular machine dissipates energy as it does something specific.
5. A molecular machine 'gains' information by selecting between two or more after states.
6. Molecular machines are isothermal engines.

Further reading

Schneider TD (1991) Theory of molecular machines. I. Channel capacity of molecular machines. *J. Theor. Biol.* 148: 83–123.

See also Molecular Efficiency, Molecular Machine Capacity, Molecular Machine Operation

MOLECULAR MACHINE CAPACITY

Thomas D. Schneider

The maximum information, in bits per molecular operation, that a molecular machine can handle is the molecular machine capacity. When translated into molecular biology, Shannon's channel capacity theorem states that:

'By increasing the number of independently moving parts that can interact cooperatively to make decisions, a molecular machine can reduce the error frequency (rate of incorrect choices) to whatever arbitrarily low level is required for survival of the organism, even when the machine operates near its capacity and dissipates small amounts of power.'

This theorem explains the precision found in molecular biology, such as the ability of the restriction enzyme EcoRI to recognize 5'-GAATTC-3′ while ignoring all other sites. The derivation is in Schneider (1991).

Further reading

Schneider TD (1991) Theory of molecular machines. I. Channel capacity of molecular machines. *J. Theor. Biol.* 148: 83–123. http://alum.mit.edu/www/toms/papers/ccmm/.

See also Channel Capacity, Molecular Efficiency

MOLECULAR MACHINE OPERATION

Thomas D. Schneider

A molecular machine operation is the thermodynamic process in which a molecular machine changes from the high energy before state to a low energy after state. There are four standard examples:

- Before DNA hybridization the complementary strands have a high relative potential energy; after hybridization the molecules are non-covalently bound and in a lower energy state.
- The restriction enzyme EcoRI selects 5'-GAATTC-3' from all possible DNA duplex hexamers. The operation is the transition from being anywhere on the DNA to being at a GAATTC site.
- The molecular machine operation for rhodopsin, the light sensitive pigment in the eye, is the transition from having absorbed a photon to having either changed configuration (in which case one sees a flash of light) or failed to change configuration.
- The molecular machine operation for actomyosin, the actin and myosin components of muscle, is the transition from having hydrolyzed an ATP to having either changed configuration (in which the molecules have moved one step relative to each other) or failed to change configuration.

Further reading

Schneider TD (1991) Theory of molecular machines. I. Channel capacity of molecular machines. *J. Theor. Biol.*, 148: 83–123.

See also Molecular Efficiency

MOLECULAR MECHANICS

Roland Dunbrack

Representation of molecules as mechanical devices with potential energy terms such as harmonic functions, torsional potentials, electrostatic potential energy, and van der Waals energy. Molecular mechanics programs for macromolecules usually include routines for energy minimization as a function of atomic coordinates and molecular dynamics simulations.

Further reading

Kollman P, Massova I, Reyes C, et al. (2000) Calculating structures and free energies of complex molecules: Combining molecular mechanics and continuum models. *Accts. Chem. Res.* 33, 889–897.

See also Electrostatic Potential Energy, van der Waals, Energy Minimization, Molecular Dynamics Simulations

MOLECULAR NETWORK, *SEE* NETWORK.

MOLECULAR REPLACEMENT

Liz Carpenter

Molecular replacement is a technique used in X-ray crystallography to solve the structures of macromolecules for which a related structure is known. The X-ray diffraction experiment gives a set of indexed diffraction spots each of which has a measured intensity, I. The observed structure factor amplitude, $|F_{obs}|$, is proportional to the square root of the intensity.

A diffraction pattern can be calculated from any molecule placed in the unit cell in any position using the equation:

$$F(hkl)_{j=1}^{allatoms} = \sum f_j \exp(2\pi i[hx + ky + lz])j = 1$$

where

F(hkl) is the structure factor for the diffraction spot with indices hkl

f_j is the atomic scattering factor for the jth atom

x, y, z are the coordinates of the jth atom in the unit cell.

The atomic scattering factor is the scattering from a single atom in a given direction and it can be calculated from the atomic scattering factor equation. In general:

$$f_{j1}^{4}(\sin\theta/\lambda) = \sum a_i . \exp(-b_i(\sin\theta/\lambda)^2) + c$$

where a_i, b_i and c are known coefficients.

Molecular replacement is a trial and error method in which a related structure is moved around inside the unit cell, the diffraction pattern is calculated using the equation above for each trial position and compared to the observed diffraction pattern. If the correlation between the observed and calculated pattern is above the background this indicates that a solution has been found. The phases calculated from the correctly positioned model can be combined with the observed diffraction pattern to give an initial map of the electron density of the unknown structure.

The search for a molecular replacement solution requires testing both different orientations of the model and different translations of the model within the unit cell. This would be a 6-dimensional search, covering three rotation parameters and three translational parameters. Fortunately this problem can be broken down into two consecutive three

M

dimensional searches. Firstly suitable rotation angles are tested using a rotation function search and then the top solutions from the rotation function search are tested in a translation function search to identify the position of the molecule within the unit cell.

Further reading

Crowther RA (1972) The fast rotation function. In *The Molecular Replacement Method* ed. M.G. Rossman, *Int. Sci. Rev. Ser.*, no. 13, pp. 173–178. Navaza J (1994) *AMoRe*: an automated package for molecular replacement. *Acta Cryst.* A50, 157–163.

Read RJ (2001) Pushing the boundaries of molecular replacement with maximum likelihood. *Acta Cryst.* D57: 1373–1382.

MONOPHYLETIC GROUP, *SEE* CLADE.

MONTE CARLO SIMULATION

Roland Dunbrack

A method for randomly sampling from a probability distribution, used in many fields including statistics, economics, and chemistry. Monte Carlo simulations must begin with an initial configuration of the system. A new configuration is drawn randomly from a proposal distribution, and is either accepted or rejected depending on whether the new configuration is more or less likely than the current configuration. If the new configuration is more probable, the new configuration is always accepted. If the new configuration lowers the probability, the new state is accepted with probability $p_{new}/p_{current}$. For molecular systems, the moves can be changes in Cartesian atom positions or in internal coordinates such as bond lengths, bond angles, and dihedral angles. The energy of the system is used to calculate the probabilities of each move. If the energy decreases, the move is always accepted. If the energy is increased, then the probability of accepting the move is calculated with the change in energy of the system, $\exp(-(E_{new}-E_{current})/k_B T)$, where T is the temperature of the system and k_B is the Boltzmann constant.

Related website

MCPRO	http://zarbi.chem.yale.edu/

Further reading

Jorgensen WL (1998) *The Encyclopedia of Computational Chemistry 3*, 1754–1763 Athens, GA USA Wiley.

See also Markov Chain Monte Carlo

MOTIF

Teresa K. Attwood and Dov Greenbaum

Motifs are short, conserved regions of peptide or nucleic acid sequence. In proteins, motifs can be delineated at the level of the primary, secondary and tertiary structure.

At the primary structure level (i.e., in the linear amino acid sequence), a motif can be defined as a consecutive string of residues whose general character is repeated, or conserved, at a particular position, in all sequences with which it is aligned. In the motif illustrated below, five residues are completely conserved (bold highlights); in addition, the polar character of the constituent residues is conserved at positions 4 and 8, while non-polar character is conserved at positions 3, 7, 11 and 13:

RYLT**W**TFS**TP**MIL
RYLT**W**TLS**TP**MIL
RYLT**W**ALS**TP**MIL
RYLD**W**VVT**TP**LLV
RYVD**W**IVT**TP**LLV
RYVD**W**VVT**TP**LLV
RYID**W**LVT**TP**ILV
RYAD**W**LFT**TP**LLL
RYVD**W**LLT**TP**LNV

When building a multiple alignment, as more distantly related sequences are included, it is often necessary to insert gaps to bring equivalent parts of adjacent sequences into the correct register. As a result of this process, islands of conservation tend to emerge within a sea of mutational change – these conserved regions, or motifs (like the one illustrated here), are typically 10–20 residues in length, and tend to correspond to the core structural and functional elements of the protein. This renders them scientifically valuable, allowing them to be used to diagnose the same or similar structures or functions in uncharacterised sequences via a range of analysis techniques: e.g., consensus (regex) patterns may be used to encode single motifs (the basis of the PROSITE database); fingerprints may be used to encode multiple motifs (the basis of the PRINTS database). None of the available analytical techniques should be regarded as the best: each exploits sequence motifs in a different way – each therefore offers a different perspective and a different (complementary) diagnostic opportunity. The best strategy is therefore to use them all.

At the secondary structure level, motifs (minimally comprising two or three substructures – i.e., coil, helix or sheet) are somewhat arbitrary, in that they can describe a whole fold or only a small segment of a peptide. A simple motif could be the meander, which comprises three up-down β-helices; many such motifs together might form a larger β-barrel structure. A common structural motif in DNA-binding proteins is the zinc finger, which is characterized by finger-like loops of amino acids stabilised by zinc atoms.

By contributing to a 'biological parts' list, motifs are helpful both in characterising proteomes and in comparing widely different genomes, providing a common ground for proteome-proteome and genome-genome comparisons.

Related websites

ScanProsite	http://prosite.expasy.org/scanprosite/
FingerPRINTScan	http://www.bioinf.manchester.ac.uk//dbbrowser/fingerPRINTScan/

Further reading

Attwood TK (2000) The quest to deduce protein function from sequence: the role of pattern databases. *Int. J. Biochem. Cell Biol.* 32(2): 139–155.

Attwood TK, Parry-Smith DJ (1999) *Introduction to Bioinformatics.* Addison Wesley Longman, Harlow, Essex, UK.

Higgs P, Attwood TK (2005) *Bioinformatics and Molecular Evolution.* Blackwell Publishing, Oxford, UK.

Mulder NJ, Apweiler R (2002) Tools and resources for identifying protein families, domains and motifs. *Genome Biol.* 3: REVIEWS2001

See also Alignment, multiple, Consensus Pattern, Fingerprint, Gap, Sequence Motifs, prediction and modeling, PRINTS, PROSITE, Cis-regulatory Motif

MOTIF DISCOVERY

Jacques van Helden

Extraction of one or several motifs characterizing a sequence set, without prior knowledge of these motifs. The criteria for considering a motif as characteristic can vary depending on the applications and algorithmic approaches, e.g. over-representation, positional biases, under-representation.

Motif discovery approaches are commonly applied to peptide sequences in order to find signatures of functional domains, or to nucleic sequences, to detect signals involved in regulatory processes (e.g. transcription factor binding, RNA post-processing). The following discussion is restricted to the context of motifs found in nucleic sequences, such as cis-regulatory motifs representing the binding specificity of transcription factors.

Many algorithms have been specifically developed for the discovery of transcription factor binding motifs (TFBM) in regulatory sequences, relying on a variety of approaches: word statistics (over-representation, position biases), expectation-maximization, greedy algorithms, Gibbs sampling, to cite a few.

Motif discovery is commonly applied to extract motifs, and infer potentially binding factors, from promoters of co-expressed genes (transcriptome analysis), or from genome-wide location analysis (peak collections obtained by ChIP-on-chip or ChIP-seq peaks), or to identify conserved motifs in promoters of orthologous genes (phylogenetic footprints).

Motif discovery differs from *motif search*, which starts from a motif of interest known in advance (e.g. built from a collection of experimentally validated sites for the factor if interest), and from *motif enrichment analysis*, where candidate motifs are selected among a wide collection of predefined motifs (e.g. a database of all known transcription factor binding motifs).

Related websites

Tool	Link and description
AlignACE	Gibbs sampling algorithm, initially developed to detect subtle motifs in protein equences. http://arep.med.harvard.edu/mrnadata/mrnasoft.html
Consensus	Greedy algorithm http://stormo.wustl.edu/consensus/
MEME	Multiple Expectation-maximization for Motif Elicitation. http://meme.nbcr.net/meme/
RSAT suite	Regulatory Sequence Analysis Tools includes a variety of word-based approaches to detect over-represented or positionally biased motifs: *oligo-analysis*, *dyad-analysis*, *position-analysis*, *local-word-analysis*, *info-gibbs*). http://rsat.ulb.ac.be/rsat/
Weeder	Word statistics http://www.pesolelab.it/

Further reading

Tompa, et al. (2005) Assessing computational tools for the discovery of transcription factor binding sites. *Nat Biotechnol* 23 (1): 137–144.

See also Motif Enrichment Analysis, Motif Search

MOTIF ENRICHMENT ANALYSIS

Jacques van Helden

Detection of motifs found more frequently than expected by chance in a given set of sequences.

Motif enrichment relies on scanning the sequences (see Motif Search) with a large collection of predefined motifs (e.g. a database of previously known transcription factor binging motifs).

Motif enrichment analysis is distinct from *motif discovery* in that it relies on a pre-existing collection of motifs (there is thus an input of biological knowledge about motifs). Motif enrichment is not suited for discovering novel motifs, but presents the advantage to associate each significant motif with some prior knowledge (e.g. transcription factor binding this motif).

An important criterion for enrichment analysis is the choice of the background model (defining the random expectation for motif occurrences) and the scoring function (used to compare occurrences between the query sequence and the background). CisTargetX (Aerts et al., 2010; Pottier et al., 2012) ranks promoters according to scores of motif occurrences, and compares the ranking distributions between a positive set (e.g. cluster of co-expressed genes) and a negative set (e.g. all the other genes of the genome).

Motif-wise enrichment analysis can be complemented by additional criteria to restrict the search space and return more relevant motifs or combinations thereof. The Web tool DIRE

M

(previously named CREME) restricts the analysis to phylogenetically conserved intergenic regions. The program further predicts cis-regulatory modules by detecting combinations of motifs co-occurring in restricted regions (Sharane t al, 2004).

Related websites

CistargetX	Target and Enhancer prediction in *Drosophila* using gene co-expression sets. http://med.kuleuven.be/lcb/i-cisTarget/
DIRE/CREME	Distant Regulatory Elements of Co-regulated genes http://dire.dcode.org/
HOMER	Motif enrichment and de novo motif discovery in a set of ChIP-seq peaks or other sets of genomic sequences. http://biowhat.ucsd.edu/homer/chipseq/peakMotifs.html

Further reading

Aerts S, Quan XJ, Claeys A, Naval Sanchez M, Tate P, Yan J, Hassan BA (2010) Robust target gene discovery through transcriptome perturbations and genome-wide enhancer predictions in Drosophila uncovers a regulatory basis for sensory specification. *PLoS Biol* 8(7): e1000435.

Potier D, Atak ZK, Sanchez MN, Herrmann C, Aerts S. 2012. Using cisTargetX to predict transcriptional targets and networks in Drosophila. *Methods Mol. Biol.* 786: 291–314.

Sharan R, et al. (2004) CREME: Cis-Regulatory Module Explorer for the human genome. *Nucleic Acids Res.* 32 (Web Server issue): W253–256.

See also Motif Search, Motif Discovery

MOTIF SEARCH

Jaap Heringa

A sequence motif is a nucleotide or amino acid pattern that is recurring in a collection of sequences or within a given sequence. Motifs are usually confined to a local segment as certain regions of a protein molecular structure are less liable to change than others. Knowledge of this kind can be used to elucidate certain characteristics of a protein's architecture such as buried *versus* exposed location of a segment or the presence of specific secondary structural elements. The most salient aspect however involves the protein's functionality: The most conserved protein region very often serves as a ligand binding site, a target for posttranslational modification, the enzyme catalytic pocket, and the like. Detecting such sites in newly determined protein sequences could save immense experimental effort and help characterise functional properties. Moreover, using the conserved regions of a protein only rather than its whole sequence for databank searching can reduce background noise and help considerably to establish distant relationships.

Sequence motifs are usually considered in the context of multiple sequence comparison, when certain contiguous sequence spans are shared by a substantial number of proteins. Staden (1988) gave the following classification for protein sequence motifs:

- exact match to a short defined sequence
- percentage match to a defined short sequence
- match to a defined sequence, using a residue exchange matrix and a cut-off score
- match to a weight matrix with cutoff score
- direct repeat of fixed length
- a list of allowed amino acids for each position in the motif.

Protein sequence motifs are often derived from a multiple sequence alignment, so that the quality of the alignment determines the correctness of the patterns. Defining consensus motifs from a multiple alignment is mostly approached using majority and plurality rules, and a consensus sequence is generally expressed as a set of heterogeneous rules for each alignment position. The stringency of the restraint imposed on a particular position should depend on how crucial a given feature is for the protein family under scrutiny and can range from an absolutely required residue type to no requirement at all. Often an initial alignment of sequences is generated on the basis of structural information available (e.g., superposition of C_α atoms) to ensure reliability (Taylor, 1986b). After creating the template from the initial alignment, it can then be extended to include additional related proteins. This process is repeated iteratively until no other protein sequence can be added without giving up on essential features. In the absence of structural information, methods have been developed based on pairwise sequence comparison as a starting point for determining sequence patterns, after which a multiple alignment is built by aligning more and more sequences to the pattern, which is allowed to develop at the same time (*e.g.* Patthy, 1987; Smith and Smith, 1990). Indirect measures of conservation have also been attempted as criteria for motif delineation, such as the existence of homogeneous regions in a protein's physical property profiles (*e.g.* Chappey and Hazout, 1992).

Frequently, a sequence motif is proven experimentally to be responsible for a certain function. A collection of functionally related sequences should then yield a discriminating pattern occurring in all the functional sequences and not occurring in other unrelated sequences. Such patterns often consist of several sequentially separated elementary motifs, which for example join in the three-dimensional structure to form a functional pocket. Such discriminating motifs have been studied extensively early on for particular cases such as helix-turn-helix motifs (Dodd and Egan, 1990) or G-protein coupled receptor fingerprints (Attwood and Findlay, 1993).

A consistent semi-manual methodology for finding characteristic protein patterns was developed by Bairoch (1993). The aim of the approach was to make the derived patterns as short as possible but still sensitive enough to recognize the maximum number of related sequences and also sufficiently specific to reject, in a perfect case, all unrelated sequences. A large collection of motifs gathered in this way is available in the PROSITE databank (Hofmann *et al.*, 1999). Associated with each motif is an estimate of its discriminative power. For many PROSITE entries, published functional motifs serve as a first approximation of the pattern. In other cases, a careful analysis of a multiple alignment for the protein family under study is performed. The initial motif is searched against the SWISS-PROT sequence databank and search hits are analysed. The motif is then empirically refined by extending or shrinking it in sequence length to achieve the minimal amount of false positives and maximal amount of true positives.

In the PROSITE database, motifs are given as regular expressions. For example, the putative AMP-binding pattern (entry PS00455) is given as

[LIVMFY] – {E} – {VES} – [STG] – [STAG] – G – [ST] – [STEI] – [SG] – x – [PASLIVM] – [KR],

where each position is separated by a hyphen, amino acids between square brackets are optional, residues between curly brackets are to be avoided and 'x' denotes no constraint. The AMP-binding motif is found 746 times in 682 different sequences from the UniProtKB/SWISS-PROT databank (Release 02); 723 hits in 659 sequences are true positives, 23 hits in 23 sequences are known false positives, while there are 72 known missed hits (false negatives). The PROSITE databank and related software is an invaluable and generally available tool for detecting the function of newly sequenced and uncharacterised proteins. The PROSITE database also includes the extended profile formalism of Bucher *et al.* (1996) for sensitive profile searching.

A description of sequence motifs in a regular expression such as in the PROSITE database, involves enumeration of all residue types allowed at particular alignment positions or by less stringent rules previously described. This necessarily leads to some loss of information, because particular sequences might not be considered. Also, the formalism is not readily applicable to all protein families. Exhaustive information about a conserved sequence span can be stored in the form of an ungapped sequence block. Henikoff and Henikoff (1991) derived a comprehensive collection of such blocks (known as the BLOCKS databank) from groups of related proteins as specified in the PROSITE databank (Hofmann *et al.*, 1999) using the technique of Smith *et al.* (1990). Searching protein sequences against the BLOCKS library of conserved sequence blocks or an individual block against the whole sequence library (Wallace and Henikoff, 1992) provides a sensitive way of establishing distant evolutionary relationships between proteins. The BLOCKS database was extended (Henikoff and Henikoff., 1999) by using, in addition to PROSITE, non-redundant information from four more databases: the PRINTS protein motifs database (Attwood and Beck, 1994) and the protein domain sequence databases Pfam-A (Bateman *et al.*, 1999), ProDom (Corpet *et al.*, 1999), and Domo (Gracy and Argos, 1998). The ELM database (Gould et al., 2010) is a repository of linear motifs, which are short segments of multidomain proteins that provide regulatory functions independently of protein tertiary structure.

With the availability of fast and reliable sequence comparison tools, methods have been developed to compile automatically as many patterns as possible from the full protein sequence databank. These methods often rely on automatic clustering and multiple alignment of protein sequence families, followed by application of motif extraction tools. Recent programs include the methods DOMAINER (Sonnhammer and Kahn, 1994), MKDOM (Gouzy *et al.*, 1997) and DIVCLUS (Park and Teichman, 1998). Pattern matching methods based upon machine learning procedures have also been attempted; in particular neural networks (*e.g.* Frishman and Argos, 1992) as these have the ability to learn an internal non-linear representation from presented examples. The method GIBBS (Lawrence et al., 1993) detects conserved regions with a residual degree of similarity from unaligned sequences and is based on iterative statistical sampling of individual sequence segments. The motifs found are multiply aligned by the method. Also the program MEME (Bailey Elkan, 1994) employs unsupervised motif-searching, but using an expectation maximization (EM) algorithm. Although MEME motifs are ungapped, the program can find multiple occurrences in individual sequences, which do not need to be encountered within each input sequence.

The evolution of pattern derivation methods and resulting motifs in turn has triggered the development of tools of varying degrees of complexity to search individual sequences and whole sequence databanks with user-specified motifs. Particularly, many programs are available for searching sequences against the PROSITE databank; a full list of publicly available and commercial programs for this purpose is supplied together with PROSITE (Hofmann *et al.*, 1999) and ELM (Gould et al., 2010).

Related websites

Prosite	http://prosite.expasy.org/
ELM	http://elm.eu.org/
MAST (Meme suite)	http://meme.nbcr.net/meme/ Supports Markov models for the background models.
matrix-scan (RSAT suite)	http://www.rsat.eu/ Supports Markov models for the background models.
patser (consensus suite)	Stand-alone version: http://stormo.wustl.edu/resources.html Web version: http://stormo.wustl.edu/consensus/ Bernoulli background models (assume independence between successive nucleotides).

Further reading

Attwood TK, Beck ME (1994) PRINTS – a protein motif fingerprint database. *Protein Engineering* 7: 841–848.

Attwood TK, Findlay JBC (1993) Design of a discriminating fingerprint for G-protein-coupled receptors. *Protein Engineering* 6: 167–176.

Bailey TL, Elkan C (1994) Fitting a mixture model by expectation maximization to discover motifs in biopolymers. *In:* Proceedings of the second international conference on Intelligent Systems for Molecular Biology. pp. 28–36. AAAI Press.

Bailey TL, Gribskov M (1998) Combining evidence using p-values: application to sequence homology searches. *Bioinformatics* 14(1): 48–54.

Bairoch A (1993) The PROSITE dictionary of sites and patterns in proteins, its current status. *Nucleic Acid. Res.* 21: 3097–3103.

Bateman A, Birney E, Durbin R, Eddy SR, Finn RD, Sonnhammer ELL (1999) Pfam 3.1: 1313 multiple alignments and profile HMMs match the majority of proteins. *Nucleic Acids Res.* 27: 260–262.

Bucher P, Karplus K, Moeri N, Hofmann K (1996) A flexible motif search technique based on generalized profiles. *Computers and Chemistry* 20: 3–24.

Chappey C, Hazout S (1992) A method for delineating structurally homogenious regions in protein sequences. *Comput. Appl. Biosci.* 8: 255–260.

Corpet F, Gouzy J, Kahn D (1999) Recent improvements of the ProDom database of protein domain families. *Nucleic Acids Res.* 27: 263–267.

Dodd IB, Egan JB (1990) Improved detection of helix-turn-helix DNA-binding motifs in protein sequences. *Nucleic Acids Res.* 18: 5019–5026.

M

Frishman DI, Argos P (1992) Recognition of distantly related protein sequences using conserved motifs and neural networks. *J. Mol. Biol.* 228: 951–962.

Gould CM, Diella F, Via A, et al. (2009) ELM: the status of the 2010 eukaryotic linear motif resource. *Nucleic Acids Res.* 38(Database issue) D167–D180.

Gouzy J, Eugène P, Greene EA, Kahn D., Corpet F (1997) XDOM, a graphical tool to analyse domain arrangements in protein families. *Comput. Appl. Biosci.* 13: 601–608.

Gracy J, Argos P (1998) Automated protein sequence database classification. II. Delineation of domain boundaries from sequence similarities. Bioinformatics 14: 174–187.

Henikoff S, Henikoff JG (1991) Automated assembly of protein blocks for database searching. *Nucleic. Acids. Res.* 19: 6565–6572.

Henikoff JG, Henikoff S (1999) New features of the Blocks database server. *Nucl. Acids Res.* 27: 226–228.

Hertz GZ, Stormo GD (1999) Identifying DNA and protein patterns with statistically significant alignments of multiple sequences. *Bioinformatics* 15(7–8): 563–577.

Hofmann K, Bucher P, Falquet L, Bairoch A (1999) The PROSITE database, its status in 1999. *Nucleic Acids Res.* 27: 215–219.

Lawrence CE, Altschul SF, Boguski MS, Liu JS, Neuwald AF, Wootton JC (1993) Detecting subtle sequence signals: a Gibbs sampling strategy for multiple alignment. *Science* 262: 208–214.

Park J, Teichmann SA (1998) DIVCLUS: an automatic method in the GEANFAMMER package that finds homologous domains in single- and multi-domain proteins. *Bioinformatics* 14: 144–150.

Patthy L (1987) Detecting homology of distantly related proteins with consensus sequences. *J. Mol. Biol.* 198: 567–577.

Smith HO, Annau TM, Chandrasegaran S (1990) Finding sequence motifs in groups of functionally related proteins. *Proc. Natl. Acad. Sci. USA* 87: 826–830.

Smith RF, Smith TF (1990) Automatic generation of primary sequence patterns from sets of related sequences. *Proc. Natl. Acad. Sci. USA* 87: 118–122.

Sonnhammer ELL, Kahn D (1994) Modular arrangement of proteins as inferred from analysis of homology. *Protein Sci.* 3: 482–492.

Staden R (1988) Methods to define and locate patterns of motifs in sequences. *Comput. Appl. Biosci.* 4: 53–60.

Staden R (1989) Methods for calculating the probabilities of finding patterns in sequences. *Comput. Appl. Biosci* 5(2): 89–96.

Taylor WR (1986a) The classification of amino acid conservation. *J. Theor. Biol.* 119: 205–218.

Taylor WR (1986b) Identification of protein sequence homology by consensus template alignment. *J. Mol. Biol.* 188: 233–258.

Turatsinze JV, Thomas-Chollier M, Defrance M, van Helden J (2008) Using RSAT to scan genome sequences for transcription factor binding sites and cis-regulatory modules. *Nat. Protoc.* 3(10): 1578–1588.

Wallace JC, Henikoff S (1992) PATMAT: a searching and extraction program for sequence, pattern and block queries and databases. *Comput. Appl. Biosci.* 8: 249–254.

See also Homology Searching, Profile Searching

MOUSE GENOME DATABASE, *SEE* MOUSE GENOME INFORMATICS.

MOUSE GENOME INFORMATICS (MGI, MOUSE GENOME DATABASE, MGD)

Dan Bolser

The Mouse Genome Informatics project is a database resource for the laboratory mouse, providing genetic, genomic, and biological data. It is an integrated resource, composed primarily of data from the Mouse Genome Database (MGD), the mouse Gene Expression Database (GXD), the Mouse Tumor Biology database (MTB), and MouseCyc. MGD catalogues gene function, nomenclature, phylogenomics, genetic variation and mouse strain and phenotype information. GXD provides expression data in various conditions of health and disease, including development. MTB provides data on the frequency, incidence, genetics, and pathology of neoplastic disorders in different strains of mice.

Related websites

MGD	http://www.informatics.jax.org/
Wikipedia	http://en.wikipedia.org/wiki/Mouse_Genome_Database
MetaBase	http://metadatabase.org/wiki/Mouse_Genome_Database

Further reading

Blake JA, Bult CJ, Kadin JA, Richardson JE, Eppig JT (2011) The Mouse Genome Database (MGD) premier model organism resource for mammalian genomics and genetics. *Nucleic Acids Res.* 39: D842–848.

See also EMA, RGD

MOUSE TUMOR BIOLOGY (MTB) DATABASE, *SEE* MOUSE GENOME INFORMATICS.

MOUSECYC, *SEE* MOUSE GENOME INFORMATICS.

MPROMDB (MAMMALIAN PROMOTER DATABASE)

Obi L. Griffith and Malachi Griffith

MPromDb is a curated database of annotated gene promoters identified by ChIP-seq experiments. Datasets are obtained primarily from published and unpublished studies made available through the NCBI GEO repository. Users can search the database by Entrez gene id/symbol or by tissue/cell specific activity and apply various filters.

M

Related website

Mammalian Promoter Database	http://mpromdb.wistar.upenn.edu/

Further reading

Gupta R, Bhattacharyya A, et al. (2011) MPromDb update 2010: An integrated resource for annotation and visualization of mammalian gene promoters and ChIP-seq experimental data. *Nucleic Acids Res.* 39: D92–97.

Sun H, et al. (2006) MPromDb: an integrated resource for annotation and visualization of mammalian gene promoters and ChIP-chip experimental data. *Nucleic Acids Res.* 34: D98–D103.

MRBAYES

Michael P. Cummings

A program for Bayesian phylogenetic analysis of binary, 'standard' (morphology), nucleotide, doublet, codon, amino-acid data and mixed models.

Among the most sophisticated programs for phylogenetic analysis, its use of mixed models allows different data partitions to be combined in the same model, with parameters linked or unlinked across partitions. Additional models include relaxed clocks for dated phylogenies, model averaging across time-reversible substitution models, and support for various types of topological constraints. Inclusion of the Bayesian estimation of species trees (BEST) algorithms allows for inference of species trees from gene trees. The program features automatic optimization of tuning parameters, which improves convergence for many problems. Convergence diagnostics and allows multiple analyses to be run in parallel with convergence progress monitored during analysis. Marginal model likelihoods for Bayes factor tests can be estimated accurately across the entire model space using the stepping-stone method. Output includes various summaries of tree and parameters statistics, as well as samples of ancestral states, site rates, site dN/dS ratios, branch rates, and node dates. A wide range of statistics on tree parameters can also be output for visualization in FigTree and compatible software.

The program is used through a command line interface, and a user can execute a series of individual commands without interaction by including appropriate commands in the MrBayes block (NEXUS file format). Support for Streaming SIMD Extensions (SSE), related processor supplementary instruction sets and the BEAGLE library, speed likelihood calculations. The program also supports check-pointing. The programs are written in ANSI C, and source code is freely available under the GNU General Public License version 3.0. Executables for some platforms is also available.

Related website

MrBayes web site	http://mrbayes.sourceforge.net/

Further reading

Huelsenbeck JP, and Ronquist F (2001) MRBAYES: Bayesian inference of phylogenetic trees. *Bioinformatics* 17: 754–755.

Ronquist F, Huelsenbeck JP (2003) MrBayes 3: Bayesian phylogenetic inference under mixed models. *Bioinformatics* 12: 1572–1574.

Ronquist F, Teslenko M, van der Mark P, et al. MrBayes 3.2: Efficient Bayesian phylogenetic inference and model choice across a large model space. *Syst Biol* 61: 539–542.

See also BAMBE, BEAGLE, FigTree

MULTIDOMAIN PROTEIN

Laszlo Patthy

Proteins that are larger than approx. 200–300 residues usually consist of multiple protein domains. The individual structural domains of multidomain proteins are compact, stable units with a unique three-dimensional structure. The interactions within one domain are more significant than with other domains. The presence of distinct structural domains in multidomain proteins indicates that they fold independently, i.e. the structural domains are also folding domains. Frequently, the individual structural and folding domains of a multidomain protein perform distinct functions.

Many multidomain proteins contain multiple copies of a single type of protein domain, indicating that internal duplication of gene segments encoding a domain has given rise to such proteins. Some multidomain proteins contain multiple types of domain of distinct evolutionary origin. Such chimeric proteins that arose by fusion of two or more gene-segments are frequently referred to as mosaic proteins or modular proteins, and the constituent domains are usually referred to as protein modules.

See also Protein Domains, Mosaic Proteins, Modular Proteins, Protein Modules

MULTIFACTORIAL TRAIT (COMPLEX TRAIT)

Mark McCarthy and Steven Wiltshire

Phenotypes (traits or diseases) for which individual susceptibility is governed by the action of (and interaction between) multiple genetic and environmental factors.

Most of the major diseases of mankind (e.g. diabetes, asthma, many cancers, schizophrenia) are multifactorial traits. Whilst they can be seen to cluster within families, such traits manifestly fail to show classical patterns of Mendelian segregation. Individual susceptibility, it is assumed, reflects variation at a relatively large number of polymorphic sites in a variety of genes, each of which has a relatively modest influence on risk. Depending on the assumptions about the number of genes involved, this may be termed polygenic or oligogenic inheritance. In addition, the phenotypic expression of this genetic inheritance in any given person will, typically, depend on their individual history of exposure to relevant environmental influences (for example, food availability in the case of obesity). For such multifactorial traits, no single variant is likely to be either necessary or sufficient for the development of disease. This incomplete correspondence between genotype and phenotype (contrasting with the tight correlation characteristic of Mendelian diseases) provides one of the major obstacles to gene mapping in such traits, because of the consequent reduction in power for both linkage and linkage disequilibrium (association) analyses.

M

Further reading

Bennett ST, Todd JA (1996) Human Type 1 diabetes and the insulin gene: principles of mapping polygenes. *Ann. Rev. Genet.* 30: 343–370.

Cardon LR, Bell JI (2001) Association study designs for complex disease. *Nat. Rev. Genet.* 2:91–99.

Lander ES, Schork NJ (1994) Genetic dissection of complex traits. *Science* 265: 2037–2048.

Risch N (2000) Searching for genetic determinants in the new millennium. *Nature* 405: 847–856.

Risch N (2001) Implications of multilocus inheritance for gene-disease association studies. *Theor Popul Biol* 60: 215–220.

See also Mendelian Disease, Linkage Analysis, Penetrance, Polygenic Inheritance, Oligogenic Inheritance

MULTIFURCATION (POLYTOMY)

Aidan Budd and Alexandros Stamatakis

An internal node of a phylogenetic tree is described as a multifurcation if (i) it is in a rooted tree and is linked to three or more daughter subtrees or (ii) it is in an unrooted tree and is attached to four or more branches. A tree that contains any multifurcations can be described as a multifurcating tree. See Figure M.4.

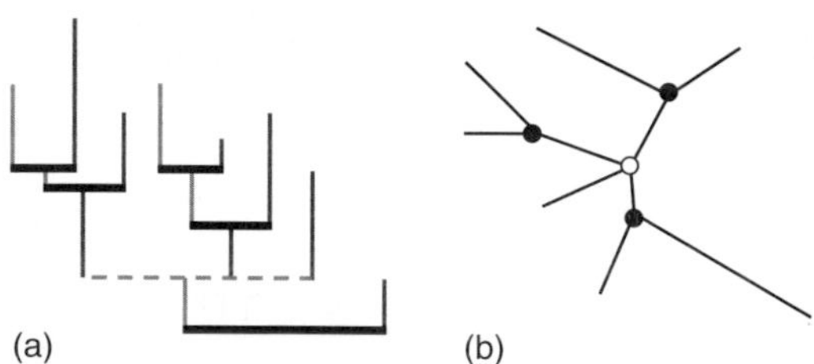

Figure M.4 The rooted tree, (a), contains five bifurcations (thick horizontal lines) and one multifurcation (dashed horizontal line). The unrooted tree, (b), contains 3 bifurcations (solid dots) and one multifurcation (hollow dot).

See also Bifurcation, Node

MULTILABEL CLASSIFICATION

Concha Bielza and Pedro Larrañaga

In *multi-label* classification, as opposed to *single-label* classification, the learning examples are associated with a set of labels. It is usually confused with multiclass classification, where examples are categorized into more than two classes. Real applications are text categorization, medical diagnosis, protein function classification, music categorization and semantic scene classification.

A common way of tackling these problems is by employing problem transformations into a set of binary classification problems, like in the binary relevance, where one binary classifier is trained per label. Other transformation methods include RAkEL, Chain

Classifiers and Ml-kNN, a variant of the k-nearest neighbours lazy classifiers. Other multi-label classification methods adapt the usual single-label algorithms to directly perform multi-label classification.

Performance measures for multi-label classification are inherently different from those used in single-label classification. Thus, they can account for the percentage of examples that have all their labels classified correctly or just look at some sets of labels.

Java implementations are available in the Mulan and Meka software packages, both built on top of Weka.

Related websites

Mulan website	http://mulan.sourceforge.net/datasets.html
Meka website	http://meka.sourceforge.net/

Further reading

Chen B, Gu W, Hu J (2010) An improved multi-label classification method and its application to functional genomics. *International Journal of Computational Biology and Drug Design* 3(2) 133–145.

Jin B, Muller B, Zhai C, Lu X (2008) Multi-label literature classification based on the Gene Ontology graph. *BMC Bioinformatics* 9: 525.

Tsoumakas G, Katakis I (2007) Multi-label classification: An overview. *International Journal of Data Warehousing & Mining* 3(3): 1–13.

MULTILAYER PERCEPTRON, *SEE* NEURAL NETWORK.

MULTIPLE ALIGNMENT

Jaap Heringa

A multiple sequence alignment is an alignment of three or more biomolecular sequences, generally protein, DNA or RNA.

The automatic generation of an accurate multiple alignment is potentially a daunting task. Ideally, one would make use of an in-depth knowledge of the evolutionary and structural relationships within the homologous family members to be aligned, but this information is often lacking or difficult to use. General empirical models of protein evolution (Benner *et al.*, 1992; Dayhoff, 1978; Henikoff & Henikoff, 1992) are widely used instead but these can be difficult to apply when the sequences are less than 30% identical (Sander & Schneider, 1991). Further, mathematically sound methods for carrying out alignments, using these models, can be extremely demanding in computer resources for more than a handful of sequences (Carrillo & Lipman, 1988; Wang & Jiang, 1994). To be able to cope with practical dataset sizes, heuristics have been developed that are used for all but the the smallest data sets.

The most commonly used heuristic methods are based on the tree-based progressive alignment strategy (Hogeweg & Hesper, 1984; Feng & Doolittle, 1987; Taylor, 1988) with ClustalW (Thompson *et al.*, 1994) being the most widely used implementation. The idea is to establish an initial order for joining the sequences, and to follow this order in gradually building up the alignment. Many implementations use an aproximation of a phylogenetic tree between the sequences as a guide tree that dictates the alignment order. Although appropriate for many alignment problems, the progressive strategy suffers from its greediness. Errors made in the first alignments during the progressive protocol cannot be corrected later as the remaining sequences are added in. Attempts to minimize such alignment errors have generally been targeted at global sequence weighting (Altschul *et al.*, 1989; Thompson *et al.*, 1994), where the contribution of individual sequences are weighted during the alignment process. However, such global sequence weighting schemes carry the risk of propagating rather than reducing error when used in progressive multiple alignment strategies (Heringa, 1999).

The main alternative to progressive alignment is the simultaneous alignment of all the sequences. Two such implementations are available, MSA (Lipman *et al.*, 1989) and DCA (Stoye *et al.*, 1997). Both methods are based on the Carillo and Lipman algorithm (Carrillo & Lipman, 1988) to limit computations to a small area in the multi-dimensional search matrix. They nonetheless remain an extremely CPU- and memory-intensive approach, applicable only to about 9 sequences of average length for the fastest implementation (DCA). Iterative strategies (Hogeweg & Hesper, 1984; Gotoh, 1996; Notredame & Higgins, 1996; Heringa, 1999, 2002) are an alternative to optimise multiple alignments by reconsidering and correcting those made during preceding iterations. Although such iterative strategies do not provide any guarantees about finding optimal solutions, they are reasonably robust and much less sensitive to the number of sequences than their simultaneous counterparts.

All of these techniques perform *global* alignment and match sequences over their full lengths. Problems with this approach can arise when highly dissimilar sequences are compared. In such cases global alignment techniques might fail to recognize highly similar internal regions because these may be overshadowed by dissimilar regions and high gap penalties normally required to achieve proper global matching. Moreover, many biological sequences are modular and show shuffled domains (Heringa and Taylor, 1997), which can render a global alignment of two complete sequences meaningless. The occurrence of varying numbers of internal sequence repeats (Heringa, 1998) can also severely limit the applicability of global methods. In general, when there is a large difference in the lengths of two sequences to be compared, global alignment routines become unwarranted. To address these problems, Smith and Waterman (1981) early on developed a so-called *local* alignment technique in which the most similar regions in two sequences are selected and aligned. For multiple sequences, the main automatic local alignment methods include the Gibbs sampler (Lawrence *et al.*, 1993), MEME (Bailey and Elkan, 1994) and Dialign2 (Morgenstern, 1999). These programs often perform well when there is a clear block of ungapped alignment shared by all of the sequences. However, performance on representative sets of test cases is poor when compared with global methods (Thompson *et al.*, 1999; Notredame *et al.*, 2000).

Some recent methods have been developed in which global and local alignment is combined to optimise multiple sequence alignment (Notredame *et al.*, 2000; Heringa, 2002), where, the popular method T-Coffee (Notredame *et al.*, 2000) and the Praline technique (Heringa, 2002) are robust and sensitive implementations. Figure M.5 shows an example of an alignment of 13 flavodoxin sequences and the divergent sequence of the chemotaxis protein cheY. Other popular methods with a good balance between speed and accuracy

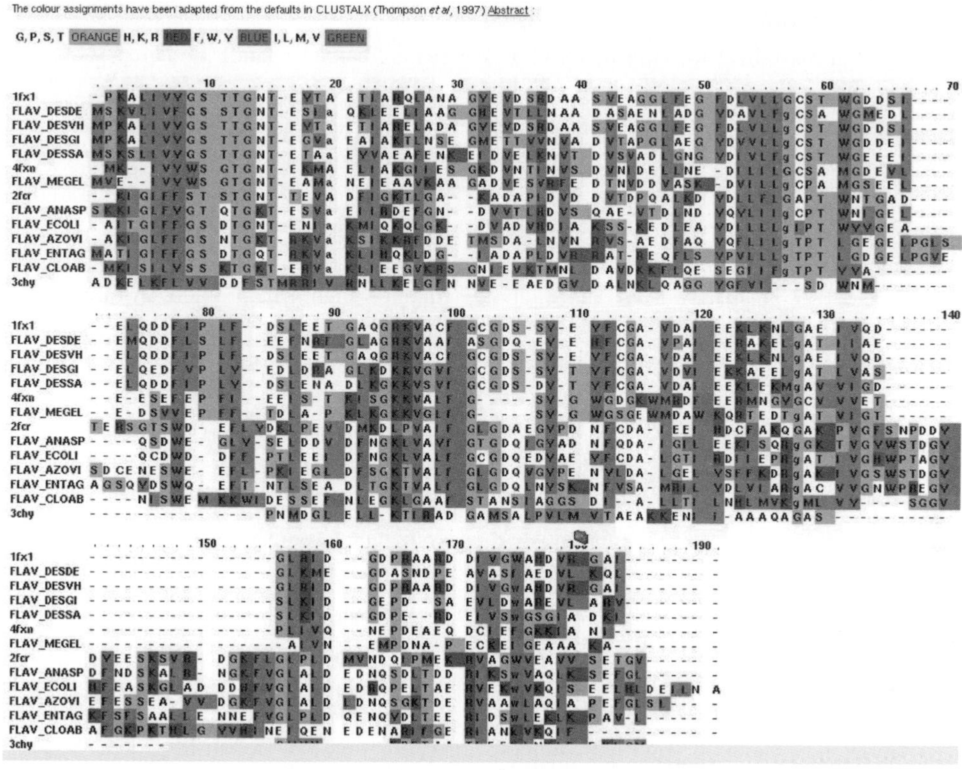

Figure M.5 Multiple Sequence Alignment of flavodoxin and cheY sequences. The alignment is comprised of 13 flavodoxin and the distantly related cheY sequence (bottom sequence). The alignment was generated using Praline (Heringa, 2002).

include MUSCLE (Edgar, 2004), Kalign (Lassmann and Sonnhammer, 2005), Probcons (Do et al, 2005) and Clustal Omega (Sievers et al., 2011). The latter method is able to align extremely large sequence sets with high accuracy.

Related websites

ClustalW2, v2.1	http://www.clustal.org
DIALIGN 2.2.1	http://dialign.gobics.de/
FSA 1.15.5	http://sourceforge.net/projects/fsa/
Kalign 2.04	http://msa.sbc.su.se/cgi-bin/msa.cgi
MAFFT 6.857	http://mafft.cbrc.jp/alignment/software/source.html
MSAProbs 0.9.4	http://sourceforge.net/projects/msaprobs/files/
MUSCLE version 3.8.31	http://www.drive5.com/muscle/downloads.htm
PRALINE P2	http://www.ibi.vu.nl/programs/pralinewww/
PROBCONS version 1.12	http://probcons.stanford.edu/download.html
T-Coffee Version 8.99	http://tcoffee.crg.cat/

M

Further reading

Altschul SF, Carrol RJ, Lipman DJ (1989) Weights for data related by a tree. *J. Mol. Biol.* 207: 647–653.

Bailey TL, Elkan C (1994) Fitting a mixture model by expectation maximization to discover motifs in biopolymers. In: *Proceedings of the second international conference on Intelligent Systems for Molecular Biology*. pp. 28–36. AAAI Press.

Benner SA, Cohen MA, Gonnet GH (1992) Response to Barton's letter: computer speed and sequence comparison. *Science* 257: 609–610.

Carillo H, Lipman DJ (1988) The multiple sequence alignment problem in biology. *SIAM J. Appl. Math*. 48: 1073–1082.

Dayhoff MO, Schwartz RM, Orcutt BC (1978) A model of evolutionary change in proteins. In: *Atlas of Protein Sequence and Structure*, Dayhoff MO (ed.), vol.5, suppl.3, Washington, DC (Natl.Biomed.Res.Found.), pp. 345–352.

Do CB, Mahabhashyam MSP, Brudno M, Batzoglou S (2005) ProbCons: probabilistic consistency-based multiple sequence alignment. *Genome Res* 15: 330–340.

Edgar RC (2004) MUSCLE: multiple sequence alignment with high accuracy and high throughput. *Nucleic Acids Res* 32: 1792–1797.

Feng DF, Doolittle RF (1987) Progressive sequence alignment as a prerequisite to correct phylogenetic trees. *J. Mol. Evol.* 21: 112–125.

Gotoh O (1996) Significant improvement in accuracy of multiple protein sequence alignments by iterative refinement as assessed by reference to structural alignments. *J. Mol. Biol.* 264: 823–838.

Henikoff S, Henikoff JG (1992) Amino acid substitution matrices from protein blocks. *Proc. Natl. Acad. Sci. USA* 89: 10915–10919.

Heringa J (1998) Detection of internal repeats: how common are they? *Curr. Opin. Struct. Biology* 8: 338–345.

Heringa J (1999) Two strategies for sequence comparison: profile-preprocessed and secondary structure-induced multiple alignment. *Comput. Chem.* 23: 341–364.

Heringa J (2002) Local weighting schemes for protein multiple sequence alignment. *Comput. Chem.* 26: 459–477.

Heringa J, Taylor WR (1997) Three-dimensional domain duplication, swapping and stealing. *Curr. Opin. Struct. Biol.* 7: 416–421.

Hogeweg P, Hesper B (1984) The alignment of sets of sequences and the construction of phyletic trees: an integrated method. *J. Mol. Evol.* 20: 175–186.

Lassmann T, Sonnhammer ELL (2005) Kalign – an accurate and fast multiple sequence alignment algorithm. *BMC Bioinformatics* 6: 298.

Lawrence CE, Altschul SF, Boguski MS, Liu JS, Neuwald AF, Wootton JC (1993) Detecting subtle sequence signals: a Gibbs sampling strategy for multiple alignment. *Science* 262: 208–214.

Lipman DJ, Altschul SF, Kececioglu JD (1989) A tool for multiple sequence alignment. *Proc. Natl. Acad. Sci. USA* 86: 4412–4415.

Morgenstern B (1999) DIALIGN 2: improvement of the segment-to-segment approach to multiple sequence alignment. *Bioinformatics* 15: 211–218.

Notredame C, Higgins DG (1996) SAGA: sequence alignment by genetic algorithm. *Nucleic Acids Res.* 24: 1515–1524.

Notredame C, Higgins DG, Heringa J (2000) T-Coffee: a novel method for fast and accurate multiple sequence alignment, *J. Mol. Biol.* 302: 205–217.

Sander C, Schneider R (1991) Database of homology derived protein structures and the structural meaning of sequence alignment. *Proteins* 9: 56–68.

Sievers F, Wilm A, Dineen DG, et al. (2011) Fast, scalable generation of high-quality protein multiple sequence alignments using Clustal Omega. *Mol. Syst. Biol.* 7:539.

Stoye J, Moulton V, Dress AWM (1997) DCA: An efficient implementation of the divide-and-conquer approach to simultaneous multiple sequence alignment. *Comput. Appl. Biosci.* 13: 625–626.

Taylor WR (1988) A flexible method to align large numbers of biological sequences. *J. Mol. Evol.* 28: 161–169.

Thompson JD, Higgins DG, Gibson TJ (1994) CLUSTAL W: improving the sensitivity of progressive multiple sequence alignment through sequence weighting, positions-specific gap penalties and weight matrix choice. *Nucleic Acids Res.* 22: 4673–4680.

Thompson JD, Plewniak F, Poch O (1999) A comprehensive comparison of multiple sequence alignment programs. *Nucl. Acids Res.* 27: 2682–2690.

See also Alignment, Tree-based Progressive Alignment, Global Alignment, Local Alignment

MULTIPLE HIERARCHY (POLYHIERARCHY)

Robert Stevens

A hierarchy in which, unlike a tree, a concept may have multiple parents. A multiple hierarchy may be seen in Figure D.3 under Directed Acyclic Graph.

See also Directed Acyclic Graph

MULTIPLEX SEQUENCING

Matthew He

Multiplex sequencing is an approach to high-throughput sequencing that uses several pooled DNA samples run through gels simultaneously and then separated and analysed.

MULTIPOINT LINKAGE ANALYSIS

Mark McCarthy and Steven Wiltshire

The determination of evidence for linkage which simultaneously takes into account the information arising from multiple typed loci.

Multipoint linkage analysis can be applied to the analysis of multiple marker loci alone, as in the case in marker map construction; or to the mapping of a putative disease-susceptibility locus onto a map of markers, in the case in linkage analysis for Mendelian disease and multifactorial/complex traits. The two most commonly used algorithms in linkage analysis – the Elston-Stewart and Lander-Green algorithms – calculate the likelihood of the data at each position on a marker map, conditional on the genotypes of all the markers in the map, although the L-G algorithm can handle substantially more markers than the E-S algorithm. A multipoint LOD score is obtained at each position on the map, and the peak in the resulting multipoint linkage profile identifies the position of the maximum evidence for linkage. By combining evidence from several markers, each of which may, in isolation, provide limited information, multipoint linkage analysis can provide enhanced evidence for, and localisation of, susceptibility loci when compared to two-point analyses. .

Examples: Vionnet and colleagues (2000) report the results of a genome scan for type 2 diabetes loci in French pedigrees that employs a multipoint linkage analysis.

Related websites
Programs for linkage analysis:

LINKAGE	ftp://linkage.rockefeller.edu/software/linkage/
GENEHUNTER	http://www-genome.wi.mit.edu/ftp/distribution/software/genehunter
MERLIN	http://www.sph.umich.edu/csg/abecasis/merlin/reference.html
Jurg Ott's Linkage page	http://linkage.rockefeller.edu/

Further reading

Ott J (1999) *Analysis of human genetic linkage*. Baltimore: The John's Hopkins University Press pp114–150.

Vionnet N, et al. (2000) Genomewide search for type 2 diabetes-susceptibility genes in French Whites: evidence for a novel susceptibility locus for early-onset diabetes on chromosome 3q27-qter and independent replication of a type 2-diabetes locus on chromosome 1q21-q24. *Am. J. Hum. Genet.* 67: 1470–1480.

See also Linkage Analysis, Multifactorial Trait, Genome Scans

MURCKO FRAMEWORK, *SEE* BEMIS AND MURCKO FRAMEWORK.

MUTATION MATRIX, *SEE* AMINO ACID EXCHANGE MATRIX.

N

N-terminus (amino terminus)

Naïve Bayes, *see* Bayesian Classifier.

National Center for Biotechnology Information, *see* NCBI.

Natural Selection

NCBI (National Center for Biotechnology Information)

NDB, *see* Nucleic Acid Database.

Nearest Neighbor Methods

Nearly Neutral Theory, *see* Neutral Theory.

Needleman-Wunsch Algorithm

Negative Selection, *see* Purifying Selection.

Negentropy (Negative Entropy)

Neighbor-Joining Method

Network (Genetic Network, Molecular Network, Metabolic Network)

Neural Network (Artificial Neural Network, Connectionist Network, Backpropagation Network, Multilayer Perceptron)

Neutral Theory (Nearly Neutral Theory)

Newton-Raphson Minimization, *see* Energy Minimization.

Next Generation DNA Sequencing

Next Generation Sequencing, De Novo Assembly, *see* De Novo Assembly in Next Generation Sequencing.

Nit

NMR (Nuclear Magnetic Resonance)

Node, *see* Phylogenetic Tree.

Noise (Noisy Data)

Non-Crystallographic Symmetry, *see* Space Group.

Non-Parametric Linkage Analysis, *see* Allele-Sharing Methods.

Non-Synonymous Mutation

NOR, *see* Nucleolar Organizer Region.

NoSQL, *see* Database.

Nuclear Intron, *see* Intron.

Nuclear Magnetic Resonance, *see* NMR.

Nucleic Acid Database (NDB)

Nucleic Acid Sequence Databases

Nucleolar Organizer Region (NOR)

Nucleotide Base Codes, *See* IUPAC-IUB Codes.

Concise Encyclopaedia of Bioinformatics and Computational Biology, Second Edition.
Edited by John M. Hancock and Marketa J. Zvelebil.

N-TERMINUS (AMINO TERMINUS)

Roman Laskowski and Tjaart de Beer

In a polypeptide chain the N-terminus is the end-residue that has a free amino group (NH_3). The other end is referred to as the C-terminus. The direction of the chain is defined as running from the N- to the C-terminus. N-terminal refers to being of or relating to, the N-terminus.

See also Polypeptide, Residue, C-terminus

NAÏVE BAYES, *SEE* BAYESIAN CLASSIFIER.

NATIONAL CENTER FOR BIOTECHNOLOGY INFORMATION, *SEE* NCBI.

NATURAL SELECTION

A.R. Hoelzel

Natural selection is the non-random, differential survival and reproduction of phenotypes in a population of organisms of a given species, whereby organisms best suited to their environment contribute a higher proportion of progeny to the next generation.

The idea of natural selection was first mentioned by Charles Darwin in his notebooks in 1838, and later described in detail as the motive force behind evolutionary change in his 1859 volume, *The Origin of Species by Means of Natural Selection or the Preservation of Favoured Races in the Struggle for Life.*

This process assumes that phenotypically expressed traits that can be selected in natural populations have a heritable basis, later established to be genes encoded by DNA. Selection acts by changing allele frequencies in natural populations, and for a single locus, two-allele system in Hardy-Weinberg equilibrium for a diploid organism this change can be expressed as:

$$\Delta_s p = \frac{pq[p(w_{11} - w_{12}) + q(w_{12} - w_{22})]}{p^2 w_{11} + 2pq w_{12} + q^2 w_{22}}$$

where w_{11}, w_{12} and w_{22} represent the viabilities of A_1A_1, A_1A_2 and A_2A_2 genotypes, respectively, and p is the allele frequency of A_1, while q is the allele frequency of A_2. These changes lead to the adaptation of organisms to their environment. Selection can be directional when one allele is favoured, balanced when selection maintains polymorphism in a population (e.g. by overdominance or frequency dependent selection), or disruptive when homozygotes are favoured over the heterozygote condition.

N

Further reading

Darwin C (1859) *The Origin of Species by Means of Natural Selection, or The Preservation of Favoured Races in the Struggle for Life.*

Ridley M (1996) *Evolution*, Blackwell Science, Oxford.

See also Kin Selection

NCBI (NATIONAL CENTER FOR BIOTECHNOLOGY INFORMATION)

Guenter Stoesser, Malachi Griffith and Obi L. Griffith

The National Center for Biotechnology Information (NCBI) develops and provides information systems for molecular biology, conducts research in computational biology and develops software tools for analysing genome data.

Analysis resources include BLAST, Clusters of Orthologous Groups (COGs) database, Consensus CDS, Conserved Domain Database, Database of Expressed Sequence Tags (dbEST), Database of Genomic Structural Variation (dbVar), Electronic PCR, Entrez Genomes, Gene Expression Omnibus (GEO), HomoloGene, Human Genome Sequencing, Human MapViewer, Human Mouse Homology Map, Molecular Modeling Database (MMDB), Online Mendelian Inheritance in Man (OMIM), OrfFinder, RefSeqGene, SAGEmap, Sequence Read Archive (SRA), Single Nucleotide Polymorphisms (dbSNP) database, and UniGene. Retrieval resources include Entrez Gene, GenBank, PubMed, PubChem, LocusLink and the Taxonomy Browser.

NCBI's mission is to develop information technologies to aid in the understanding of fundamental molecular and genetic processes that control health and disease. NCBI is located and maintained at the National Institute of Health (NIH) in Bethesda, USA.

Related websites

NCBI homepage	http:www.ncbi.nlm.nih.gov
GenBank	http://www.ncbi.nlm.nih.gov/Genbank/
Literature Databases	http://www.ncbi.nlm.nih.gov/Literature/
Genome Biology	http://www.ncbi.nlm.nih.gov/Genomes/
Data Mining Tools	http://www.ncbi.nlm.nih.gov/Tools/

Further reading

Altschul SE, et al. (1990) Basic local alignment search tool *J Mol Biol.* 215: 403–410.

Edgar R, et al. (2002) Gene Expression Omnibus: NCBI gene expression and hybridization array data repository. *Nucleic Acids Res.* 30: 207–210.

Pruitt K, Maglott D (2001) RefSeq and LocusLink: NCBI gene-centered resources. *Nucleic Acids Res.* 29: 137–140.

Sayers EW, et al. (2012) Database resources of the National Center for Biotechnology Information: 2012 update *Nucleic Acids Res.* 40: D13–25.

Schuler GD, et al. (1996) Entrez: molecular biology database and retrieval system. *Methods Enzymol.* 266: 141–162.

NDB, *SEE* NUCLEIC ACID DATABASE.

NEAREST NEIGHBOR METHODS

Patrick Aloy

The *nearest neighbor* is a classification algorithm that assigns a test instance to the class of a 'nearby' example whose class is known. When applied to secondary structure prediction, this means that the secondary structure state of the central residue of a test segment is assigned according to the secondary structure of the closest homolog of known structure.

A key element in any nearest-neighbor prediction algorithm is the choice of a scoring system for evaluating segment similarity. For such a purpose, a variety of approaches have been used (e.g. similarity matrices, neural networks) with different success.

Some methods that use this approach (e.g. NNSSP) have achieved three-state-per-residue accuracies of 72.2%.

Related website

NNSSP	http://bioweb.pasteur.fr/seqanal/interfaces/nnssp-simple.html

Further reading

Salamov AA, Solovyev VV (1995) Prediction of protein secondary structure by combining nearest-neighbor algorithms and multiple sequence alignments. *J Mol Biol* 247: 11–15.

Yi TM, Lander ES (1993) Protein secondary structure prediction using nearest-neighbor methods. *J Mol Biol* 232: 1117–1129.

See also Secondary Structure Prediction, Neural Network

NEARLY NEUTRAL THEORY, *SEE* NEUTRAL THEORY.

NEEDLEMAN-WUNSCH ALGORITHM

Jaap Heringa

The technique to calculate the highest scoring or optimal alignment is generally known as the dynamic programming (DP) technique. While the physicist Richard Bellman first conceived DP and published a number of papers on the topic between 1955 and 1975, Needleman and Wunsch (1970) introduced the technique to the biological community and their paper remains among the most cited in the area. The technique falls in the class of global alignment techniques, which align protein or nucleotide sequences over their full lengths. The original Needleman-Wunsch algorithm penalizes the inclusion of gaps in the alignment through a single gap penalty for each gap, irrespective of how many amino acids are spanned by the gap, and also penalizes end gaps.

N

Further reading

Needleman SB, Wunsch CD (1970) A general method applicable to the search for similarities in the amino acid sequence of two proteins. *J. Mol. Biol.* 48: 443–453.

NEGATIVE SELECTION, *SEE* PURIFYING SELECTION.

NEGENTROPY (NEGATIVE ENTROPY)

Thomas D. Schneider

The term negentropy was defined by Brillouin (Brillouin 1962, p. 116) as 'negative entropy', $N = -S$. Supposedly living creatures feed on negentropy from the sun. However it is impossible for entropy to be negative, so negentropy is always a negative quantity. The easiest way to see this is to consider the statistical-mechanics (Boltzmann) form of the entropy equation:

$$S \equiv -k_B \sum_{i=1}^{\Omega} P_i \ln P_i \left(\frac{joules}{K \bullet microstate} \right)$$

where k_B is Boltzmann's constant, Ω is the number of microstates of the system and P_i is the probability of microstate i. Unless one wishes to consider imaginary probabilities it can be proven that S is positive or zero. Rather than saying 'negentropy' or 'negative entropy', it is clearer to note that when a system dissipates energy to its surroundings its entropy decreases. So it is better to refer to $-\Delta S$ (a negative change/decrease in entropy).

Further reading

Brillouin L (1962) *Science and Information Theory*. Academic Press, Inc., New York, second edition.

NEIGHBOR-JOINING METHOD

Sudhir Kumar and Alan Filipski

A method for inferring phylogenetic trees based on the minimum evolution principle.

Neighbor-joining is a stepwise method that begins with a star tree and successively resolves the phylogeny by joining pairs of taxa in each step such that the selected pairing minimizes the sum of branch lengths. Thus, the ME principle is applied locally at each step and a final tree is produced along with the estimates of branch lengths. This method does not require the assumption of molecular clock and is known to be computationally efficient and phylogenetically accurate for realistic simulated data.

Software

Felsenstein J (1993) *PHYLIP: Phylogenetic Inference Package*. Seattle, WA, University of Washington.

Kumar S, et al. (2001) MEGA2: molecular evolutionary genetics analysis software. *Bioinformatics* 17: 1244–1245.

Swofford DL (1998) *PAUP*: Phylogenetic Analysis Using Parsimony (and other methods)*. Sunderland, MA., Sinauer Associates.

Further reading

Saitou N, Nei M (1987) The Neighbor-Joining Method – a New Method for Reconstructing Phylogenetic Trees. *Mol Biol Evol* 4: 406–425.

Studier JA, Keppler KJ (1988) A note on the neighbor-joining algorithm of Saitou and Nei. *Mol Biol Evol.* 5: 729–31.

See also Minimum Evolution Principle

NETWORK (GENETIC NETWORK, MOLECULAR NETWORK, METABOLIC NETWORK)

Denis Thieffry

Biological macromolecules (DNA fragments, proteins, RNAs, other (bio)chemical compounds) interact with each other, affecting their respective concentrations, conformations or activities. A collection of interacting molecular components thus forms a (often intertwined) molecular network, endowed with emerging dynamical properties.

One speaks of a genetic interaction occurring between two genes when the perturbation (deletion, over-expression) of the expression of the first gene affects that of the second, something which can be experimentally observed at the phenotypic or the molecular level. Such interactions can be direct or indirect, i.e. involving intermediate genes. Turning to the molecular level, one can further distinguish between different types of molecular interaction: Protein-DNA interactions (e.g., transcriptional regulation involving a regulatory protein and a DNA cis-regulatory region near the regulated gene), Protein-RNA interactions (as in the case of post-transcriptional regulatory interactions), and Protein-Protein interactions (e.g., phosphorylation of a protein by a kinase, formation of multimeric complexes, etc.).

Molecular interactions between biological macromolecules are often represented as a graph, with genes as vertices and interactions between them as edges. Information on gene interactions can also be obtained directly from genetic experiments (mutations, genetic crosses, etc.). Regardless of the types of experimental evidence, one often refers to genetic networks, though the corresponding interactions usually involve various types of molecular components (e.g., proteins, RNA species, etc.). The genetic networks described in the literature predominantly encompass transcription regulatory interactions. At the level of the organism, the complete set of regulatory interactions forms the regulome.

Regulatory interactions of various types (protein-protein interactions, transcriptional regulation, etc.) form so called regulatory cascades. Signaling molecule are sensed by membrane receptors, leading to (cross-) phoshorylation by kinases (or other types of protein-protein interactions), ultimately leading to the translocation or the change of activity of transcription factors, which will in turn affect the transcription of specific subsets of genes in the nucleus. Different regulatory cascades (each forming a graph, typically with a tree topology) can share some elements, thus leading to further regulatory integration, something, which can be represented in terms of acyclic graphs. When regulatory cascades ultimately affect the expression of genes coding for some of their crucial components (e.g., membrane receptors), the system is then better represented by (cyclic) directed graphs.

In the context of metabolism or (to a lesser extent) gene regulation, one speaks of regulatory feedback when the product of a reaction participates in the regulation of the activity or of the expression of some of the enzymes at work in the reaction (or pathway). Such regulatory feedback can be inhibitory (negative feedback, e.g. the inhibition of a reaction by its product) or activatory (positive feedback). This notion of regulatory feedback has to be distinguished from that of a regulatory (feedback) circuit defined in the context of regulatory graphs. A feedback regulatory circuit often counts a regulatory feedback among its edges.

The term module is used in various contexts with varying meanings in biology and bioinformatics. In the context of metabolic or genetic networks, it corresponds to the delineation of interactive structures associated with specific functional processes. From a graph theoretical perspective, in contrast to regulatory cascades - corresponding to a tree graph topology (devoid of feedback circuits), regulatory modules correspond to (strongly) connected components, thus often to sets of intertwined feedback circuits.

Cell metabolism involves numerous reactions sharing some of their reactants, products and catalytic enzymes, thus also forming complex networks, including cycles such as the Krebbs cycle. The full set of metabolic reactions occurring in one organism is called the metabolic network of the organism, or metabolome.

Information on genetic and metabolic networks is scarcely available in general molecular databases. Furthermore, when the information is present, it lacks the appropriate higher-level integration needed to allow queries dealing with multigenic regulatory modules or pathways. As a consequence, a series of research groups or consortia are developing dedicated databases to integrate genetic regulatory data or metabolic information. See Table N.1 for a list of the prominent (publicly available) genetic databases and metabolic databases.

Molecular, metabolic and genetic (sub)graphs may be characterized by different network structures, depending on the pattern of interactions between the genes or molecular species involved. Formally, one can make rigorous distinctions on the basis of graph theory (see the entry on graph topology).

When facing a metabolic or genetic network, various biological questions can be phrased in terms of comparison of interactive structures. One can for example check whether a set of genes, each paralogous to a gene of another set, also present a conserved interaction pattern. When comparing orthologous or paralogous sets of genes across different organisms, one can evaluate the conservation of regulatory networks or metabolic pathways beyond the conservation of the structure or sequence of individual components (genes or molecular species). Formally, these comparisons can be expressed in terms of graph theoretic notions such as (sub)graph isomorphism (similar interactive structures) or homeomorphism (similar topologies).

Table N.1 Examples of Databases Integrating Genetic, Molecular or Metabolic Interaction Data

Name	Types of data	URL
Amaze	Biochemical pathways	http://www.ebi.ac.uk/research/pfmp/
Ecocyc/Metacyc	Metabolic pathways	http://ecocyc.panbio.com/
KEGG	Metabolic pathways	http://www.genome.ad.jp/kegg/
TransPath	Signal transduction pathways	http://transpath.gbf.de/
BIND	Protein interactions and complexes	http://www.bind.ca/
GeneNet	Gene networks	http://wwwmgs.bionet.nsc.ru/mgs/systems/genenet/
CSNDB	Cell signaling networks	http://geo.nihs.go.jp/csndb/

A molecular, metabolic or genetic network can be displayed and analysed on the basis of the use of various types of representations. The most intuitive description consists simply in using diagrams such as those found in many biological articles. But such a graphical description does not usually conform to strictly defined standards, making difficult any computer implementation or analysis. When turning to more standardized and formal approaches, one can distinguish three main (related) formal representations: the first referring to graph theory (see interaction graph), the second referring to the matrix formalism (i.e. the use of matrix to represent interactions between genes, each corresponding to a specific column and a specific row), and the last referring to dynamical system theory and using different types of (differential) equations (see mathematical modeling).

Molecular, metabolic and genetic networks can be visualized via dedicated Graphical User Interfaces (GUIs). These network visualization tools offer various types of functions, such as zooming or editing (selection, deletion, insertion, and displacement of objects). Various software packages are under development for this.

The variation of metabolic activity or of gene expression across time constitutes what is called network dynamics. The term is often used to designate the full set of temporal behavior associated with a network, depending on different signals or perturbations. Formally, this behavior can be represented by time plots (see Ordinary Differential Equations) or dynamical graphs. Network state usually refers to the levels of activity or concentration of all components of a network at a given time.

In contrast with the derivation of the temporal behavior from prior knowledge of a regulatory (metabolic or genetic) network, the notion of reverse engineering refers to the derivation of the network or of parts of its regulatory structure on the basis of dynamical data (i.e., a set of characterized states of the network). Reverse engineering methods can be developed in the context of various formal approaches, for example by fitting a differential model to dynamical data using computer optimization methods (see also Bayesian Networks).

N

Further reading

Bower JM, Bolouri H (2001) *Computational Modeling of Genetic and Biochemical Networks*. MIT Press.

Goldbeter A (1997) *Biochemical Oscillations and Cellular Rhythms: The Molecular Bases of Periodic and Chaotic Behaviour*. Cambridge University Press.

Voit EO (2000) *Computational Analysis of Biochemical Systems: A Practical Guide for Biochemists and Molecular Biologists*. Cambridge University Press.

See also Graph Representation of Genetic, Molecular and Metabolic Networks, Mathematical Modeling of Molecular/Metabolic/Genetic Networks

NEURAL NETWORK (ARTIFICIAL NEURAL NETWORK, CONNECTIONIST NETWORK, BACKPROPAGATION NETWORK, MULTILAYER PERCEPTRON)

Nello Cristianini

A general class of algorithms for machine learning, originally motivated by analogy with the structure of neurons in the brain (see Figure N.1).

A Neural Network can be described as a parametrized class of functions, specified by a weighted graph (the network's architecture). The weights associated with the edges of the graph are the parameters, each choice of weights identifying a function.

For directed graphs, we can distinguish recurrent architectures (containing cycles) and feed-forward architectures (acyclic). A very important special case of feed-forward networks is given by layered networks, in which the nodes of the graph are organized into an ordered series of disjoint classes (the layers), such that connections are possible only between elements of two consecutive classes and following the natural order. The weight between the unit k and the unit j of a network is indicated with w_{kj} and we assume that all elements of a layer are connected to all elements of the successive layer (fully connected architecture). In this way, the connections between two layers can be represented by a weight matrix W, whose entry jk corresponds to the connection between node j and node k in successive layers (see Figure N.1). It is customary to call input and output layers the first and the last ones, and hidden layers all the remaining.

In a layered network, the function is computed sequentially, assigning the value of the argument to the input layer, then calculating the activation level of the successive layers as described below, until the output layer is reached. The output of the function computed by the network is the activation value of the output unit.

All units in a layer are updated simultaneously, all the layers are updated sequentially, based on the state of the previous layer. Each unit k calculates its value y_k by a linear combination of the values at the previous layer, x, followed by a nonlinear transformation $t{:}\mathrm{R} \to \mathrm{R}$, as follows: $y_k = t(Wx)$ where t is called the transfer function and for which a common choice is the logistic function $f(z) = \frac{1}{1+e^{-z}}$.

Notice that the input/output behavior of the network is determined by the weights (or, in other words, each neural network represents a class of nonlinear functions parametrized by the weights), and training the network amounts to automatically choosing the values of the weights. A Perceptron can be described as a network of this type with no hidden units.

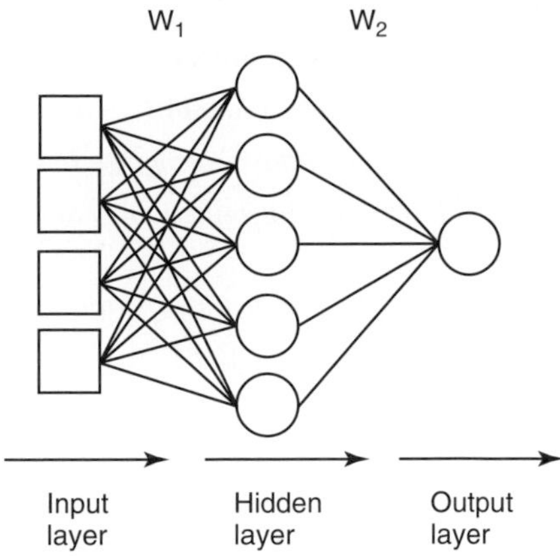

Figure N.1 A feed-forward neural network or multi-layer perceptron.

It can also be seen as the building block of complex networks, in that each unit can be regarded as a Perceptron (if instead of the transfer function one uses a threshold function, returning Boolean values). Layered Feed Forward Neural Networks are often referred to as Multilayer Perceptrons.

Given a (labeled) training set of data and a fixed error function for the performance of the network, the training of a neural network can be done by finding those weights that minimize the network's error on such sample (i.e. by fitting the network to the data). This can be done by gradient descent, if the error function is differentiable, by means of the Back Propagation algorithm.

In the parameter space, one can define a cost function that associates each configuration with a given performance on the training set. Such function is typically non-convex, so that it can only be locally minimized by gradient descent. Back propagation provides a way to compute the necessary gradients, so that the network finds a local minimum of the training error with respect to network weights.

The chain rule of differentiation is used to compute the gradient of the error function with respect to the weights. This gives rise to a recursion proceeding from the output unit to the input, hence the name back-propagation.

If y_i is the value of the *i-th* unit, for each w_{kj} connecting it to the previous layer's units one can write the partial derivative of the error function as $\frac{\partial E}{\partial w_{ij}} = \frac{\partial E}{\partial y_i}\frac{\partial y_i}{\partial w_{ij}} = \frac{\partial E}{\partial y_i} t'(x_i)x_i = \varepsilon_i$ where $t()$ is the transfer function defined before, and hence the update for such weight will be $\Delta w_{ij} = -\eta\varepsilon_i y_i$, where □ is a parameter known as the learning rate. Any differentiable error function can be used.

Among the main problems of neural networks are that this training algorithm is only guaranteed to converge to a local minimum, and the solution is affected by the initial conditions, since the error function is in general non convex. Also problematic is the design of the architecture, often chosen as the result of trial and error. Some such problems have been overcome by the introduction of the related method of Support Vector Machines.

Further reading

Other types of networks arise from different design choices. For example Radial Basis Function networks use a different transfer function; Kohonen networks are used for clustering problems, and

N

Hopfield networks for combinatorial optimization problems. Also different training methods exist. Good starting points for a literature search are the following books.

Baldi P, Brunak S (2001) *Bioinformatics: A Machine Learning Approach*. MIT Press.

Bishop C (1996) *Neural Networks for Pattern Recognition*. Oxford University Press.

Cristianini N, Shawe-Taylor J (2000) *An Introduction to Support Vector Machines* Cambridge University Press.

Duda RO, Hart PE, Stork DG (2001) *Pattern Classification*, John Wiley & Sons, USA.

Mitchell T *Machine Learning*, McGraw Hill, 1997.

NEUTRAL THEORY (NEARLY NEUTRAL THEORY)

A.R. Hoelzel

Theory of evolution suggesting that most evolutionary change at the molecular level is caused by the random genetic drift of mutations that are selectively neutral or nearly neutral.

Proposed by Kimura (1968) the neutral theory was controversial from the start and has undergone numerous modifications over the years. Although it has been called 'non-Darwinian' because of its emphasis on random processes, it does not actually contradict Darwin's theory. It proposes that most substitutions have no influence on survival, but does not deny the importance of selection in shaping the adaptive characteristics of many morphological, behavioural and life history characteristics.

The evidence in support of the neutral theory included the observation that the rate of amino acid substitution in vertebrate lineages was remarkably constant for different loci, which in turn led to the proposal of a molecular clock. The observed rate of change by substitution was also noted to be similar to the estimated mutation rate, consistent with theoretical expectations. One modification to the theory, especially to account for discrepancies between the expected and observed relationship between amino acid substitution rate and generation time, suggested that most mutations are slightly deleterious, as opposed to neutral (e.g. Ohta, 1973).

Further reading

Kimura M (1968) Evolutionary rate at the molecular level. *Nature*, 217: 624–626.

Ohta T (1973) Slightly deleterious mutant substitutions in evolution. *Nature*, 246: 96–98.

Ohta T (1992) The nearly neutral theory of molecular evolution. *Annu Rev Ecol Systematics* 23: 263–286.

See also Evolution, Natural Selection, Evolutionary Clock

NEWTON-RAPHSON MINIMIZATION, *SEE* ENERGY MINIMIZATION.

NEXT GENERATION DNA SEQUENCING

Stuart Brown

The determination of the sequence of a molecule of DNA has been a fundamental technology for biology research since the method was invented by Frederick Sanger in 1975. Improvements in sequencing were slow and incremental for more than a quarter of a century, despite substantial investment in efforts such as the Human Genome Project. The technologies used for DNA sequencing were revolutionized in the period from 2004 to 2008 by machines developed by the private companies 454 Life Sciences, Solexa, and Applied Biosystems. These Next Generation DNA Sequencers (NGS) use micro-fluidics, nanotechnology, and computer power to increase the production of DNA sequence data by 3 to 5 orders of magnitude compared to methods that use the traditional Sanger technology. Additional competitors with new variations on this technology continue to enter the marketplace. In every year since 2004, the cost per million bases of DNA sequenced on these machines has fallen more than fourfold per year. This influx of new DNA sequence has created data storage challenges for research institutions and public databases.

There are substantial differences in the details of technologies used by the competing manufacturers, but NGS technologies all use methods that simultaneously determine the sequence of DNA bases from many thousands (or millions) of different DNA templates in a single biochemical reaction volume. Each template molecule is affixed to a solid surface in a spatially separate location, and then clonally amplified to increase signal strength. Then the template molecules are copied, and the sequence is determined as each new base is added to the copy. Early versions of NGS machines produced short reads from each DNA sequence fragment (25–100 bp). These short reads made it more difficult to assemble *de novo* genomes and more difficult to collect any form of sequence data from repetitive regions of well known genomes. Read lengths have steadily increased with newer versions of NGS machines. Several manufacturers have introduced methods for paired-end sequencing, which reads sequence from both ends of each DNA fragment, allowing for improved computational methods for both *de novo* assembly and mapping of sequence reads to a reference genome.

There are many applications for NGS that range from sequencing single molecules of DNA to entire genomes. Methods for breaking a genome into many small DNA fragments and simultaneously sequencing all of the fragments have been optimized to the point that the sequence of an entire human genome can be determined on a single machine in less than a week. The use of clonally amplified single molecule templates also allows for the detection and quantitation of rare variant sequences in a mixed population, such as detecting a rare drug resistance mutation in an HIV sample from an infected patient. It is also possible to sequence an entire collection of mRNA from a tissue sample (RNA-seq), allowing for a direct quantitation of gene expression by counting RNA molecules from each gene. DNA extracted from environmental samples can be sequenced in bulk to identify genes from all of the organisms present (Metagenomics). NGS has also been applied to the study of protein-DNA interactions and epigenetics, by sequencing DNA fragments bound to proteins that are selected by immunoprecipitation with a specific antibody (ChIP-seq).

One of the most common medical applications of NGS is to find disease causing mutations. These mutations might be related to a heritable disease in an infant, or somatic mutations in the tumors of a cancer patient. In either case, the use of NGS methods creates a challenge related to coverage and accuracy. When the genome is broken into random fragments and sequenced, not all positions on all chromosomes will be covered equally well. If the fragments are distributed in a purely random fashion (i.e. a Poisson distribution), then, the total length of sequenced fragments needs to be much greater than the length of the genome to be certain that all positions are covered. NGS machines are not perfectly accurate. In fact, the most popular machines produce 0.5 to 1% errors. Therefore, every difference from the standard reference genome discovered in a DNA fragment from a patient must be confirmed with several additional overlapping fragments. Even with very deep coverage, some false discovery of mutations and false negatives (mutations that fail to be discovered) may occur.

NEXT GENERATION SEQUENCING, DE NOVO ASSEMBLY, *SEE* DE NOVO ASSEMBLY IN NEXT GENERATION SEQUENCING.

NIT

Thomas D. Schneider

Natural units for information or uncertainty are given in nits.

If there are M messages, then $\ln(M)$ nits are required to select one of them, where ln is the natural logarithm with base e (=2.71828...). Natural units are used in thermodynamics where they simplify the mathematics. However nits are awkward to use because results are almost never integers. In contrast the bit unit is easy to use because many results are integer (e.g. log2 32 = 5) and these are easy to memorize. Using the relationship:

$$\frac{\ln(x)}{\ln(2)} = \log_2(x)$$

allows one to present all results in bits.

Related website

The appendix in the Primer on Information Theory gives a table of powers of two that is useful to memorize	http://alum.mit.edu/www/toms/paper/primer

Further reading

Schneider TD (1995) *Information Theory Primer*. http://alum.mit.edu/www/toms/paper/primer/

NMR (NUCLEAR MAGNETIC RESONANCE)

Liz Carpenter

Nuclear magnetic resonance (NMR) is a method for solving the three dimensional structures of macromolecules and small molecules in solution.

The nuclei of atoms have associated magnetic fields which are referred to as spin states. The spin state of an atom depends on the number of protons and neutrons in the nucleus. ^{12}C and ^{16}O both have I = 0, where I is the spin number. ^{1}H, ^{13}C and ^{15}N have I = ½ whereas ^{14}N and ^{2}H have I = 1. When nuclei with spin greater than zero are placed in a magnetic field the spin direction aligns with the magnetic field and an equilibrium state is obtained. If radio frequency pulses are then applied to the samples, higher energy states are produced and when these revert to the equilibrium state, radio frequency radiation is emitted. The exact frequency of the radiation emitted depends on the environment around the nucleus, so each nucleus will emit at a different frequency (unless the nuclei are exactly chemically equivalent).

In order to solve the structure of a macromolecule the types of radio frequency pulses used are varied and nature of the nuclei probed. Two useful types of NMR experiment involve COSY (correlation spectroscopy) which gives information about nuclei that are connected through bonds and nuclear Overhauser effect (NOE) experiments which give information about the through space interactions. Since these spectra on individual isotopes of a particular nucleus often give rise to overlapping peaks, it is necessary to perform two, three and four dimensional experiments using different isotopes of various nuclei. The isotope of hydrogen, carbon and nitrogen present in a macromolecule can be varied by producing the protein in bacteria that are grown in media rich in the required isotope. For example ^{12}C is replaced with ^{13}C and/or ^{13}N is replaced with ^{14}N.

Once spectra have been collected it is necessary to assign all the peaks to individual atoms and interactions between atoms. This is done using through-bond experiments such as COSY, TOCSY and 3- and 4-dimensional experiments. The NOEs (through space) provide information about which parts of the protein that are distant in the sequence are adjacent in the folded structure. Newer techniques exist that can provide longer-range interactions based on partial alignment of the molecules in solution, e.g. by use of lipid solutions. Once these distances have been measured, they can be used to generate a series of distance restraints to apply to the protein to form a folded structure. Usually structures are generated by various molecular mechanics techniques, obtaining an ensemble of structures that represent the experimental data.

Related websites

NMR teaching pages by David Gorenstein.	http://www.biophysics.org/btol/NMR.html
Gary Trammell's teaching web pages on NMR	http://www.uis.edu/~trammell/che425/nmr_theory/
Joseph Hornak's web pages	http://www.cis.rit.edu/htbooks/nmr/inside.htm

NODE, *SEE* PHYLOGENETIC TREE.

NOISE (NOISY DATA)

Thomas D. Schneider

Noise is a physical process that interferes with transmission of a message.

Shannon pointed out that the worst kind of noise has a Gaussian distribution. Since thermal noise is always present in practical systems, received messages will always have some probability of having errors.

Further reading

Shannon CE (1949) Communication in the presence of noise. *Proc. IRE*, 37: 10–21.

NON-CRYSTALLOGRAPHIC SYMMETRY, *SEE* SPACE GROUP.

NON-PARAMETRIC LINKAGE ANALYSIS, *SEE* ALLELE-SHARING METHODS.

NON-SYNONYMOUS MUTATION

Laszlo Patthy

Nucleotide substitution occurring in translated regions of protein-coding genes that alters the amino acid are called amino acid changing or non-synonymous mutations.

NOR, *SEE* NUCLEOLAR ORGANIZER REGION.

NOSQL, *SEE* DATABASE.

NUCLEAR INTRON, *SEE* INTRON.

NUCLEAR MAGNETIC RESONANCE, *SEE* NMR.

NUCLEIC ACID DATABASE (NDB)

Guenter Stoesser

The Nucleic Acid Database (NDB) assembles and distributes information about the crystal structures of nucleic acids.

The *Atlas of Nucleic Acid Containing Structures* highlights the special aspects of each structure in the NDB. Archives contain nucleic acid standards, nucleic acid summary information, software programs, coordinate files, and structure factor data.

Data for the crystal structures of nucleic acids are deposited using the AutoDep Input Tool. NDB maintains the mmCIF Web site; mmCIF (macromolecular Crystallographic Information File) is the IUCr-approved data representation for macromolecular structures. Databases for monomer units and ligands have been created by á la mode, which is 'A Ligand And Monomer Object Data Environment' for building models.

NDB is located in the Department of Chemistry and Chemical Biology at Rutgers University in Piscataway (NJ), USA.

Related websites

NDB Home Page	http://ndbserver.rutgers.edu/NDB/
NDB Search	http://ndbserver.rutgers.edu/NDB/structure-finder/ndb/
NDB Atlas	http://ndbserver.rutgers.edu/NDB/NDBATLAS/
Structure Deposition	http://ndbserver.rutgers.edu/NDB/deposition/

Further reading

Berman HM, et al. (1992) The Nucleic Acid Database: A comprehensive relational database of three-dimensional structures of nucleic acids. *Biophys. J.* 63: 751–759.

Berman HM, et al. (1996) The Nucleic Acid Database: present and future. *J. Res. Natl. Inst. Stand. Technol.* 101: 243–257.

Berman HM, et al. (1999) The Nucleic Acid Database: a research and teaching tool. In: *Handbook of Nucleic Acid Structure*, S. Neidle, Editor. Oxford University Press, pp. 77–92.

Olson WK, et al. (2001) a standard reference frame for the description of nucleic acid base-pair geometry *J Mol Biol* 313: 229–237.

NUCLEIC ACID SEQUENCE DATABASES

Guenter Stoesser

The principal nucleic acid sequence databases, DDBJ/EMBL/GenBank, constitute repositories of all published nucleotide sequences. These comprehensive archival databases are primary factual databases containing data, which can be mined and analysed by computer analysis. In addition to these primary sequence databases, there exists a large variety of organism- or molecule-specific secondary sequence databases.

While data in primary databases originates from original submissions by experimentalists, sequence data in secondary databases are typically derived from a primary database and are complemented by further human curation, computational annotation, literature research, personal communications etc. For primary databases the ownership of data lies with the submitting scientist, while secondary databases have editorial power over contents of the database.

Related websites

EBI Database Resources	http://www.ebi.ac.uk/Databases/
NCBI Database Resources	http://www.ncbi.nlm.nih.gov/Database/
DDBJ	http://www.ddbj.nig.ac.jp/

Further reading

Baxevanis AD (2002) The Molecular Biology Database Collection: 2002 update *Nucleic Acids Res.* 30: 1–12.

NUCLEOLAR ORGANIZER REGION (NOR)

Katheleen Gardiner

Sites of tandem repeats of ribosomal RNA genes.

In metaphase chromosomes, NORs appear as secondary constrictions. In interphase chromosomes, NORs show attached spherical structures, called nucleoli, where the ribosomal RNAs are transcribed. In the human genome, NORs are located on the short arms of the acrocentric chromosomes 13, 14, 15, 21 and 22.

Further reading

Miller OJ, Therman E (2000) *Human Chromosomes*. Springer.

Wagner RP, Maguire MP, Stallings RL (1993) *Chromosomes – a synthesis*. Wiley-Liss.

NUCLEOTIDE BASE CODES, *SEE* IUPAC-IUB CODES.

OBF (The Open Bioinformatics Foundation)

Object, *see* Data Structure.

Object-Relational Database

OBO-Edit

OBO Foundry

Observation, *see* Feature.

Occam's Razor, *see* Parsimony.

ODB (Operon DataBase)

Offspring Branch (Daughter Branch/Lineage)

OKBS, *see* Open Knowledge Base Connectivity.

Oligo Selection Program, *see* OSP.

Oligogenic Effect, *see* Oligogenic Inheritance.

Oligogenic Inheritance (Oligogenic Effect)

Omics

OMIM (Online Mendelian Inheritance in Man)

Online Mendelian Inheritance in Man, *see* OMIM.

Ontology

Open Biological and Biomedical Ontologies, *see* OBO Foundry.

Open Reading Frame (ORF)

Open Reading Frame Finder, *see* ORF Finder.

Open Regulatory Annotation Database, *see* OregAnno.

Operon DataBase, *see* OPD.

Open Source Bioinformatics Organizations

Operating System

Operational Taxonomic Unit (OTU)

OPLS

Optimal Alignment

Oral Bioavailability

ORF, *see* Open Reading Frame.

ORegAnno (Open Regulatory Annotation Database)

ORFan, *see* Orphan Gene.

Organelle Genome Database, *see* GOBASE.

Organism-Specific Database, *see* MOD.

Organismal Classification, *see* Taxonomic Classification.

Orphan Gene (ORFan)

Ortholog (Orthologue)

Outlier, *see* Outlier Mining.

Outlier Mining (Outlier)

Overdominance

Overfitting (Overtraining)

Overtraining, *see* Overfitting.

OWL, see Web Ontology Language.

Concise Encyclopaedia of Bioinformatics and Computational Biology, Second Edition.
Edited by John M. Hancock and Marketa J. Zvelebil.

O

OBF (THE OPEN BIOINFORMATICS FOUNDATION)

Pedro Fernandes

The O|B|F is an umbrella group that provides a central resource for a variety of open source bioinformatics projects.

The O|B|F grew out of the BioPerl project, which was officially organised in 1995. The project was an international association of developers of open source Perl tools for bioinformatics. The O|B|F acts as a distribution point for a number of similar or related projects including biojava.org, biopython.org, Distributed Annotation Server (DAS), bioruby.org, biocorba.org, Ensembl and EMBOSS.

The O|B|F underwrites and supports the BOSC conferences and developer-centric 'Bio-Hackathon' events in cooperation with several research organizations.

Related websites

O|B|F Website	http://open-bio.org/
BioJava	http://biojava.org
BioPerl	http://bioperl.org
BioCorba	http://www.bioperl.org/wiki/BioCORBA
BioPython	http://biopython.org
BioRuby	http://bioruby.org
Distributed Annotation Server	http://biodas.org
EMBOSS	http://www.emboss.org/
Ensembl	http://www.ensembl.org/

OBJECT, *SEE* DATA STRUCTURE.

OBJECT-RELATIONAL DATABASE

Matthew He

Object-relational databases combine the elements of object orientation and object-oriented programming languages with database capabilities. They provide more than persistent storage of programming language objects. Object databases extend the functionality of object programming languages (e.g., C++, Smalltalk, or Java) to provide full-featured database programming capability. The result is a high level of congruence between the data model for the application and the data model of the database. Object-relational databases are used in Bioinformatics to map molecular biological objects (such as sequences, structures, maps and pathways) to their underlying representations (typically within the rows and columns of relational database tables.) This enables the user to deal with the biological objects in a more intuitive manner, as they would in the laboratory, without having to worry about the underlying data model of their representation.

O

OBO-EDIT

John M. Hancock

OBO-Edit is an easy-to-use, GUI-based application for the creation and editing of ontologies. Although optimized to make use of the OBO biological ontology file format it can also handle OWL and some other file formats. OBO-Edit incorporates a simple reasoner and the ability to search within an ontology.

Relevant website

OBO-Edit website	http://oboedit.org/

Further reading

Day-Richter J, Harris MA, Haendel M; Gene Ontology OBO-Edit Working Group, Lewis S (2007) OBO-Edit – an ontology editor for biologists. *Bioinformatics*. 23: 2198–2200.

OBO FOUNDRY

John M. Hancock

The OBO Foundry site aims to collect together a set of biological and biomedical ontologies that conform to, or aim to conform to, a set of principles of ontology construction. Constituent ontologies should make use of relations contained in the Relation Ontology, have definitions that conform to a standard structure, and construct compound terms using terms from an OBO ontology if possible.

Related website

OBO Foundry	http://www.obofoundry.org/

Further reading

Smith B, Ashburner M, Rosse C, et al. The OBI Consortium (2007) The OBO Foundry: coordinated evolution of ontologies to support biomedical data integration. *Nature Biotechnology* 25: 1251–1255.

OBSERVATION, *SEE* FEATURE.

OCCAM'S RAZOR, *SEE* PARSIMONY.

ODB (OPERON DATABASE)

Obi L. Griffith and Malachi Griffith

ODB is a curated collection of operons from sequenced genomes. 'Known' operons are obtained directly from descriptions in the literature. 'Conserved' operons are identified where genes in a known operon have orthologous genes in other genomes that are consecutively located on the same strand of the genome. Each operon record includes a unique operon identifier, its species, the operon name, the gene names, the definition of the operon, PubMed identifiers, and the source of the operon information. Users can query the database by species and can examine the structure and peripheral genes of each operon by a web-based graphical viewer. The complete database can also be downloaded in tab-delimited text format.

Related website

Operon Database	http://operondb.jp/

Further reading

Okuda S, Yoshizawa AC (2011) ODB: a database for operon organizations, 2011 update. *Nucleic Acids Res. 39*(Database issue): D552–555.

Okuda S, et al. (2006) ODB: a database of operons accumulating known operons across multiple genomes. *Nucleic Acids Res. 34*(Database issue): D358–362.

OFFSPRING BRANCH (DAUGHTER BRANCH/LINEAGE)

Aidan Budd and Alexandros Stamatakis

Branches that link a specified internal node within a phylogenetic tree to those taxonomic units (i.e., nodes) that are offspring (or 'daughters') of this specified node. The branches linking this specified node to a taxonomic unit that is ancestral to the specified node can be described as an 'ancestral branch'.

In an unrooted tree, as the location of the root is unknown or unspecified, it is not possible to distinguish daughter branch lineages from ancestral lineages.

OKBS, *SEE* OPEN KNOWLEDGE BASE CONNECTIVITY.

OLIGO SELECTION PROGRAM, *SEE* OSP.

OLIGOGENIC EFFECT, *SEE* OLIGOGENIC INHERITANCE.

O

OLIGOGENIC INHERITANCE (OLIGOGENIC EFFECT)

Mark McCarthy and Steven Wiltshire

A genetic architecture whereby variation in a trait is determined by variation in several genes, each of which has substantial, additive, effects on the phenotype.

The oligogenic architecture is applicable, as with polygenic inheritance, to both quantitative traits and discrete traits under the liability threshold model. Oligogenicity is to be distinguished from polygenic (many genes) and Mendelian – single gene – inheritance patterns, though the boundaries between these are, in practice, often indistinct.

Further reading

Falconer D, McKay TFC (1996) *Introduction to Quantitative Genetics*. Harlow: Prentice Hall, pp 100–183.

Hartl DL, Clark AG (1997) *Principles of Population Genetics*. Sunderland: Sinauer Associates, pp 397–481.

See also Mendelian Disease, Multifactorial Traits, Polygenic Inheritance, QuantitativeTrait

OMICS

John M. Hancock

Catch-all for fields of study attempting to integrate the study of some feature of an organism.

Recent years have seen an explosion in terms ending in the suffix -omics. These all derive from the term genome (hence genomics), a term invented by Hans Winkler in 1920, although the use of -ome is older, signifying the 'collectivity' of a set of things (see web article by Lederberg). The oldest of these terms, and one that seems due to come back into fashion, may be biome. Thus although the explosion in the use of this terminology may appear fatuous, it does signify a widespread interest in moving towards an integrative, rather than reductionist, approach to biology following from the early successes of genomics and functional genomics.

Related websites

Article by Lederberg	http://www.the-scientist.com/?articles.view/articleNo/13313/title/-Ome-Sweet--Omics---A-Genealogical-Treasury-of-Words/
The Omes Page	http://bioinfo.mbb.yale.edu/what-is-it/omes/omes.html

Further reading

Winkler H (1920) *Verbreitung und Ursache der Parthenogenesis im Pflanzen- und Tierreiche* Jena: Verlag Fischer.

See also Systems Biology, Networks

OMIM (ONLINE MENDELIAN INHERITANCE IN MAN)

Obi L. Griffith and Malachi Griffith

OMIM is a comprehensive, frequently updated, catalogue of human genes and their genetic phenotypes, providing full-text, referenced overviews of all known mendelian disorders and over 12,000 genes. OMIM focuses specifically on the relationship between phenotype and genotype with a particular focus on clinically relevant disease phenotypes. The database began in the early 1960s as a catalog of mendelian traits and disorders, entitled Mendelian Inheritance in Man (MIM). Multiple book editions have been published and the online version (OMIM) was developed in 1987 as a collaboration between the National Library of Medicine and Johns Hopkins University and then made available for the World Wide Web by the NCBI. Each gene-based record provides detailed information such as phenotype-gene relationships, clinical features (associated diseases), clinical management, inheritance patterns, cytogenetics, genetic mapping, diagnosis, molecular genetics, pathogenesis, available animal models, and relevant references.

Related website

Online Mendelian Inheritance in Man	http://www.ncbi.nlm.nih.gov/omim http://omim.org/

Further reading

Amberger J, et al. (2011) A new face and new challenges for Online Mendelian Inheritance in Man (OMIM®). *Hum Mutat* 32(5): 564–567.

ONLINE MENDELIAN INHERITANCE IN MAN, *SEE* OMIM.

ONTOLOGY

Robert Stevens

In computer science an ontology is a way of capturing how a particular community thinks about its field. An ontology attempts to capture a community's knowledge or understanding of a domain of interest. Like any representation, an ontology is only a partial representation of the community's understanding. Ontologies are used to share a common understanding between both people and computer systems. Most ontologies consist of at least concepts, the terms that name those concepts and relationships between those concepts. Ideally, the concepts have definitions and the collection of terms (lexicon) provides a vocabulary by which a community can talk about its domain. The ontology provides a structure for the domain and constrains the interpretations of the terms the ontology provides. The goal of an ontology is to create an agreed-upon vocabulary and semantic structure for exchanging information about that domain.

Within molecular biology ontologies have found many uses. The most common use is to provide a shared, controlled vocabulary. For example, the Gene Ontology provides a common understanding for the three major attributes of gene products. The ontologies in RiboWeb and EcoCyc form rich and sophisticated schema for their knowledge bases and offer inferential support not normally seen in a conventional database management system.

Further reading

Bechhofer S (2002) Ontology Language Standardisation Efforts; OntoWeb Deliverable D4.0; http://ontoweb.aifb.uni-karlsruhe.de/About/Deliverables/d4.0.pdf.

Uschold M et al. (1998). The Enterprise Ontology.*Knowledge Eng. Rev.*13: 32–89.

OPEN BIOLOGICAL AND BIOMEDICAL ONTOLOGIES, *SEE* OBO FOUNDRY.

OPEN READING FRAME (ORF)

Niall Dillon

Set of translation codons spanning the region between a start and a stop codon. In prokaryotes, which have no introns, finding open reading frames forms the basis of gene finding. In eukaryotes introns make this approach inapplicable, but defining the largest ORF in a cDNA sequence is an important component of identifying the likely protein product of a gene, and of identifying pseudogenes.

OPEN READING FRAME FINDER, *SEE* ORF FINDER.

OPEN REGULATORY ANNOTATION DATABASE, *SEE* OREGANNO.

OPERON DATABASE, *SEE* OPD.

OPEN SOURCE BIOINFORMATICS ORGANIZATIONS

John M. Hancock

Bioinformatics organizations dedicated to the open source model of software development and dissemination.

Although bioinformatics originated as an academic discipline, its increasing importance has led to a plethora of commercial products, some of which have become almost indispensable to their users. The open source model, familiar from the activities of organizations such as GNU, Linux and SourceForge, asserts that software and data should be freely available and distributed under non-restrictive licensing conditions.

Related websites

Open Bioinformatics Foundation	http://open-bio.org/
GNU	http://www.gnu.org/
Linux	http://www.linux.org/
Open Source Initiative	http://www.opensource.org/
SourceForge	http://sourceforge.net/

OPERATING SYSTEM

Steve Pettifer, James Marsh and David Thorne

An operating system (OS) is a layer of software that provides an interface between application programs and a computer's hardware. At a bare minimum, an OS will provide functions that allow a single program to execute: such functions include managing memory, and controlling access to backing store and other devices such as keyboards and screens. Modern operating systems, such as GNU/Linux, Microsoft Windows or Apple's OS X and their mobile variants provide substantially more functionality than this.

Operating systems on early computers closely modeled the underlying hardware, allowing a single program to execute at any one time on a single CPU, with no explicit support for the concept of users. The OS would typically provide a mode in which a program could be loaded and run (sometimes referred to as 'monitor mode'). Once running, a program would execute its own instructions (directly on the CPU), periodically passing control to functions provided by the OS in order to allocate memory or access devices. Once complete, the system would return to monitor mode.

Over time, operating systems became more sophisticated, providing an increasingly abstract view of the underlying hardware. Systems introduced the concept of multiple users, and thus were required to model their privileges and permissions. Multi-processing operating systems developed scheduling mechanisms whereby several programs could apparently execute as simultaneous processes: here the OS became responsible for giving each program a small slice of CPU time and for swapping between programs in a manner that appeared to give each sole access to the underlying hardware. Broadly speaking, most contemporary operating systems consist of a 'kernel' (providing low level access to hardware as per historical systems), and a set of system processes scheduled by the kernel that provide higher level features such as filesystems.

Whether code is part of an operating system or not is becoming increasingly ambiguous as systems and hardware evolve. For example, both OS X and Windows include functionality for managing a single style of graphical user interface; in essence the GUI is part of the OS. By contrast, GNU/Linux allows for numerous different GUIs, which consist of a window manager (of which several varieties exist), and a subsystem for displaying graphics on-screen (typically X-Windows). In this case, the GUI is logically separate from the underlying operating system, however is so commonly included in GNU/Linux distributions as to be arguably a part of the OS in some senses. A similar argument applies to utility programs such as process/activity monitors that are commonly distributed along with an OS: in some

O

senses these are 'application software' much as any other program; in other senses because they are essential to the everyday operation of a machine, and always available, they can be considered as an OS component. Thus the distinction between 'applications', 'utilities' and 'operating system components' is somewhat blurred.

Traditionally, operating systems have been classified according to various criteria such as 'real-time', 'multi-user', 'multi-tasking', 'distributed' and 'embedded'. Except in highly specialist areas, such classifications are now for the most part redundant, as modern operating systems can be configured to suit most purposes, and the hardware restrictions that forced the creation of bespoke embedded and real-time systems are less often an issue. GNU/Linux, for example is frequently used as an 'embedded' system in devices such as set-top boxes, media players and routers, but is equally suitable for use in mobile devices, desktop computers and servers. Modern mobile operating systems such as iOS, Android and Windows Phone are sophisticated OSs with GUIs tailored for small screens, and power management regimes suitable for devices with limited battery life, and have more in common with fully-fledged operating systems than with the embedded systems of previous times.

Numerous projects have assembled operating system distributions or virtual machine images that are pre-configured with common bioinformatics tools.

OPERATIONAL TAXONOMIC UNIT (OTU)

John M. Hancock

In a phylogenetic tree, the terminal nodes, or leaves, corresponding to observable genes or species.

The term is used to distinguish these observable nodes from internal nodes which are presumed to have existed earlier in evolution and not to be observable. A further implication of the term is that, in the context of taxonomy, these nodes may correspond to species, subspecies or some other defined unit of taxonomy. They may, however, correspond to individuals within a population or members of a gene family within an individual species.

See also Phylogenetic Tree

OPLS

Roland Dunbrack

A molecular mechanics potential energy function for organic molecules developed by William Jorgensen and colleagues at Yale University

Related website

Jorgensen website	http://zarbi.chem.yale.edu/

Further reading

Jorgensen WL (1998) *The Encyclopedia of Computational Chemistry 3.*, 3281–3285 Athens, GA, USA, Wiley.

OPTIMAL ALIGNMENT

Jaap Heringa

The optimal alignment of two sequences refers to the highest scoring alignment out of very many possible alignments, as assessed using a scoring system consisting of a residue exchange matrix and gap penalty values. The most widely applied technique for obtaining an optimal global alignment or local alignment is the dynamic programming algorithm (Needleman and Wunsch, 1970; Smith and Waterman, 1981). The fast homology search methods BLAST and FASTA comprise individual heuristics to generate approximate optimal alignments, the scores of which are assessed by statistical evaluations, such as E values.

Further reading

Needleman SB, Wunsch CD (1970) A general method applicable to the search for similarities in the amino acid sequence of two proteins. *J. Mol. Biol.* 48: 443–453.

Smith TF, Waterman MS (1981) Identification of common molecular subsequences. *J. Mol. Biol.* 147: 195–197.

See also Alignment, Residue Exchange Matrix, Gap Penalty, Global Alignment, Local Alignment, Homology Search, Sequence Sinilarity, BLAST, FASTA, E Value

ORAL BIOAVAILABILITY

Bissan Al-Lazikani

A measure reflecting the fraction of a compound or drug, when administered orally, to become available in the blood stream, and thus theoretically be able to reach its intended place of action in the body. This requires that the compound possesses the correct balance of physciochemical properties required for it to pass through the gut wall and enter the blood stream with minimal metabolism and clearance from the blood.

See also LogP

ORF, *SEE* OPEN READING FRAME.

OREGANNO (OPEN REGULATORY ANNOTATION DATABASE)

Obi L. Griffith and Malachi Griffith

ORegAnno is an open-access, open-source database for the curation of known regulatory elements from scientific literature including regulatory regions (e.g., promoters or enhancers), transcription factor binding sites, and regulatory polymorphisms. Annotation is collected from users worldwide for various biological assays and is automatically cross-referenced against PubMED, Entrez Gene, Ensembl, dbSNP, the eVOC: Cell type ontology, and the Taxonomy database, where appropriate, and includes information regarding the original experimentation performed using its own evidence ontology. Oreganno also provides the ability to incorporate well-established datasets or existing databases and to date has incorporated the Drosophila DNase I Footprint Database, Stanford ENCODE Promoters, Regulatory Element Database for Drosophila (REDfly), VISTA Enhancers, NFIRegulomeDB, phylogenetically conserved non-coding elements (PCNEs), and many ChIP-chip or ChIP-seq datasets for such factors as STAT1, REST, CTCF, ESR1, RELA, Foxa2, and more. Oreganno was conceived as an alternative to other closed and commercial models (e.g., Transfac). For related projects, see PAZAR and JASPAR.

Related websites

OregAnno	http://www.oreganno.org
PAZAR	http://www.pazar.info/
JASPAR	http://jaspar.cgb.ki.se/

Further reading

Griffith OL, Montgomery SB, et al. Open Regulatory Annotation Consortium. (2008) ORegAnno: an open-access community-driven resource for regulatory annotation. *Nucleic Acids Res. 36* (Database issue): D107–113.

Montgomery SB, Griffith OL, et al. (2006) ORegAnno: an open access database and curation system for literature-derived promoters, transcription factor binding sites and regulatory variation. *Bioinformatics.* 22(5): 637–640.

ORFAN, *SEE* ORPHAN GENE.

ORGANELLE GENOME DATABASE, *SEE* GOBASE.

ORGANISM-SPECIFIC DATABASE, *SEE* MOD.

ORGANISMAL CLASSIFICATION, *SEE* TAXONOMIC CLASSIFICATION.

ORPHAN GENE (ORFAN)

Jean-Michel Claverie

A (often putative) protein sequence (Open Reading Frame) apparently unrelated to any other previously identified protein sequence from any other organism.

In practice, this notion depends on a similarity threshold (e.g. percentage identity or BLAST score) and/or a given significance level (e.g. Blast E-value). Orphan genes (often nicknamed ORFans) are more precisely defined as protein sequences which cannot be *reliably* aligned with any other, i.e. with similarity levels in the twilight zone. This definition is time-dependent, as it depends on the content of the sequence database. Thus the fraction of ORFans tends to diminish as more genomes get sequenced. The ORFan definition is adapted to the presence of very close species (or bacterial strains) in the databases, by been used in an evolutionary distance-dependent fashion, meaning 'only found in'. For instance, '*E. coli* ORFans' will have clear homologues in the other *E. coli* strain sequences, but not elsewhere. The relative proportion of ORFans corresponding to truly unique genes *vs.* genes that have simply diverged too far from their relative in other genomes is unknown. The determination of the 3-D structure of ORFan proteins is one way of addressing this question.

Further reading

Fischer D, Eisenberg D. (1999) Finding families for genomic ORFans. *Bioinformatics,* 15: 759–762.

Monchois V, et al. (2001) *Escherichia coli* ykfE ORFan gene encodes a potent inhibitor of C-type lysozyme. *J. Biol. Chem.,* 276: 18437–18441.

ORTHOLOG (ORTHOLOGUE)

Dov Greenbaum

Orthologs are homologous sequences that are separated by a speciation event: the copies of the gene, one in each of the resulting species, sharing the same ancestor. The term 'ortholog' was coined in 1970 by Walter Fitch.

In particular, orthologs are defined as genes derived from a single ancestral gene in the last common ancestor of the compared species.

With much of ancestry of genes somewhat uncertain given different genomic evolutionary events, evidence that genes are actually orthologs sharing ancestry is often found by finding shared function among the sequences or doing phylogenic analyses.

Orthologs are valuable in determining taxonomic classification and can be used to track relatedness between genomes: closely related genomes have very similar sequences between their respective genes in an ortholog pair; genomes that are not closely related will have more dissimilarity between their orthologs' sequences.

O

Pseudoorthologs, a sub category are genes that actually are paralogous but nevertheless appear to be orthologous due to differential, lineage-specific gene loss.

Related websites

COG	http://www.ncbi.nlm.nih.gov/COG/
OrthologID	http://nypg.bio.nyu.edu/orthologid/
eggNOG	http://eggnog.embl.de/version_3.0/
OrthoDB	http://cegg.unige.ch/orthodb5
OrthoMCL	http://www.orthomcl.org/cgi-bin/OrthoMclWeb.cgi

Further reading

Fitch WM (1970) Distinguishing homologous from analogous proteins. *Syst. Zool.* 19: 99–106.

Forslund K, Schreiber F, Thanintorn N, Sonnhammer EL (2011) OrthoDisease: tracking disease gene orthologs across 100 species. *Brief Bioinform.* 12(5): 463–73.

Gharib WH, Robinson-Rechavi M (2011) When orthologs diverge between human and mouse. *Brief Bioinform.* 12(5): 436–41.

Koonin EV (2005) Orthologs, paralogs, and evolutionary genomics. *Annu Rev Genet.* 39: 309–338.

Kuzniar A, van Ham RC, Pongor S, Leunissen JA (2008) The quest for orthologs: finding the corresponding gene across genomes. *Trends Genet.* 24(11): 539–551.

O'Brien KP, Remm M, Sonnhammer ELL (2005) Inparanoid: a comprehensive database of eukaryotic orthologs. *Nucleic Acids Research 33* (Database Issue): D476.

OUTLIER, *SEE* OUTLIER MINING.

OUTLIER MINING (OUTLIER)

Feng Chen and Yi-Ping Phoebe Chen

Outlier represents a kind of data which possesses a different model or feature from the majority of data. Outlier mining is an interesting and significant task due to the fact that one's noise could be another one's signal. Outlier mining generally has two steps: a definition about what kind of data outliers are in database; and an effective method to discover outliers. The mining strategies can be categorized as: statistic distribution-based method, distance-based method, density-based method and deviation-based method.

Firstly, a distribution or probability model for the given dataset such as Poisson or Gaussian is used. Then a discordance test is used to discover the outliers which do not conform to the given distribution. Subsequently, the method detects one outlier with the parameters of *pct* and *dmin* if the distance between it and *pct* data points are larger than *dmin*. Finally, the dataset in which the data are not distributed evenly are considered because these are good for discovering local outliers. One local outlier is such data point if it is apart from its

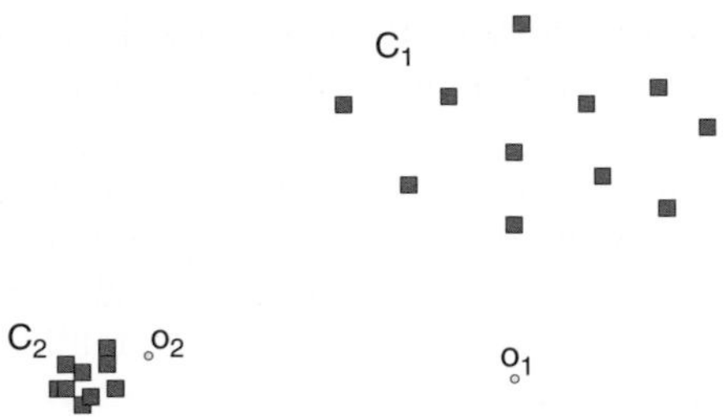

Figure O.1 Density-based outlier detection.

neighbourhood. Figure O.1 illustrates an example of density based method and clarifies the difference between density based and distance based. In this figure, C_1 and C_2 represent two clusters with different density. o_1 and o_2 are outliers. o_2 is very easy to be detected as outlier no matter which method is used. But o_1 cannot be discovered correctly if we use distance based method because it is close to the data points in cluster C_2. The fourth generalizes the main features of a dataset. Outliers are determined if they deviate from such main features.

How to find outliers in bioinformatics dataset is based on how one defines outliers. For example, a gene could be an outlier if it is far away from others; a protein structure might be special due to its dissimilarity with other structures. Outlier detection can imply a unique phenomenon, probably leading to a unique biological process.

Further reading

Barnett V, Lewis, T. (1994) *Outliers in Statistical Data*. John Wiley & Sons., 3rd edition, New York.

Cho HJ, Kim Y, Jung HJ, Lee S-W, Lee JW (2008) *OutlierD: an R package for Outlier Detection Using Quantile Regression on Mass Spectrometry Data. Bioinformatics*, 24(6): 882–884.

Motulsky HJ, Brown RE (2006) Detecting outliers when fitting data with nonlinear regression – a new method based on robust nonlinear regression and the false discovery rate. *BMC Bioinformatics*, 7: 123.

OVERDOMINANCE

A.R. Hoelzel

Overdominance occurs when the heterozygote condition at a locus is selectively more fit than the homozygotes at that locus.

Considering the relative fitness of genotypes for a two-allele system at a locus in a diploid organism, the following relationship holds:

Genotype:	A_1A_1	A_1A_2	A_2A_2
Relative fitness:	1	1-*hs*	1-*s*

O

Where s is the selection coefficient and h is the heterozygous effect. A_1 is dominant when h=0, and A_2 is dominant when h=1, but overdominance occurs when $h < 0$.

Overdominance is also referred to as 'heterozygote advantage'. A famous example is the relationship between sickle cell anemia and malarial resistance (Allison 1954). In this example a mutation for the sickle cell haemoglobin (HbS) causes serious anemia in homozygotes, but the heterozygote exhibits only mild anemia. Because *Plasmodium sp.* (the malarial parasite) are less able to infect the red blood cells of individuals heterozygous for this condition, they have a selective advantage in locations where malaria is prevalent.

Further reading

Allison AC (1954) Protection afforded by sickle-cell trait against subtertian malaria. *British Medical Journal*, 6 Feb.

Hartl DL (2000) *A Primer of Population Genetics*. Sinauer Associates

See also Natural Selection

OVERFITTING (OVERTRAINING)

Nello Cristianini

In machine learning, the phenomenon by which a learning algorithm identifies relations in the training data that are due to noise or chance, and do not reflect the underlying laws governing the given dataset.

It occurs mostly in the presence of small and/or noisy training samples, when too much flexibility is allowed to the learning algorithm. When the learning algorithm is overfitting, its performance on the training set appears to be good, but the performance on the test data or in cross-validation is poor. It is addressed by reducing the capacity of the learning algorithm. The size of a neural network or a decision tree, the number of centers in k-means, are all rough measures of capacity of a learning algorithm. A precise formal definition is given within the field of Statistical Learning Theory, and is at the basis of the design of a new generation of algorithms explicitly aimed at counteracting overfitting.

Further reading

Duda RO, Hart PE, Stork DG (2001) *Pattern Classification*, John Wiley & Sons, USA, 2001.

Mitchell T (1997) *Machine Learning*, McGraw Hill, Maidenhead, UK.

See also Neural Networks, Capacity, Support Vector Machines, Cross-Validation

OVERTRAINING, *SEE* OVERFITTING.

OWL, *SEE* WEB ONTOLOGY LANGUAGE.

P

Pairwise Alignment
PAM Matrix (of Amino Acid Substitutions), *see* Dayhoff Amino Acid Substitution Matrix, Amino Acid Exchange Matrix.
PAM Matrix of Nucleotide Substitutions (Point Accepted Mutations)
PAML (Phylogenetic Analysis by Maximum Likelihood)
Paralinear Distance (LogDet)
Parallel Computing in Phylogenetics
Paralog (Paralogue)
Parameter
Parametric Bootstrapping, *see* Bootstrapping.
Paraphyletic Group, *see* Cladistics.
Parent, *see* Template.
Parity Bit
Parsimony
Partition Coefficient, *see* LogP.
Pattern
Pattern Analysis
Pattern Discovery, *see* Motif Discovery.
Pattern of Change Analysis, *see* Phylogenetic Events Analysis.
Pattern Recognition, *see* Pattern Analysis.
PAUP* (Phylogenetic Analysis Using Parsimony (and Other Methods))
PAZAR
Pearson Correlation, *see* Regression Analysis.
Penalty, *see* Gap Penalty.
Penetrance
Peptide
Peptide Bond (Amide Bond)
Peptide Mass Fingerprint
Peptide Spectrum Match (PSM)
PeptideAtlas
Percent Accepted Mutation Matrix, *see* Dayhoff Amino Acid Substitution Matrix.
Petri Net
Pfam
PFM, *see* Position-Frequency Matrix.
Phantom Indel (Frame Shift)
Pharmacophore
Phase (Sensu Linkage)
PheGenI, *see* dbGAP.
PHRAP
PHRED
PHYLIP (PHYLogeny Inference Package)
Phylogenetic Events Analysis (Pattern of Change Analysis)
Phylogenetic Footprint

Concise Encyclopaedia of Bioinformatics and Computational Biology, Second Edition.
Edited by John M. Hancock and Marketa J. Zvelebil.

Phylogenetic Footprint Detection
Phylogenetic Placement of Short Reads
Phylogenetic Profile
Phylogenetic Reconstruction, *see* Phylogenetic Tree.
Phylogenetic Shadowing, *see* Phylogenetic Footprinting.
Phylogenetic Tree (Phylogeny, Phylogeny Reconstruction, Phylogenetic Reconstruction)
Phylogenetic Trees, Distance, *see* Distances Between Trees.
Phylogenetics
Phylogenomics
Phylogeny, Phylogeny Reconstruction, *see* Phylogenetic Tree.
Piecewise-Linear Models
PIPMAKER
PlantsDB
Plesiomorphy
Point Accepted Mutations, *see* PAM Matrix of Nucleotide Substitutions.
Polar
Polarization
Polygenic Effect, *see* Polygenic Inheritance.
Polygenic Inheritance (Polygenic Effect)
Polymorphism (Genetic Polymorphism)
Polypeptide
Polyphyletic Group, *see* Cladistics.
Polytomy, *see* Multifurcation.
PomBase
Population Bottleneck (Bottleneck)
Position-Specific Scoring Matrix, *see* Profile.
Position Weight Matrix
Position Weight Matrix of Transcription Factor Binding Sites
Positional Candidate Approach
Positive Classification
Positive Darwinian Selection (Positive Selection)
Positive Selection, *see* Positive Darwinian Selection.
Post-Order Tree Traversal, *see* Tree Traversal.
Posterior Error Probability (PEP)
Potential of Mean Force
Power Law (Zipf's Law)
Prediction of Gene Function
Predictive ADME (Absorption, Distribution, Metabolism, and Excretion), *see* Chemoinformatics.
PRIDE
PRIMER3
Principal Components Analysis (PCA)
PRINTS
Probabilistic Network, *see* Bayesian Network.
ProDom
Profile (Weight Matrix, Position Weight Matrix, Position-Specific Scoring Matrix, PSSM)
Profile, 3D
Profile Searching
Programming and Scripting Languages
Promoter
The Promoter Database of Saccharomyces cerevisiae (SCPD)
Promoter Prediction
PROSITE
Protégé
Protein Array (Protein Microarray)
Protein Data Bank (PDB)
Protein Databases
Protein Domain

Protein Family

Protein Family and Domain Signature Databases

Protein Fingerprint, *see* Fingerprint.

Protein Inference Problem

Protein Information Resource (PIR)

Protein Microarray, *see* Protein Array.

Protein Module

Protein-Protein Coevolution

Protein-Protein Interaction Network Inference

Protein Sequence, *see* Sequence of Protein.

Protein Sequence Cluster Databases

Protein Structure

Protein Structure Classification Databases, *see* Structure 3D Classification.

Proteome

Proteome Analysis Database (Integr8)

Proteomics

Proteomics Standards Initiative (PSI)

Proteotypic Peptide

Pseudoparalog (pseudoparalogue)

Pseudogene

PSI BLAST

PSSM, *see* Profile.

Purifying Selection (Negative Selection)

PAIRWISE ALIGNMENT

Jaap Heringa

Co-linear matching of two molecular sequences, which derive from a common ancestor, such that the evolutionary relationships between the residues is reproduced optimally. Alignment of two sequences involves matching of residues, and insertion of gap characters in one of the two sequences, which represent insertion or deletion events during molecular evolution.

Many methods have been developed for the calculation of sequence alignments, of which implementations of the dynamic programming algorithm (Needleman and Wunsch, 1970; Smith and Waterman, 1981) are considered the standard in yielding the most biologically relevant alignments. The dynamic programming (DP) algorithm requires a scoring matrix, which is an evolutionary model in the form of a symmetrical 4×4 nucleotide or a 20×20 amino acid exchange matrix. Each matrix cell approximates the evolutionary propensity for the mutation of one nucleotide or amino acid type into another. The DP algorithm also relies on the specification of gap penalties, which model the relative probabilities for the occurrence of insertion/deletion events. Normally, a gap opening and extension penalty is used for creating a gap and each extension respectively (affine gap penalties), so that the chance for an insertion/deletion depends linearly upon the length of the associated fragment. Given an exchange matrix and gap penalty values, which together are commonly called the scoring scheme, the DP algorithm is guaranteed to produce the highest scoring alignment of any pair of sequences; the optimal alignment. Two types of alignment are generally distinguished: global and local alignment. Global alignment (Needleman and Wunsch, 1970) denotes an alignment over the full length of both sequences, which is an appropriate strategy to follow when two sequences are similar or have roughly the same length. However, some sequences may show similarity limited to a motif or a domain only, while the remaining sequences stretches may be essentially unrelated. In such cases, global alignment may well misalign the related fragments, as these become overshadowed by the unrelated sequence portions that the global method attempts to align, possibly leading to a score that would not allow the recognition of any similarity. If similarity between one complete sequence and part of the other sequence is suspected, for example, when aligning a gene against a homologous genome, or a single-domain protein with a homologous multi-domain protein, the semi-global alignment technique should be attempted. This technique facilitates the inclusion of gaps at either flank of the shorter sequence, leading to an alignment where the shorter sequence is not interrupted by spurious gaps.

If not much knowledge about the relationship of two sequences is available, it is usually better to align selected fragments of either sequence. This can be done using the local alignment technique (Smith and Waterman, 1981). The first method for local alignment, often referred to as the Smith-Waterman (SW) algorithm, is in fact a minor modification of the dynamic programming algorithm for global alignment. It basically selects the best scoring sub-sequence from each sequence and provides their alignment, thereby disregarding the remaining sequence fragments. Later elaborations of the algorithm include methods to generate a number of sub-optimal local alignments in addition to the optimal pair-wise alignment (e.g. Waterman and Eggert, 1987).

Further reading

Needleman SB, Wunsch CD (1970) A general method applicable to the search for similarities in the amino acid sequence of two proteins. *J. Mol. Biol.* 48: 443–453.

Smith TF, Waterman MS (1981) Identification of common molecular subsequences. *J. Mol. Biol.* 147: 195–197.

Waterman MS, Eggert M (1987) A new algorithm for best subsequences alignment with applications to the tRNA-rRNA comparisons. *J. Mol. Biol.* 197: 723–728.

See also Alignment, Pairwise Alignment, Global Alignment, Local Alignment, Optimal Alignment, Residue Exchange Matrix, Gap Penalty

PAM MATRIX (OF AMINO ACID SUBSTITUTIONS), *SEE* DAYHOFF AMINO ACID SUBSTITUTION MATRIX, AMINO ACID EXCHANGE MATRIX.

PAM MATRIX OF NUCLEOTIDE SUBSTITUTIONS (POINT ACCEPTED MUTATIONS)

Laszlo Patthy

PAM matrices for scoring DNA sequence alignments incorporate the information from mutational analyses which revealed that point mutations of the transition type (A↔G or C↔T) are more probable than those of the transversion type (A↔C, A↔T, G↔T, G↔C).

Further reading

Li WH, Graur D (1991) *Fundamentals of Molecular Evolution*, Sinauer Associates, Sunderland, Massachusetts.

States DJ, et al. (1991) Improved sensitivity of nucleic acid database searches using application specific scoring matrices. *Methods: A Companion to Methods in Enzymology* 3: 66–70.

PAML (PHYLOGENETIC ANALYSIS BY MAXIMUM LIKELIHOOD)

Michael P. Cummings

A package of computer programs for maximum likelihood-based analysis of DNA and protein sequence data.

The strengths of the programs are in providing sophisticated likelihood models at the level of nucleotides, codons or amino acids. Specific programs allow for simulating sequences, analyses incorporating rate variation among sites, along branches, and combinations of sites and branches, inferring ancestral sequences, and estimating synonymous and non-synonymous substitution rates using codon-based models. The programs also allow for likelihood ratio-based hypothesis tests including those for rate constancy among sites, evolutionary homogeneity among among different genes, and molecular clock.

The programs are written in ANSI C, and are available as source code and executables for some platforms.

Related website

PAML home page	http://abacus.gene.ucl.ac.uk/software/paml.html

Further reading

Bielawski JP, Yang Z (2005) Maximum likelihood methods for detecting adaptive protein evolution In *Statistical Methods in Molecular Evolution*, Nielsen R (ed.) Springer-Verlag: New York. pp. 103–124.

Yang Z (1997) PAML: a program package for phylogenetic analysis by maximum likelihood. *Comput. Appl. Biosci.* 15: 555–556.

Yang Z, Nielsen R (2000) Estimating synonymous and nonsynonymous substitution rates under realistic evolutionary models. *Mol. Biol. Evol.* 17: 32–43.

Yang Z, Nielsen R (2002) Codon-substitution models for detecting molecular adaptation at individual sites along specific lineages. *Mol. Biol. Evol.* 19: 908–917.

Yang Z, Swanson WJ (2002) Codon-substitution models to detect adaptive evolution that account for heterogeneous selective pressures among site classes. *Mol. Biol. Evol.* 19: 49–57.

Yang Z, Wong WSW, Nielsen R (2005) Bayes empirical Bayes inference of amino acid sites under positive selection. *Mol. Biol. Evol.* 22: 1107–1118.

Yang Z, et al. (2000) Codon-substitution models for heterogeneous selection pressure at amino acid sites. *Genetics* 155: 431–449.

Zhang J, Nielsen R, Yang Z (2005) Evaluation of an improved branch-site likelihood method for detecting positive selection at the molecular level. *Mol. Biol. Evol.* 22: 2472–2479.

See also Seq-gen

PARALINEAR DISTANCE (LOGDET)

Dov Greenbaum

An additive measure of the genealogical distance between two sequences.

Paralinear distance, also known as LogDet (invented simultaneously by Mike Steel and James Lake) is based on the most general models of nucleotide substitution. A paralinear distance is an additive measure of the genealogical distance between two sequences i and j (either DNA or protein). It assumes that all sites in the sequence can vary equally. This assumption of equal variation at all sites is one of the drawbacks of the model.

$$d_{ij} = -\log_e \frac{\det J_{ij}}{(\det D_i)^{1/2} (\det D_j)^{1/2}}$$

where det is the determinent of the matrix.

This method does not suffer, to the same extent as other methods, in dealing with unequal rate effects. These effects group together diverse species that, although they are not evolutionarily related are evolving at a fast pace. In general it has been found that there is significant heterogeneity in the substitution patterns in different taxa; in the paralinear model base substitutions can vary throughout the tree.

Further reading

Lake JA (1994) Reconstructing evolutionary trees from DNA and protein sequences: paralinear distances. *Proc Natl Acad Sci USA* 91: 1455–1459.

Székely L, et al (1993) Fourier inversion formula for evolutionary trees. *Appl Math Lett* 6: 13–17.

PARALLEL COMPUTING IN PHYLOGENETICS

Aidan Budd and Alexandros Stamatakis

The growing size of datasets and the development of more elaborate and complex models of sequence evolution has generated a large amount of activity with respect to the parallelization of phylogenetic inference codes since the mid 90ies. Most widely-used phylogeny programs have been parallelized in various ways and adapted to different parallel hardware architectures, such as GPUs, multi-core processors, programmable/reconfigurable hardware devices, or supercomputer systems. One can typically distinguish between three levels of parallelism in phylogenetic inference: 1. embarrassing parallelism (e.g., executing independent tree searches in parallel) 2. Inference parallelism (e.g., parallelizing the steps of a tree search algorithm) 3. fine-grain parallelism (e.g., computing the likelihood of a single tree concurrently on several processors by splitting up the alignment evenly among processors).

For likelihood-based models, fine-grain parallelism is particularly important for accommodating the memory requirements of super-matrix analyses, because the memory requirements can be split up across a larger number of nodes with more RAM.

Parallelized software

RAxML, GARLI, PhyloBayes, MrBayes, PHYML, IQPNNI, POY, TNT.

Further reading

Ayres DL, Darling A, Zwickl, et al. (2012) BEAGLE: an Application Programming Interface and High-Performance Computing Library for Statistical Phylogenetics. *Systematic Biology* 61(1): 170–173.

Stamatakis A (2010) Orchestrating the Phylogenetic Likelihood Function on Emerging Parallel Architectures. In: B Schmidt (ed.) *Bioinformatics: High Performance Parallel Computer Architectures*, pp 85–115, CRC Press, Taylor & Francis.

See also Supermatrix Approach.

PARALOG (PARALOGUE)

Austin L. Hughes, Dov Greenbaum and Laszlo Patthy

Two genes are said to be paralogous (or paralogues) if they are descended from a common ancestral gene with one or more intervening gene duplication events.

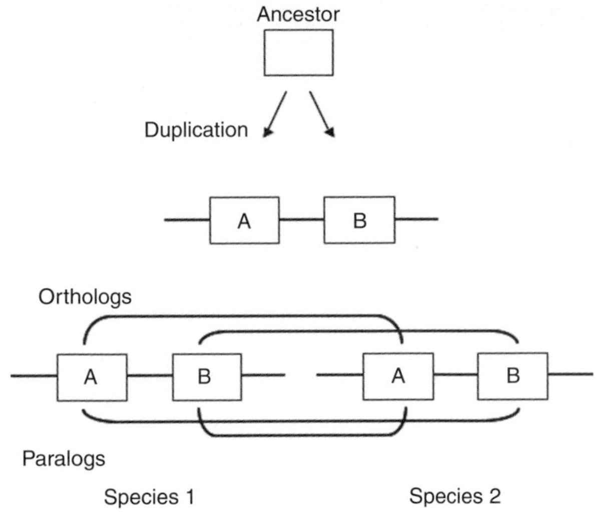

Figure P.1 Illustration of paralogous genes.

When there is a gene duplication event followed by independent mutations and sequence changes within the duplicate genes, the result is a set of paralogous genes. That is, a set of genes within one organism with similar sequence but differing functions. Because they arise from gene duplication events, there is less evolutionary pressure to maintain the sequence in all paralogous genes in the set, thus in some instances, the encoded protein in a paralogous gene may have a totally different function or may decay into pseudogenes.

A commonly cited example of paralogous genes is the HOX cluster of transcription factor genes that is found in many organisms. In both mice and humans there are four paralogous clusters of 13 genes, each on different chromosomes.

In Figure P.1, gene A in species 1 is paralogous to gene B (in either species 1 or species 2). Likewise gene B is paralogous to gene A.

See also Homology, Ortholog, Pseudoparalog, Xenolog

PARAMETER

Matthew He

Parameters are user-selectable values, typically experimentally determined, that govern the boundaries of an algorithm or program. For instance, selection of the appropriate input parameters governs the success of a search algorithm. Some of the most common search parameters in bioinformatics tools include the stringency of an alignment search tool, and the weights (penalties) provided for mismatches and gaps.

PARAMETRIC BOOTSTRAPPING, *SEE* BOOTSTRAPPING.

PARAPHYLETIC GROUP, *SEE* CLADISTICS.

PARENT, *SEE* TEMPLATE.

PARITY BIT

Thomas D. Schneider

A parity bit determines a code in which one data bit is set to either 0 or 1 so as to always make a transmitted binary word contain an even or odd number of 1s.

The receiver can then count the number of 1's to determine if there was a single error. This code can only be used to detect an odd number of errors but cannot be used to correct any error. Unfortunately for molecular biologists, the now-universal method for coding characters, 7 bit ASCII words, assigns to the symbols for the nucleotide bases A, C and G only a one bit difference between A and C and a one bit difference between C and G:

A: $101_8 = 1000001_2$
C: $103_8 = 1000011_2$
G: $107_8 = 1000111_2$
T: $124_8 = 1010100_2$

For example, this choice could cause errors during transmission of DNA sequences to the international sequence repository, GenBank. If we add a parity bit on the front to make an even parity code (one byte long), the situation is improved and more resistant to noise because a single error will be detected when the number of 1s is odd:

A: $101_8 = 01000001_2$
C: $303_8 = 11000011_2$
G: $107_8 = 01000111_2$
T: $324_8 = 11010100_2$

Related website

GenBank	http://www.ncbi.nlm.nih.gov/Genbank/

PARSIMONY

Dov Greenbaum

A general principle of evolutionary reconstruction aimed at constructing scenarios, particularly with regard to phylogenic trees, with the minimal number of events required to account for the available data.

Parsimony uses a matrix of discrete phylogenetic characters to infer phylogenetic trees for a set of set of species, or in some instances, an isolated populations of a single species.

Typically parsimony methodologies evaluate potential phylogenetic trees using a scoring system. With the number of potential trees for each species or isolated population often running into the millions, determining phylogenies becomes non-trivial particularly considering homoplasy wherein two distinct species may share a trait that was nevertheless absent from their branch point on the phylogenic tree.

Maximum parsimony analysis is a method often used to calculate phylogentic trees. The method scores individual phylogenic trees according to the degree to which they imply a parsimonious distribution: the most parsimonious tree represents the best fit of the potential relationships within the population.

Typically, a simple algorithm may be used to score the maximum parsimony, calculated the number of evolutionary steps that are necessary to explain the data distribution.

In some examples, it may be the case that an analysis will return more than one most-parsimonious trees (MPTs), although this may reflect more on the data inputed, or the lack thereof, then the methodologies themselves.

Related websites

MetaPIGA 2	http://www.metapiga.org/welcome.html
Algorithms for Parsimony Phylogeny	http://ftp.cse.sc.edu/bioinformatics/notes/020321fengwang.pdf

Further reading

Hill T, Lundgren A, Fredriksson R, Schiöth HB (2005) Genetic algorithm for large-scale maximum parsimony phylogenetic analysis of proteins. *Biochim Biophys Acta.* 30;1725(1): 19–29.

Kannan L, Wheeler WC (2012) Maximum Parsimony on Phylogenetic Networks. *Algorithms Mol Biol.* 2;7(1): 9.

Stebbing ARD (2006) Genetic parsimony: a factor in the evolution of complexity, order and emergence. *Biological Journal of the Linnean Society* 88(2): 295–308.

Whelan S (2008) Inferring trees. *Methods Mol Biol.* 452: 287–309.

PARTITION COEFFICIENT, *SEE* LOGP.

PATTERN

Matthew He

A pattern is a type of theme of recurring events or objects, sometimes referred to as elements of a set of objects. Molecular biological patterns usually occur at the level of the characters making up the gene or protein sequence.

See also Motif

P

PATTERN ANALYSIS

Nello Cristianini

The detection of relations within datasets.

In data analysis, by 'pattern' one means any relation that is present within a given dataset. Pattern analysis or discovery therefore deals with the problem of detecting relations in datasets. Part of this process is to ensure that (with high probability and under assumptions) the relations found in the data are not the product of chance, but can be trusted to be present in future data.

Statistical and computational problems force one to select a priori the type of patterns one is looking for, for example linear relations between the features (independent variables) and the labels (dependent variables).

Further reading

Duda RO, et al. (2001) *Pattern Classification*, John Wiley & Sons, USA.

See also Machine Learning

PATTERN DISCOVERY, *SEE* MOTIF DISCOVERY.

PATTERN OF CHANGE ANALYSIS, *SEE* PHYLOGENETIC EVENTS ANALYSIS.

PATTERN RECOGNITION, *SEE* PATTERN ANALYSIS.

PAUP* (PHYLOGENETIC ANALYSIS USING PARSIMONY (AND OTHER METHODS))

Michael P. Cummings

A program for phylogenetic analysis using parsimony, maximum likelihood and distance methods.

The program features an extensive selection of analysis options and model choices, and accommodates DNA, RNA, protein and general data types. Among the many strengths of the program is the rich array of options for dealing with phylogenetic trees including importing, combining, comparing, constraining, rooting and testing hypotheses.

Versions for Mac OS, and to a lesser extent Windows, have a graphical user interface; others versions are command line driven. In all versions a user can execute a series of individual commands without interaction by including appropriate commands in the PAUP block (NEXUS file format). The program accommodates several input and output formats for data. Written in ANSI C the program is available as executable files for several platforms.

Related website

PAUP* home page	http://paup.csit.fsu.edu/

Further reading

Swofford DL. (2002) *PAUP* Phylogenetic Analysis Using Parsimony (*and Other Methods, version 4).* Sinauer Associates, Sunderland, Massachusetts.

PAZAR

Obi L. Griffith and Malachi Griffith

PAZAR is an open-access, open-source software framework for the construction and maintenance of regulatory sequence annotations that allows multiple 'boutique' databases to function independently within a larger system (or information mall). It aims to be the public repository of all regulatory data. Annotation is collected from users worldwide under a model in which curators own their data and can release it according to their own interests. Users can query the database by transcription factor, gene, sequence or transcription factor binding profile. Through its model of collecting existing datasets, PAZAR has incorporated more than 85 other datasets or databases with a strong focus recently placed on large-scale ChIP-seq studies. PAZAR was conceived as an alternative to other closed and commercial models (e.g., Transfac). For related projects, see Oreganno and JASPAR.

Related websites

PAZAR	http://www.pazar.info/
OregAnno	http://www.oreganno.org
JASPAR	http://jaspar.cgb.ki.se/

Further reading

Portales-Casamar E, et al. (2007) PAZAR: a framework for collection and dissemination of cis-regulatory sequence annotation. *Genome Biol.* 8(10): R207.

Portales-Casamar E, et al. (2009) The PAZAR database of gene regulatory information coupled to the ORCA toolkit for the study of regulatory sequences. *Nucleic Acids Res.* 37(Database issue): D54–60.

PEARSON CORRELATION, *SEE* REGRESSION ANALYSIS.

PENALTY, *SEE* GAP PENALTY.

P

PENETRANCE

Mark McCarthy and Steven Wiltshire

The conditional probability that an individual with a given genotype expresses a particular phenotype.

The concept of penetrance is most readily understood in relation to Mendelian diseases. Such a disease may, for example, be described as fully penetrant if susceptibility genotypes are invariably associated with disease development (in other words, the probability associated with those genotypes is one). In multifactorial traits, where any given variant will characteristically be neither necessary nor sufficient for the development of disease, the penetrance of susceptibility genotypes will be less than one ('incomplete penetrance'), and the penetrance of non-susceptibility genotypes, greater than zero (see phenocopies). In either situation, penetrance may vary with age (most markedly so, for diseases of late onset), or gender, or pertinent environmental exposures.

Examples: The penetrance of Huntington's chorea, even in individuals known to have inherited a disease-causing allele, is effectively zero until middle-age, and climbs to 100% during late middle-life.

Further reading

Ott J (1999) *Analysis of Human Genetic Linkage*. (Third Edition) Baltimore: Johns Hopkins University Press, pp 151–170.

Peto J, Houlston RS (2001) Genetics and the common cancers. *Eur J Cancer* 37 (Suppl 8): S88–96.

Sham P (1998) *Statistics in Human Genetics*. London: Arnold, pp 88–91.

See also Multifactorial Trait, Phenocopy, Polygenic Inheritance, Oligogenic Inheritance

PEPTIDE

Roman Laskowski and Tjaart de Beer

A compound of two or more amino acids joined covalently by a peptide bond between the carboxylic acid group (–COOH) of one and the amino group (–NH) of the other. The amino acids forming a peptide are referred to as amino acid residues as their formation removes an element of water.

Long chains of amino acids joined together in this way are termed polypeptides. All proteins are polypeptides.

Further reading

Branden C, Tooze J (1998) *Introduction to Protein Structure*. Garland Science, New York.

See also Amino Acid, Peptide Bond, Residue, Polypeptides.

PEPTIDE BOND (AMIDE BOND)

Roman Laskowski and Tjaart de Beer

The covalent bond joining two amino acids, between the carboxylic (-COOH) group of one and the amino group (-NH_2) of the other, to form a peptide.

The bond has a partial double bond character and so the atoms shown below tend to lie in a plane and act as a rigid unit.

Because of this, the dihedral angle *omega*, defined about the bond, tends to be close to 180°.

Further reading

Branden C, Tooze J (1998) *Introduction to Protein Structure*. Garland Science, New York.

PEPTIDE MASS FINGERPRINT

Simon Hubbard

The mass spectrum produced by digesting a candidate protein (often from a 2D or 1D protein gel) with a proteolytic enzyme and analysing the peptide mixture. This generates a 'fingerprint' of the resulting peptide ions that is hopefully diagnostic of the protein in question, and enables its identification from a protein database using an appropriate database search engine, such as Mascot or Protein Prospector.

PEPTIDE SPECTRUM MATCH (PSM)

Simon Hubbard

In mass spectrometry-based proteomics, proteins can be characterized by putative identifications of their peptides generated by endoprotelytic digestion and analysed in the mass spectrometer. Commonly, trypsin is the proteolytic enzyme used to generate the peptides. The subsequent tandem mass spectra generated from analysis of these peptides are searched using *database search engines* such as Mascot or Sequest, to achieve candidate matches between experimental and theoretical product ion spectra. Usually, the top-scoring PSM is considered and some statistical confidence assigned via False Discovery Rate statistics or a Posterior Error Probability.

See also False Discovery Rate, Posterior Error Probability

PEPTIDEATLAS

Simon Hubbard

A popular publicly accessible database containing peptides identifications from tandem mass spectrometry based proteomic experiments, containing builds for multiple species.

The database curators search the spectra against protein databases using standard search engines, standardizing the identifications and presenting them all in a consistent format and with high quality confidence scores derived from the Trans Proteomic Pipeline (TPP).

Related website

Peptide Atlas	http://www.peptideatlas.org/

Further reading

Deutsch EW, Lam H, Aebersold R (2008) PeptideAtlas: a resource for target selection for emerging targeted proteomics workflows. *EMBO Rep.* 9: 429–434.

PERCENT ACCEPTED MUTATION MATRIX, *SEE* DAYHOFF AMINO ACID SUBSTITUTION MATRIX.

PETRI NET

Anton Feenstra

Petri nets are a formalism geared towards modeling and analysis of concurrent systems. A Place-Transition (PT) Petri net is a bipartite graph of Places and Transitions connected by arcs. It is denoted as a quadruple (P, T, F, m), where P is a set of places, T a set of transitions, F the weights of arcs, and m a marking (number of tokens in all places) of the network, which represents its state.

In its application to the operational modeling of biological systems, places represent genes, protein species and complexes, while transitions represent reactions or transfer of a signal. Arcs represent reaction substrates and products, and tokens the availability of the species represented by the place.

The intuitiveness of the graphical representation of Petri nets, tied with its strict formal definition, contributes greatly to establish a common ground for 'biologists' and 'computer scientists' in modeling the complex processes in living cells.

Further reading

Bonzanni N, Feenstra KA, Fokkink WJ, Krepska E (2009) What can formal methods bring to systems biology? *Proc. FM'09. LNCS* 5850: 16–22.

Bonzanni N, Krepska E, Feenstra KA, et al. (2009) Executing Multicellular Development: Quantitative Predictive Modelling of *C. elegans* vulva, *Bioinformatics*, 25: 2049–2056.

Chaouiya C (2007) Petri net modelling of biological networks. *Brief Bioinform* 8: 210–219.

Krepska E, Bonzanni N, Feenstra KA, et al. (2008) Design issues for qualitative modelling of biological cells with Petri nets. *Proc. FMSB'08*, volume 5054 of LNBI, pp. 48–62.

Murata T (1989) Petri nets: Properties, analysis and applications. *Proceedings of the IEEE* 77(4): 541–580.

Petri CA (1962) Kommunikation mit Automaten. Bonn: Institut für Instrumentelle Mathematik, *Schriften des IIM* Nr. 2.

Pinney, et al. (2003) *Biochem Soc Trans.* 31(Pt 6): 1513–155.

PFAM

Teresa K. Attwood

A database of hidden Markov models (HMMs) encoding a wide range of protein and domain families. A member of the InterPro consortium, Pfam specializes in characterizing domains and superfamilies that are highly divergent. As such, it complements PROSITE, which excels in the identification of functional sites, and PRINTS, which concentrates on the hierarchical diagnosis of gene families, from superfamily, through family, down to subfamily levels.

The resource has two main components: Pfam-A and Pfam-B. Entries in Pfam-A are created from seed alignments from which profile HMMs are constructed using the HMMER3 software; the profile HMMs are then searched against UniProtKB, and family-specific thresholds are applied to select sequences to contribute to final full family alignments. By contrast, Pfam-B is derived fully automatically from clusters produced by the ADDA algorithm; entries in Pfam-B are hence neither manually verified nor annotated. Annotation of entries in Pfam-A tends to be culled from existing sources (such as PROSITE, PRINTS and InterPro) or from the community via Wikipedia.

Pfam is hosted at the Wellcome Trust Sanger Institute, UK. Its entries have unique accession numbers and database identifiers, and include, where possible, functional descriptions, literature references, links to related databases and annotation resources (SCOP, PROSITE, PRINTS, InterPro, Wikipedia, etc.), some technical details of how the HMMs were derived, together with both seed and final alignments. The database may be interrogated with simple keywords, or searched with query sequences.

Related websites

Pfam	http://pfam.sanger.ac.uk/
HMMER3	http://hmmer.janelia.org/

Further reading

Finn RD, Clements J, Eddy SR (2011) HMMER web server: interactive sequence similarity searching. *Nucleic Acids Res.* (Web Server Issue) 39: W29–W37.

Heger A, Holm L (2003) Exhaustive enumeration of protein domain families. *J. Mol. Biol.* 328: 749–767.

Punta M, Coggill PC, Eberhardt RY, et al. (2012). The Pfam protein families database. *Nucleic Acids Res.* (Database Issue) 40: D290–D301.

See also Domain Family, Hidden Markov Models, InterPro, PRINTS, PROSITE, Protein Family

PFM, *SEE* POSITION-FREQUENCY MATRIX.

P

PHANTOM INDEL (FRAME SHIFT)

Jaap Heringa

A phantom indel corresponds to a spurious insertion or deletion in a nucleotide sequence that arises when physical irregularities in a sequencing gel cause the reading software either to call a base too soon, or to miss a base altogether. The occurrence of a phantom indel then leads to a frameshift, which is an alteration in the reading sense of DNA resulting from an inserted or deleted base, such that the reading frame for all subsequent codons is shifted with respect to the number of changes made. For example, if a sequence should read GUC-AGC-GAG-UAA (translated into amino acids Val-Ser-Asp as UAA is the stop codon), and a single A is added to the beginning, the new sequence would then read AGU-CAG-CGA-CUA, and give rise to a completely different protein product Ser-Gln-Arg-Leu-. . . as also the stop codon would be changed so that translation would proceed until the next spurious stop codon is encountered. Because the codon table differs between species, such that the same mRNA can lead to different proteins across species, the effects of frameshifts can also be different.

PHARMACOPHORE

Bissan Al-Lazikani

A pharmacophore is an abstraction of the geometric, steric and electronic features representing a ligand, ligand class or receptor binding pocket. These features are the most important parts of the ligand or the site for binding. Typical pharmacophore features include hydrophobic centers with defined radii, hydrogen bond donors or acceptors, cations, and anions.

Pharmacophores are a useful tool in virtual screening.

PHASE (SENSU LINKAGE)

Mark McCarthy and Steven Wiltshire

Phase describes the relationship between alleles at a pair of loci, most relevantly when an individual is heterozygous at both. For the phase of the alleles in such double heterozygotes to be determined, the haplotypes of their parents need to be known.

Consider the three-generation family in Figure P.2. The child in the third generation has genotype Aa at one locus and Bb at a second (linked) locus. Based on this information alone, it is not possible to determine the haplotype arrangement of this individual (they could be AB and ab; or Ab and aB). However, in the presence of parental (and, as in this case, grandparental) genotypes, which allow the parental haplotypes to be unambiguously defined, the phase of the parental meioses is clear, and the haplotypes of the child can be deduced. Such information, allows us, in this example, to see that (under the assumption of no mutation) there has been a paternal recombination between the two loci (marked with a cross). Linkage analysis relies on the detection of recombinant events, and when phase is known, transmitted haplotypes can readily be separated into those which are recombinant and those which are non-recombinant. When phase is not known, linkage analysis

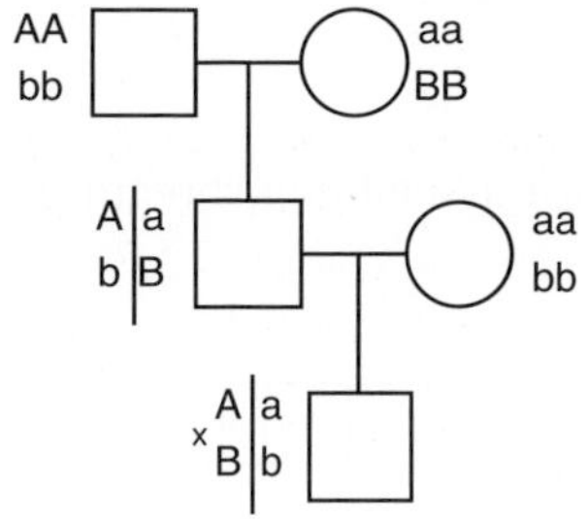

Figure P.2 Three-generation family types for two loci. See text for explanation.

is still possible and is typically implemented through maximum-likelihood methods using appropriate software. The development of statistical tools (including software listed below) to facilitate phase determination (haplotype construction) in phase-ambiguous individuals (such as unrelated cases and control subjects in whom there is no genetic information from relatives) is important for linkage disequilibrium analyses to exploit the power of haplotype analyses based on dense maps of single nucleotide polymorphisms.

Related websites

Programs for haplotype (and phase) determination:

PHASE	http://www.stats.ox.ac.uk/mathgen/software.html
HAPLOTYPER	http://www.peoplc.fas.harvard.edu/~junliu/
SNPHAP	http://www-gene.cimr.cam.ac.uk/clayton/software/
SIMWALK	http://watson.hgen.pitt.edu/docs/simwalk2.html

Further reading

Niu T, et al. (2002) Bayesian haplotype inference for single-nucleotide polymorphisms. *Am J Hum Genet* 70: 157–169.

Ott J (1999) *Analysis of Human Genetic Linkage* (Third Edition). Baltimore: Johns Hopkins University Press, pp 7 et seq.

Schaid DJ, et al. (2002) Score tests for association between traits and haplotypes when linkage phase is ambiguous. *Am J Hum Genet* 70: 425–434.

Sham P (1998) *Statistics in Human Genetics*. London: Arnold pp 62–67.

Stephens M, et al (2001) A new statistical method for haplotype reconstruction from population data. *Am J Hum Genet* 68: 978–989.

See also Haplotype, Linkage Analysis

PHEGENI, *SEE* DBGAP.

P

PHRAP
Rodger Staden

A widely used program for assembling shotgun DNA sequence data.

Related website

PHRAP	http://www.phrap.org/phredphrapconsed.html

PHRED
Rodger Staden

The base-calling program which gave its name to the PHRED scale of base call confidence values. Defines the confidence $C = -10\log(P_{error})$ where P_{error} is the probability that the base-call is erroneous.

Further reading

Ewing B, Green P (1998) Base-calling of automated sequencer traces using Phred. II. Error Probabilities. *Genome Res.* 8: 186–194.

Related website

PHRAP site	http://www.phrap.org/phredphrapconsed.html

PHYLIP (PHYLOGENY INFERENCE PACKAGE)
Michael P. Cummings

A set of modular programs for performing numerous types of phylogenetic analysis.

Individual programs are broadly grouped into several categories: molecular sequence methods; distance matrix methods; analyses of gene frequencies and continuous characters; discrete characters methods; and tree drawing, consensus, tree editing, tree distances. Together the programs accommodate a broad range of data types including, DNA, RNA, protein, restriction sites, and general data types. The programs encompass a broad variety of analysis types including parsimony, compatibility, distance, invariants and maximum likelihood, and also include both jackknife and bootstrap re-sampling methods. The output from one program often forms the input for other programs within the package (e.g., *dnadist* generates a distance matrix from a file of DNA sequences, which is then used as input for neighbor to generate a neighbor-joining tree and the tree is then viewed with *drawtree*). Therefore for a typical analysis the user makes choices regarding each aspect of an analysis and chooses specific programs accordingly. Programs are run interactively via a text-based interface that provides a list of choices and prompts users for input.

The programs are available as ANSI C source code. Executables are also available for several platforms.

Related website

PHYLIP home page	http://evolution.genetics.washington.edu/phylip.html

Further reading

Felsenstein J (1981) Evolutionary trees from DNA sequences: a maximum likelihood approach. *J. Mol. Evol.* 17: 368–376.

Felsenstein J (1985) Confidence limits on phylogenies: an approach using the bootstrap. *Evolution* 39: 783–791.

Felsenstein J (1985) Phylogenies from gene frequencies: a statistical problem. *Syst. Zool.* 34: 152.

PHYLOGENETIC EVENTS ANALYSIS (PATTERN OF CHANGE ANALYSIS)

Jamie J. Cannone and Robin R. Gutell

As noted in the Covariation Analysis section (See **Covariation Analysis**), the traditional method for determining the columns in an alignment with similar patterns of variation is to calculate the overall frequencies for each base pair type for each pairwise comparison of columns. However this simpler method that usually utilizes a chi-square, mutual information (or similar) metric does not identify the total number of changes or variations that have occurred at each position during the evolution of the RNA, our ultimate objective. The Phylogenetic Event Counting (PEC) method does estimate the number of changes based on the relationships for all of the organisms on a phylogenetic tree. These two

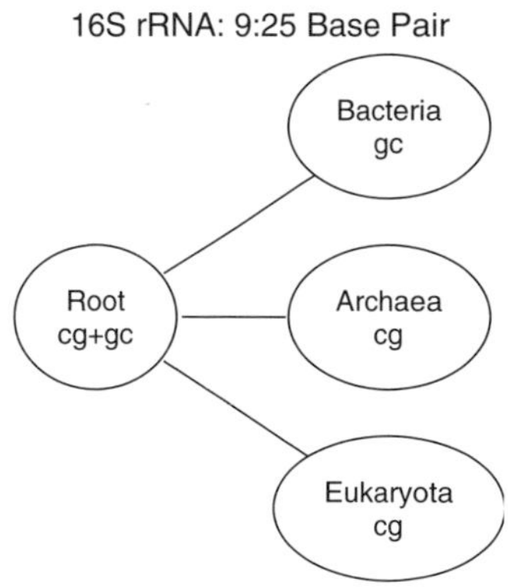

Figure P.3 Examples of Phylogenetic Event Counts for two base pairs in 16S rRNA. Position numbers are based on the *Escherichia coli* 16S rRNA.
• 9:25 base pair. Nearly all *Bacteria* have a G:C base pair at this location while nearly all of the *Archaea* and *Eukaryota* have a C:G base pair.
• 245:283 base pair. This base pair exchanges frequently between C:C and U:U within the *Bacteria*, *Archaea* and *Eukaryota*.

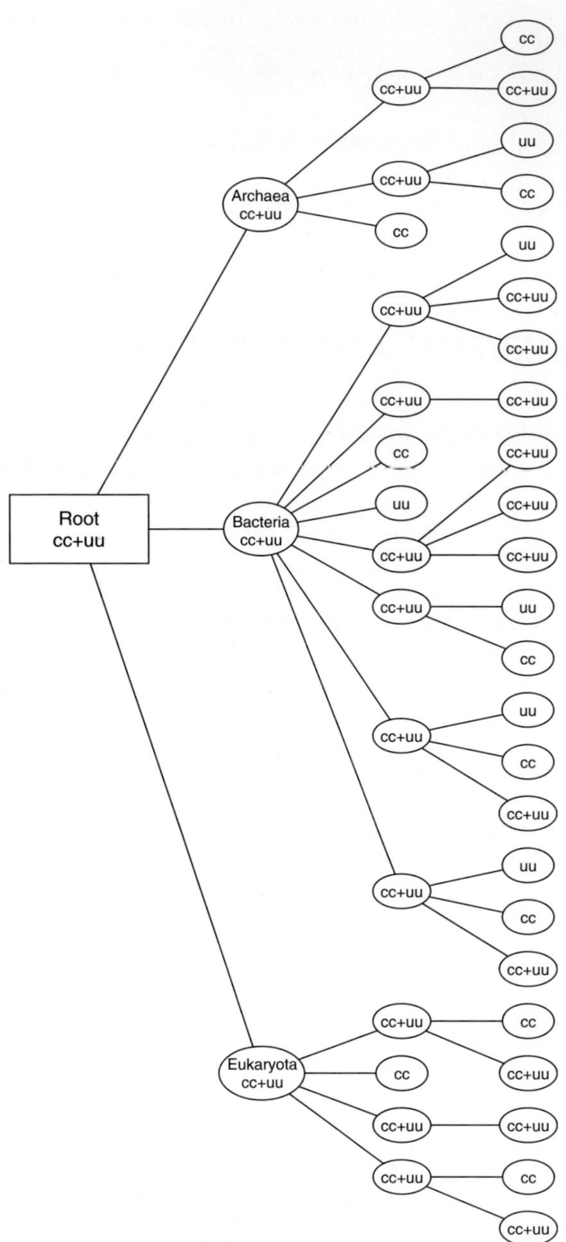

Figure P.3 (*Continued*)

different metrics are exemplified by three different base pairs in 16S rRNA: 9:25, 502:543, and 245:283 (*E. coli* numbering). The 9:25 base-pair has approximately 67% G:C and 33% C:G in the nuclear-encoded rRNA genes in the three primary forms of life, the Eucarya, Archaea, and the Bacteria. The minimal number of times or events these base pairs evolved (phylogenetic events) on the phylogenetic tree is approximately 1. In contrast the 245:283 base-pair has 38% C:C and 62% U:U in the same set of 16S rRNA sequences, with a minimum of 25 phylogenetic events (see Figure P.3).

Analysis published in 1986 revealed that a PEC covariation metric has the potential to be a more sensitive gauge for positional covariation. The number of mutual changes during the evolution of that base pair is directly proportional to the greater confidence for the authenticity of a base pair. These analyses determined the number of phylogenetic events after the base pair was tentatively predicted with traditional methods.

However, the search for covariation with PEC methods requires more sophisticated computational methods and compute resources. Recently it has been possible to either estimate the number of phylogenetic events based on models or map the variations directly onto each node of a phylogenetic tree. The results, as expected, do reveal more positional covariations are identified when the number of mutual changes are measured during the evolution of the RNA, in contrast with a simpler count of the frequency for each base pair type.

Further reading

Dutheil J, Pupko T, Jean-Marie A, Galtier N (2005) A model-based approach for detecting coevolving positions in a molecule. *Mol. Biol. Evol.* 22: 1919–1928.

Gutell RR, Noller HF, Woese CR (1986) Higher order structure in ribosomal RNA. *The EMBO Journal* 5: 1111–1113.

Shang L, Xu W, Ozer S, Gutell RR (2012) Structural constraints identified with Covariation Analysis in 16S ribosomal RNA. *PLoS ONE* 7: e39383.

Yeang CH, Darot JF, Noller HF, Haussler D (2007) Detecting the coevolution of biosequences – an example of RNA interaction prediction. *Mol. Biol. Evol.* 24: 2119–2131.

PHYLOGENETIC FOOTPRINT

Jacques van Helden

Non-coding genomic region conserved between distant species, presumably because it contains binding sites for transcription factors, and, consequently, mutations that would disrupt these binding sites were counter-selected during evolution.

The term ‘footprint’ was used by reference to the experimental method called ‘DNAse footprinting’, classically used to characterize the precise binding locations of transcription factors. Sequencing gels shows ‘footprints’ (i.e. missing bands) at the places where DNA is bound by a transcription factor, and thereby protected from the cleavage action of DNAse. Similarly, ‘phylogenetic footprints’ reveal positions where a selective pressure supposedly eliminated variations that would disrupt some cis-regulatory element.

P

PHYLOGENETIC FOOTPRINT DETECTION

Jacques van Helden

Detection of phylogenetic footprints, i.e. conserved motifs in regulatory regions of orthologous genes.

Various approaches have been applied to detect phylogenetic footprints. The detection of non-coding conserved blocks relies on the alignment of several genomes onto a genome of reference (e.g. all insect genomes against drosophila melanogaster). This approach relies on an assumption that the cis-regulatory region is conserved as a whole (e.g. a few hundreds base pairs corresponding to a full enhancer), and may thus miss effective cis-regulatory elements and modules due to site shuffling or degradation of the inter-site sequences. The tools DIRE and CREME detect combinations of known motifs co-occurring in conserved blocks (see Motif Enrichment Analysis).

Several motif discovery algorithms have been adapted to analyse cis-regulatory regions of orthologous genes: *Footprinter* (Blanchette et al., 2003), *PhyloCon* (Wang and Stormo, 2003), *PhyloGibbs* (Siddhartan et al., 2005), *footprint-discovery* (Janky and van Helden, 2008), to cite a few. These approaches do not rely on the assumption of conservation of whole regulatory regions (cis-regulatory modules). These methods can also detect sites without assuming conservation of their position, and are thus more robust to the shuffling of regulatory sites.

Related websites

UCSC genome browser	http://genome.ucsc.edu/
Monkey	http://labs.csb.utoronto.ca/moses/monkey.html http://rana.lbl.gov/monkey
Footprinter	http://genome.cs.mcgill.ca/cgi-bin/FootPrinter3.0/ FootPrinterInput2.pl
footprint-discovery (RSAT suite)	http://www.rsat.eu/
DIRE/CREME	http://dire.dcode.org/
MEME suite	http://meme.nbcr.net/meme/
PhyloCon	http://stormo.wustl.edu/PhyloCon/
PhyloGibbs	http://www.phylogibbs.unibas.ch/cgi-bin/ phylogibbs.pl

Further reading

Blanchette M, Tompa M (2003) FootPrinter: A program designed for phylogenetic footprinting. *Nucleic Acids Res* 31(13): 3840–3842.

Janky R, van Helden J (2008) Evaluation of phylogenetic footprint discovery for predicting bacterial cis-regulatory elements and revealing their evolution. *BMC Bioinformatics* 9: 37.

Siddharthan R, Siggia ED, van Nimwegen E. 2005. PhyloGibbs: a Gibbs sampling motif finder that incorporates phylogeny. *PLoS Comput Biol* 1(7): e67.

Wang T, Stormo GD (2003) Combining phylogenetic data with co-regulated genes to identify regulatory motifs. *Bioinformatics* 19(18): 2369–2380.

See also Phylogenetic Footprint, Motif Discovery, Motif Enrichment Analysis

PHYLOGENETIC PLACEMENT OF SHORT READS

Aidan Budd, Erick Matsen and Alexandros Stamatakis

Algorithms/methods for assigning collections of short reads (e.g. from metagenomics samples) to the branches of a given reference tree based on a full-length reference sequence alignment. Initially, all reads are aligned with respect to the reference alignment. Thereafter, for each short read a separate and independent (of all other reads) optimal insertion into the reference tree is computed. Such assignments can be computed by using parsimony, likelihood or posterior probability. The output is a reference tree and a distribution of reads over that reference phylogeny. This can serve to determine evolutionary read provenance and to compare different metagenomic samples, without using a species or OTU concept, by using the same reference tree as a fixed scaffold. There also exist dedicated alignment methods for aligning short reads to the joint signal of the reference alignment and reference phylogeny.

Software

Stark M, Berger S, Stamatakis A, von Mering C (2010) MLTreeMap-accurate Maximum Likelihood placement of environmental DNA sequences into taxonomic and functional reference phylogenies. *BMC Genomics* 11(1): 461.

Berger SA, Krompass D, Stamatakis A (2011) Performance, accuracy, and web server for evolutionary placement of short sequence reads under maximum likelihood. *Systematic Biology* 60(3): 291–302.

Berger SA, Stamatakis A (2011) Aligning short reads to reference alignments and trees. *Bioinformatics* 27(15): 2068–2075.

Matsen FA, Kodner RB, Armbrust EV (2010) pplacer: linear time maximum-likelihood and Bayesian phylogenetic placement of sequences onto a fixed reference tree. *BMC Bioinformatics* 11: 538.

Further reading

Evans SN, Hoffman NG, Matsen FA (2012) Edge principal components and squash clustering: using the special structure of phylogenetic placement data for sample comparison. Submitted, http://arxiv.org/abs/1107.5095

Evans SN, Matsen FA (2012) The phylogenetic Kantorovich-Rubinstein metric for environmental sequence samples. Accepted to the Journal of the Royal Statistical Society (B), preprint: http://arxiv.org/abs/1005.1699.

PHYLOGENETIC PROFILE

Patrick Aloy

Phylogenetic profiles are based on the presence or absence of a certain number of genes in a set of genomes. Systematic presence or absence of a set of proteins (e.g. similarity of phylogenetic profiles) might indicate a common functionality of these proteins. Sometimes this reflects physical interactions between proteins forming a complex, but, most often implies a related function (e.g. they are components of the same metabolic pathway).

P

This approach can only be applied to entire genomes (i.e. not individual pairs of proteins) and it cannot be used to test interactions between essential proteins present in most organisms (e.g. ribosomal proteins).

Further reading

Pellegrini M, Marcotte EM, Thompson MJ, Eisenberg D, Yeates TO (1999) Assigning protein functions by comparative genome analysis: protein phylogenetic profiles. *Proc Natl Acad Sci USA* 96: 4285–4288.

PHYLOGENETIC RECONSTRUCTION, *SEE* PHYLOGENETIC TREE.

PHYLOGENETIC SHADOWING, *SEE* PHYLOGENETIC FOOTPRINTING.

PHYLOGENETIC TREE (PHYLOGENY, PHYLOGENY RECONSTRUCTION, PHYLOGENETIC RECONSTRUCTION)

Jaap Heringa

A phylogenetic tree depicts the evolutionary relationships of a target set of sequences. It contains interior and exterior (terminal) nodes. Normally, the input sequences are contemporary and referred to as the operational taxonomic units (OTUs). They correspond with the exterior nodes of the evolutionary tree, whereas the internal nodes represent ancestral sequences which have to be guessed from the OTUs and the tree topology. The length of each branch connecting a pair of nodes may correspond to the estimated number of substitutions between two associated sequences. The minimal evolution hypothesis is that the 'true' phylogenetic tree is the rooted tree; i.e., contains a node ancestral to all other nodes, which has the shortest overall length and thus comprises the lowest cumulative number of mutations.

The computation of a phylogenetic tree typically involves four steps: (i) selection of a target set of orthologous sequences; (ii) multiple alignment of the sequences; (iii) construction of a pairwise sequence distance matrix by calculating the pairwise distances using the multiple alignment; and (iv) compilation of the tree by applying a clustering technique to the distance matrix. As a main alternative to the last two steps, parsimony or maximum likelihood methods may also be used to infer a tree from a multiple alignment (see below).

The construction of phylogenetic trees from protein and DNA data was pioneered more than three decades ago by Edwards and Cavalli-Sforza (1963) who reconstructed phylogenies by exploring the concept of minimum evolution based on gene frequency data.

Camin and Sokal (1965) first implemented this concept on discrete morphological characters, which they called the parsimony method, a now standard term. Following Camin and Sokal, Eck and Dayhoff (1966) then introduced the parsimony method for molecular sequence data. Other early attempts to reconstruct evolutionary pathways from present-day sequences were based on distance methods which explore a matrix containing all pairwise distances of a set of multiply aligned sequences (Sokal and Sneath, 1963; Cavalli-Sforza and Edwards, 1967; Fitch and Margoliash, 1967). All of these methods, each with different emphasis, try to reconstruct the past using a minimalist approach; i.e., they use as few evolutionary changes as possible. More recently, another important class of computer methods appeared for sequence data based on a stochastic model of evolution and was named the maximum likelihood technique (Felsenstein, 1981a).

A widely used principle early on for constructing evolutionary or phylogenetic trees from protein sequence data is that of maximum parsimony (Farris, 1970) based on a deterministic model of sequence evolution and requiring as few mutations as possible. Unlike distance methods, parsimony methods are character based and examine each sequence position (character) separately. Multiple alignments of the subject sequences (OTUs) are screened for the occurrence of *phylogenetically informative* alignment positions where at least two residue types are represented more than once. From the phylogenetically informative columns, the number of mutations between each sequence (including hypothetical ancestors) is determined.

The first step of a parsimony method deals with the generation of the interior nodes of a tree. Fitch (1971) proved that it is possible, given an initial alignment of sequences and an initial tree or dendrogram, to construct maximum parsimony ancestral sequences associated with the interior nodes of the tree such that the minimum number of mutations is obtained. So, for every given tree topology it is possible to infer the ancestral sequences associated with the internal nodes together with all the branch lengths, yielding the minimum overall length for that tree. The second step is computationally prohibitive since it involves, for a given set of OTUs, exploring all possible tree topologies to find the one with minimum cost. There is no global strategy to infer the most parsimonous tree, which therefore can only be found by exhaustive searching. Unfortunately, the number of trees to be inspected explodes with the number of sequences such that much more than 10 sequences is not feasible. The number of possible rooted tree topologies for N sequences is $(2n-3)!/(2^{n-2}(n-2)!)$, yielding about 34.5 × 106 different topologies for only 10 sequences. Therefore, heuristic search approaches are needed.

Felsenstein (1978, 1981b) and Saitou and Nei (1986) demonstrated that parsimony methods cannot deal well with many biological phenomena, including the occurrence of mutation rates that differ through various lineages, or the fact that not all alignment positions are equally informative given different tertiary structural constraints. Felsenstein (1981b) therefore suggested applying different weights to different alignment positions. Generally, parsimony methods are effective for closely related sequences resulting from divergent evolution.

Another issue is the multiple sequence alignment required as input to infer mutations. As Hogeweg and Hesper (1984) and Feng and Doolittle (1987) demonstrated, multiple alignment accuracy and derivation of a phylogenetic tree are closely interrelated. These authors devised phylogenetic methods that utilize this relationship using distance methods. The construction of a multiple alignment is dependent on an amino acid exchange matrix and gap penalties, especially when distant sequences are aligned which are likely to acquire many insertions and deletions. An accurate multiple alignment in fact requires an *a priori* knowledge of evolutionary relationships (see Tree-based Progressive Alignment).

Many different parsimony programs have been described over the years, generally operating on nucleotide sequences. Moore *et al.* (1973) extended the parsimony method of Fitch (1971) for nucleotide sequences to protein sequences. Also the PHYLIP package (Felsenstein, 1989, 1990) and PAUP (Swofford, 1992) offer various parsimony algorithms for protein sequences. A further class of methods related to maximum parsimony techniques is that of compatibility analysis (Le Quesne, 1974) where tree branching is made compatible with observed mutations at a maximum number of given multiple alignment sites; sites that are not compatible are ignored. Maximum likelihood methods (Felsenstein, 1981b) present an alternative way to derive the phylogeny of a set of sequences. Here the tree construction and assignment of branch lengths is performed using evolutionary probabilities of nodal connections from which statistical significance is inferred. The probabilities are based on a stochastic model of sequence evolution. Furthermore, alternative tree topologies can be readily evaluated using the associated likelihoods. The method attempts to maximize the probability that the data will fit well onto a tree under a given evolutionary model. Explicit assumptions of sequence evolution are required, such as equal mutation rates, independent evolution of alignment sites, and the like. They can be tailored to a particular data set under study. Golding and Felsenstein (1990) used maximum likelihood calculations to detect the occurrence of selection in the evolution of restriction sites in *Drosophila*. Unfortunately, likelihood methods are also computer intensive, difficult to apply to large data sets and can fall into local traps (Golding and Felsenstein, 1990).

Although most maximum likelihood methods deal with nucleotide sequences (e.g., Felsenstein, 1989, 1990; Olsen et al., 1994), the DNAML program from the PHYLIP package (Felsenstein, 1990) allows codon positions to be independently weighted for protein sequence analysis. Adachi and Hasegawa (1992) devised a maximum likelihood program for protein phylogeny, which can handle larger sets of sequences using a fast star-decomposition approach to search through tree topologies (Saitou and Nei, 1987).

Closely related to likelihood methods are the so-called invariant techniques (e.g. Lake, 1987), which compare phylogenetically trees only based on their topology, thereby ignoring branch lengths. Their advantage is insensitivity to different branch lengths and different evolutionary rates at various (alignment) sites but to date, they can only operate on four-species trees and nucleotide data.

Distance methods derive a tree from a distance matrix which compares pairwise all tree constituents. Distances can be obtained from sequence identities (Fitch and Margoliash, 1967) or pairwise sequence alignment scores (Hogeweg and Hesper, 1984). An agglomerative cluster criterion is utilized reflecting the evolutionary information in the best possible way. One advantage of distance methods over parsimony methods is that they usually are less CPU intensive as they employ a fixed strategy to arrive at a final tree without the need to sample the complete tree space. Many clustering criteria have been introduced over the years, each having an underlying assumption of evolutionary dynamics.

The first cluster method used in molecular sequence phylogeny was the UPGMA or group averaging method (Sokal and Sneath, 1963). It takes the average value over all inter-group distances to measure the evolutionary distance between two groups of sequences and has the underlying assumption of identical mutation rates in all lineages. It is an appropriate technique only if the so-called 3-point condition holds and the distance matrix is ultrametric, which implies a uniform evolutionary clock with equal mutation rates in all lineages. For not too large differences in evolutionary clock speeds in different lineages, the method of Fitch and Margoliash (1967) is more appropriate since it infers the branch length for each nearest-neighbor pair with information from the next closest sequence. The Fitch-Margoliash method yields a tree topology identical to that from the UPGMA technique, but the branch lengths could be different.

The method of Saitou and Nei (1987), called neighbor-joining (NJ), is acclaimed by many workers in phylogenetic analysis. The neighbor-joining method is able to reconstruct evolution properly under a variety of parameters such as different mutation rates, back mutations, and the like (Saitou and Imanishi, 1989; Czelusniak *et al*., 1990; Huelsenbeck and Hillis, 1993). The method relies on a protocol of progressive pairwise joining of nearest sequences (each sequence being represented by a node) such that each time two nodes are joined, they are represented by an internal node; the two nodes selected at each step for joining keep the overall tree length at a minimum. This process is iterated until all nodes are joined. The NJ method departs from the UPGMA and Fitch–Margoliash techniques in that it does not use the evolutionary (dis)similarity among groups but merely is a strategy to join sequences and calculate branch lengths. For the NJ method to work well, the distance matrix should comply with the four-point condition, meaning that all pairwise sequence distances are additive. However, it has been shown to be reasonably robust in more complex cases where the four-point condition does not hold.

The simplest way to calculate the distance between sequences is by the percent divergence, which for two aligned sequences involves counting the number of non-identical matches (ignoring positions with gaps) divided by the number of positions considered. When a multiple alignment is used, all positions containing a gap in any of the sequences are commonly ignored. The real evolutionary time between the divergence of two sequences depends on the speed of the evolutionary clock, a matter of ongoing controversy. Even under a uniform clock, sequence identity as a measure of distance underestimates the real number of mutations. Certainly in diverged sequences there is an increasing chance that multiple substitutions have occurred at a site. The greater the divergence, the more the evolutionary times are underestimated. For example, Dayhoff *et al*. (1978) estimated that a sequence identity of 20% would correspond to about 2.5 mutations per amino acid. The Dayhoff *et al*. relationship between observed percent identity and percent accepted mutation shows that the estimated evolutionary time rises much faster than the observed sequence distances; the curve is not defined for sequence differences higher than 93%. Kimura (1983) corrected for this effect by curve fitting such that the corrected evolutionary time from the distance K (percent divergence divided by 100) is given by $K_{\text{corrected}} = -\ln(1.0 - K - K^2/5.0)$. The formula applies to cases with a reasonably uniform evolutionary clock and sequence identities from 15% and fits the data well from identities higher than 35%. It is good practice to start tree-building routines from such corrected sequence distances. The Dayhoff *et al*. relationship is valid for a fixed amino acid composition of the data. Whenever deviating compositions occur, the user can resort to the PHYLIP package (Felsenstein, 1989, 1990) to calculate evolutionary distances based on the actual amino acid composition.

Because evolutionary trees may be a result of local traps in the search space, it is important to estimate the significance of a particular tree topology and associated branch lengths. Felsenstein (1985) introduced the concept of bootstrapping, which involves resampling of

P

the data such that the multiple alignment positions are randomly selected and placed in some order and then a tree is generated by the original method. This process is repeated a statistically significant number of times. Comparison of frequencies at which the $N-3$ internal branches (N is the number of sequences) occur in the original tree and those from bootstrapping allow probability estimates of significance. It is common practice to consider frequencies higher than or equal to 95% as supportive for the occurrence of an original tree branch in the bootstrapped trees. In this way, bootstrapping tests the stability of groupings given the data set and the method, thus lessening the chance of incorrect tree structures due to conservative and/or back mutations. Bootstrapping can be performed for any method that generates a tree from a multiple alignment, albeit the biological significance of a particular tree is not addressed.

A molecular biologist wishing to derive a phylogeny from a given set of protein sequences, should first seek a sensitive multiple alignment routine. If secondary or tertiary structures are known, the information should be applied to achieve the most accurate alignments. A variety of phylogenetic methods should then be used (preferably including maximum likelihood methods) in conjunction with bootstrapping and the results compared carefully for consistency. It should be kept in mind that many wrong trees can be derived from a particular sequence set, often with more seemingly interesting phylogenics than that of the one correct tree.

Related websites

PHYLIP	http://evolution.genetics.washington.edu/phylip.html
PAUP	http://paup.csit.fsu.edu/

Further reading

Adachi J, Hasegawa M (1992) MOLPHY: programs for molecular phylogenetics, I. PROTML: maximum likelihood inference of protein phylogeny. *Computer Science Monographs* no. 27, Institute of Statistical Mathematics, Tokyo, Japan.

Camin JH, Sokal RR (1965) Computer comparison of new and existing criteria for constructing evolutionary trees from sequence data. *J. Mol. Evol.* 19: 9–19.

Cavalli-Sforza LL, Edwards AWF (1967) Phylogenetic analysis: models and estimation procedures. *Am. J. Hum. Genet.* 19: 233–257.

Czelusniak J, Goodman M, Moncrief ND, Kehoe SM (1990) Maximum parsimony approach to construction of evolutionary trees from aligned homologous sequences. *Methods Enzym.* 183: 601–615.

Dayhoff MO (1965) *Atlas of Protein Sequence and Structure*. Maryland: National Biomedical Research Foundation.

Eck RV, Dayhoff, MO (1966) *Atlas of Protein Sequence and Structure 1966*. Silver Spring, Maryland: Natl. Biomed. Res. Found.

Edwards AFW, Cavalli-Sforza LL (1963) The reconstruction of evolution. *Ann. Hum. Genet.* 27: 105.

Farris JS (1970) Methods for computing Wagner trees. *Syst. Zool.* 19: 83–92.

Felsenstein J (1978) Cases in which parsimony or compatibility methods will be positively misleading. *Syst. Zool.* 27: 401–410.

Felsenstein J (1981a) Evolutionary trees from DNA sequences: a maximum likelihood approach. *J. Mol. Evol.* 17: 368–376.

Felsenstein J (1981b) A likelihood approach to character weighting and what it tells us about parsimony and compatibility. *Biol. J. Linn. Soc.* 16: 183–196.

Felsenstein J (1985) Confidence limits on phylogenies: an approach using the bootstrap. *J. Evolution* 39: 783–791.

Felsenstein J (1989) PHYLIP – phylogeny inference package (version 3.2). *Cladistics* 5: 164–166.

Felsenstein J (1990) *PHYLIP Manual version 3.3.* University Herbarium, University of California, Berkeley, CA.

Feng DF, Doolittle RF (1987) Progressive sequence alignment as a prerequisite to correct phylogenetic trees. *J. Mol. Evol.* 21: 112–125.

Fitch WM (1971) Toward defining the course of evolution: minimum change for a specified tree topology. *Syst. Zool.* 20: 406–416.

Fitch WM, Margoliash E (1967) Construction of phylogenetic trees. *Science* 155: 279–284.

Golding B, Felsenstein J (1990) A maximum likelihood approach to the detection of selection from a phylogeny. *J. Mol. Evol.* 31: 511–523.

Hogeweg P, Hesper B (1984) The alignment of sets of sequences and the construction of phyletic trees: an integrated method. *J. Mol. Evol.* 20: 175–186.

Huelsenbeck JP, Hillis DM (1993) Success of phylogenetic methods in the four-taxon case. *Syst. Biol.* 42: 247–264.

Kimura M (1983) *The Neutral Theory of Molecular Evolution.* Cambridge University Press, Cambridge, England.

Lake JA (1987) A rate-independent technique for analysis of nucleic acid sequences. *Mol. Biol. Evol.* 4: 167.

Le Quesne WJ (1974) The uniquely evolved character concept and its cladistic application. *Syst. Zool.* 23: 513–517.

Moore GW, Barnabas J, Goodman M (1973) A method for constructing maximum parsimony ancestral amino acid sequences on a given network. *J. Theor. Biol.* 38: 459–485.

Olsen GJ, Matsuda H, Hagstrom R, Overbeek R (1994) fastDNAml: a tool for construction of phylogenetic trees of DNA sequences using maximum likelihood. *CABIOS* 10: 41–48.

Saitou N, Imanishi T (1989) Relative efficiencies of the Fitch-Margoliash, maximum parsimony, maximum-likelihood, minimum-evolution, and neighbor-joining methods of phylogenetic tree construction in obtaining the correct tree. *Mol. Biol. Evol.* 6: 514–525.

Saitou N, Nei M (1986) The number of nucleotides required to determine the branching order of three species with special reference to human-chimpanzee-gorilla divergence. *J. Mol. Evol.* 24: 189–204.

Saitou N, Nei M (1987) The neighbor-joining method: a new method for reconstructing phylogenetic trees. *Mol. Biol. Evol.* 4: 406–425.

Sokal RR, Sneath PHA (1963) Principles of numerical taxonomy. W.H. Freeman, San Fransisco.

Swofford DL (1992) PAUP: Phylogenetic Analysis Using parsimony, Version 3.0s, Illinois Natural History Survey, Champaign, IL.

See also Alignment, Multiple

PHYLOGENETIC TREES, DISTANCE, *SEE* DISTANCES BETWEEN TREES.

P

PHYLOGENETICS

Aidan Budd and Alexandros Stamatakis

The study of the phylogenetic relationships between biological taxa, i.e., the patterns of transfer of genetic information between taxa. These studies focus on both (i) methods used to estimate these patterns, i.e., phylogenetic trees and (ii) the description of the processes of transformation of characters of taxonomic units along the branches of phylogenetic trees.

PHYLOGENOMICS

Jean-Michel Claverie

An area of research making use of phylogenetic methods (hierarchical dendrogram building) in a large-scale context.

Apparently introduced by Eisen to designate the use of evolutionary relationship to complement straightforward similarity between homologues to more accurately predict gene functions. The term has then been extended to include a wide range of studies with nothing in common but the computation of trees from a large number of sequences. The database of Clusters of Orthologous Groups of proteins (COGs) is one successful example of phylogenomic study. More recently, this term has been used to qualify the phylogenetic classification of organisms based on properties computed from their whole gene content. Accordlingly, phylogenomics is part of comparative genomics. Finally, this term is also used to designate whole *vs.* all other genome analyses to establish relationships between genes of known and unknown functions such co-evolution profiling, 'Rosetta stone' gene-fusion detection, or genomic context studies.

Related websites

Clusters of Orthologous Groups	http://www.ncbi.nlm.nih.gov/COG/
UCLA-DOE Institute for Genomics and Proteomics	http://www.doe-mbi.ucla.edu/

Further reading

Eisen JA (1998) Phylogenomics: improving functional predictions for uncharacterized genes by evolutionary analysis. *Genome Res.* 8: 163–167.

Huynen M, et al. (2000) Predicting protein function by genomic context: quantitative evaluation and qualitative inferences. *Genome Res.* 10: 1204–1210.

Marcotte EM et al. (1999) Detecting protein function and protein-protein interactions from genome sequences. *Science* 285: 751–753.

Pellegrini M, et al. (1999) Assigning protein functions by comparative genome analysis: protein phylogenetic profiles. *Proc. Natl. Acad. Sci. USA.* 96: 4285–4288.

Sicheritz-Ponten T, Andersson SG (2001) A phylogenomic approach to microbial evolution. *Nucleic Acids Res.* 29: 545–552.

Snel B, et al. (1999) Genome phylogeny based on gene content. *Nat. Genet.* 21: 108–110.

Tatusov RL, et al. (1997) A genomic perspective on protein families. *Science* 278: 631–637.

See also Phylogenetic Tree

PHYLOGENY, PHYLOGENY RECONSTRUCTION, *SEE* PHYLOGENETIC TREE.

PIECEWISE-LINEAR MODELS

Luis Mendoza

Piecewise-linear models have been proposed for the modeling of regulatory networks. These equations, originally proposed in (Glass and Kauffman, 1973) have been amply studied (Glass and Pasternack, 1978; Mestl et al., 1995a; 1995b) from a theoretical point of view. These models were developed under the assumption that the switch-like behavior of a gene can be approximated by discontinuous step functions. Variables in the piecewise-linear approach represent the concentrations of proteins, while the differential equations describe their rate of production as a function of their regulatory inputs minus a decay rate. Each differential equation contains two terms, namely the activation part consisting of a weighted sum of products of step functions and the decay rate. The mathematical form of these equations divides the state space into n-dimensional hyperboxes. The main advantage of this approach is that inside each of these multidimensional boxes, the equations are reduced to linear ODEs, making the behavior of the system inside a given volume straightforward to analyse. Despite the clear advantage of studying multiple regions of local linear behavior, the global behaviour of picewise-linear differential equations can be quite complex, and is not well understood in general. Indeed, it has been shown that chaotic behaviour is rather common in spaces where $n \geq 4$ (Lewis and Glass, 1991; Mestl et al., 1996; 1997). However, piecewise-models have been used to analyse several regulatory networks of biological interest (De Jong et al., 2004; Ropers et al., 2006; Viretta and Fussenegger, 2004).

Further reading

De Jong H, Gouzé JL, Hernandez C, Page M, Sari T, Geiselmann J (2004) Qualitative simulation of genetic regulatory networks using piecewise-linear models. *Bull. Math. Biol.* 66: 301–340.

Glass L, Kauffman SA (1973) The logical analysis of continuous non-linear biochemical control networks. *J. Theor. Biol.* 39: 103–129.

Glass L, Pasternack JS (1978) Prediction of limit cycles in mathematical models of biological oscillations. *Bull. Math. Biol.* 40: 27–44.

Lewis JE, Glass L (1991) Steady states, limit cycles, and chaos in models of complex biological networks. *Int. J. Bifurcat. Chaos* 1: 477–483.

Mestl T, Plahte E, Omholt SW (1995a) A mathematical framework for describing and analysing gene regulatory networks. *J. Theor. Biol.* 176: 291–300.

Mestl T, Plahte E, Omholt SW (1995b) Periodic solutions in systems of piecewise-linear differential equations. *Dynam. Stabil. Syst.* 10: 179–193.

Mestl T, Bagley RJ, Glass L (1997) Common chaos in arbitrarily complex feedback networks. *Phys. Rev. Lett.* 79: 653–656.

Mestl T, Lemay C, Glass L (1996) Chaos in high-dimensional neural and gene networks. *Physica D* 98: 33–52.

P

Ropers D, de Jong H, Page M, Schneider D, Geiselmann, J (2006) Qualitative simulation of the carbon starvation response in Escherichia coli. *Biosystems* 84: 124–152.

Viretta AU, Fussenegger M (2004) Modeling the quorum sensing regulatory network of human-pathogenic *Pseudomonas aeruginosa. Biotechnol. Prog.* 20: 670–678.

PIPMAKER

John M. Hancock

A program to identify evolutionarily conserved regions between genomes.

PIPMAKER carries out alignment between two long genomic sequences using BLASTZ, a variant of the BLAST algorithm. BLASTZ generates a series of local alignments which form the raw material of PIPMAKER. The main novel feature of PIPMAKER is the display of these alignments in the form of a Percentage Identity Plot, or PIP. This displays the local alignments in the context of the first sequence, which may have been annotated to show positions of exons and is processed with RepeatMasker to show positions of repeated sequences. CpG islands are also identified. The segments of sequence 2 that align with sequence 1 are represented as horizontal lines below sequence 1 and are placed on a vertical axis corresponding to the percentage match in the alignment. In this way regions of strong similarity between genomes are readily visualised. Matches can be constrained so that they have to be in the same order and orientation in the two genomes. MULTIPIPMAKER takes a similar approach but allows comparison of more than two sequences.

Related website

PIPMAKER website	http://bio.cse.psu.edu/pipmaker/

Further reading

Schwartz S, et al. (2000) PipMaker – A web server for aligning two genomic DNA sequences. *Genome Res.* 10: 577–586.

See also BLAST

PLANTSDB

Dan Bolser

PlantsDB is a public database of plant genomic information run by the MIPS plant genomics group. Starting with Arabidopsis, the group now contributes to the analysis of maize, *Medicago truncatula*, Lotus, rice, tomato, sorghum, barley and other plants. PlantsDB provides a platform for integrative and comparative plant genome research, covering monocots and dicots. The tools provided include the CrowsNest comparative map viewer.

Related website

PlantsDB	http://mips.helmholtz-muenchen.de/plant/genomes.jsp

Further reading

Spannagl M, Noubibou O, Haase D, et al. (2007) MIPSPlantsDB – plant database resource for integrative and comparative plant genome research. *Nucleic Acids Res.* 35: 834–40.

See also Ensembl Plants, Gramene, MaizeGDB, SGN

PLESIOMORPHY

A.R. Hoelzel

An ancestral or unchanged state of an evolutionary character.

As a hypothesis about the pattern of evolution among operational taxonomic units (OTUs), a phylogenetic reconstruction can inform us about the relationship between character states. We can identify ancestral and descendent states, and therefore identify primitive and derived states. A plesiomorphy (meaning 'near-shape' in Greek) is the primitive state (shown as open circles in Figure P.4).

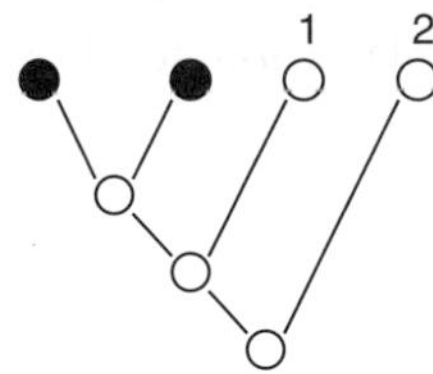

Figure P.4 Illustration of plesiomorphy. Open circles (1 & 2) represent OTUs with an ancestral state of the character.

A shared primitive state between OTUs, as between 1&2 in Figure P.4, is known as a synplesiomorphy (*c.f.* synapomorphy).

Further reading

Maddison DR, Maddison WP (2001) *MacClade 4: Analysis of Phylogeny and Character Evolution.* Sinauer Associates.

See also Apomorphy, Synapomorphy, Autapomorphy

POINT ACCEPTED MUTATIONS, *SEE* PAM MATRIX OF NUCLEOTIDE SUBSTITUTIONS.

P

POLAR
Roman Laskowski and Tjaart de Beer

A molecule which has an uneven distribution of electrons and a consequent dipole moment is said to be polar. Perhaps the best-known polar molecule is that of water.

In proteins, the amino acid residues having polar side chains are: serine, threonine, cysteine, asparagine, tyrosine, histidine, tryptophan and glutamine.

See also Amino Acid, Side Chain

POLARIZATION
Roland Dunbrack

Changes in electron distribution in a molecule due to intermolecular and intramolecular interactions. Intermolecular interactions include solvent and ions, as well as other kinds of molecules. Intramolecular polarization may occur as a flexible molecule changes conformation. Polarization is often not represented explicitly in molecular mechanics potential energy functions, but rather is accounted for by partial charges that are averaged over likely interactions with solvent.

Further reading

Halgren TA, Damm W (2001) Polarizable force fields. *Curr Opin Struct Biol* 11, 236–242.

POLYGENIC EFFECT, *SEE* POLYGENIC INHERITANCE.

POLYGENIC INHERITANCE (POLYGENIC EFFECT)
Mark McCarthy and Steven Wiltshire

A genetic architecture whereby variation in a trait is determined by variation in many genes, each of which has individually small, additive, effects on the phenotype.

The term owes its origin to RA Fisher who demonstrated in 1918 that continuous characteristics could be explained in terms of the small additive effects of many genes, each inherited according to Mendel's Law of Segregation. The aggregate of such genes is sometimes referred to as a polygene. Models for quantitative traits, variance components analysis and segregation analysis routinely feature polygenic components, often in addition to genes of individually larger, measurable, effect termed quantitative trait loci. However, use of the term also extends to describing a possible genetic architecture for discrete complex, or multifactorial, traits. Such traits can be thought of as having an underlying, continuously-distributed, disease liability to which the many genetic and environmental risk factors contribute. An individual manifests the disease when his/her genetic and environmental risk factors together exceed a certain threshold. Many apparently 'discrete' traits have underlying, measurable, quantitative characteristics, as demonstrated for example, by the relationship between blood pressure (a continuous trait) and hypertension (a discrete phenotype).

Further reading

Falconer D, McKay TFC (1996) *Introduction to Quantitative Genetics*. Harlow: Prentice Hall, pp100–183.

Hartl DL, Clark AG (1997) *Principles of Population Genetics*. Sunderland: Sinauer Associates, pp 397–481.

See also Quantitative Trait, Oligogenic Inheritance, Variance Components, Multifactorial Traits

POLYMORPHISM (GENETIC POLYMORPHISM)

Mark McCarthy and Steven Wiltshire

Any genomic position (or locus) at which the DNA sequence shows variation between individuals.

Polymorphic sites are of importance for several reasons. First, by definition, inherited variation in traits of biological and medical importance is due to polymorphic variation. Second, polymorphic loci, if their genomic location is known, represent markers that can be used in linkage and linkage disequilibrium analyses to follow chromosomal segregation, define sites of recombination, and thereby aid susceptibility gene identification. Third, highly polymorphic loci are used in a wide range of legal and forensic situations. Several types of DNA polymorphism exist but the most frequently used in genetic analysis are single nucleotide polymorphisms (SNPs) and microsatellites (these are usually di-, tri- or tetra-nucleotide repeats). The degree of polymorphism at a locus is described in terms of either the heterozygosity (the probability that a random individual is heterozygous for any two alleles at the locus) or a related measure termed the polymorphism information content (PIC). Historically, the term polymorphism was restricted to sites at which the minority allele frequency exceeded one percent, and carried some imputation that the variant was subject to either neutral or balancing selection. However, in the light of contemporary insights into the development and maintenance of genomic variation in man, these restrictions have become less relevant.

Related websites

Databases of polymorphic sites in man:

Online Mendelian Inheritance in Man	http://www.ncbi.nih.gov/Omim/
dbsNP:	http://www.ncbi.nlm.nih.gov/SNP/
GWAS central	http://www.gwascentral.org/index

Utility programs (including determination of PIC values for polymorphisms)

Jurg Ott's Linkage Utility programs	http://linkage.rockefeller.edu/ott/linkutil.htm

P

Further reading
Balding DJ, Donnelly P (1995) Inferring identity from DNA profile evidence. *Proc Natl Acad Sci USA* 92: 11741–11745.

Gray IC, et al. (2000) Single nucleotide polymorphisms as tools in human genetics. *Hum Mol Genet* 9: 2403–2408.

Sachidanandam R, et al. (2001) A map of human genome sequence variation containing 1.42 million single nucleotide polymorphisms. *Nature* 409: 928–933.

Tabor HK, et al. (2002) Candidate-gene approaches for studying complex genetic traits: practical considerations. *Nat Rev Genet* 3: 1–7.

Wang D, et al. (1998) Large-scale identification, mapping, and genotyping of single-nucleotide polymorphisms in the human genome. *Science* 280: 1077–1082.

See also Marker, SNP, Linkage Analysis, Linkage Disequilibrium, Association Analysis

POLYPEPTIDE

Roman Laskowski and Tjaart de Beer

As defined in the IUPAC Compendium of Chemical Terminology, a polypeptide is any peptide containing ten or more amino acids. By convention its direction runs from the N-terminus to the C-terminus.

Relevant website

IUPAC	http://goldbook.iupac.org

Further reading
Branden C, Tooze J (1998) *Introduction to Protein Structure*. Garland Science, New York.

POLYPHYLETIC GROUP, *SEE* CLADISTICS.

POLYTOMY, *SEE* MULTIFURCATION.

POMBASE

Dan Bolser

PomBase is a database for the fission yeast *Schizosaccharomyces pombe*, providing curated structural and functional gene annotation. Users can download, search, or browse chromosomes, gene and peptide sequences, sequence library information, and functional annotations. The site acts as a community portal, allowing curation and deposition of data.

Related website

PomBase	http://www.pombase.org/

Further reading

Wood V, Harris MA, McDowall MD, et al. (2011) PomBase: a comprehensive online resource for fission yeast. *Nucleic Acids Res.* 28.

See also SGD, SCPD

POPULATION BOTTLENECK (BOTTLENECK)

A.R. Hoelzel

A population bottleneck occurs when a large population quickly reduces in size.

This can impact genetic diversity in the reduced population through the random loss of alleles (*c.f.* founder effect), and through the loss of heterozygosity. The latter is a consequence of the increased chance of recombining alleles that are identical by descent in small populations. If the pre-bottleneck population was large and highly polymorphic, recessive alleles may have been maintained in the heterozygous state. The expression of these as homozygotes through more frequent consanguineous mating in the bottlenecked population can lead to inbreeding depression (a reduction in fitness resulting from inbreeding). The degree of loss in genetic diversity depends on the size of the bottlenecked population and the intrinsic rate of growth for that population, whereby smaller bottlenecks and lower rates of growth both lead to a greater loss of diversity. The time required to regain lost variation through mutation can be approximated by the reciprocal of the mutation rate (sometimes longer than the lifespan of the species).

Another possible consequence of a population bottleneck is an impact on quantitative characters expressed as phenotype. Given that there are many genes interacting for a given (polygenic) character, and that the variance resulting from this interaction has both additive and non-additive components, a disruption of this interaction during a bottleneck event can lead to both an increase in the diversity of that character, and the disruption of developmental stability (resulting in increased fluctuating asymmetry).

Further reading

Hartl DL (2000) *A Primer Of Population Genetics*. Sinauer Associates Inc.

Hoelzel AR (1999) Impact of a population bottleneck on genetic variation and the importance of life history; a case study of the northern elephant seal. *Biol. J. Linn. Soc.*, 68, 23–39.

See also Evolution, Founder Effect

POSITION-SPECIFIC SCORING MATRIX, *SEE* PROFILE.

P

POSITION WEIGHT MATRIX

Jacques van Helden

Particular type of position-specific scoring matrix, where the values indicate weights, i.e. log-odds of residue frequencies versus prior probabilities. A PWM is typically used to scan DNA sequences for predicting transcription factor binding sites.

Note that assigning a weight score to each position of a matrix assumes independence between prior frequencies of successive residues (Bernoulli model), whereas for DNA sequences the background is better modelled by Markov chains.

See also Profile

POSITION WEIGHT MATRIX OF TRANSCRIPTION FACTOR BINDING SITES

James Fickett

A tool to describe the DNA binding specificity of a transcription factor or, equivalently, to describe a prototypical transcription factor binding site.

The most common description of specificity in everyday writing is the consensus sequence, which gives the most commonly occurring nucleotide at each position of the binding site. However the consensus sequence loses a great deal of information, as it does not distinguish between a highly informative position where T, say, always occurs, and an essentially random one where T occurs 28% of the time, and A, C, and G each occur 24% of the time. The Position Weight Matrix (PWM) has one column for each position in the binding site, and one row for each possible nucleotide. The entries are derived from the relative frequency f_{ij} of each nucleotide i at each position j. If one measures the background frequency p_i of each nucleotide as well, and defines a PWM entry as the logarithm of the f_{ij}/p_i, then the matrix can provide statistically and physically meaningful numerical models. Define the PWM score of a potential site as the sum of the corresponding matrix entries. Then, under certain simplifying assumptions, the PWM score of a site is proportional both to the probability that the site is bound by the cognate factor, and to the free energy of binding (Stormo & Fields, 1998). The sequence logo (Schneider, 2001) is a graphical device to represent the information in a PWM intuitively. The height of a character at a particular position is proportional to the relative frequency of the corresponding nucleotide, while the overall height of the column of letters is proportional to the information content of that position, in the sense of Shannon (Cover & Thomas, 1991). The PWM score does typically predict *in vitro* binding fairly well; its greatest limitation is that *in vitro* binding does not necessarily correlate with biological activity (Tronche et al., 1997). Less frequently, it may be an unrealistic assumption that the different positions in the binding site make independent contributions to the free energy of binding (Bulyk et al., 2002).

The TRANSFAC database covers some of the known transcription factors, binding sites, and derived PWMs (Wingender et al., 2001). To date, binding sites have been laboriously determined one at a time, and only a very small fraction of the total is known (Tupler et al., 2001). However genome-wide localization of DNA-binding proteins is now possible through formaldehyde cross-linking of the protein to the DNA, followed by fragmentation, chromatin immunoprecipitation, and finally hybridization of the resulting fragments to a microarray of genomic fragments (Ren et al., 2000).

The first step in constructing a PWM is alignment of known sites. More generally, one may attempt to discover, and align, a set of binding sites common to a particular transcription factor from a set of regulatory regions thought to contain such sites. For example, putative promoters may be extracted from a set of coordinatedly up-regulated genes (as determined by expression array analysis), and these promoters searched for the binding sites of common regulating factors (Hughes et al., 2000). For both these variations on the alignment problem, an iterative algorithm, such as Expectation Maximization or Gibbs Sampling, is often used (Lawrence et al., 1993). If the upstream regions of genes are extensive, with widely placed regulatory regions, as in human, phylogenetic footprinting must be used to limit the search to regions of high promise (Wasserman et al., 2000).

Related websites

BIOBASE	http://www.biobase-international.com/
Wadsworth Center Bayesian Bioinformatics	http://www.wadsworth.org/resnres/bioinfo

Further reading

Bulyk ML, et al. (2002) *Nucleic Acids Research* 30, 1255–1261.

Cover TM, Thomas JA (1991) *Elements of Information Theory*. Wiley.

Hughes JD, et al. (2000) *Journal of Molecular Biology* 296, 1205–1214.

Lawrence CE, et al. (1993) *Science* 262, 208–214.

Ren B, et al. (2000) *Science* 290, 2306–2309.

Schneider TD (2001) *Nucleic Acids Research* 29, 4881–4891.

Stormo GD, Fields DS (1998) *Trends in Biochemical Sciences* 23, 109–113.

Tronche F, et al. (1997) *Journal of Molecular Biology* 266, 231–245.

Tupler R, et al. (2001) *Nature* 409, 832–833.

Wasserman WW, et al. (2000) *Nature Genetics* 26, 225–228.

Wingender E, et al. (2001) *Nucleic Acids Research* 29, 281–283.

See also Transcription Factor, Transcription Factor Binding Site, Consensus Sequence, Sequence Logo, TRANSFAC, Expectation Maximization, Phylogenetic Foortprinting

POSITIONAL CANDIDATE APPROACH

Mark McCarthy and Steven Wiltshire

The dominant strategy for susceptibility gene discovery in multifactorial traits, this is a hybrid approach that seeks to move towards susceptibility gene discovery using information derived from both positional and biological information.

The former will generally be derived through a genome scan for linkage, which may be expected, if adequately powered, to define the approximate genomic position of the most significant susceptibility effects. Further refinement of location may be possible through fine-mapping within these regions, using both linkage and linkage disequilibrium analyses. In parallel, transcripts known to map to the region are reviewed to identify those positional candidate genes with the strongest prior claims, on biological grounds, for involvement in the pathogenesis of the disease concerned.

Examples: Two recent examples of this approach have been the identification of NOD2 as a susceptibility gene for Crohn's disease (Hugot et al., 2001; Ogura et al., 2001), and of ADAM33 for asthma (van Eerdewegh et al., 2002).

Further reading

Collins FS (1995) Positional cloning moves from the perditional to traditional. *Nat Genet* 9: 347–350.

Hugot J-P, et al (2001) Association of NOD2 leucine-rich repeat variants with susceptibility to Crohn's disease. *Nature* 411: 599–603.

Ogura Y, et al. (2001) A frameshift mutation in NOD2 associated with susceptibility to Crohn's disease. *Nature* 411: 603–607.

van Eerdewegh P, et al. (2002) Association of the ADAM33 gene with asthma and bronchial hyperresponsiveness. *Nature* 418: 426–430.

See also Genome-Wide Association, Candidate Gene, Genome Scan, Multifactorial Trait

POSITIVE CLASSIFICATION

Pedro Larrañaga and Concha Bielza

In the genomics setting, an increasingly common data configuration consists of a small set of sequences possessing a targeted property (positive instances) amongst a large set of sequences for which class membership is unknown (unlabeled instances). This constitutes an example of the so called positive classification (or positive and unlabelled classification).

Adaptations of some of the standard supervised classification methods have been proposed in the literature. Bayesian classifier such as naïve Bayes and tree augmented naïve Bayes have been adapted by Denis et al. (2002) and Calvo et al. (2007), respectively. Also, approaches based on logistic regression (Lee et al. 2003) and the EM algorithm (Ward et al., 2009) have been developed.

Application in Bioinformatics include the prioritization of candidate cancer genes in oncogenomic studies (Furney et al., 2008) and the prediction of HLA binding and alternative splicing conservation between human and mouse (Xiao and Segal, 2008).

Further reading

Calvo B, et al. (2007) Learning Bayesian classifiers from positive and unlabeled examples. *Pattern Recognition Letters* 28(16): 2375–2384.

Denis F, et al. (2002) Text classification from positive and unlabeled examples. In *Proceedings of the 9th International Conference on Information Processing and Management of Uncertainty*, pp. 1927–1934.

Furney S, et al. (2008) Prioritization of candidate cancer genes. An aid to oncogenomic studies. *Nucleic Acids Research* 36(18): e115.

Lee WS, et al. (2003) Learning with positive and unlabeled examples using weighted logistic regression. In *Proceedings of the 20th International Conference on Machine Learning*.

Ward G, et al. (2009) Presence-only data and the EM algorithm. *Biometrics* 65(2): 554–563.

Xiao Y, Segal MR (2008) Biological sequence classification utilizing positive and unlabeled data. *Bioinformatics* 24(9): 1198–1205.

P

POSITIVE DARWINIAN SELECTION (POSITIVE SELECTION)

Austin L. Hughes

Positive Darwinian selection is natural selection that acts to favor a selectively advantageous mutation. There is evidence that this type of selection is a relatively rare phenomenon over the course of evolution. When we have a biological reason for predicting that natural selection may have favored multiple amino acid replacements in a given protein domain, we can test this prediction by comparing the number of synonymous nucleotide substitutions per synonymous site (d_S) with the number of nonsynonumous (i.e, amino acid-altering) nucleotide substitutions per nonsynonymous site (d_N). In most comparisons between homologous protein-coding genes, d_S exceeds d_N because most nonsynonymous substitutions are deleterious and thus are eliminated by purifying selection. On the other hand, we may find that d_N is greater than d_S in the domain on which we predict positive selection to be acting. Such a finding supports the hypothesis that natural selection has actually favored change at the amino acid level in this domain.

Several statistical packages have been developed that claim to provide a test for positive selection by searching for individual codons at which $d_N/d_S > 1$ (so-called 'codon-based' tests). Often such programs use likelihood ratio tests to compare a model of codon evolution including a category of codons at which $d_N/d_S > 1$ with a model lacking such a category. If the former model is significantly better that the latter model by the likelihood ratio test, that result is taken as evidence of positive selection. But this conclusion is not really valid because many data sets contain one or more codons at which $d_N/d_S > 1$ even under strong purifying selection, simply as a result of the random nature of mutation. Codon-based tests thus cannot distinguish between positive selection and the relaxation of purifying selection. Codon-based tests are also very sensitive to alignment errors and sequencing errors.

In general, there is no such thing as a 'signature of positive selection' that can be detected computationally. The hypothesis that a given amino acid replacement or other sequence change has occurred as a result of positive selection needs to be tested empirically, with evidence that the sequence change in fact confers a fitness benefit.

Further reading

Hill RE, Hastie ND (1987) Accelerated evolution in the reactive centre regions of serine protease inhibitors. *Nature* 326: 96–99.

Hughes AL, Nei M (1988) Pattern of nucleotide substitution at MHC class I loci reveals overdominant selection. *Nature* 335: 167–170.

Hughes AL (1999) *Adaptive Evolution of Genes and Genomes.* Oxford University Press, New York.

Hughes AL (2007) Looking for Darwin in all the wrong places: the misguided quest for positive selection at the nucleotide sequence level. *Heredity* 99: 364–373.

Hughes AL, Friedman R (2008) Codon-based tests of positive selection, branch lengths, and the evolution of mammalian immune system genes. *Immunogenetics* 60: 495–506.

Schneider A, Souvorov A, Sabath N, Landan G, Gonnet GH, Graur D (2009) Estimates of positive Daowinian selection are inflated by errors in sequencing, annotation, and alignment. *Genome Biol Evol* 5: 114–118.

See also Purifying Selection

POSITIVE SELECTION, *SEE* POSITIVE DARWINIAN SELECTION.

POST-ORDER TREE TRAVERSAL, *SEE* TREE TRAVERSAL.

POSTERIOR ERROR PROBABILITY (PEP)

Simon Hubbard

In high-throughout proteomics experiments, the *Posterior Error Probability* is estimated as a local *False Discovery Rate* measure as whether a single *Peptide Spectrum Match* is incorrect for a given score or e-value, based on the global distributions of those scores for correct and incorrect *PSMs*. It measures the probability for an error of a single PSM, as opposed to the error rate for a whole collection of PSMs.

Further reading

Käll L, Storey JD, MacCoss MJ, Noble WS. Posterior error probabilities and false discovery rates: two sides of the same coin. *J Proteome Res.* 2008, 7: 40–44.

POTENTIAL OF MEAN FORCE

Roland Dunbrack

A potential energy function derived from molecular dynamics simulations or other sources of conformational sampling for a system that represents the energy of some component of the system (for instance, the interaction of a pair of ions) with the effects of the remainder of the system (for instance, solvent water) averaged out. This term is often used to describe statistical potential functions.

See also Statistical Potential Energy

P

POWER LAW (ZIPF'S LAW)

Dov Greenbaum

A frequency distribution of text strings in which a small number of strings are very frequent.

The most famous example of the power law is that of the usage of words in texts. Zipf's law states that some words – e.g. and/or/the – are used frequently while most other words are used infrequently. When the size of each group of words is plotted against the usage of that word, the distribution can be approximated to a power-law function; that is, the number of words *(N)* with a given occurrence *(F)* decays according to the equation $N = aF^{-b}$.

In molecular biology it has been found that short strings of DNA, DNA words, in non-coding DNA followed a similar behaviour, suggesting that non-coding DNA resembled a natural language. The power-law has also been applicable in contexts such as the connectivity within metabolic pathways, the occurrences of protein families, folds or protein-protein interactions, and the occurrence of pseudogenes and pseudomotifs.

The finding of power-law behaviour is non-trivial, as it provides a mathematical description of a biological feature: the dominance, within a larger population, of a few members.

Further reading

Mantegna, RN, et al. (1994) Linguistic features of noncoding DNA sequences. *Phys Rev Lett* 73: 3169–3172.

Qian J, et al. (2001) Protein family and fold occurrence in genomes: power-law behavior and evolutionary model. *J Mol Biol* 313: 673–681.

Zipf, GK (1949) *Human Behavior and the Principle of Least Effort*. Addison-Wesley.

PREDICTION OF GENE FUNCTION

Jean-Michel Claverie

The process of interpreting a newly determined sequence in order to classify it into one of the previously characterized families of proteins.

Functional prediction mostly relies on the detection of similarity with proteins of known function. There are numerous levels of increasing sophistication in the recognition of these similarities, leading to functional predictions of decreasing accuracy and confidence. These various levels are:

- Significant global pairwise similarity, associated with the expected positioning of the new sequence in a phylogenetic tree. This is used to define orthologues, i.e. genes with the exact same role in two different species.
- Significant global pairwise similarity, and reciprocal best match throughout whole genome comparison is used to extend the above definition of orthology.
- Significant global pairwise similarity, associated with the strict conservation of a set of residues at positions defining a function specific motif, is used to define paralogues.
- Significant local pairwise similarity, encompassing a function-specific motif, indicates evolutionarily more distant, but functionally related genes (possibly associated with unrelated domains).

P

- Significant local similarity solely detected by sequence family-specific scoring schemes (Psi-BLAST, HMM, etc.), indicating an even more distant functional relationship (usually biochemical, e.g. 'kinase').
- The occurrence of a set of conserved positions/residues (e.g. PROSITE regular expression) previously associated with a function.
- No sequence similarity, but the evidence of a link with a gene of known function provided by one of the guilt-by-association methods, or experimental protein-protein interaction data (e.g. two hybrid system).
- No sequence similarity, but the evidence of a link with another gene (e.g. co-expression) or a disease, organ, tissue, or physiological condition (e.g. differential expression) from gene expression measurements and profiling.

3-D structure similarity can also be used to infer functional relationship between proteins with absolutely no significant sequence similarity.

It should be noted that the operational definitions of orthology and paralogy used here differ from stricter definitions used in the discipline of molecular evolution. For example, in molecular evolution orthology is not taken to imply that two genes play the same role in different species.

Related websites

NCBI BLAST page	http://www.ncbi.nlm.nih.gov/BLAST/
PROSITE home page	http://www.expasy.ch/prosite/
InterPro home page	http://www.ebi.ac.uk/interpro/
Pfam home page	http://pfam.sanger.ac.uk/
PRINTS home page	http://bioinf.man.ac.uk/dbbrowser/PRINTS/
ProDom home page	http://prodom.prabi.fr/prodom/current/html/home.php
SMART home page	http://smart.embl-heidelberg.de/

Further reading

Altschul SF, et al. (1997) Gapped BLAST and PSI-BLAST: a new generation of protein database search programs. *Nucleic Acids Res.* 25: 3389–3402.

Apweiler R, et al. (2000) InterPro-an integrated documentation resource for protein families, domains and functional sites. *Bioinformatics.* 16: 1145–1150.

Bateman A, et al. (2002) The Pfam protein families database. *Nucleic Acids Res.* 30: 276–280.

Claverie JM (1999) Computational methods for the identification of differential and coordinated gene expression. *Hum. Mol. Genet.* 8: 1821–1832.

Eisen JA (1998) Phylogenomics: improving functional predictions for uncharacterized genes by evolutionary analysis. *Genome Res.* 8: 163–167.

Enright AJ, et al. (1999) Protein interaction maps for complete genomes based on gene fusion events. *Nature* 402: 86–90.

Marcotte EM, et al. (1999) Detecting protein function and protein-protein interactions from genome sequences. *Science* 285: 751–753.

Tatusov RL, et al. (2001) The COG database: new developments in phylogenetic classification of proteins from complete genomes. *Nucleic Acids Res.* 29: 22–28.

See also Motif, Gene Family, Functional Signature, Ortholog, Paralog

PREDICTIVE ADME (ABSORPTION, DISTRIBUTION, METABOLISM, AND EXCRETION), *SEE* CHEMOINFORMATICS.

PRIDE

Simon Hubbard

The PRIDE database (Proteomics IDEntifications) is a popular repository for experimental proteomics data from spectra through to peptide and protein identifications, based at the EBI in Hinxton, UK. The resource is open source and freely available, and contains many useful features and views, as well as over 20,000 experiments. Importantly, it supports the latest community driven reporting standards developed by the HUPO-sponsored Proteomics Standards Initiative (PSI), including mzML (for spectra) as well as its own XML format, PRIDE-XML.

Related website

PRIDE repository of proteomics experimental data	http://www.ebi.ac.uk/pride/

Further reading

Vizcaíno JA, Côté R, Reisinger F, et al. (2009) A guide to the Proteomics Identifications Database proteomics data repository. *Proteomics.* 9: 4276–4283.

See also Proteomics Standards Initiative

PRIMER3

John M. Hancock

A program for the design of PCR primers.

PRIMER3 takes a DNA sequence as input and, given user defined constraints, provides a list of the highest quality pairs of PCR primers that will amplify from that sequence consistent with those constraints. Constraints include position and length of the region to be amplified, preferred experimental conditions (of the PCR reaction), product GC content, and potential of the primers to self anneal or anneal to repetitive elements.

P

Related websites

PRIMER3 web site	http://primer3.sourceforge.net/
PRIMER3PLUS Web Form	http://primer3plus.com/cgi-bin/dev/primer3plus.cgi

Further reading

Rozen S, Skaletsky HJ (2000) Primer3 on the WWW for general users and for biologist programmers. In: Krawetz S, Misener S (eds) *Bioinformatics Methods and Protocols: Methods in Molecular Biology*. Humana Press, Totowa, NJ, pp 365–386.

Untergasser A, Cutcutache I, Koressaar T, et al. (2012) Primer3 – new capabilities and interfaces. *Nucleic Acids Res.* 40(15): e115.

PRINCIPAL COMPONENTS ANALYSIS (PCA)

John M. Hancock

A statistical method for identifying a limited set of uncorrelated (orthogonal) variables which explain most of the variability in a sample.

The variables derived by the method are termed the principal components. These are linear combinations of the original variables so that it may be possible to ascribe meaning to them, although this is not always easy to achieve.

Related website

Wikipedia	http://en.wikipedia.org/wiki/Principal_component_analysis

Further reading

Hilsenbeck SG. et al. (1999) Statistical analysis of array expression data as applied to the problem of tamoxifen resistance. *J Natl Cancer Inst.* 91: 453–459.

Misra J, et al. (2002) Interactive exploration of microarray gene expression patterns in a reduced dimensional space. *Genome Res.* 12: 1112–1120.

PRINTS

Teresa K. Attwood

A database of 'fingerprints' that encode a range of gene and domain families. A founding member of the InterPro consortium, PRINTS specialises in the hierarchical diagnosis of gene families, from superfamily, through family, down to subfamily levels. As such, it complements PROSITE, which excels in the identification of functional sites, and Pfam, which concentrates on characterising highly divergent domains and superfamilies.

Entries in PRINTS are created from hand-crafted seed alignments from which the most conserved motifs are selected manually – each group of conserved motifs constitutes a fingerprint (the fingerprint concept incorporates the number of motifs, the order in which they occur within an alignment and the relative distances between them). Using an algorithm that converts the motifs into residue frequency matrices, these are used to search UniProtKB iteratively. Sequences that match all the motifs, but were not in the seed alignment, are assimilated into the process, and the database is searched again until the scans converge – i.e., until no more new matches can be found that contain all the motifs. The resulting fingerprint is then manually annotated prior to deposition in PRINTS.

The database was originally maintained as a single flat file, but has also been migrated to a relational database management system, in order to facilitate maintenance and to support more complex queries – the streamlined version is known as PRINTS-S. To augment the resource, an automatically generated supplement, prePRINTS, was also created using sequence clusters in ProDom and automatically annotated using the PRECIS software.

PRINTS is hosted at the University of Manchester, UK. Its entries have unique accession numbers and database identifiers, and include high-level and fine-grained functional descriptions of the encoded families, where possible including disease and/or structural information, literature references, links to related databases (PROSITE, Pfam, InterPro, PDB, OMIM, etc.), technical details of how the fingerprints were derived, together with the initial and final motifs, and match-lists against UniProtKB (including true-positives, and false-positives if they exist); the seed alignments used to create the fingerprints are also made available from the companion resource, ALIGN. PRINTS may be interrogated with simple keywords, or searched with query sequences via the FingerPRINTScan suite.

Related websites

PRINTS	http://www.bioinf.manchester.ac.uk/dbbrowser/PRINTS/
PRINTS-S	http://www.bioinf.manchester.ac.uk/dbbrowscr/sprint/
pre-PRINTS	http://www.bioinf.manchester.ac.uk/dbbrowser/prePRINTS/
ALIGN	http://www.bioinf.manchester.ac.uk/dbbrowser/ALIGN/
FingerPRINTScan	http://www.bioinf.manchester.ac.uk//dbbrowser/fingerPRINTScan/
PRECIS	http://www.bioinf.manchester.ac.uk/cgi-bin/dbbrowser/precis/precis.cgi
PRINTS Manual	http://www.bioinf.manchester.ac.uk/dbbrowser/PRINTS/printsman.php

Further reading

Attwood TK, Beck ME (1994) PRINTS – a protein motif fingerprint database. *Protein Eng.* 7(7): 841–848.

Attwood TK, Bradley P, Flower DR, et al. (2003) PRINTS and its automatic supplement, prePRINTS. *Nucleic Acids Res.* 31(1): 400–402.

Attwood TK, Croning MDR, Flower DR, et al. (2000) PRINTS-S: the database formerly known as PRINTS. *Nucleic Acids Res.* 28(1): 225–227.

Mitchell AL, Reich JR, Attwood TK (2003) PRECIS: protein reports engineered from concise information in SWISS-PROT. *Bioinformatics* 19(13): 1664–1671.

Scordis P, Flower DR, Attwood TK (1999) FingerPRINTScan: intelligent searching of the PRINTS motif database. *Bioinformatics* 15(10): 799–806.

See also Fingerprint, Domain Family, Gene Family, InterPro, Motif, PROSITE, Protein Family

PROBABILISTIC NETWORK, *SEE* BAYESIAN NETWORK.

P

PRODOM

Teresa K. Attwood

A database of automatically created sequence clusters encoding a wide range of protein domain families. A member of the InterPro consortium, ProDom specialises in providing graphical interpretations of protein domain arrangements. As such it complements the manual, discriminator-based approaches embodied in databases like PROSITE, which excels in the identification of functional sites, and PRINTS, which specialises in the hierarchical diagnosis of gene families, from superfamily, through family, to subfamily levels.

Because entries in ProDom are generated automatically from UniProtKB, its philosophical basis is rather different from the discriminator-based databases: the populations of a given domain family will depend on the parameters used by the clustering algorithm rather than on the expert knowledge of a biologist or bioinformatician.

ProDom is hosted at the Universite Claude Bernard, France. Each entry has a unique accession number and entry name, representing a single domain alignment of all sequences detected in UniProtKB with a level of similarity at or above a defined clustering threshold. Being automatically derived, ProDom is more comprehensive than that of the discriminator-based databases, but consequently includes minimal annotation. Domain architectures may be browsed via ProDom entry name or accession number; the database may also be searched with query sequences using BLASTP or BLASTX.

Related websites

ProDom	http://prodom.prabi.fr/
BLAST ProDom	http://prodom.prabi.fr/prodom/current/html/form.php

Further reading

Bru C, Courcelle E, Carrère S, Beausse Y, Dalmar S, Kahn D (2005) The ProDom database of protein domain families: more emphasis on 3D. *Nucleic Acids Res.* 33: D212–D215.

See also BLAST, Domain Family, InterPro, PRINTS, PROSITE

PROFILE (WEIGHT MATRIX, POSITION WEIGHT MATRIX, POSITION-SPECIFIC SCORING MATRIX, PSSM)

Teresa K. Attwood

A position-specific scoring table derived from a conserved region of a sequence alignment, and used to characterise and classify sequences according to their membership of different gene or domain families. By contrast with motif-based diagnostic approaches, which focus on local regions of high similarity, profiles typically span gene families or domains over their entire length, including gapped regions. The scoring system defines which residues are allowed at given positions; which positions are conserved and which degenerate; and which regions can tolerate insertions. Variable gap costs are used to penalise insertions and deletions in highly conserved regions. In addition to data implicit in the alignment, the scores may include evolutionary weights and results from structural studies. A more complete technical description of profile structure and syntax can be found in the `profile.txt` file made available with the PROSITE database.

Powerful discriminators, profiles are particularly effective in the diagnosis of highly divergent protein domains and superfamilies, where few residues are well conserved; they therefore provide a useful complement to more selective discriminators like fingerprints and consensus patterns.

Profiles are used in PROSITE both to supplement consensus patterns whose diagnostic power for particular protein families is not optimal, and to offer diagnostic tools for families that are too divergent to be able to derive effective patterns. They may be harnessed to characterise user-defined query sequences using the ScanProsite tool; a variety of other profile collections may also be searched via the MotifScan suite hosted at the Swiss Institute of Bioinformatics.

Related websites

PROSITE Manual	http://prosite.expasy.org/prosuser.html#meth2
MotifScan	http://hits.isb-sib.ch/cgi-bin/motif_scan
ScanProsite	http://prosite.expasy.org/scanprosite/
PROSITE	http://prosite.expasy.org/

Further reading

Bucher P, Bairoch A (1994) A generalized profile syntax for biomolecular sequence motifs and its function in automatic sequence interpretation. *Proc. Int. Conf. Intell. Syst. Mol. Biol.* 2: 53–61.

Sigrist CJ, Cerutti L, de Castro E, et al. (2010) PROSITE, a protein domain database for functional characterization and annotation. *Nucleic Acids Res.* 38(Database issue): D161–166.

See also Alignment, multiple, Domain Family, Gene Family, Profile Searching, PROSITE, Sequence Motifs, prediction and modeling

PROFILE, 3D

Andrew Harrison and Christine Orengo

A 3D profile of template encodes information about the structural conservation of positions in the 3D structure or fold adopted by a group of related structures. These structures are typically adopted by proteins related by a common evolutionary ancestor (homologues) or proteins adopting the same fold because of the constraints on the packing of secondary structures. This latter is described as convergent evolution and related structures are then described as analogues.

Different approaches have been developed for capturing information on the structural characteristics of a protein fold and how these are conserved across a structural family. The aim of generating a 3D profile is usually to recognize the most highly conserved positions for a group of structures in order to use this information to recognize other relatives sharing the same structural characteristics for these conserved positions. A 3D profile can therefore also be thought of as a pattern and are analogous to 1D-profiles used in sequence alignment, as their performance similarly depends on matching those positions which have been most highly conserved during evolution and will therefore correspond even between very distant relatives.

In constructing 3D-profiles, residue positions are usually considered and information on variation in 3D coordinates of the Cα or Cβ atom may be determined and described in the profile. However, in order to compare the positions it is first necessary to multiply align the structures against each other or superpose them in 3D. Thus equivalent positions can be identified across the set and variation in co-ordinates can be easily calculated. Some approaches also consider variation in the structural environments of residues, where the structural environment may be described by the distances or vectors to all other residues or neighboring residues in the protein. Information on those subsets of residues which have conserved contacts in the fold can also be included. The algorithm for generating the 3D profile usually employs some mechanism for preventing the profile becoming biased towards any subsets of structures in the group which are over-represented.

When using 3D-profiles to recognize relatives adopting a similar 3D structure or fold it is usual to employ a dynamic programming algorithm which can identify where insertions or deletions occur in the query structure to obtain an optimal alignment of the query against the template. Weights are employed to amplify scores associated with matching highly conserved positions in the 3D profile.

Relevant website

Conserved Domains Database (CDD)	http://www.ncbi.nlm.nih.gov/Structure/cdd/cdd.shtml

PROFILE SEARCHING

Jaap Heringa

A powerful technique for characterizing the putative structure and function of a sequence. Profile searching is a sequence comparison method for finding and aligning distantly related sequences. The comparison uses a scoring matrix and an existing optimal alignment of two or more homologous protein sequences. The group or 'family' of similar sequences are first aligned together to create a multiple sequence alignment. The information in the multiple sequence alignment is then represented as a table of position-specific symbol comparison values and gap penalties. This table is called a profile. The similarity of new sequences to an existing profile can be tested by comparing each new sequence to the profile using a modification of the Smith-Waterman algorithm.

If multiple query sequences representing a homologous family are available, a profile representing the query family can be built and compared with existing profiles, involving profile-to-profile comparison.

Profile searching is commonly done in three different ways:

1. Aligning a query sequence optimally against an annotated database of profiles, each representing a family of homologous sequences (sequence-to-profile comparison)
2. Aligning a profile representing a query of homologous sequences against an annotated sequence database (sequence-to-profile comparison)
3. A combination of these two, where a query profile is compared against a database of family profiles (profile-to-profile comparison).

Profile searching techniques use information over an entire sequence alignment of a certain protein family to find additional related family members. The earliest conceptually clear technique of this kind of sequence searching was called profile analysis (Gribskov *et al.*, 1987), and combined a full representation of a sequence alignment with a sensitive searching algorithm.

A profile is constructed from a multiple sequence alignment and is an alignment-specific scoring table which comprises the likelihood of each residue type to occur in each position of the multiple alignment. A basic profile has L*20 elements, where L is the total length of the alignment and 20 the number of amino acid types. In addition to 20 columns, profiles may contain further columns to describe the number of gaps occurring at each alignment position, or to specify a gap penalty weight for each alignment position. Many profile techniques normalize the observed amino acid frequencies at each alignment position using expected frequencies (or background frequencies); for example, amino acid frequencies observed in a large sequence database, in a genome, in some particular protein family, or in the alignment the profile is derived from. The thus normalized scores are then logarithmically scaled to convert the amino acid frequencies occurring at each alignment position into probabilities: $P_{i,j} = \log(f_{i,j}/b_i)$, where $f_{i,j}$ is the frequency of amino acid type i at alignment position j and b_i the background frequency of amino acid type i.

An issue with logarithmic scoring schemes is the case of residue types not occurring at a given position in the multiple sequence alignment from which the profile is derived, which leads to undefined logarithmic scores. A way to alleviate this problem is the application of so-called pseudo-counts, which denotes the addition of small fractions to each of the amino acid frequencies to avoid zero scores: $f'_{i,j} = (\alpha f_{i,j} + \beta g_{i,j})/(\alpha + \beta)$, where α and β are scaling factors and $g_{i,j}$ is the amino acid pseudo frequency, which can be a background frequency derived from a large sequence database. In its simplest form, α is set to the number of amino acids observed at a given alignment position, $\beta = 1$ and each $g_{i,j}$ is set to 0.05 (equal amino acid frequencies); i.e., the number of each observed amino acid type is added by 1.

Many profile implementations reserve a profile column for so-called position specific gap penalty weights, which are determined using the number of gaps observed across each alignment position. The advantage of positional gap penalties is that multiple alignment regions with gaps (loop regions) will be assigned lowered gap penalties, hence will be more likely than core regions to attract gaps in a target sequence (or profile) during profile alignment, consistent with structural considerations. After calculation of the profile, it is aligned with a target sequence (or another profile) by means of the Smith and Waterman (1981) dynamic programming procedure, which finds the best local alignment, using a residue exchange matrix and appropriate gap penalty values. The aforementioned position specific gap weights are applied to the gap penalties used at each alignment position, and hence are lowered locally depending on the gap weights in the profile. The PSI-BLAST method (Altschul *et al.*, 1997) (see BLAST) is a widely-used heuristic and iterative local sequence-to-profile comparison technique. The method starts with comparing a given query sequence against a large annotated sequence database, after which local alignments scoring beyond a specified threshold are collected to calculate a profile. The profile is than iteratively compared against the sequence database to find distantly related family members of the initial query sequence. PSI-BLAST uses a profile-formalism without gap weights to describe a set of aligned sequence motifs, which is coined a position specific scoring matrix (PSSM).

More recently, Profile hidden Markov models (profile HMMs) have been developed to perform sensitive database searching using statistical descriptions of a sequence family's consensus. Profile HMMs have a formal probabilistic basis and consistent theory behind gap and insertion scores, in contrast to standard profile methods which use heuristic methods. HMMs apply a statistical method, using so-called emission and transition probabilities, to estimate the true frequency of a residue at a given position in the alignment from its observed frequency, while standard profiles use the observed frequency itself to assign the score for that residue. This means that a profile HMM derived from only 10 to 20 aligned sequences can be of equivalent quality to a standard profile created from 40 to 50 aligned sequences. Often, producing good profile HMMs requires less skill and manual intervention than producing good standard profiles. However, the technique is computationally intensive, which is an issue when integrated in genomic analysis pipelines.

After constructing a profile HMM, the alignment of the model with a database sequence corresponds to following a path through the model to generate a sequence consistent with the model. The probability of any sequence that is generated depends on the transition and emissions probabilities at each node in the profile HMM. The most sensitive methods to date employ profile-to-profile alignments using profile HMMs. Some widely used HMM-based profile searching tools include SAM-T2K (Karplus *et al.*, 1998), HMMER3 (Finn et al., 2011), PRC (Madeira, 2008) and HHsearch (Hildebrandt *et al.*, 2009).

Further reading

Altschul SF, Lipman DJ (1990) Protein database searches for multiple alignments. *Proc. Natl. Acad. Sci. USA* 87: 5509–5513.

Barton GJ, Sternberg MJE (1990) Flexible protein sequence patterns: a sensitive method to detect weak structural similarities. *J. Mol. Biol.* 212: 389–402.

Eddy SR (1998) Profile hidden Markov models. *Bioinformatics* 14: 755–763.

Finn RD, Clements J, Eddy SR (2011) HMMER web server: interactive sequence similarity searching. *Nucleic Acids Research* (Web Server Issue) 39: W29–W37.

Gribskov M, McLachlan AD, Eisenberg, D (1987) Profile analysis: detection of distantly related proteins. *Proc. Natl. Acad. Sci. USA* 84: 4355–4358.

Hildebrand A, Remmert M, Biegert A, Soding J (2009) Fast and accurate automatic structure prediction with HHpred. *Proteins* 77 Suppl 9: 128–132.

Karplus K, Barrett C, Hughey R (1998) Hidden Markov models for detecting remote protein homologies. *Bioinformatics* 14: 846.

Madera M (2008) Profile Comparer: a program for scoring and aligning profile hidden Markov models. *Bioinformatics* 24: 2630–2631.

See also Alignment, multiple, Local Alignment, Sequence Similarity

PROGRAMMING AND SCRIPTING LANGUAGES

David Thorne, Steve Pettifer and James Marsh

A programming language (or *scripting* language depending on context) is a high-level language used by developers to write software. Every such language consists of both grammar and syntax easily understood by a human, and which is structured strictly enough for a compiler (another piece of software) to be able to translate it into machine code, the low level instructions that then run on a computer's processor.

Languages can follow a number of paradigms, such as *functional programming* or *logic programming*, but by far the most prolific is that of *imperative programming*, in which the user writes a sequence of statements that are then executed in order to control the state and flow of the software. The most common languages for application and server development are imperative in nature: C/C++, Python, Ruby, Java, Perl, and PHP name a few. They vary in how their instructions are executed, whether they're *compiled* directly into machine code, or *interpreted* on-the-fly by an interpreter or a *virtual machine*, but they nevertheless all provide the same core facilities as each other: variables, arithmetic, functions, conditional and iterative control structures (e.g. if-then-else branches and for loops), object orientation, pattern matching, and many more besides. For a number of these languages there exist useful libraries that provide standard bio-related functionality that can greatly help bioinformaticians; the imaginatively named BioPython, BioPerl, BioRuby, and BioJava libraries are common examples of these that provide, amongst other features, the parsing and translation of biological data formats, simple analysis of biologically derived data, and access to standard web services.

Other languages exist, such as R and SAS, that are tailored instead to the task of statistical analysis, incorporating such features as modeling, distribution, regression, and significance, and as such have become invaluable in the field of bioinformatics.

Any bioinformatics solution, given the globally distributed and complex nature of the field, will almost certainly incorporate a number of different languages, fulfilling their roles according to their strengths and the legacy of their involvement.

PROMOTER

Niall Dillon

Sequence that specifies the site of initiation of transcription.

The term promoter is an operational definition describing sequences located close to the transcription start site that specify the site of transcriptional initiation. Polymerase I and II genes have their promoter sequences located immediately upstream and downstream from the transcription initiation site. RNA polymerase III promoters are located upstream from the transcription start site in some genes (snRNA genes) and mainly within the transcribed region in others (5SRNA and tRNA genes). Promoters for all three polymerases have spatially constrained sequences that bind specify the site or sites of initiation of transcription by binding multi-protein intitiation complexes.

Further reading

Dvir A, et al. (2001) Mechanism of transcription initiation and promoter escape by RNA polymerase II. *Curr. Opin. Genet. Dev.*11: 209–214.

Geiduschek EP, Kassavetis GA (2000) The RNA polymerase III transcription apparatus. *J. Mol. Biol.* 310: 1–26.

Grummt I (1999) Regulation of mammalian ribosomal gene transcription by RNA polymerase I. *Prog. Nucleic Acid Res. Mol. Biol.* 62: 109–154.

P

THE PROMOTER DATABASE OF *SACCHAROMYCES CEREVISIAE* (SCPD)

Dan Bolser

SCPD provides information on and sequences of the regulatory regions of selected yeast genes. Information includes transcription factor binding affinities for a handful of genes collected from the literature. The database contains useful but sparse information, including 580 experimentally mapped transcription factor binding sites and 425 transcriptional start sites.

Related website

SCPD	http://rulai.cshl.org/SCPD/

Further reading

Zhu J, Zhang MQ (1999) SCPD: a promoter database of the yeast *Saccharomyces cerevisiae*. *Bioinformatics.* 15: 607–611.

See also TRANSFAC, SGD, Transcription Factor Binding Site

PROMOTER PREDICTION

James Fickett

Computational methods to localize the transcription start site (TSS) in genomic DNA (Ohler & Niemann, 2001), and to interpret transcription factor binding sites to predict under what conditions the gene will be expressed.

Definitions of a promoter differ. The essential thing is that the promoter contains signals necessary to both localize and time the initiation of transcription. If a gene is alternatively spliced, there may be multiple TSS and, correspondingly, multiple promoters. In prokaryotes, all transcriptional regulation is typically in the promoter. In vertebrates, although the promoter does usually serve both functions of localizing the TSS and transducing control signals, most control is typically elsewhere, in so-called enhancers. Together, promoters and enhancers constitute regulatory regions. See Predicting Regulatory Regions for the computational methods intended to predict the context of transcription.

In eukaryotes, computational prediction of the TSS is a difficult and unsolved problem (Fickett & Hatzigeorgiou, 1997). At the same time, it is a key component of gene identification. The core motif in use to recognize eukaryotic promoters is the so-called TATA box (Bucher, 1990), which binds a central protein (TATA binding protein) of the trancription initiation complex. Not all promoters contain a TATA box and, in fact, the great difficulty in promoter recognition is that no single set of motifs is characteristic. The most common components of prediction tools are:

- scoring for common motifs (like the TATA box)
- similarity of oligonucleotide frequencies to those of known promoters
- density of putative transcription factor binding sites
- phylogenetic footprinting
- scoring (in vertebrates) for CpG islands in the vicinity.

A CpG island is a region of high C+G content and a high frequency of CpG dinucleotides relative to the rest of the genome (Gardiner-Garden & Frommer, 1987). One recent algorithm gains accuracy by redefining the problem as first exon prediction, searching for pairs of promoters and donor splice sites. This and other algorithms seem to perform with acceptable accuracy when a gene occurs in conjunction with a CpG island, but not otherwise (Davuluri et al., 2001). Development of new algorithms depends on a collection of genes with well-characterized TSS (Praz et al., 2002).

The problem in bacteria is somewhat different. Because operons are common, the Ribosome Binding Site rather than the TSS is used in gene identification to mark the beginning of the gene.

Related websites

EPD	http://www.epd.isb-sib.ch
Zhang Lab, CSH	http://rulai.cshl.edu/software/index1.htm

Further reading

Abeel T, Van de Peer Y, Saeys Y (2009) Toward a gold standard for promoter prediction evaluation. *Bioinformatics.* 25(12): i313–i320.

Bucher P (1990) *Journal of Molecular Biology* 212, 563–578.

Davuluri, RV, et al. (2001) *Nature Genetics* 29, 412–417.

Fickett JW, Hatzigeorgiou A. (1997) *Genome Research* 7, 861–878.

Gardiner-Garden M, Frommer M. (1987) *Journal of Molecular Biology* 196, 261–282.

Ohler U, Niemann H (2001) *Trends in Genetics* 17, 56–60.

Praz V, et al. (2002) *Nucleic Acids Research* 30, 322–324.

Suzuki Y. et al. (2002) *Nucleic Acids Research* 30, 328–331.

Zeng J, Zhu S, Yan H (2009) Towards accurate human promoter recognition: a review of currently used sequence features and classification methods. *Brief Bioinform.* 10(5): 498–508.

PROSITE

Teresa K. Attwood

A database of consensus patterns (or regular expressions (regexs)) and profiles encoding a range of gene and domain families. A founding member of the InterPro consortium, PROSITE characterizes both highly conserved functional sites (via its consensus patterns) and divergent domains and superfamilies (via its profile collection). As such, it complements PRINTS, which specialises in hierarchical diagnosis of gene families, from superfamily, through family, to subfamily levels, and Pfam, which also focuses on divergent domains and superfamilies.

Consensus patterns in PROSITE are derived from hand-crafted seed alignments from which the most conserved motif is selected manually. The patterns are searched against UniProtKB:Swiss-Prot, and iteratively refined to determine the full extent of the family. Where a single motif cannot capture the full family, additional patterns are derived until an optimal set is produced that characterizes all, or most, family members. Profile entries are also hand-crafted to produce a discriminator with optimal diagnostic power – they are included in the resource to counterbalance the relatively poor diagnostic performance of some of its patterns. All patterns and profiles are manually annotated prior to deposition in PROSITE.

PROSITE is hosted at the Swiss Institute of Bioinformatics, Switzerland. Each entry has a database identifier and two accession numbers: one for the data file and the other for the associated documentation file. The data file lists the consensus pattern(s) or profile, various technical and numerical details relating to their diagnostic performance, their matches to a given version of UniProtKB:Swiss-Prot (including true-positives, false-positives and false-negatives), together with links to related PDB entries, where relevant. The documentation file contains high-level functional descriptions of the encoded families, where possible including disease and/or structural information, details of representative family members, literature references, and so on. PROSITE may be interrogated with simple keywords, or searched with query sequences via the ScanProsite tool.

Related websites

PROSITE	http://prosite.expasy.org/
ScanProsite	http://prosite.expasy.org/scanprosite/

Further reading

Sigrist CJ, Cerutti L, de Castro E, et al. (2010) PROSITE, a protein domain database for functional characterization and annotation. *Nucleic Acids Res.* 38(Database issue): D161–166.

See also Consensus Pattern, Consensus Pattern Rule, InterPro, Pfam, PRINTS, Profile, Motif

PROTÉGÉ

Robert Stevens and John M. Hancock

Protégé (latest version Protégé 4) is an extensible tool that enables users to build ontologies and use them for developing applications. Protégé uses a frame-based language including classes and metaclasses, arranged into hierarchies, with slots representing relationships between these classes, and constraining axioms. From such an ontology, the system can automatically generate forms, which can be used to acquire information about instances of the classes in the ontology and thereby to construct a knowledge base. Protégé can be extended by adding modular plug-ins from a centrally maintained library to enhance the behaviour of their knowledge-acquisition systems. Developers can easily make new plug-ins to provide new functionality.

Related websites

Protégé	http://protege.stanford.edu/index.html
Web Protégé	http://protegewiki.stanford.edu/wiki/WebProtege

PROTEIN ARRAY (PROTEIN MICROARRAY)

Jean-Michel Claverie

A technique for proteomics and functional genomics based on the immobilization of protein molecules or the *in situ* synthesis of peptides on a solid support.

The explosion of data on gene sequences and expression patterns has generated interest in the function and interactions of the corresponding proteins. This has triggered the demand for advanced technology offering greater throughput and versatility than 2D gel electrophoresis combined with mass spectrometry. Protein microarrays incorporate advances in surface chemistries, developments of new capture agents, and new detection methods. As for DNA, the technologies differ by the nature of the physical support, and the protocol of protein/peptide production and immobilization. Protein arrays make possible the parallel multiplex screening of thousands of interactions, encompassing protein-antibody, protein-protein, protein-ligand or protein-drug, enzyme-substrate screening and multianalyte diagnostic assays. In the microarray or chip format, such determinations can be carried out with minimum use of materials while generating large amounts of data. Moreover, since most proteins are made by recombinant methods, there is direct connectivity between results from protein arrays and DNA sequence information.

Related website

Protein Arrays Resource Page	http://www.functionalgenomics.org.uk/sections/resources/protein_arrays.htm

Further reading

Cahill DJ (2001) Protein and antibody arrays and their medical applications. *J. Immunol. Methods* 250: 81–89.

Fung ET, et al. (2001) Protein biochips for differential profiling. *Curr. Opin. Biotechnol.* 12: 65–69.

Jenkins RE, Pennington SR (2001) Arrays for protein expression profiling: towards a viable alternative to two-dimensional gel electrophoresis? *Proteomics* 1: 13–29.

Mitchell P (2002) A perspective on protein microarrays. *Nature Biotechnol.* 20: 225–229.

Vaughan CK, Sollazzo M (2001) Of minibody, camel and bacteriophage. *Comb. Chem. High. Throughput Screen.* 4: 417–430.

Walter G, et al. (2000) Protein arrays for gene expression and molecular interaction screening. *Curr. Opin. Microbiol.* 3: 298–302.

Weinberger SR, et al. (2000) Recent trends in protein biochip technology. *Pharmacogenomics* 1: 395–416.

Wilson DS, Nock S (2002) Functional Protein Biochips. *Curr. Opin. Chem. Biol.* 6: 81–85.

Zhou H, et al. (2001) Solution and chip arrays in protein profiling. *Trends Biotechnol.* 19: S34–39.

P

PROTEIN DATA BANK (PDB)

Eric Martz

The single internationally recognized primary repository of all published three-dimensional biological macromolecular structure data, including nucleic acids and carbohydrates as well as proteins.

The Protein Data Bank archives atomic coordinate files, sometimes accompanied by supporting empirical data (crystallographic structure factors or NMR restraints). Each data file represents a fragment of a molecule, a whole molecule, or a complex of fragments or molecules. Each data file is assigned a unique four-character identification code ('PDB ID'). The first character is always a numeral (1–9) while the remaining three characters may be numerals or letters. The data files are freely available through the Internet. In some cases, authors do not release data until a year or more after publication of the structure; however, some prominent journals require that the data become publically available upon the date of publication.

The Protein Data Bank (PDB) was founded at Brookhaven National Laboratory (Long Island, New York, USA) in 1971, at a time when about a dozen protein structures had been solved by crystallography. It was maintained at Brookhaven until 1999, when management changed to a consortium of three geographically separated groups under the umbrella of the Research Collaboratory for Structural Bioinformatics. The number of entries had reached about 15,000 by year 2000. In 2002, about 83% are crystallographic results, 15% are from nuclear magnetic resonance, and 2% are theoretical. About 90% are proteins, while about 10% are nucleic acids or protein-nucleic acid complexes.

There is high redundancy in the PDB dataset, however, so the number of sequence-dissimilar proteins is only a few thousand (depending on the stringency of the criteria employed). Furthermore, most of these are single domains rather than the intact naturally expressed forms. Proteins difficult to crystallize, notably integral membrane proteins, are under-represented. Altogether, empirical structure data are available for less than a few percent of the human genome as of 2002.

Related websites

Protein Data Bank	http://www.pdb.org
Research Collaboratory for Structural Bioinformatics	http://www.rcsb.org

Further reading

Berman HM, et al. (2000) The Protein Data Bank. *Nucleic Acids Research*, 28: 235–242.

Bernstein FC, et al. (1977) The Protein Data Bank: a computer-based archival file for macromolecular structures. *J. Mol. Biol.* 112: 535–542.

Meyer EF (1997) The First Years of the Protein Data Bank. *Protein Science* 6: 1591–1597.

Westbrook J, et al. (2002) The Protein Data Bank: Unifying the archive. *Nucleic Acids Research*, 30: 245–248.

See also Atomic Coordinate Files, Crystallography, Models – molecular, visualization, Molecular, nuclear magnetic resonance

PROTEIN DATABASES

Rolf Apweiler

Protein databases are a whole range of databases containing information on various protein properties.

The various protein databases hold different types of information on proteins like sequence data; bibliographical references, description of the biological source of a protein, function(s) of a protein, post-translational modifications, protein families, domains and sites, secondary, tertiary, and quaternary structure, disease(s) associated with deficiencie(s) in a protein, 2D-gel data, and various other information depending on the type of protein database. The different kinds of protein databases include protein sequence databases, protein structure databases, protein sequence cluster databases, protein family and domain signature databases, 2D-gel protein databases, and specialized protein database resources like the proteome analysis database.

Further reading

Apweiler R. (2000) Protein sequence databases. In: Richards FM, Eisenberg DS, Kim PS (eds) *Advances in Protein Chemistry*. 54 31–71, Academic Press, New York.

Zdobnov EM, Lopez R, Apweiler R, Etzold T (2001) Using the Molecular Biology Data. In: *Biotechnology*, vol. 5b: *Genomics and Bioinformatics*, CW Sernsen (ed.); pp. 281–300, Wiley-VCH.

PROTEIN DOMAIN

Laszlo Patthy

A structurally independent, compact spatial unit within the three-dimensional structure of a protein.

Distinct structural domains of multidomain proteins usually interact less extensively with each other than do structural elements within the domains. A protein domain of a multidomain protein folds into its unique three-dimensional structure irrespective of the presence of other domains. Different domains of multidomain proteins may be connected by flexible linker segments.

Related websites

Structure databases of protein domains:

PDB	http://www.rcsb.org/pdb/
Dali	http://www2.embl-ebi.ac.uk/dali/domain/

Sequence databases of protein domains:

INTERPRO	http://www.ebi.ac.uk/interpro
PFAM	http://www.sanger.ac.uk/Pfam
PRODOM	http://protein.toulouse.inra.fr/prodom.html
SMART	http://smart.embl-heidelberg.de
SBASE	http://www3.icgeb.trieste.it/~sbasesrv/

See also Multidomain Protein

P

PROTEIN FAMILY

Andrew Harrison, Christine Orengo, Dov Greenbaum, and Laszlo Patthy

Given the variety of proteins in nature, it is helpful to pigeonhole proteins into a categorical system. The most common system is that of hierarchical families, akin to a taxonomic classification used for organisms. Protein families were originally conceived as comprising proteins related through sequence identity. Part of the Dayhoff classification scheme, protein families are more closely related evolutionarily and share a closer ancestor than the more broadly characterized grouping of superfamilies. By contrast closely related homologous proteins may be grouped into protein subfamilies. Dayhoff originally defined a protein family as a group of proteins of similar function whose amino acid sequences are more than 50% identical; this cutoff value is still used by the Protein Information Resource (PIR). Other databases may use lower cut-off values and may not use functional criteria for the definition of a protein family.

This type of homology-based protein classification is straightforward in the case of proteins consisting of a single protein domain: in this case the term 'protein family' is identical with the term 'protein domain family'.

Multidomain proteins containing multiple types of protein domains of distinct evolutionary origin pose a special problem for homology-based protein classification: a single protein may belong to several different domain families. Another special problem with multidomain proteins is that in many cases homologous multidomain proteins display only partial homology or local homology, limited to a domain-type present in both proteins. Moreover, the homology of some multidomain proteins may not be assigned to any of the main subcategories of homology (orthology, paralogy, xenology): if homology of two multidomain proteins is due only to the independent acquisition of the same domain types they are termed as epaktologs.

There are now several protein domain family databases, including Pfam, PRINTS, ProDom, SMART and PANTHER. Pfam is the most comprehensive database, to date, containing 14831 protein families (version 27.0). Most protein domain family databases, generate sequence profiles, patterns or regular expressions from a set of clear relatives within the family and use these to scan the sequence databases, such as SP-TREMBL, GenBank, to recognize further relatives to include in the classification. The comprehensive Pfam database builds Hidden Markov models for each family, whilst the PRINTS database uses a fingerprint of regular expressions encoding conserved patterns at different positions in a multiple alignment of relatives. The PRINTS database contains extensive functional information for each protein domain family and is manually validated. By contrast, protein families in the ProDom database are identified using completely automated protocols.

Related websites

Structural classification of protein domain families:

SCOP	http://scop.mrc-lmb.cam.ac.uk/scop
CATH	http://www.cathdb.info/

Structural annotations of genomes:

Gene3D	http://gene3d.biochem.ucl.ac.uk/Gene3D/
SUPERFAMILY	http://supfam.cs.bris.ac.uk/SUPERFAMILY/

Further reading

Mulder NJ, Apweiler R (2002) Tools and resources for identifying protein families, domains and motifs. *Genome Biol* 3: REVIEWS2001.

See also Hidden Markov Models, Protein Domain

PROTEIN FAMILY AND DOMAIN SIGNATURE DATABASES

Rolf Apweiler

Protein family and domain signature databases provide signatures diagnostic for certain protein families or domains.

Protein family and domain signature databases use different sequence-motif methodologies and a varying degree of biological information on well-characterized protein families and domains to derive signatures diagnostic for certain protein families or domains. The different protein family and domain signature databases provide varying degrees of annotation for the protein families or domains. These signature databases have become vital tools for identifying distant relationships in novel sequences and hence for inferring protein function. Diagnostically, the signature databases have different areas of optimum application owing to the different strengths and weaknesses of their underlying analysis methods (e.g., regular expressions, profiles, Hidden Markov Models, fingerprints).

Relevant websites

InterPro	http://www.ebi.ac.uk/interpro/
PROSITE	http://www.expasy.org/prosite/
Pfam	http://pfam.sanger.ac.uk/
PRINTS	http://www.bioinf.manchester.ac.uk/dbbrowser/PRINTS/index.php
SMART	http://SMART.embl-heidelberg.de

Further reading

Apweiler R, Attwood TK, Bairoch A, et al. (2001) The InterPro database, an integrated documentation resource for protein families, domains and functional sites. *Nucl. Acids Res.* 29: 37–40.

Kriventseva EV, Biswas M, Apweiler R (2001) Clustering and analysis of protein families. *Current Opinion in Structural Biology* 11: 334–339.

PROTEIN FINGERPRINT, *SEE* FINGERPRINT.

P

PROTEIN INFERENCE PROBLEM

Simon Hubbard

A well-recognized problem in mass spectrometry based proteomics, where degeneracy in protein amino acid sequences leads to challenges in assigning the unique 'true' parent of a peptide sequence identified in a proteomics experiment. This is often an issue when peptides from paralogous proteins or protein isoforms are identified that are found in one or more protein species, such as those produced from the same genomic locus by alternate splicing in eukaryotic organisms. Clearly, this has implications for quantitative proteomics experiments seeking to measure the abundance of these proteins.

Further reading

Li YF, Arnold RJ, Li Y, Radivojac P, Sheng Q, Tang H (2009) A bayesian approach to protein inference problem in shotgun proteomics. *J. Comput. Biol.* 2009, 16: 1183–1193.

Nesvizhskii AI, Aebersold R (2005) Interpretation of shotgun proteomic data: the protein inference problem. *Mol. Cell. Proteomics.* 4: 1419–1440.

PROTEIN INFORMATION RESOURCE (PIR)

Rolf Apweiler

The Protein Information Resource is a universal annotated protein sequence database covering proteins from many different species.

The National Biomedical Research Foundation (NBRF) established PIR in 1984 as a successor of the original NBRF Protein Sequence Database. Since 1988 the database has been maintained by PIR-International, a collaboration between the NBRF, the Munich Information Center for Protein Sequences (MIPS), and the Japan International Protein Information Database (JIPID).

In 2002 PIR, along with its international partners, EBI (European Bioinformatics Institute) and SIB (Swiss Institute of Bioinformatics), were awarded a grant from NIH to create UniProt, a single worldwide database of protein sequence and function, by unifying the PIR-PSD, Swiss-Prot, and TrEMBL databases.

Relevant website

PIR	http://pir.georgetown.edu/

Further reading

Wu CH, Huang H, Arminski L, et al. (2002) The Protein Information Resource: an integrated public resource of functional annotation of proteins. *Nucl. Acids Res.* 30: 35–37.

See also SWISS-PROT

PROTEIN MICROARRAY, *SEE* PROTEIN ARRAY.

PROTEIN MODULE

Laszlo Patthy

A protein module is a structurally independent protein domain that has been spread by module shuffling and may occur in different multidomain proteins with different domain combinations.

Further reading

Patthy L (1985) Evolution of the proteases of blood coagulation and fibrinolysis by assembly from modules. *Cell*, 41: 657–663.

Patthy L (1991) Modular exchange principles in proteins. *Curr Opin Struct Biol* 1: 351–361.

Patthy L (1996) Exon shuffling and other ways of module exchange. *Matrix Biol.* 15: 301–310.

See also Protein Domain, Module Shuffling, Multidomain Protein

PROTEIN-PROTEIN COEVOLUTION

Laszlo Patthy

Coevolution is a reciprocally induced inherited change in a biological entity in response to an inherited change in another with which it interacts.

Specific protein-protein interactions essential for some biological function (e.g. binding of a protein ligand to its receptor, binding of a protein inhibitor to its target enzyme) are ensured by a specific network of inter-residue contacts between the partner proteins. During evolution, sequence changes accumulated by one of the interacting proteins are usually compensated by complementary changes in the other. As a corollary, positions where changes occur in a correlated fashion in the interacting molecules tend to be close to the protein-protein interfaces, permitting the prediction of contacting pairs of residues of interacting proteins.

Coevolution of protein-protein interaction partners is also reflected by congruency of their cladograms. Thus, comparison of phylogenetic trees constructed from multiple sequence alignments of interacting partners (e.g. ligand-families and the corresponding receptor families) show that protein ligands and their receptors usually co-evolve so that each subgroup of ligand has a matching subgroup of receptors.

Bioinformatics approaches have been developed for the prediction of protein-protein interactions based on the correspondence between the phylogenetic trees of interacting proteins and detection of correlated mutations between pairs of proteins.

Further reading

Goh CS, et al. (2000) Co-evolution of proteins with their interaction partners. *J Mol Biol* 299: 283–293.

Lovell SC, Robertson DL (2010) An integrated view of molecular coevolution in protein-protein interactions. *Mol Biol Evol.* 27(11): 2567–2575.

Pazos F, Valencia A (2001) Similarity of phylogenetic trees as indicator of protein-protein interaction. *Protein Eng* 14: 609–614.

See also Coevolution, Alignment – Multiple

P

PROTEIN-PROTEIN INTERACTION NETWORK INFERENCE

Hedi Peterson

Prediction of physical interactions between two proteins based on a variety of experimental and computational evidence.

Protein-protein interactions (PPI) are inferred from biological experiments and computational predictions. Experimentally validated interactions are collected into specific databases, e.g. DIP and Intact. Softwares like String combine known and predicted interactions into integrated networks. Connections in protein-protein networks are depicted by undirected edges, because the directionality of interactions is hard to measure with current methods. Edges can be assigned weights to express interaction confidence based on the type of conducted experiments.

Experimentally the interactions can be extracted from a variety of high-throughput methods. Most widely are used affinity-purification methods followed by mass spectrometry, yeast two-hybrid screens or protein microarrays. Different experimental conditions and sensitivity of methods lead to small overlap between inferred protein-protein pairs. Results from different protocols should be taken into account and combined as complementary tools to describe the best possible interaction space.

Most widely used computational methods for inferring protein-protein interactions are evolutionally conserved co-expressions between interactors, co-localization of proteins in the cell, functioning in the same biological process. The latter is mostly based on proteins belonging to the same Gene Ontology term. Often knowledge from one organism is transferred to other less studied organisms.

Further reading

Aranda B, et al. (2010) The IntAct molecular interaction database in 2010 *NAR* 38(database issue): D525–531.

Bader JS, et al. (2004) Gaining confidence in high-throughput protein interaction networks. *Nature Biotechnology* 22(1): 78–85.

Ewing RL, et al. (2007) Large-scale mapping of human protein-protein interactions by mass spectrometry. *Molecular Systems Biology* 3: 89.

Kaushansky A et al. (2010) Quantifying protein-protein interactions in high throughput using protein domain microarrays. *Nature Protocols* 5(4): 773–790.

Rual JF, et al. (2005) Towards a protein-scale map of the human protein-protein interaction network. *Nature* 437(7062): 1173–1178.

Szklarczyk D, et al. (2011) The STRING database in 2011: functional interaction networks of proteins, globally integrated and scored. *NAR* 39(Database issue): D561–568.

Xenarios I, et al. (2002) DIP: The Database of Interacting Proteins. A research tool for studying cellular networks of protein interactions. *NAR* 30(1): 303–5.

PROTEIN SEQUENCE, *SEE* SEQUENCE OF PROTEIN.

PROTEIN SEQUENCE CLUSTER DATABASES

Rolf Apweiler

Protein sequence cluster databases group related proteins together and are derived automatically from protein sequence databases using different clustering algorithms.

Protein sequence cluster databases use different clustering algorithms to group related proteins together. Since they are derived automatically from protein sequence databases without manual crafting and validation of family discriminators, these databases are relatively comprehensive although the biological relevance of clusters can be ambiguous and can sometimes be an artefact of particular thresholds.

Related websites

ProDom	http://prodom.prabi.fr/prodom/current/html/home.php
SYSTERS	http://systers.molgen.mpg.de/
ProtoMap	http://protomap.stanford.edu/
CluSTr	http://www.ebi.ac.uk/clustr/
Clusters of Orthologous Groups of proteins (COGs)	http://www.ncbi.nlm.nih.gov/COG

Further reading

Kriventseva EV, et al. (2001) Clustering and analysis of protein families. *CurrentOpin Struct Biol* 11: 334–339.

PROTEIN STRUCTURE

Roman Laskowski, Andrew Harrison, Christine Orengo and Tjaart AP de Beer

The final, folded 3D conformation of the atoms making up a protein, which describes the geometry of that protein in 3D space. The term is loosely applied to the models of the protein structure that are obtained as the result of experiment (for example, by means of X-ray crystallography or NMR spectroscopy). Such models aim to give as good an explanation for the experimental data as possible but are merely representations of the molecule in question; they may be accurate or inaccurate models, precise or imprecise. Models from X-ray crystallography tend to be static representations of a time-averaged structure, whereas proteins frequently exhibit dynamic behaviour.

Nevertheless, knowledge of a protein's structure, as from such an experimentally determined model, can be crucial for understanding its biological function and mechanism. Comparisons of structures can provide insights into general principles governing these complex molecules, the interactions they make and their evolutionary relationships.

The details of the protein's active site can not only help explain how it achieves its biological function but may also assist the design of drugs to target the protein and block or inhibit its activity.

Protein structure is usually represented as a set of 3D coordinates for each atom in the protein. From these coordinates other geometrical relationships and properties can be calculated. For example the distances between all Cα atoms within the protein or specific angles e.g. dihedral angles which are used to describe the curvature of the polypeptide chain. Protein structures are deposited in the Protein Databank (PDB) currently held at the Research Collaboratory of Structural Biology in Rutgers University in the States. There is also a European node, the PDBe, where structures are also deposited, held at the European Bioinformatics Institute in Cambridge, UK.

Relevant websites

RCSB PDB	http://www.pdb.org
PDBe	http://www.ebi.ac.uk/pdbe

Further reading

Branden C, Tooze J (1998) *Introduction to Protein Structure.* Garland Science, New York.

See also Dihedral Angle, Protein Data Bank

PROTEIN STRUCTURE CLASSIFICATION DATABASES, *SEE* STRUCTURE 3D CLASSIFICATION.

PROTEOME

Dov Greenbaum

Groups of proteins sharing homology, structure or function.

Given the variety of proteins in nature, it is helpful to pigeonhole proteins into a categorical system. The most common system is that of hierarchical families, akin to a taxonomic classification used for organisms. Protein families were originally conceived as comprising proteins related through sequence identity, although the colloquial use has extended to include also the grouping together of proteins related via similar function. Part of the Dayhoff classification scheme, protein families are more closely related evolutionarily and share a closer ancestor than the more broadly characterized grouping of superfamilies. By contrast closely related homologous proteins may be grouped into protein subfamilies. Dayhoff originally defined a protein family as a group of proteins of similar function whose amino acid sequences are more than 50% identical; this cutoff value is still used by the Protein Information Resource (PIR). Other databases may use lower cut-off values and may not use functional criteria for the definition of a protein family. In many cases there are specific sequences within each member of the family that are required for the functional or structural features of that protein. The analysis of these conserved portions of the sequence allow researchers to analyze and derive defining signature sequences.

There are now several protein family databases, including Pfam, PRINTS, ProDom, SMART. Pfam is the most comprehensive database, to date, containing 3360 protein families (version 7.0). There is usually considerable similarity in the functions of relatives classified within the same protein family. Most protein family databases generate sequence profiles, patterns or regular expressions from a set of clear relatives within the family and use these to scan the sequence databases, such as SP-TREMBL, GenBank, to recognize further relatives to include in the classification. The comprehensive Pfam database builds Hidden Markov models for each family, whilst the PRINTS database uses a fingerprint of regular expressions encoding conserved patterns at different positions in a multiple alignment of relatives. The PRINTS database contains extensive functional information for each protein family and is manually validated. By contrast, protein families in the ProDom database are identified using completely automated protocols.

Related websites

Sequence-based classification of protein families:

PROCLASS	http://www-nbrf.georgetown.edu/gfserver/proclass.html
PFAM	http://www.sanger.ac.uk/Pfam/
PIR	http://www-nbrf.georgetown.edu/pirwww/pirhome.shtml
PRINTS	http://bioinf.man.ac.uk/dbbrowser/PRINTS/
SMART	http://smart.embl-heidelberg.de/
ProtFam List	http://www.hgmp.mrc.ac.uk/GenomeWeb/prot-family.html
MetaFam	http://metafam.ahc.umn.edu/
ProDom	http://prodes.toulouse.inra.fr/prodom/doc/prodom.html

Structural classification of protein families:

SCOP	http://scop.mrc-lmb.cam.ac.uk/scop
FSSP	http://www2.embl-ebi.ac.uk/dali/fssp/fssp.html
CATH	http://www.biochem.ucl.ac.uk/bsm/cath_new/index.html
PROSITE	http://www.expasy.ch/prosite/
TRANSFAC	http://transfac.gbf.de/TRANSFAC/

Further reading

Mulder NJ, Apweiler R (2002) Tools and resources for identifying protein families, domains and motifs. *Genome Biol 3*: REVIEWS2001.

Rappsilber J, Mann M (2002) Is mass spectrometry ready for proteome-wide protein expression analysis? *Genome Biol 3*: COMMENT2008.

Silverstein KTE, Shoop JE, Johnson A, et al. (2001) The MetaFam Server: a comprehensive protein family resource, *Nucleic Acids Research*, 29: 49–51.

Zhu H, et al. (2001) Global analysis of protein activities using proteome chips. *Science*. 293: 2101–2105.

PROTEOME ANALYSIS DATABASE (INTEGR8)

Marketa J. Zvelebil

The Proteome Analysis Database is a specialized protein database resource integrating information from a variety of sources that together facilitate the classification of proteins in complete proteome sets.

Relevant website

Proteome Analysis Database	http://www.ebi.ac.uk/proteome/

Further reading

Kersey P, Bower L, Morris L, et al. (2004) Integr8 and Genome Reviews: integrated views of complete genomes and proteomes. *Nucleic Acids Research* 33, D297–D302.

PROTEOMICS

Jean-Michel Claverie

A research area, and the set of approaches, dealing with the study of the proteome as a whole (or the components of large macromolecular complexes).

Beside the large-scale aspect, an essential difference with traditional protein biochemistry is that the proteins of interest are not defined prior to the experiment.

The critical pathway of proteome research includes:

1. Sample collection, handling and storage
2. Protein separation (2-D Electrophoresis)
3. Protein identification (peptide mass fingerprinting and mass spectrometry)
4. Protein characterisation (amino acid sequencing)
5. Bioinformatics (cross reference of protein informatics with genomic databases).

Main topics in proteomics are:

1. listing the parts: the detailed identification of all the components of an organism, organ, tissue, organelle, or other intracellular structures
2. profiling the expression: the analysis of protein abundance across different cell types, organs, tissue, physiological or disease conditions
3. finding the function: submitting the proteome (a large subset) to parallel functional assays such as ligand binding, interaction with a macromolecule, etc.

Topics 1 and 2 require good separation and highly sensitive detection, quantitation and identification techniques. Topic 3 requires protein mass production, robotics, arraying (e.g. protein arrays) and sensitive interaction assays.

Related websites

ExPASy Proteomics tools page	http://us.expasy.org/tools/
Proteomics Interest Group	http://proteome.nih.gov/

Further reading

Bakhtiar R, Nelson RW (2001) Mass spectrometry of the proteome. *Mol. Pharmacol.* 60: 405–415.

P

Griffin TJ, Aebersold R (2001) Advances in proteome analysis by mass spectrometry. *J. Biol. Chem.* 276: 45497–45500.

Issaq HJ (2001) The role of separation science in proteomics research. *Electrophoresis* 22: 3629–3638.

Jenkins RE, Pennington SR (2001) Arrays for protein expression profiling: towards a viable alternative to two-dimensional gel electrophoresis? *Proteomics* 1: 13–29.

Nelson RW, et al. (2000) Biosensor chip mass spectrometry: a chip-based proteomics approach. *Electrophoresis* 21: 1155–1163.

Patton WF (2000) A thousand points of light: the application of fluorescence detection technologies to two-dimensional gel electrophoresis and proteomics. *Electrophoresis* 21: 1123–1144.

Rappsilber J, Mann M (2002) What does it mean to identify a protein in proteomics? *Trends Biochem. Sci.* 27: 74–78.

PROTEOMICS STANDARDS INITIATIVE (PSI)

Simon Hubbard

The Proteomics Standards Initiative is a HUPO subgroup aiming to develop community driven standards for reporting, dissemination and comparison of proteomics data, ranging from gels, columns, and spectra, through peptide and protein identifications and on in to protein interactions and protein quantitation information. Different workgroups have been established to deal with these different aspects, each of which has reached different levels of maturity. For example, there are now well established XML standards for molecular interaction data (PSI-MI XML), recent releases for mass spectra (mzML) and proteome informatics (mzIdentML), and nascent candidate releases for quantitation (mzQuantML).

The PSI group is open to anyone wishing to contribute and is active with regular international meetings, often timetabled around major conferences.

Related website

Proteomics Standards Initiative wiki	http://www.psidev.info/

See also PRIDE

PROTEOTYPIC PEPTIDE

Simon Hubbard

This is a peptide that is usually produced from a proteolytic digestion of a given protein and is therefore a good characteristic ‘marker’ that that particular protein is present in a given proteomic experiment. Because they are usually present when the protein is, these peptides are good choices for surrogates in quantitative proteomic experiments seeking to measure the level of the parent protein via a proxy molecule. A variety of machine learning based prediction algorithms have been published for prediction and selection of such peptides for proteomics experiments.

P

Further reading

Eyers CE, Lawless C, Wedge DC, Lau KW, Gaskell SJ, Hubbard SJ. CONSeQuence: prediction of reference peptides for absolute quantitative proteomics using consensus machine learning approaches. *Mol Cell Proteomics*. 10: M110.003384.

Mallick P, Schirle M, Chen SS, et al. (2011) Computational prediction of proteotypic peptides for quantitative proteomics. *Nat Biotechnol*. 2007, 25: 125–31.

PSEUDOPARALOG (PSEUDOPARALOGUE)

Laszlo Patthy

Two homologous genes within a genome are said to be pseudoparalogous (or pseudoparalogues) if they are not the result of gene duplication but evolved through a combination of vertical inheritance and horizontal gene transfer. Pseudoparalogs are common in genomes affected by endosymbiosis.
See also Homology, Ortholog, Paralog, Epaktolog

Further reading

Koonin EV (2005) Orthologs, paralogs, and evolutionary genomics. *Annu Rev Genet* 39: 309–338.

Makarova KS, Wolf YI, Mekhedov SL, Mirkin BG, Koonin EV (2005) Ancestral paralogs and pseudoparalogs and their role in the emergence of the eukaryotic cell. *Nucleic Acids Res*. 33(14): 4626–4638. Print 2005.

PSEUDOGENE

Dov Greenbaum

Pseudogenes are sequences of genomic DNA which although they have similarity to normal genes, are non-functional.

Pseudogenes are interesting in genomics as they provide a picture of how the genomic sequence has mutated over evolutionary time, and can also be used in determining the underlying rate of nucleotide insertion, deletion and substitution.

Although some pseudogenes may have introns or promoters, some do not. Those that do not are termed processed pseudogenes and are likely copied from mRNA and incorporated into the chromosome.

Most pseudogenes some gene-like features such as promoters. Nevertheless, pseudogenes as a class are considered to be non-functional given their inability to code protein. This inability to code proteins may be from various genetic disablements, including: premature stop codons or frameshifts.

Pseudogenes are caused by one of two possible processes. (i) Duplication to produce non processed pseudogenes. During duplication any number of modifications – mutations, insertions, deletions or frame shifts – can occur to the primary genomic sequence. Subsequent to these modifications a gene may lose its function at either, or both of, the translational or transcriptional level. (ii) Processed pseudogenes. These arise when a mRNA transcript is reverse transcribed and integrated back into the genome. Mutations and other disablements may then occur to this sequence over the course of evolution. As they are reverse transcription copies of a mRNA transcript, they do not contain introns.

In addition, recently another class of pseudogenes, i.e., disabled genes or unitary pseudogenes represent genes where varied mutations stopped the gene from being successfully transcribed and or translated leading the gene to become non-functional, and in some examples, deactivated particularly when the problematic mutations become fixed in the population.

Pseudogenes are identified through the use of sequence alignment programs. Initially annotated genes are collected into paralog families. These families are then used to survey the entire genome. Potential homologs are determined to be pseudogenes if there is evidence for one of the two methods mentioned above. For each potential pseudogene there are a number of validation steps including: checking for over-counting and repeat elements, overlap on the genomic DNA with other homologs, and cross-referencing with exon assignments from genome annotations.

Related website

Gerstein Lab Pseudogenes Page	http://bioinfo.mbb.yale.edu/genome/pseudogene/

Further reading

Cooper DN (1999) *Human Gene Evolution*, Bios Scientific Publishers, pp 265–285.

Gerstein M, Zheng D (2006) The Real Life of Pseudogenes. *Scientific American* August, 49–55.

Harrison PM, Gerstein M (2002) Studying genomes through the acons: protein families, pseudogenes and proteome evolution. *J Mol Biol* 318: 1155–1174.

Mighell AJ, Smith NR, Robinson PA, Markham AF (2000) Vertebrate pseudogenes. *FEBS Lett* 468: 109–114.

PSI BLAST

Dov Greenbaum

A version of the BLAST program employing iterative improvements of an automatically generated profile to detect weak sequence similarity.

BLAST is a heuristic that attempts to optimize a specific similarity measure. It permits a tradeoff between speed and sensitivity, with the setting of a ‘threshold’ parameter, T. A higher value of T yields greater speed, but also an increased probability of missing weak similarities.

Previous versions of BLAST could find only local alignments between proteins or DNA sequences that did not contain gaps. These programs were able to give a value, the E value, that gave some idea of the probability of the results occurring wholly by chance. It is thought, although not mathematically proven, that E values are reliable for gapped sequence alignments as well.

PSI BLAST, Position Specific Iterated Basic Local Alignment Search Tool is based on the Gapped Blast program, but after alignments have been found, a profile is constructed from

P

the significant matches. This profile is then compared and locally aligned to proteins in the available protein databases. This is then iterated, if desired, an arbitrarily number of times with the new profiles that are discovered in each successive search.

PSI BLAST is one of the most powerful and popular sequence homology programs available – the papers describing the algorithms for this and related BLASTS are some of the most heavily cited ever.

Related website

BLAST	http://www.ncbi.nlm.nih.gov/blast/

Further reading

Altschul SF, et al. (1990) Basic local alignment search tool. *J. Mol. Biol.* 215: 403–410.

Altschul SF, et al. (1997) Gapped BLAST and PSI-BLAST: a new generation of protein database search programs. *Nucleic Acids Res.* 25: 3389–3402.

Bhagwat M, Aravind L (2007) PSI-BLAST tutorial. *Methods Mol Biol.*395: 177–186.

See also BLAST

PSSM, *SEE* PROFILE.

PURIFYING SELECTION (NEGATIVE SELECTION)

Austin L. Hughes

Natural selection that acts to eliminate mutations that are harmful to the fitness of the organism is known as purifying selection.

When we have evidence that purifying selection has acted on a given genomic region (whether coding or non-coding), this in turn is evidence that the region in question plays an important function and thus is subject to functional constraint. In protein-coding regions, purifying selection eliminates the majority of non-synonymous (amino acid-altering) mutations. As evidence that such selection has occurred, the number of synonymous nucleotide substitutions per synonymous site (d_S) exceeds the number of nonsynonymous nucleotide substitutions per nonsynonymous site (d_N) in the vast majority of protein-coding genes.

Single nucleotide polymorphisms often show evidence of ongoing purifying selection against slightly deleterious variants. For example, in many species, it has been found that non-synonymous variants tend to occur at lower frequencies than synonymous variants in the same genes. This result is predicted if most nonsynonymous variants are slightly deleterious and are thus in the process of being eliminated from the population by purifying selection.

Further reading

Kimura M (1977) Preponderance of synonymous changes as evidence for the neutral theory of molecular evolution. *Nature* 267: 275–276.

Kimura M (1983) *The Neutral Theory of Molecular Evolution*. Cambridge University Press, Cambridge.

Li W-H, et al. (1985) Evolution of DNA sequences, pp 1–94, In: MacIntyre RJ (ed.) *Molecular Evolutionary Genetics*. Plenum Press, New York.

P

Q

Qindex (Qhelix; Qstrand; Qcoil; Q3)
QM/MM Simulations
QSAR (Quantitative Structure Activity Relationship)
Qualitative and Quantitative Databases used in Systems Biology
Qualitative Differential Equations
Quantitative Proteomics
Quantitative Trait (Continuous Trait)
Quartets, Phylogenetic
Quartet Puzzling, *see* Quartet, Phylogenetic.
Quaternary Structure

Concise Encyclopaedia of Bioinformatics and Computational Biology, Second Edition.
Edited by John M. Hancock and Marketa J. Zvelebil.

Q

Q

QINDEX (QHELIX; QSTRAND; QCOIL; Q3)

Patrick Aloy

Per-residue prediction accuracy is the simplest, although not the best, measure of secondary structure prediction accuracy. It gives the percentage of residues correctly predicted as helix, strand, coil or for all three conformational states.

For a single conformational state:

$$Q_i = \frac{\text{number of residues correctly predicted in state } i}{\text{number of residues observed in state } i} * 100$$

where i is helix, strand or coil.

For all three states:

$$Q_3 = \frac{\text{number of residues correctly predicted}}{\text{number of residues}} * 100$$

Q_3 is a very generous way of measuring prediction accuracy, as it only considers correctly predicted residues and does not penalize wrong predictions.

Further reading

Rost B, Sander C (2000) Third generation prediction of secondary structures. *Methods Mol Biol* 143, 71–95.

See also Secondary Structure Prediction

QM/MM SIMULATIONS

Roland Dunbrack

Molecular mechanics simulations in which part of the system is treated with *ab initio* or semi-empirical quantum mechanics while the rest of the system is treated with standard empirical molecular mechanics energy functions. QM/MM simulations are needed to study inherently quantum-mechanical processes such as breaking and forming covalent bonds. In this case, the active site of an enzyme is treated quantum mechanically, and the rest of the protein and the solvent is treated with empirical energy functions.

Further reading

Gao J, Truhlar DG (2002) Quantum mechanical methods for enzyme kinetics. *Annu Rev Phys Chem* 53, 467–505.

QSAR (QUANTITATIVE STRUCTURE ACTIVITY RELATIONSHIP)

Marketa J. Zvelebil and Bissan Al-Lazikani

The method of QSAR aims to derive a function connecting biological activity of a set of similar chemical compounds with parameters that describe a structural feature of these molecules which reflects properties of the binding cavity on a protein target. The derived function is used as a guide to select best candidates for drug design.

Basically, the QSAR method puts together statistical and graphical models of biological activity based on molecular structures. The models are then used to make predictions for the activity of untested compounds.

This is a broad field and encompasses data-mining, 2D and 3D analyses. It is applied at different stages of the drug discovery process including drug-design and lead optimization.

See also SAR

Further reading

Klebe G (1998) Comparative molecular similarity indices: CoMSIA. In: *3D QSAR in Drug Design*, H Kubinyi, G Folkers, YC Martin (eds), Kluwer Academic Publishers, UK.

Kim KH, Greco G, Novellino E (1998) A critical review of recent CoMFA applications. In: *3D QSAR in Drug Design*, H Kubinyi, G Folkers, YC Martin (eds), Kluwer Academic Publishers, UK.

QUALITATIVE AND QUANTITATIVE DATABASES USED IN SYSTEMS BIOLOGY

Pascal Kahlem

With the constant development of faster and more reliable high-throughput biotechnologies, the scientific community is presented with a growing collection of biological information, some qualitative, some quantitative. Formal databases include catalogs of genes (EnsEMBL) proteins (UniProt) enzymes and their substrates (BRENDA) and molecular reactions (Reactome), many from multiple species. Quantitative data resulting from large-scale experiments are also collected in databases; some of them are public, including gene expression (ArrayExpress), protein interactions (IntAct), reaction kinetics (SABIO-RK), cellular phenotypes (MitoCheck) and whole organism phenotypes (e.g. EuroPhenome) amongst others. A list of relevant resources is given below:

	Type of data	**Name of the database**	**URL**
Knowledge databases	Small chemical entities	ChEBI	www.ebi.ac.uk/chebi
	Gene ontology	GO	www.geneontology.org
	Taxonomy	Uniprot taxonomy	www.uniprot.org/taxonomy
	Genetic diseases	OMIM	www.ncbi.nlm.nih.gov/omim
	Literature	CiteXplore	www.ebi.ac.uk/citexplore
	Genomes	Ensembl	www.ensembl.org
	Nucleotide sequences	EMBL-bank	www.ebi.ac.uk/embl
	Proteins	UniProt	www.uniprot.org
	Enzymes	IntEnz	www.ebi.ac.uk/intenz
	Structure	MSD	www.ebi.ac.uk/msd
	Enzymes	BRENDA	www.brenda-enzymes.info
	Pathways	Reactome	www.reactome.org
	Pathways	KEGG	www.genome.jp/kegg
	Pathways	Chillibot	www.chilibot.net
	Pathways	iHOP	www.ihop-net.org/UniPub/iHOP
	Pathways	Panther	www.pantherdb.org/panther
	Pathways	BioCyc	www.biocyc.org
	Genome-scale metabolic models	Thermodynamics SEED models	http://seed-viewer.theseed.org/
	Biological models	BioModels	www.ebi.ac.uk/biomodels
	Transporters	TransportDB	www.membranetransport.org
	Thermodynamics	NIST	xpdb.nist.gov/enzyme_thermodynamics/
Quantitative databases	Gene expression	Array express	www.ebi.ac.uk/aerep
	Protein interaction	IntAct	www.ebi.ac.uk/intact
	Reaction kinetics	SABIO-RK	sabiork.villa-bosch.de
	Cellular phenotypes	MitoCheck	www.mitocheck.org
	Growth phenotypes	ASAP	asap.ahabs.wisc.edu/asap/home.php
	Metabome, fluxome	Escherichia coli multi-omics database	ecoli.iab.keio.ac.jp
	Kinetic parameters	KMedDB	sysbio.molgen.mpg.de/KMedDB

Further reading

Aranda B, Achuthan P, et al. (2010) The IntAct molecular interaction database in 2010. *Nucleic Acids Research* 38(Database issue): D525–531.

D'Eustachio P (2011) Reactome knowledgebase of human biological pathways and processes. *Methods in Molecular Biology* 694: 49–61.

Erfle H, Neumann B, et al. (2007) Reverse transfection on cell arrays for high content screening microscopy. *Nat Protoc* 2(2): 392–399.

Flicek P, Amode MR, et al. (2011) Ensembl 2012. *Nucleic Acids Research*.

Li C, Donizelli M, et al. (2010) BioModels Database: An enhanced, curated and annotated resource for published quantitative kinetic models. *BMC Systems Biology* 4: 92.

Magrane M, Consortium U (2011) UniProt Knowledgebase: a hub of integrated protein data. *Database: the journal of biological databases and curation* 2011: bar009.

Morgan H, Beck T, et al. (2010) EuroPhenome: a repository for high-throughput mouse phenotyping data. *Nucleic Acids Res* 38(Database issue): D577–585.

Parkinson H, Sarkans U, et al. (2011) ArrayExpress update – an archive of microarray and high-throughput sequencing-based functional genomics experiments. *Nucleic Acids Research* 39(Database issue): D1002–1004.

Scheer M, Grote A, et al. (2011) BRENDA, the enzyme information system in 2011. *Nucleic Acids Research* 39(Database issue): D670–676.

QUALITATIVE DIFFERENTIAL EQUATIONS

Luis Mendoza

With qualitative differential equations, the dependent variable takes a qualitative value composed of a qualitative magnitude and a direction. Here, the qualitative magnitude is a discretization of a continuous variable, while the qualitative direction is the sign of its derivative. Furthermore, each equation is actually a set of constraints that restrict the possible qualitative values of the variable. To solve the system, it is necessary to create a tree with all possible sequences of state transitions from the initial state. Now, this characteristic makes the methodology difficult to apply for large biological systems, since the trees describing the dynamical behavior rapidly grow out of bounds due to the weak nature of qualitative constraints. This scalability problem has restricted the application of qualitative equations to a small number of models (Akutsu et al., 2000; de Jong et al., 2001; Heidtke and Schulze–Kremer, 1998; Trelease et al., 1999).

Further reading

Akutsu T, Miyano S, Kuhara S (2000) Inferring qualitative relations in genetic networks and metabolic pathways. *Bioinformatics* 16: 727–734.

de Jong H, Page M, Hernandez C, Geiselmann J. (2001) Qualitative simulation of genetic regulatory networks: Method and Application. In: Nebel B (ed.), *Proc. 17th Int. Joint Conf. Artif. Intell. (IJCAI–01)*, 67–73. San Mateo, CA. Morgan Kaufmann.

Heidtke KR, Schulze–Kremer S (1998) Design and implementation of a qualitative simulation model of lambda phage infection. *Bioinformatics* 14: 81–91.

Trelease RB, Henderson RA, Park JB (1999) A qualitative process system for modeling NF–kappaB and AP-1 gene regulation in immune cell biology research. *Artif. Intell. Med.* 17: 303–321.

Q

QUANTITATIVE PROTEOMICS

Simon Hubbard

The field of *proteomics* is now quantitative and offers a range of methods and techniques which offers means to determine the relative or absolute abundance of a given protein species in a complex proteome sample. The approaches usually rely on gel electrophoresis and/or mass spectrometry as the analytical techniques from which quantitative information can be obtained, and can be divided as follows:

1. Gel-based methods, including Differential in Gel Electrophoresis (DIGE) where individual protein samples and labelled with a different fluorescent probe prior to analysis on a 2D gel, image analysis and data processing.
2. Label-free methods. Protein quantitation is obtained either from spectral counting using methods such as *emPAI*, or by signal intensity label free methods comparing features in a 2D retention time vs. mass-to-charge space to each other to obtain relative quantitation, integrating over multiple peptide species to obtain protein level abundances.
3. Label-mediated methods. Quantitation is obtained through stable isotope labelling of proteins/peptides, and comparing values across samples that are subsequently differentiated by mass shifts from the labelling. In isobaric tagging approaches, such as iTRAQ™ or Tandem Mass Tags (TMT), the proteolytic peptides are labelled and quantification is obtained from different reporter ions specific to each tag type (4-plex and 8-plex versions of the technology are available). In metabolic labelling strategies, proteins are labelled in cell culture to generate 'heavy' and 'light' versions in different cell populations prior to mixing and analysis. Aternatively, surrogate heavy-labelled peptides can be spiked in against which endogenous peptides can be compared, using techniques such as *Selected Reaction Monitoring* for high sensitivity.

This is a fast moving field with new methods appearing on a regular basis and no clear 'winners' have yet emerged. It is possible using spiked-in standards to calculate absolute quantitative values, such as concentrations or copies-per-cell estimates and the leaders in the field are able to quantify proteins down to 50 copies per cell in the yeast *Saccharomyces cerevisiae*.

Related website

PaxDB – collected and normalized datasets of quantitative proteomics experiments across several species	http://pax-db.org

See also MaxQuant, Selected Reaction Monitoring, SILAC, Isobaric Tagging

QUANTITATIVE TRAIT (CONTINUOUS TRAIT)

Mark McCarthy and Steven Wiltshire

A trait or character whose values are measured on a metric or continuous scale.

The quantitative phenotype (P_i) of a particular individual i, will reflect the combined effects of those genetic (G_i) and environmental factors (E_i) acting on that individual:

$$P_i = G_i + E_i$$

The architecture of the genetic component, G_i, can vary in terms of the number of genes involved, their frequencies and the magnitude of their effects on the trait. Any gene influencing the value of a quantitative trait can be termed a quantitative trait locus (or QTL), although this term is usually reserved for genes with individually measurable effects on the trait. A trait under the influence of several such genes is often termed oligogenic, whereas a genetic architecture comprising many genes each with individually small, additive, effects is termed polygenic, and the aggregate of such genes often referred to as the polygene. Many biomedically significant traits (such as obesity, blood pressure) are thought to be under the control of one or several QTLs acting on a polygenic background. The environmental component, E, encompasses all the non-genetic factors acting on the phenotype, and, by definition, has a mean of 0 in the population as a whole.

The mean phenotypic value of individuals with a genotype at a given QTL is referred to as the genotype mean, so for a trait influenced by a biallelic QTL (with alleles A and B) there will be three genotypic means $-\mu_{AA}$, μ_{AB} and μ_{BB} The overall population mean, μ, represents the weighted sum of these three genotypic means, the weights being the population frequencies of the respective genotypes. An analogous situation applies when the effects of multiple QTLs are considered. Continuous traits are often, although not always, distributed normally in humans.

The phenotype P_i of individual i can be modelled in an alternative fashion. If G_i and E_i, each with mean of zero, are now both defined as deviations from the overall population mean, μ, the phenotype of individual i can be expressed thus:

$$P_i = \mu + G_i + E_i$$

This model is the basis for partitioning the total phenotypic variance into its genetic and environmental variance components.

It is important to note that the term quantitative trait is also widely-used to refer to meristic traits – those in which the trait is ordinally distributed – and to discrete traits, in which it is assumed that an underlying continuously distributed liability exists with the disease phenotype expressed when individual liability exceeds some threshold.

Further reading

Almasy L, Blangero J (1998) Multipoint quantitative-trait linkage analysis in general pedigrees. *Am J Hum Genet* 62: 1198–1211.

Almasy L, et al. (1999) Human pedigree-based quantitative-trait-locus mapping: localization of two genes influencing HDL-cholesterol metabolism. *Am J Hum Genet* 64: 1686–1693.

Falconer D, McKay TFC (1996) *Introduction to Quantitative Genetics*. Harlow: Prentice Hall.

Flint J, Mott R (2001) Finding the molecular basis of quantitative traits: successes and pitfalls. *Nat Rev Genet* 2: 437–445.

Hartl DI, Clark AG (1997) *Principles of Population Genetics*. Sunderland: Sinauer Associates, pp 397–481.

Sabatti C, Service SK, Hartikainen A-L, et al. (2009) Genomewide association analysis of metabolic phenotypes in a birth cohort from a founder population. *Nature Genetics*, 41: 35–46.

See also Polygenic Inheritance, Oligogenic Inheritance, Variance Components, Heritability, Haseman-Elston Algorithm

QUARTETS, PHYLOGENETIC

Sudhir Kumar and Alan Filipski

A phylogenetic quartet is a tree of four taxa or groups of taxa.

Quartets are important in phylogenetic theory because four is the smallest number of taxa that can have alternative unrooted topologies. For three taxa, there only exists one possible unrooted tree topology whereas for four taxa there exist three alternative, distinct unrooted tree topologies. For this reason four taxon trees (quartets) are often used for mathematical and theoretical arguments about models and properties of phylogenies.

If the correct Maximum Likelihood configurations for all possible quartets of a set of taxa are computed, then a phylogenetic tree over the entire set of taxa may be created by a process known as quartet puzzling, that is, by merging such quartets. This is essentially one possible heuristic approach for inferring Maximum Likelihood trees. However, the number of possible quartets increases rapidly with the number of taxa n such that not all quartets can be computed for large trees.

Software

Schmidt HA, et al. (2002) TREE-PUZZLE: maximum likelihood phylogenetic analysis using quartets and parallel computing. *Bioinformatics* 18: 502–504.

Further reading

Rzhetsky A, et al. (1995) Four-cluster analysis: a simple method to test phylogenetic hypotheses. *Mol Biol Evol* 12: 163–167.

Strimmer K, von Haeseler A (1996) Quartet puzzling: A quartet maximum-likelihood method for reconstructing tree topologies. *Mol Biol Evol* 13: 964–969.

See also Consensus Tree, Maximum Likelihood Phylogeny Reconstruction, Minimum Evolution Principle, Distances Between Trees

See also Distances Between Trees (Quartet Distance)

QUARTET PUZZLING, *SEE* QUARTET, PHYLOGENETIC.

QUATERNARY STRUCTURE

Roman Laskowski and Tjaart de Beer

Describes the arrangement of two or more interacting macromolecules: for example, how the chains of a multimeric protein pack together and the extent of the contact regions between these chains. Many proteins can only achieve their biological function in concert with one or more other proteins, the functional units ranging in complexity from simple dimers to large multi-protein and RNA assemblies such as the ribosome. The assemblies are held together by hydrogen bonds, van der Waals and coulombic forces.

Related websites

Quaternary structure predictor	http://www.mericity.com/
PDBePISA	http://www.ebi.ac.uk/msd-srv/prot_int/pistart.html

See also Hydrogen Bonds

R

R-Factor
r8s
Ramachandran Plot
Random Forest
Random Trees
Rat Genome Database (RGD)
Rate Heterogeneity
Rational Drug Design, *see* Structure-based Drug Design.
RDF
RDF Database, *see* Database.
readseq
Reasoning
Recombination
RECOMBINE, *see* LAMARC.
Record, *see* Data Structure.
Recursion
Reference Genome (Type Genome)
Reference Sequence Database, *see* RefSeq.
Refinement
RefSeq (the Reference Sequence Database)
Regex, *see* Regular Expresssion.
Regression Analysis
Regression Tree
Regular Expression (Regex)
Regularization (Ridge, Lasso, Elastic Net, Fused Lasso, Group Lasso)
Regulatory Motifs in Network Biology
Regulatory Network Inference
Regulatory Region
Regulatory Region Prediction
Regulatory Sequence, *see* Transcriptional Regulatory Region.
Regulome
Relational Database
Relational Database Management System (RDBMS)
Relationship
REPEATMASKER
Repeats Alignment, *see* Alignment.
Repetitive Sequences, *see* Simple DNA Sequence.
Residue, *see* Amino Acid.
Resolution in X-Ray Crystallography
Response, *see* Label.
RESTful Web Services
Restriction Map
Retrosequence
Retrotransposon
Reverse Complement
Rfam
Rfrequency
RGD, *see* Rat Genome Database.
Ri
Ribosomal RNA (rRNA)

Concise Encyclopaedia of Bioinformatics and Computational Biology, Second Edition.
Edited by John M. Hancock and Marketa J. Zvelebil.

Ribosome Binding Site (RBS)
RMSD, *see* Root Mean Square Deviation.
RNA (General Categories)
RNA Folding
RNA Hairpin
RNA-seq
RNA Splicing, *see* Splicing.
RNA Structure
RNA Structure Prediction (Comparative Sequence Analysis)
RNA Structure Prediction (Energy Minimization)
RNA Tertiary Structure Motifs
Robustness
ROC Curve
Role
Root Mean Square Deviation (RMSD)
Rooted Phylogenetic Tree, *see* Phylogenetic Tree. Contrast with Unrooted Phylogenetic Tree.
Rooting Phylogenetic Trees
Rosetta Stone Method
Rotamer
Rough Set
rRNA, *see* Ribosomal RNA.
Rsequence
Rule
Rule Induction
Rule of Five (Lipinski Rule of Five)

R-FACTOR

Liz Carpenter

The R-factor is the principle measure of how well the refinement of a crystallographic structure is proceeding. It is a measure of the difference between the observed structure factors, $|F_{obs}|$, (proportional to the square root of the measured intensity) and the structure factors calculated from the model of the macromolecular structure $|F_{calc}|$.

$$R = \frac{\sum_{hkl} ||F_{obs}| - |F_{calc}||}{\sum_{hkl} |F_{obs}|}$$

Values of R-factors are generally expressed as percentages. For a macromolecule a completely random distribution of atoms in the unit cell would give an R-factor of 59%. Values below 24 % suggest that the model is generally correct. Near atomic resolution structures (1.3 Å and above) can have R factors below 15 %. The R-factor is sensitive to over-refinement, particularly at lower resolution, worse than say 2.8 Å. Since the refinement programs used to perfect the structure are designed to minimise the R-factor, it tends to decrease even when no real improvement in the structure is occurring. The free R-factor (see free R-factor) was introduced to overcome this problem.

Further reading

Glusker JP, Lewis M, Rossi M (1994) *Crystal Structure Analysis for Chemists and Biologists*, VCH Publishers.

See also X-ray Crystallography for Structure Determination, Diffraction of X-Rays, Crystals

R8S

Michael P. Cummings

A program for estimating divergence times and rates of evolution based on a phylogenetic tree with branch lengths using maximum likelihood, semi- and non-parametric methods.

The program accommodates local, global and relaxed clock assumptions, as well as one or more calibration points.

The program can be used interactively or without interaction by including appropriate commands in the r8s block (NEXUS file format). Written in ANSI C the program is available as source code and executable for Linux (x86).

Related website

r8s web page	http://loco.biosci.arizona.edu/r8s/

R

Further reading

Sanderson MJ (2002) Estimating absolute rates of molecular evolution and divergence times: a penalized likelihood approach. *Mol. Biol. Evol.*, 19: 101–109.

Sanderson MJ, Doyle JA (2001) Sources of error and confidence intervals in estimating the age of angiosperms from rbcL and 18S rDNA data. *Amer. J. Bot.*, 88: 1499–1516.

RAMACHANDRAN PLOT

Roman Laskowski and Tjaart de Beer

A 2D plot of the permitted combinations of *phi* and *psi* values in protein structures, where *phi* and *psi* are the main chain dihedral angles. The favored regions were originally computed by Ramakrishnan and Ramachandran in 1965 by taking side chains to be composed of hard-sphere atoms and deriving the regions of *phi-psi* space that were accessible and those that were 'disallowed' on the basis of steric hindrance. Nowadays, the favourable and disallowed regions of the plot are determined empirically using the known protein structures deposited in the PDB.

Different residues exhibit slightly different Ramachandran plots due to the differences in their side chains, the most different being glycine which, due to its lack of a proper side chain can access more of the *phi-psi* space than can other side chains, and proline whose *phi* dihedral angle is largely restricted to the range -120° to 30°.

The plot is primarily used as a check on the 'quality' of a protein structure. That is, to check whether a given model of a protein structure (whether derived by experiment or other means) appears to be a reasonable one, or whether it might be unreliable. By plotting the *phi-psi* combination for every residue in the protein structure on the Ramachandran plot one sees whether the values cluster in the favoured regions, as expected, or whether any stray into the 'disallowed' regions. While it is possible for individual residues to have such disallowed *phi-psi* values due to strain in the structure, any model of a protein having many such residues has probably been incorrectly determined and is not a reliable representation of the protein in question.

Relevant websites

Ramachandran plot generators	http://eds.bmc.uu.se/ramachan.html
IISc Ramachandran plot	http://dicsoft1.physics.iisc.ernet.in/rp/

Further reading

Kleywegt GJ, Jones TA (1996) Phi/psi-chology: Ramachandran revisited. *Structure*, 4: 1395–1400.

Ramachandran GN, Ramakrishnan C, Sasisekharan V (1963) Stereochemistry of polypeptide chain configurations. *J. Mol. Biol.*, 7: 95–99.

Ramachandran GN, Sasisekharan V (1968) Conformations of polypeptides and proteins. *Adv. Prot. Chem*, 23: 283.

See also Main Chain, Side Chains, Dihedral Angles

RANDOM FOREST

Concha Bielza and Pedro Larrañaga

Originally developed for improving the predictive accuracy of decision trees, random forests is an ensemble classifier that can also be used for improving regression trees. A (big) number of decision trees is grown depending on the values of a random vector sampled independently and with the same distribution for all trees in the forest. Thus, for growing each tree, take as learning data a bootstrap sample of the original data set (i.e. sample with replacement n cases at random from the n original cases), select a relatively small subset of randomly chosen candidate attributes and the best split on these is used to split the node (random selection of features). For classifying a new example, the random forest decides the final prediction by letting the trees vote for the most popular class.

The random forest method is robust as it reduces the variance of tree-based algorithms. The error rate of this classifier depends on the correlation between any two trees in the forest (the error increases if the correlation increases) and on the error rate of each individual tree (the forest error decreases if the errors of the trees decrease). The size of the subset of candidate attributes in the random selection of features influences the forest error rate. Increasing this size increases the correlation and also decreases the individual tree errors. Thus, somewhere in between is a good choice for this size, usually quite wide.

Related website

Commercial implementation	http://www.salford-systems.com/

Further reading

Breiman L (2001) Random forests. *Machine Learning*, 45(1): 5–32.

Díaz-Uriarte R, Alvarez de Andrés S (2006) Gene selection and classification of microarray data using random forest. *BMC Bioinformatics*, 7: 3.

Jiang R, Tang W, Wu X, Fu W (2009) A random forest approach to the detection of epistatic interactions in case–control studies. *BMC Bioinformatics*, 30;10 Suppl 1: S65.

Pang H, Lin A, Holford M, et al. (2006) Pathway analysis using random forests classification and regression. *Bioinformatics*, 22(16): 2028–2036.

Qi Y (2012) Random forest for bioinformatics. *In: Ensemble Machine Learning: Methods and Applications*, Zhang C, Ma Y (eds.), Springer, 307–323.

See also Random Trees, Bagging, Decision Trees

RANDOM TREES

Concha Bielza and Pedro Larrañaga

A classification method that builds a decision tree by considering a given number of random features at each node. A random tree on its own tends to be too weak (poor performance). This is why it is usually used as 'building block' with random forests, i.e. by bagging several random trees we have a random forest. Also, we can have boosted random trees to make the resulting classifier strong enough.

R

Further reading

Breiman L (2001) Random forests. *Machine Learning*, 45 (1): 5–32.

See also Random Forests, Bagging, Decision Trees

RAT GENOME DATABASE (RGD)

Guenter Stoesser

The Rat Genome Database (RGD) collects and integrates data generated from ongoing rat genetic and genomic research efforts.

RGD develops and provides informatics tools that allow researchers to further explore the data. Besides rat-centric tools in the context of genetic mapping and gene prediction, RGD also provides algorithms which are of use to researchers working with other organisms, particularly human and mouse. RGD is based at the Medical College of Wisconsin (MCW) and is a direct collaboration between the MCW, the Mouse Genome Database (MGD) and the National Center for Biotechnology Information (NCBI).

Related website

RGD homepage	http://rgd.mcw.edu

Further reading

Petri V, et al; RGD Team (2011) *The Rat Genome Database pathway portal*. Database (Oxford). 2011:bar010.

See also Model Organism Database

RATE HETEROGENEITY

Aidan Budd and Alexandros Stamatakis

In statistical substitution models for phylogenetic inference the term is used to describe the phenomenon that different sites of the alignment (e.g., different areas of a gene) evolve at different evolutionary rates/speeds because of distinct evolutionary pressures. Thus, the rate changes and is heterogeneous as we move along an alignment. Rate heterogeneity is typically incorporated into likelihood models by integrating the per-site likelihood over a discrete approximation of the Γ function. The parameter that is actually estimated (or integrated in Bayesian frameworks) is the α shape parameter of the Γ function. In phylogenetics software the default number of discrete rates for approximating the integral of the likelihood over the Γ function at each site is set to 4, because this represents a good trade-off between speed and accuracy. However, computing the likelihood on a phylogeny under the Γ model of rate heterogeneity with 4 discrete rates leads to a four-fold increase of required numerical operations and memory with respect to a model that does not incorporate the Γ model of rate heterogeneity.

Another, simpler, approach is to estimate the proportion of invariable sites. Finally, mostly for computational reasons, it has also been proposed to use a certain number of per-site rate categories, that is, each site, only evolves under a specific evolutionary rate that has previously been estimated and assigned to this site.

Software
Almost all current likelihood-based tree inference programs offer this capability, e.g. IQPNNI, GARLI, PHYML, RAxML, MrBayes, PhyloBayes.

Further reading
Stamatakis A (2006) Phylogenetic models of rate heterogeneity: a high performance computing perspective. Proceedings of 20th Parallel and Distributed Processing Symposium, IEEE.

Yang Z (1993) Maximum-likelihood estimation of phylogeny from DNA sequences when substitution rates differ over sites. *Molecular Biology and Evolution*, 10(6): 1396–1401.

Yang Z (1996) Among-site rate variation and its impact on phylogenetic analyses. *Trends in Ecology and Evolution*, 11(9): 367–372.

See also Mixture Models, Heterotachy, Bayesian Inference

RATIONAL DRUG DESIGN, *SEE* STRUCTURE-BASED DRUG DESIGN.

RDF

Carole Goble and Katy Wolstencroft

The Resource Description Framework (RDF) is a language for representing information about web resources. Although initially intended for the representation of metadata about resources, RDF is increasingly also being used for the representation of data. RDF is the foundation of 'Linked Data': that is the exchange and publishing of data on the Web.

The basic building block of RDF is the triple; a statement that asserts a relationship between a *subject* and *object* using a *predicate*. Subjects, predicates and objects are identified using URIs, while the object of a triple may also be a literal value. A collection of triples can make statements using the same URIs. In this way, they can be considered as a *graph*. *Blank nodes* (nodes that do not have universal identities) can also occur in RDF graphs, allowing for the representation of structured data.

RDF has a number of different serialisations that can be used to express, communicate and store RDF graphs. The most frequently used is RDF/XML, an XML based format, although text-based formats such as N3, Turtle and N-Triples are also in common usage.

An RDF *triplestore* provides storage of RDF graphs along with mechanisms supporting query of those graphs, in particular through the use of the SPARQL Query Language for RDF. A wide range of triplestore implementations are available, including research, open source and commercial products.

R

Related websites

RDF	http://www.w3.org/RDF/
Data	http://linkeddata.org/

See also Linked Data

RDF DATABASE, *SEE* DATABASE.

READSEQ

Michael P. Cummings

A program for converting sequence files from one format to another.

The program reads and writes a very broad range of file formats. The most recent version of the program is written in Java and is available as source code.

Related website

readseq web page	http://iubio.bio.indiana.edu/soft/molbio/readseq/java/

REASONING

Robert Stevens

Reasoning is the process by which new facts are produced from those already existing. This process is known as inference. Computer-based reasoners can process knowledge encoded in a formal Knowledge Representation Language to produce new facts. In a Description Logic environment, a reasoner is used to infer the subsumption hierarchy that forms the taxonomy of the ontology. The reasoner will also check that the statements made within the encoding are logically consistent. A computer-based reasoner used for classification is sometimes known as a classifier.

RECOMBINATION

Mark McCarthy and Steven Wiltshire

Recombination describes the rearrangement—or recombining—of alleles between haplotypes on homologous chromosomes during meiosis.

Assume that individual i inherits haplotype A_pB_p from its father and A_mB_m from its mother. If meiosis in individual *i* produces gametes (and subsequently offspring) with haplotypes A_pB_m or A_mB_p, a recombination event is said to have occurred between loci A and B, and the resulting gametes (and offspring) are recombinants for the two loci. Such offspring, as a result, carry haplotypes of loci A and B that were not inherited by individual i from its parents, i.e. there has been a recombining of alleles in the parental haplotypes inherited by individual *i*. However, if the gametes produced by individual *i* are the same as the parental types that it inherited, that is, A_pB_p and A_mB_m, no recombination has been detected between loci A and B: such gametes (and offspring) contain haplotypes of loci A and B that were inherited by individual *i* from its parents. (The fact that no recombination has been detected does not necessarily mean that no recombination has occurred, since an even number of recombination events between the loci will also result in non-recombinant haplotypes). Recombination occurs when the chromatids of a pair of homologous chromosomes physically cross over at chiasmata during meiosis and exchange DNA. The probability of a recombination between any two loci is termed the recombination fraction (θ) and lies between 0 and 0.5. Two loci for which the recombination fraction is < 0.5 are said to be in genetic linkage or linked. Recombination fractions may be converted into genetic (or map) distances using map functions. Detecting linkage and estimating the recombination fractions between loci form the basis of linkage analysis.

Example: In a genome-wide parametric linkage analysis of a large Australian Aboriginal pedigree for loci influencing susceptibility to type 2 diabetes, Busfield et al. (2002) detected linkage to marker D2S2345 with a maximum two-point LOD score of 2.97 at a recombination fraction of 0.01.

Related websites

Programs for estimating recombination fractions between loci, for identifying recombinations in individual pedigrees and/or for performing linkage analysis include:

LINKAGE	ftp://linkage.rockefeller.edu/software/linkage/
GENEHUNTER2	www-genome.wi.mit.edu/ftp/distribution/software/genehunter/
SIMWALK2	http://watson.hgen.pitt.edu/docs/simwalk2.html
MERLIN	http://www.sph.umich.edu/csg/abecasis/Merlin/

Further reading

Busfield F, et al. (2002) A genomewide search for type 2 diabetes-susceptibility genes in indigenous Australians. *Am J Hum Genet*, 70: 349–357.

Ott J (1999) *Analysis of Human Genetic Linkage*. Baltimore: The Johns Hopkins University Press pp 1–23.

Sham P (1998) *Statistics in Human Genetics*. London: Arnold pp 51–58.

See also Linkage Map Function, LOD Score, Phase, Linkage Analysis

RECOMBINE, *SEE* LAMARC.

RECORD, *SEE* DATA STRUCTURE.

R

RECURSION
Matthew He

A recursion is an algorithmic procedure whereby an algorithm calls on itself to perform a calculation until the result exceeds a threshold, in which case the algorithm exits. Recursion is a powerful procedure with which to process data and is computationally quite efficient.

REFERENCE GENOME (TYPE GENOME)
John M. Hancock

In genomics, and particularly in Next Generation Sequencing, a genome sequence against which another sequence is compared or assembled. Standard reference genomes are the type genome sequences for the relevant species but other references may be used under particular circumstances. In some cases the reference may derive from a different but closely related species, for example for speedy assembly of a novel type genome.

See also Next Generation DNA Sequencing

REFERENCE SEQUENCE DATABASE, *SEE* REFSEQ.

REFINEMENT
Liz Carpenter

When an initial structure is obtained from an experimental electron density map, there will be many inaccuracies in the structure, due to errors in obtaining the intensities, the phases and in positioning the atoms in the electron density map. Refinement techniques are used to improve the crystal structure so that the model agrees with the data more closely while also having reasonable bond lengths and angles.

A number of mathematical techniques are used to improve the agreement between data and the model including least squares minimization, simulated annealing and maximum likelihood methods.

In general to define a particular parameter such as the position x,y,z or B factor of an atom it is necessary to have at least three experimental measurements for each parameter. The resolution of the data will define how many measurements are obtained for each dataset. In macromolecular crystallography there are often an insufficient number of measurements to properly define all the parameters. This problem can be overcome by including prior knowledge of the structure such as information on bond lengths and angles that are known from the structures of small molecules. This information can be included in the form of constraints (where a parameter is set to a certain value) or as restraints (where a parameter is set to vary within a range).

Refinement adjusts the positions of atoms within the unit cell. It can make changes over a small distance, fractions of an Ångstrom. Refinement protocols will not make large movements, for example re-positioning an entire loop of a protein that is not in density. This has to be done manually on the graphics or with a structure building program. Several rounds of refinement and rebuilding of a model are necessary to obtain a final structure.

Related websites

CNS	http://cns.csb.yale.edu/v1.1/
Refmac	http://www.ccp4.ac.uk/dist/html/refmac5.html

Further reading

Murshudov GN, Vagin AA, Dodson EJ (1997) Refinement of macromolecular structures by the maximum-likelihood method. *Acta Cryst.*, D53: 240–255.

REFSEQ (THE REFERENCE SEQUENCE DATABASE)

Malachi Griffith and Obi L. Griffith

The reference sequence (RefSeq) database is a collection of high quality records describing naturally occurring DNA, RNA and protein sequences from multiple species including human. Unlike GenBank and other sequence collections, RefSeq entries are designed to be non-redundant. The purpose of these sequences is to provide a reference standard for a variety of downstream applications and experiments. Each RefSeq record contains a collection of curated information synthesized from multiple primary data sources. Each record explicitly links genome sequences to corresponding transcript and protein sequences. Over time, RefSeq entries are manually curated by one or more 'editors'. These records will be marked as 'reviewed'. Sequence records are tracked by systematic version numbers and record history is maintained indefinitely. The RefSeq staff are contributing to the CCDS project. RefSeq is maintained by the NCBI and all sequence data used by RefSeq are derived from submissions to the International Nucleotide Sequence Database Collection (INSDC).

Related websites

RefSeq homepage	http://www.ncbi.nlm.nih.gov/RefSeq/
RefSeq manual	http://www.ncbi.nlm.nih.gov/books/NBK21091/

Further reading

Pruitt K, et al. (2011) The Reference Sequence (RefSeq) Database. *The NCBI Handbook [Internet]*. Chapter 18. McEntyre J, Ostell J, Eds. Bethesda (MD): National Center for Biotechnology Information.

Pruitt KD, et al. (2009) NCBI Reference Sequences: current status, policy, and new initiatives. *Nucleic Acids Res.*, 37(Database issue): D32–6.

R

REGEX, *SEE* REGULAR EXPRESSSION.

REGRESSION ANALYSIS

Nello Cristianini

The statistical methodology for predicting values of one or more response (dependent) variables from a collection of predictor (independent) variable values.

In machine learning literature the responses are often called 'labels', and the predictors are called 'features'. This method can also be used to assess the effect of the predictor variables on the responses. When there is more than one predictor variable (feature), it is termed 'multivariate regression'. When there is more than one response variable (label) it is called 'multiple regression'.

Further reading

Kleinbaum DG, Kupper LL, Muller KE (1997) *Applied Regression Analysis and Multivariable Methods*. Wadsworth.

REGRESSION TREE

Concha Bielza and Pedro Larrañaga

Similarly to decision trees, regression trees consist of internal nodes corresponding to attributes, edges corresponding to the values of the attributes and terminal nodes or leaves corresponding to a constant value that estimates the expected value of the response (continuous) variable. Conditions (pairs of attributes and corresponding values) evaluated in each internal node are conjunctively connected. The attributes can be either continuous or discrete. The basic algorithm for growing regression trees is similar to that for growing decision trees. The key factor is how to choose the best attribute for the next node. For this purpose, measures like the change of variance in the response variable is frequently used. For the prediction of the response value of a new example, from the root node we traverse along the edges according to the example's attribute values until we reach a leaf. Each leaf contains the response value for the examples falling in the leaf.

Reliability of the response value estimations is highly dependent on the number of learning examples. Post-pruning techniques can be used.

When a linear function of a subset of variables is used in the leaves rather than a constant value, the (more complex) regression tree is called a model regression tree.

Further reading

Breiman L, Friedman JH, Olshen RA, Stone CJ (1984) *Classification and Regression Trees*. Wadsworth, Inc.

Nepomuceno-Chamorro IA, Aguilar-Ruiz JS, Riquelme JC (2010) Inferring gene regression networks with model trees. *BMC Bioinformatics*, 11: 517.

Phuong TM, Lee D, Lee KH (2004) Regression trees for regulatory element identification. *Bioinformatics*, 20(5): 750–757.

See also Regression, Regression Analysis, Decision Trees

REGULAR EXPRESSION (REGEX)

Teresa K. Attwood

A single consensus expression derived from a conserved region (motif) of a sequence alignment, and used as a characteristic signature of family membership. Regular expressions (regexs) thus discard sequence data, retaining only the most conserved residue information. Typically, they are designed to be family-specific, and encode motifs of 15–20 residues. An example regex is shown below:

[LIVMW]-[PGC]-x(3)-[SAC]-K-[STALIM]-[GSACNV]-[STACP]-x(2)-[DENF]-[AP]-x(2)-[IY]

Within this expression, the single-letter amino acid code is used: residues within square brackets are allowed at the specified position in the motif; the symbol *x* denotes that any residue is allowed at that position; individual residues denote completely conserved positions; and numbers in parentheses indicate the number of residues of the same type allowed sequentially at that location in the motif. Regexs can also be used to disqualify particular residues from occurring at any position—conventionally, these are grouped in curly brackets (e.g., {PG} indicates that proline and glycine are disallowed).

Regexs are tremendously simple to derive and use in database searching. However, they suffer from various diagnostic drawbacks. In particular, in order to be recognized, expressions of this type require sequences to match them exactly. This means that any minor change, no matter whether it is conservative (e.g., V in position 8 of the above motif), will result in a mismatch even if the rest of the sequence matches the pattern exactly. Thus many closely related sequences may fail to match a regex if its allowed residue groups are too strict. Conversely, if the groups are relaxed to include additional matches, the expression may make too many chance matches. Thus a match to a pattern is not necessarily true, and a mismatch is not necessarily false!

Regexs underpin the PROSITE database. In the former, expressions are derived exactly from observed motifs in sequence alignments; in the latter, to counteract the limitations of binary (match/no-match) diagnoses, expressions are derived in a more tolerant manner, using a set of prescribed physicochemical groupings (DEQN, HRK, FYW, etc.). Both approaches have drawbacks, hence caution (and biological intuition) should be used when interpreting matches.

Relevant website

ScanProsite	http://www.expasy.org/tools/scanprosite/

Further reading

Attwood TK, Parry-Smith DJ (1999) *Introduction to Bioinformatics*. Addison Wesley Longman, Harlow, Essex, UK.

Attwood TK (2000) The quest to deduce protein function from sequence: the role of pattern databases. Int. J. Biochem. Cell Biol., 32(2), 139–155.

See also Motif, Alignment, Amino Acid, Conservation

R

REGULARIZATION (RIDGE, LASSO, ELASTIC NET, FUSED LASSO, GROUP LASSO)

Concha Bielza and Pedro Larrañaga

Regularization is a way of introducing additional information to solve an ill-posed problem or to prevent overfitting. This information is usually introduced as a penalty for complexity. Its theoretical justification is to try to impose Occam's razor (parsimony) on the solution. From a Bayesian point of view, many regularization techniques correspond to imposing certain prior distributions on model parameters.

A general class of regularization problems has the form

$$min_{f\in\mathcal{H}}\left[\sum_{i=1}^{N} \mathrm{L}\left(\mathrm{y_i}, \mathrm{f}\left(\mathbf{x}_\mathrm{i}\right)\right) + \lambda \mathrm{J(f)}\right]$$

where $\mathrm{L(y_i, f(\mathbf{x}_i))}$ is a loss function of the observed values $\mathrm{y_i}$ of the response variable Y and their prediction $\mathrm{f}(\mathbf{x}_\mathrm{i})$ for a general prediction function f on the observed values $\mathbf{x}_\mathrm{i}$ of the predictor variables $\mathbf{X}=(\mathrm{X}_1,\ldots,\mathrm{X}_\mathrm{p})$, J(f) is a penalty functional, $\mathcal{H}$ is a space of functions on which J(f) is defined and N is the number of learning examples. $\lambda > 0$ is the penalty or regularization parameter and controls the amount of regularization. The larger the λ, the stronger its influence.

Typical examples of regularization in linear regression include ridge regression, lasso and elastic net. These methods impose a penalty on the size of the regression coefficients, trying to shrink them towards zero. Therefore, regularized estimators are restricted maximum likelihood estimators, since they maximize the likelihood function subject to restrictions on the regression parameters. Equivalently, they minimize a penalized residual sum of squares:

$$\left[min_{\beta}\sum_{i=1}^{N}\left(\mathrm{y_i} - \beta_0 - \sum_{\mathrm{j}=1}^{\mathrm{p}}\mathrm{x_{ij}}\beta_\mathrm{j}\right)^2 + \lambda\sum_{\mathrm{j}=1}^{\mathrm{p}}\psi(\beta_\mathrm{j})\right]$$

where p is the number of predictor variables, and β_j are the regression coefficients. When $\lambda=0$, the solution is the ordinary maximum likelihood estimators, whereas if $\lambda \to \infty$, the β_j all tend to 0. λ is usually chosen by cross-validation. The error, BIC or AIC can be used as the criteria to be optimized.

A typical choice for the penalization function is $\psi(\beta_\mathrm{j}) = |\beta_\mathrm{j}|^\mathrm{q}, \mathrm{q} > 0$. Thus, the quadratically-regularized approach, i.e. q = 2, is called ridge regression, which seeks maximum likelihood estimators subject to spherical restrictions on the parameters. When q=1 it results in lasso (Least Absolute Shrinkage and Selection Operator), which is interesting since its penalty encourages some estimators to be exactly zero, thereby automatically performing feature selection. The elastic net combines linearly the ridge and the lasso penalties. As a consequence, it encourages a grouping effect, where strongly correlated predictors tend to be in or out of the model together. Other variants of lasso include the group lasso, which is able to do variable selection on (predefined) groups of variables, and the fused lasso, which penalizes the coefficients and their successive differences (assuming the features are ordered) obtaining sparsity of both types of coefficients.

The little bias allowed in these approaches provides more stable estimates with smaller variance. Regularization methods are more continuous than usual discrete processes of retaining-or-discarding features thereby not suffering as much from high variability.

In the specific literature on DNA microarray classification which is the most representative example of 'the large p, small *N*' problem, regularization methods are very useful.

Further reading

Li J, Das K, Fu G, Li R, Wu R (2011) The Bayesian lasso for genome-wide association studies. *Bioinformatics*, 27 (4): 516–523.

Li C, Li H (2008) Network-constrained regularization and variable selection for analysis of genomic data. *Bioinformatics*, 24 (9): 1175–1182.

Tibshirani R (1996) Regression shrinkage and selection via the Lasso. *Journal of the Royal Statistical Society, Series B*, 58 (1): 267–288.

Tibshirani R, Saunders M, Rosset S, Zhu J, Knight K (2005) Sparsity and smothness via the fused lasso. *Journal of the Royal Statistical Society, Series B*, 67: 91–108.

Waldron L, Pintilie M, Tsao M-S, Shepherd FA, Huttenhower C, Jurisica I (2011) Optimized application of penalized regression methods to diverse genomic data. *Bioinformatics*, 27 (24): 3399–3406.

See also Regression Analysis

REGULATORY MOTIFS IN NETWORK BIOLOGY

Denis Thieffry

In the context of the analysis of genetic networks, regulatory motifs generally denote relatively small, oriented, connected subgraphs associated with a specific function, and thereby potentially preferentially conserved across evolution. The notion of regulatory circuit or feedback loop emerged in the field of bacterial genetics from the late 1950s onwards. A regulatory circuit can be simply defined as a simple oriented circular path, which, in the simplest case, may amount to an auto-regulation, i.e. a loop in the graph terminology.

Following René Thomas, regulatory circuits can be classified into a positive versus negative circuit, depending on the parity of the number of negative interactions. Involving an even number of negative interactions, positive circuits are required to generate multiple gene expression states, in other words to enable cell differentiation.

In contrast, involving an odd number of negative interactions, negative circuits are needed to generate sustained, robust oscillations or homeostatic gene expression (as in the case of a thermostatic device). As set by Thomas in the 1980s, these rules have been recently translated into theorems within the differential and logical mathematical frameworks. Strongly connected components involving single or several intertwined circuits may be considered as functional regulatory modules, and are indeed found recurrently at the core of many specific gene network models. Already simple combinations of negative and positive circuits can result into exquisitely sophisticated dynamical behaviours.

Recently, other types of regulatory motifs have been found as particularly over-represented and potentially conserved across evolution. In particular, feed-forward loops (FFL) group three component motifs where one component (input) regulates two other components, while only one of these targets regulates the other one (output). Depending on the signs of the different interactions involved, one may consider different variants. Coherent FFL may filter transient noises, whereas the an incoherent FFL may constitute pulse generators.

Further reading

Alon U (2007) Network motifs: theory and experimental approaches. *Nature Reviews—Genetics*, 8(6): 450–461.

Thieffry D (2007) Dynamical roles of biological regulatory circuits. *Briefings in Bioinformatics*, 8(4): 220–225.

Thomas R, D'Ari R (1990) *Biological Feedback*. CRC Press.

REGULATORY NETWORK INFERENCE

Hedi Peterson

Process of identifying gene interactions from experimental data through computational analysis.

Gene regulation networks can be inferred from high-throughput measurements, such as microarray experiments measuring mRNA expression or chromatin immunoprecipitation (ChIP) experiments identifying DNA binding regulatory proteins.

Gene expression measurements originating from microarray experiments are widely used to reverse engineer gene regulatory networks. Both time course and perturbation experiments allow to infer pairwise gene-gene relationships. Gene expression changes in perturbation experiments lead to identification of connections between regulators and their targets. Such relationships can be direct or indirect. In the latter case, regulatory signal is mediated by other proteins, noncoding RNAs or metabolites. Regarding regulatory relationships from time course experiments, data can be extracted by identifying a correlated time-lag between gene expression profiles.

The physical interactions measured by ChIP experiments lead mainly to direct regulation events where the transcription factor binding to a promoter region positively regulates the target gene. However, some of the bound regulatory proteins may, to the contrary, negatively regulate target genes by blocking the relevant promoter region. However, in other cases, regulatory proteins can bind indirectly to DNA through intermediate physical interaction with co-factors. In addition, pattern matching of known regulatory motifs and discovery of *de novo* motifs on known promoter regions are used to identify putative regulatory connections.

Combinations of different network interference methods and data sources are usually applied in order to describe specific biological questions under investigation.

Further reading

Bansal M, et al. (2007) How to infer gene networks from expression profiles. *Molecular Systems Biology*, 3: 78.

De Smet R, Marchal K (2010) Advantages and limitations of current network inference methods. *Nature Reviews Microbiology*, 8(10): 717–729.

Marbach D, et al. (2010) Revealing strengths and weaknesses of methods for gene network inference. *PNAS*, 107(14): 6286–6291.

Margolis AA, et al. (2006) ARACNE: an algorithm for the reconstruction of gene regulatory networks in a mammalian cellular context. *BMC Bioinformatics*, 7 Suppl 1: S7.

REGULATORY REGION

Jacques van Helden

Genomic region exerting a cis-regulatory effect on the transcription of a given gene. In bacterial and fungal genomes, cis-regulatory elements are typically found in the non-coding sequences located upstream the regulated gene, and are restricted to a few hundred base pairs per gene. In metazoan genomes, cis-regulatory elements can be found in upstream regions, introns, downstream regions. They can be located in close proximity to the gene (proximal regions) or at larger distances (several kb away from the transcription start site). In some cases, a cis-regulatory element can act on genes located further away than the nearest neighbour genes (e.g. cis-regulation of the *achaete-scute* complex in *Drosophila melanogaster*).

See also Cis-regulatory Element, Cis-regulatory Module (CRM)

REGULATORY REGION PREDICTION

James Fickett

The prediction of sequence regions responsible for the regulation of gene prediction.

In eukaryotes, transcriptional regulatory regions (TRRs) are either promoters, at the transcription start site, or enhancers, elsewhere. Enhancers, as well as the regulatory portion of promoters, are typically made up of functionally indivisible cis-regulatory modules (CRM) in which a few transcription factors have multiple binding sites (Yuh and Davidson, 1996). The common computational problem is to recognize functional clusters of transcription factor binding sites. Clusters of sites for a single factor may be part of the key to localizing functional sites for that factor (Wagner, 1999), while coordinate action of multiple factors may hold the key to highly specific expression patterns (Berman et al., 2002). A number of computational methods have been used to characterize the degree of clustering. It is possible that the quality of the position weight matrices used to describe binding sites, and the use of phylogenetic footprinting to eliminate false positive matches, are more important, in practical terms, than the particular mathematical treatment of clusters.

There is almost certainly some structure in TRRs beyond mere clustering of the binding sites. It is well known that certain combinations of transcription factors act synergistically (one collection of such cases may be found in the database COMPEL (Kel-Margoulis et al., 2002)). It has been shown that physical interaction between pairs of transcription factors is, in at least some cases, reflected in constraints on spacing of their binding sites (Fickett, 1996). Some success has been had, in simple cases, in describing the overall structure of a class of CRMs (Klingenhoff et al., 1999). However in most cases a formal characterization of any spacing and order constraints remains elusive.

In eukaryotes, algorithms giving practical prediction of regulatory specificity are available for only a few cases. Those developed for, for example, muscle (Wasserman and

R

Fickett, 1998), liver (Krivan and Wasserman, 2001), and anterior-posterior patterning in early Drosophila development (Berman et al., 2002), were only developed following years of extensive work, in a broad community, determining the biological function and specificity of the transcription factors involved. It will require further work to deduce the set of relevant transcription factors (or their binding specificities), and develop a prediction algorithm, directly from a set of co-expressed genes.

In bacteria, regulatory regions are much simpler, often consisting of the binding site of a single regulator. Because intergenic distances are typically small, the localization problem is one of determining operon groupings. Many bacterial genomes are now available, and comparison of gene and putative transcription element order across many species can often provide insight into both operon structure and likely regulatory mechanism (Gelfand et al., 2000).

Related websites

Cister	http://sullivan.bu.edu/~mfrith/cister.shtml
Regulatory Sequence Analysis Tools	http://rsat.ulb.ac.be/rsat/
Wyeth Wasserman	http://www.cmmt.ubc.ca/research/investigators/wasserman/lab

Further reading

Berman BP, et al. (2002) Exploiting transcription factor binding site clustering to identify cis-regulatory modules involved in pattern formation in the Drosophila genome *Proc Natl Acad Sci USA*, 99: 757–762.

Fickett J (1996) *Gene*, 172: GC19–32.

Gelfand MS, et al. (2000) Prediction of transcription regulatory sites in Archaea by a comparative genomic approach *Nucleic Acids Res*, 28: 695–705.

Kel-Margoulis O, et al. (2002) TRANSCompel: a database on composite regulatory elements in eukaryotic genes *Nucleic Acids Res*, 30: 332–334.

Klingenhoff A, et al. (1999) Coordinate positioning of MEF2 and myogenin binding sites *Bioinformatics*, 15: 180–186.

Krivan W, Wasserman WW (2001) A predictive model for regulatory sequences directing liver-specific transcription *Genome Res*, 11: 1559–1566.

Wagner A (1999) Genes regulated cooperatively by one or more transcription factors and their identification in whole eukaryotic genomes. *Bioinformatics*, 15: 776–784.

Wasserman WW and Fickett JW (1998) Identification of regulatory regions which confer muscle-specific gene expression *J Mol Biol*, 278: 167–181.

Yuh C-H, Davidson EH. (1996) Modular cis-regulatory organization of Endo16, a gut-specific gene of the sea urchin embryo *Development*, 122: 1069–1082.

See also Promoter, Transcription Start Site, Enhancer, Cis-regulatory Module, Transcription Factor Binding Site, Phylogenetic Footprinting, Software Suites for Regulatory Sequences

REGULATORY SEQUENCE, *SEE* TRANSCRIPTIONAL REGULATORY REGION.

REGULOME

John M. Hancock

Variously defined as the genome-wide regulatory network and the combined transcription factor activities of the cell.

For future studies of systems biology it will be essential to understand the interactions that take place within a given cell-type. Characterization of the regulome (effectively the set of interactions taking place between individual gene pairs in a cell-type) will provide a necessary catalogue of interactions to be built into any such model.

See also Networks, Systems Biology, Interactome

RELATIONAL DATABASE

Feng Chen and Yi-Ping Phoebe Chen

A relation database is a series of tables, each of which has a variety of attributes to store tuples. In one table, every tuple is an object described by the related attributes.

See also Database, Data Warehouse, Non-relational Database, Relational Database Management System

RELATIONAL DATABASE MANAGEMENT SYSTEM (RDBMS)

Matthew He

Relational Database Management System (RDBMS) is **a** software system that includes a database architecture, query language, and data loading and updating tools and other ancillary software that together allow the creation of a relational database application. An RDBMS stores data in a database consisting of one or more tables of rows and columns. The rows correspond to a record (tuple); the columns correspond to attributes (fields) in the record. In an RDBMS, a view, defined as a subset of the database that is the result of the evaluation of a query, is a table. RDBMSs use Structured Query Language (SQL) for data definition, data management, and data access and retrieval. Relational and object-relational databases are used extensively in bioinformatics to store sequence and other biological data.

RELATIONSHIP

Robert Stevens

A relationship is a type of association between two concepts. A relation is the nature of the association and a relationship is the relation plus the concepts it associates together. For example membership, causality, part/whole are all types of relation that form relationships within ontologies.

R

Common relations are:

- is-kind-of relates subcategories to more general categories. This is one of the most important relationships used to relate organise categories into a taxonomy.
- is-instance-of relates individuals to categories
- part-of describes the relationship between a part and its whole.

Relations themselves can have properties. A relation can have an inverse; so, 'part-of' has an inverse 'has-part'. Instead of an inverse, a relation may be symmetric; so that 'homologous-to' has the same relation in the other direction. This enables the ontology to have the relationships 'Protein homologous-to Protein' with the same relation in each direction. Relations may have different kinds of quantification: Existential quantification indicates that there is *some* concept filling the relationship. Similarly, there can be precise cardinality constraints indicating the numbers or range of concepts filling the relation. Universal quantification indicates what type of concept can possibly form the relationship. Similarly, a relation can have domain and range constraints, which determine what is logically allowed to be used on either end of the relation. Relations may also be said to be transitive; this means that the effect of the relation is transferred along a chain of relations of the same type. An 'Amino acid' is 'part-of' an 'Alpha helix', and that class is also 'part-of' the class 'Protein'. If the 'part-of' relation is made transitive, the ontology is also stating that the 'Amino acid' is 'part-of' the 'Protein'.

REPEATMASKER

John M. Hancock

A program for detecting repeated and repetitive sequences in DNA sequences.

Repeatmasker compares input sequences against a library including both interspersed and internally repetitive sequences. It returns a detailed annotation of the repeats found and a version of the input sequence in which repeats are replaced by Ns (a so-called masked sequence). The program uses a modified Smith-Waterman algorithm to align library sequences against the input sequence. Sequence masking by repeatmasker is commonly used before comparing DNA sequences against databases to reduce the chances of spurious hits.

Related website

Repeatmasker Home Page	http://www.repeatmasker.org/

See also Smith-Waterman

REPEATS ALIGNMENT, *SEE* ALIGNMENT.

REPETITIVE SEQUENCES, *SEE* SIMPLE DNA SEQUENCE.

RESIDUE, *SEE* AMINO ACID.

RESOLUTION IN X-RAY CRYSTALLOGRAPHY

Liz Carpenter

An important parameter in determining the quality of an X-ray crystallographic structure is the resolution of the data. The resolution of a diffraction spot observed in X-ray crystallography is dependant on the spacing between the planes that gave rise to the diffraction spot. Bragg's law states that for diffraction to occur the following equation must be correct:

$$2d \sin \theta = n\lambda.$$

The planes in the crystal are separated by a distance d. For a fixed wavelength, the smaller the value of d, the larger is the angle of diffraction θ. With larger values of θ the intensity of the diffraction decreases, until eventually it becomes indistinguishable from the background. This fall-off in intensity with Bragg angle is due to thermal motion of the atoms in the crystal. On an X-ray diffraction image the spots in the middle of the image with small θ values are strong and those near the edge of the detector, with high θ values are weak or undistinguishable. The maximum resolution of a dataset is the smallest value of d for which diffraction spots can be detected in the background.

By analogy with the light microscopy this relates to the smallest distance between two objects that can be separated. So if there are diffraction spots to 2 Å then atoms 2 Å apart are seen as separate objects, whereas if the data only extends to 3 Å, then only atoms 3 Å apart are separate. This means that in datasets where there are reflections to a resolution of 4 Å only the overall shape of a molecule and a few alpha helices can be distinguished. A 3Å dataset would allow the backbone of a protein to be traced, the helices and sheets to be identified and some side-chains to be positioned but the detail of interactions would not be visible. At 2 Å the side-chains and water structure would be clearly visible. At 1 Å resolution each individual atom is in its own separate density and atoms can be assigned six thermal parameters instead of the usual one B factor parameter, which indicate the thermal motion of the atoms.

Further reading

Blundell TL, Johnson LN (1976) *Protein Crystallography*, Academic Press.

Drenth J, *Principles of Protein X-ray Crystallography*, Springer Verlag (1994).

Glusker JP, Lewis M, Rossi M (1994) *Crystal Structure Analysis for Chemists and Biologists*, VCH Publishers.

See also X-ray Crystallography for Structure Determination, Diffraction of X-rays, Crystals

R

RESPONSE, *SEE* LABEL.

RESTFUL WEB SERVICES

Carole Goble and Katy Wolstencroft

REpresentational State Transfer (REST) is a format and web architecture for describing and executing Web Services. Conforming to this architecture is referred to as being RESTful.

The RESTful approach uses a standard URI (Uniform Resource Identifier) to make a call to a web service like https://www.mycompany.com/program/method?Parameters=xx. The approach is simple to understand and can be executed on any client or server that has HTTP/HTTPS support. The command can be executed using the HTTP Get method. REST-style web services are increasing in popularity and are used in greater numbers than WSDL/SOAP interfaces. They are considered to be easier to implement, light-weight and more in line with the protocols of the Web because they use the existing Web infrastructure. The approach works well for situations with: limited bandwidth and resources, totally stateless operations and caching situations.

See also Web Services, WSDL/SOAP Web Services

RESTRICTION MAP

Matthew He

A physical map or depiction of a gene (or genome) derived by ordering overlapping restriction fragments produced by digestion of the DNA with a number of restriction enzymes.

RETROSEQUENCE

Austin L. Hughes

A retrosequence is any genomic sequence that originates from the reverse transcription of any RNA molecule and its subsequent integration into the genome, excluding sequences that encode reverse transcriptase.

Usually, the RNA template is the RNA transcript of a gene. Most of the retrosequences derived from protein-coding genes are derived from processed mRNAs which thus lack introns. A retrosequence that is functional may be called a retrogene, while one that is non-functional is called a retropseudogene. Retropseudogenes derived from processed mRNAs are often called processed pseudogenes.

Further reading

Weiner AM, et al. (1986) Nonviral retroposons: genes, pseudogenes, and transposable elements generated by the reverse flow of genetic information. *Annu Rev Biochem*, 55: 631–661.

See also Retrotransposon

R

RETROTRANSPOSON

Dov Greenbaum

Transposable elements that spread by reverse transcription of mRNAs.

Retrotransposons likely make up more than 40% of the human genome, with many of the most common retrptransposons, including, L1, Alu, SVA, and HERVs remaining active in their transpositioning.

Retrotransposons resemble retroviruses in many ways. They contain long terminal repeats that flank two or three reading frames: gag-like, env-like and pol-like. Unlike retroviruses, they are not infectious. They are very abundant within many genomes.

Retrotransposons are thought to exert mutagenic pressures on genes iand genomes through a number of distinct mechanisms including via insertion mutations that lead to muations and genomic instability and changes in of gene expression.

There are many different families of retrotransposons. The major types of retrotransposons in the human genome include Long Terminal Repeat retrotransposons, characterized by the presence of the two long-terminal repeats (LTRs), and the non-LTR retrotransposons such as ALU.

Retrotransposons move by first being transcribed into mRNA. These pieces of RNA are transcribed back into DNA and are inserted into the genome. Retrotransposons require the help of the enzymes integrase (for insertion) and reverse transcriptase to convert the mRNA transcript into a DNA sequence that can be inserted back into the genome.

There are two classes of retrotransposons. Viral and Non-viral. Viral transposons include retrovirus-like transposons such as Ty in Saccharomyces, Bs1 in maize, copia in Drosophila and LINE-like elements. Non-viral retrotransposons include SINEs and processed pseudogenes.

Related websites

dRIPs	http://dbrip.brocku.ca/
RetrOryza	http://retroryza.fr/

Further reading

Brosius J (1991) Retroposons—Seeds of evolution. *Science*, 251: 753.

Liang P, Tang W (2012) Database documentation of retrotransposon insertion polymorphisms. *Front Biosci*, 1;4: 1542–55.

Patience C, et al. (1997) Our retroviral heritage. *Trends Genet.*, 13: 116–120.

REVERSE COMPLEMENT

John M. Hancock

As the DNA double helix is made up of two base-paired strands, the sequences of the two strands are not identical but have a strict relationship to one another. For example, a sequence 5′-ACCGTTGACCTC-3′ on one strand pairs with the sequence 5′-GAGGTCAACGGT-3′. This second sequence is the known as the reverse complement of the first. Converting a sequence to its complement is a simple operation and may be required, for example, if a sequence has been read off one strand but a coding region lies on the other strand.

See also DNA Sequence, Complement

RFAM

Sam Griffiths-Jones

Rfam is a database of structured non-protein-coding RNA families, run by the Bateman group at the Wellcome Trust Sanger Institute. Each family is represented by a multiple sequence alignment of a set of homologous RNA sequences. Each family also has an associated base-paired secondary structure, usually derived from experimental data or comparative sequence analysis. Rfam aims to provide structure-annotated multiple sequence alignments of all functional RNA families, methods and models (called covariance models) to annotate homologs of known structured RNAs in complete genomes, textual annotation, links to other databases, and literature references associated with the sequences. Textual annotation in Rfam is provided via the community RNA Wikipedia project. The sequence alignments and homology search functions of Rfam are provided using the INFERNAL software. Release 10.1 of Rfam (June 2011) contains 1973 RNA families.

Related website

Rfam	http://rfam.sanger.ac.uk

Further reading

Daub J, Gardner PP, Tate J, et al. (2008) The RNA WikiProject: community annotation of RNA families. *RNA.*, 14: 2462–2464.

Gardner PP, Daub J, Tate J, et al. (2011) Rfam: Wikipedia, clans and the 'decimal' release. *Nucleic Acids Res*, 39: D141–145.

Griffiths-Jones S, Bateman A, Marshall M, Khanna A, Eddy SR. Rfam: an RNA family database. *Nucleic Acids Res.*, 31: 439–441.

Nawrocki EP, Kolbe DL, Eddy SR (2009) Infernal 1.0: inference of RNA alignments. *Bioinformatics.*, 25: 1335–1337.

RFREQUENCY

Thomas D. Schneider

Rfrequency is the amount of information needed to find a set of binding sites out of all the possible sites in the genome. If the genome has G possible binding sites and γ binding sites, then Rfrequency $= \log_2\left(\frac{G}{\lambda}\right)$ bits per site. Rfrequency predicts the expected information in a binding site, Rsequence.

Related website

Rfrequency Paper	http://alum.mit.edu/www/toms/paper/schneider1986

Further reading

Schneider TD (2000) Evolution of biological information. *Nucleic Acids Res.*, 28: 2794–2799. http://alum.mit.edu/www/toms/paper/ev/

Schneider TD, et al. (1986) Information content of binding sites on nucleotide sequences. *J. Mol. Biol.*, 188: 415–431. http://alum.mit.edu/www/toms/paper/schneider1986/

RGD, *SEE* RAT GENOME DATABASE.

RI

Thomas D. Schneider

Ri is a short hand notation for individual information. Following Claude Shannon, the 'R' stands for a 'rate of information transmission'. For molecular biologists this is usually bits per base or bits per amino acid. The 'i' stands for 'individual'.

Further reading

Schneider TD (1997) Information content of individual genetic sequences. *J. Theor. Biol.*, 189: 427–441. http://alum.mit.edu/www/toms/paper/ri/

RIBOSOMAL RNA (rRNA)

Robin R. Gutell

Ribosomal RNAs (rRNAs) are the structural and catalytic core of the ribosome, the macromolecule that is the site for protein synthesis in all living cells. The ribosomal RNA is directly involved in the decoding of the triplet codes in the mRNA into the amino acids in proteins and the peptidyl transferase activity.

The three major forms of rRNA are the 5S, small subunit (SSU, 16S, or 16S-like), and large subunit (LSU, 23S, or 23S-like) rRNAs. rRNA may comprise up to 90% of a cell's RNA. The ribosomal RNAs are organized into operons in the genome (except for a few very bizarre organisms). The typical Bacterial and Archaeal rRNA operon contains the 16S (SSU), 23S (LSU), and 5S rRNA genes. The rDNA are separated by internal spacers. Transfer RNA genes can be located between these rRNAs. The typical higher Eukaryotic nuclear-encoded operon contains the 18S (SSU) and the LSU that is typically broken into the 5.8S at the 5′ end of the LSU and a 25S- 28S rRNA. The number of operons in Bacteria and Archaea range from one to approximately 10 per cell. In Eukaryotes, the number of nuclear encoded operons can be 300–400 per cell. Plant chloroplasts encode for Bacterial-like rRNA operons, while mitochondria also encode for rRNAs.

With the general exception of the LSU rRNA composed of the 5.8S/25–28S in the higher eukaryotes, the SSU and LSU rRNAs are frequently composed of a single unfragmented

RNA molecule. However many examples of fragmented rRNAs have been documented. The most unusual rRNA documented to date, the *Plasmodium* mitochondrial rRNA is comprised of approximately 30 fragments, ranging from 23 to 190 nucleotides in length, with the SSU and LSU fragments intermixed in the mitochondrial genome, not encoded in the the order they occur on a typical rRNA sequence, and encoded on both strands.

While the SSU and LSU rRNAs can be fragmented into smaller RNAs, the rRNA genes can also be fragmented with the exons (the regions of the gene that code for rRNA) separated by introns. And while the frequency of the total number of sequenced rRNA genes that have introns spanning the entire tree of life is small, introns do frequently occur on specific phylogenetic branches (see http://www.rna.icmb.utexas.edu/SAE/2C/rRNA_Introns/).

The lengths of the rRNAs vary, most notably for the SSU and LSU rRNAs for organisms that span the entire tree of life. The *Escherichia coli* rRNAs, which were the first to be sequenced across their entire length, serve as one of the primary reference organisms for comparative analysis of rRNA. Their lengths are 120, 1542, and 2904 nucleotides for the 5S, SSU, and LSU rRNAs, respectively.

The quest to determine phylogenetic relationships for organisms that span the entire tree of life is one of the grand challenges in biology. Since the ribosomal RNAs are present in all organisms and have one of the most conserved primary and secondary structures, rRNA sequences have been determined for more than 1,000,000 organisms. The analyses of these sequences have determined the phylogenetic relationships for all of the major forms of life, including the discovery of the third phylogenetic domain—the Archaea. For the same reasons that SSU rRNAs are one of the ideal genetic sequences to determine the phylogenetic relationships for organisms that span the entire tree of life, the bacterial 16S ribosomal RNA is one of the primary gene sequences determined for the microbiome (the totality of microbes) studies.

Secondary structure models for the ribosomal RNAs have been proposed using comparative sequence analysis. The majority of the base-pairs predicted by these methods are canonical (G:C, A:U, and G:U) base-pairs that are consecutive and antiparallel with each other. And the vast majority of these form nested secondary structure helices. Many tertiary structure interactions were also proposed with comparative analysis. These interactions include base triples, non-canonical base-pairs, pseudoknots, and many RNA motifs (see 'RNA Tertiary Structure, Motifs'). The most recent versions of the comparative models for the *Escherichia coli* 16S and 23S rRNAs are shown in Figure R.1.

Size variation in ribosomal RNA: Approximate ranges of size (in nucleotides) for complete sequences are shown in Table R.1. Unusual sequences (of vastly different) length are excluded.

Table R.1 Length ranges of ribosomal RNAs, in nucleotides

rRNA Molecule Phylogenetic Domain/Cell Location	**5S**	**SSU**	**LSU**
Bacteria	105–128	1470–1600	2750–3200
Archaea	120–135	1320–1530	2900–3100
Eukaryota Nuclear	115–125	1130–3725	2475–5450
Eukaryota Chloroplast	115–125	1425–1630	2675–3200
Eukaryota Mitochondria	115–125	685–2025	940–4500
Overall	105–135	685–3725	940–5450

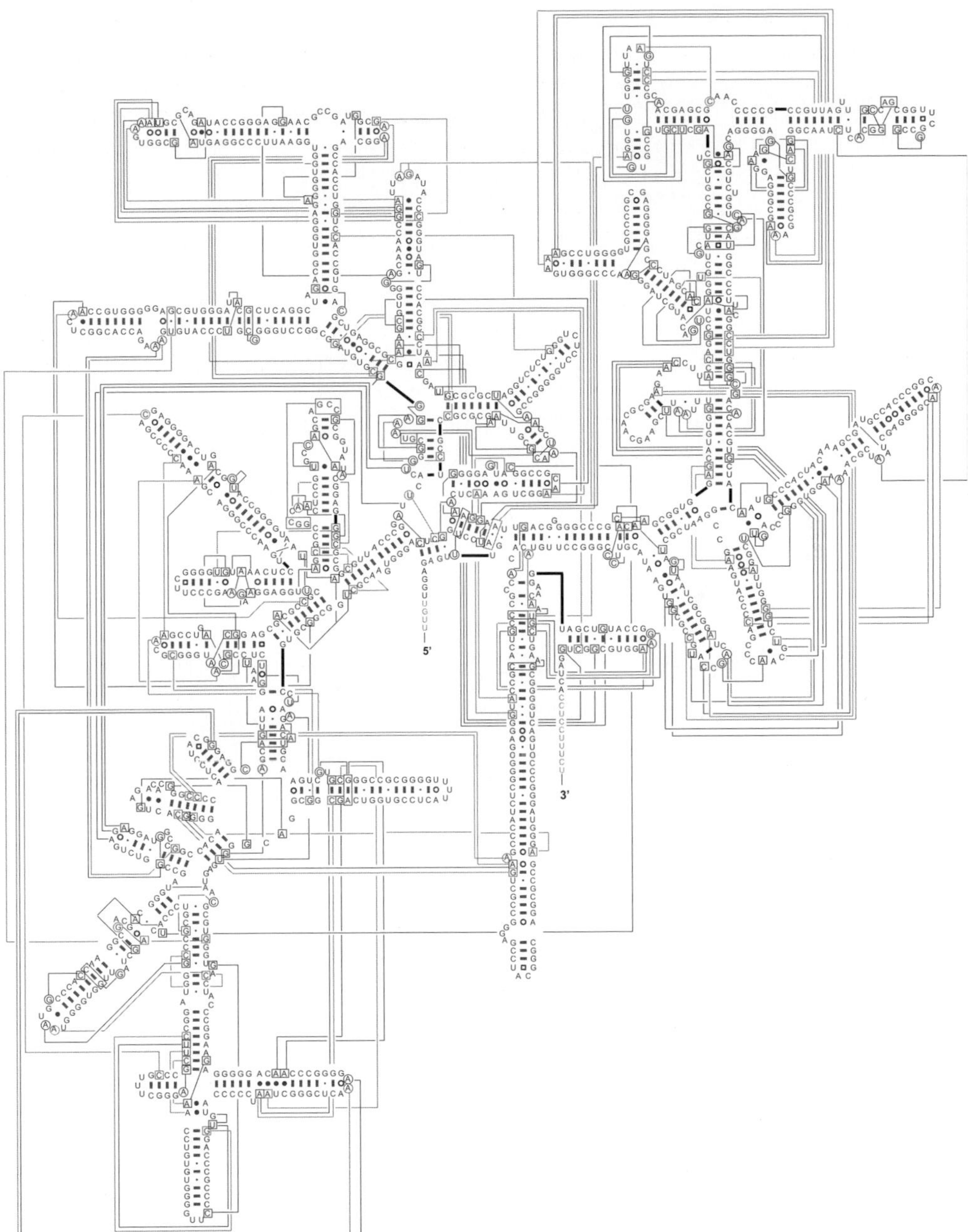

Figure R.1 The bacterial *Thermus thermophilus* 16S rRNA secondary structure, as initially determined with comparative methods and confirmed with the high-resolution crystal structure of the 30S ribosomal subunit. Tertiary structure base-base and base-backbone interactions are shown with the thin lines connecting nucleotides in the secondary structure. (Source: Gutell RR, Lee JC, Cannone JJ (2002) The Accuracy of Ribosomal RNA Comparative Structure Models. *Current Opinion in Structural Biology*, 12(3):301–310. Reproduced with permission from Elsevier). (See Colour plate R.1)

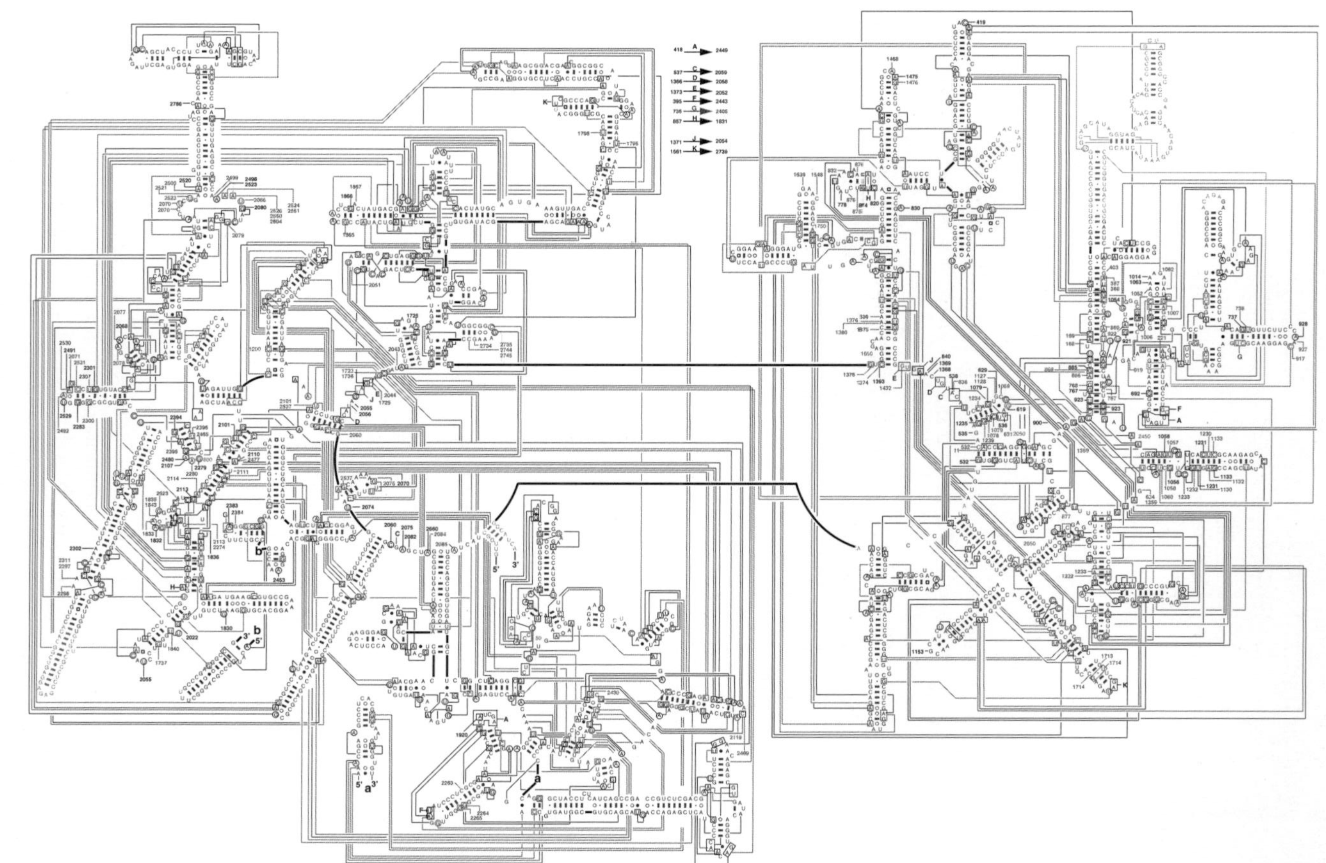

Figure R.2 The archaeal *Haloarcula marismortui* 23S rRNA secondary structure, as initially determined with comparative methods and confirmed with the high-resolution crystal structure of the 50S ribosomal subunit. Tertiary structure base-base and base-backbone interactions are shown with the thin lines connecting nucleotides in the secondary structure. (Source: Gutell RR, Lee JC, Cannone JJ (2002) The accuracy of ribosomal RNA comparative structure models. *Current Opinion in Structural Biology*, 12(3): 301–310. Reproduced with permission from Elsevier). (See Colour plate R.2)

More than twenty years after the first 16S and 23S rRNA comparative structure models were proposed, the most recent comparative structure models (see Figures R.2 and R.3) were evaluated against the high-resolution crystal structures of both the small (Wimberly *et al.* 2000) and large (Ban *et al.* 2000) ribosomal subunits that were determined in 2000. The results were affirmative; 97–98% of the 16S and 23S rRNA base-pairs, including nearly all of the tertiary structure base-pairs, predicted with covariation analysis were present in these crystal structures. However the majority of the tertiary interactions in the high-resolution crystal structures were not identified with comparative methods since nearly all of the sets of nucleotides involved in tertiary structure base pairs do not covary with one another.

The majority of the secondary structure in the nuclear-encoded ribosomal RNA genes is conserved in the three primary phylogenetic domains. However regions of the Archaea, Bacteria and Eucarya rRNA secondary structure are unique within each of these three major phylogenetic groups. Usually homologous structural elements between the phylogenetic groups can be readily identified due to very similar structural features. The high-resolution crystal structure of a eukaryotic ribosome revealed that a few regions of the rRNA structure, while different in secondary structure, are similar in their three-dimensional structure.

Analysis of the comparative secondary structure models and the high-resolution crystal structures have facilitated the identification and characterization of new structure motifs (see RNA Tertiary-Structure Motifs).

Related websites

Wikipedia: Ribosomal RNA	http://en.wikipedia.org/wiki/Ribosomal_RNA
Wikipedia: 16S rRNA	http://en.wikipedia.org/wiki/16S_ribosomal_RNA
Wikipedia: 18S rRNA	http://en.wikipedia.org/wiki/18S_ribosomal_RNA
Wikipedia: 23S rRNA	http://en.wikipedia.org/wiki/23S_ribosomal_RNA
Wikipedia: 28S rRNA	http://en.wikipedia.org/wiki/28S_ribosomal_RNA
Wikipedia: 5S rRNA	http://en.wikipedia.org/wiki/5S_ribosomal_RNA
Wikipedia: 30S ribosomal subunit	http://en.wikipedia.org/wiki/30S
Wikipedia: 40S ribosomal subunit	http://en.wikipedia.org/wiki/40S
Wikipedia: 50S ribosomal subunit	http://en.wikipedia.org/wiki/50S
Wikipedia: 60S ribosomal subunit	http://en.wikipedia.org/wiki/60S
Wikipedia: Ribosomal RNA spacer	http://en.wikipedia.org/wiki/Dna,_ribosomal_spacer

Further reading

Ban N, et al. (2000) The complete atomic structure of the large ribosomal subunit at 2.4 Å resolution. *Science*, 289: 905–920.

Feagin JE, Harrell MI, Lee JC, et al (2012) The Fragmented Mitochondrial Ribosomal RNAs of *Plasmodium falciparum*. *PLoS One*, 7(6): e38320.

Gray MW, Schnare MN (1996) Evolution of rRNA gene organization. *In:* Ribosomal RNA: Structure, Evolution, Processing and Function in Protein Biosynthesis. pp. 49–69. Editors: Zimmermann RA, Dahlberg A.E. CRC Press. Boca Raton, FL.

R

Gutell RR, et al (2002) The Accuracy of Ribosomal RNA Comparative Structure Models. *Curr Opin Struct Biol*, 12: 301–310.

Harms J, et al. (2001) High resolution structure of the large ribosomal subunit from a mesophilic eubacterium. *Cell*, 107: 679–688.

Jackson SA, Cannone JJ, Lee JC, Gutell RR, Woodson SA (2002) Distribution of rRNA introns in the three-dimensional structure of the ribosome. *Journal of Molecular Biology*, 323: 35–52.

Klinge S, Voigts-Hoffmann F, Leibundgut M, Ban N (2012) Atomic structures of the eukaryotic ribosome. *Trends Biochem Sci.*, 37: 189.

Lee JC, Gutell RR (2012) A *Comparison of the Crystal Structures of Eukaryotic and Bacterial* SSU ribosomal RNAs reveals common structural features in the hypervariable regions. *PLoSOne.*, 7: e38203.

Melnikov S, Ben-Shem A, Garreau de Loubresse N, Jenner L, Yusupova G, Yusupov M (2012) One core, two shells: bacterial and eukaryotic ribosomes. *Nat Struct Mol Biol.*, 19: 560–567.

Schluenzen F, et al. (2000) Structure of functionally activated small ribosomal subunit at 3.3 Å resolution. *Cell*, 102: 615–623.

Shang L, Xu W, Ozer S, Gutell RR (2012) Structural constraints identified with covariation analysis in ribosomal RNA. *PLoS ONE*, 7(6): e39383.

Wimberly BT et al. (2000). Structure of the 30 S ribosomal subunit. *Nature*, 407: 327–339.

Woese CR (1987). Bacterial evolution. *Microbiol. Rev.*, 51: 221–271.

RIBOSOME BINDING SITE (RBS)

John M. Hancock

The site on the mRNA that directs its binding to the ribosome.

The identification of an RBS is an important indicator of the position of the true translation start site. The sequence nature of these sites differs between prokaryotes and eukaryotes. The prokaryotic RBS is known as the Shine-Dalgarno sequence; it has the consensus AGGAGG and is located 4–7 bp 5′ of the translation start site. In eukaryotes the equivalent is the Kozac Sequence.

Further reading

Kozak M (1987) An analysis of 5′-noncoding sequences from 699 vertebrate messenger RNAs. *Nucleic Acids Res.* 15: 8125–8148.

Kozak M (1997) Recognition of AUG and alternative initiator codons is augmented by G in position +4 but is not generally affected by the nucleotides in positions +5 and +6. *EMBO J.*, 16: 2482–2492.

Shine J, Dalgarno L (1974) The 3′-terminal sequence of E. coli 16s ribosomal RNA: Complementarity to nonsense triplets and ribosome binding sites. *PNAS*, 71: 1342–1346.

RMSD, *SEE* ROOT MEAN SQUARE DEVIATION.

RNA (GENERAL CATEGORIES)

Robin R. Gutell

Ribonucleic acid (RNA) has a most prominent role in the structure, function and regulation of the cell. RNA has characteristics of DNA and proteins. RNA and DNA both form base pairs with similar but not identical rules. While the most prominent base pairs in RNA (G:C and A:U(T)) also form in DNA, the higher-order structure in RNA also contains many non-canonical (not G:C and A:U) base pairs, including G:U (wobble), A:G, A:A, G:G, U:U, C:C, A:C, and U:C. Consecutive and antiparallel canonical base pairs form helices in RNA and DNA. And RNA, like protein, can form a three-dimensional structure capable of catalysing chemical reactions. It is now generally accepted that RNA was a precursor to DNA and proteins, and thus essential for the origin of life.

RNA like DNA is composed of a ribose sugar, a phosphate and a cyclic base. RNA has four primary bases: adenine, guanine, cytosine, and uracil although other bases and modified forms of these bases occur. The two major differences in the structure of RNA and DNA are: 1) DNA has T (thymine) in place of the U (uracil) in RNA, 2) The sugar in DNA is deoxyribose while RNA has ribose. The ribose sugar increases RNA's susceptibility to degradation compared to DNA. As a consequence DNA is the better repository for long-term genetic inheritance while RNA's labiality is preferred for functions that have a shorter half-life.

The three major forms of RNA, messenger RNA (mRNA), ribosomal RNA (rRNA) and transfer RNA (tRNA) are directly involved in protein synthesis. The mRNA sequence codes for the amino acids in proteins. Three consecutive nucleotides form a codon that encodes for an amino acid, as specified by the genetic code for each organism. During transcription, mRNA is transcribed from DNA, and then translated into proteins. The mRNAs can contain untranslated regions (UTRs) flanking the protein coding sequence at their 5′ and 3′ ends. These UTRs are involved in gene regulation and expression. The mRNA is quickly synthesized and degraded as part of the regulation of protein synthesis. The rRNA is a major part of the ribosome, the site of protein synthesis. Approximately 2/3 of the mass of the bacterial ribosome is rRNA while the remaining 1/3 is ribosomal proteins. In other organisms the mass of the rRNA could be closer to 1/2. The anticodon sequence in the tRNA specifies the amino acid that is attached to the 3′ end of tRNA. During protein synthesis the tRNA with the anticodon that is complementary to the codon binds to the mRNA, and facilitates the addition of the correct amino acid to the growing peptide chain. The rRNA, mRNA, and tRNA were once the most studied and characterized RNA types. Their role in protein synthesis dominated and biased our perspective of RNA. Now different RNA types have been identified and characterized that are also involved in protein synthesis, regulation, post-transcriptional modification, catalysis, DNA replication, RNA genomes, parasitic RNAs, Archaea and Bacterial immune/genome defense response, X-chromosome inactivation, diseases, and a growing number of other tantalizing functions.

RNAs that are functional and do not code for proteins are called noncoding RNAs (ncRNA). The first non-coding RNAs identified were the tRNAs and rRNAs. A newer term – functional RNA (fRNA) is used by some to categorize RNAs that do not code for protein and have a specific biological function. A brief description of some of the ncRNA/fRNA are:

- SRP RNA (Signal Recognition RNA) is complexed with proteins to transport newly-translated secretory proteins to the cytosol. The SRP binds to a signal at the N-terminus of a protein as the protein is synthesized.
- The bacterial tmRNA (Transfer Messenger RNA) is a chimeric molecule with tRNA-like and mRNA-like characteristics. It binds to the ribosome to terminate translation when the abundance of tRNAs is low. This RNA is also known as SsrA and 10Sa RNA.
- RNase P is complexed with protein. The RNA component catalyzes the cleavage and maturation of tRNA.
- Small nucleolar RNAs (snoRNAs) typically contain 60–300 nucleotides. They are abundant in the nucleolus of different eukaryotes. The snoRNAs associate with proteins to form small nucleolar ribonucleoproteins (snoRNPs). snoRNAs function is to guide the chemical modification of other RNAs, primarily the rRNAs, tRNAs, and other sno RNAs.
- MicroRNAs are 22 or so nucleotices in length, present in eukaryotes, and regulate genes post-transcriptionally. MicroRNAs have associated with the regulation of many different genes, and implicated in multiple diseases, including numerous cancers.
- RNAi (RNA Interference) are short, approximately 20 nucleotides in length with multiple functions in eukaryotic cells. RNAi can protect eukaryotic cells from foreign nucleic acids. RNAi can up and down regulate the expression of genes.
- Guide RNAs (gRNAs) facilitate the insertion or deletion of uridines (RNA editing) in the mitochondrial RNA in kinetoplastid protists.
- Piwi RNAs interact with piwi proteins for gene silencing and epigenetic modifications of retrotransposons and nucleic acids in animal germ cells.
- CRISPR (**C**lustered **R**egularly **I**nterspaced **S**hort **P**alindromic **R**epeats) are present in some but not all Bacteria and Archaea. This nucleic-acid based immune system protects cells and their genomes from nucleic acids attacks from other cells.
- snRNAs (small nuclear RNAs) are associated with proteins to form the spliceosome, a macromolecule involved in the removal of introns from transcribed RNA. The splicesome is composed of RNAs U1, U2, U4, U5 and U6. Other U RNA types occur less frequently and occur in 'minor spliceosomes'.
- Group I and II introns are two types of introns that catalyse their own excision from the flanking exon.
- Telomerase RNA (or Telomerase RNA Component – TERC) is complexed with proteins to form Telomerase, a ribonucleoprotein polymerase that extends the end of the chromosomes with the sequence TTAGGG.
- A Riboswitch is a segment of the mRNA that binds small molecules and influences gene activity.
- Ribozymes are RNAs that form a three-dimensional structure capable of catalyzing a chemical reaction.
- Long non-coding RNAs is a general category for RNAs that do not code for proteins, contain at least 200 nucleotides, and do not parse into the other larger noncoding RNA types (e.g. rRNA, group I intron). These RNAs have been associated with the regulation of eukaryotic transcription, post-transcriptional processing of the RNA, epigenetic regulation, X-chromosome inactivation, and numerous diseases.

R

Related websites

Wikipedia: tmRNA	http://en.wikipedia.org/wiki/TmRNA
Wikipedia: SRP RNA	http://en.wikipedia.org/wiki/Signal_recognition_particle_RNA
Wikipedia: RNase P	http://en.wikipedia.org/wiki/RNase_P
Wikipedia: snoRNA	http://en.wikipedia.org/wiki/SnoRNAs
Wikipedia: Guide RNAs	http://en.wikipedia.org/wiki/Guide_RNA
Wikipedia: Piwi RNAs	http://en.wikipedia.org/wiki/Piwi-interacting_RNA
Wikipedia: CRISPR	http://en.wikipedia.org/wiki/CRISPR
Wikipedia: RNAi	http://en.wikipedia.org/wiki/RNAI
Wikipedia: microRNA	http://en.wikipedia.org/wiki/MicroRNA
Wikipedia: Spliceosome	http://en.wikipedia.org/wiki/Spliceosome
Wikipedia: small nuclear RNA	http://en.wikipedia.org/wiki/Small_nuclear_RNA
Wikipedia: Introns	http://en.wikipedia.org/wiki/Introns
Wikipedia: Group I Introns	http://en.wikipedia.org/wiki/Group_I_catalytic_intron
Wikipedia: Group II Introns	http://en.wikipedia.org/wiki/Group_II_intron
Wikipedia: Telomerase	http://en.wikipedia.org/wiki/Telomerase
Wikipedia: Telomerase RNA	http://en.wikipedia.org/wiki/Telomerase_RNA_component
Wikipedia: Riboswitch	http://en.wikipedia.org/wiki/Riboswitch
Wikipedia: Ribozymes	http://en.wikipedia.org/wiki/Ribozyme
Wikipedia: Long Noncoding RNAs	http://en.wikipedia.org/wiki/Long_noncoding_RNA

Further reading

Bachellerie JP, Cavaillé J (1998) Small nucleolar RNAs guide the ribose methylations of eukaryotic rRNAs. In: *Modification and Editing of RNA*. Grosjean H, Benne R, editors. ASM Press, Washington, DC.

Estévez AM, Simpson L (1999) Uridine insertion/deletion RNA editing in trypanosome mitochondria – a review. *Gene*, 240: 247–260.

Guthrie C, Patterson B (1988) Spliceosomal snRNAs. *Ann Rev Genet*, 22: 387–419.

Hutvagner G, Zamore PD (2002) RNAi: nature abhors a double-strand. *Curr Opin Genet Dev*, 12: 225–232.

Kiss T (2002) Small nucleolar RNAs: an abundant group of noncoding RNAs with diverse cellular functions. *Cell*, 109: 145–148.

Mattick JS (2001) Non-coding RNAs: the architects of eukaryotic complexity. *EMBO Reports*, 2: 986–991.

Reinhart BJ, et al. (2002) MicroRNAs in plants. *Genes Dev*, 16: 1616–1626.

Samarsky DA, Fournier MJ (1999) A comprehensive database for the small nucleolar RNAs from *Saccharomyces cerevisiae*. *Nucleic Acids Res*, 27: 161–164.

R

RNA FOLDING

Jamie J. Cannone and Robin R. Gutell

The RNA Folding problem is one of the grand challenges in Biology. The goal is to predict the correct secondary and three-dimensional structure from one or a set of sequences. Comparative sequence analysis (see RNA Structure Prediction (Comparative Sequence Analysis)) has, for select RNAs, accurately identified that RNA secondary structure that is common to the set of analyzed sequences. The prediction of RNA secondary structure and other higher-order structures with energy minimization is the second major method used for RNA folding (see RNA Structure Prediction (Energy Minimization)). The accuracies vary significantly with this method.

See also Covariation Analysis, RNA Structure Prediction: Energy Minimization

RNA HAIRPIN

Antonio Marco and Sam Griffiths-Jones

An RNA hairpin is a simple base-paired secondary structure consisting of a helix stem and a loop, often called a stem-loop or foldback structure.

Single-stranded RNA molecules often form secondary structures by intra-molecular base-pairing. The simplest stable structure is produced by folding an RNA sequence back on itself to produce a hairpin or stem-loop structure. MicroRNA precursors adopt a simple RNA hairpin structure. RNA hairpins are recognized by some endonucleases, including in microRNA biogenesis.

The computational detection of microRNAs often starts with the prediction of hairpin structures in a given sequence. Hairpins can be easily predicted using dynamic-programming approaches such as those implemented in mfold (http://mfold.rna.albany.edu/) and the ViennaRNA package (http://www.tbi.univie.ac.at/RNA/).

Related websites

Mfold	http://mfold.rna.albany.edu/
ViennaRNA	http://www.tbi.univie.ac.at/RNA/

Further reading

Berezikov E, Cuppen E, Plasterik RHA (2006) Approaches to microRNA discovery. *Nat Genet*, 38: S2–S7.

Washietl S (2010) Sequence and structure analysis of noncoding RNAs. *Methods Mol Biol*, 609: 285–306.

See also MicroRNA Discovery, RNA Folding Energy

RNA-SEQ

Stuart Brown

The bulk sequencing of RNA from biological samples (RNA-seq) is an application of Next Generation DNA Sequencing (NGS). RNA molecules are reverse transcribed into cDNA, fragmented, and primers are attached to create libraries suitable for cluster amplification in NGS protocols. After sequencing, reads must be mapped to a set of known genes in order to produce count data that can be used to quantify gene expression. This mapping and counting is computationally challenging, since most alignment algorithms optimized for NGS data (to align millions or reads to a large eukaryotic genome) do not accurately map RNA reads across the gaps created by introns. This problem is further complicated by alternative splicing and multiple gene isoforms. Some solutions to this problem rely on mapping RNA reads to a database of splice junction sequences, which are created from the known gene models in a specific database such as RefSeq or ENSEMBL. Other methods create a set of *de novo* splice junction sequences from the RNA-seq data itself, identifying reads that map to two adjacent genomic positions, likely to fall within a single gene, that also contain known splice consensus sequences.

There are several advantages of measuring gene expression with RNA-seq compared with hybridization of RNA to a microarray. RNA-seq directly counts mRNA molecules per gene, so there is no bias based on hybridization artifacts such as GC/AT content or sequence similarity to another gene. RNA-seq with deep coverage provides greater sensitivity to study low expressed genes. RNA-seq is more linear across a wide dynamic range (from very high to very low expressing genes). As NGS methods have become cheaper and the amount of sequence data obtained per sample has increased, RNA-seq has become competitive or less expensive than microarray as a method to measure gene expression.

The sequence data from RNA-seq can be used for additional applications that are not possible with microarrays. RNA-seq can identify expression from new genes or new parts of genes that have not previously been identified and annotated in public databases. It is also possible to use RNA-seq to study the transcriptome of a speices for which genome sequence has not been determined (of for which no gene expression microarray has been developed). RNA-seq can be used to identify mutations in the coding regions of genes. RNA-seq can be used to explore and quantify alternative splicing and allele-specific expression.

One challenge for RNA-seq is optimizing methods for the removal of very high levels of mitochondrial and ribosomal RNA (rRNA) in total RNA. Methods that depend on selection of poly-A tailed molecules lead to 3' bias in the final sequence data.

Another challenge is the development of appropriate statistical techniques that can handle a wide variety of different sample types and experimental designs. In order to measure changes in gene expression due to an experimental treatment or biological condition, it is necessary to compare RNA-seq read counts per gene for different samples. Statistical methods are needed to normalize counts across different samples with different total numbers of reads, and then to identify differentially expressed genes between biological treatments. Methods developed for microarray data may not be appropriate for RNA-seq data, since gene expression is measured as a continuous variable (fluorescent intentisy) on microarrays, but as discrete read counts in RNA-seq. In addition, RNA-seq gene expression represents a zero sum, where high expression of one (or a few genes) can consume most of the counts, forcing other genes to have lower expression measurements.

RNA SPLICING, *SEE* SPLICING.

RNA STRUCTURE

Jamie J. Cannone and Robin R. Gutell

For this discussion, secondary structure is defined as two or more consecutive and canonical base-pairs that are nested and antiparallel with one another. All other structure, including non-canonical or non-nested base-pairs and RNA motifs, is considered to be tertiary structure. Paired nucleotides are involved in interactions in the structure model; all other nucleotides are considered to be unpaired.

Base-Pair (canonical and non-canonical)

A base-pair is the set of bases in two nucleotides that form hydrogen bonds with one another. While the two paired bases can adopt 15 or so different conformations, the 'canonical' Watson–Crick conformation occurs most frequently. Most RNA base pairs are oriented with the backbones of the two nucleotides in an antiparallel configuration.

The canonical base-pairs, G:C and A:U, plus the G:U ('wobble') base-pair, were originally proposed by Watson and Crick. All other base pair types (e.g. A:A, A:C, A:G, C:C, C:U, G:G, U:U) are non-canonical. While 'non-canonical' connotes an unusual or unlikely combination of two nucleotides, a significant number of non-canonical RNA base pairs has been proposed in rRNA comparative structure models and substantiated by the ribosomal subunit crystal structures. However a much larger number of non-canonical base pairs, that were not predicted with comparative analysis, are present in the high-resolution crystal structures, indicating that the majority of the sets of positions that form a tertiary structure base pairs do not have similar patterns of variation (i.e. no covariation).

Stem

A stem (or simple helix) is a set of base pairs that are arranged adjacent to and antiparallel with one another, without any irregularity (e.g. bulge loop, non-canonical base pair). A set of stems or simple helices that are connected with internal loops or bulges is a compound helix.

The base pairs that comprise a stem are nested; that is, drawn graphically, each base-pair either contains or is contained within its neighbours. For two nested base-pairs, a:a′ and b:b′, where $a < a'$, $b < b'$, and $a < b$ in the 5′ to 3′ numbering system for a given RNA molecule, the statement $a < b < b' < a'$ is true. Figure R.3 shows nesting for tRNA in two different formats that represent the global arrangement of base-pairs. Nesting arrangements can be far more complicated in a larger RNA molecule. Most base-pairs are nested, and most helices are also nested. Base-pairs that are not nested are pseudoknots (see below).

R

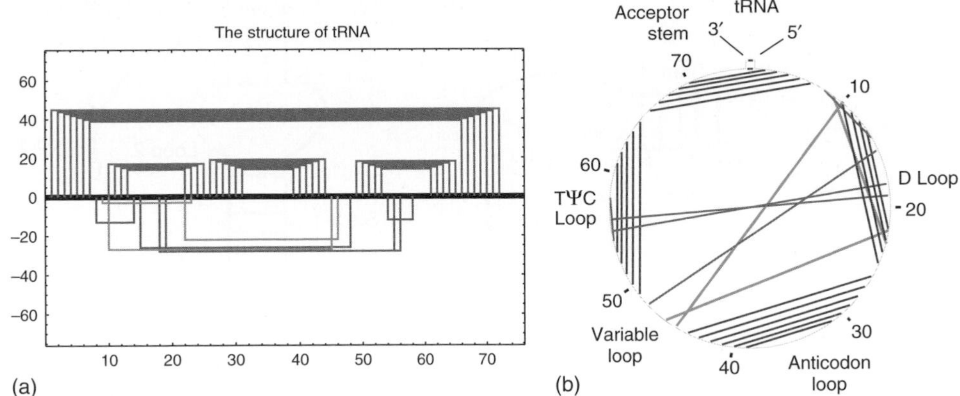

Figure R.3 Nested and non-nested base-pairs. The secondary and tertiary structure of tRNA is shown in two formats that highlight the helical and nesting relationships. All of the secondary structure base pairs are nested while some of the tertiary base-pairs are nested. Tertiary base pairs in both formats that cross other lines are not nested. A. Histogram format, with the tRNA sequence shown as a 'baseline' from left to right (5′ to 3′). Secondary structure elements are shown above the baseline and tertiary structure elements are shown below the baseline. The distance from the baseline to the interaction line is proportional to the distance between the two interacting positions within the RNA sequence. B. Circular format, with the sequence drawn clockwise (5′ to 3′) in a circle, starting at the top and base-base interactions shown as lines traversing the circle. The tRNA structural elements are labeled.

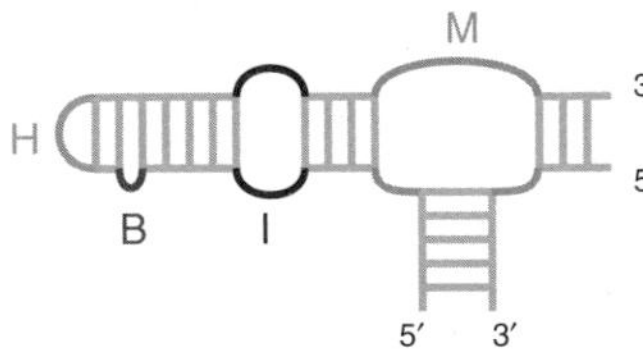

Figure R.4 Loop types. This schematic RNA secondary structure contains one example of each of the four loop types. B, bulge loop, H, hairpin loop, I, internal loop, M, multi-stem loop. (See Colour plate R.4)

Loop

Unpaired nucleotides in a secondary structure model are commonly referred to as loops. Many of these loops close one end of an RNA stem and are called 'hairpin loops;' phrased differently, the nucleotides in a hairpin loop are flanked by a single stem. Loops that are flanked by two stems come in several forms. A 'bulge loop' occurs only in one strand in a stem; the second strand's nucleotides are all forming base-pairs. An 'internal loop' is formed by parallel bulges on opposing strands, interrupting the continuous base-pairing of the stem. Finally, a 'multi-stem' loop forms when three or more stems intersect. Figure R.4 is a schematic RNA that shows each of these types of loop. The sizes of each type of loop can vary; certain combinations of loop size and nucleotide composition have been shown to be more energetically stable.

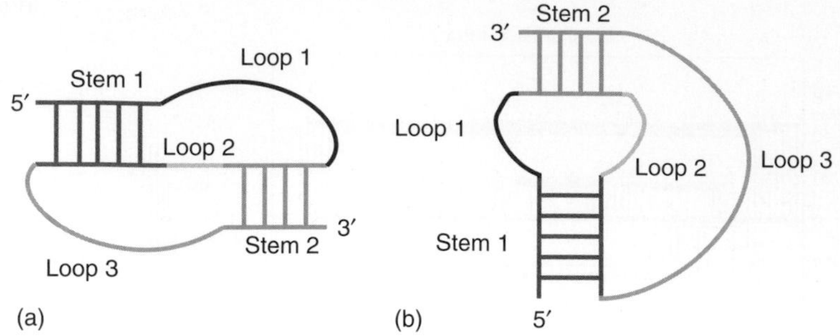

Figure R.5 Schematic drawings of a single pseudoknot. (a) Standard format. (b) Hairpin loop format (Source: after Hilbers et al. 1998). (See Colour plate R.5)

Pseudoknots

A pseudoknot is an arrangement of helices and loops where the helices are not nested with respect to each other. Pseudoknots are so named due to the optical illusion of knotting evoked by secondary and tertiary structure representations. A simple pseudoknot is represented in Figure R.5. For two non-nested (pseudoknot) base-pairs, $a{:}a'$ and $b{:}b'$, where $a < a'$ and $b < b'$ in the 5′ to 3′ numbering system for a given RNA molecule, the following statement will be true: $a < b < a' < b'$. Contrast this situation with a set of nested base-pairs (see Stems), where $a < b < b' < a'$. Another descriptive explanation of pseudoknot formation is when the hairpin loop nucleotides from a stem-loop structure form a helix with nucleotides outside the stem-loop. The colour version of Figure R.5 shows pseudoknot interactions in green; note how these lines cross the blue lines that represent nested helices.

Related websites

The Comparative RNA Web (CRW) Site	http://www.rna.ccbb.utexas.edu/
Definitions of structural elements in RNA secondary structure	http://www.rna.ccbb.utexas.edu/CAR/1C/
Base pairing conformations, other three-dimensional/ chemical structure	http://www.rna.ccbb.utexas.edu/SIM/4A/CRWStructure/

Further reading

Chastain M, Tinoco I Jr (1994) Structural elements in RNA. *Progr Nucleic Acids Res Mol Biol*, 44: 131–177.

Gutell RR, Larsen N, Woese CR (1994) Lessons from an evolving ribosomal RNA: 16S and 23S rRNA structure from a comparative perspective. *Microbiological Reviews*, 58(1): 10–26.

Hendrix DK, Brenner SE, Holbrook SR (2005) RNA structural motifs: building blocks of a modular biomolecule. *Quarterly Reviews of Biophysics*, 38: 221–243.

Hilbers CW, et al. (1998) New developments in structure determination of pseudoknots. *Biopolymers*, 48: 137–153.

Lee JC, Gutell RR (2004) Diversity of base-pair conformations and their occurrence in rRNA structure and RNA structural motifs. *Journal of Molecular Biology*, 344(5): 1225–1249.

Moore PB (1999) Structural Motifs in RNA. *Ann. Review of Biochemistry.*, 68: 287–300.

Pleij CWA (1994) RNA pseudoknots. *Curr Opin Struct Biol*, 4: 337–344.

ten Dam E, et al. (1992) Structural and functional aspects of RNA pseudoknots. *Biochemistry*, 31: 11665–11676.

RNA STRUCTURE PREDICTION (COMPARATIVE SEQUENCE ANALYSIS)

Jamie J. Cannone and Robin R. Gutell

Comparative Analysis is based on a very important discovery in molecular biology. The same RNA secondary and tertiary structure can form with an extraordinary large number of different RNA sequences. The structure and function for specific RNA molecules is maintained during the evolutionary process. Natural selection acts on genetic mutations to create and then maintain the same higher-order structure. The overall objective for the prediction of RNA structure with comparative analysis is to identify that higher-order structure that is conserved in the sequences in the same RNA family.

In practice homologous base-pairs that occur at the same positions in all of the sequences in the data set are identified. Covariation analysis identifies those base pairs that have similar patterns of variation at each of the paired nucleotides (see Covariation Analysis). Comparative analysis can also identify RNA structural motifs (see Motifs in RNA Tertiary Structure). While these motifs can be composed of base pairs identified with covariation analysis, the majority of these structural motifs are composed of elements that do not covary. Instead they are composed of specific sets of nucleotides that have a unique arrangement of paired and unpaired nucleotides. These occur at unique locations in the RNAs higher-order structure.

The comparative method has been used to identify higher-order structure models for many RNA molecules, including tRNA, the three rRNAs (5S, 16S and 23S), group I and II introns, RNase P, tmRNA, SRP RNA, telomerase RNA, ITS and IVS rRNAs, 5′ and 3′ UTRs for different protein coding genes, and many other short and long non-coding RNAs. The accuracy of these methods can be very high when the number of sequences is large and the diversity among these sequences is high. Approximately 97–98% of the base pairs in the prokaryotic and eukaryotic comparative rRNA structure models are in the high resolution crystal structures.

Related websites

Rfam	http://rfam.sanger.ac.uk/
Comparative RNA Web (CRW) Site	http://www.rna.ccbb.utexas.edu/METHODS/

R

Further reading

Cannone JJ, Subramanian S, Schnare MN, et al. (2002) The Comparative RNA Web (CRW) Site: An Online Database of Comparative Sequence and Structure Information for Ribosomal, Intron, and other RNAs. *BioMed Central(BMC) Bioinformatics*, 3: 2.

Gutell RR, et al. (1994) Lessons from an evolving rRNA: 16 S and 23 S rRNA structures from a comparative perspective. *Microbiol Rev*, 58: 10–26.

Gutell RR, et al. (2002) The accuracy of ribosomal RNA comparative structure models. *Curr Opin Struct Biol*, 12: 301–310.

Michel F, Westhof E (1990) Modelling of the three-dimensional architecture of group I catalytic introns based on comparative sequence analysis. *J. Mol. Biol.*, 216 (3): 585–610.

Michel F, et al. (2000) Modeling RNA tertiary structure from patterns of sequence variation. *Methods Enzymol*, 317: 491–510.

Woese CR, Pace NR (1993) Probing RNA structure, function, and history by comparative analysis. In *The RNA World* (Gesteland RF, Atkins JF, editors), pp. 91–117, Cold Spring Harbor Laboratory Press, Cold Spring Harbor, NY.

RNA STRUCTURE PREDICTION (ENERGY MINIMIZATION)

Jamie J. Cannone and Robin R. Gutell

RNA secondary structure models are predicted with comparative sequence analysis (see RNA Structure Prediction (Comparative Sequence Analysis)) or with programs that attempt to determine the most thermodynamically stable structure, using energy minimization techniques. Three of the fundamental premises for the energy minimization approach are: 1) the total free-energy of a predicted RNA structure is the sum total of the energy values for the smaller structural elements, 2) the energy values determined experimentally for small structural elements in isolation will be similar for structural elements that are assembled into the complete RNA structure, 3) energy values have been determined for all of the necessary structural elements required for the accurate prediction of a secondary structure for any given RNA. Energetic values have been determined for consecutive base pairs within a helix, focusing initially on canonical base pairs (i.e. G:C, A:U and G:U) and more recently noncanonical base pairs. The energy values for several non-helical structural elements have also been determined with experimental methods. These include the dangling ends of a helix, and hairpin, internal, and multi-stem loops, coaxial stacking, and other structural motifs, including the UAA/GAN motif.

To improve the accuracy in the prediction of RNA secondary structure with minimum free energy-based folding programs, the energy values for a larger number of structural elements is necessary. Recently, pseudo-energies (statistical potentials) have been determined for more structural elements than possible with experimental methods. The different approaches utilized to determine the statistical potentials are based on the concept that the frequency for each structural element is approximately proportional to its stability. Analysis is revealing that the prediction accuracies are higher with statistical potentials than experimentally determined free energy values. While this latter result reveals that structural potentials for a larger set of structural elements could improve the prediction accuracies, other factors such as kinetics are known to be involved in the process of folding RNA into its secondary and other higher order structure.

Related websites

Wikipedia: Nucleic acid structure prediction	http://en.wikipedia.org/wiki/RNA_structure_prediction
Wikipedia: List of RNA structure prediction software	http://en.wikipedia.org/wiki/List_of_RNA_structure_prediction_software

Further reading

Andronescu M, Condon A, Hoos HH, Mathews DH, Murphy K P (2010) Computational approaches for RNA energy parameter estimation. *RNA*, 16: 2304–2318.

Dima RI, Hyeon C, Thirumalai D (2005) Extracting stacking interaction parameters for RNA from the data set of native structures. *J. Mol. Biol.*, 347: 53–69.

Do CB, Woods DA, Batzoglou S (2006) CONTRAfold: RNA secondary structure prediction without physics-based models. *Bioinformatics.*, 22(14): e90–8.

Doshi KJ, Cannone JJ, Cobaugh CW, Gutell RR (2004) Evaluation of the suitability of free energy minimization using nearest-neighbor energy parameters for RNA secondary structure prediction. *BMC Bioinformatics*, 5: 105.

Gardner DP, Ren P, Ozer S, Gutell RR (2011) Statistical potentials for hairpin and internal loops improve the accuracy of the predicted RNA structure. *Journal of Molecular Biology*, 413: 473–483.

Hajiaghayi M, Condon A,Hoos HH (2012) Analysis of energy-based algorithms for RNA secondary structure predictin. *BMC Bioinformatics*, 13: 22.

Mathews DH, et al. (1999) Expanded sequence dependence of thermodynamic parameters provides robust prediction of RNA secondary structure. *J Mol Biol*, 288: 911–940.

Rivas E., Lang R, Eddy SR (2012) A range of complex probabilistic models for RNA secondary structure prediction that includes the nearest-neighbor model and more. *RNA*, 18: 193–212.

Turner DH, Mathews DH (2010) NNDB: the nearest neighbor parameter database for predicting stability of nucleic acid secondary structure. *Nucleic Acids Research*, 38: D280–D282.

Zuker M (1989) On finding all suboptimal foldings of an RNA molecule. *Science*, 244: 48–52.

Zuker M (2000) Calculating nucleic acid secondary structure. *Current Opinion in Structural Biology*, 10: 303–310.

See also Covariation Analysis

RNA TERTIARY STRUCTURE MOTIFS

Jamie J. Cannone and Robin R. Gutell

Underlying the complex and elaborate secondary and tertiary structures for different RNA molecules is a collection of different RNA building blocks, or structural motifs.

Beyond the abundant G:C, A:U, and G:U base-pairs in the standard Watson-Crick conformation that are arranged into regular secondary structure helices, structural motifs are usually composed of non-canonical base-pairs (*e.g.* A:A) with non-standard base-pair conformations that are usually not consecutive and antiparallel with one another. RNA structural motifs have been identified with a variety of different experimental and computational methods; these motifs are listed in alphabetical order below.

- 2′-OH-Mediated Helical Interactions: extensive hydrogen bonding between the backbone of one strand and the minor groove of another in tightly-packed RNAs.
- A Story: unpaired adenosines in the covariation-based structure models.
- A-Minor: the minor groove faces of adenosines insert into the minor groove of another helix, forming hydrogen bonds with the 2′-OH groups of C:G base-pairs.
- AA.AG@helix.ends: A:A and A:G oppositions exchange at ends of helices.
- Adenosine Platform: two consecutive adenosines form a pseudo-base-pair that allows for additional stacking of bases.
- Base Triple: a base-pair interacts with a third nucleotide.
- Bulge-Helix-Bulge: Archaeal internal loop motif that is a target for splicing.
- Bulged-G: links a cross-strand A stack to an A-form helix.
- Coaxial Stacking of Helices: two neighboring helices are stacked end-to-end.
- Cross-Strand Purine Stack: consecutive adenosines from opposite strands of a helix are stacked.
- Dominant G:U Base-Pair: G:U is the dominant base-pair (50% or greater), exchanging with canonical base-pairs over a phylogenetic group, in particular structural locations.
- E Loop/S Turn: an asymmetric internal or multi-stem loop (with consensus sequence 5′-AGUA/RAA-3′) forms three non-canonical base-pairs.
- E-like Loop: a symmetric internal loop that resembles an E Loop (with consensus sequence 5′-GHA/GAA-3′) forms three non-canonical base pairs.
- G-quadruplex: while initially characterized in DNA, this structural motif is now identified in RNA, in particular in the 5′ UTRs. Four guanines base pair with one another to form a four-stranded G-quartet structure.
- Kink-Turn: named for its kink in the RNA backbone; two helices joined by an internal loop interact via the A-minor motif.
- Kissing Hairpin Loop: two hairpin loops interact to form a pseudocontinuous, coaxially stacked three-stem helix.
- Lone Pair: a base pair that has no consecutive, adjacent base-pair neighbors.
- Lonepair Triloop: a base pair with no consecutive, adjacent base-pair neighbors encloses a three-nucleotide hairpin loop. 'T-loop' motif is a subset of this lonepair triloop motif.
- Metal-Binding: guanosine and uracil residues can bind metal ions in the major groove of RNA.
- Metal-Core: specific nucleotide bases are exposed to the exterior to bind specific metal ions.
- Pseudoknots (see RNA Structure)
- Ribose Zipper: as two helices dock, the ribose sugars from two RNA strands become interlaced.
- Tandem G:A Opposition: two consecutive G:A oppositions occur in an internal loop.
- Tetraloop: four-nucleotide hairpin loops with specific sequences.
- Tetraloop Receptor: a structural element with a propensity to interact with tetraloops, often involving another structural motif.
- Triplexes: stable 'triple helix' observed only in model RNAs.
- tRNA D-Loop:T-Loop: conserved tertiary base pairs between the D and T loops of tRNA.
- U-turn: a loop with the sequences UNR or GNRA contains a sharp turn in its backbone, often followed immediately by other tertiary interactions.

R

Related websites

The Comparative RNA Web (CRW) Site (descriptions and images of structural elements; motif-related publications)	http://www.rna.icmb.utexas.edu/
CRW site—3D structure	http://www.rna.icmb.utexas.edu/SIM/4A/CRWStructure
Wikipedia: G-guadruplex	http://en.wikipedia.org/wiki/G-quadruplex
Wikipedia: Nucleic Acid secondary structure	http://en.wikipedia.org/wiki/Nucleic_acid_secondary_structure
Wikipedia	http://en.wikipedia.org/wiki/Pseudoknot
Wikipedia: RNA Tertiary Structure	http://en.wikipedia.org/wiki/RNA_Tertiary_Structure
Wikipedia: Stem-loop	http://en.wikipedia.org/wiki/Hairpin_loop
Wikipedia: Tetraloops	http://en.wikipedia.org/wiki/Tetraloop
Wikipedia: U-turn	http://en.wikipedia.org/wiki/U-turn

Further reading

Batey RT, et al. (1999) Tertiary motifs in RNA structure and folding. *Angewandte Chemie* (International ed. in English), 38: 2326–2343.

Butcher SE and Pyle AM (2011) The molecular interactions that stabilize RNA tertiary structure: RNA motifs, patterns, and networks. *Accounts of Chemical Research.*, 44: 1302–1311.

Cate JH, Doudna JA (1996) Metal-binding sites in the major groove of a large ribozyme domain. *Structure*, 4: 1221–1229.

Cate JH (1996) RNA tertiary structure mediation by adenosine platforms. *Science*, 273: 1696–1699.

Cate JH, et al. (1996) Crystal structure of a group I ribozyme domain: principles of RNA packing. *Science*, 273: 1678–1685.

Cate JH, et al. (1997) A magnesium ion core at the heart of a ribozyme domain. *Nature Structural Biology*, 4: 553–558.

Chang KY, Tinoco I Jr (1997) The structure of an RNA 'kissing' hairpin complex of the HIV TAR hairpin loop and its complement. *J Mol Biol*, 269: 52–66.

Correll CC, et al. (1997) Metals, motifs, and recognition in the crystal structure of a 5S rRNA domain. *Cell*, 91: 705–712.

Costa M, Michel F (1995) Frequent use of the same tertiary motif by self-folding RNAs. *EMBO J*, 14: 1276–1285.

Costa M, Michel F (1997) Rules for RNA recognition of GNRA tetraloops deduced by in vitro selection: comparison with in vivo evolution. *EMBO J*, 16: 3289–3302.

R

Diener JL, Moore PB (1998) Solution structure of a substrate for the archaeal pre-tRNA splicing endonucleases: the bulge-helix-bulge motif. *Mol Cell*, 1: 883–894.

Dirheimer G, et al. (1995) Primary, secondary, and tertiary structures of tRNAs. In: *tRNA: Structure, Biosynthesis, and Function*, Söll D, RajBhandary U (editors). American Society for Microbiology, Washington, DC, pp. 93–126.

Doherty EA, et al. (2001) A universal mode of helix packing in RNA. *Nature Struct Biol*, 8: 339–343.

Elgavish T, et al. (2001) AA.AG@Helix.Ends: A:A and A:G base-pairs at the ends of 16 S and 23 S rRNA helices. *J Mol Biol*, 310: 735–753.

Gautheret D, et al. (1994). A major family of motifs involving G-A mismatches in ribosomal RNA. *J Mol Biol*, 242: 1–8.

Gautheret D, et al. (1995). Identification of base-triples in RNA using comparative sequence analysis. *J Mol Biol*, 248: 27–43.

Gautheret D, et al. (1995). GU base pairing motifs in ribosomal RNAs. *RNA*, 1: 807–814.

Gutell RR (1994) Lessons from an evolving ribosomal RNA: 16S and 23S rRNA structure from a comparative perspective. *Microbiol Rev*, 58: 10–26.

Gutell RR, et al. (2000) A story: unpaired adenosine bases in ribosomal RNA. *J Mol Biol*, 304: 335–354.

Gutell RR, et al. (2000) Predicting U-turns in ribosomal RNA with Comparative Sequence Analysis. *J Mol Biol*, 300: 791–803.

Ippolito JA, Steitz TA (1998) A 1.3-A resolution crystal structure of the HIV-1 trans-activation response region RNA stem reveals a metal ion-dependent bulge conformation. *Proc Natl Acad Sci USA*, 95: 9819–9824.

Jaeger L, et al. (1994) Involvement of a GNRA tetraloop in long-range RNA tertiary interactions. *J Mol Biol*, 236: 1271–1276.

Klein DJ, et al. (2001) The kink-turn: a new RNA secondary structure motif. *EMBO J*, 20: 4214–4221.

Lee JC, et al. (2003) The Lonepair Triloop: A New Motif in RNA Structure. *J Mol Biol*, 325: 65–83.

Leonard GA, et al. (1994) Crystal and molecular structure of r(CGCGAAUUAGCG): an RNA duplex containing two G(anti).A(anti) base pairs. *Structure*, 2: 483–494. [2′-OH-Mediated Helical Interactions]

Leontis NB, Westhof E (1998) A common motif organizes the structure of multi-helix loops in 16 S and 23 S ribosomal RNAs. *J Mol Biol*, 283: 571–583.

Lietzke SE. et al. (1996) The structure of an RNA dodecamer shows how tandem U-U base pairs increase the range of stable RNA structures and the diversity of recognition sites. *Structure*, 4: 917–930.

Massoulié J (1968) [Associations of poly A and poly U in acid media. Irreversible phenomenon] (French). *Eur J Biochem*, 3: 439–447.

Moore PB (1999) Structural motifs in RNA. *Ann Rev Biochem*, 68: 287–300.

Nissen P, et al. (2001) RNA tertiary interactions in the large ribosomal subunit: the A-minor motif. *Proc Natl Acad Sci USA*, 98: 4899–4903.

Pleij CWA (1994) RNA pseudoknots. *Curr Opin Struct Biol*, 4: 337–344.

SantaLucia J Jr, et al. (1990) Effects of GA mismatches on the structure and thermodynamics of RNA internal loops. *Biochemistry*, 29: 8813–8819.

Tamura M, Holbrook SR (2002) Sequence and structural conservation in RNA ribose zippers. *J Mol Biol*, 320: 455–474.

Traub W, Sussman JL (1982) Adenine-guanine base pairing ribosomal RNA. *Nucleic Acids Res*, 10: 2701–2708.

Wimberly B (1994) A common RNA loop motif as a docking module and its function in the hammerhead ribozyme. *Nature Struct Biol*, 1: 820–827.

Wimberly B, et al. (1993) The conformation of loop E of eukaryotic 5S ribosomal RNA *Biochemistry*, 32: 1078–1087.

Woese CR, et al. (1983) Detailed analysis of the higher-order structure of 16S-like ribosomal ribonucleic acids. *Microbiol Rev*, 47: 621–669.

Woese CR, et al. (1990). Architecture of ribosomal RNA: constraints on the sequence of 'tetra-loops.' *Proc Natl Acad Sci USA*, 87: 8467–8471.

ROBUSTNESS

Dov Greenbaum and Marketa J. Zvelebil

Robustness, in terms of system biology, can be defined as the maintenance of certain functions despite variability in components or the environment.

There can be four possible types of mechanisms used to achieve robust behavior: System control, redundancy, structural stability and modularity. Some papers suggest that robustness is achieved through complexity an in the *E.coli* example:

Robustness also refers to an organism's ability to be resilient against various genetic perturbations. In addition an organism may be genetically robust when it has redundant systems, or repair systems, or thermodynamic stability at the RNA level as well as at the protein level. In other instances an organism may be robust via regulatory mechanisms. Further, epistatic effects across a genome further contribute to robustness.

Robustness allows for complexity of a system as well as its fragility. Minimal cellular life such as Mycoplasm, survives with only a few hundred genes but it can only live under specific environmental conditions and is very sensitive to any fluctuations. However, even a slightly more complex life, such as the *E. coli*, have approximately 4000 genes of which only about 300 are classified as essential. The presence of complex regulatory networks for robustness that accommodate various stress situations is thought to be the reason for this additional gene complexity. The *E. coli* organism can live happily in fluctuating environments.

At the single gene level robustness may be seen to be the result of a dominant rather than recessive trait. Thus a dominant trait will be robust in the face of a mutation that may effect the corresponding gene from the other parent.

Organisms can be both robust and at the same time plastic. Thus they have the ability to evolve and adapt, while retaining many particular characteristics.

Robustness of an organism can occur at various biological levels: (1) gene expression; (2) metabolic flux; (3) protein folding; (4) physiological homeostasis and others.

From an evolutionary perspective robustness can be seen to be a method of increasing the fitness of a desired trait by decreasing the phenotypic effects of a mutation.

Further reading

Bateson P, Gluckman P. (2012) Plasticity and robustness in development and evolution. *Int J Epidemiol*, 41(1): 219–223.

R

Kitano H (2002) Systems biology: A brief overview. *Science*, 295: 1662–1664.

Lauffenburger D (2000) Cell signaling pathways as control modules: Complexity for simplicity? *Proceedings of the National Academy of Science, USA*, 97(10): 5031–5033.

Wagner, A. (2005) *Robustness and Evolvability of Living Systems*. Princeton Univ. Press, Princeton.

Nijhout HF (2002) The nature of robustness in development. *Bioessays*, 24: 553–563.

Gerhar, J, Kirschner M (1997) *Cells, Embryos and Evolution*. Blackwell Science, Malden.

Arjan J, de Visser Joachim Hermisson JM, et al. Perspective: evolution and detection of genetic robustness workshop. *Detection and Evolution of Genetic Robustness*, April 23–25, 2002 at the Santa Fe Institute, Santa Fe, NM.

ROC CURVE

Pedro Larrañaga and Concha Bielza

A receiver operating characteristic (ROC), or simply ROC curve, is a two-dimensional plot which illustrates the performance of a binary classifier system as its discrimination threshold is varied. (See Figure R.6.) It is created by plotting the fraction of true positives out of the

positives (TPR = true positive rate) in the Y axis vs. the fraction of false positives out of the negatives (FPR = false positive rate) in the X axis, at various threshold settings. TPR is also known as sensitivity, and FPR is one minus the specificity (true negative rate). The area under the ROC curve (AUC), when using normalized units, is equal to the probability that a classifier will rank a randomly chosen positive instance higher than a randomly chosen negative one (assuming 'positive' ranks higher than 'negative'). It can be shown that AUC is closely related to the Mann–Whitney U statistic, Wilcoxon test of ranks and the Gini coefficient.

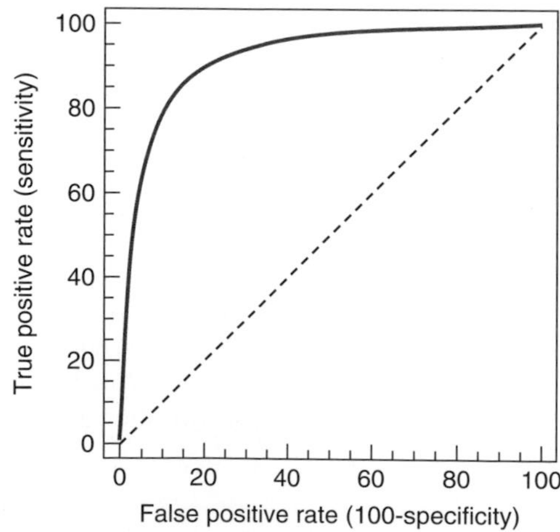

Figure R.6 A ROC curve.

The ROC curve was first developed by electrical engineers and radar engineers during World War II for detecting enemy objects in battle fields and was soon introduced to psychology, medicine (Hanley and McNeil, 1982, 1983) and other areas for many decades and is increasingly used in machine learning research (Fawcett, 2004). See for example Parodi et al. (2008) for identifying differentially expressed genes, based on a generalization of the ROC curve.

Further reading

Fawcett T (2004) ROC graphs: Notes and practical considerations for researchers. *Pattern Recognition Letters*, 27(8): 882–891.

Hanley JA, McNeil BJ (1982) The meaning and use of the area under a receiver operating characteristic (ROC) curve. *Radiology*, 143 (1): 29–36.

Hanley A, McNeil BJ (1983) A method of comparing the areas under receiver operating characteristic curves derived from the same cases. *Radiology*, 148 (3): 839–843.

Parodi S, et al. (2008) Not proper ROC curves as new tool for the analysis of differentially expressed genes in microarray experiments. *BMC Bioinformatics*, 9: 410.

ROLE

Robert Stevens

In description logics, the word ‘role’ is used for ‘relationships type’ or ‘property’. Other Knowledge Representation and Object Oriented Programming Languages use the word in different ways.

ROOT MEAN SQUARE DEVIATION (RMSD)

Roman Laskowski and Tjaart de Beer

The root-mean-square deviation. In Statistics it is defined as the square root of the sum of squared deviations of the observed values from the mean value, divided by the number of observations. The units of rmsd are the same as the units used for the observed values.

$$R(x) = \sqrt{\langle x^2 \rangle}$$

In protein crystallography it is common to cite the rmsd of bond lengths and bond angles in the final, refined model of the protein structure, these giving the root-mean-square deviations of the final bond lengths and angles about the ‘target values’ used during refinement.

Also used as a measure of the similarity of two molecular conformations. The two molecules are first superimposed and the distances between equivalent atoms in the two structures calculated. The square root of the sum of the squares of these distances, divided by the number of distances, gives an indication of how similar or dissimilar the two conformations are. When citing such an rmsd it is necessary to specify which atoms

R

were used in its calculation. For example, the rmsd between two protein structures might be given as 1.4Å for equivalenced C-alpha atoms.

In structure determination by NMR spectroscopy it is usual to give an rmsd of all conformers in the ensemble from an averaged set of coordinates.

Relevant website

PDBeFold	http://www.ebi.ac.uk/msd-srv/ssm/

Further reading

Norman GR, Streiner DL. (2000) *Biostatistics; the Bare Essentials*. BC Decker Inc. (2000).

See also Conformation, C-alpha

ROOTED PHYLOGENETIC TREE, *SEE* PHYLOGENETIC TREE. CONTRAST WITH UNROOTED PHYLOGENETIC TREE.

ROOTING PHYLOGENETIC TREES

Sudhir Kumar and Alan Filipski

Rooting is the process of determining placement of the root in a phylogenetic tree.

The standard method is to use an outgroup. An outgroup is a taxon, or clade, in a phylogenetic tree that lies outside of (or is a sister group to) the clade containing the taxa under study (the ingroup). An outgroup is commonly used for rooting phylogenetic trees by placing the root onto the branch that leads from the outgroup to the remaining taxa. Choosing an appropriate outgroup can be a difficult task, because it may not be possible to accurately and reliably place/connect too distant outgroups to the ingroup.

If outgroup knowledge is unavailable, one alternative may be to apply a mid-point rooting algorithm to an unrooted binary tree. In mid-point rooting, the root can be placed at the midpoint of the longest path between any pair of terminal taxa. If all taxa evolve at the same rate, this will give the correct root, but it may yield misleading results in the presence of deviations from a strict molecular clock. Note that, there also exist alternative algorithms/definitions for mid-point rooting.

Another alternative is to use non time-reversible statistical models of sequence evolution for finding a root. Unlike for time reversible models, where one assumes that evolution occurred in the same way if followed forward or backward in time, each distinct virtual root that is placed into the tree for executing the Felsenstein pruning algorithm to calculate the likelihood will yield a distinct score. Hence, under non time-reversible models, one can evaluate the likelihood for all possible rootings of a tree and root the phylogeny at *the* branch yielding the best likelihood score.

Further reading

Boussau B, Gouy M (2006) Efficient likelihood computations with nonreversible models of evolution. *Systematic Biology*, 55(5): 756–768.

de La Torre-Barcena JE, Kolokotronis S, Lee EK, et al. (2009) The impact of outgroup choice and missing data on major seed plant phylogenetics using genome-wide EST data. *PloS One*, 4(6): e5764.

Hess PN, De Moraes Russo CA (2007) An empirical test of the midpoint rooting method. *Biological Journal of the Linnean Society*, 92(4): 669–674.

See also Phylogenetic Tree

ROSETTA STONE METHOD

Jean-Michel Claverie

A sequence analysis method attempting to predict interactions (and functional relationships) between the products of genes *A* and *B* of a given organism (*e.g.* E. coli) by the systematic search of gene fusions such as A_{ortho} -B_{ortho} (or B_{ortho} - A_{ortho}) in another species (where A_{ortho} and B_{ortho} are orthologous to *A* and *B*, respectively).

The name 'Rosetta' refers to the 'stone of Rosette', the discovery of which allowed a crucial breakthrough in the deciphering of Egyptian hieroglyphs. This basalt slab, found in July 1799 in the small Egyptian village Rosette (Raschid), displays three inscriptions that represent a single text in three different variants of script: hieroglyphs (script of the official and religious texts), Demotic (everyday Egyptian script), and in Greek. A 3-way comparison allowed the French scholar Jean Francois Champollion (1790–1832) to decipher the hieroglyphs in 1822.

The molecular rationale behind the approach is that protein sequences merging A and B sequences will only be possible if they can fold and form a stable 3-D structure where domain A and B interact closely. By extension, isolated proteins constituted of separate A-like and B-like domains are predicted to form a bi-molecular complex. The A-B or B-A gene fusion sequence, from which the potential A/B interaction is inferred, is called the 'Rosetta' sequence. The approach is easier to implement for genomes and or genes without introns. The logic can be extended from a strictly structural interaction to a functional relationship by considering that two enzymes A and B catalysing different steps of a biochemical pathway could become a bi-functional enzyme where A-like and B-like domains, connected by a flexible linker, might not be required to stably interact. The method, introduced in 1999 has become part of the classical phylogenomics tool chest. The results obtained from the Rosetta stone analysis need to be carefully examined to remove the contribution of pseudogenes and/or split genes found in some bacterial genomes.

Related website

UCLA-DOE Institute for Genomics and Proteomics	http://www.doe-mbi.ucla.edu/

Further reading

Date SV (2008) The Rosetta stone method. *Methods Mol Biol.*, 453: 169–180.

R

Eisenberg D, et al. (2000) Protein function in the post-genomic era. *Nature*, 405: 823–826.

Enright AJ, et al. (1999) Protein interaction maps for complete genomes based on gene fusion events. *Nature*, 402: 86–90.

Huynen M, et al. (2000) Predicting protein function by genomic context: quantitative evaluation and qualitative inferences. *Genome Res.*, 10: 1204–1210.

Marcotte EM, et al. (1999) Detecting protein function and protein-protein interactions from genome sequences. *Science*, 285: 751–753.

Ogata H, et al. (2001) Mechanisms of evolution in *Rickettsia conorii* and *R. prowazekii*. *Science*, 293: 2093–2098.

Tsoka, S, Ouzounis CA (2000) Prediction of protein interactions: metabolic enzymes are frequently involved in gene fusion. *Nat Genet.*, 26: 141–142.

ROTAMER

Roland Dunbrack

Short for 'rotational isomer', a single protein side-chain conformation represented as a set of values, one for each dihedral-angle degree of freedom. Since bond angles and bond lengths in proteins have rather small variances, they are usually not included in the definition of a rotamer.

A rotamer is usually thought to be a local minimum on a potential energy map, or an average conformation over some region of dihedral angle space. However, broad distributions of side-chain dihedral angles (such as amides) may be represented by several rotamers, which may not all be local minima or population maxima or means. *Non-rotameric* is sometimes used to describe side chains that have dihedral angles far from average values or far from a local energy minimum on a potential energy surface.

Further reading

Bower MJ, Cohen FE, Dunbrack RL Jr (1997) Prediction of protein side-chain rotamers from a backbone-dependent rotamer library: a new homology modeling tool. *J Mol Biol*, 267: 1268–1282.

Dunbrack RL Jr, Cohen FE (1997) Bayesian statistical analysis of protein sidechain rotamer preferences. *Prot. Sci.*, 6: 1661–1681.

See also Conformation, Amino acids, Modeling

ROUGH SET

Feng Chen and Yi-Ping Phoebe Chen

In reality, the data cannot be always partitioned in exact classes by the available attributes. Therefore, rough set is constructed for describing such 'approximate' situation. Rough set is based on the equivalence classes in the training data. All the data tuples that forms equivalence classes are indifferent, which means the samples are equivalent for the attributes describing the data. For a given class C, rough set defines two sets: C's lower approximation and C's upper approximation. Lower approximation consists of data tuples that

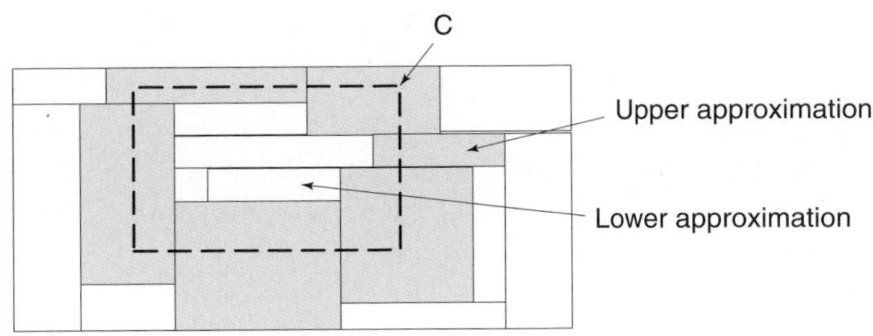

Figure R.7 An example of upper approximation and lower approximation.

definitely belong to C according to their attribute values. Upper approximation is composed of all the tuples which may belong to C. Figure R.7 illustrates an example of two kinds of approximations, in which every rectangle is an equivalent class.

Rough set is always used in but not limited to the following aspects: 1). Classification. Is especially for discovering the structure relation in incorrect data or data with noise; 2). The selection of an attribute set. It can be employed to filter the attributes that do not contribute to classify training data; and 3). Correlation analysis (see Correlation Analysis). It can evaluate the importance of every attribute based on classification criterion effectively and efficiently.

RRNA, *SEE* RIBOSOMAL RNA.

RSEQUENCE

Thomas D. Schneider

Rsequence is the total amount of information conserved in a binding site, represented as the area under the sequence logo.

Related website

Schneider, et al. (1986)	http://alum.mit.edu/www/toms/paper/schneider1986

Further reading

Schneider TD, et al. (1986) Information content of binding sites on nucleotide sequences. *J. Mol. Biol.*, 188: 415–431. http://alum.mit.edu/www/toms/paper/schneider1986/

Schneider, TD (2000) Evolution of biological information. *Nucleic Acids Res.*, 28: 2794–2799. http://alum.mit.edu/www/toms/paper/ev/

See also Rfrequency

R

RULE

Teresa K. Attwood

A short regular expression (typically 3–6 residues in length) used to identify generic (non-family-specific) patterns in protein sequences. Rules tend to be used to encode particular functional sites: *e.g.*, sugar attachment sites, phosphorylation, glycosylation, hydroxylation, sulphation sites, etc. Example rules are shown below:

Functional site	Rule
Protein kinase C phosphorylation site	[ST]-x-[RK]
N-glycosylation site	N-{P}-[ST]-{P}

Their small size means that rules do not provide good discrimination—in a typical sequence database, matches to expressions of this type number in the thousands. They therefore cannot be used to show that a particular functional site exists in a protein sequence, but rather give a guide as to whether such a site *might* exist. Biological knowledge must be used to confirm whether such matches are likely to be meaningful.

Rules are used in PROSITE to encode functional sites, such as those mentioned above and many others. For routine searches of the database, however, the ScanProsite tool provides an option to exclude patterns with a high probability of occurrence, in order that outputs should not be flooded with spurious matches.

Relevant website

ScanProsite	http://www.expasy.org/tools/scanprosite/

Further reading

Attwood TK, Parry-Smith DJ (1999) *Introduction to Bioinformatics*. Addison Wesley Longman, Harlow, Essex, UK.

Attwood TK (2000) The quest to deduce protein function from sequence: the role of pattern databases. *Int. J. Biochem. Cell Biol.*, 32(2), 139–155.

See also Regular Expression

RULE INDUCTION

Pedro Larrañaga and Concha Bielza

The rules induced from examples are represented as logical expressions of the following form:

IF (*conditions*) *THEN* (*decision class*);

where conditions are conjunctions of elementary tests on values of attributes, and decision class indicates the assignment of an object (which satisfies the conditions part) to a given decision class. A number of algorithms for inducing rules of this kind have been proposed, e.g. *OneR* and *Ripper*.

OneR or 'One Rule' (Holte, 1993) is a simple algorithm that builds one rule for each attribute in the training data and then selects the rule with the smallest error rate as its 'one rule'. To create a rule for an attribute, the most frequent class for each attribute value must be determined. The most frequent class is simply the class that appears most often for that attribute value. A rule is simply a set of attribute values assigned to their majority class. *OneR* selects the rule with the lowest error rate. In the event that two or more rules have the same error rate, the rule is chosen at random.

Ripper (Cohen, 1995) builds a ruleset by repeatedly adding rules to an empty ruleset until all positive examples are covered. Rules are formed by greedily adding conditions to the antecedent of a rule (starting with an empty antecendent) until a criterion is not improved. After a ruleset is constructed, a heuristic is used to refine it so as to reduce its size and improve its fit to the training data.

Applications of rule induction algorithms in Computational Biology include the prediction of transcription factors binding to DNA (Huss and Nordstrom, 2006), and the finding of microRNA regulatory modules (Tran et al., 2008).

Further reading

Cohen WW (1995) Fast effective rule induction. In: *Proceedings of the Twelfth International Conference on Machine Learning*, pp. 115–123.

Holte RC (1993) Very simple classification rules perform well on most commonly used datasets. *Machine Learning*, 11: 63–91.

Huss M, Nordstrom K (2006) Prediction of transcription factor binding to DNA using rule induction methods. *Journal of Integrative Bioinformatics*, 3(2): 42.

Tran DH, Satou K, Ho TB (2008) Finding microRNA regulatory modules in human genome using rule induction. *BMC Bioinformatics*, 9(Suppl 12): S5.

RULE OF FIVE (LIPINSKI RULE OF FIVE)

Bissan Al-Lazikani

Also known as Lipinski Rule-of-Five is a rule-of-thumb that can be used to predict whether a compound is likely to be orally bioavailable using its 2D chemical structure. The rule dictates that if a compound has the following properties, it is likely to have the solubility and permeability consistent with being orally bioavailable:

- molecular weight $\leq$500
- $\leq$5 hydrogen-bond donors
- $\leq$10 hydrogen-bond acceptors
- ClogP$\leq$5

This rough rule is a rule for solubility and permeability, and not, as frequently and incorrectly used, a measure of *drug-likeness*.

R

Further reading

Lipinski CA, Lombardo F, Dominy BW, Feeney PJ (2001) Experimental and computational approaches to estimate solubility and permeability in drug discovery and development settings. *Adv Drug Deliv Rev.*, 46(1–3): 3–26.

S

Saccharomyces Genome Database, *see* SGD.
Safe Zone
SAGE (Serial Analysis of Gene Expression)
SAR (Structure–Activity Relationship)
Scaffold
Scaled Phylogenetic Tree, *see* Branch. Contrast with Unscaled Phylogenetic Tree.
Schematic (Ribbon, Cartoon) Models
Scientific Workflows
Score
Scoring Matrix (Substitution Matrix)
SCWRL
SDF
Search by Signal, *see* Sequence Motifs: prediction and modeling.
Second Law of Thermodynamics
Secondary Structure of Protein
Secondary Structure Prediction of Protein
Secretome
Segmental Duplication
Segregation Analysis
Selected Reaction Monitoring (SRM)
Selenoprotein
Self-Consistent Mean Field Algorithm
Self-Organizing Map (SOM, Kohonen Map)
Semantic Network
Semi-Global Alignment, *see* Global Alignment.
Seq-Gen
SeqCount, *see* ENCprime.
Sequence Alignment
Sequence Assembly
Sequence Complexity (Sequence Simplicity)
Sequence Conservation, *see* Conservation.
Sequence Distance Measures
Sequence Logo
Sequence Motif, *see* Motif.
Sequence Motifs: Prediction and Modeling (Search by Signal)
Sequence of a Protein
Sequence Pattern
Sequence Read Archive (SRA, Short Read Archive)
Sequence Retrieval System, *see* SRS.
Sequence Similarity
Sequence Similarity-based Gene Prediction, *see* Gene Prediction, homology-based.
Sequence Similarity Search
Sequence Simplicity, *see* Sequence Complexity.

Concise Encyclopaedia of Bioinformatics and Computational Biology, Second Edition.
Edited by John M. Hancock and Marketa J. Zvelebil.

Sequence Tagged Site (STS)
Sequence Walker
Serial Analysis of Gene Expression, *see* SAGE.
SGD (Saccharomyces Genome Database)
Shannon Entropy (Shannon Uncertainty)
Shannon Sphere
Shannon Uncertainty, *see* Shannon Entropy.
Short-Period Interspersion, Short-Term Interspersion, *see* Interspersed Sequence.
Short Read Archive, *see* Sequence Read Archive.
Shuffle Test
Side Chain
Side-Chain Prediction
Signal-to-Noise Ratio, *see* Noise.
Signature, *see* Fingerprint.
SILAC, *see* Stable Isotope Labelling with Amino Acids in Cell Culture.
Silent Mutation, *see* Synonymous Mutation.
Similarity Index, *see* Distance Matrix.
SIMPLE (SIMPLE34)
Simple DNA Sequence (Simple Repeat, Simple Sequence Repeat)
Simple Repeat, *see* Simple DNA Sequence.
Simple Sequence Repeat, *see* Simple DNA Sequence.
SIMPLE34, *see* SIMPLE.
Simulated Annealing
Simultaneous Alignment and Tree Building
Single Nucleotide Polymorphism (SNP)
Sippl Test, *see* Ungapped Threading Test B.
Sister Group
Site, *see* Character.
SITES
Small Sample Correction
SMILES
Smith-Waterman
SNP, *see* Single Nucleotide Polymorphism.
Software Suites for Regulatory Sequences
Solanaceae Genomics Network (SGN)
Solvation Free Energy
SOV
Space-Filling Model
SPARQL (SPARQL Protocol and RDF Query Language)
Spliced Alignment
Splicing (RNA Splicing)
Split (Bipartition)
Spotted cDNA Microarray
SQL (Structured Query Language)
SRA, *see* Sequence Read Archive.
SRS (Sequence Retrieval System)
Stable Isotope Labelling with Amino Acids in Cell Culture (SILAC)
STADEN
Standard Genetic Code, *see* Genetic Code.
Standardized Qualitative Dynamical Systems
Stanford HIV RT and Protease Sequence Database (HIV RT and Protease Sequence Database)
Start Codon, *see* Genetic Code.
Statistical Mechanics
Statistical Potential Energy
Steepest Descent Method, *see* Gradient Descent.
Stem-loop, *see* RNA hairpin.

Stochastic Process
Stop Codon, *see* Genetic Code.
Stream Mining (Time Series, Sequence, Data Stream, Data Flow)
Strict Consensus Tree, *see* Consensus Tree.
Structural Alignment
Structural Genomics
Structural Motif
Structurama
Structure
Structure–3D Classification
Structure-Activity Relationship, *see* SAR.
Structure-based Drug Design (Rational Drug Design)
STS, *see* Sequence Tagged Site.
Subfunctionalization
Substitution Process
Subtree
Superfamily
Superfold
Supermatrix Approach
Supersecondary Structure
Supertree, *see* Consensus Tree.
Supervised and Unsupervised Learning
Support Vector Machine (SVM, Maximal Margin Classifier)
Surface Models
Surprisal
SVM, *see* Support Vector Machine.
Swiss-Prot, *see* UniProt.
SwissModel
Symmetry Paradox
Synapomorphy
Synonymous Mutation (Silent Mutation)
Synteny
Systems Biology

SACCHAROMYCES GENOME DATABASE, *SEE* SGD.

SAFE ZONE

Patrick Aloy

The *safe zone* is the region of the sequence-structure space where sequence alignments can unambiguously distinguish between protein pairs of similar and different structure. This is for high values of the pairwise sequence identity (i.e. > 40%).

The *safe zone* is occupied by homologous proteins that have evolved from a common ancestor. If the alignment of a sequence of unknown structure and one with known 3D conformation falls into the *safe zone*, homology modeling techniques can be successfully applied to obtain accurate models for our sequence of interest.

Further reading

Rost B (1999) Twilight zone of protein sequence alignments. *Protein Eng* 12: 81–94.

See also Alignment – Multiple, Identity, Homology

SAGE (SERIAL ANALYSIS OF GENE EXPRESSION)

Jean-Michel Claverie

Method for the detection and quantitation of transcripts.

The SAGE method is based on the isolation of unique sequence tags from individual transcripts and concatenation of tags serially into long DNA molecules. Rapid sequencing of concatemer clones reveals individual tags and allows quantitation and identification of cellular transcripts. Two principles underlie the SAGE methodology: 1) A short sequence tag (10–14bp) contains enough information to uniquely identify a transcript provided that that the tag is obtained from a unique position within each transcript; 2) counting the number of times a particular tag is observed provides an estimate of the expression level of the corresponding transcript. Sequencing one insert (from a SAGE library) provides information on the expression of up to 30 different genes rather than on a single one with standard cDNA libraries used to generate ESTs. The same number of sequencing reactions thus provides a much larger set of tags that can then be used, like ESTs, in the 'gene discovery' or 'gene profiling' modes. While average EST sequencing projects are set up to generate 5000 tags, SAGE projects routinely generate of the order of 50,000 tags. The same statistical analysis (i.e. tag counting) applies to both methods, but SAGE experiments allow more accurate and more sensitive expression estimates to be made. SAGE studies have been particularly successful in the characterization of various cancers.

Recent developments include the use of longer tags in the SuperSAGE protocol (Matsumura et al., 2006).

Related website

SAGE page	http://www.sagenet.org/

S

Further reading

Audic S, Claverie J-M (1997) The significance of digital gene expression profiles. *Genome Research* 7: 981–995.

Claverie JM (1999) Computational methods for the identification of differential and coordinated gene expression. *Hum. Mol. Genet.* 8: 1821–1832.

Lal A, et al. (1999) A public database for gene expression in human cancers. *Cancer Res.* 59: 5401–5407.

Matsumura H, Bin Nasir K H, et al. (2006) 'SuperSAGE array: the direct use of 26-base-pair transcript tags in oligonucleotide arrays'. *Nature Methods* 3(6): 469.

Velculescu VE, et al. (1995) Serial analysis of gene expression. *Science* 270: 481–487.

Velculescu VE, et al. (2000) Analysing uncharted transcriptomes with SAGE. *Trends Genet.* 16: 421–425.

SAR (STRUCTURE–ACTIVITY RELATIONSHIP)

Bissan Al-Lazikani

Structure-Activity-Relationship. The relationship between the details of a chemical structure and its biological activity. SAR is a fundamental part of the drug discovery process as it allows the understanding of which components of a compound are most important for its biological activity. For example, a compound identified through high-throughput screening to show some inhibitory effect on a specific enzyme, can be modified systematically. The daughter compounds generated are then screened in the same assay and the inhibitory effect measured. This allows identification of the most important chemical groups in the compound for the activity and identified parts of the compound that can be modified to improve affinity.

See also QSAR

SCAFFOLD

Bissan Al-Lazikani

The core structure of a compound that orientates the peripheral side chains. It is a useful abstraction of compounds that allows the grouping of compounds into related series or scaffold families. The scaffolds can be defined in different ways including Bemis and Murcko frameworks, scaffold trees and maximal common substructres.

Further reading

Bemis GW, Murcko MA (1996) The properties of known drugs. 1. Molecular frameworks. *J Med Chem* 39(15): 2887–2893.

Langdon SR, Brown N, Blagg J (2011) Scaffold diversity of exemplified medicinal chemistry space. *J Chem Inf Model* 51(9): 2174–2185.

Schuffenhauer A, Ertl P, Roggo S, Wetzel S, Koch MA, Waldmann H (2007) The scaffold tree – visualization of the scaffold universe by hierarchical scaffold classification. *J Chem Inf Model* 47(1): 41–58.

See also Bemis and Murcko Framework

S

SCALED PHYLOGENETIC TREE, *SEE* BRANCH. CONTRAST WITH UNSCALED PHYLOGENETIC TREE.

SCHEMATIC (RIBBON, CARTOON) MODELS

Eric Martz

Smoothed backbone models that indicate secondary structure schematically. (see illustration at Models, molecular.)

Alpha helices are represented as helical ribbons, or further simplified to cylinders. Beta strands are represented as relatively straight ribbons, with arrowheads designating the carboxy ends. In some cases, the cylinders representing alpha helices may have their carboxy ends pointed. Backbone regions that have neither alpha nor beta secondary structure may be represented as thin ribbons or ropes, following a smoothed backbone trace.

Hand-drawn schematic depictions first appeared in the mid-1960s. Schematic ribbon drawings and were perfected and popularized by Jane Richardson in the early 1970s. At that time, some visualization software packages were already being adapted to display schematic models. Schematic models are designated 'cartoons' in the popular visualization freeware RasMol and its derivatives (see visualization).

Further reading

Lesk AM, Hardman KD (1982) Computer-generated schematic diagrams of protein structures. *Science* 216: 531–40.

Lesk AM, Hardman KD (1985) Computer-generated pictures of proteins. *Meth. Enzymol.* 115: 381–390.

Richardson JS (1981) The anatomy and taxonomy of protein structure. *Adv. Protein Chem.* 34: 161–339.

Richardson JS (1985) Schematic drawings of protein structures. *Meth. Enzymol.* 115: 351–80.

Richardson JS (2000) Early ribbon drawings of proteins. *Nature Structural Biol.* 7: 621–625.

See also Models, molecular, Backbone Models, Secondary Structure, Visualization Molecular

Eric Martz is grateful for help from Eric Francoeur, Peter Murray-Rust, Byron Rubin, and Henry Rzepa.

SCIENTIFIC WORKFLOWS

Carole Goble and Katy Wolstencroft

Workflows are sophisticated, automated pipelines that allow scientists to analyse their data by linking together a series of tools and resources. Typically, workflows link resources using Web Services (either WSDL or REST), which means that workflows can link local and distributed resources. This makes them a powerful tool in data integration, management and analysis, allowing scientists to perform computationally intensive investigations from their own desktops.

Workflow management systems, like Taverna or Galaxy, have large user communities in the Life Sciences. Workflows are particularly suited to the analysis of high throughput 'omics data (e.g. transcriptomics, metabolomics, next generation sequencing analysis) because they are re-usable, reproducible and executable informatics protocols. As data and analysis methods become larger and more complex, the necessity for adopting methods to ensure reproducible research increases.

Related websites

Taverna	http://www.taverna.org.uk
Galaxy	http://usegalaxy.org

Further reading

Romano P (2008) Automation of in-silico data analysis processes through workflow management systems, *Briefings in Bioinformatics*, 9(1):51–68.

See also Web Services, WSDL, REST

SCORE

Thomas D. Schneider

Many methods in bioinformatics give results as scores.

From an information theory perspective, it is worth noting that scores can be multiplied by an arbitrary constant and remain comparable. In contrast, information is measured in bits, and this cannot be multiplied by an arbitrary constant and still retain the same units of measure. Scores cannot be compared between different binding sites, whereas it is reasonable to compare bits. For example, it is interesting that donor splice sites do not have the same information as acceptor splice sites (Stephens & Schneider 1992).

Related website

Splice	http://alum.mit.edu/www/toms/paper/splice/

Further reading

Stephens RM, Schneider TD (1992) Features of spliceosome evolution and function inferred from an analysis of the information at human splice sites. *J. Mol. Biol.*, 228: 1124–1136. http://alum.mit.edu/www/toms/paper/splice/

See also Sequence Conservation

SCORING MATRIX (SUBSTITUTION MATRIX)

Teresa K. Attwood

A table of pairwise values encoding relationships between either nucleotides or amino acid residues, used to calculate alignment scores in sequence comparison methods. In the simplest type of scoring matrix (a unitary or identity matrix), the pairwise value for identities (*e.g.*, guanine pairing with guanine, or leucine pairing with leucine) is 1, while the value given for non-identical pairs (say, guanine with adenine, or leucine with valine) is 0. The unitary matrix for nucleotides is illustrated below:

	A	**C**	**T**	**G**
A	1	0	0	0
C	0	1	0	0
T	0	0	1	0
G	0	0	0	1

The unitary matrix is sparse, meaning that most of its elements are 0. Its diagnostic power is therefore relatively poor, because all identical matches carry equal weight. To improve diagnostic performance, the aim is to enhance the scoring potential of weak, but biologically significant signals without also amplifying noise. To this end, scoring matrices have been derived for protein sequence comparisons that, for example, weight matches between non-identical residues according to observed substitution rates across large evolutionary distances. The most commonly used series are the BLOSUM matrices, based on blocks of aligned sequences in the BLOCKS database, and the Dayhoff amino acid substitution matrices, based on the concept of the Point Accepted Mutation (PAM)–the widely used log-odds matrix for 250 PAMs is illustrated below. Here, the greater the value for an amino acid pair, the more likely it would be expected to occur in related sequences than random chance would predict (*e.g.*, tyrosine pairing with phenylalanine scores +7); conversely, the smaller the value, the less likely the pair would be expected to occur in related sequences than chance would predict (*e.g.*, glycine pairing with tryptophan scores −7).

	C	S	T	P	A	G	N	D	E	Q	H	R	K	M	I	L	V	F	Y	W
C	**12**																			
S	0	**2**																		
T	-2	1	**3**																	
P	-3	1	0	**6**																
A	-2	1	1	1	**2**															
G	-3	1	0	-1	1	**5**														
N	-4	1	0	-1	0	0	**2**													
D	-5	0	0	-1	0	1	2	**4**												
E	-5	0	0	-1	0	0	1	3	**4**											
Q	-5	-1	-1	0	0	-1	1	2	2	**4**										
H	-3	-1	-1	0	-1	-2	2	1	1	3	**6**									
R	-4	0	-1	0	-2	-3	0	-1	-1	1	2	**6**								
K	-5	0	0	-1	-1	-2	1	0	0	1	0	3	**5**							
M	-5	-2	-1	-2	-1	-3	-2	-3	-2	-1	-2	0	0	**6**						
I	-2	-1	0	-2	-1	-3	-2	-2	-2	-2	-2	-2	-2	2	**5**					
L	-6	-3	-2	-3	-2	-4	-3	-4	-3	-2	-2	-3	-3	4	2	**6**				
V	-2	-1	0	-1	0	-1	-2	-2	-2	-2	-2	-2	-2	2	4	2	**4**			
F	-4	-3	-3	-5	-4	-5	-4	-6	-5	-5	-2	-4	-5	0	1	2	-1	**9**		
Y	0	-3	-3	-5	-3	-5	-2	-4	-4	-4	0	-4	-4	-2	-1	-1	-2	7	**10**	
W	-8	-2	-5	-6	-6	-7	-4	-7	-7	-5	-3	2	-3	-4	-5	-2	-6	0	0	**17**

S

When using scoring matrices, it should be appreciated that they are inherently noisy, because they indiscriminately weight relationships that may be inappropriate in the context of a particular sequence comparison–in other words, scores of random matches are boosted along with those of weak biological signals. The significance of matches derived from searches using scoring matrices should therefore be evaluated with this in mind; this is especially important for matches that lie in the Twilight Zone.

Further reading

Dayhoff MO, Schwartz RM, Orcutt BC (1978) A model of evolutionary change in proteins. In: *Atlas of Protein Sequence and Structure*, Vol.5, Suppl. 3, Dayhoff MO (ed.) NBRF, Washington, DC, pp.341–352.

Henikoff S, Henikoff JG (1992) Amino acid substitution matrices from protein blocks. *Proc. Natl. Acad. Sci. USA* 89(22): 10911–10919.

Schwartz RM, Dayhoff MO (1978) Matrices for detecting distant relationships. In: *Atlas of Protein Sequence and Structure*, Vol.5, Suppl. 3, Dayhoff MO (ed.), NBRF, Washington, DC, pp.353–358.

See also Alignment, multiple, Substitution Process, Dayhoff Amino Acid Substitution Matrix, BLOSUM Matrix

SCWRL

Roland Dunbrack

A program for predicting side-chain conformations given a backbone model and sequence, used primarily in comparative modeling. SCWRL uses a backbone-dependent rotamer library to build the most likely conformation for each side-chain for the backbone conformation. Steric conflicts with the backbone and between side-chains are removed with a combination of dead-end elimination and branch-and-bound algorithms. SCWRL allows the user to specify conserved side-chains whose Cartesian coordinates can be preserved from the input structure. SCWRL also allows the user to provide a second input file of atomic coordinates for ligands (small molecules, ions, DNA, other proteins) that can act as a background against which the predicted side-chains must fit without significant steric conflicts.

Related website

SCWRL4	http://dunbrack.fccc.edu/scwrl4/

Further reading

Bower MJ, et al (1997) Prediction of protein side-chain rotamers from a backbone-dependent rotamer library: a new homology modeling tool. *J. Mol. Biol.* 267: 1261–1282.

Dunbrack RL Jr, Cohen FE (1997) Bayesian statistical analysis of protein sidechain rotamer preferences. *Prot. Science* 6: 1661–1681.

Dunbrack RL Jr (1999) Comparative modeling of CASP3 targets using PSI-BLAST and SCWRL. *Proteins Suppl.* 3: 81–87.

See also Comparative Modeling, Rotamer Library

SDF

Bissan Al-Lazikani

Structure Data Format (SDF) file is a chemical file format to store multiple chemical structures. Originally developed by MDL (now part of Accelrys), it concatenates MOL files into a single file, separated by '$$$$'. It is a useful and lossless format to store large virtual libraries.

See also Mol file format, InChI, SMILES

SEARCH BY SIGNAL, *SEE* SEQUENCE MOTIFS: PREDICTION AND MODELING.

SECOND LAW OF THERMODYNAMICS

Thomas D. Schneider

The Second Law of Thermodynamics is the principle that the disorder of an isolated system (entropy) increases to a maximum.

The Second Law appears in many surprisingly distinct forms. Transformations between these forms were described by Jaynes (1988). The relevant form for molecular information theory is

$$E_{\min} = K_b T \ln(2) = -q/R \text{ (joules per bit)},$$

where K_b is Boltzmann's constant, T is the absolute temperature, ln(2) is a constant that converts to bits, -q is the heat dissipated away from the molecular machine and R is the information gained by the molecular machine (Schneider 1991).

Supposedly Maxwell's demon violates the Second Law, but if we approach the question from the viewpoint of modern molecular biology, the puzzles go away (Schneider, 1994).

Related websites

Probability Theory As Extended Logic	http://bayes.wustl.edu/
A light-hearted introduction to the 'Mother of all Murphy's Laws' which clarifies an issue that is often incorrectly presented in discussions about entropy	http://www.secondlaw.com http://jchemed.chem.wisc.edu/Journal/Issues/Current/abs1385.html
An Equation for the Second Law of Thermodynamics	http://alum.mit.edu/www/toms/paper/secondlaw/html/index.html
Information Is Not Entropy, Information Is Not Uncertainty!	http://alum.mit.edu/www/toms/information.is.not.uncertainty.html
Rock Candy: An Example of the Second Law of Thermodynamics	http://alum.mit.edu/www/toms/rockcandy.html

Further reading

Jaynes ET (1988) The evolution of Carnot's principle. In: GJ Erickson, CR Smith, editors, *Maximum-Entropy and Bayesian Methods in Science and Engineering, volume 1*, pages 261–281, Dordrecht, The Netherlands. Kluwer Academic Publishers. Lambert FL (1999) Shuffled cards, messy desks, and disorderly dorm rooms–examples of entropy increase? Nonsense! *J. Chem. Educ.* 76: 1381–1387.

Schneider TD (1991) Theory of molecular machines. II. Energy dissipation from molecular machines. *J. Theor. Biol*, 148:125–137. http://alum.mit.edu/www/toms/paper/edmm/

Schneider TD (1994) Sequence logos, machine/channel capacity, Maxwell's demon, and molecular computers: a review of the theory of molecular machines. *Nanotechnology*, 5: 1–18. http://alum.mit.edu/www/toms/paper/nano2/

See also Molecular Efficiency

SECONDARY STRUCTURE OF PROTEIN

Roman Laskowski and Tjaart de Beer

Describes the local conformation of a protein's backbone. The secondary structure can be regular or irregular. The regular conformations are stabilized by hydrogen bonds between main chain atoms. The most common of these conformations are the alpha-helix and beta-sheet first proposed by Linus Pauling in 1951. Others include the 3_{10}, or collagen helix, and various type of turn. Regions of backbone that have an irregular conformation are said to have random coil, or simply coil, conformation.

The regular conformations are characterized by specific combinations of the backbone torsion angles, *phi* and *psi* and map onto clearly defined regions of the Ramachandran plot. In secondary structure prediction the different types are usually simplified into three categories: helix, sheet and coil.

Various computer algorithms exist for automatically computing the secondary structure of each part of a protein, giving the 3D coordinates of the protein structure. The best known is the DSSP algorithm of Kabsch and Sander.

Related website

DSSP	http://mrs.cmbi.ru.nl/hsspsoap/

Further reading

Kabsch W, Sander C. (1983) Dictionary of protein secondary structure: pattern recognition of hydrogen-bonded and geometrical features. *Biopolymers*, 22: 2571–2637.

See also Backbone, Hydrogen Bonds, Main Chain, Alpha-Helix, Beta-Sheet, Dihedral Angle, Ramachandran Plot.

SECONDARY STRUCTURE PREDICTION OF PROTEIN

Patrick Aloy

Protein *secondary structure prediction* is the determination of the regions of secondary structure in a protein (e.g. α-helix, β-strand, coil) from its amino acid sequence.

The first automated methods for protein secondary structure prediction appeared during the seventies and were mainly based on residue preferences derived from very limited databases of known 3D structures. They reached three-state-per-residue accuracies of about 50%.

During the 1980s a second generation of prediction algorithms appeared, the main improvements coming from the use of segment statistics (putting single residues in context) and larger databases. These methods achieved three-state-per-residue accuracies slightly higher than 60%.

A third generation of secondary structure prediction methods emerged during the 1990s. They combined evolutionary information extracted from multiple sequence alignments with 'intelligent' algorithms, such as neural networks, to predict at 75% accuracy.

Analyses of 3D structures of close homologues and NMR experiments suggest an upper limit for secondary structure prediction accuracy of 80%. Recent studies have estimated the accuracy of current methods to be 76 ± 10%, so we might have already reached this upper limit and only more entries in structural databases can improve the accuracies of the current methods.

Related websites

PredictProtein server	http://cubic.bioc.columbia.edu/predictprotein/
Jpred Server	http://www.compbio.dundee.ac.uk/~www-jpred/submit.html
A review of methods	http://cmgm.stanford.edu/WWW/www_predict.html

Further reading

Rost B, Sander C (2000) Third generation prediction of secondary structures. *Methods Mol. Biol.* 143: 71–95.

Zvelebil MJ, Barton GJ, Taylor WR, Sternberg MJE (1987) Prediction of protein secondary structure and active site using the alignment of homologous sequences. *J. Mol. Biol.*, 195: 951–967.

See also Web-based Prediction Programs

SECRETOME

Dov Greenbaum

The secretome is the population of protein products that are secreted from the cell.

First used in reference to the secreted proteins of the *Bacillus subtilis*. The analysis of the secretome covers the protein transport system within the cell and specifically the secretory pathways. The secretome also represents a subsection of the proteome/translatome, i.e. those proteins that are exported.

The secretome can be measured experimentally, via biochemical identification of all secreted proteins. High throughput proteomic methods such as two-dimensional electrophoresis to separate the proteins in the extracellular environment of the organism followed by mass spectrometry identification of each of the individual proteins are often used. The secretome can also be determined computationally via dedicated algorithms designed to determine which protein sequences signal the cell to excrete that specific protein. Presently computational methods suffer from the inability to detect lipoproteins and those proteins which, although they contain no recognizable secretion signals, are nevertheless secreted from the cell.

In particular, the exploration of secretomes is of commercial use with regard to the exploitation of so-called industrial bacteria such as *Bacillus subtilis* that are used to produce, through their secretory pathways industrially important products such as synthetics and drugs. Furthermore the secretome provides an easily and biochemically well described population which is informative with regard to inter-cellular interaction both within and between organisms.

Further reading

Antelmann H, et al. (2001) A proteomic view on genome-based signal peptide predictions. *Genome Res* 11: 1484–1502.

Tjalsma H, et al. (2000) Signal peptide-dependent protein transport in Bacillus subtilis: a genome-based survey of the secretome. *Microbiol Mol Biol Rev* 64: 515–547.

See also Proteome, Translatome

SEGMENTAL DUPLICATION

Dov Greenbaum

Segmental duplications (duplicons) are segments of DNA with near-identical DNA sequences, i.e., nucleotide sequence similarity of at least 90% and at least 1kb in length cover around 5% of the genome at issue.

In some examples, segmental duplications may give rise to low copy repeats (LCRs). Segmental duplications may also play a role in creating new primate genes; thus genomic duplications allow for functional divergence of genes and may contribute to the general complexity of genomes.

In humans, chromosomes Y and 22 have been found to have the greatest proportion of segmental duplications: 50.4% of Y and 11.9% on chromosome 22.

Segmented duplications have been found principally through two methods: (1) whole genome assembly comparison (WGAC) and (2) whole genome shotgun sequence detection (WSSD).

Segmental duplications are thought to arise through mechanisms such as non-allelic homologous recombination (NAHR) and non-homologous end joining (NHEJ). Some specific sites in the human genome have been shown to be hot spots (e.g., predisposition sites) for the occurrence of non-allelic homologous recombination or unequal crossing-over or and non-homologous end joining leading in some instances to duplications.

Related website

Human Genome Segmental Duplication Database	http://humanparalogy.gs.washington.edu/

S

Further reading

Kahn CL, Mozes S, Raphael BJ (2010) Efficient algorithms for analyzing segmental duplications with deletions and inversions in genomes. *Algorithms Mol Biol*. 4;5(1): 1.

Marques-Bonet T, Girirajan S, Eichler EE (2009) The origins and impact of primate segmental duplications. *Trends Genet*. 25(10): 441–454.

Marques-Bonet T, Eichler EE (2009) The evolution of human segmental duplications and the core duplicon hypothesis. *Cold Spring Harb Symp Quant Biol*. 74: 351–62

SEGREGATION ANALYSIS

Mark McCarthy and Steven Wiltshire

Analysis of the distribution and segregation of phenotypes within pedigrees designed to characterize the genetic architecture of the trait concerned.

In its simplest form, segregation analysis of Mendelian traits in simple nuclear pedigrees amounts to nothing more than comparing the observed segregation ratio of affected/unaffected offspring arising from a particular parental mating type with that expected under different possible disease models (dominant, recessive, co-dominant and sex-linked). For large pedigrees, and traits displaying incomplete penetrance, computer-implemented likelihood ratio based methods are used. The parameters of the genetic model that need to be specified and/or estimated from the data include putative disease allele frequency, penetrance probabilities, ascertainment correction and transmission probabilities. The more challenging problem of segregation analysis of complex traits–both qualitative and quantitative–usually proceeds by one of two likelihood-ratio based approaches. Mixed model segregation analysis explicitly models the quantitative trait in terms of a major gene effect, polygenic effect and environmental deviation. For discrete traits, a continuous underlying liability function is specified, and the model includes penetrance probabilities rather than genotypic means. Regressive models instead express the quantitative trait in terms of the effects of a segregating major gene together with a residual familial correlation, achieved by regressing each offspring on its predecessors in the pedigree. For both methods, hypothesis testing proceeds by comparing model likelihoods to identify the most parsimonious, statistically significant model. The value of segregation analysis methods for the dissection of multifactorial trait architecture is limited by two main factors: first, difficulty in describing the appropriate correction to account for the ascertainment scheme by which the families were collected; and second, data sets are rarely large enough to allow discrimination between the large number of alternative models involved in any realistic description of the architecture of a multifactorial trait (which is likely to involve several interacting susceptibility loci).

Examples: An et al. (2000) used mixed model segregation analysis to detect the presence of a recessive major gene influencing baseline resting diastolic blood pressure as part of the HERITAGE study. Thein et al. (1994) used regressive model segregation analysis to detect the presence of a dominant or codominant major gene influencing hereditary persistence of foetal haemoglobin. Rice et al. (1999) used mixed model segregation analysis to examine body mass index in Swedish pedigrees: though they found evidence for transmission for a major gene effect, transmission probabilities were not consistent with Mendelian segregation.

S

Related websites

Programs for performing mixed model or regressive model segregation analysis include:

PAP	http://hasstedt.genetics.utah.edu/
POINTER	http://cedar.genetics.soton.ac.uk/pub/PROGRAMS/pointer
SAGE	http://darwin.cwru.edu/pub/sage.html

Further reading

An P, et al. (2000) Complex segregation analysis of blood pressure and heart rate measured before and after a 20-week endurance exercise training program: the HERITAGE Family Study. *Am J Hypertens* 13: 481–497.

Ott J (1990) Cutting a Gordian knot in the linkage analysis of complex human traits. *Am J Hum Genet* 46: 211–221.

Rice T, et al. (1999) Segregation analysis of body mass index in a large sample selected for obesity: the Swedish Obese Subjects study. *Obes Res* 7: 241–255.

Sham P (1998) *Statistics in Human Genetics*. London: Arnold, pp 11–50: 251–261.

Thein SL, et al. (1994) Detection of a major gene for heterocellular hereditary persistence of fetal hemoglobin after accounting for genetic modifiers. *Am J Hum Genet* 54: 211–228.

Weiss KM (1995) *Genetic Variation and Human Disease*. Cambridge: Cambridge University Press, pp 51–116.

See also Mendelian Disease, Multifactorial Trait, QuantitativeTrait, Variance Components

SELECTED REACTION MONITORING (SRM)

Simon Hubbard

This is a mass spectrometry based technique where a peptide ion is fragmented in the mass spectrometer to generate selected product ions, chosen by the researcher, to increase sensitivity. The resulting extracted ion chromatogram corresponds solely to signal derived from ions which have a given precursor ion mass and yield specific product ion transitions (the 'selected reaction'). This mass spectrometry technique is well suited to quadrupole based instruments such as a triple quad, and is often used in quantitation experiments in proteomics using proteotypic peptides. Databases of high quality selected reactions are available, and are considered to be reproducible for use by other researchers on other instrument platforms. SRM and MRM are often used interchangeably where Multiple (M) replaces Selected (S).

Related websites

SRMatlas–collection of reactions for peptides across a range of species	http://www.srmatlas.org/
MRMaid–design tool for selecting SRMs for use in proteomics experiments	http://www.mrmaid.info/

Further reading

Picotti P, Bodenmiller B, Mueller LN, Domon B, Aebersold R. Full dynamic range proteome analysis of S. cerevisiae by targeted proteomics. *Cell*. 2009, 138: 791–806.

Picotti P, Lam H, Campbell D, Deutsch EW, Mirzaei H, Ranish J, Domon B, Aebersold R. A database of mass spectrometric assays for the yeast proteome. *Nat Methods*. 2008, 5: 911–4.

SELENOPROTEIN

Enrique Blanco and Josep F. Abril

Selenoproteins are proteins that incorporate the amino acid selenocysteine, a cysteine analog in which a selenium atom is found in place of sulfur.

During selenoprotein synthesis, the UGA codon is recoded into a selenocysteine through the influence of SECIS (the selenocysteine insertion sequence), an mRNA secondary/tertiary structure found in the 3′-untranslated region of genes.

Selenoproteins are essential for viability and play a key role in fundamental biological processes like development, reproduction, immune function, ageing, cancer, viral infections and cardiovascular disorders. Most selenoproteins act as oxidoreductases that prevent damage to cellular components, repair this damage, regulate redox signaling state of proteins or have other antioxidant functions.

Canonical gene identification approaches obtain modest success in the prediction of selenoproteins mostly due to the dual function of the UGA codon. However, more sophisticated pipelines of protein homology-based methods, SECIS structural evaluation and *in silico* gene predictors able to deal with in-frame TGA codons have shown to be very effective for this purpose. In fact, whole genome analysis in multiple species have reveled the existence of a selenoproteome conserved throughout evolution, being recognized up to 25 selenoproteins in humans.

Further reading

Brown KM, Arthur JR (2001) Selenium, selenoproteins and human health: a review. *Public Health Nutr*. 4: 591–599.

Castellano S, Gladyshev VN, Guigó R, Berry MJ (2008) Seleno DB 1.0: a database of selenoprotein genes, proteins and SECIS elements. *Nucl. Acids Res*. 36: D332–D338.

Driscoll DM, Chavatte L (2004) Finding needles in a haystack. *In silico* identification of eukaryotic selenoprotein genes. *EMBO reports* 5: 141–141.

Kryukov GV, Castellano S, Novoselov SV, Lobanov AV, Zehtab O, Guigó R, Gladyshev VN (2003) Characterization of mammalian selenoproteomes. *Science* 300: 1431–1443.

See also Gene Prediction, non-canonical

SELF-CONSISTENT MEAN FIELD ALGORITHM

Roland Dunbrack

An algorithm for protein side-chain placement developed by Patrice Koehl and Marc Delarue. Each rotamer of each side chain has a certain probability, $p(r_i)$. The total energy is a weighted sum of the interactions with the backbone and interactions of side chains with each other:

$$E_{tot} = \sum_{i=1}^{N} \sum_{ri=1}^{n_{rot}(i)} p(r_i)E_{bb}(r_i) + \sum_{i=1}^{N-1} \sum_{ri=1}^{n_{rot}(i)} \sum_{j=i+1}^{N} \sum_{rj=1}^{n_{rot}(j)} p(r_i)p(r_j)E_{SC}(r_i, r_j)$$

In this equation, $p(r_i)$ is the density or probability of rotamer r_i of residue i, $E_{bb}(r_i)$ is the energy of interaction of this rotamer with the backbone, and $E_{SC}(r_i,r_j)$ is the interaction energy (van der Waals, electrostatic) of rotamer r_i of residue i with rotamer r_j of residue j. Some initial probabilities are chosen for the p's, and the energies calculated. New probabilities $p'(r_i)$ can then be calculated with a Boltzmann distribution based on the energies of each side chain and the probabilities of the previous step:

$$E(r_i) = E_{bb}(r_i) + \sum_{j=1, j\neq i}^{N} \sum_{r_j=1}^{n_{rot}(j)} p(r_j)E_{SC}(r_i, r_j)$$

$$p(r_i) = \frac{\exp\left(\frac{-E(r_i)}{kT}\right)}{\sum_{r_i=1}^{n_{rot}(i)} \exp\left(\frac{-E(r_i)}{kT}\right)}$$

Alternating steps of new energies and new probabilities can be calculated from these expressions until the changes in probabilities and energies in each step become smaller than some tolerance.

Further reading

Koehl P, Delarue M (1996) Mean-field minimization methods for biological macromolecules. *Curr. Opin. Struct. Biol.* 6: 221–226.

See also Rotamer

SELF-ORGANIZING MAP (SOM, KOHONEN MAP)

John M. Hancock

Self-Organizing Maps are unsupervised machine learning algorithms which are designed to learn associations between groups of inputs.

The classical output of an SOM algorithm is a one- or two-dimensional representation. Because of this SOMs can also be considered as a means of visualizing complex data, as they effectively project a set of data onto a lower-dimension representation.

A SOM takes as input a set of sample vectors. It then carries out an iterative matching between these vectors and a set of weight vectors. The weight vectors carry values corresponding to the values in the sample vectors plus coordinates that represent their position on the map. After initializing the weight vectors (which may be carried out randomly or in some other way) the algorithm selects a sample vector and identifies the weight vector most similar to it. Once this best fit is identified, weight vectors adjacent to the winning weight vector are adjusted to make them more similar to the winning matrix. Over tens of thousands of iteration, this process should result in a map in which adjacent weight vectors link a cluster of sample vectors.

Related website

Introduction to SOMs	http://davis.wpi.edu/~matt/courses/soms/

Further reading

Wang J, et al. (2002) Clustering of the SOM easily reveals distinct gene expression patterns: results of a reanalysis of lymphoma study. *BMC Bioinformatics*. 3: 36.

See also Clustering

SEMANTIC NETWORK

Robert Stevens

Any representation of knowledge based on nodes representing concepts and arcs representing binary relations between those concepts. Early semantic networks were relatively informal and intended to indicate associations. Formalization led to Knowledge Representation Languages and Conceptual Graphs.

SEMI-GLOBAL ALIGNMENT, *SEE* GLOBAL ALIGNMENT.

SEQ-GEN

Michael P. Cummings

A program that simulates nucleotide or amino acid sequences for a phylogenetic tree using Markov models.

The program allows for a choice of models and model parameter values. Written in ANSI C the program is available as source code and executable files for some platforms.

Related website

Seq-Gen web page	http://tree.bio.ed.ac.uk/software/seqgen/

S

Further reading

Rambaut A, Grassly NC (1997) Seq-Gen: An application for the Monte Carlo simulation of DNA sequence evolution along phylogenetic trees. *Comput. Appl. Biosci.* 13: 231–238.

See also PAML

SEQCOUNT, *SEE* ENCPRIME.

SEQUENCE ALIGNMENT

Jaap Heringa

A linear comparison of amino or nucleic acid sequences in which insertions are made in order to bring equivalent position in the sequences into the correct register. Alignments are the basis of sequence analysis methods, and are used to pinpoint the occurrence of conserved motifs.

See also Alignment – Multiple, Alignment – Pairwise, Dynamic Programming, Sequence Similarity

SEQUENCE ASSEMBLY

Rodger Staden

The process of arranging a set of overlapping sequence readings (or reads) into their correct order along the genome.

SEQUENCE COMPLEXITY (SEQUENCE SIMPLICITY)

Katheleen Gardiner

Information content of a DNA sequence related to repeat sequence content.

Complex sequences lack repeat motifs, and are also considered as unique sequence (one copy per genome) and as having high information content. A pool of all the coding regions of the human genome thus represents a much more complex sequence than a similar pool of all intronic sequences, which contain repetitive sequence such as Alus and L1s.

Originally measured by reassociation kinetics of hybridization, complexity or unique sequence content can now be identified from direct analysis of genomic DNA sequence.

Further reading

Cooper DN (1999) *Human Gene Evolution* Bios Scientific Publishers, pp 261–285.

Ridley M (1996) *Evolution* Blackwell Science, Inc pp 261–276.

Wagner RP, Maguire MP, Stallings RL (1993) *Chromosomes – a synthesis.* Wiley-Liss.

See also SIMPLE, Simple DNA Sequence, Low Complexity Region

SEQUENCE CONSERVATION, *SEE* CONSERVATION.

SEQUENCE DISTANCE MEASURES

Sudhir Kumar and Alan Filipski

A sequence distance measure is a numerical representation of the phylogenetic dissimilarity or divergence between two molecular (DNA or protein) sequences.

Sequence distance measures can be used for instance to calculate a pair-wise distance matrix that is provided as input to a neighbor-joining method.

The simplest example of a sequence distance measure is the *p*-distance, which is simply the fraction of sites that differ between two aligned DNA or protein sequences. A Poisson correction may be applied to account for multiple substitutions at a site. In this case the corrected distance is given by $d = -log(1\text{-}p)$.

More complex distance measures are available for DNA sequences. The Jukes-Cantor distance is a simple distance measure for DNA which corrects for multiple substitutions under the assumption that there are exactly 4 states (nucleotides) at any site. It also assumes that the probability of change from any nucleotide to another is the same and that all sites evolve at the same rate. The Kimura 2-parameter distance is based on an extension of the Jukes-Cantor model to allow for different rates for transitional and transversional substitutions.

Sequence distances may be calculated using all sites in a set of sequences or, for protein coding DNA sequences, using exclusively synonymous or non-synonymous sites, in which case synonymous or non-synonymous sequence distances are obtained. Non-synonymous distances can be obtained by codon-by-codon or protein sequence analysis. Synonymous distances can be estimated by using codon-by-codon approaches or by extracting neutral or fourfold degenerate sites and conducting site-by-site analysis.

Software

Felsenstein J (1993) *PHYLIP: Phylogenetic Inference Package*. Seattle, WA, University of Washington.

Kumar S, et al. (2001). MEGA2: molecular evolutionary genetics analysis software. *Bioinformatics* 17: 1241–1245.

Further reading

Felsenstein J (2004) Inferring Phylogenies. Sinauer Associates.

Johnson NL, Kotz S (1970) *Continuous Univariate Distributions*. New York, Hougton Mifflin.

Li W-H (1997) *Molecular Evolution*. Sunderland, Mass., Sinauer Associates.

Nei M, Kumar S (2000) *Molecular Evolution and Phylogenetics*. Oxford, Oxford University Press.

Yang Z (1996) Among-site rate variation and its impact on phylogenetic analyses. *Trends Ecol Evol* 11: 361–372.

See also Evolutionary Distance Measures

SEQUENCE LOGO

Thomas D. Schneider

A sequence logo is a graphic representation of an aligned set of binding sites (see Figure S.1). A logo displays the frequencies of bases at each position as the relative heights of letters, along with the degree of sequence conservation as the total height of a stack of letters, measured in bits of information. Subtle frequencies are not lost in the final product as they would be in a consensus sequence. The vertical scale is in bits, with a maximum of 2 bits possible at each position. Note that sequence logos are an average picture of a set of binding sites (which is why logos can have several letters in each stack) while sequence walkers are the individuals that make up that average (which is why walkers have only one letter per position). More examples are in the Sequence Logo Gallery in the discussion of binding site symmetry.

Related websites

Enologos	http://biodev.hgen.pitt.edu/cgi-bin/enologos/enologos.cgi
Sequence logos	http://alum.mit.edu/www/toms/logoprograms.html
Web based server	http://weblogo.berkeley.edu
Structure logos	http://www.cbs.dtu.dk/gorodkin/appl/slogo.html

Further reading

Blom N, et al. (1999) Sequence and structure-based prediction of eukaryotic protein phosphorylation sites. *J. Mol. Biol.*, 294: 1351–1362.

Gorodkin J, et al. (1997) Displaying the information contents of structural RNA alignments: the structure logos. *Comput Appl Biosci*, 13: 581–586.

Schneider TD (1996) Reading of DNA sequence logos: Prediction of major groove binding by information theory. *Meth. Enzym.*, 274: 441–455. http://alum.mit.edu/www/toms/paper/oxyr/

Schneider TD, Stephens RM (1990) Sequence logos: A new way to display consensus sequences. *Nucleic Acids Res.*, 18: 6091–6100. http://alum.mit.edu/www/toms/paper/logopaper/

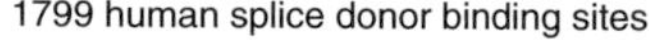

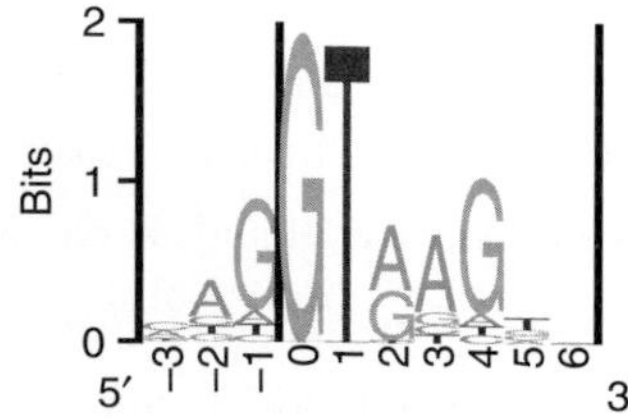

Figure S.1 Sequence logo of human splice donor sites (Schneider & Stephens, 1990). (See Colour plate S.1)

SEQUENCE MOTIF, *SEE* MOTIF.

SEQUENCE MOTIFS: PREDICTION AND MODELING (SEARCH BY SIGNAL)

Roderic Guigó

Approach to identifying the function of a sequence by searching for patterns within it.

Given an alphabet, (for instance that of nucleic acids A,C,G,T), a motif is an object denoting a set of sequences on this alphabet, either in a deterministic or probabilistic way. Given a sequence *S* and a (deterministic) motif *m*, we will say that the motif *m* occurs in S if any of the sequences denoted by *m* occurs in *S*. We will use here interchangeably the terms motif, pattern, signal, etc. although these terms may be used with different meaning (see Motif Searching and Pattern). The simplest motif is just one string or sequence in the alphabet, often a so-called exact word. Exact words may encapsulate biological functions, often when in the appropiate context. For instance, the sequence 'TAA' denotes, under the appropiate circumstances, a translation stop codon.

Often, however, biological functions are carried out by related, but not identical, sequences. Usually these sequences can be aligned. For instance the sequences below

```
CTAAAAATAA
TTAAAAATAA
TTTAAAATAA
CTATAAATAA
TTATAAATAA
CTTAAAATAG
TTTAAAATAG
```

are all known to bind the MEF2 (*Myocyte enhancer factor 2*) transcription factor (Yu et al., 1992, Fickett, 1996). Sets of functionally related aligned sequences can be described by consensus sequences. The simplest form of a consensus sequence is obtained by picking the most frequent base at each position in the set of aligned sequences. More information on the underlying sequences can be captured by extending the alphabet with additional symbols, that allow alternative possibilities that may occur at a given position to be denoted. For instance using the IUB (International Union of Biochemistry and Molecular Biology) nucleotide codes, the sequences above could be represented by the motif YTWWAAATAR, where Y=[CT], W=[AT] and R=[AG].

Regular expressions extend the alphabet further. For instance, the pattern C..?[STA]..C[STA][^P]C denotes the following amino acid sequences: Cysteine (C); any amino acid (.); any amino acid, that may or may not be present (.?); either Serine, Threonine or Alanine ([STA]); any amino acid (.); Cysteine (C); either Serine, Threonine or Alanine ([STA]); any amino acid but Proline ([^P]); and Cysteine (C). This pattern is the iron-sulfur binding region signature of the 2Fe-2S ferredoxin, from the PROSITE database (Falquet et al., 2002). In its early days PROSITE was a database mostly of regular expressions.

Information about the relative occurrence of each symbol at each position is lost in the motifs above. For instance in the alignment of MEF2 sites, both A and G are possible at the last position, but A appears in five sequences, while G appears only in two. This may reflect some underlying biological feature, for instance that the affinity of the binding is increased when adenosine instead of guanine appears at this position. We can capture explicitally this information by providing the relative frequency or probability of each symbol at each position in the alignment. These probabilities conform the so-called Position Weight Matrices (PWMs) or Position Specific Scoring Matrices (PSSMs). Below is the PSSM derived from a set of aligned canonical vertebrate donor sites (see DONOR SITES):

	-3	-2	-1	+1	+2	+3	+4	+5	+6
A	35.1	59.6	8.7	0.0	0.0	50.7	72.1	7.0	15.8
C	34.8	13.3	2.7	0.0	0.0	2.8	7.6	4.7	17.2
G	18.5	13.2	80.9	100.0	0.0	43.9	12.2	83.1	18.8
T	11.6	13.9	7.7	0.0	100.0	2.5	8.1	5.2	48.3
	C/A	A	G	G	T	A	A	G	T

This matrix, M, allows us to compute the probability of a given query under the hypothesis that it is a donor site. Indeed, the probability of sequence $s=s_1 \ldots s_l$ if s is a donor site, is $P(s) = \prod_{i=1}^{l} M_{s_i i}$, where M_{ij} is the probability of symbol i at position j. (In the case of the example, l=9). We can compute as well the background probability Q(s) of the sequence s. The logarithm of the ratio of these two probabilities, $R(s) = log\frac{P(s)}{Q(s)}$ is often used to score the query sequence. If the score is positive, the sequence occurs in donor sites more often than at random, while if the score is negative the sequence occurs less often in donor sites than randomly expected. The score *R(s)* can be easily computed if the individual coefficients of the matrix are themselves converted into log-likelihood ratios: $L_{ij} = log\frac{M_{ij}}{f_i}$ where f_i is the background probability of nucleotide i. Assuming equiprobability as the background probability of individual nucleotides, the frequency matrix above becames

	-3	-2	-1	+1	+2	+3	+4	+5	+6
A	0.34	87	-1.05	-inf	-inf	0.71	1.06	-1.27	-0.46
C	0.33	-0.63	-2.22	-inf	-inf	-2.17	-1.19	-1.68	-0.38
G	-0.30	-0.64	1.17	1.39	-inf	0.56	-0.72	1.20	-0.29
T	-0.77	-0.59	-1.18	-inf	1.39	-2.29	-1.13	-1.58	0.66
	C/A	A	G	G	T	A	A	G	T

Now, the score *R(s)* is simply $\sum_{i=1}^{l} L_{s_i i}$.

While a regular expression m denotes a subset of all sequences in the range of lengths l_1, l_2, a PSSM m is indeed a probability distribution over the set of all sequences of length l. Thus, while finding matches to regular expression m in sequence S is finding all subsequences of S of length $l_1 \leq l \leq l_2$ that belong to the set denoted by m, finding matches to the PSSM m is scoring each subsequence of S of length l according to matrix *m*; in practice, sliding a window of length *l* along the sequence. Usually, however, only subsequences scoring over a pre-defined threshold are considered matches to the motif described by the PSSM.

Sequence logos are useful graphical representations of PSSMs, while Sequence Walkers are useful to visually scan a sequence with a PSSM.

In PSSMs, adjacent positions are assumed to be independent. This is often unrealistic; for instance, in the case of the donor sites above. The frequencies of the donor site PSSM reflect the complementariety between the precursor RNA molecule at the donor site and the 5′ end of the U1 snRNP. The stability of this interaction is affected by the stacking energies, which depend on nearest neighbour arrengements along the nucleic acid sequences. Positions along the donor site sequence, thus, do not appear to be independent. If this is the case, the donor site motif would be better described by the conditional probability of each nucleotide at each position, depending on the nucleotide at the precedent position. Below are these conditional probabilities on the tree exon positions for the vertebrate donor sites above.

	position -3				position -2				position -1			
	A	C	G	T	A	C	G	T	A	C	G	T
A	29.2	31.9	25.5	13.4	62.4	9.5	15.2	12.9	7.0	1.7	86.2	5.1
C	48.6	32.5	6.2	12.7	69.2	11.6	6.4	12.8	19.1	7.1	55.2	18.5
G	38.8	36.2	17.7	7.3	62.6	15.8	12.3	9.3	12.3	2.4	79.1	6.2
T	16.4	41.3	29.5	12.9	17.7	25.6	29.5	27.2	2.9	3.3	84.4	9.4
	35.1	34.8	18.5	11.6	59.6	13.3	13.2	13.9	8.7	2.7	80.9	7.7

As it is possible to see, although the overall probability of G at position -3 is 0.18, this probability is 0.30 if the nucleotide at -4 is T, while it is only 0.06 if this nucleotide is C. These conditional (or transition) probabilities conform a first order non-homogeneous Markov Model. These models are called non-homogeneous because the transition probabilities may change along the sequence. Higher order Markov Models can also also be used (see Markov Models).

PSSMs and non-homogeneous Markov Models are particularly useful to model protein-protein and protein-nucleic acid interactions (Berg and von Hippel, 1987; Schneider, 1997), and thus have been used to describe a variety of elements: splice sites, translation initiation codons, promoter elements, amino acid motifs etc.

One limitation of Markov Models is that they can model only interactions between adjacent positions. In some cases, however, strong dependencies appear to exist between non-adjacent positions. Burge and Karlin (1997) introduced Maximal Dependence Decomposition, a decision tree like method to model distant dependencies in donor sites. These dependencies can be interpreted in terms of the thermodynamics of RNA duplex formation between U1 small nuclear RNA (snRNA) and the 5′ splice site region of the pre-mRNA. A number of methods have been further developed to infer all relevant dependencies between positions in a set of aligned sequences (see for instance Agarwal and Bafna, (1998), Cai et al., (2000)). More complex patterns, for instance including gaps or heterogeneous domains, can be modelled with more complex structures (see Neural Networks, Hidden Markov Models).

S

Related websites (splice site prediction)

Here we list a number of www severs for splice site prediction. Splice site motifs have been modeled using a wide variety of methods and they are, thus, a good case study. Interested readers can obtain information on the particular methods at the addresses below.

GENESPLICER	http://www.tigr.org/tdb/GeneSplicer/gene_spl.html
SPLICEPREDICTOR	http://bioinformatics.iastate.edu/cgi-bin/sp.cgi
BRAIN	http://metaponto.sci.unisannio.it:8080/brain/index.jsp
NETPLANTGENE	http://www.cbs.dtu.dk/services/NetPGene/
NETGENE	http://www.cbs.dtu.dk/services/NetGene2/
SPL	http://www.softberry.com
NNSPLICE	http://www.fruitfly.org/seq_tools/splice.html
GENIO/splice	http://genio.informatik.uni-tuttgart.de/GENIO/splice/
SPLICEVIEW	http://biogenio.com/splice/

Further reading

Agarwal P, Bafna V (1998) Detecting non-adjoining correlations within signals in DNA. In *RECOMB 98: Proceedings of the Second Annual International Conference on Computational Molecular Biology*.

Berg O, von Hippel P (1987) Selection of DNA binding sites by regulatory proteins, statistical-mechanical theory and application to operators and promoters. *J Mol Biol* 193: 721–750.

Bucher P, et al. (1994). A flexible search technique based on generalized profiles. *Comput Chem* 20: 1–24.

Burge CB, Karlin S (1997) Prediction of complete gene structures in human genomic DNA *J Mol Biol* 268: 71–94.

Burge CB (1998) Modeling dependencies in pre-mRNA splicing signals. In: Salzberg S, Searls D, Kasif S (eds.) *Computational Methods in Molecular Biology*, Elsevier Science, Amsterdam, pp. 121–163.

Cai D, et al. (2000). Modeling splice sites with bayes networks. *Bioinformatics*, 16: 151–158.

Falquet L, et al. (2002) The prosite database, its status in 2002. *Nucleic Acids Res.* 30: 231–238.

Schneider T (1997) Information content of individual genetic sequences. *J Theor Biol* 189: 421–441.

Yu Y, et al. (1992) Human myocyte-specific enhancer factor 2 comprises a group of tissue-restricted mads box transcription factors. *Genes Dev* 6: 1781–1798.

See also Pattern, Motif Search, Sequence Logos, Sequence Walkers, Neural Network, Markov Models, Hidden Markov Models, Splice Sites

SEQUENCE OF A PROTEIN

Roman Laskowski and Tjaart de Beer

The order of neighbouring amino acids in a protein, listed from the N- to the C-terminus. This is the primary structure of the protein. It is most simply represented by a string of letters, such as 'MTAPISALTYDREGC', where each letter codes for one of the 20 amino acids.

The sequence uniquely determines how the protein folds into its final three-dimensional conformation.

The sequence is relatively straightforward to determine experimentally, either directly using automated techniques, or by derivation, via the genetic code, from regions of DNA known to encode for the protein. Protein sequence databases nowadays contain the sequences of many hundreds of thousands of proteins from a wide variety of organisms.

Many databases exist that contain protein sequence data, such as UniProt and NCBI.

Related websites

UniProt	http://www.uniprot.org/
NCBI	http://www.ncbi.nlm.nih.gov/guide/proteins/

See also N-terminus, C-terminus, Amino Acid, Conformation

SEQUENCE PATTERN

Roman Laskowski, Tjaart de Beer and Thomas D. Schneider

A sequence pattern is defined by the nucleotide sequences of a set of aligned binding sites or by a common protein structure.

In contrast, consensus sequences, sequence logos and sequence walkers are only models of the patterns found experimentally or in nature. Models do not capture everything in nature. For example, there might be correlations between two different positions in a binding site (Stephens and Schneider, 1992; Gorodkin et al, 1997). A more sophisticated model might capture these but still not capture three-way correlations. It is impossible to make the more detailed model if there is not enough data.

Further reading

Gorodkin J, et al. (1997) Displaying the information contents of structural RNA alignments: the structure logos. *Comput. Appl. Biosci.*, 13: 581–586.

Stephens RM, Schneider TD (1992) Features of spliceosome evolution and function inferred from an analysis of the information at human splice sites. *J. Mol. Biol.*, 228: 1121–1136. http://alum.mit.edu/www/toms/paper/splice/

SEQUENCE READ ARCHIVE (SRA, SHORT READ ARCHIVE)

Obi L. Griffith and Malachi Griffith

The Sequence Read Archive (SRA) is the public repository of next-generation sequence data. It is jointly operated by the International Nucleotide Sequence Database Collaboration comprised of the NCBI (USA), ENA/EBI (Europe) and DDBJ (Japan). Increasing numbers of journals and funding agencies require that next-generation sequencing data be deposited into the SRA. A number of different file formats are or have been accepted for submission including SRA, SFF, SRF, native vendor formats, and fastq. It is expected that the BAM file format will replace SRF/SFF as the preferred submission format. The explosion in next-generation sequence data, at a rate exceeding the growth of hard drive storage, has posed serious challenges for the SRA. Storage of image and signal data has been limited and improved compression strategies are being investigated.

Related website

Sequence Read Archive	http://www.ncbi.nlm.nih.gov/sra

Further reading

Leinonen R et al; International Nucleotide Sequence Database Collaboration (2011) The sequence read archive. *Nucleic Acids Res.* 39(Database issue): D11–21.

SEQUENCE RETRIEVAL SYSTEM, *SEE* SRS.

SEQUENCE SIMILARITY

Jaap Heringa

Sequence similarity is a measure to assess the relatedness of protein or nucleotide sequences, normally by using a sequence alignment. Numerous studies into DNA or protein sequence relationships evaluate sequence alignments using a simple binary scheme of matched positions being identical or non-identical. Sequence identity is normally expressed as the percentage identical residues found in a given alignment, where normalization can be performed using the length of the alignment or that of the shorter sequence. The scheme is simple and does not rely on an amino acid exchange matrix. However, if two proteins are said to share a given percentage in sequence identity, this is based on a sequence alignment which will have been almost always constructed using an amino acid exchange matrix and gap penalty values, so that sequence identity cannot be regarded independent from sequence similarity.

Using sequence identity as a measure, Sander and Schneider (1991) estimated that if two protein sequences are longer than 80 residues, they could relatively safely be assumed to be homologous whenever their sequence identity is 25% or more. Despite its popularity, using sequence identity percentages is not optimal in homology searches (Abagyan and Batalov, 1997) and as a result, no major sequence analysis method employs sequence identity scores in deriving statistical significance estimates.

Sequence alignment methods are essentially pattern searching techniques, leading to an alignment with a similarity score even in case of absence of any biological relationship. In some database search engines, such as PSI-Blast (Altschul *et al.*, 1997), sequences are scanned for the presence of so-called low-complexity regions (Wooton and Federhen, 1996), which are then excluded from alignment. Although similarity scores of unrelated sequences are essentially random, they can behave like 'real' scores and, for example, like the latter are correlated with the length of the sequences compared. Particularly in the context of database searching, it is important to know what scores can be expected by chance and how scores that deviate from random expectation should be assessed. If within a rigid statistical framework a sequence similarity is deemed statistically significant, this provides confidence in deducing that the sequences involved are in fact biologically related.

As a result of the complexities of protein sequence evolution and distant relationships observed in nature, any statistical scheme will invariably lead to situations where a sequence is assessed as unrelated while it is in fact homologous (false negative), or the inverse, where a sequence is deemed homologous while it is in fact biologically unrelated (false positive). The derivation of a general statistical framework for evaluating the significance of sequence similarity scores has been a major task. However, a rigid framework has not been established for global alignment, and has only partly been completed for local alignment.

Further reading

Abagyan RA, Batalov S (1997) Do aligned sequences share the same fold? *J. Mol. Biol.* 273: 351–368.

Altschul SF, et al. (1997) Gapped BLAST and PSI-BLAST: a new generation of protein database search programs. *Nucl. Acids. Res.* 25: 3381–3402.

Sander C, Schneider R (1991) Database of homology derived protein structures and the structural meaning of sequence alignment. *Proteins Struct. Func. Evol.* 9: 51–68.

Wootton JC, Federhen S (1993) Statistics of local complexity in amino acid sequences and sequence databases. *Comput Chem* 17: 141–163.

See also Sequence Conservation

SEQUENCE SIMILARITY-BASED GENE PREDICTION, *SEE* GENE PREDICTION, HOMOLOGY-BASED.

SEQUENCE SIMILARITY SEARCH

Laszlo Patthy

Related genes, proteins, protein domains usually retain a significant degree of sequence similarity that can be measured by comparing their aligned sequences. Sequence similarity, however, does not necessarily reflect homology since it may also result from convergence or may simply occur by chance.

Various types of sequence similarity searches may be performed with a query sequence to identify genes or proteins that share sequence similarity with a significance that may imply common ancestry. The search attempts to align the query sequence with all sequences in the database and calculates the similarity scores. The search provides a list of database sequences that can be aligned with the query sequence, ranked in the order of decreasing similarity scores.

Sequence similarity searches may be performed with a single query sequence, with multiple alignments, patterns, motifs or position-specific scoring matrices.

S

Related websites

FASTA	http://fasta.bioch.virginia.edu/fasta/
MAST	http://meme.sdsc.edu/meme/website/mast.html
BLAST	http://www.ncbi.nlm.nih.gov/BLAST/
PROSITE	http://www.expasy.ch/prosite
INTERPRO	http://www.ebi.ac.uk/interpro
CDD/IMPALA	http://www.ncbi.nlm.nih.gov/Structure/cdd/cdd.shtml
PSI-BLAST	http://www.ncbi.nlm.nih.gov/BLAST/
PROFILESEARCH	ftp.sdsc.edu/pub/sdsc/biology
PFAM	http://www.sanger.ac.uk/Pfam
FASTA	http://www.ebi.ac.uk/fasta33/
SW	http://www.ebi.ac.uk/bic_sw/

See also Homology

SEQUENCE SIMPLICITY, *SEE* SEQUENCE COMPLEXITY.

SEQUENCE TAGGED SITE (STS)

John M. Hancock and Rodger Staden

A unique sequence-defined landmark in a genome which can be detected by a specific PCR reaction.

An STS is a short (200–500 bp) DNA sequence that occurs uniquely in a genome and whose location and base sequence are known. STSs are useful for localizing and orienting mapping and sequence data and serve as landmarks on the physical map of a genome.

SEQUENCE WALKER

Thomas D. Schneider

A sequence walker is a graphic representation of a single possible binding site, with the height of letters indicating how bases match the individual information weight matrix at each position. Bases that have positive values in the weight matrix are shown right-side up; bases that have negative values are shown upside down and below the 'horizon'. As in a sequence logo, the vertical scale is in bits; the maximum is 2 bits and the minimum is negative infinity. Bases that do not appear in the set of aligned sequences are shown negatively and in a black box. Bases that have negative values lower than can fit in the space available have a purple box. The zero coordinate is inside a rectangle which (in this

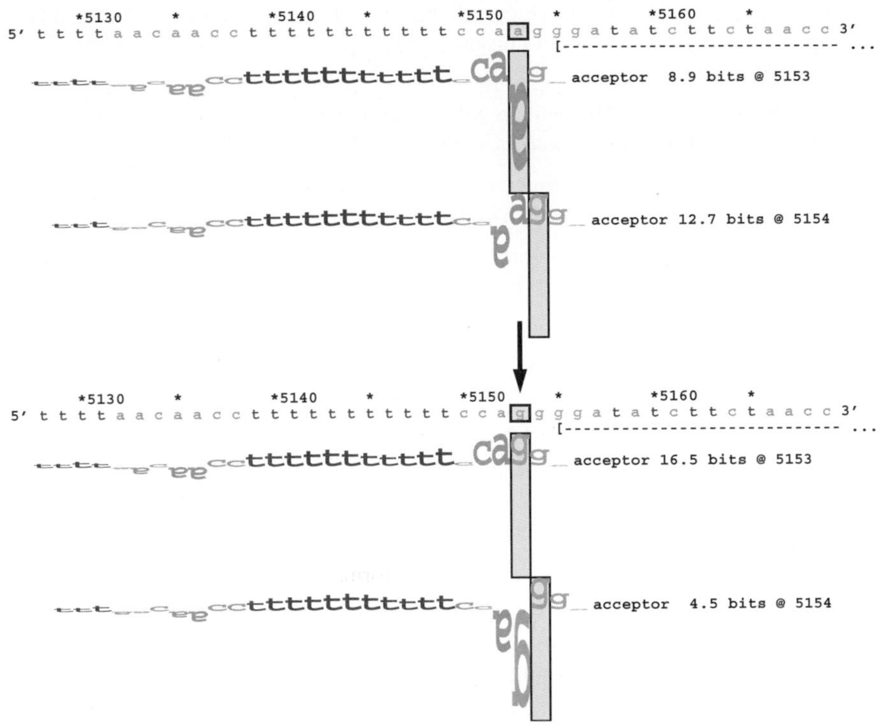

Figure S.2 Sequence walkers for human acceptor splice sites at intron 3 of the iduronidase synthetase gene (IDS, L35485). (See Colour plate S.2)

case) runs from −3 to +2 bits in height. If the background of the rectangle is light green, the sequence has been evaluated as a binding site, while if it is pink it is not a binding site. Figure S.2 shows sequence walkers for human acceptor splice sites at intron 3 of the iduronidase synthetase gene (IDS, L35485). An A to G mutation decreases the information content of the normal site while simultaneously increasing the information content of a cryptic site, leading to a genetic disease (Rogan, et al. 1998).

Related websites

Ri	http://alum.mit.edu/www/toms/paper/ri/
Walker	http://alum.mit.edu/www/toms/paper/walker/
Web page on walkers	http://alum.mit.edu/www/toms/walker/
Try sequence walkers yourself with the Delila Server	http://alum.mit.edu/www/toms/delilaserver.html

Further reading

Allikmets R, et al. (1998) Organization of the ABCR gene: analysis of promoter and splice junction sequences. *Gene*, 215: 111–122. http://alum.mit.edu/www/toms/paper/abcr/

Arnould I, et al. (2001) Identifying and characterizing a five-gene cluster of ATP-binding cassette transporters mapping to human chromosome 17q24: a new subgroup within the ABCA subfamily. *GeneScreen*, 1: 151–164.

Emmert S, et al. (2001) The human XPG gene: Gene architecture, alternative splicing and single nucleotide polymorphisms. *Nucleic Acids Res.*, 29: 1441–1452.

Hengen PN, et al. (1997) Information analysis of Fis binding sites. *Nucleic Acids Res.*, 25: 4991–5002. http://alum.mit.edu/www/toms/paper/fisinfo/

Kahn SG, et al. (1998) Xeroderma Pigmentosum Group C splice mutation associated with mutism and hypoglycinemia–A new syndrome? *J. Invest Dermatol*, 111: 791–796.

Rogan PK, et al. (1998) Information analysis of human splice site mutations. *Hum Mut*, 12: 151–171. http://alum.mit.edu/www/toms/paper/rfs/

Schneider TD (1997) Information content of individual genetic sequences. *J. Theor. Biol.*, 189: 421–441. http://alum.mit.edu/www/toms/paper/ri/

Schneider TD (1997) Sequence walkers: a graphical method to display how binding proteins interact with DNA or RNA sequences. *Nucleic Acids Res.*, 25: 4401–4415. http://alum.mit.edu/www/toms/paper/walker/

Schneider TD (1999) Measuring molecular information. *J. Theor. Biol.*, 201: 81–92. http://alum.mit.edu/www/toms/paper/ridebate/

Schneider TD, Rogan PK (1999) Computational analysis of nucleic acid information defines binding sites, United States Patent 5867402.

Shultzaberger RK, Schneider TD (1999) Using sequence logos and information analysis of Lrp DNA binding sites to investigate discrepancies between natural selection and SELEX. *Nucleic Acids Res.*, 27: 881–887. http://alum.mit.edu/www/toms/paper/lrp/

Shultzaberger RK, et al. (2001). Anatomy of *Escherichia coli* Ribosome Binding Sites. *J. Mol. Biol.*, 313: 211–228. http://alum.mit.edu/www/toms/paper/flexrbs/

Stephens RM, Schneider TD (1992) Features of spliceosome evolution and function inferred from an analysis of the information at human splice sites. *J. Mol. Biol.*, 228: 1121–1136. http://alum.mit.edu/www/toms/paper/splice/

Svojanovsky SR, et al. (2000) Redundant designations of BRCA1 intron 11 splicing mutation; c. 4211–2A>G; IVS11–2A>G; L78833, 37698, A>G. Hum Mut, 16: 264. http://www3.interscience.wiley.com/cgi-bin/abstract/73001161/START

Wood TI, et al. (1999). Interdependence of the position and orientation of SoxS binding sites in the transcriptional activation of the class I subset of *Escherichia coli* superoxide-inducible promoters. *Mol. Microbiol.*, 34: 411–430.

Zheng M, et al. (1999) OxyR and SoxRS regulation of fur. *J. Bacteriol*, 181: 4631–4643. http://alum.mit.edu/www/toms/paper/oxyrfur/

Zheng M, et al. (2001) Computation-Directed Identification of OxyR-DNA Binding Sites in *Escherichia coli*. *J. Bacteriol.*, 183: 4571–4579.

SERIAL ANALYSIS OF GENE EXPRESSION, *SEE* SAGE.

SGD (SACCHAROMYCES GENOME DATABASE)

Obi L. Griffith and Malachi Griffith

The Saccharomyces Genome Database project provides and maintains comprehensive information on the genome and molecular biology of the yeast *Saccharomyces cerevisiae*. It maintains and annotates the reference genome assemblies and the *S. cerevisiae* gene name registry. Data are accessed through the SGD website via a genome browser (based on GMOD/GBrowse). Other analysis tools include a built-in BLAST aligner, GO term finder, primer designer, restriction enzyme mapper, and more.

Related website

Saccharomyces Genome Database	http://www.yeastgenome.org/

Further reading

Skrzypek MS, Hirschman J (2011) Using the Saccharomyces Genome Database (SGD) for analysis of genomic information. *Curr Protoc Bioinformatics.* Chapter 1: Unit 1.20.1–23.

See also Model Organism Database

SHANNON ENTROPY (SHANNON UNCERTAINTY)

Thomas D. Schneider

Information theory measure of uncertainty.

The story goes that Shannon didn't know what to call his measure and so asked the famous mathematician von Neumman. Von Neumann said he should call it the entropy because nobody knows what that is and so Shannon would have the advantage in every debate (the story is paraphrased from Tribus and McIrvine, 1971.). This has led to much confusion in the literature because entropy has different units than uncertainty. It is the latter which is usually meant. Recommendation: when making computations from symbols always use the term uncertainty, with recommended units of bits per symbol. If referring to the entropy of a physical system, then use the term entropy, which has units of joules per kelvin (energy per temperature). It is also important to note that information is not entropy and that information is not uncertainty!

Related websites

Reference	http://alum.mit.edu/www/toms/information.is.not.uncertainty.html
Pitfalls	http://alum.mit.edu/www/toms/pitfalls.html

Further reading

Tribus M, McIrvine EC (1971). Energy and information. *Sci. Am.*, 225: 171–188. (Note: the table of contents in this volume incorrectly lists this as volume 224).

See also Molecular Efficiency

S

SHANNON SPHERE

Thomas D. Schneider

A sphere in a high dimensional space which represents either a single message of a communications system (after sphere) or the volume that contains all possible messages (before sphere) could be called a Shannon sphere, in honour of Claude Shannon who recognized its importance in information theory.

The radius of the smaller after spheres is determined by the ambient thermal noise, while that of the larger before sphere is determined by both the thermal noise and the signal power (signal-to-noise ratio), measured at the receiver. The log of the number of small spheres that can fit into the larger sphere determines the channel capacity (Shannon, 1949). The high-dimensional packing of the spheres is the coding of the system.

There are two ways to understand how the spheres come to be. Consider a digital message consisting of independent voltage pulses. The independent voltage values specify a point in a high dimensional space since independence is represented by coordinate axes set at right angles to each other. Thus three voltage pulses correspond to a point in a 3-dimensional space and 100 pulses correspond to a point in a 100-dimensional space. The first, 'non-Cartesian' way to understand the spheres is to note that thermal noise interferes with the initial message during transmission of the information such that the received point is dislocated from the initial point. Since noisy distortion can be in any direction, the set of all possible dislocations is a sphere. The second, 'Cartesian' method is to note that the sum of many small dislocations to each pulse, caused by thermal noise, gives a Gaussian distribution at the receiver. The probability that a received pulse is disturbed a distance from the initial voltage is of the form $p(x) \approx e^{-x^2}$. Disturbance of a second pulse will have the same form, $p(y) \approx e^{-y^2}$. Since these are independent, the probability of both distortions is multiplied $p(x,y) = p(x)p(y)$. Combining equations,

$$p(x, y) \approx e^{-(x^2+y^2)} = e^{-r^2},$$

where r is the radial distance. If $p(x,y)$ is a constant, the locus of all points enscribed by r is a circle. With more pulses the same argument holds, giving spheres in high dimensional space. Shannon used this construction in his channel capacity theorem.

For a molecular machine containing n atoms there can be as many as $3n$-6 independent components (degrees of freedom) so there can be $3n$-6 dimensions. The velocity of these components corresponds to the voltage in a communication system and they are disturbed by thermal noise. Thus the state of a molecular machine can also be described by a sphere in a high dimensional velocity space.

Further reading

Schneider TD (1991) Theory of molecular machines. I. Channel capacity of molecular machines. *J. Theor. Biol.*, 148: 81–123. http://alum.mit.edu/www/toms/paper/ccmm/

Schneider TD (1994). Sequence logos, machine/channel capacity, Maxwell's demon, and molecular computers: a review of the theory of molecular machines. *Nanotechnology*, 5: 1–18. http://alum.mit.edu/www/toms/paper/nano2/

See also Before Sphere, After Sphere, Gumball Machine, Molecular Efficiency

SHANNON UNCERTAINTY, *SEE* SHANNON ENTROPY.

SHORT-PERIOD INTERSPERSION, SHORT-TERM INTERSPERSION, *SEE* INTERSPERSED SEQUENCE.

SHORT READ ARCHIVE, *SEE* SEQUENCE READ ARCHIVE.

SHUFFLE TEST

David Jones

A method for estimating the significance of matches in fold recognition, though most commonly for threading methods.

One major problem in fold recognition is how to determine the statistical significance of matches. In the absence of any theoretical methods for calculating the significance of matches in a fold recognition search, empirical methods are typically used. The most commonly used approach is to make use of sequence shuffling, which is often used to assess the significance of global sequence alignments. The test is fairly straightforward to implement. The score for a test sequence threaded onto a particular template is first calculated. Then the test sequence is shuffled i.e. the order of the amino acids is randomised. This shuffling ensures that random sequences are generated with identical amino acid composition to the original protein. This shuffled sequence is then also threaded onto the given template and the energy for the shuffled sequence compared to the energy for the un-shuffled sequence. In practice a large number of shuffled threadings are carried out (typically 100–1000) and the mean and standard deviation of the energies calculated. The unshuffled threading score can then be transformed into a Z-score by subtracting the mean and then dividing by the standard deviation of the shuffled scores. As a rule of thumb, Z-scores of 4 or more correspond to significant matches.

See also Threading

SIDE CHAIN

Roman Laskowski and Tjaart de Beer

In a peptide or protein chain, all atoms other than the backbone (or main chain) atoms are termed side chain atoms. For most of the amino acids, the side chain atoms spring off from the backbone Cα the exception being glycine, which has no side chain (other than a single hydrogen atom), and proline whose side chain links back onto its main chain nitrogen.

In proteins, interactions between side chains help determine the structure and stability of the folded state. Hydrophobic-hydrophobic contacts are instrumental in forming the folded protein's core. Side chains on the surface of the protein play a crucial role in recognition of, and interaction with, other molecules (*i.e* DNA, other proteins, metabolites, metals, and so on) and for performing the protein's biological function (for example, the side chains that carry out the biochemical reactions in enzymes–one example being the catalytic triad).

Related website

Atlas of Protein Side-Chain Interactions	http://www.biochem.ucl.ac.uk/bsm/sidechains

Further reading

Singh J, Thornton JM (1992) *Atlas of Protein Side-Chain Interactions*, Vols. I & II, IRL Press, Oxford.

See also Backbone, Main Chain, Side Chain

SIDE-CHAIN PREDICTION

Roland Dunbrack

A step in comparative modeling of protein structures in which side-chains for the residues of the target sequence are built onto the backbone of the template or parent structure. For residues that are conserved in the target-parent alignment, the Cartesian coordinates of the parent side-chain can be preserved in the model. But residues that are different in the two sequences require new side-chain coordinates in the model.

Most side-chain prediction methods are based on choosing conformations from rotamer libraries and an optimization scheme to choose conformations with low energy, as defined by some potential energy function. Side-chains are usually built with standard bond lengths and bond angles, and dihedral angles are obtained from the rotamer library used. However, Cartesian coordinate rotamer libraries can also be used to account for side-chain flexibility. Alternatively, flexibility around rotamer dihedral angles can improve results in some cases.

Related website

SCWRL	http://dunbrack.fccc.edu/scwrl4/

Further reading

Bower MJ, et al. (1997) Prediction of protein side-chain rotamers from a backbone-dependent rotamer library: a new homology modeling tool. *J. Mol. Biol.* 267: 1261–1282.

Holm L, Sander C (1991) Database algorithm for generating protein backbone and sidechain coordinates from a Cα trace: Application to model building and detection of coordinate errors. *J. Mol. Biol.* 218: 181–194.

Liang S, Grishin NV (2002) Side-chain modeling with an optimized scoring function. *Protein Sci* 11: 321–331.

Mendes J. et al. (2001). Incorporating knowledge-based biases into an energy-based side-chain modeling method: application to comparative modeling of protein structure. *Biopolymers* 59: 71–86.

See also Target, Template, Rotamer Libraries

S

SIGNAL-TO-NOISE RATIO, *SEE* NOISE.

SIGNATURE, *SEE* FINGERPRINT.

SILAC, *SEE* STABLE ISOTOPE LABELLING WITH AMINO ACIDS IN CELL CULTURE.

SILENT MUTATION, *SEE* SYNONYMOUS MUTATION.

SIMILARITY INDEX, *SEE* DISTANCE MATRIX.

SIMPLE (SIMPLE34)

John M. Hancock

Program for detecting simple sequence regions in DNA and protein sequences.

SIMPLE was originally developed to analyse DNA sequences for potential substrates for DNA slippage that were not obviously tandemly repeated. The program detects both tandem repeats and these less obvious 'cryptically simple' sequences by a heuristic algorithm that counts the frequency of repeats of the central motif of a 65 bp window and compares it to frequencies observed in random simulated copies of that sequence with the same length and base or (latterly) dinucleotide composition. It thus attempts to distinguish repetition that is beyond the chance expectation for the composition of a given test sequence. More recent versions of the program identify repeated motifs and apply the algorithm to protein sequences.

Related website

SIMPLE server	http://www.biochem.ucl.ac.uk/bsm/SIMPLE/index.html

Further reading

Albà MM, et al. (2002) Detecting cryptically simple protein sequences using the SIMPLE algorithm. *Bioinformatics* 18: 671–688.

Hancock JM, Armstrong JS (1994) SIMPLE34: an improved and enhanced implementation for VAX and SUN computers of the SIMPLE algorithm for analysis of clustered repetitive motifs in nucleotide sequences. *Comput. Appl. Biosci.* 10: 61–70.

Tautz D, et al. (1986) Cryptic simplicity in DNA is a major source of genetic variation. *Nature* 322: 651–656.

See also Low Complexity Region, Simple DNA Sequence

S

SIMPLE DNA SEQUENCE (SIMPLE REPEAT, SIMPLE SEQUENCE REPEAT)

John M. Hancock and Katheleen Gardiner

DNA sequence composed of repeated identical or highly similar short motifs.

Simple sequences may be polymorphic in length. The term encompasses both minisatellites and microsatellites and may also include non-tandem arrangements of motifs. Sequences containing overrepresentations of short motifs that are not tandemly arranged are known as cryptically simple sequences. Simple repeats are widely dispersed in the human genome and account for at least 0.5% of genomic DNA.

Further reading

Cooper DN (1999) *Human Gene Evolution*. Bios Scientific Publishers, pp 261–285.

Ridley M (1996) *Evolution*. Blackwell Science, Inc pp 261–276.

Rowold DJ, Herrera RJ (2000) Alu elements and the human genome Genetics 108: 51–72.

Tautz D, Trick M, Dover GA (1986) Cryptic simplicity in DNA is a major source of genetic variation. *Nature* 322: 651–656.

See also SIMPLE, Microsatellite, Minisatellite

SIMPLE REPEAT, *SEE* SIMPLE DNA SEQUENCE.

SIMPLE SEQUENCE REPEAT, *SEE* SIMPLE DNA SEQUENCE.

SIMPLE34, *SEE* SIMPLE.

SIMULATED ANNEALING

Roland Dunbrack

A Monte Carlo simulation used to find the global minimum energy of a system, in which the temperature is slowly decreased in steps, with the system equilibrated at each temperature step. The name is derived from the analogous process of heating and slowly cooling a substance to produce a crystal. If the cooling is done too quickly, the substance will freeze into a disordered glass-like state. But if cooling is slow, the substance will crystallize.

Further reading

Kirkpatrick S, et al. (1983) Optimization by simulated annealing. *Science* 220: 671–680.

See also Monte Carlo Simulation, Energy Minimization

SIMULTANEOUS ALIGNMENT AND TREE BUILDING

Aidan Budd and Alexandros Stamatakis

The process of trying to simultaneously infer/compute an alignment as well as an evolutionary tree for a set of unaligned molecular sequences.

Computational methods and models have been proposed for parsimony and likelihood-based methods. Maximum Likelihood methods can deploy a simulated annealing search technique while Bayesian alignment methods extend the MCMC search mechanisms by another variable (the alignment) over which they integrate/sample.

Software

Wheeler W (2006) Dynamic homology and phylogenetic systematics: a unified approach using POY. American Museum of Natural History New York.

Fleissner R, Metzler D, Von Haeseler A. (2005) Simultaneous statistical multiple alignment and phylogeny reconstruction. *Systematic Biology* 54(4): 541–561.

Suchard MA, Redelings BD (2006) BAli-Phy: simultaneous Bayesian inference of alignment and phylogeny. *Bioinformatics* 22(16): 2041–2048.

Further reading

Lunter G, Miklos I, Drummond A, Jensen J, Hein J (2005) Bayesian coestimation of phylogeny and sequence alignment. *BMC Bioinformatics* 6(1): 83.

Thorne JL, Kishino H, Felsenstein J (1992) Inching toward reality: an improved likelihood model of sequence evolution. *Journal of Molecular Evolution* 34(1): 1–16.

SINGLE NUCLEOTIDE POLYMORPHISM (SNP)

Dov Greenbaum

SNPs or Single Nucleotide Polymorphisms are DNA variations, i.e. differences in a single base within a genomic sequence, within a population., i.e., when a single nucleotide (A,T,C,or G) in the genome sequence is altered.

There is 99.9 % identity between most humans; the remaining variation can be partly attributed to SNPs. In other words, SNPs that are responsible for over 80–90% of the variation between two individuals, they are ideal for the task of hunting for correlations between genotype and phenotype.

They are evolutionarily stable and occur frequently in both coding and non-coding regions–approximately every 100 to 300 bases within the human genome. Studies have shown that the occurrence of SNPs is lowest in exons (i.e. coding regions) and higher in repeats, introns and pseudogenes. The majority of these polymorphisms have a cytosine replaced by a thymine.

These variations do not, for the most part, produce phenotypic differences. SNPs are important in biomedical research as they can be used as markers for disease-associated mutations and they allow population geneticists to follow the evolution of populations. Additionally these minimal variations within the genome may have significant effects to an organism's response to environment, bacteria, viruses, toxins, disease and drugs.

To mine the human genome for SNPs, the SNP consortium, a unique association of both commercial and academic institutions, was set up in 1999. SNPs can be found through random sequencing using consistent and uniform methods.

SNP analysis has become popular throughout the life sciences and is now used throughout the life sciences including molecular diagnostics, agriculture, food testing, identity testing, pathogen identification, drug discovery and development, and pharmacogenomics.

Related websites

HAPMAP	http://snp.cshl.org/
NCBI SNP Page	http://www.ncbi.nlm.nih.gov/SNP/

Further reading

Gray IC, Campbell DA, Spurr NK (2000) Single nucleotide polymorphisms as tools in human genetics *Hum. Mol. Genet.* 9: 2401–2408.

Kwok P-Y, Gu Z (1999) Single nucleotide polymorphism libraries: why and how are we building them? *Mol. Med. Today* 5: 531–543.

SIPPL TEST, *SEE* UNGAPPED THREADING TEST B.

SISTER GROUP

Aidan Budd and Alexandros Stamatakis

In a rooted bifurcating tree, any internal node represents an ancestor of two subtrees. These two subtrees are sometimes referred to as 'sister groups' of each other, i.e., subtree A is the sister group of subtree B (and accordingly subtree B is the sister group of subtree A). As this definition depends on knowing the direction of genetic transmission along the branches of the tree (i.e., which nodes/branches are ancestral and which are descendant), it is only possible to identify sister groups in the context of a rooted phylogenetic tree; in unrooted trees the direction of transmission of genetic information is not specified, so it is not possible to identify which of the subtrees linked to an internal node are ancestral and which are descendant. To overcome this problem, the concept of adjacent groups was developed for referring to subtrees linked to the same internal node in unrooted trees.

SITE, *SEE* CHARACTER.

SITES

Michael P. Cummings

A program for the comparative analysis of polymorphism in DNA sequence data.

SITES is typically applied to sets of similar sequences (i.e., multiple alleles of the same locus within a species). The program provides summary statistics for synonymous sites, replacement sites, non-coding sites, codons, transitions, transversions, GC content and other features. The program also estimates several population genetic parameters (θ [$=4N\mu$], π, γ, $4Nc$, population size, linkage disequilibrium), and provides hypothesis tests for departures from neutral theory expectations.

The program can be used in an interactive fashion whereby the program provides prompts and lists of choices. Alternatively the program can be invoked with command line options to perform analyses. Written in ANSI C, the program is available as source code and as executables for several platforms.

Related website

SITES home page	http://astro.temple.edu/~tuf29449/software/software.htm

Further reading

Hey J, Wakeley J (1997) A coalescent estimator of the population recombination rate. *Genetics* 145: 831–846.

Wakeley J, Hey J (1997) Estimating ancestral population parameters. *Genetics* 145: 841–855.

SMALL SAMPLE CORRECTION

Thomas D. Schneider

A correction to the Shannon uncertainty measure to account for the effects of small sample sizes.

Related website

The Small Sample Correction for Uncertainty and Information Measures	http://alum.mit.edu/www/toms/small.sample.correction.html

Further reading

Basharin GP (1959) On a statistical estimate for the entropy of a sequence of independent random variables. *Theory Probability Appl.*, 4: 331–336.

Miller GA (1955) Note on the bias of information estimates. In H. Quastler, editor, *Information Theory in Psychology*, pp 91–100, Glencoe, IL. Free Press.

Schneider TD, et al. (1986) Information content of binding sites on nucleotide sequences. *J. Mol. Biol.*, 188: 411–431. http://alum.mit.edu/www/toms/paper/schneider1986/

S

SMILES
Bissan Al-Lazikani

Simplified Molecular Input Line Entry System (SMILES) is a proprietary ASCII line representation of molecules and reactions from Daylight Chemical Information Systems Inc. It is a simple representation that is useful for rapid entry into chemical sketchers or database searching. However, it is not a lossless format. The SMILES string cannot always be unambiguously converted back to the original structure. However, it is a popular format as it is lighter and more human readable than InChIs.

benzene	c1ccccc1
L-Proline	C1C[C@@H](NC1)C(=O)O

(lower case atom symbols represent aromatic atoms)

Further reading

Details of SMILES can be found here:

http://www.daylight.com/dayhtml/doc/theory/theory.smiles.html

See also InChI, InChI Key, Mol File Format

SMITH-WATERMAN
Jaap Heringa

Inventors of a technique for local alignment based upon the dynamic programming algorithm (Smith and Waterman, 1981). The Smith-Waterman technique is a minor modification of the global dynamic programming algorithm (Needleman and Wunsch, 1970). The algorithm selects and aligns similar segments of two sequences and needs a residue exchange matrix with negative values and appropriate gap penalty values; if a non-negative exchange matrix is used, the algorithm will mostly function as a global alignment method.

Further reading

Needleman SB, Wunsch CD (1970) A general method applicable to the search for similarities in the amino acid sequence of two proteins. *J. Mol. Biol.* 48: 441–453.

Smith TF, Waterman MS (1981) Identification of common molecular subsequences. *J. Mol. Biol.* 147: 191–197.

See also Local Alignment, Dynamic Programming, Global Alignment

SNP, *SEE* SINGLE NUCLEOTIDE POLYMORPHISM.

S

SOFTWARE SUITES FOR REGULATORY SEQUENCES

Jacques van Helden

Many software tools have been developed for the analysis of regulatory sequences, which are generally oriented towards a specific task: motif discovery, motif enrichment, pattern matching, CRM prediction, phylogenetic footprints, motif comparison, analysis of microarray clusters end such. A few websites provide a more generic environment enabling to apply a variety of algorithms to treat these diverse questions.

Related websites

Database	Description
MEME	The MEME suite. Includes motif discovery, pattern matching, CRM detection, functional enrichment, motif comparison. http://meme.nbcr.net/meme/
RSAT	Regulatory Sequence Analysis Tools. Includes motif discovery, pattern matching, CRM detection, motif comparison, random generators for sequences, motifs and sites, distributions of matrix scores. http://www.rsat.eu/

SOLANACEAE GENOMICS NETWORK (SGN)

Dan Bolser

The SGN is a clade-oriented database for plant genomes in the Euasterid clade, including the families *Solanaceae* (tomato, potato, eggplant, pepper and petunia) and *Rubiaceae* (coffee). The database contains comparative genetic, genomic, phenotypic, and taxonomic information for these genomes. The database is community curated, meaning that anyone can contribute annotations to genes and phenotypes.

Related websites

SGN	http://solgenomics.net/
MetaBase	http://metadatabase.org/wiki/SGN

Further reading

Bombarely A, Menda N, Tecle IY, et al. (2011) The Sol Genomics Network (solgenomics.net): growing tomatoes using *Perl*. *Nucleic Acids Res*. 39: 1141–55.

See also Ensembl Plants, Gramene MaizeGDB, PlantsDB

S

SOLVATION FREE ENERGY

Roland Dunbrack

The free energy change involved in moving a molecule from vacuum to solvent. The solute is assumed to be in a fixed conformation, and the free energy change includes terms for interaction of the solute with the solvent (electrostatic and dispersion forces including polarization of the solvent and solute), averaged over time, and changes in solvent-solvent interactions. The contribution of hydrophobicity to solvation free energy is usually obtained from terms that are proportional to the exposed surface area of solute atoms.

Further reading

Lazaridis T, Karplus M (1999) Effective energy function for proteins in solution. *Proteins* 35: 131–152.

Wesson L, and Eisenberg D (1992) Atomic solvation parameters applied to molecular dynamics of proteins in solution. *Prot. Science* 1: 221–235.

SOV

Patrick Aloy

Segment overlap (SOV) is a measure for the evaluation of secondary structure prediction methods based on secondary structure segments rather than individual residues.

SOV is a more stringent measure of prediction accuracy than Q_3 and discriminates reasonably well between good and mediocre or bad predictions, as it penalizes wrongly predicted segments.

For a single conformational state:

$$S_{ov}(i) = 100 * \frac{1}{N(i)} \sum_{S(i)} \left[\frac{minov(S_1, S_2) + \delta(S_1, S_2)}{maxov(S_1, S_2)} * len(S_1) \right]$$

with the normalization value N(i) defined as:

$$N(i) = \sum_{S(i)} len(S_1) + \sum_{S'(i)} len(S_1)$$

S_1 and S_2 correspond to the secondary structure segments being compared. S(i) is the set of all the overlapping pairs of segments (S_1,S_2). S'(i) is the set of all segments S_1 for which there is no overlapping segment S_2 in state i. len(S_1) is the number of residues in the segment S_1, minov(S_1,S_2) is the length of the overlap between S_1 and S_2 for which both segments have residues in state i and maxov(S_1,S_2) is the total extent for which either of the segments S_1 or S_2 has a residue in state i. δ(S_1,S_2) is defined as:

$$\delta(S_1, S_2) = \min\{(maxov(S_1, S_2) - minov(S_1, S_2)); minov(S_1, S_2); int(len(S_1)/2); int(len(S_2)/2)\}$$

where min{$x1$,$x2$; ... ;xn} is the minimum of n integers.

For all three states:

$$S_{ov} = 100 * \left[\frac{1}{N} \sum_{i \in (H,E,C)} \sum_{S(i)} \frac{minov(S_1, S_2) + \delta(S_1, S_2)}{maxov(S_1, S_2)} * len(S_1) \right]$$

where the normalization value N is the sum of N(i) over all three conformational states

$$N = \sum_{i \in \{H,E,C\}} N(i)$$

The normalization ensures that all S_{ov} values are within the 1–100 range and thus can be used as percentage accuracy and compared to other measures, such as Q3.

Related website

Accuracy estimators	http://predictioncenter.llnl.gov/local/local.html

Further reading

Zemla A, Venclovas C, Fidelis K, Rost B (1999) SOV, a segment-based measure for protein secondary structure prediction assessment. *Proteins* 34: 221–223.

See also Secondary Structure Prediction, Qindex

SPACE-FILLING MODEL

Eric Martz

A space-filling model (or surface model) depicts the molecular volume into which other atoms cannot move, barring conformational or chemical changes. Space-filling models are three-dimensional depictions of molecular structures in which atoms are rendered as spheres with radii proportional to van der Waals radii, or less frequently, covalent, or ionic radii. (See illustration at models, molecular.)

Space-filling models originated with physical models designed to represent steric constraints–the atoms in the model cannot come unrealistically close together, occupying the same space. In the 1950s, Corey and Pauling used hard wood and plastic models at a scale of one inch or 0.5 inches per Ångstrom for their work on the folding of the polypeptide chain. In the 1960's, a committee of the US National Institutes of Health improved the design of these models, especially through a connector designed by Walter L. Koltun, increasing their suitability for macromolecules. Commercial production of these Corey-Pauling-Koltun, or CPK, models was implemented by the American Society of Biological Chemists, with financial support from the US National Science Foundation. CPK models are associated with a commonly used colour scheme for the elements (see models, molecular).

Most atomic coordinate files for proteins lack hydrogen atoms. In these cases, some visualization programs (e.g. RasMol and its derivatives, see visualization) increase the spherical radii of carbon, oxygen, nitrogen and sulfur atoms to 'united atom' radii approximating the sums of the volumes of these atoms with their bonded hydrogens. Because of the ease and speed with which they can produce accurate space-filling models, computer visualization programs have supplanted the construction of physical space-filling models for most purposes.

Related websites

Physical Molecular Models	http://www.netsci.org/Science/Compchem/feature14b.html
Radii used in space-filling models of RasMol and its derivatives	http://www.umass.edu/microbio/rasmol/rasbonds.htm

Further reading

Corey RB, Pauling L (1953) Molecular models of amino acids, peptides and proteins. *Rev. Sci. Instr.* 24: 621–627.

Koltun WL (1965) Precision space-filling atomic models. *Biopolymers* 3: 661–679.

Platt JR (1960) The Need for Better Macromolecular Models. *Science* 131: 1301–1310.

Yankeelov JA Jr, Coggins, JR (1971) Construction of space-filling models of proteins using dihedral angles. *Cold Spring Harbor Symposia on Quantitative Biology* 36: 581–587.

See also Models, molecular, Surface Models, Visualization, Molecular

Eric Martz is grateful for help from Eric Francoeur, Peter Murray-Rust, Byron Rubin and Henry Rzepa.

SPARQL (SPARQL PROTOCOL AND RDF QUERY LANGUAGE)

John M. Hancock

SPARQL is the semantic web equivalent of SQL and is the standard query language for RDF triple stores. It enables remote querying of data stored in RDF format.

SPARQL 1.0 became an official W3C Recommendation in January 2008.

Related website

SPARQL	http://www.w3.org/TR/rdf-sparql-query/

SPLICED ALIGNMENT

Roderic Guigó

Given a genomic sequence, the spliced alignment algorithms finds a legal chain of exons predicted along this sequence with the best fit to a related target protein, or cDNA.

The program PROCRUSTES (Gelfand et al., 1996) introduced the spliced alignment concept. In PROCRUSTES the set of candidate exons is constructed by selection of all blocks between candidate acceptor and donor sites and further gentle statistical filtration. PROCRUSTES considers all possible chains of candidate exons in this set and finds a chain with the maximum global similarity to a given target protein sequence. In the Las-Vegas version (Sze and Pevzner, 1997), PROCRUSTES generates a set of suboptimal spliced alignments and uses it to assess the confidence level for the complete predicted gene or individual exons. Other examples of spliced alignment algorithms are GENEWISE (Birney and Durbin, 1997) for DNA/protein alignment and SIM4 (Florea et al., 1998) and EST_GENOME (Mott, 1997) for DNA/cDNA alignment.

Most spliced alignment algorithms use dynamic programming to obtain the optimal chain of exons in the query sequence. Therefore, running them in database search mode to identify potential genes in large anonymous genomic sequences may become computationally prohibitive.

Spliced algorithms are quite accurate in recovering the correct set of exons when the target protein is very similar to the protein encoded in the genomic sequence, but the accuracy drops as the similarity between query and target sequences decreases (Guigó et al., 2000).

Related websites

SIM4 documentation	http://globin.cse.psu.edu/html/docs/sim4.html
EST_GENOME Home Page	http://www.well.ox.ac.uk/~rmott/ESTGENOME/est_genome.shtml

Further reading

Birney E, Durbin R (1997) Dynamite: a flexible code generating language for dynamic programming methods used in sequence comparison. *Ismb*, 5: 51–64.

Florea L, et al. (1998) A computer program for aligning a cDNA sequence with a genomic DNA sequence. *Genome Res.* 8: 961–974.

Gelfand MS, et al. (1996) Gene recognition via spliced alignment. *Proc Natl Acad Sci USA* 93: 9061–9066.

Guigó R, et al. (2000) Sequence similarity based gene prediction. In Suhai S, editor, *Genomics and Proteomics: Functional and Computational Aspects*, pages 91–105. Kluwer Academic / Plenum Publishing.

Mott R (1997) EST_GENOME: a program to align spliced DNA sequences to unspliced genomic DNA. *Comput Appl Biosci* 13: 471–478.

Sze SH, Pevzner P (1997) Las vegas algorithms for gene recognition: suboptimal and error-tolerant spliced alignment. *J. Comput. Biol.* 4: 291–309.

See also Gene Prediction homology-based, GeneWise, Gene Prediction, accuracy, Dynamic Programming

S

SPLICING (RNA SPLICING)

John M. Hancock

The process of removal of introns from RNAs to produce the mature RNA molecule.

The term is commonly used to describe the removal of nuclear introns from the primary transcript to produce the mature mRNA but it may also be applied to other types of RNA processing.

Further reading

Lewin B (2000) *Genes VII*. Oxford University Press.

See also Intron

SPLIT (BIPARTITION)

Aidan Budd and Alexandros Stamatakis

When a branch is removed from a phylogenetic tree to yield two unconnected subtrees, each of these two subtrees contains a subset of the operational taxonomic units (OTUs) present in the initial tree. These two sets (i.e., the sets of OTUs present in the two subtrees) are non-overlapping, and the union of the two sets is the same as the set of OTUs in the initial tree. Such a pair of sets of OTUs is described as a split or bipartition. All terminal branches of a tree describe a 'trivial' or 'non-informative' bipartition (i.e., one set contains the OTU associated with that terminal branch, the other set contains all other OTUs). In contrast, the bipartitions specified by internal branches are described as 'non-trivial' or 'informative'. The list of all the informative bipartitions specified by a tree is sufficient to reconstruct the topology of the tree and represents an equivalent mathematical representation of a tree.

See also Consensus Tree, Distances Between Trees

SPOTTED CDNA MICROARRAY

Stuart Brown

Microarray constructed by spotting DNA onto glass microscope slides.

The microarray technique is an extension of the pioneering work done by Southern on the hybridization of labeled probes to nucleic acids attached to a filter. This technology evolved into filter-based screening of clone libraries (dot blots), but a key innovation by Pat Brown, Mark Schena, and colleagues was to use automated pipetting robots to spot tens of thousands of DNA probes onto glass microscope slides. The probes were derived from PCR amplification of inserts from plasmids containing known cDNA sequences. After developing methods for printing and hybridizing microarrays, Brown and colleagues published detailed plans for constructing array spotters on the Internet as the 'Mguide.'

These 'spotted arrays' generally use full-length cDNA clones as probes and hybridize each array with a mixture of two cDNA samples (an experimental and a control) labeled with two different fluorescent dyes. The hybridized chip is then scanned at the two fluorescent wavelengths and ratios of signal intensities are calculated for each gene. By using full-length cDNA probes, this system has the advantage of high sensitivity and the ability to detect virtually all alternatively spliced transcripts. Its disadvantages include the potential for cross-hybridization from transcripts of genes with similar sequences (members of multi-gene families) and the logistical difficulty of maintaining large collections of cDNA clones and doing large scale PCR amplifications

Related website

Pat Brown's MGuide to cDNA microarrays	http://cmgm.stanford.edu/pbrown/mguide/index.html

Further reading

Phimister B (1999) Going global. *Nat. Genet. Suppl.* 21: 1.

Schena, M et al. (1995) Quantitative monitoring of gene expression patterns with a complementary DNA microarray. *Science* 270: 461–470.

Southern EM (1975) Detection of specific sequences among DNA fragments separated by gel electrophoresis. *J. Mol. Biol.* 98: 501–517.

SQL (STRUCTURED QUERY LANGUAGE)

Matthew He

SQL is a standard language for accessing databases. It is designed for managing data in relational database management systems (RDBMS). Many database products support SQL with proprietary extensions to the standard language. Queries take the form of a command language that lets you select, insert, update, find out the location of data, and so forth.

SRA, *SEE* SEQUENCE READ ARCHIVE.

SRS (SEQUENCE RETRIEVAL SYSTEM)

Guenter Stoesser

The Sequence Retrieval System (SRS) has become an integration system for both data retrieval and applications for data analysis. It provides capabilities to search multiple databases by shared attributes and to query across databases. The SRS server at the EBI integrates and links a comprehensive collection of specialized databanks along with the main nucleotide and protein databases. This server contains more than 130 biological databases and integrates more than 10 data analysis applications. Complex querying and linking across all available databanks can be executed.

Related website

Sequence Retrieval System (SRS)	http://srs.ebi.ac.uk/.

S

Further reading

Etzold T, et al. (1996) SRS: information retrieval system for molecular biology data banks. *Methods Enzymol.* 266: 114–128.

Zdobnov EM, et al. (2002) The EBI SRS server—recent developments. *Bioinformatics* 18: 361–373.

See also Nucleic Acid Sequence Databases, Protein Databases

STABLE ISOTOPE LABELLING WITH AMINO ACIDS IN CELL CULTURE (SILAC)

Simon Hubbard

A labelling technique used in quantitative proteomic experiments that incorporates stable isotope labelled amino acids metabolically into intracellular proteins, so that newly synthesized proteins become 'heavy' or 'light' labelled. Labelling is usually done for several rounds of cell division and/or uses auxotrophic strains lacking the ability to synthesize a given amino acid, in order to achieve very high levels of label incorporation. Ideally, this should be 100%. Cell populations can then be mixed in equal amounts and quantitation is performed using mass spectrometry, calculating changes in protein expression from differences in the signal between heavy and light labelled peptides from the same protein. Software tools such as MaxQuant and the TPP are able to process the mass spectrometry data.

See also Quantitative proteomics, MaxQuant

STADEN

John M. Hancock and M.J. Bishop

Package of software for the treatment of sequencing data and for sequence analysis.

The STADEN package is one of the longest established sequence analysis packages and has gone through a series of revisions. Its functions can be classified under the following general headings: sequence trace and reading file manipulation, sequence assembly (including sequencing trace viewing and preparation and contaminant screening); mutation detection; and sequence analysis. The sequence analysis tool *spin* carries out a wide variety of analyses including base and dinucleotide composition, codon usage, finding open reading frames, mapping restriction sites, searching for subsequences or motifs and gene finding using a number of statistics. It also provides sequence comparison functions.

Related website

STADEN home page	http://staden.sourceforge.net/

Further reading

Staden R, et al. (1998) The Staden Package, 1998. *Computer Methods in Molecular Biology, vol. 132: Bioinformatics Methods and Protocols* Misener S, Krawetz SA (eds). The Humana Press Inc., Totowa, NJ, pp. 111–130.

See also Open Reading Frame, Sequence Assembly, Chimeric DNA, DNA Sequencing Trace

STANDARD GENETIC CODE, *SEE* GENETIC CODE.

STANDARDIZED QUALITATIVE DYNAMICAL SYSTEMS

Luis Mendoza

Boolean models are the simplest implementation of regulatory networks as dynamical systems. While Boolean models approximate a continuous sigmoid with a discontinuous step function, the standardized qualitative dynamical systems (SQUAD) method does the opposite by approximating a Boolean network with the use of a set of ordinary differential equations. In this method, variables representing the state of activation are normalized, so that they are constrained in the range [0,1]. This feature enables a direct comparison of the attractors obtained with a continuous model against the attractors of a purely binary model. The equations used in the SQUAD method are constructed with the rate of change of activation (or synthesis, or transcription) as a sigmoid function of the state of activation of the input variables, minus a decay rate. Importantly, the sigmoid function has a standard form; this is, it is not network- specific. Because of this characteristic, it is possible to construct the dynamical system from the topological information of a regulatory network in a fully automated way. This can be achieved because the SQUAD method was developed as a qualitative modeling tool, and hence there is no need to incorporate stoichiometric or rate constants. Moreover, by using default values for certain parameters in the equations, like the steepness of the sigmoid response curve, it is possible to perform a completely automated dynamical analysis of the network just by providing as input the topology of a network (Mendoza and Xenarios, 2006; Di Cara et al. , 2007). Given that this qualitative method was recently developed, it has been used to model only a small number of biological networks (Mendoza and Xenarios, 2006; Sánchez-Corrales et al. , 2010; Mendoza and Pardo, 2010; Sankar et al. , 2011). The standardized qualitative dynamical systems methodology has been useLd for the development of a software package for the automated analysis of signaling networks (http://www.enfin.org/squad).

Further reading

Di Cara A, Garg A, De Micheli G, Xenarios I, Mendoza L (2007) Dynamic simulation of regulatory networks using SQUAD. *BMC Bioinformatics* 8: 462.

Mendoza L Pardo F (2010) A robust model to describe the differentiation of T-helper cells. *Theory in Biosciences* 129: 283–293.

Mendoza L, Xenarios I (2006). A method for the generation of standardized qualitative dynamical systems of regulatory networks. *Theoretical Biology and Medical Modelling* 3: 13.

Sánchez-Corrales YE, Álvarez-Buylla ER, Mendoza L (2010) The Arabidopsis thaliana flower organ specification gene regulatory network determines a robust differentiation process. *Journal of Theoretical Biology* 264: 971–983.

Sankar M, Osmont KS, Rolcik J, et al. (2011) A qualitative continuous model of cellular auxin and brassinosteroid signaling and their crosstalk. *Bioinformatics* 27: 1404–1412.

S

STANFORD HIV RT AND PROTEASE SEQUENCE DATABASE (HIV RT AND PROTEASE SEQUENCE DATABASE)

Guenter Stoesser, Malachi Griffith, Obi L. Griffith

The HIV RT and Protease Sequence Database is a curated database containing a compilation of most published HIV RT and protease sequences, including submissions from International Collaboration databases (DDBJ/EMBL/GenBank), and sequences published in journal articles. The database catalogs evolutionary and drug-related sequence variation in HIV reverse transcriptase (RT) and protease enzymes, which are the molecular targets of antiretroviral drugs. Sequence changes in the genes coding for these enzymes are directly responsible for phenotypic resistance to RT and protease inhibitors. The database project constitutes a major resource for studying evolutionary and drug-related variation in the molecular targets of anti-HIV therapy. The website divides content into three categories: genotype-treatment correlations, genotype-phenotype correlations, and genotype-clinical outcome correlations. The HIV RT and Protease Sequence Database is maintained at the Stanford University Medical Center.

Related website

Stanford HIV RT and Protease Sequence Database homepage	http://hivdb.stanford.edu/

Further reading

Gifford RJ, et al. (2009) The calibrated population resistance tool: standardized genotypic estimation of transmitted HIV-1 drug resistance. *Bioinformatics*. 25(9): 1191–8.

Liu TF, Shafer RW (2006) Web Resources for HIV type 1 Genotypic-Resistance Test Interpretation. *Clin Infect Dis*. 42(11): 1601–18.

Rhee SY et al; International Non Subtype B HIV-1 Working Group (2006) HIV-1 pol mutation frequency by subtype and treatment experience: extension of the HIVseq program to seven non-B subtypes. *AIDS*. 20(5): 641–651

Shafer RW (2006) Rationale and Uses of a Public HIV Drug-Resistance Database. *J. Infect Dis*. 194(Suppl 1): S51–8.

START CODON, *SEE* GENETIC CODE.

STATISTICAL MECHANICS

Patrick Aloy

Statistical mechanics is the branch of physics in which statistical methods are applied to the microscopic constituents of a system in order to predict its macroscopic properties.

The earliest application of this method was Boltzmann's attempt to explain the thermodynamic properties of gases on the basis of the statistical properties of large assemblies of molecules. Since then, statistical mechanics theories have been applied to a variety of biological problems with great success. For example, statistical mechanics theories applied to protein folding rely on a series of major simplifications about the energy as a function of conformation and/or simplifications of the representation of the polypeptide chain, such as one point per residue on a cubic lattice, that allow us to obtain a tractable model of a very complex system.

STATISTICAL POTENTIAL ENERGY

Roland Dunbrack

Any potential energy function derived from statistical analysis of experimentally determine structures. Such energy functions are usually of the form $E(x) = -k\log(p(x))$ where x is some conformational degree of freedom and k is some constant. When $p(x) = 0$, E is technically infinity, but often some large non-infinite value is arbitrarily chosen. The value of k chosen is often k_BT where k_B is the Boltzmann constant and T is the temperature in degrees Kelvin. However, there is no theoretical justification for this choice, since the ensemble of structural features used to derive the statistical energy function is not a statistical mechanical ensemble, such as the canonical ensemble. A value of k can be chosen by plotting energy differences from the statistical analysis, as $\log(p(x_1))$-$\log(p(x_2))$, versus the same energy differences, $(E(x_2)$ - $E(x_1))$, from a different source, such as molecular mechanics. The slope of the line is $-k$. Statistical potential energy functions are used in side-chain prediction and in threading.

Further reading

Dunbrack RL, Jr. (1999) Comparative modeling of CASP3 targets using PSI-BLAST and SCWRL. *Proteins Suppl.* 3: 81–87.

Sippl MJ (1990) Calculation of conformational ensembles from potentials of mean force: An approach to the knowledge-based predictions of local structures in globular proteins. *J. Mol. Biol.* 213: 851–883.

Thomas PD, Dill KA (1996) Statistical potentials extracted from protein structures: how accurate are they? *J. Mol. Biol.* 257: 451–469.

Thomas PD, Dill KA (1996) An iterative method for extracting energy-like quantities from protein structures. *Proc Natl Acad Sci USA* 93: 11621–11633.

See also Side-Chain Prediction, Threading

STEEPEST DESCENT METHOD, *SEE* GRADIENT DESCENT.

STEM-LOOP, *SEE* RNA HAIRPIN.

S

STOCHASTIC PROCESS

Matthew He

A stochastic process is the counterpart to a deterministic process (or deterministic system). Instead of dealing with only one possible reality of how the process might evolve under time (as is the case, for example, for solutions of an ordinary differential equation), in a stochastic or random process there is some indeterminacy in its future evolution described by probability distributions. This means that even if the initial condition (or starting point) is known, there are many possibilities the process might go to, but some paths may be more probable and others less so.

STOP CODON, *SEE* GENETIC CODE.

STREAM MINING (TIME SERIES, SEQUENCE, DATA STREAM, DATA FLOW)

Feng Chen and Yi-Ping Phoebe Chen

A data stream, which is an important mining strategy in data mining, goes in or out of a computer system by distinct speed. The data are of very huge volume and are based on time. It is impossible to store or analyse the whole data stream due to too large a volume and fast changes. Generally speaking, stream mining includes time series and sequence patterns.

A database of time series is composed of values or events measured at the different time points. The intervals between these time points are usually fixed and unchangeable, such as a minute, hour or day. Time series can be applied widely in a lot of fields, like stock analysis, economy prediction and so forth. Sequence is a database in which the data are ordered by a certain sequence. This kind of sequence may be not related to time, such as gene detection in DNA.

No matter which kind of data is analysed, stream mining generally includes the following tasks: 1). Tendency prediction. The future tendency can be speculated based on the current data; 2). Period detection. It is very effective for detecting periodical changes on data stream; 3). Subsequence matching. Given the whole sequence X and a requirement sequence S, this method is used to find all the similar sequences with S in X; 4). Whole sequence matching. This is an effective method for discovering all the similar subsequences in X; and 5). Sequence Alignment. This kind of alignment is categorized as pairwise sequence alignment and multiple sequence alignment. In bioinformatics, sequence alignment is widely used on homologous DNA or RNA for gene detection, unknown sequence fragment prediction and so on. Apart from that, other techniques, such as outlier detection and correlation analysis, can be also employed in stream mining.

Stream mining is significant in bioinformatics because a lot of datasets, such as DNA sequence, Protein primary structure, microarray, cDNA sequence and so forth, can be regarded as a data stream. So a majority of techniques of mining streams can be used in bioinformatics. For instance, the comparison with the whole DNA sequences can find the function of an unknown DNA fragment; while finding the clusters on a retrovirus insertion sequence can imply where the oncogenes are located. In summery, the current stream mining methods are generally based on statistics or theories from experiments. In addition to known biological features, stream mining is a powerful tool for biological streams.

Further reading

Gaber MM, Zaslavsky A, Krishnaswamy S (2005) *Mining Data Streams: A Review*. SIGMOD Record, 34,2.

Gama J, Gaber MM (2007) *Learning from Data Streams*. Springer.

See also Hidden Markov Model, HMM

STRICT CONSENSUS TREE, *SEE* CONSENSUS TREE.

STRUCTURAL ALIGNMENT

Andrew Harrison and Christine Orengo

A structural alignment identifies equivalent positions in the 3D structures of the proteins being compared. More than 40 different methods have been developed for aligning protein structures although most are based on comparing similarities in the properties and/or relationships of the residues or secondary structures. One of the earliest approaches was pioneered by Rossman and Argos who developed a method for superposing two protein structures in 3D by translating both proteins to a common coordinate frame and then rotating one protein relative to the other to maximise the number of equivalent positions superposed within a minimum distance from each other.

Other approaches (e.g. COMPARER, STRUCTAL, SSAP, CATHEDRAL) usually employ dynamic programming algorithms to optimise the alignment by efficiently determining where insertions or deletions of residues occur between the structures being compared. Structural environments of residues can be compared to identify equivalent positions, comprising, for example, information on distances from a residue to neighbouring residues. Some approaches (DALI, COMPARER) use sophisticated optimisation protocols like simulated annealing and Monte Carlo optimisation. Some multiple structure alignment methods have also been developed (MAMMOTH, CORA) for aligning groups of related structures.

Once the set of equivalent positions have been determined, many approaches superpose the structures on these equivalent positions to measure the root mean square deviation (RMSD). This is effectively the average distance between the superposed positions calculated by taking the square root of the average squared distances between all equivalent positions measured in angstroms. Related protein structures possessing similar folds typically possess an RMSD less than 4 angstroms, though this depends on the size of the domain. Some researchers prefer to calculate the RMSD over those positions which are very structurally similar and superpose within 3 angstroms. In which case it is usual to quote the number of positions over which the RMSD has been measured.

S

Related website

DALI server	http://ekhidna.biocenter.helsinki.fi/dali_server/
CATHEDRAL server	http://www.cathdb.info/
CE server	http://source.rcsb.org/jfatcatserver/ceHome.jsp

Further reading

Brown, NP, Orengo CA, Taylor WR (1996) A protein structure comparison methodology. *Computers Chem.* 20: 351–380.

Gu J, Bourne PE (2009) *Structural Bioinformatics.* Wiley-Blackwell; 2nd Edition.

Holm L, Rosenström P (2010) Dali server: conservation mapping in 3D. *Nucl. Acids Res.* 38: W541–549.

Redfern OC, Harrison A, Dallman T, Pearl FM, Orengo CA (2007) CATHEDRAL: a fast and effective algorithm to predict folds and domain boundaries from multidomain protein structures. *PLoS. Comput. Biol.* 3(11): e232.

Taylor WR, Orengo CA (1989) Protein Structure Alignment. *Journal of Molecular Biology*, 208: 1–22.

See also Alignment–Multiple, Alignment–Pairwise, Dynamic Programming, Indel, Simulated Annealing, Monte Carlo Simulations

STRUCTURAL GENOMICS

Jean-Michel Claverie

A research area, and the set of approaches, dealing with the determination of protein 3-D structures in a high throughput fashion.

The concept of Structural Genomics was first introduced in reference to the project of determining the 3-D structure of all genes products in a given (small) bacterial genome (hence the use of 'genomics'). Given the high redundancy of protein structures (i.e. many different protein sequences fold in similar ways) such projects are now seen as a waste of resources and effort. Structural Genomics projects are now developing in two different directions: *fundamental* projects aiming at determining the complete set of basic protein folds (estimates range from 1000 to 40,000), and *applied* Structural Genomics aiming at determining the structure of many proteins of a given family, or sharing common properties usually of biomedical interest (e.g. evolutionary conserved pathogen proteins of unknown function). In the latter context, Structural Genomics can be seen as part of Functional Genomics. Given that protein structures evolve much slower than sequences, 3-D structure similarities allow a fraction of proteins of unknown function to be linked with previously described families, when no sequence relationships can be established with confidence (i.e. in the 'twilight zone' of sequence similarity). The knowledge of a representative 3-D structure is also invaluable for the interpretation of conservation patterns identified by multiple sequence alignments. Purely computational studies attempting to predict protein folds and perform homology modeling from all protein-coding sequences in a given genome have also been referred to as 'Structural Genomics'.

Structural Genomics projects are made possible by a combination of recent advances in:

- recombinant protein expression (in vivo and in cell-free extract robotized screening)
- crystallogenesis (robotized large-scale screening of conditions),
- X-ray beam of higher intensity and adjustable wavelength (synchrotron radiation facilities) allowing single-crystal based 'phasing protocols',
- crystal handling (cryo-cooling)
- software for structure resolution (noise reduction, phasing, density map interpretation)
- NMR technology

However, Structural Genomics projects are presently limited to proteins that are not associated with membranes, due to the extreme difficulty to express (and crystallize) non soluble proteins.

Relevant websites

Midwest Center for Structural Genomics	http://www.mcsg.anl.gov/
New York Structural Genomics Research Consortium	http://www.nysgrc.org/
Joint Center for Structural Genomics	http://www.jcsg.org/scripts/prod/home.html

Further reading

Baker D, Sali A. (2001) Protein structure prediction and structural genomics. *Science* 294: 91–96.

Eisenstein E, et al. (2000) Biological function made crystal clear–annotation of hypothetical proteins via structural genomics. *Curr. Opin. Biotechnol.* 11: 21–30.

Kim, SH (1998) Shining a light on structural genomics. *Nat. Struct. Biol.* 5 Suppl: 641–645.

Moult J, Melamud E. (2000) From fold to function. *Curr. Opin. Struct. Biol.* 10: 381–389.

Orengo CA, et al. (1999) From protein structure to function. *Curr. Opin. Struct. Biol.* 9: 371–382.

Simons KT, et al. (2001) Prospects for *ab initio* protein structural genomics. *J. Mol. Biol.* 306: 1191–1199.

Skolnick J, et al. (2000) Structural genomics and its importance for gene function analysis. *Nat. Biotechnol.* 18: 281–287.

Zarembinski TI, et al. (1998) Structure-based assignment of the biochemical function of a hypothetical protein: a test case of structural genomics. *Proc. Natl. Acad. Sci. USA* 95: 15181–15193.

STRUCTURAL MOTIF

Andrew Harrison and Christine Orengo

A structural motif typically refers to a small fragment of a 3D structure, usually smaller than a domain, which has been found to recur either within the same structure or within other structures possessing different folds and architectures. Most structural motifs which

S

have been characterized comprise fewer than five secondary structures and are common to a particular protein class. For example the $\alpha\beta$ motif, split $\alpha\beta$ motif and $\alpha\beta$-plait are found within many structures in the alpha-beta class. Whilst the β-meander, β-hairpin, greek roll and jelly roll motifs are found in the mainly-β class. Some structural motifs are found to recur in different protein classes. For example the α-hairpin which occurs in some mainly-α structures and also in some alpha-beta structures. Recurrent structural motifs may correspond to particularly favoured folding arrangements of the polypeptide chain, due to thermodynamic or kinetic factors. Some researchers have suggested that some structural motifs may correspond to evolutionary modules which have been extensively duplicated and combined in different ways with themselves and other modules to give rise to the variety of different folds observed. The PROMOTIF database contains information on the structural characteristics of many structural motifs reported in the literature.

Related website

PROMOTIF	http://www.biochem.ucl.ac.uk/~gail/promotif/promotif.html

See also Protein Domain, Secondary Structure of Proteins

STRUCTURAMA

Michael P. Cummings

A program for inferring population structure using multi-locus genotype data.

An assumption is that the sampled loci are in linkage equilibrium and that the allele frequencies for each population are drawn from a Dirichlet probability distribution. The program transcends Structure in that it allows the number of populations to be a random variable and can estimate the number of populations under the Dirichlet process prior.

Related website

Structurama home page	http://cteg.berkeley.edu/~structurama/index.html

Further reading

Huelsenbeck JP, Andolfatto P, Huelsenbeck ET (2011) Structurama: Bayesian inference of population structure. *Evol. Bioinform. Online* 7: 51–59.

STRUCTURE

Michael P. Cummings

A software package to infer population structure using multi-locus genotype data.

Most often used for inferring the presence of distinct populations and assigning individuals to populations. It accommodates data including single nucleotide polymorphisms (SNPs), microsatellites, restriction fragment length polymorphisms (RFLPs) and amplified length polymorphisms (AFLPs).

The program can be used via the command line or graphical user interface. It is available as source code as well as executables for Mac OS, Windows, Linux and Solaris.

Related website

Structure home page	http://pritch.bsd.uchicago.edu/structure.html

Further reading

Falush D, Stephens M, Pritchard JK (2003) Inference of population structure using multilocus genotype data: linked loci and correlated allele frequencies. *Genetics* 164: 1561–1587.

Falush D, Stephens M, Pritchard JK (2007) Inference of population structure using multilocus genotype data: dominant markers and null alleles. *Mol. Ecol. Notes* 7: 571–578.

Hubisz MJ, Falush D, Stephens M, Pritchard JK (2009) Inferring weak population structure with the assistance of sample group information. *Mol. Ecol. Resour.* 9: 1321–1332.

Pritchard JK, Stephens M, Donnelly P (2000) Inference of population structure using multilocus genotype data. *Genetics* 155: 941–959.

STRUCTURE–3D CLASSIFICATION

Andrew Harrison and Christine Orengo

Structure classifications were established in the mid 1990s. In these, protein structures are classified according to phylogenetic relationships and phonetic (that is purely geometric) relationships. Although multidomain structures can be classified into evolutionary families and superfamilies, most structural classification also distinguish the individual domain structures within them and domain structures are grouped into further hierarchical levels (see below). The largest, most complete classifications accessible over the web are the SCOP, CATH and DALI Domain databases (see below). Other resources such as HOMSTRAD, CAMPASS and 3Dee are also valuable web-accessible resources.

Although the various resources employ different algorithms and protocols for recognizing related structures, most group domain structures into the following levels: At the top of the hierarchy, Class, groups domains depending on the percentage and packing of α-helices and β-strands in the structure. Within each class, architecture groupings are sometimes used to distinguish structures having different arrangements of the secondary structures in 3D. Whilst topology or fold group clusters structures having similar topologies regardless of evolutionary relationships. That is similarities in both the orientations of the secondary structures in 3D space and their connectivity. Within each fold group, domains are clustered according to evolutionary relationships into homologous superfamilies. Families are groups of closely related domains within each superfamily, usually sharing significant sequence and functional similarity.

Related websites

SCOP	http://scop.mrc-lmb.cam.ac.uk/scop/
DALI	http://ekhidna.biocenter.helsinki.fi/dali_server/
CATH	http://www.cathdb.info/
HOMSTRAD	http://tardis.nibio.go.jp/homstrad/

S

Further reading

Murzin AG, Brenner SE, Hubbard T, Chothia C. (1995) SCOP: a structural classification of proteins database for the investigation of sequences and structures. *J. Mol. Biol.* 247: 531–540.

Orengo CA, Michie AD, Jones S, Jones DT, Swindells MB, Thornton JM (1997) CATH—a hierarchic classification of protein domain structures. *Structure*. 5(8):1091–108.

See also Phylogeny, Superfamily, Protein Domain

STRUCTURE-ACTIVITY RELATIONSHIP, *SEE* SAR.

STRUCTURE-BASED DRUG DESIGN (RATIONAL DRUG DESIGN)

Marketa J. Zvelebil

Structure based drug design is based on the identification of candidate molecules that will bind selectively to a specific enzyme and thus (usually) inhibit its activity. The identification and docking of such ligand is one of the most important steps in rational drug design. Once a binding or active site has been identified it is possible to model a known ligand of interest into the binding site or to find a potential ligand from a database of small molecules. The ligand is then fitted into its binding site. Analysing the interactions between ligand and protein leads to a much better understanding of the protein's function and therefore to ways that the function could be modified. Binding is generally determined by the shape, chemical properties of both the binding site and substrate and the orientation of the ligand.

Related websites

DOCK	http://dock.compbio.ucsf.edu/
Network Science	http://www.netsci.org/Science/Compchem/feature14i.html

Further reading

Ladbury JE, Connelly PR (eds) *Structure-Based Drug Design: Thermodynamics, Modeling and Strategy* Springer-Verlag, New York.

See also Docking, Dock

STS, *SEE* SEQUENCE TAGGED SITE.

SUBFUNCTIONALIZATION

Austin L. Hughes

Subfunctionalization is hypothesized to occur when, after duplication of an ancestral gene with broad tissue expression, mutations occur causing each of the daughter genes to be expressed in a subset of the tissues in which the ancestral gene was expressed.

This process is most easily visualized by considering a gene with two different regulatory elements (A and B in Figure S.3) which control expression in two different tissues. After gene duplication, a mutation in one gene copy knocks out regulatory element A, while in the other copy a mutation knocks out regulatory element B. As a result, each of the daughter genes is necessary if the organism is to continue to express the gene product in all tissues in which it was expressed before gene duplication. A similar process might occur with regard to protein functions if the ancestral gene encoded a protein performing two separate functions, which could be performed by separate genes after gene duplication.

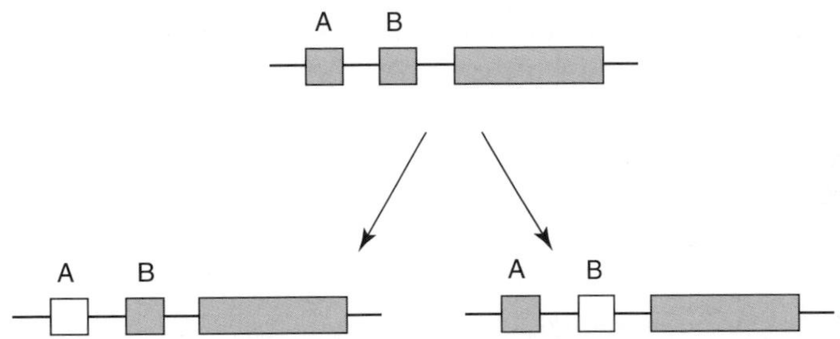

Figure S.3 Illustration of subfunctionalization. Of the regulatory elements A and B, a different one is inactivated in two copies of the gene, resulting in differential expression in different tissues.

Further reading

Force A, et al. (1999) Preservation of duplicate genes by complementary degenerative mutations. *Genetics* 151: 1531–1545.

Hughes AL (1994) The evolution of functionally novel proteins after gene duplication. *Proc R Soc Lond B* 256: 111–124.

Lynch M, Force A (2000) The probability of duplicate gene preservation by subfunctionalization. *Genetics* 154: 451–473.

Orgel LE (1977) Gene-duplication and the origin of proteins with novel functions. *J Theor Biol.* 67: 773.

See also Gene Sharing

SUBSTITUTION PROCESS

Sudhir Kumar and Alan Filipski

This refers to the stochastic parameters of the process by which one nucleotide or amino acid replaces another in evolutionary time.

Usually this substitution process is modeled as Markov process, *i.e.*, the transition probabilities only depend on the current state and the prior substitution history of the state at a specific site.

The homogeneity of a substitution process refers to the presence of the same process of substitution (at DNA or protein sequence level) over time in a given lineage (but see, Heterotachy). The term stationarity may refer to composition or process; in the former case, it refers to the property of transition probabilities not changing over time; in the latter case, it refers to a constant base composition (except for stochastic variation) over time.

Substitution pattern refers to the pattern of relative probabilities of state changes. For example, if we accelerate the process by doubling all substitution rates of one nucleotide to another, the substitution pattern itself stays unchanged.

Further reading

Kumar S, Gadagkar SR (2001) Disparity index: a simple statistic to measure and test the homogeneity of substitution patterns between molecular sequences. *Genetics* 158: 1321–1327.

Nei M, Kumar S (2000) *Molecular Evolution and Phylogenetics*. Oxford University Press.

Tavare S (1986) Some probabilistic and statistical problems on the analysis of DNA sequences. *Lectures in Mathematics in the Life Sciences*. Providence, RI, American Mathematical Society. 17: 51–86.

Yang Z (1994) Estimating the pattern of nucleotide substitution. *J Mol Evol* 39: 101–111.

SUBTREE

Aidan Budd and Alexandros Stamatakis

A tree obtained by detaching a branch from a larger phylogenetic tree. See Figure S.4.

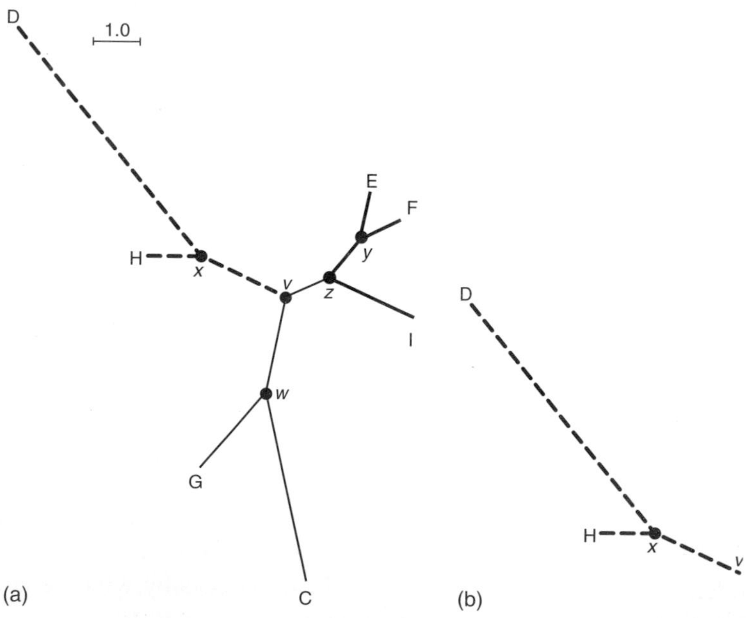

Figure S.4 Tree (b) shown in the figure above is a subtree of tree (a), as it can be obtained by detaching the branch that links internal nodes x and v in tree (a) from the rest of tree (a).

See also Phylogenetic Tree

S

SUPERFAMILY

Dov Greenbaum

Superfamilies fit within the Dayhoff protein classification system that uses a hierarchical structure with reference to the protein relatedness.

While originally coined to refer to evolutionarily related proteins, (the term is traditionally used only in reference to full sequences and not only domains) it has also come to signify groups of proteins related through functional or structural similarities independent of evolutionary relatedness.

The PIR (International Protein Sequence Database), classifying over 33,000 superfamilies as of January 2002, divides proteins into distinct superfamilies using an automated process that takes into account many criteria including domain arrangement and percent identity between sequences. In general proteins in the same superfamily do not differ significantly in length and have similar numbers and placements of domains. SCOP, (Structural Classification of Proteins) defines 'superfamily' more rigidly as a classification grouping the most distantly related proteins having a common evolutionary ancestor.

Homeomorphic superfamilies are defined as those families containing proteins that can be lined up end-to-end, i.e. containing the same overall domain architecture in the same order. All members of the family are deemed to have shared a similar evolutionary history.

Related websites

Superfamily	http://supfam.mrc-lmb.cam.ac.uk/SUPERFAMILY/index.html
PIR	http://pir.georgetown.edu/ PIR

Further reading

Dayhoff MO (1976) The origin and evolution of protein superfamilies, *Fed. Proc.* 35: 2131–2138.

Gough J, Chothia C (2002) SUPERFAMILY: HMMs representing all proteins of known structure. SCOP sequence searches, alignments and genome assignments. *Nucleic Acids Res* 30: 261–272.

Lo Conte L, et al. (2002) SCOP database in 2002: refinements accommodate structural genomics. *Nucleic Acids Res* 30: 261–267.

Orengo CA, Thornton JM (2005) Protein families and their evolution-a structural perspective. *Annu. Rev. Biochem.* 74: 861–900.

SUPERFOLD

Andrew Harrison and Christine Orengo

Superfold is the name given to those frequently occurring domain structures which have been found to recur in proteins which may or may not be related by divergent evolution. Although analyses of structural classifications have shown that most fold groups contain only proteins related by a common evolutionary ancestor (homologues), some folds are adopted by proteins which are not homologues. For example 29 different homologous superfamilies adopt the TIM barrel fold. Superfolds have been shown to account for nearly one third of protein domain structures; current analyses of CATH show that the ten most highly populated folds make up 36% of the non-redundant structures in the database. Although, these may be very distantly related proteins where all trace of sequence or functional similarity has disappeared during evolution, they may also be the consequence of convergence of different superfamilies to a favoured folding arrangement. Superfolds may have been selected due to the thermodynamic stability of the fold or kinetic factors associated with the speed of folding.

Related websites

DALI server	http://ekhidna.biocenter.helsinki.fi/dali_server/
SCOP	http://scop.mrc-lmb.cam.ac.uk/scop/
CATH server	http://www.cathdb.info/

Further reading

Baker D (2000) A surprising simplicity to protein folding. *Science.* 405: 31–42.

Holm L, Sander C (1998) Dictionary of recurrent domains in protein structures. *Proteins: Struct. Funct. Genet.* 33: 81–96.

Murzin AG, Brenner SE, Hubbard T, Chothia C (1995) SCOP: a structural classification of proteins database for the investigation of sequences and structures. *J. Mol. Biol.* 247: 531–540.

Orengo CA, Jones DT, Thornton JM (1994) Protein superfamilies and domain superfolds. *Nature.* 372: 631–634.

See also Homologous Genes

SUPERMATRIX APPROACH

Aidan Budd and Alexandros Stamatakis

An approach for inferring the evolutionary history for a set of species for which data from several genes is available. In the supermatrix approach one infers a single joint tree across all genes, by concatenating the per-gene multiple sequences alignments into one large multi-gene alignment (a super-matrix). Methods and tools have been developed for assembling super-matrices and for selecting a subset of informative genes. Because of their sheer size, the phylogenetic analysis of super-matrics, in particular under likelihood-based methods (Maximum Likelihood, Bayesian) is characterized by significant computational requirements (memory and CPU time) and increasingly requires access to supercomputers. A problem that is frequently discussed is that of missing data, that is, sequence data typically is not available from all species for each gene under study.

Software

Mare software: http://mare.zfmk.de/

Jones M, Koutsovoulos G, Blaxter M (2011) *iPhy: an integrated phylogenetic workbench for supermatrix analyses, BMC Bioinformatics* 12(1): 30.

Further reading

Narechania A, et al. (2011) Random addition concatenation analysis: a novel approach to the exploration of phylogenomic signal reveals strong agreement between core and shell genomic partitions in the Cyanobacteria. *Genome Biology and Evolution*: http://dx.doi.org/10.1093/gbe/evr121

Queiroz, A., Gatesy, J (2007) The supermatrix approach to systematics. *Trends in Ecology and Evolution* 22(1): 31–41.

Sanderson MJ, McMahon MM, Steel M (2011) Terraces in Phylogenetic Tree Space. *Sciences* 333(6041): 448.

See also Parallel Computing in Phylogenetics

SUPERSECONDARY STRUCTURE

Roman Laskowski and Tjaart de Beer

Assemblies of commonly occurring secondary structure elements which constitute a higher level of structure than secondary structure but a lower level than structural domains.

Examples include beta-alpha-beta motfis, Greek keys, alpha- and beta-hairpins, helix-turn-helix motifs, and EF-hands.

Further reading

Branden C, Tooze J (1998) *Introduction to Protein Structure*. Garland Science.

See also Secondary Structure of Proteins, Motif.

SUPERTREE, *SEE* CONSENSUS TREE.

SUPERVISED AND UNSUPERVISED LEARNING

Nello Cristianini

The two general classes of machine learning.

One can distinguish two main kinds of machine learning settings in data analysis applications: supervised and unsupervised. In the first case, the data (independent features) are 'labeled' by specifying to which category they belong (dependent feature), and the algorithm is requested to predict the labels on new unseen data. In the second case they are not labeled, and the algorithm is required to discover patterns (such as cluster structures) in the data.

Further reading

Duda RO, Hart PE, Stork DG (2001) *Pattern Classification*, John Wiley & Sons, USA.

Mitchell T (1997) *Machine Learning*, McGraw Hill.

See also Labels, Features, Machine Learning, Data Analysis, Clustering, Classification, Regression

S

SUPPORT VECTOR MACHINE (SVM, MAXIMAL MARGIN CLASSIFIER)

Nello Cristianini

Algorithms for learning complex classification and regression functions, belonging to the class of kernel methods.

In the binary classification case, SVMs work by embedding the data into a feature space by means of kernels, and by separating the two classes with a hyperplane in such a space. For statistical reasons, the hyperplane sought has maximal margin, or maximal distance from the nearest data point. The presence of noise in the data, however, requires a slightly adapted algorithm.

The maximal margin hyperplane in the feature space can be found by solving a convex (quadratic) optimization problem, and this means that the training of Support Vector Machines is not affected by local minima. This hyperplane in the feature space can correspond to a complex (non linear) decision function in the original input domain.

The final decision function can be written as

$$f(x) = \langle w, \phi(x) \rangle + b = \sum y_i \alpha_i K(x_i, x) + b$$

where the pairs (x_i, y_i) are labeled training points, x is a generic test point, K is a kernel function, and the α_i are parameters tuned by the training algorithm. There is one such parameter per training point, and only points whose parameter α is non-zero affect the solution. It turns out that only points lying nearest to the hyperplane have non-zero coefficient, and those are called 'support vectors' (see Figure S.5).

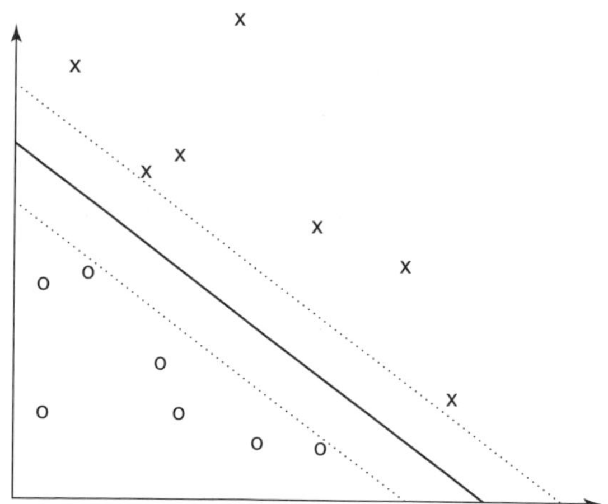

Figure S.5 A maximal margin.

Generalizations of this algorithm to deal with noisy data, regression problems, unsupervised learning and a number of other important cases have been proposed in recent years. First introduced in 1992, Support Vector Machines are now one of the major tools in pattern recognition applications, mostly due to their computational efficiency and statistical stability.

Related website

Kernel Machines website	www.kernel-machines.org

Further reading

Cristianini N, Shawe-Taylor J (2000) *An Introduction to Support Vector Machines* Cambridge University Press.

Vapnik V (1995) *The Nature of Statistical Learning Theory* Springer Verlag.

SURFACE MODELS

Eric Martz

Models that represent the surfaces of molecules; that is, the volume into which other atoms cannot penetrate, barring conformational or chemical changes. (See illustration at models, molecular.)

Most commonly, the surface is a solvent-accessible surface, and is generated by rolling a spherical probe over the space-filling model of the molecule. For a probe representing a water molecule, a sphere of radius 1.4 Å is most commonly used. The concept of the solvent-accessible surface was introduced by Lee and Richards in 1971 to quantitate the burial of hydrophobic moieties in folded proteins. Electron-density maps of molecules, resulting from X-ray crystallography, are another representation of molecular surfaces. For early low-resolution electron density maps, physical models were constructed with layers of balsa wood that represented the molecular surface.

The surface can be defined either by the path of the probe's center (the original Lee and Richards 'solvent-accessible' surface), or by the path of the probe's proximal surface (Richmond and Richards, 1978, the 'contact surface'). The volume enclosed by the solvent-accessible surface is larger than the unified van der Waals radius by the radius of the probe, and has been termed 'solvent-excluded volume' by some (Richmond, 1984), while others use that term to designate the volume of the contact surface (Connolly, 1985). The volume enclosed by the contact surface represents the portion of the van der Waals surface that can be contacted by solvent, and includes the van der Waals volume plus the interstitial volume, the latter being the spaces between atoms that are too small to admit solvent.

Molecular surfaces are often colored to represent molecular electrostatic potential, or molecular lipophilicity potential.

Related website

Molecular Surfaces	http://www.netsci.org/Science/Compchem/feature14.html

Further reading

Connolly ML (1983) Solvent-accessible surfaces of proteins and nucleic acids, *Science* 221: 701–713.

Connolly ML (1985) Computation of molecular volume. *J. Am. Chem. Soc.* 107: 1111–1124.

See also Models, molecular, Visualization, molecular, Space-Filling Model, van der Waals radius.

Eric Martz is grateful for help from Eric Francoeur, Peter Murray-Rust, Byron Rubin and Henry Rzepa.

SURPRISAL

Thomas D. Schneider

The surprisal is how surprised one would be by a single symbol in a stream of symbols. It is computed from the probability of the *i*th symbol, P_i, as $u_i = -\log_2 P_i$. For example, late at night as I write this, the phone rarely rings so the probability of silence is close to 1 and the surprisal for silence is near zero. (If the probability of silence is 99% then $u_{silence} = -\log_2(0.99) = 0.01$ bits per second, where the phone can ring only once per second.) On the other hand a ring is rare so the surprisal for ringing is very high. (For example, if the probability of ringing is 1% per second then $u_{ring} = - log_2(0.01) = 6.64$ bits per second). The average of all surprisals over the entire signal is the uncertainty ($0.99 * 0.01 + 0.01 * 6.64 = 0.08$ bits per second in this example). The term comes from Myron Tribus' book.

Further reading

Tribus M (1961) *Thermostatics and Thermodynamics*. D. van Nostrand Company, Inc., Princeton, N. J.

SVM, *SEE* SUPPORT VECTOR MACHINE.

SWISS-PROT, *SEE* UNIPROT.

SWISSMODEL

Roland Dunbrack

A program for comparative modeling of protein structure developed by Wolfgang Peitsch and colleagues. SwissModel is intended to be a complete modeling procedure accessible via a web server that accepts the sequence to be modeled, and then delivers the model by electronic mail. SwissModel follows the standard protocol of homologue identification, sequence alignment, determining the core backbone, and modeling loops and side chains. SwissModel determines the core backbone from the alignment of the target sequence to the parent sequence(s) by averaging the structures according to their local degree of sequence identity with the target sequence. The program builds new segments of backbone for loop regions by a database scan of the PDB using anchors of 4 Cα atoms on each end. Side chains are now built for those residues without information in the parent structure by using the most common (backbone-independent) rotamer for that residue type. If a side chain can not be placed without steric overlaps, another rotamer is used. Some additional refinement is performed with energy minimization with the GROMOS program.

Related website

SwissModel	http://swissmodel.expasy.org/

Further reading

Guex N, Peitsch MC (1997) SWISS-MODEL and the Swiss-PdbViewer: an environment for comparative protein modeling. *Electrophoresis* 18: 2711–2723.

See also Comparative Modeling

SYMMETRY PARADOX

Thomas D. Schneider

Specific individual sites may not be symmetrical (i.e. completely self-complementary) even though the set of all sites are bound symmetrically.

This raises an experimental problem: how do we know that a site is symmetric when bound by a dimeric protein if each individual site has variation on the two sides? A solution if we assume that the site is symmetrical, is to write Delila instructions for both the sequence and its complement. The resulting sequence logo will, by definition, be symmetrical. If, on the other hand, we write the instructions so as to take only one orientation from each sequence, perhaps arbitrarily, then by definition the logo will be asymmetrical. That is, one gets the output of what one puts in. This is a serious philosophical and practical problem for creating good models of binding sites. One solution would be to use a model that has the maximum information content (Schneider and Mastronarde, 1996), although this may be difficult to determine in many cases because of small sample sizes.

See Figure S.6.

Related websites

Delila	http://alum.mit.edu/www/toms/delila/instshift.html
Delila inst	http://alum.mit.edu/www/toms/delilainstructions.htmlsymmetry

Further reading

Schneider TD, Mastronarde D (1996) Fast multiple alignment of ungapped DNA sequences using information theory and a relaxation method. *Discr. Appl. Math* 71: 251–268. http://alum.mit.edu/www/toms/paper/malign

Schneider TD, Stephens RM (1990) Sequence logos: A new way to display consensus sequences. *Nucleic Acids Res*, 18: 6091–6100.

Schneider TD, Stormo GD (1989) Excess information at bacteriophage T7 genomic promoters detected by a random cloning technique. *Nucleic Acids Res*, 17: 659–674.

See also Sequence Logo, Delila Instructions

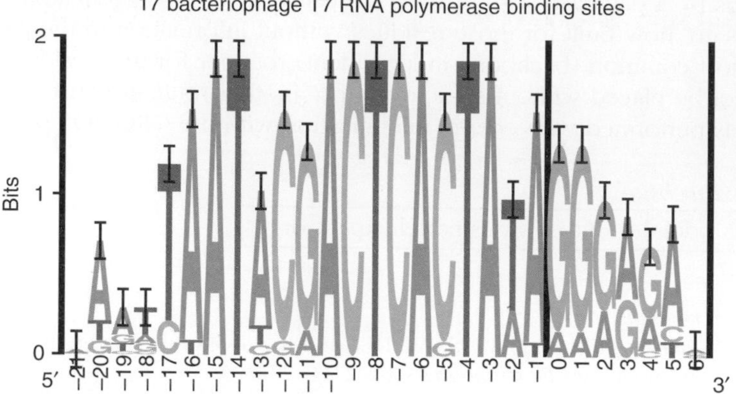

Figure S.6 A sequence logo for T7 RNA polymerase binding sites (Schneider & Stormo 1989, Schneider & Stephens, 1990). (See Colour plate S.6)

SYNAPOMORPHY

A.R. Hoelzel

Two evolutionary characters sharing a derived state.

As a hypothesis about the pattern of evolution among operational taxonomic units (OTUs), a phylogenetic reconstruction can inform us about the relationship between character states. We can identify ancestral and descendent states, and therefore identify primitive and derived states. An apomorphy (meaning 'from-shape' in Greek) is a derived state (shown as filled circles in Figure S.7):

A shared derived state between OTUs, as between 1&2 in Figure S.7, is known as a synapomorphy. This is an instance of homology, since 1&2 share derived characters from a common ancestor.

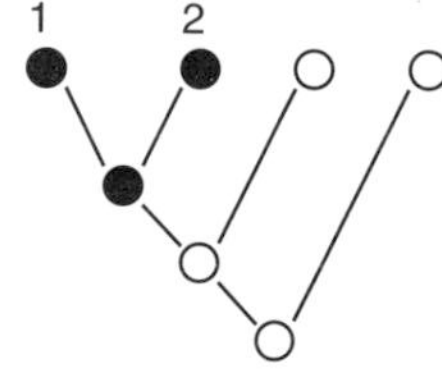

Figure S.7 Example of synapomorphy. The two OTUs 1 and 2 share a common derived state.

Further reading

Maddison DR, Maddison WP (2001) *MacClade 4: Analysis of Phylogeny and Character Evolution.* Sinauer Associates.

See also Apomorphy, Plesiomorphy, Autapomorphy

SYNONYMOUS MUTATION (SILENT MUTATION)

Laszlo Patthy

Nucleotide substitutions occurring in translated regions of protein-coding genes are synonymous (or silent) if they cause no amino acid change. This results as the altered codon codes for the same amino acid as the original one.

SYNTENY

Dov Greenbaum

Synteny is the physical co-localization of genetic loci on the same chromosome within an individual or species. Shared synteny, or conserved synteny represents the preserved co-localization of genes on chromosomes of different species.

Evolutionary pressures may make or break synteny: rearrangements and translocations may separate two loci apart or join two previously separate pieces of chromosomes together.

Stronger than expected conserved synteny may reflect evolutionary selection for functional relationships between syntenic genes and/or shared regulatory mechanisms.

Shared synteny may be able to establish the orthology of genomic regions in different species.

Patterns of shared synteny may be a tool in determining phylogenetic relationships among several species,

A Syntenic block may refer to a set of syntenic genomic regions, typically on a single chromosome usually comprised of syntenic gene sets.

Researchers may distinguish between preservation of synteny in large portions of a genome, (i.e., macrosynteny) and microsynteny representing the preservation of synteny for only a few genes.

Related websites

Synteny Database	http://teleost.cs.uoregon.edu/synteny_db/
Determine conserved synteny between the genomes of two organisms	http://www.ncbi.nlm.nih.gov/guide/howto/find-syntenic-regions/

Further reading

Catchen JM, Conery JS, Postlethwait JH (2009) Automated identification of conserved synteny after whole-genome duplication. *Genome Res.* 19(8): 1491–505.

Muffato M, Louis A, Poisnel CE, Crollius HR (2010) Genomicus: a database and a browser to study gene synteny in modern and ancestral genomes, *Bioinformatics* 15; 26(8): 1119–1121.

S

SYSTEMS BIOLOGY

Pascal Kahlem

Systems Biology is a synthetic approach to the study of a given biological system, which comes as an alternative to the classical analytical biology. Since the beginning of the 21st century, the development of high-throughput micro- and nano-technologies has enabled the acquisition of full molecular profiles of biological systems in given states of interest. The studies often cover genome, transcriptome and proteome because the corresponding biotechnologies are widespread, but they can also tackle lower levels of granularity, such as cellular or organism phenotypes or the physiology of organs, amongst others. Using statistics to compensate the unavoidable existence of noise and errors in these large-scale datasets, one can obtain global representations of a system, which can be used for predicting discrete protein function or reconstructing gene networks. Systems-level modeling is the ultimate level of large-scale data integration that can be used to study complex molecular networks properties and derive biological targets that could not be identified by classical hypothesis-driven approaches.

Further reading

Birney EA, Ciliberto, et al. (2005) Report of an EU projects workshop on systems biology held in Brussels, Belgium on 8 December 2004. *Systems Biology* 152(2): 51–60.

Meyer PLG. Alexopoulos S, et al. (2011) Verification of systems biology research in the age of collaborative competition. *Nature Biotechnology* 29(9): 811–815.

Westerhoff HV (2011) Systems biology left and right. *Methods in Enzymology* 500: 1–11.

See also Robustness

T

Tandem Mass Spectrometry (MS)
Tandem Repeat
Tanimoto Distance
Target
TATA BOX
Taxonomic Classification (Organismal Classification)
Taxonomic Unit
Taxonomy
Telomere
Template (Parent)
Template Gene Prediction, *see* Gene Prediction, ab initio.
Term
Terminology
Text Mining (Information Retrieval, IR)
Thermal Noise
Thesaurus
Thousand Genomes Project
THREADER, *see* Threading.
Threading
TIM-barrel
Trace Archive
Tracer
Trans-Proteomic Pipeline (TPP)
Transaction Database (Data Warehouse)
Transcription
Transcription Factor
Transcription Factor Binding Motif (TFBM)
Transcription Factor Binding Site
Transcription Factor Database
Transcription Start Site (TSS)
Transcriptional Regulatory Region (Regulatory Sequence)
Transcriptome
TRANSFAC
Transfer RNA (tRNA)
Translation
Translation End Site
Translation Start Site
Translatome
Transposable Element (Transposon)
Transposon, *see* Transposable Element.
Tree, *see* Phylogenetic Tree.
Tree of Life
Tree-based Progressive Alignment
Tree-Puzzle, *see* Quartets, Phylogenetic.
TreeStat
Tree Topology
Tree Traversal
TreeView X

Concise Encyclopaedia of Bioinformatics and Computational Biology, Second Edition.
Edited by John M. Hancock and Marketa J. Zvelebil.

Trinucleotide Repeat
Triple Store, *see* Database.
tRNA, *see* Transfer RNA.
Turn
Twilight Zone
Two-Dimensional Gel Electrophoresis (2DE)
Type Genome *see* Reference Genome

TANDEM MASS SPECTROMETRY (MS)

Simon Hubbard

The combination of two stages of mass spectrometry, frequently used in high-throughput proteomics experiments, whereby precursor peptide ions are analysed in the first mass spectrometer, then fragmented in to ion series and analysed again in a second spectrometer. The two stages of MS can be coupled in either space or time, in different types of instrument, but the effect is the same: generation of characteristic product ions from which amino acid sequence can be inferred and putative peptide identifications can be obtained.

TANDEM REPEAT

Katheleen Gardiner

Identical DNA sequences immediately adjacent to each other and in the same orientation. Identical adjacent sequences in opposite orientations are termed inverted repeats.

Further reading

Cooper DN (1999) *Human Gene Evolution*. Bios Scientific Publishers, pp 265–285.

Ridley M (1996) *Evolution* Blackwell Science, Inc pp 265–276.

See also Gene Family, Microsatellite, Minisatellite

TANIMOTO DISTANCE

John M. Hancock

Tanimoto Distance is a distance measure based on arrays containing binary presence/absence data. Given such a pair of arrays the Tanimoto Distance is:

$$T_d(X, Y) = -\log_2(T_s(x, Y))$$

where T_s the Tanimoto similarity ratio is:

$$T_s(X, Y) = \frac{\sum_i (X_i \wedge Y_i)}{\sum_i (X_i \vee Y_i)}$$

The Tanimoto Distance is sometimes confused with the Jaccard Distance because of the similar forms of their similarity measures.

Relevant website

Wikipedia	http://en.wikipedia.org/wiki/Jaccard_index

See also Compound Similarity and Similarity Searching, Jaccard Distance

T

T

TARGET

Roland Dunbrack

In comparative modeling, the protein of unknown structure for which a model will be constructed, based on a homologous protein of known structure, referred to as the template (or parent). In some situations such as sequence database searching, the target is also referred to as the query.

See also Comparative Modeling, Template

TATA BOX

Niall Dillon

Consensus sequence located 30–35 bases upstream from the transcription start site of many RNA polymerase II transcribed genes.

Two different elements have been identified that are involved in specifying the transcriptional initiation site of polymerase II transcribed genes in eukaryotes. These are the TATA box and the initiator. Promoters can contain one or both of these elements and a small number have neither. The TATA box is located 30–35 bp upstream from the site of transcriptional initiation and binds the the multi-protein complex TFIID, which is involved in specifying the initiation site.

Further reading

Roeder RG (1996) The role of general initiation factors in transcription by RNA polymerase II. *Trends Biochem. Sci.*, 21: 327–335.

TAXONOMIC CLASSIFICATION (ORGANISMAL CLASSIFICATION)

Dov Greenbaum

A hierarchically structured terminology for organisms.

There are seven nested taxa, (plural of taxon) or levels, in the present classification (or Phylogeny or Organismal Diversity) system:

1. kingdom, of which there are conventionally five; proposed (in 1969) by Robert Whittaker
   a. Plantae: multicellular eukaryotes which derive energy from photosynthesis and whose cells are enclosed by cellulose cell walls
   b. Animalia: multicellular, mobile and heterotrophic eukaryotes whose cells lack walls
   c. Fungi: eukaryotic, heterotrophic and (usually) multicellular organisms that have multinucleated cells enclosed in cells with cell walls.
   d. Protista: eukaryotes that are not plants, fungi, or animals
   e. Monera: unicellular prokaryotic organisms lacking membrane bound organelles but surrounded by a cell wall.

It has been suggested that this be compressed into three Archaea, Bacteria, and Eukarya.

2. hylum or division (In the animal kingdom, the term phylum is used instead of division.)
3. class
4. order
5. family
6. genus
7. species.

The hierarchical format has kingdom as the largest and most encompassing group and species as the most narrow. As such the closer a set of organisms are evolutionarily the more groups they share.

The species name, designed by Carolus Linnaeus, an 18th-century Swedish botanist and ratified by the International Congress of Zoologists, is by convention, binomial. Each species is given a two-part Latin name, formed by appending a specific epithet to the genus name, for example *Homo sapiens* for the human species. Other general rules of classification include the fact that all taxa must belong to a higher taxonomic group.

Related websites

NCBI taxonomy	http://www.ncbi.nlm.nih.gov/Taxonomy/taxonomyhome.html/index.cgi)
Taxon Lift	http://www.ucmp.berkeley.edu/help/taxaform.html
International Society of Zoological Sciences	http://www.globalzoology.org/
Ensembl Species List	http://www.ensembl.org/info/about/species.html
Species 2000 (Global Species List)	http://www.sp2000.org/index.php

Further reading

Cavalier-Smith T (1993) Kingdom protozoa and its 18 phyla. *Microbiol. Rev.* 57: 953–994.

Gerlach W, Jünemann S, Tille F, Goesmann A, Stoye J (2009) WebCARMA: a web application for the functional and taxonomic classification of unassembled metagenomic reads. *BMC Bioinformatics.* 18;10:430.

Kotamarti RM, Hahsler M, Raiford D, McGee M, Dunham MH (2010) Analyzing taxonomic classification using extensible Markov models. *Bioinformatics.* 26(18): 2235–2241.

Lake JA (1991) Tracing origins with molecular sequences: metazoan and eukaryotic beginnings. *Trends Biochem. Sci.* 16: 46–50.

T

Rosen GL, Reichenberger ER, Rosenfeld AM (2011) NBC: the Naive Bayes Classification tool web-server for taxonomic classification of metagenomic reads. *Bioinformatics*. 27(1): 127–129.

Whittaker RH, Margulis L (1978) Protist classification and the kingdoms of organisms BioSystems 10: 3–18.

TAXONOMIC UNIT

Aidan Budd and Alexandros Stamatakis

Phylogenetic trees can be used to describe patterns of genetic transmission between different kinds of entities, for example: different species, different individuals within a population of the same species, different genes within a gene family. The term 'taxonomic unit' is used to refer to the entities between which patterns and paths of genetic transfer are described. Thus, for some trees, the taxonomic units will be individuals within a population, in other trees they will be different species.

See also Operational Taxonomic Unit, Hypothetical Taxonomic Unit

TAXONOMY

Robert Stevens

1. Generically, the grouping of concepts into hierarchies based solely on the subtype relationships.
2. A classification of organisms, which may correspond to their phylogenetic relationships
3. The science and art of developing 2) above.

See also Taxonomic Classification

TELOMERE

Katheleen Gardiner

Specialized DNA sequences found at the ends of eukaryotic (linear) chromosomes.

Telomeres are composed of simple repeats, TTAGGG is common in vertebrates, reiterated several hundred times. A telomere stabilizes the chromosome ends, prevents fusion with other DNA molecules and allows DNA replication to proceed to the end of the chromosome without loss of DNA material.

Further reading

Blackburn EH (2001) Switching and signaling at the telomere Cell 106:661–673.

Miller OJ, Therman E (2000) *Human Chromosomes* Springer.

Wagner RP, Maguire MP, Stallings RL (1993) *Chromosomes – a synthesis* Wiley-Liss.

TEMPLATE (PARENT)

Roland Dunbrack

In comparative modeling, the protein of known structure used as a template for building a model of a protein of unknown structure, referred to as the target. The template and target proteins are usually homologous to one another. The template is sometimes referred to as the parent structure.

See also Comparative Modeling, Target

TEMPLATE GENE PREDICTION, *SEE* GENE PREDICTION, AB INITIO.

TERM

Robert Stevens

A label or name that is used to refer to a concept in some language.

The concept '*Protein*' may have terms such as 'Polypeptide', 'Protein', 'Folded protein', etc. The concept and its label are formally independent but often confused, because it is impossible to refer to concepts without the use of terms to name them. Ontology developers should adopt conventions for their terms. For instance it is usual to use the singular form of a noun and use initial capitals. As far as is possible the terms used within an ontology should follow a community's conventional understanding of a domain, as well as assist in controlling usage of terms within a domain.

TERMINOLOGY

Robert Stevens

The concepts of an ontology, together with their lexicon, form a terminology.

See also Ontology, Lexicon

TEXT MINING (INFORMATION RETRIEVAL, IR)

Feng Chen and Yi-Ping Phoebe Chen

Text mining (information retrieval), which is designed for knowledge discovery in text or documents, usually includes the following mining tasks: first, text classification and clustering; second, relevant text recognition; and third, important text extraction, and so forth.

In text mining, two significant definitions are *precision* and *recall*. For example, in Figure T.1 the documents relevant to an inquiry are called Relevant while the document

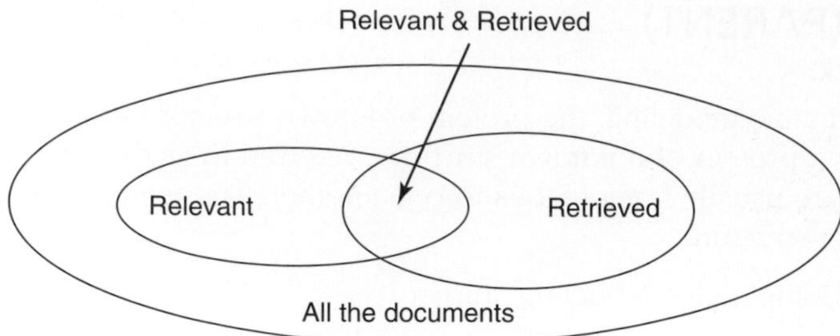

Figure T.1 The relation of relevant and retrieved.

set that we find is denoted as Retrieved. Then the formulas about precision and recall are as follows in Figure T.1:

$$Precision = \frac{|\{Relevant\} \cap \{Retrieved\}|}{|\{Retrieved\}|}$$

$$Recall = \frac{|\{Relevant\} \cap \{Retrieved\}|}{|\{Relevant\}|}$$

Generally speaking, information retrieval is trying to find a balance between precision and recall, given by the F-score:

$$F\_score = \frac{recall \times precision}{(recall + precision)/2}$$

Text mining has a special application in bioinformatics. Nowadays, a great deal of documents have been published, almost every one of which includes unique biological information. For example, we have to search too many papers when we want to see if our results conform to the established research. Consequently, text mining provides a series of tools, like text classification and clustering, document extraction, the analysis between documents and so forth that allow for a comprehensive analysis.

Further reading

Blaschke C, Hirschman L, Valencia A, Yeh A (2004) A critical assessment of text mining methods in molecular biology. *BMC Bioinformatics*, 6.

Feldman R, Sanger J (2006) *The Text Mining Handbook*. New York: Cambridge University Press.

Kao A, Poteet S (2004) *Natural Language Processing and Text Mining*. Springer, 2006.

THERMAL NOISE

Thomas D. Schneider

Thermal noise is caused by the random motion of molecules at any temperature above absolute zero Kelvin.

Since the third law of thermodynamics prevents one from extracting all heat from a physical system, one cannot reach absolute zero and so cannot entirely avoid thermal noise. In 1928 Nyquist worked out the thermodynamics of noise in electrical systems and in a back-to-back paper Johnson demonstrated that the theory was correct.

Further reading

Johnson JB (1928) Thermal agitation of electricity in conductors. *Phys Rev*, 32: 97–109.

Nyquist H (1928) Thermal agitation of electric charge in conductors. *Phys Rev*, 32: 110–113.

See also: Molecular Efficiency

THESAURUS

Robert Stevens

A thesaurus is a hierarchy or multi-hierarchy of index terms organised on the relationships of 'broader than' (parents) and 'narrower than' for children: 'The vocabulary of a controlled indexing language, formally organized so that the a priori relationships between concepts (for example as 'broader' and 'narrower') are made explicit' (ISO 2788, 986:2).

The terms within a thesaurus are traditionally used to index a classification of documents, usually in a library: 'A controlled set of terms selected from natural language and used to represent, in abstract form, the subjects of documents' (ISO 2788, 1986:2). Thus, the terms of a thesaurus are used to find documents within a classification.

THOUSAND GENOMES PROJECT

Andrew Collins

The 1000 genomes project has as its target the sequencing of the genomes of a large number of people. The outcome of this work is a comprehensive catalogue of human genetic variation which describes the majority of variants with a frequency of at least 0.01 in the populations studied. The 1000 genome catalogue is invaluable as a resource for gene mapping by describing patterns of genetic polymorphism not associated with specific disease. These data therefore annotate polymorphic variation that might be usefully filtered out of disease genome data to remove predominantly 'neutral' polymorphisms from further consideration. The complete project plan is to sequence 2500 samples with four-fold sequencing depth coverage. Because high depth sequencing is so costly the project employs genotype imputation to infer variants and genotypes in regions with well defined haplotype structure. The first set of samples sequenced includes 1167 samples from 13 populations with samples of 633 and 700 individuals to be sequenced in subsequent phases.

Further reading

The 1000 Genomes Project Consortium (2011) A map of human genome variation from population-scale sequencing. *Nature* 467: 1061–1073.

T

Related website

1000 genomes	http//www.1000genomes.org/

See also Haplotype, Next Generation Sequencing, Genotype Imputation

THREADER, *SEE* THREADING.

THREADING

David Jones

The process of replacing the side chains of one protein with the side chains of another protein whilst keeping the main chain conformation constant.

The side chain replacement is carried out according to a specific sequence-to-structure alignment. Threading is typically used as one of a number of approaches to recognizing protein folds (see Fold Recognition). For fold recognition, threading methods consider the 3-D structure of a protein and evaluate the compatibility of the target sequence with the template by means of pair potentials and solvation potentials. A variety of algorithms can be used to find the optimum threading alignment given a particular energy or scoring function, including double dynamic programming, simulated annealing, branch-and-bound searching and Gibbs sampling.

Related website

Phyre2	http://www.sbg.bio.ic.ac.uk/phyre2/html/page.cgi?id=index

Further reading

Bryant SH, Lawrence CE (1993) An empirical energy function for threading protein sequence through the folding motif. *Proteins*, 16(1): 92–112.

Jones DT, Miller RT, Thornton JM (1995) Successful protein fold recognition by optimal sequence threading validated by rigorous blind testing. *Proteins*, 23(3): 387–97.

Miller RT, Jones DT, Thornton JM (1996) Protein fold recognition by sequence threading: tools and assessment techniques. *Faseb J*, 10(1): 171–178.

TIM-BARREL

Roman Laskowski and Tjaart de Beer

A beta barrel consisting of eight parallel beta strands with each pair of adjacent strands connected by a loop containing an alpha helix springing from the end of the first strand and looping around the outside of the barrel before connecting to the start of the next strand. The name comes from the name of the first protein in which such a barrel was observed: triose phosphate isomerase (TIM). It is one of the most common tertiary folds observed in high resolution protein crystal structures

Related website

DATE: a database of TIM barrel enzymes	http://www.mrc-lmb.cam.ac.uk/genomes/date/

See also Beta Barrel, Beta Strand, Alpha Helix, Fold

TRACE ARCHIVE

Obi L. Griffith and Malachi Griffith

The Trace Archive is a permanent repository of DNA sequence chromatograms (traces), base calls, and quality estimates for single-pass reads from various large-scale sequencing projects. The Trace Archive is a project of the International Nucleotide Sequence Database Collaboration (INSDC) comprised of NCBI (USA), DDBJ (Japan) and EBI/ENA (Europe). The archive contains traces of nucleotide sequences produced primarily by traditional Sanger sequencing projects. With the advent of next-generation sequencing, the production volume of comparable 'trace data' has surpassed the ability to economically store it. Therefore, the Trace Archive has become a legacy database with declining new submissions and importance.

Related websites

NCBI Trace Archive	http://www.ncbi.nlm.nih.gov/Traces/
DDBJ Trace Archive	http://trace.ddbj.nig.ac.jp/dta/

Further reading

Sayers EW, et al. (2012) Database resources of the National Center for Biotechnology Information. *Nucleic Acids Res.* 40(Database issue):D13–25.

TRACER

Michael P. Cummings

A program for analysing results from Bayesian MCMC programs such as BEAST, LAMARC, MrBayes.

The continuous values sampled from the MCMC are used to generate various summary statistics, estimates of effective sample size (ESS), some of these statistics accompanied by some graphics.

Related website

Tracer home page	http://beast.bio.ed.ac.uk/Main_Page

See also BEAST, LAMARC, MrBayes

T

TRANS-PROTEOMIC PIPELINE (TPP)

Simon Hubbard

The Trans-Proteomics Pipeline (TPP) is a suite of software tools installed on a local computer for analysis and processing of a large variety of proteomics associated tasks linked to MS/MS data. This includes data format conversion, spectral processing, assignment of statistical significance scores to peptides and proteins in high throughput proteomics experiments, and quantitative analysis via several techniques. The software is made freely available, through sourceforge, and is available for most platforms including linux, windows and Mac OSX.

Related website

The TPP wiki	http://tools.proteomecenter.org/wiki/index.php?title=Software:TPP

Further reading

Deutsch EW, Mendoza L, Shteynberg D, et al. (2010) A guided tour of the Trans-Proteomic Pipeline. *Proteomics*. 10: 1150–1159.

TRANSACTION DATABASE (DATA WAREHOUSE)

Feng Chen and Yi-Ping Phoebe Chen

A transaction database is composed of a file, in which every record is a transaction. Generally, a transaction includes an ID and a list of items in this transaction. For example, one transaction represents a customer who goes shopping in supermarket and the items in it means the goods that he buys.

TRANSCRIPTION

John M. Hancock

The process of generating an RNA molecule from a gene.

Transcription is initiated at the transcription start site under the control of the promoter and other regulatory elements. It proceeds until termination occurs, which may take place at a specific site or may be probabilistic in nature. The product of transcription is an immature RNA which then undergoes a variety of processing reactions to produce a mature message.

Further reading

Lewin B (2000) *Genes VII*. Oxford University Press.

See also hnRNA, mRNA

TRANSCRIPTION FACTOR

Jacques van Helden

Protein affecting the level of transcription of a specific set of genes. A transcription factor is qualified as *activator* or *repressor* depending on whether it increases of represses the expression of its target gene(s). It has to be noted that the *activator/repressor* qualifier applies to the interaction between the TF and a given gene rather than on the TF itself, since a factor can activate some genes and repress other ones. Transcription factors are qualified of *specific* or *global* depending on whether they act on a restricted or a large number of genes (the boundary between specific and global factors is somewhat arbitrary).

Mechanisms: DNA-binding transcription factors act by binding to specific genomic locations, called transcription factor binding sites. Transcription factor may also act indirectly on the expression of their target genes by interacting with other transcription factors. For example, the yeast repressor Gal80p does not bind DNA, but interacts with the DNA-binding transcription factor Gal4p and prevents it from activating its target genes.

Further reading

Kuras L, Struhl K (1999) Binding of TBP to promoters in vivo is stimulated by activators and requires Pol II holoenzyme *Nature* 399: 609–613.

Lee TI, et al. (2000) Redundant roles for the TFIID and SAGA complexes in global transcription *Nature* 405: 701–704.

See also Transcription Factor Binding Site (TFBS), Transcription Factor Binding Motif (TFBM), Transcription Factor Database

TRANSCRIPTION FACTOR BINDING MOTIF (TFBM)

Jacques van Helden

Representation of the binding specificity of a transcription factor, generally obtained by summarizing the conserved and variable positions of a collection of binding sites. Several modes or representation can be used to describe TFBM: consensus, position-specific scoring matrices, Hidden Markov models.

Further reading

D'haeseleer P (2006) What are DNA sequence motifs? *Nat Biotechnol* 24(4): 423–425.

See also Consensus, Position-Specific Scoring Matrix (PSSM), Transcription factor binding site (TFBS)

T

TRANSCRIPTION FACTOR BINDING SITE

Jacques van Helden

Position on a DNA molecule where a transcription factor (TF) specifically binds. By extension, the sequence of the bound DNA segment. Note that there is a frequent confusion in the literature between the concepts of binding site and binding motif. We recommend to reserve the term 'site' to denote the particular sequence (genomic or artificial) where a factor binds, and the term 'motif' for the generic description of the binding specificity, obtained by synthesizing the information provided by a collection of sites.

The sites bound by a given transcription factor show similarities in their nucleotide sequences, some positions showing strong constrains on residues (sometimes called 'conserved' positions, although the different sites of a same genome are generally not homologous), whereas other positions can be modified without affecting the TF binding (sometimes called 'degenerate', with the same semantic remark). As an illustration, Table T.1 shows some examples of binding sites bound by the yeast transcription factor Pho4p. Transcription factor binding sites are generally short (between 5 and 20 base pairs), with a small number of constrained positions (typically 5–7).

Table T.1 A Collection of Sites Bound by the Yeast Transcription Factor Pho4p

Regulated gene	Site left	Site right	Site strand	Site sequence
PHO5	−370	−347	D	TAAATTAG**CACGTTT**TCGCATAGA
PHO5	−262	−239	D	TGGCACTCA**CACGTGGG**ACTAGCA
PHO8	−540	−522	R	ATCGCTG**CACGTGGC**CCGA
PHO81	−350	−332	R	TTATTCG**CACGTGCC**ATAA
PHO84	−421	−403	D	TTTCCAG**CACGTGGG**GCGG
PHO84	−442	−425	D	TAGTTC**CACGTGGA**CGTG
PHO84	−267	−250	D	TAATACG**CACGTTTT**TAA

Each site is described by its sequence, and its position (left, right) relative to the start codon of the regulated gene. Bold letters indicate the core of the binding sites, i.e. the residues entering in direct contact with the transcription factor.

Various experimental methods have been used to characterize transcription factor binding sites, with different levels of precision: crystal structures (atom-level characterization of TF/DNA interactions); DNAse protection (footprinting, nucleotide-level resolution); Electro-Mobility Shift Assays (EMSA, short oligonucleotides); SELEX (collections of oligonucleotides, with specificity depending on the number of amplification/selection cycles); Chromatin Immuno-Precipitation (ChIP, and the derived ChIP-on-chip and ChIP-seq methods) return large genomic regions (a few hundred base pairs) containing one or several binding sites.

See also Consensus, Motif Search, Transcription Factor, Transcription Factor Binding Motif, Transcription Factor Binding Site Prediction Transcription Factor Database

TRANSCRIPTION FACTOR DATABASE

Jacques van Helden

Several databases provide access to information concerning the binding and regulatory activity of transcription factors. Some databases provide detailed information on each piece of information (evidences for binding, references to the literature) whereas other databases are focused on pragmatic objects (e.g. matrices that can be used for TFBS prediction).

Related websites

Database	Description
DBTBS	Promoters, TFBS, operons in the bacteria *Bacillus subtilis*. http://dbtbs.hgc.jp/
FlyFactorSurvey	Binding sites for the fruit fly Drosophila melanogaster. http://pgfe.umassmed.edu/TFDBS/
FlyReg	Binding sites and cis-regulatory modules in the genome of Drosophila melanogaster. http://www.flyreg.org/
JASPAR	Position-specific scoring matrices and collections of sites having served to build them (individual sites are not documented by links to the literature). http://jaspar.cgb.ki.se/
ORegAnno	Community-based annotation of cis-regulatory elements in various model organisms. http://www.oreganno.org/oregano/
RegulonDB	Binding sites, position-specific scoring matrices, promoters, operons for the bacteria *Escherichia coli K12*. http://regulondb.ccg.unam.mx/
SCPD	Binding sites and position-specific scoring matrices for the yeast *Saccharomyces cerevisiae*. http://rulai.cshl.edu/SCPD/
TRANSFAC	Commercial database. Binding sites (detailed information on experimental evidence for each individual site) + position-specific scoring matrices. http://www.gene-regulation.com/pub/databases.html
Yeastract	Binding sites and position-specific scoring matrices for the yeast *Saccharomyces cerevisiae*. http://www.yeastract.com/

See also Transcription Factor, Transcription Factor Binding Site, Transcription Factor Binding Motif, Cis-Regulatory Module

T

TRANSCRIPTION START SITE (TSS)

John M. Hancock

The point at the 5' end of a gene at which RNA polymerase initiates transcription.

The TSS will be adjacent to the promoter and is the place at which the RNA polymerase complex associates with the DNA. Details of the context of the TSS will depend on the gene, what organism it is in, what type of polymerase is involved, and other factors.

Related website

dbTSS	http://dbtss.hgc.jp/

Further reading

Valen E, Sandelin A (2011) Genomic and chromatin signals underlying transcription start-site selection. *Trends Genet.* 27(11): 475–485.

TRANSCRIPTIONAL REGULATORY REGION (REGULATORY SEQUENCE)

Niall Dillon and James Fickett

Sequence that is involved in regulating the expression of a gene (eg. promoter, enhancer, LCR).

In eukaryotes, transcriptional regulatory regions (TRRs) are either promoters, at the transcription start site, or enhancers, elsewhere. Enhancers, as well as the regulatory portion of promoters, are typically made up of functionally indivisible cis-regulatory modules (CRM) in which a few transcription factors have multiple binding sites (Yuh and Davidson, 1996).

Further reading

Krebs JE (ed.) (2009) *Lewin's Genes X*. Jones and Bartlett Publishers, Inc.

See also Promoter, Transcription Start Site, Enhancer, Transcription Factor, Transcription Factor Binding Site

TRANSCRIPTOME

Dov Greenbaum and Jean-Michel Claverie

The complete set of all transcripts that can be generated from a given genome. This includes all alternative transcripts (splice variant, alternative 3' and 5' UTR). In an experimental context, it is also used to designate the subset of transcripts found in a given cell, organ or tissue. In a more quantitative way, the 'transcriptome' can also refer to the set of all transcripts indexed by (associated to) a measurement of their abundance level.

The transcriptome connects the genome to the translatome and is measured on a high throughput scale. Until recently transcriptomics have employed cDNA microarray,

Affymetrix GeneChip or SAGE technologies to determine transcript levels but increasingly this is now being carried out by RNA-seq. While there has been an effort to quantify the protein content of the cell, this does not diminish the importance of quantifying and understanding the mRNA population in the cell. While the genome is generally static, the ability of the transcription machinery to produce alternatively processed forms of transcripts provides much of the cellular heterogeneity and adaptability.

Transcriptome analysis has been used, among other uses, to compare disease states, predict cancer types, diagnose diseases and to cluster genes into functional categories for functional annotation transfer. In addition to comparing cell states, whole organisms can be compared with each other based on the degree of similarity or specific differences in their mRNA populations.

There still remain many uncertainties and inaccuracies in the mRNA data, due to differences in experimental techniques between labs and also to the delicate and sensitive nature of the experimental procedures. As such, researchers must use caution whenever analyzing an mRNA dataset. A more robust approach might include creating comprehensive datasets assembling multiple similar experiments into a more reliable data set.

Related websites

Yeast transcriptome analysis	http://bioinfo.mbb.yale.edu/expression/transcriptome/
SGD Expression Connection	http://spell.yeastgenome.org/
NCBI Gene Expression Omnibus	http://www.ncbi.nlm.nih.gov/geo/
ArrayExpress	http://www.ebi.ac.uk/arrayexpress/

Further reading

Saha S, et al. (2002) Using the transcriptome to annotate the genome. *Nat Biotechnol.* 20: 508–512.

Velculescu VE, et al. (1997) Characterization of the yeast transcriptome *Cell.* 88: 243–251

See also RNA-seq, Affymetrix GeneChip™ Oligonucleotide Microarray, Spotted cDNA Microarray, Serial Analysis of Gene Expression

TRANSFAC

Obi L. Griffith and Malachi Griffith

TRANSFAC began in the late 1980s and early 1990s as a database to collect information about transcription factors (TFs), transcription factor binding sites (TFBS), and the binding specificity of individual TFs or TF classes with the ultimate goal of constructing a regulatory map for all genes and genomes. This effort led to the development of tools for de novo prediction of TFBSs by, for example, consensus or matrix searches. The core TRANSFAC database was subsequently complemented by the creation of TRANSCompel, a database cataloguing the importance of interactions between TFBSs (composite elements). Similarly,

TRANSPATH integrated information about transcription factors and the signal transduction pathways to which they belong. This allows the identification of candidate master regulators for specific biological events. TRANSFAC and related databases are accessible to academic users after registration. Commercial enterprises are required to license the database and accompanying programs from BIOBASE GmbH. The perceived disadvantages of TRANSFAC's closed and commercial model have been cited as motivation for the development of more open databases with similar goals or functions (see JASPAR, ORegAnno, and PAZAR).

Related website

TRANSFAC (BIOBASE)	http://www.gene-regulation.com

Further reading

Matys V, et al. (2006) TRANSFAC and its module TRANSCompel: transcriptional gene regulation in eukaryotes. *Nucleic Acids Res.* 34(Database issue):D108–110.

Wingender E (2008) The TRANSFAC project as an example of framework technology that supports the analysis of genomic regulation. *Brief Bioinform.* 9(4):326–32.

TRANSFER RNA (tRNA)

Robin R. Gutell

Transfer RNAs (tRNAs) are usually 70–90 nt in length in nuclear and chloroplast genomes, and are involved in protein synthesis.

The carboxyl-terminus of an amino acid is specifically attached to the 3′ end of the tRNA (aminoacylated). These aminoacylated tRNAs are substrates in translation, interacting with a specific mRNA codon to position the attached amino acid for catalytic transfer to a growing polypeptide chain. Transfer RNAs decode (or translate) the nucleotide sequence during protein synthesis.

tRNAs have a characteristic 'cloverleaf' secondary structure that was initially determined with comparative analysis and substantiated with X-ray crystallography. High resolution crystal structures of tRNA also revealed that different tRNAs formed very similar tertiary structures The 'variable loop' of tRNA is primarily responsible for length variation among tRNAs; some of the mitochondrial tRNAs are smaller than the typical tRNA, shortening or deleting the D or T ψ C helices. Over fifty modified nucleotides have been observed in different tRNA molecules. See Figure T.2.

Related website

Wikipedia: Transfer RNA	http://en.wikipedia.org/wiki/Transfer_RNA

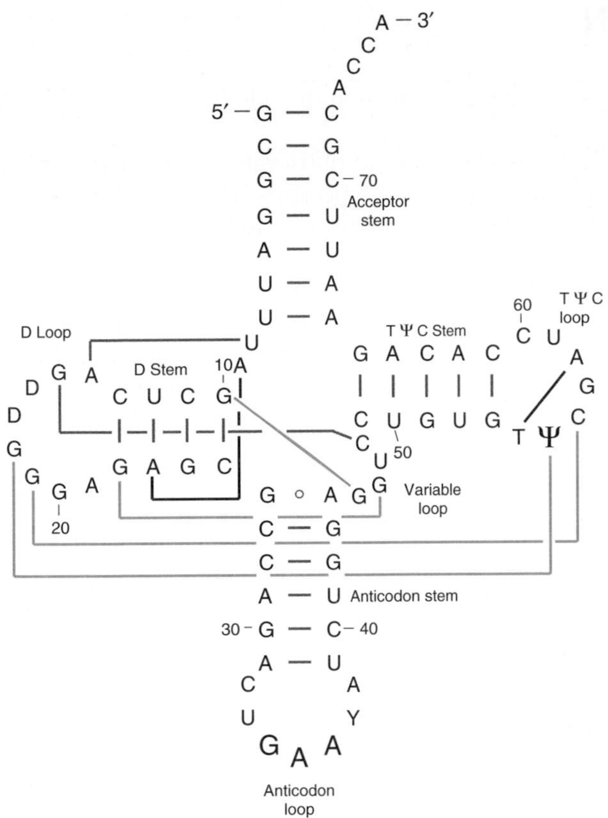

Figure T.2 tRNA secondary structure (*Saccharomyces cerevisiae* phenylalanine tRNA). Structural features are labelled (Wikipedia http://en.wikipedia.org/wiki/Trna). (See Colour plate T.2)

Further reading

Holley RW, et al. (1965) Structure of a ribonucleic acid. *Science* 147: 1462–1465.

Kim SH (1979) Crystal structure of yeast tRNAphe and general structural features of other tRNAs. In: *Transfer RNA: Structure, Properties, and Recognition,* Schimmel PR, Soll D, Abelson JN (eds), pp. 83–100. Cold Spring Harbor Laboratory Press, Cold Spring Harbor, New York.

Kim SH, et al. (1974) Three-dimensional tertiary structure of yeast phenylalanine transfer RNA. *Science* 185: 435–440.

Levitt M (1969) Detailed molecular model for transfer ribonucleic acid. *Nature* 224: 759–763.

Marck C, Grosjean H (2002) tRNomics: Analysis of tRNA genes from 50 genomes of Eukarya, Archaea, and Bacteria reveals anticodon-sparing strategies and domain-specific features. *RNA* 8: 1189–1232.

Quigley GJ, Rich A (1976) Structural domains of transfer RNA molecules. *Science* 194: 796–806.

Robertus JD, et al. (1974). Structure of yeast phenylalanine tRNA at 3 Å resolution. *Nature* 250: 546–551.

T

TRANSLATION

John M. Hancock

The process of reading information encoded in a mature mRNA and converting it into a protein molecule.

Translation takes place on the ribosome and involves a number of cytoplasmic factors as well as the ribosomal components (rRNAs and ribosomal proteins). It consists of a series of complex chemical processes including the selection and binding of cognate tRNA molecules, the movement of mRNA and nascent protein chains between binding sites on the ribosomal surface, and the sequential formation of peptide bonds between the growing protein chain and amino acids on incoming tRNA molecules.

Proteins may be subject to a variety of post-translational modifications including cutting by specific proteases, phosphorylation and glycosylation before the mature protein is formed. These processes my be cell type-specific in eukaryotes and give rise to additional variation seen in proteomic experiments.

Further reading

Lewin B (2000) *Genes VII*. Oxford University Press.

TRANSLATION END SITE

Enrique Blanco and Josep F. Abril

Codon on an mRNA sequence at which the translation by the ribosome into amino acid sequence is terminated.

There are three stop codons in the standard genetic code: UAG, UAA and UGA. Translation termination is initiated when a stop codon appears in the A site of the ribosome. Since no tRNA can be associated to the stop codon, a release factor occupies this location, stimulating the correct completing of the polypeptide synthesis, its release and the dissociation of the ribosomal subunits.

Alternative decoding of UGA into selenocysteine (Sec) during the synthesis of selenoproteins has been reported in presence of the SECIS element (the selenocysteine insertion sequence), a stem-loop structure around 60 nucleotides long, on the 3'UTR of the messenger RNA.

Pyrrolysine (Pyl) is another non-standard amino acid that is encoded by the UAG codon. It is mainly used by some methanogenic archaea in their methane-producing metabolism, being associated to a stem-loop structure present on the 3'UTR (the pyrrolysine insertion sequence, PYLIS). Variants of the standard genetic code can define a different translation for a stop codon, such UGA encoding Tryptophan (Trp) in *Mycoplasma*, yet in those cases specific recoding signals in the mRNA are not required.

Further reading

Brenner S, Stretton A, Kaplan S (1965) Genetic code: the 'nonsense' triplets for chain termination and their suppression. *Nature* 206: 994–998.

Brenner S, Barnett L, Katz E, Crick F (1967) UGA: a third nonsense triplet in the genetic code. *Nature* 213: 449–450.

Kisselev LL, Buckingham RH (2000) Translational termination comes of age. *Trends Biochem Sci.* 25: 561–566.

Zhang Y, Baranov PV, Atkins JF, Gladyshev VN (2005) Pyrrolysine and selenocysteine use dissimilar decoding strategies. *J. Biol. Chem.* 280(21): 20740–20751.

See also Translation Start Site, Selenoproteins, Gene Prediction, non-canonical, Gene Prediction, *ab initio*

TRANSLATION START SITE

Roderic Guigó

Codon on an mRNA sequence at which the translation by the ribosome into amino acid sequence is initiated.

The start codon is usually AUG (which codes for the amino acid methionine), but in prokaryotes it can also be GUG (coding for leucine). In prokaryotes, binding of the ribosome to the start AUG is mediated by the so called Shine-Dalgarno motif (Shine and Dalgarno, 1974), a sequence complementary to the 3′ of the 16 S rRNA of the 30 S subunit of the ribosome. The consensus Shine-Dalgarno sequence is UAAGGAG, but the motif is highly variable.

In eukaryotes, ribosomes do not bind directly to the region of the messenger RNA that contains the AUG initiation codon. Instead, the 40S ribosomal subunit starts at the 5′ end and 'scans' down the message until it arrives at the AUG codon (Kozak, 1989). A short recognition sequence, the Kozak motif, usually surrounds the initiation codon. The consensus of this motif is ACCAUGG. If this sequence is too degenerate, the first AUG may be skipped, and translation is initiated at a further downstream AUG.

Related websites

NetStart Prediction Server	http://www.cbs.dtu.dk/services/NetStart/
ATGpr Prediction Server	http://www.hri.co.jp/atgpr/
AUG_EVALUATOR Prediction Server	http://www.itb.cnr.it/webgene/

Further reading

Kozak M. (1999) Initiation of translation in prokaryotes and eukaryotes. *Gene*, 234: 187–208.

Shine J, Dalgarno L. (1974). The 3′-terminal sequence of *E.Coli* 16s ribosomal RNA: Complementary to nonsense triplets and ribosome binding sites. *Proc. Natl. Acad. Sci. USA*, 71: 1342–1346.

See also Gene Prediction, Motif Search

TRANSLATOME

Dov Greenbaum

An alternative term to proteome which refers specifically to the complement of translated products in a given cell or cell type.

The translatome refers to any quantifiable population of proteins within a cell at any given moment or under specific cellular circumstances. The translatome can be measured through multiple high throughput technologies (see Proteome). Basically this requires that each protein be isolated, measured, and then identified. The isolation usually occurs through two-dimensional electrophoresis and protein identification occurs primarily through mass spectrometry methods.

Quantification can be experimentally determined, for example, through radioactive labeling or computationally, (using specific software such as Melanie, Z3 and similar programs) to quantify the pixel representation for individual proteins on a two dimensional gel image.

While there has been significant work on the elucidation of the transcriptome, much work has yet to be done with regard to the translatome, the analysis of the end product of gene expression. Although there is some correlation between mRNA expression and protein abundance, and mRNA expression clustering has been very useful in research, the translatome is still a superior picture of the cell and its processes.

Related website

Translatome	http://bioinfo.mbb.yale.edu/expression/translatome/

Further reading

Greenbaum D, et al. (2002) Analysis of mRNA expression and protein abundance data: an approach for the comparison of the enrichment of features in the cellular population of proteins and transcripts. *Bioinformatics* 18: 585–596.

Gygi SP, et al. (2000) Measuring gene expression by quantitative proteome analysis. *Curr Opin Biotechnol* 11: 396–401.

See also Proteome, Proteomics

TRANSPOSABLE ELEMENT (TRANSPOSON)

Dov Greenbaum and Katheleen Gardiner

DNA sequences that can move from one genomic site to another. First described in maize by Barbara McClintock.

This insertion of new pieces of DNA into a sequence can increase or decrease the amount of DNA and possibly cause mutations if they jump into a coding region of a gene.

There are three distinct classes of transposable element:

- Class I – Retrotransposons transcribe mRNA into DNA and insert this new piece of DNA into the genome. Retrotransposons are often flanked by long terminal repeats (LTRs) that can be over 1000 bases in length. About 40% of the entire human genome consists of retrotransposons.

- Class II are pieces of DNA that move directly from one position to another. They require the activity of an enzyme, transposase, to cut and paste them into new positions within the genome. The transposase enzyme is often encoded within the transposon itself.
- Class II transposons can be site specific, that is that they insert into specific sequences within the genome, others insert randomly.
- Class III (Miniature Inverted Repeat Transposable Elements) MITEs, which are too small to encode any protein, have been found in the genome of humans, rice, apples and Xenopus.

Transposons can be mutagens, that is they can cause mutations within the genome of a cell. This is accomplished in three ways:

- (i) insertion into a gene or its flanking regions can change either enhance or prevent gene expression depending on where the transposon is inserted.
- (ii) when a transposon leaves its original site, there may be a failure of the cell to repair the gap and this may cause mutations.
- (iii) long strings of repeats caused by transposon insertion can interfere in pairing during meiosis.
- (iv) poly(A) tails carried by some transposable elements may evolve into microsatellites, in some cases causing disease as in the case of the GAA repeat associated with an Alu element in the Friedreich's Ataxia gene.

Related website

Gypsy Database	http://gydb.org/index.php/Main_Page

Further reading

Cooper DN (1999) *Human Gene Evolution* Bios Scientific Publishers, pp 265–285.

Finnegan DJ (1992) Transposable elements. *Curr Opin Genet Dev* 2: 861–867.

Kines KJ, Belancio VP (2012) Expressing genes do not forget their LINEs: transposable elements and gene expression. *Front Biosci*. 1;17: 1329–1344.

Levin HL, Moran JV (2011) Dynamic interactions between transposable elements and their hosts. *Nat Rev Genet*. 18;12(9): 615–627.

McDonald JF. Evolution and consequences of transposable elements (1993). *Curr Opin Genet Dev* 3: 855–864.

Ridley M (1996) *Evolution* Blackwell Science, Inc pp 265–276.

Rowold DJ, Herrera RJ (2000) Alu elements and the human genome *Genetics* 108: 57–72.

See also LINE, SINE, Alu, L1

TRANSPOSON, *SEE* TRANSPOSABLE ELEMENT.

TREE, *SEE* PHYLOGENETIC TREE.

T

TREE OF LIFE

Sudhir Kumar and Alan Filipski

A phylogenetic framework depicting the evolutionary relationships among and within the three domains of extant life: Eukaryotes, Eubacteria and Archaea.

At present, many aspects of deep evolutionary relationships remain controversial or unknown as lateral gene transfer appears to have been common in the early history of life. For this reason, a traditional phylogenetic tree model may not be an appropriate representation of relationships among genomes of these primitive organisms.

The last common ancestor of all extant life forms on earth, is denoted the Last Universal Common Ancestor (LUCA). The characteristics and the time of origin of this organism is usually inferred by examining genes and biochemical mechanisms common to all extant life forms. It is possible that no such organism ever existed and that several lineages arose from pre-biotic precursors.

Further reading

Brown JR, Doolittle WF (1995) Root of the universal tree of life based on ancient aminoacyl-tRNA synthetase gene duplications. *Proc. Natl. Acad. Sci. USA*, 92: 2441–2445.

Hedges SB et al (2001) A genomic timescale for the origin of eukaryotes. *BMC Evol Biol* 1 (4), 10.

Woese C (1998) The universal ancestor. *Proc. Natl. Acad. Sci. USA*, 95: 6854–6859.

See also Phylogenetic Tree, Last Universal Common Ancestor

TREE-BASED PROGRESSIVE ALIGNMENT

Jaap Heringa

A heuristic strategy for generating a multiple sequence alignment (see Multiple Alignment). The most commonly used heuristic multiple sequence alignment methods are based on the progressive alignment strategy (Hogeweg and Hesper, 1984; Feng and Doolittle, 1987; Taylor, 1988) with ClustalW (Thompson *et al*., 1994) being the most widely used implementation. The idea is to establish an initial order for joining the sequences, and to follow this order in gradually building up the alignment. Many implementations use an approximation of a phylogenetic tree between the sequences as a so-called guide tree which dictates the alignment order. The advantage of a guide tree is that the order of the sequences to become aligned, as dictated by the tree, leads to the easiest alignment problems to be dealt with first, before more distant and error-prone alignments need to be made.

Although appropriate for many alignment problems, the progressive strategy suffers from its greediness. Errors made in the first alignments during the progressive protocol cannot be corrected later as the remaining sequences are added in ('Once a gap, always a gap'; Feng and Doollitle, 1987). Another problem with the progressive alignment strategy is incomplete use of the sequence information; the first step uses only the information of two sequences to be aligned, and only at the last alignment step of the progressive protocol, when most sequences have been matched, the sequence information is used to the full extent. Some alignment methods employ consistency-based strategies to include information from all sequences at each alignment step during progressive alignment, including T-Coffee (Notredame et al., 2000), Praline (Heringa, 1999; Simossis et al., 2005) and ProbCons (Do et al., 2005).

Further reading

Do CB, Mahabhashyam MSP, Brudno M, Batzoglou S (2005) ProbCons: probabilistic consistency-based multiple sequence alignment. *Genome Res.* 15: 330–340.

Feng DF, Doolittle RF (1987) Progressive sequence alignment as a prerequisite to correct phylogenetic trees. *J. Mol. Evol.* 21: 112–125.

Heringa J (1999) Two strategies for sequence comparison: Profile-preprocessed and secondary structure-induced multiple alignment. *Comp. Chem.* 23: 341–364.

Hogeweg P, Hesper B (1984) The alignment of sets of sequences and the construction of phyletic trees: an integrated method. *J. Mol. Evol.* 20: 175–186.

Notredame C, Higgins DG, Heringa J (2000) T-Coffee: a novel method for fast and accurate multiple sequence alignment. *J. Mol. Biol.* 302: 205–217.

Simossis VA, Kleinjung J, Heringa J (2005) Homology-extended sequence alignment. *Nucleic Acids Res.* 33(3): 816–824.

Taylor WR (1988) A flexible method to align large numbers of biological sequences. *J. Mol. Evol.* 28: 161–169.

Thompson JD, Higgins DG, Gibson TJ (1994) CLUSTAL W: improving the sensitivity of progressive multiple sequence alignment through sequence weighting, positions-specific gap penalties and weight matrix choice. *Nucleic Acids Res.* 22: 4673–4680.

See also Alignment – multiple, Sequence Similarity

TREE-PUZZLE, *SEE* QUARTETS, PHYLOGENETIC.

TREESTAT

Michael P. Cummings

A program to calculate summary statistics from a set of phylogenetic.

Input is PHYLIP or NEXUS format tree file. Output is a tab-delimited file for analysis in Tracer or statistics packages. Among the summary statistics include tree-balance statistics, cherry count, tree height, node heights, tree length, and others.

Related website

TreeStat home page	http://tree.bio.ed.ac.uk/software/treestat/

See also PHYLIP, Tracer

T

TREE TOPOLOGY

Aidan Budd and Alexandros Stamatakis

The pattern of linkage between the nodes in a tree.

A given tree topology can be represented in many different ways. Consider, for example, the tree topology in which the following list describes all pairs of nodes that are directly linked to each other: AE, ED, DB, EF, and DC. Three different representations of this topology are shown in Figure T.3.

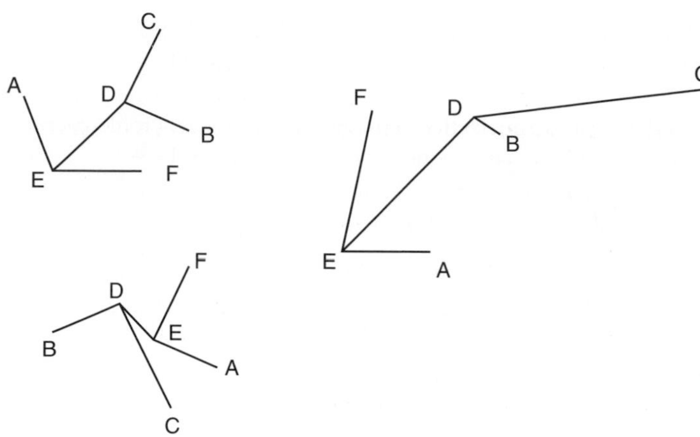

Figure T.3 Three distinct representations of the same, identical tree topology.

In Figure T.4, however, the two trees have different topologies i.e. in Tree A the list of nodes directly linked to each other is HD, DE, EF, EC and DG, while for Tree B it is HD, DE, EF, EG, and DC.

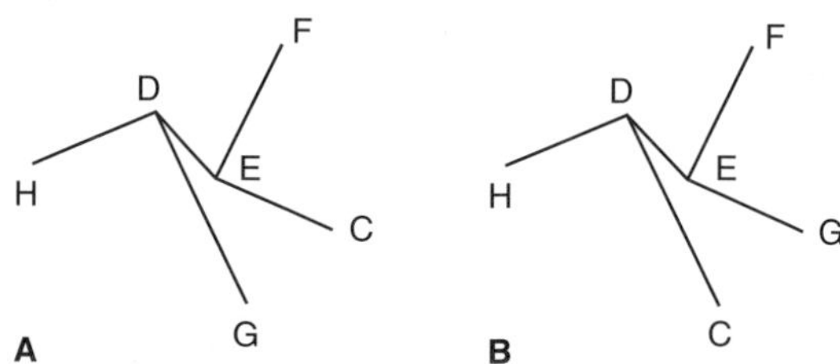

Figure T.4 Two similar looking representations of different tree topologies

Note that the pattern of links between nodes, and hence the tree topology, is independent of the lengths of the branches.

See also Phylogenetic Tree

TREE TRAVERSAL

Aidan Budd and Alexandros Stamatakis

The process of carrying out an action (computation) on each node of a tree, such that the action is carried out for each node exactly once. Phylogenetic computations often make use of a tree traversal procedure known as 'post-order tree traversal', which is applied in the context of a rooted tree structure and operates on the nodes in a particular order. The process can be carried out using a simple recursive function that takes a node from the tree as its input, and which is initially applied to the root node of the tree. The function:

- calls the function recursively for each of its descendant node (which results in the function being called also by all nodes in the descendant subtrees of these nodes)
- once all descendant nodes and their associated descendant subtrees have been traversed, carry out an action based on some value associated with the current node.

This approach is described as 'post-order' as the evaluation/action step occurs after the application of the same action to all descendant nodes.

Many phylogenetic calculations, for example calculating the likelihood of a tree (also known as Felsenstein pruning algorithm), use functions that are applied to each node of a tree, and require that the function has already been applied to all of its descendant nodes before it can return a value; this is the sequence of events in which actions are carried out in a post-order traversal, hence such traversals are often used in phylogenetic algorithms.

TREEVIEW X

Michael P. Cummings

A program that displays and print phylogenetic trees.

There are several options for the form of displayed trees (rectangular cladogram, slanted cladogram, phylogram, radial). The program accommodates a number of commonly used input file formats.

Executables are available for several platforms.

Related website

TreeView X home page	http://code.google.com/p/treeviewx/

Further reading

Page RDM (1996) TreeView: an application to display phylogenetic trees on personal computers. *Comput. Appl. Biosci.* 12: 357–358.

See also FigTree, PHYLIP, MacClade

T

TRINUCLEOTIDE REPEAT

Katheleen Gardiner

A tandem repeat of three nucleotides.

Repeats of this class have become of special interest because many trinucleotide repeats starting with C and ending with G (CAG, CGG, CTG, CCG) and present normally in ~5–~50 copies tend to be intergenerationally unstable. Repeats of this kind have been found to be associated with human genetic diseases when they expand beyond a critical copy number. About twenty neurological disease associations have been identified, including Fragile X Syndrome, Huntington's Disease and Myotonic Dystrophy. Repeats may be located within coding regions (e.g. polyglutamine tracts) or in regulatory regions where they may alter methylation status and disrupt gene expression.

Further reading

Brouwer JR, Willemsen R, Oostra BA (2009) Microsatellite repeat instability and neurological disease. *BioEssays* 31: 71–83.

Ferro P, dell'Eva R, Pfeffer U (2001) Are there CAG repeat expansion-related disorders outside the central nervous system? *Brain Res Bull* 56:259–264.

Grabczyk E, Kumari D, Usdin K (2001) Fragile X syndrome and Friedreich's ataxia: two different paradigms for repeat induced transcript insufficiency *Brain Res Bull* 56:367–373.

TRIPLE STORE, *SEE* DATABASE.

TRNA, *SEE* TRANSFER RNA.

TURN

Roman Laskowski and Tjaart de Beer

A reversal in the direction of the backbone of a protein that is stabilized by hydrogen bonds between backbone NH and CO groups and which is not part of a regular secondary structure region such as an alpha helix.

Turns are classified into various types, the most common being beta- and gamma-turns, which themselves have further subclasses. In a beta-turn, which consists of 4 residues, the CO group of residue i is usually, but not always, hydrogen-bonded to the NH group of residue $i+3$. A gamma-turn consists of 3 residues, and has a hydrogen bond between residues i and $i+2$.

Related websites

BTPERD	http://www.biochem.ucl.ac.uk/bsm/btpred/index.html
GammaPred	http://www.imtech.res.in/raghava/gammapred/

Further reading

Chan AWE, Hutchinson EG, Harris D, Thornton JM (1993) Identification, classification, and analysis of beta-bulges in proteins. *Protein Science* 2(10): 1574–1590.

See also Backbone, Secondary Structure of Proteins, Alpha Helix

TWILIGHT ZONE

Teresa K. Attwood

A zone of identity (in the range ~10–25%) within which sequence alignments may appear plausible to the eye but are not statistically significant (in other words, the sequence correspondence, and hence the alignment, could have arisen by chance).

Many different analysis techniques have been devised to analyse evolutionary relationships within the Twilight Zone, essentially to determine to what extent sequence similarities within this region reflect shared structural features, and are hence biologically significant. Extending the sensitivity of pairwise sequence comparisons, some of these methods use characteristics of multiple alignments (*e.g.*, consensus pattern and fingerprint approaches), others use complex weighting schemes that exploit mutation and/or structural information (*e.g.*, profile approaches). Examples of some of these techniques and their sensitivity ranges are shown in Figure T.5.

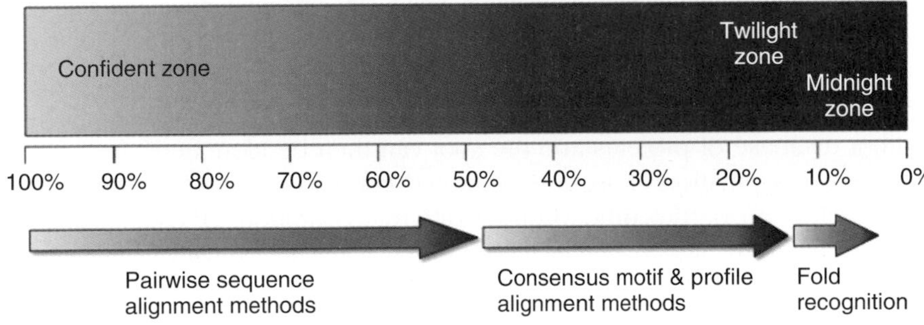

Figure T.5 Illustration showing a variety of sequence analysis techniques and their sensitivity ranges. Relationships can be confidently assigned to sequences using pairwise methods, down to around 50% identity; more sensitive consensus and/or profile methods are required to identify relationships down into the Twilight Zone; structure-based approaches are required to penetrate the Midnight Zone.

Each method offers a slightly different perspective, depending on the type of information used in the search; none should be regarded as giving the right answer, or the full picture – none is infallible. For best results, a combination of approaches should be used.

It should be noted that there is a theoretical limit to the effectiveness of sequence analysis techniques, because some sequences have diverged to such an extent that their relationships are only apparent at the level of shared structural features. These cannot be detected even using the most sensitive sequence comparison methods, because there is no significant sequence similarity between them. To detect relationships in this 'Midnight Zone', fold-recognition algorithms tend to be employed to determine whether a particular sequence is compatible with a given fold.

Further reading

Doolittle RF (1986) *In Of URFs and ORFs: A primer on how to analyse derived amino acid sequences.* University Science Books, Mill Valley, CA.

Rost B (1998) Marrying structure and genomics. *Structure* 6(3): 259–263.

Rost B (1999) Twilight zone of protein sequence alignments. *Protein Eng.* 12(2): 85–94.

See also Consensus Pattern, Fingerprint, Fold Recognition, Midnight Zone, Profile

TWO-DIMENSIONAL GEL ELECTROPHORESIS (2DE)

Dov Greenbaum

Technique used in proteomics to separate protein molecules.

Two dimensional gel-electrophoresis is a relatively old technology. Essentially a protein population is run through a gel in two dimensions: by charge (pI) through isoelectric focusing (this is done using a pH gradient) and by size via Sodium dodecyl sulphate Polyacrylamide Gel Electrophoresis (SDS-PAGE). SDS-PAGE uses SDS as a detergent to denature the proteins. Finally the proteins are visualized either through radioactive labeling or through staining (e.g. Coomasie blue).

Spots/proteins of interest are excised from the gel, digested with trypsin and then run though mass spectrometry. The masses resulting from this fragmentation can be compared with a database of proteins and the spot can then be identified.

Once a population of spots is identified gels can be compared and relative abundances of proteins can be determined under different conditions. Presently there are many programs that are designed to measure and compare the intensity of a spot within an image of the 2D gel.

The method allows for the separation of possibly thousands of proteins, each one can be seen as an individual spot on the gel. Additionally the procedure is fast, relatively easy and cheap.

Still there are many problems with the method as it does not resolve low mass and rare proteins easily. Additionally some protein types, such as membrane or hydrophobic proteins, have been shown to be problematic.

Related website

2-DE data	http://www.expasy.org/proteomics/mass_spectrometry_and_2-DE_data

Further reading

Görg A, et al. (2000) The current state of two-dimensional electrophoresis with immobilized pH gradients. *Electrophoresis*, 21: 1037–1053.

TYPE GENOME *SEE* REFERENCE GENOME

U

UCSC Genome Browser

Uncertainty

Ungapped Threading Test B (Sippl Test)

Unigene

UniProt

Universal Genetic Code, *see* Genetic Code.

Unsupervised Learning, *see* Supervised and Unsupervised Learning.

Unrooted Phylogenetic Tree, *see* Phylogenetic Tree.

Unscaled Phylogenetic Tree, *see* Branch.

UPGMA

Upstream

Concise Encyclopaedia of Bioinformatics and Computational Biology, Second Edition.
Edited by John M. Hancock and Marketa J. Zvelebil.

UCSC GENOME BROWSER

Obi L. Griffith and Malachi Griffith

The UCSC Genome Browser is a genome sequence and annotation database including graphical views and web-searchable datasets. Together with Ensembl it represents one of the two primary genome browser systems.

The project creates and applies software systems, which allow visual and programmatic exploration of the genomes of more than 57 species. Data 'tracks' are uploaded onto each genome by annotators from UCSC and all over the world. As an example, the current version of the human genome (hg19) is annotated by more than 100 data tracks representing mapping and sequencing details, phenotype and disease associations, genes and gene predictions, mRNA and EST tracks, expression, regulation, comparative genomics, variations, repeats, and more.

In addition to genomes and their annotation tracks, a number of associated tools facilitate analyses. These include Blat for sequence alignment, the Table Browser for querying and manipulating the Genome Browser annotation tables, Gene Sorter for displaying sorted tables of genes that are related to one another, an In Silico PCR tool for testing PCR primers, VisiGene for virtual microscope views of in situ images, and a Proteome browser. The genome browser system has also been extended for other purposes such as the Cancer Genomics Browser which provides web-based tools for visualization, integration and analysis of cancer genomics and associated clinical data.

Related websites

UCSC Genome Bioinformatics	http://genome.ucsc.edu/
Genome Browser	http://genome.ucsc.edu/cgi-bin/hgGateway
Blat	http://genome.ucsc.edu/cgi-bin/hgBlat
Table Browser	http://genome.ucsc.edu/cgi-bin/hgTables
Gene Sorter	http://genome.ucsc.edu/cgi-bin/hgNear
In Silico PCR	http://genome.ucsc.edu/cgi-bin/hgPcr
VisiGene	http://genome.ucsc.edu/cgi-bin/hgVisiGene
Proteome Browser	http://genome.ucsc.edu/cgi-bin/pbGateway
Cancer Genome Browser	https://genome-cancer.ucsc.edu/

Further reading

Fujita PA, et al. (2001) The UCSC Genome Browser database: update 2011. *Nucleic Acids Res.* 39(Database issue): D876–82.

Hsu F, et al. (2005) The UCSC Proteome Browser. *Nucleic Acids Res.* 33(Database issue): D454–458.

Karolchik D, et al. (2004) The UCSC Table Browser data retrieval tool. *Nucleic Acids Res.* 32(Database issue): D493–496.

Kent WJ (2002) BLAT the BLAST-like alignment tool. *Genome Res.* 12(4): 656–664.

Kent WJ, et al. (2002) The human genome browser at UCSC. *Genome Res.* 12(6): 996–1006.

Rosenbloom KR, et al. (2010) ENCODE whole-genome data in the UCSC Genome Browser. *Nucleic Acids Res.* 38(Database issue): D620–625.

Zhu J, et al. (2009) The UCSC Cancer Genomics Browser. *Nature Methods.* 6: 239–240.

U

UNCERTAINTY

Thomas D. Schneider

Uncertainty is a logarithmic measure of the average number of choices that a receiver or a molecular machine has available.

The uncertainty is computed as:

$$H = -\sum_{i=1}^{M} P_i \log_2 P_i \text{ bits/symbol}$$

where P_i is the probability of the ith symbol and M is the number of symbols. Uncertainty is the average surprisal. The information is the difference between the uncertainty before and after symbol transmission.

Related website

Information Is Not Entropy, Information Is Not Uncertainty!	http://alum.mit.edu/www/toms/information.is.not.uncertainty.html

See also Before State, After State, Small Sample Correction, Entropy, Negentropy, Molecular Efficiency, Negentropy

UNGAPPED THREADING TEST B (SIPPL TEST)

David Jones

A method for evaluating the efficacy of a particular set of potentials or scoring function for threading.

Sometimes called the 'Sippl Test' (after Manfred Sippl who popularized the test). The idea is to try to locate the native conformation for a particular protein sequence amongst a large set of decoys. These decoys are generated by taking fragments of others proteins, where the fragment length is equal to the length of the test protein. For example, a template protein of length 105 can provide 6 different decoy structures for a 100 residue protein in an ungapped threading test. Because the length of the template fragment is equal to the length of the test protein, there is no need to insert gaps in either protein, hence the test is labelled 'ungapped'.

A minimum requirement for a useful threading potential is that the energy calculated for the native structure of a protein should be lower than the calculated energy for all of the decoy conformations. Usually potentials are tested by performing the ungapped threading test for a large number of test proteins and calculating what fraction of the test proteins pass the test.

Despite the simplicity of the test, passing it must be considered a minimum requirement for a useful set of threading potentials. Many quite poor threading potentials achieve good results on the ungapped threading test.

UNIGENE

Malachi Griffith and Obi L. Griffith

The UniGene database attempts to identify the set of transcript sequences that come from a single transcription locus. Each locus thus defined is further annotated with available information on protein similarities in related species, gene expression patterns across tissue types, and availability of cDNA clone resources. The input data for the UniGene database are sequence repositories such as dbEST and mRNAs deposited in GenBank. UniGene produces both a transcript-based and genome-based build. Detailed descriptions of these two pipelines are available on the UniGene website. Briefly, expressed sequences are clustered either by pairwise alignments to each other or by alignment of all expressed sequences to a reference genome. Each UniGene record describes a single locus for a single species and the EST and mRNA members of the defined cluster of sequences aligning to that locus. UniGene provides tools to allow browsing of EST/mRNA data by library type and tissue source and also provides a 'Digital Differential Display' (DDD) tool for comparing EST profiles between tissues to identify genes with significantly different expression levels. UniGene is maintained by the NCBI.

Related website

Unigene homepage	http://www.ncbi.nlm.nih.gov/unigene

Further reading

Pontius JU, et al. (2003) UniGene: a unified view of the transcriptome. In: *The NCBI Handbook*. Bethesda (MD): National Center for Biotechnology Information.

Wheeler DL, et al. (2003) Database Resources of the National Center for Biotechnology. *Nucl Acids Res*. 31: 28–33.

UNIPROT

Marketa J. Zvelebil

Universal Protein Resource (UniProt) is a all-inclusive resource for protein sequence and annotation data. The UniProt databases consist of the UniProt Knowledgebase (UniProtKB), the UniProt Reference Clusters (UniRef), and the UniProt Archive (UniParc). UniProt is a collaboration between the European Bioinformatics Institute (EBI), the SIB Swiss Institute of Bioinformatics and the Protein Information Resource (PIR).

EBI and SIB used to produce Swiss-Prot and TrEMBL, and PIR produced the Protein Sequence Database. These two data sets coexisted with different protein sequence coverage and annotations. While TrEMBL (Translated EMBL Nucleotide Sequence Data Library) was originally created because sequence data was being generated at a pace that surpassed Swiss-Prot's ability to keep up. In 2002 the three institutes decided to combine their resources and formed the UniProt consortium.

The 2013_02 release of UniProtKB/TrEMBL contains 297,699,71 sequence entries.

U

Related websites

Stats	http://www.ebi.ac.uk/uniprot/TrEMBLstats/
UniProt	http://www.uniprot.org/

UNIVERSAL GENETIC CODE, *SEE* GENETIC CODE.

UNSUPERVISED LEARNING, *SEE* SUPERVISED AND UNSUPERVISED LEARNING.

UNROOTED PHYLOGENETIC TREE, *SEE* PHYLOGENETIC TREE.

UNSCALED PHYLOGENETIC TREE, *SEE* BRANCH.

UPGMA

Sudhir Kumar and Alan Filipski

A simple clustering method for reconstructing a phylogeny when rates of evolution are assumed constant over different lineages.

This Unweighted Pair-Group Method using arithmetic Averaging constructs a phylogenetic tree in a stepwise manner under the molecular clock assumption. It is a distance-based hierarchical clustering method in which the closest pair of taxa is clustered together in each step. The distance between any two clusters is defined to be the average of the pairwise distances between sequences, one from each cluster. If the original distances are not ultrametric, as is the case for most real data, an incorrect phylogeny may be produced.

Nowadays, UPGMA is rarely used, because significantly more elaborate methods and models exist for inferring divergence times and deploying/testing molecular clocks.

Further reading

Nei M, (1987) *Molecular Evolutionary Genetics*. New York, Columbia University Press.

Sneath PHA, Sokal RR (1973) *Numerical Taxonomy; The Principles and Practice of Numerical Classification*. San Francisco, W. H. Freeman.

See also Molecular clock, Evolutionary Distances

UPSTREAM

Niall Dillon

Describes a sequence distal to a specific point in the direction opposite to the direction of transcription (i.e. in a 5′ direction on the strand being transcribed).

V

Validation Measures for Clustering

Variance Components (Components of Variance, VC)

Variation (Genetic)

VarioML

VAST (Vector Alignment Search Tool)

VC, *see* Variance Components.

Vector, *see* Data Structure.

Vector Alignment Search Tool, *see* VAST.

VecScreen

VectorNTI

VEGA (Vertebrate Genome Annotation Database)

Vertebrate Genome Annotation Database, *see* VEGA.

Virtual Library

Virtual Screening

Virtualization

VISTA

Visualization, Molecular

Visualization of Multiple Sequence Alignments – Physicochemical Properties

Concise Encyclopaedia of Bioinformatics and Computational Biology, Second Edition.
Edited by John M. Hancock and Marketa J. Zvelebil.

V

VALIDATION MEASURES FOR CLUSTERING

Pedro Larrañaga and Concha Bielza

The evaluation of the goodness of a clustering is a problematic and controversial issue because there is no universal definition for what is a good clustering (Bonner, 1964). The evaluation remains on the eyes of the beholder. Nevertheless, several evaluation criteria have been proposed in the literature. These criteria are usually divided into two categories: internal and external.

Internal quality criteria measure the compactness of the clusters using some similarity measure. Intra-cluster homogeneity, inter-cluster separability or a combination of these two are used for measuring the compactness of the obtained clusters. A common characteristic of these internal quality criteria measures is that they do not use any external information besides the data itself. Examples include sum of squared error, scatter criteria, Condorcet's criterion (Condorcet, 1785) and the C-criterion (Fortier and Solomon, 1996).

External quality criteria measures examine how the structure of the clusters match to some predefined classification of the instances. Examples include the mutual information based measure, the precision-recall measure, the Rand index (Rand, 1971) and the adjusted Rand index (Hubert and Arabie, 1985).

Validation of clustering measures has been applied in gene expression data (Yeung et al., 2001; Handl, et al. 2005).

Further reading

Bonner R (1961) On some clustering techniques. *IBM Journal of Research and Development* 8: 22–32.

Condorcet MJA de (1785) *Essai sur l'Application de l'Analyse à la Probabilité des Décisions Rendues à la Pluralité des Voix*. l'Imprimerie Royal, Paris.

Fortier JJ, Solomon H (1996) Clustering procedures. In: *Proceedings of Multivariate Analysis*, pp. 439–506.

Handl J (2005) Computational cluster validation in post-genomic data analysis. *Bioinformatics* 21 (15): 3201–3212.

Hubert L, Arabie P (1985) Comparing partitions. *Journal of Classifications* 5: 193–218.

Rand WM (1971) Objective criteria for the evaluation for clustering methods. *Journal of the American Statistical Association* 66: 846–850.

Yeung KY, et al. (2001) Validating clustering for gene expression data. *Bioinformatics* 17 (4): 309–318.

VARIANCE COMPONENTS (COMPONENTS OF VARIANCE, VC)

Mark McCarthy and Steven Wiltshire

Describes the decomposition of the total variance of a quantitative trait into its various genetic and environmental components.

The phenotype P of individual *i* can be described according to the following basic model:

$$P_i = \mu + G_i + E_i$$

where μ is the overall population mean for the trait, and G_i and E_i are random variables representing the independent deviations from μ caused by individual i's genotype and non-shared (micro) environment, respectively: both G_i and E_i have a mean of zero and variances σ_G^2 and σ_E^2, respectively, in the population as a whole. Optionally, the effects of common family environment, C_i, shared by relatives, with population mean of zero and variance σ_C^2, and the fixed effects of a set of measured covariates can also be incorporated into this model. The overall variance of the trait, σ_P^2, can be written as the sum of these components of variance: thus, for the basic model, above:

$$\sigma_P^2 = \sigma_G^2 + \sigma_E^2$$

The genetic variance component, σ_G^2, can in turn be further partitioned into additive, σ_A^2, dominance, σ_D^2, and interaction, σ_I^2, components:

$$\sigma_G^2 = \sigma_A^2 + \sigma_D^2 + \sigma_I^2$$

reflecting, respectively, the effects on the phenotypic variation of the trait-susceptibility alleles themselves; of interactions between alleles at the same locus; and interactions due to epistasis between different loci.

This framework is the basis of a widely-used form of quantitative trait linkage analysis in which the total phenotypic variance is modeled as the sum of the effects of a quantitative trait locus (QTL), σ_Q^2 (the location of which is to be determined), a residual genetic component, σ_{PG}^2, which contains the effects of all other genes influencing the trait, and the non-shared environment component:

$$\sigma_P^2 = \sigma_Q^2 + \sigma_{PG}^2 + \sigma_E^2$$

As above, the two genetic components may themselves be partitioned into additive, dominance and interaction components and terms for familial environment and covariates are readily included. The covariance between pedigree members' trait values is expressed in terms of the variance components attributable to the QTL and residual polygenic effect, as described above (usually restricting this to the additive components of each, for simplicity) and measures of the genetic similarity between the relatives, namely the proportion of alleles shared identical by descent at an individual genetic marker, $\widehat{\pi}_{ij}$, and across the genome as a whole, 2ϕ, respectively:

$$Cov(i,j) = \sigma_Q^2 + \sigma_{PG}^2 + \sigma_E^2, \text{ when } i = j, \text{ and}$$
$$cov(i,j) = \widehat{\pi}_{ij}\sigma_Q^2 + 2\phi\sigma_{PG}^2, \text{when } i \neq j.$$

These variances and covariances are arranged as a variance-covariance matrix, Ω. The likelihood of the pedigree is calculated from Ω, and the vectors of trait values, y, and population means, μ, assuming multivariate normality, thus:

$$\ln(L) = c - \frac{1}{2}|\Omega| - \frac{1}{2}[(y-\mu)'\Omega^{-1}(y-\mu)]$$

A test of the null hypothesis of no linkage is achieved by comparing the log-likelihood of the data given $\sigma_Q^2 = 0$ (i.e no linkage) with that given $\sigma_Q^2 > 0$ (i.e. linkage). This likelihood ratio statistic can be expressed as a LOD (or logarithm of odds) score.

Examples: Hirschhorn et al. (2002), Perola et al. (2002) and Wiltshire et al. (2002) used variance components analysis to identify evidence for QTLs influencing adult stature in Swedish, Finnish, Quebecois and British populations.

Related website

Programs for variance components linkage analysis of quantitative traits include:

SOLAR	http://txbiomed.org/departments/genetics/genetics-detail?r=37

Further reading

Almasy L, Blangero J (1998) Multipoint quantitative trait linkage analysis in general pedigrees. *Am J Hum Genet* 62: 1198–1211.

Amos CI (1994) Robust variance-components approach for assessing genetic linkage in pedigrees. *Am J Hum Genet* 54: 535–543.

Falconer DS, Mackay TFC (1996) *Introduction to Quantitative Genetics*. Harlow: Prentice Hall, pp100–159.

Hirschhorn JN, et al. (2002) Genomewide linkage analysis of stature in multiple populations reveals several regions with evidence of linkage to adult height. *Am J Hum Genet* 69: 106–116.

Perola M, et al. (2002) Quantitative-trait-locus analysis of body-mass index and of stature, by combined analysis of genome scans of five Finnish study groups. *Am J Hum Genet* 69: 117–123.

Pratt SC, et al. (2000) Exact multipoint quantitative-trait linkage analysis in pedigrees by variance components. *Am J Hum Genet* 66: 1153–1157.

Wiltshire S, et al. (2002) Evidence for linkage of stature to chromosome 3p26 in a large U.K. Family data set ascertained for type 2 diabetes. *Am J Hum Genet* 70: 543–546.

See also Quantitative Trait, Linkage Analysis, IBD, LOD Score, Multifactorial Trait

VARIATION (GENETIC)

Matthew He

Variation in genetic sequences and the detection of DNA sequence variants genome-wide allow studies relating the distribution of sequence variation to a population history. This in turn allows one to determine the density of SNPS or other markers needed for gene mapping studies.

VARIOML

John M. Hancock

VariomML is an XML format for the description of sequence variant information. It is derived from the Observ-OM object model and designed for use by Locus-Specific Databases.

The data elements captured within VarioML fall under two main headings – the data source and the variant description. The description tag subsumes information on the gene source, reference sequence, name, pathogenicity, genetic origin, location within the reference sequence and the data sharing policy that applies.

Related website

VarioML	http://www.varioml.org/

Further reading

Byrne M, Fokkema IF, Lancaster O, et al. (2012) VarioML framework for comprehensive variation data representation and exchange. *BMC Bioinformatics* 13: 254.

See also Locus-specific Database, Human Variome Project

VAST (VECTOR ALIGNMENT SEARCH TOOL)

John M. Hancock and M.J. Bishop

A program to identify structurally similar proteins based on statistical criteria.

VAST attempts to identify 'surprising' similarities between proteins structures using the common statistical approach that the similarity would be expected to occur by chance with less than a certain threshold likelihood. It compares units of tertiary structure, which are defined as pairs of secondary structural elements within a protein. It bases a measure of similarity between structural elements on the the type, relative orientation, and connectivity of these pairs of secondary structural elements and derives expectations by drawing these properties at random. Results of the analysis are precompiled and made available via the NCBI web site. *VAST Search* allows the VAST algorithm to be used to search a new protein structure against the NCBI's protein structure database MMDB.

Related websites

VAST web site	http://www.ncbi.nlm.nih.gov/Structure/VAST/vast.shtml
VAST Search page	http://www.ncbi.nlm.nih.gov/Structure/VAST/vastsearch.html

Further reading

Gibrat J-F, et al. (1996) Surprising similarities in structure comparison. *Curr. Opin. Struct. Biol.* 6: 377–385.

Madej T, et al. (1995) Threading a database of protein cores. *Protein Struct. Funct. Genet.* 23: 356–369.

See also Protein Data Bank, NCBI

VC, *SEE* VARIANCE COMPONENTS.

VECTOR, *SEE* DATA STRUCTURE.

VECTOR ALIGNMENT SEARCH TOOL, *SEE* VAST.

VECSCREEN

John M. Hancock and M.J. Bishop

A program to screen sequences for contaminating vector sequences.

VecScreen uses BLAST to screen sequences for vector sequences held in the UniVec database of vector sequences. Because contaminant sequences are likely to be identical to known vector sequences, or almost so, parameters for the BLAST search are set accordingly. VecScreen returns four types of positive match, according to the length and quality of the match: strong, moderate, weak and segment of suspect origin.

Related website

VecScreen web site	http://www.ncbi.nlm.nih.gov/VecScreen/VecScreen.html

See also BLAST, Chimeric Sequence, Vector

VECTORNTI

John M. Hancock

A desktop package that provides a set of sequence manipulation and analysis facilities.

Unlike other desktop bioinformatics packages VectorNTI is based around a database which allows data and results to be stored and shared within a work group. The package supports the following classes of function: database exploration, BLAST and Entrez searches, molecule editing, PCR primer design, sequence annotation, back translation, downloading reference genomes, multiple sequence alignment, viewing PDB files, aligning expressed sequences to genomes, simulating cloning experiments, assembling synthetic DNA sequences, and assembling new sequence contigs.

Related website

Commercial suppliers	http://www.invitrogen.com/

VEGA (VERTEBRATE GENOME ANNOTATION DATABASE)

Obi L. Griffith and Malachi Griffith

The VEGA database is a collection of manually curated gene annotations of vertebrate genomes with particular focus on the human, mouse and zebrafish genomes. The effort is carried out primarily by the HAVANA group at the Welcome Trust Sanger Institute with contributions from multiple other genome centers. It can be considered a parallel and complementary effort to the automated annotation that occurs in the Ensembl pipeline. Since VEGA only displays manually annotated genes on the most current genome assemblies, these annotations can be considered a 'gold standard' and a useful reference for predicting gene structures on low coverage genomes from other vertebrates. VEGA is also annotating all loci identified by the CCDS project to ensure maximum compatibility between annotation systems. VEGA annotations can be accessed through the VEGA website which is built on the same underlying schema as Ensembl and also through the Ensembl BioMart system. Where possible, VEGA annotations are also linked from or merged with Ensembl gene records adding significant value to both resources.

Related website

Vertebrate Genome Annotation	http://vega.sanger.ac.uk/

Further reading

Wilming LG, et al. (2008) The vertebrate genome annotation (Vega) database. *Nucleic Acid Res.* 36(Database issue): D753–760.

VERTEBRATE GENOME ANNOTATION DATABASE, *SEE* VEGA.

VIRTUAL LIBRARY

Bissan Al-Lazikani

A database of chemical structures correctly enumerated and represented in a way that allows chemical searching and clustering. The chemical structures may correspond to actual compounds that exist in physical form, or they may be theoretical compounds not yet synthesized. Virtual libraries are used for virtual screening and hypothesis generation. They can be generated from existing databases (such as bioactivity databases; chemical vendor catalogues) or designed by computational chemists. In drug discovery, organizations typically contain a combination of proprietary chemical structures as well as public or commercially available ones. A number of virtual libraries are popular in academic chemoinformatics analysis, some of which are listed below.

Related websites

ZINC	http://zinc.docking.org/
ENZO FDA approved drug library	http://www.enzolifesciences.com/BML-2841/fda-approved-drug-library/
The Cambridge Structural Database (CSD)	http://www.ccdc.cam.ac.uk/products/csd/

See also Virtual Screening

VIRTUAL SCREENING

Bissan Al-Lazikani

Computational search to identify compounds from a virtual library that are likely to bind the protein in question. There are multiple ways in which this search could be performed but ultimately all are ligand-initiated or receptor (protein)-initiated methods. Ligand initiated screening involves defining one or more queries that describe a desired ligand, for example, using pharmacophores to define the ligand.

See also Pharmacophores

VIRTUALIZATION

James Marsh, David Thorne and Steve Pettifer

Virtualization is the name given in computing to the process of making a resource appear to be real when actually emulated through some combination of hardware and software.

At the simplest level almost all computing involves some virtual devices: physical disks are divided into partitions or logical volumes that appear as distinct devices to the operating system; virtual memory presents large contiguous blocks of memory pages to software processes that the operating system maps to physical memory behind the scenes; network storage volumes appear as if they are local disks. In these cases abstracting the complexity of specific hardware to a more general model simplifies software design and system management.

Increasingly whole computer systems are virtualized. This allows a single physical machine to manage what appears to the outside world to be multiple distinct systems. The management software (often called a hypervisor) of the physical host presents a full range of devices to the guest operating systems that in many cases are oblivious that these disks, network cards, screens, input devices and so on are actually being emulated and mapped in a variety ways to the underlying physical hardware.

Virtual machines are used for many reasons such as abstracting complex systems to be more conceptually simple, flexibly allocating resources according to demand and reducing energy costs by consolidating multiple servers together.

'Cloud computing' is often used as a marketing term for the provision of remote virtual machines by services such as Amazon EC2 and Microsoft's Windows Azure. Such services are useful to dynamically adapt computing power to server tasks without the customer needing to keep additional underutilised physical hardware in reserve.

Related websites

Vbox	https://www.virtualbox.org/
Vmware	http://www.vmware.com/virtualization/virtual-machine.html
ec2	http://aws.amazon.com/ec2/

See also Operating Systems, Virtual Memory, Logical Volumes, Network Attached, Storage, Distributed Computing

V

VISTA
John M. Hancock

Package for aligning and investigating long genomic sequences.

VISTA (Visualization Tools for Alignments) consists of three core programs, AVID, mVISTA and rVISTA. AVID is a sequence alignment program designed to align long segments of genomic sequence rapidly. It approaches the problem of alignment by finding a set of maximal repeated substrings between two sequences. These are found by concatenating the two sequences and identifying repeated substrings using suffix trees. Alignment is then carried out by defining anchor sequence pairs and aligning between them using the Needleman–Wunsch algorithm. mVISTA allows visualization of AVID alignments while rVISTA identifies transcription factor binding sites that are conserved in an AVID alignment.

Related website

VISTA web site	http://genome.lbl.gov/vista/

Further reading

Bray N, et al. (2003) AVID: A global alignment program. *Genome Res.* 13: 97–102.

Loots GG, et al. (2002) rVista for comparative sequence-based discovery of functional transcription factor binding sites. *Genome Res.* 12: 832–839.

See also Multiple Sequence Alignment, Needelman–Wunsch Algorithm, Transcription Factor

VISUALIZATION, MOLECULAR
Eric Martz

Looking at a model of a molecule in order to grasp its three-dimensional structure.

Molecular visualization involves generation of three-dimensional images or physical models of molecules that can be examined from multiple perspectives. Computer renderings can be rotated and viewed from varied distances ('zoomed') with views targeted to selected focal points. Computer visualization facilitates rendering of the model in different modes (back-bone, schematic ('cartoon'), space-filled, surface, etc.) and with different color schemes (see models). Optionally, some parts may be hidden to avoid obscuring moieties of interest. Some visualization software can display animations ('movies' or morphs) of conformational changes in proteins (see Websites).

'Molecular modeling' includes visualization, but 'modeling' may also include the ability to change the conformation or covalent structure of the molecule, while visualization, strictly speaking, may be limited to displaying a model using software tools that are unable to modify its structure. Some popular modeling software packages include DeepView (free), WHAT IF, Insight, Quanta, and for crystallographic modeling, O (see Websites).

Visualization of macromolecules requires that data for a three-dimensional model be available, either from empirical or theoretical sources. Such data are typically represented as atomic coordinate files that specify the positions of each atom in space. Most empirical macromolecular structures are obtained by X-ray crystallography, which produces an electron density map. The amino acid sequence of the protein is usually known in advance, and real-time interactive visualization software is used to assist in visualizing the electron density map, and fitting the amino acids into it. One of the most popular computer systems used by crystallographers for this purpose in the 1980s was manufactured by Evans and Sutherland, and cost about a quarter of a million US dollars at that time.

By the end of the 1980s, powerful computers were much less expensive. The first widely popular software that brought macromolecular visualization to ordinary personal computers was the free program MAGE released in 1992 by David C. Richardson. This supported presentations of the authors' viewpoints called Kinemages, as well as interactive visualization. In 1993, Roger A. Sayle released RasMol, with excellent support for self-directed interactive visualization, which also became widely popular. In the mid-1990s, a team at GlaxoWellcome released a powerful visualization and modeling program called Swiss-PDBViewer (recently renamed DeepView). In 1996, RasMol's open source code was adapted to a web browser plugin, Chime ('Chemical MIME'), initially used for web-delivered presentations of the authors' viewpoints. RasMol was also used as part of the foundation of WebLab Viewer. In 1999, Chime was employed as the basis for Protein Explorer, a web-based user-interface that facilitates self-directed interactive visualization while freeing the user from having to learn the RasMol command language. About the same time, the US National Center for Biotechnology Information developed a completely independent self-directed interactive macromolecular visualization program, Cn3D. Galleries of sample images, as well as visualization software itself can be obtained through the World Index of Molecular Visualization Resources (see Websites below).

Software that allows interactive rotation of the model compromises resolution and image quality in order to achieve adequate rotation speed in real time. Most figures of three-dimensional structures published in scientific journals are generated by software that emphasizes image quality and resolution at the expense of speed of image generation. The most popular is probably Molscript, a program that emits an image description meta-language that serves as input to rendering software such as Raster3D or POVRay.

Relevant websites

World Index of Molecular Visualization Resources	http://molvisindex.org
RasMol (latest version)	http://openrasmol.org
Mage	http://kinemage.biochem.duke.edu/
Database of Molecular Motions	http://molmovdb.mbb.yale.edu/MolMovDB/
Raster3D	http://www.bmsc.washington.edu/raster3d/raster3d.html
POV-Ray	http://www.povray.org/

Further reading

Bernstein, HJ (2000) Recent changes to RasMol, recombining the variants. *Trends Biochem. Sci.* 25: 453–455.

Guex N, Diemand A, Peitsch MC (1999) Protein modelling for all. *Trends Biochem. Sci.* 24: 364–367.

Martz E (2002) Protein Explorer: easy yet powerful macromolecular visualization. *Trends Biochem. Sci.* 27: 107–109.

Richardson DC, Richardson JS (1992) The kinemage: a tool for scientific communication. *Protein Sci.* 1: 3–9.

Wang Y, Geer LY, Chappey C, Kans JA, Bryant SH (2000) Cn3D: sequence and structure views for Entrez. *Trends Biochem. Sci.* 25: 300–302.

See also Protein Data Bank, Backbone, Models, Molecular

Eric Martz is grateful for help from Eric Francoeur, Peter Murray-Rust, Byron Rubin and Henry Rzepa.

VISUALIZATION OF MULTIPLE SEQUENCE ALIGNMENTS – PHYSICOCHEMICAL PROPERTIES

Teresa K. Attwood

Multiple sequence alignments are central to protein sequence analysis: they underpin evolutionary studies, phylogenetic analysis, identification of conserved motifs and domains, and so on. How they are visualised determines what users see, and hence what information can be extracted from them.

A variety of methods can be used to depict alignments. For example, an alignment may be coloured according to the degree of conservation in each of its columns (top line in Figure V.1); local alignment blocks may be highlighted by shading identical and/or similar residues within those regions (middle of Figure V.1); or alignments may coloured according to their constituent amino acid physicochemical properties (bottom of Figure V.1). Each of these approaches gives a different view of the degree of relatedness of sequences within an alignment: here, the top view suggests a greater degree of conservation than the middle view, while the middle suggests a greater degree of conservation than the bottom view, even though the same region of the same alignment is shown in each case. In order to comprehend the information content of an alignment, it is therefore essential to understand the particular 'metaphor' implied by the colouring scheme used in the visualization.

The bottom view in Figure V.1 employs an intuitive metaphor (borrowed largely from the realms of chemistry and physics) that colours amino acid residues broadly to reflect their physicochemical properties – details of the scheme are shown below. Use of such a scheme obviates much of the cognitive effort required to interpret alignments, as their constituent residue properties can be understood at a glance: e.g., sulphur-containing residues that may participate in disulphide bond formation are coloured yellow; residues with acidic or basic side-chains, which may form salt bridges, are coloured red and blue, respectively; polar residues, which may form hydrogen bonds, are green; and so on. Grouping similar residues in this way helps to highlight conserved residues and physicochemically related residue groups (this can be useful, for example, when analysing conserved motifs and domains that characterise particular gene and domain families).

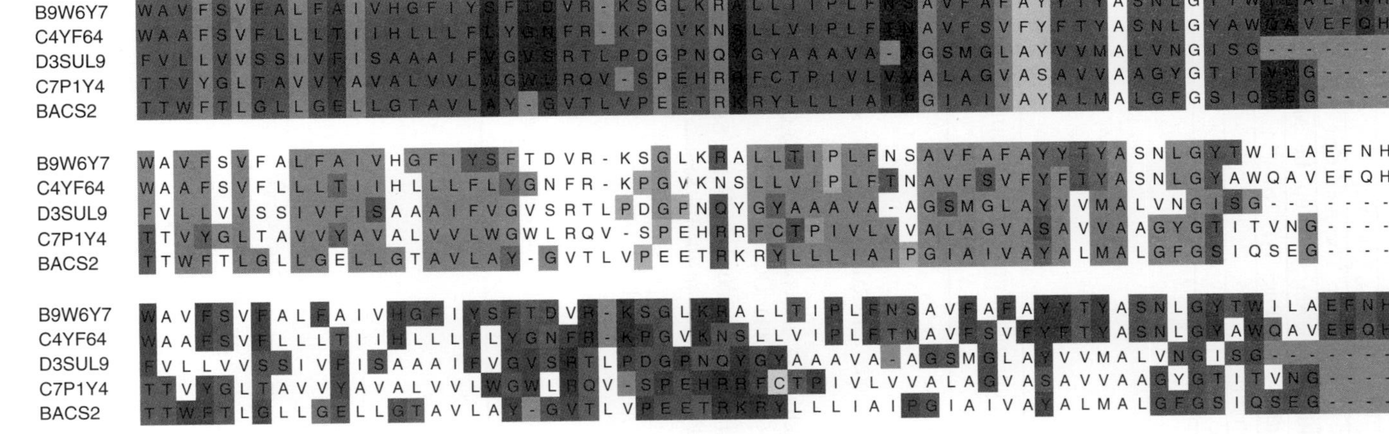

Figure V.1 Part of a multiple sequence alignment represented using different colouring schemes. The top view shows consensus colouring, where alignment columns are shaded according to the number of residues found at each position — the lighter the colour, the more conserved the position. In the middle view, regions are coloured using a hybrid scheme, in which colours may denote both general amino acid property groups and the degree of conservation in a given part of the alignment. In the bottom view, residues are coloured only according to their physicochemical properties. The different colouring approaches offer very different perspectives on the degree of conservation in this portion of the alignment, in some cases suggesting greater degrees of similarity than they really possess. (See Colour plate V.1)

Property	Residue	Colour
Acidic	Asp, Glu	Red
Basic	His, Lys, Arg	Blue
Polar uncharged	Ser, Thr, Asn, Gln	Green
Hydrophobic aliphatic	Ala, Val, Leu, Ile, Met	White
Hydrophobic aromatic	Phe, Tyr, Trp	Purple
Disulphide bond former	Cys	Yellow
Structural properties	Pro, Gly	Brown

Relevant website

CINEMA Colours	http://utopia.cs.man.ac.uk/utopia/documentation/miscellaneous-topics

Further reading

Lord PW, Selley JN, Attwood TK (2002) CINEMA-MX: a modular multiple alignment editor. *Bioinformatics* 18(10): 1402–1403.

Parry-Smith DJ, Payne AWR, Michie AD, Attwood TK (1998) CINEMA – A novel Colour INteractive Editor for Multiple Alignments. *Gene* 221(1): GC57–63.

Pettifer S, Thorne D, McDermott P, et al. (2009) Visualising biological data: a semantic approach to tool and database integration. *BMC Bioinformatics* 10: S18.

See also Alignment, multiple, Amino Acid

Web Ontology Language (OWL)
Web Services
Weight Matrix, *see* Sequence Motifs: Prediction and Modeling.
Whatcheck
WhatIf
WormBase
WSDL/SOAP Web Services

Concise Encyclopaedia of Bioinformatics and Computational Biology, Second Edition.
Edited by John M. Hancock and Marketa J. Zvelebil.

WEB ONTOLOGY LANGUAGE (OWL)

John M. Hancock

OWL is the W3C standard for the creation and exchange of ontologies. It is underpinned by an expressive Description Logic and can be submitted to a reasoning engine to infer taxonomies and check consistency.

The current version of OWL, OWL2, was introduced in 2009 and provides three sublanguages: OWL Lite, which was intended for users primarily needing a classification hierarchy and simple constraints; OWL DL, which has the form of a Description Logic, and OWL Full.

Related website

Wikipedia	http://en.wikipedia.org/wiki/Web_Ontology_Language

See also Description Logic

WEB SERVICES

Carole Goble and Katy Wolstencroft

The World Wide Web Consortium (W3C) defines a Web Service as 'a software system designed to support interoperable machine-to-machine interaction over a network.' Web Services represent the latest generation of web communication, allowing sophisticated interactive content. In the life sciences, many data analysis tools and database resources have web service interfaces. This means that they can be called from other applications programmatically (e.g. web based applications, or components of workflows). BioCatalogue.org is a registry of Web Services relevant to the Life Sciences.

Web services are largely divided into those with a WSDL/SOAP interface and those with a RESTful interface.

Related websites

BioCatalogue	http://www.biocatalogue.org
W3C Web Services Activities	http://www.w3.org/2002/ws/

See also WSDL/SOAP Web Services, RESTful Web Services, Scientific Workflows

WEIGHT MATRIX, *SEE* SEQUENCE MOTIFS: PREDICTION AND MODELING.

W

WHATCHECK
Roland Dunbrack

A program for assessing the stereochemical quality of a protein structure, developed by Gert Vriend and colleagues. Whatcheck reports deviations in covalent geometry, dihedral angles, steric clashes, and hydrogen bonding.

Related website

Whatcheck	http://swift.cmbi.ru.nl/gv/whatcheck/

Further reading

Hooft RW, et al. (1996) Errors in protein structures. *Nature* 381, 272.

Hooft RW, et al. (1997) Objectively judging the quality of a protein structure from a Ramachandran plot. *Comput Appl Biosci* 13: 425–430.

Rodriguez R, Chinea G, Lopez N, Pons T, Vriend G (1998) Homology modeling, model and software evaluation: three related resources. *CABIOS* 14: 523–528.

WHATIF
Roland Dunbrack

A multi-functional program for protein structure analysis, comparative modeling, and visualization developed by Gert Vriend and colleagues. WHATIF has a graphical interface for menu-driven commands, including access to a database of protein fragments for loop and side-chain modeling.

Related website

Whatif	http://swift.cmbi.ru.nl/whatif/

Further reading

Hekkelman ML, Te Beek TA, Pettifer SR, Thorne D, Attwood TK, Vriend G (2010) WIWS: a protein structure bioinformatics Web service collection. *NAR.* 38: W719–723.

Rodriguez R, et al. (1998) Homology modeling, model and software evaluation: three related resources. *Bioinformatics* 14: 523–528.

Vriend G (1990) WHAT IF: A molecular modeling and drug design program. *J. Mol. Graphics* 8: 52–56.

See also Comparative Modeling

WORMBASE
Obi L. Griffith and Malachi Griffith

WormBase is an international collaboration providing comprehensive information about the genetics, genomics, and biology of *C. elegans* and related nematodes. It makes available current releases of genome assemblies along with numerous annotation tracks such as gene models, alleles, transposon insertions, SNPs, and many more. Data are accessed through the WormBase master site via a genome browser (based on GMOD/GBrowse) or WormBase BioMart. Other analysis tools include built-in BLAST and BLAT aligners, a 'Synteny Browser', and tools for identifying genetic markers, or searching by cell, neuron, life stage, protein, sequence, expression pattern, and more.

Related website

WormBase homepage	http://www.wormbase.org/

Further reading

Yook K, et al. (2012) WormBase 2012: more genomes, more data, new website. *Nucleic Acids Res.* 40(D1):D735–D741.

See also Model Organism Database

WSDL/SOAP WEB SERVICES

Carole Goble and Katy Wolstencroft

The Web Services Description Language (WSDL) is the W3C standard format for describing a Web Service interface. It is an XML based language that specifies how a service can be called, what input parameters it requires, and what will be returned, in a machine-readable format. The WSDL description (also known as the WSDL file) represents a service endpoint. This endpoint can be read by client applications in order to understand what service operations are available.

WSDL files generally describe SOAP services. SOAP (originally referred to as the Simple Object Access Protocol, but now simply referred to as SOAP), is a protocol for transferring structured information usually in XML (Extensible Markup Language). It is the basic messaging framework for web services and is made up of three parts: 1) an envelope that defines what is in the message and how to process it; 2) a set of encoding rules for expressing instances of application-defined datatypes; and 3) a convention for representing procedure calls and responses.

SOAP is a mature and well-defined approach that is particularly applicable for applications that need: asynchronous processing and invocation; formal contracts; and stateful operations.

Related websites

WSDL	http://www.w3.org/TR/wsdl
SOAP	http://www.w3.org/TR/soap/

See also Web Services, RESTful Services

X-Ray Crystallography for Structure Determination

X Chromosome, *see* Sex Chromosome.

Xenbase

Xenolog (Xenologue)

XML (eXtensible Markup Language)

Concise Encyclopaedia of Bioinformatics and Computational Biology, Second Edition.
Edited by John M. Hancock and Marketa J. Zvelebil.

X-RAY CRYSTALLOGRAPHY FOR STRUCTURE DETERMINATION

Liz Carpenter

X-ray crystallography is a technique for studying the structures of macromolecules and small molecules in the crystalline state.

Crystals of the molecule under study are exposed to a beam of X-rays usually of one wavelength. X-rays are scattered by the electrons in the atoms of the crystal. The regular array of atoms in the crystal results in constructive interference in some directions. This gives rise to diffraction spots which are recorded on a detector. Since there are no lenses available for X-rays the image of the molecule in the crystal can not be obtained directly from the diffraction pattern. An image of the molecule in the crystal can however be obtained computationally. The density of electrons at each point in the repeat unit of the crystal (the unit cell) can be calculated given a knowledge of the intensities of the diffraction spots and their phase. Detectors can record intensity but not phase information, so a variety of techniques (molecular replacement, heavy atom derivatives, MAD, SAD, SIRAS) are used to obtain information on the phase of the diffraction spots. An initial electron density map can then be calculated, from which the positions of the atoms in the protein structure can be obtained. X-ray crystallography has the advantage that it can be used to obtain three dimensional structures of any molecule from a few atoms to millions of atoms. The disadvantage of X-ray crystallography is that well ordered crystals are required for data collection.

Further reading

Blundell TL, Johnson LN (1976) *Protein Crystallography*, Academic Press.

Drenth J (1999) *Principles of Protein X-ray Crystallography*, second edition. Springer, Verlag.

Glusker JP, Lewis M, Rossi M (1994) *Crystal Structure Analysis for Chemists and Biologists*, VCH Publishers.

Ladd MFC, Palmer RA (1994) *Structure Determination by X-ray Crystallography*. Plenum Press, New York.

Rhodes G (1993) *Crystallography Made Crystal Clear*, Academic Press.

See also Diffraction of X-rays, Crystal Macromolecular, Crystallization, Space Group, Phase Problem, Heavy Atom Derivative, Refinement, Electron Density Map, R-Factor

X CHROMOSOME, *SEE* SEX CHROMOSOME.

X

XENBASE

Dan Bolser

Xenbase is a model organism database providing tools, genomic data and biological information for the frogs *Xenopus laevis* and *Xenopus tropicalis*. Data include a gene catalogue, anatomical expression information and a genome browser. The resource includes a collection of papers and textbooks that can be intelligently searched via a text-mining system and a directory of people, labs and organizations in the Xenopus community.

Related websites

Xenbase	http://www.xenbase.org/
Wikipedia	http://en.wikipedia.org/wiki/Xenbase
MetaBase	http://metadatabase.org/wiki/XenBase

Further reading

Bowes JB, Snyder KA, Segerdell E, et al. (2010) Xenbase: gene expression and improved integration. *Nucleic Acids Res.* 38: 607–612.

See also FlyBase, GEISHA

XENOLOG (XENOLOGUE)

Laszlo Patthy

Homologous genes acquired through horizontal transfer of genetic material between different species are called xenologs.

See also Homology

XML (EXTENSIBLE MARKUP LANGUAGE)

Eric Martz

XML is a markup language that identifies data elements within a document. It is widely considered the universal format for structured documents and data on the Web. Its power is that the recipient of a document can recognize the contained data elements following a worldwide standard, and can therefore more easily utilize and display the data in any desired manner.

XML annotates content, while, in contrast, HTML (HyperText Markup Language) annotates primarily appearance. HTML has a fixed set of tags, while XML (the X stands for eXtensible) is designed to accomodate an ever-growing set of discipline-specific tags. XML is an emerging standard, whereas HTML is a standard already in universal use for annotating information on the web.

HTML specifies a static appearance for a document when displayed in a web browser. HTML alone does not enable the recipient web browser to identify the types of data elements contained in the document. For example, HTML can specify that a particular phrase be displayed centred in a large red bold font, but tells nothing about the kind of information in the phrase. In contrast, XML identifies the type of information. For example, XML tags could identify a certain portion of a document as the amino acid sequence for a specified protein. Presently, the content of a document that is of interest to a user must usually be selected by non-standard methods that are often inadequately automated. Once the desired information is extracted in computer-readable form, the format employed is often understood by only a subset of the programs in the world that deal with that type of information. XML provides a worldwide standard for content identification.

XML itself is set of rules for writing discipline-specific implementations. The specifications for XML are managed by the World-Wide Web consortium in the form of W3C Recommendations. Each subject discipline must develop its own a 'Document Type Definition' (DTD) or 'XML Schema' (*see* examples below). Discipline-specific content within a document is identified by a 'namespace'. A document can contain mixtures of types of content, that is, multiple namespaces. Within each namespace, the allowed content-identifying tags, and the nesting relationships of the tags, are specified by the corresponding Schema or DTD. For example, in the Chemical Markup Language (CML) Schema or DTD, one element is 'cml:molecule'. It can have child elements such as 'cml:atomArray' and 'cml:bondArray'. The cml:bondArray element has 'cml:atom' elements as its children. The XML Schema for a particular discipline, such as CML, is enforced by a validation process. Each Schema has a primary website, where the discipline-specific tags and and rules for validation are defined.

Examples of well-defined XML languages for common generic use include:

1. XHTML (a general document markup language derived from HTML)
2. SVG (scalable vector graphics) for expressing diagrams and visual information
3. MathML (Mathematical markup language).

Specific scientific, technical and medical (STM) namespaced applications of XML include:

1. STMML (for description of scientific units and quantities)
2. CML (chemical markup language) for expressing molecular content
3. CellML for computer-based biological models
4. GAME (genome annotation markup elements)
5. MAGE-ML (MicroArray and Gene Expression Markup Language)
6. Molecular Dynamics (Markup) Language (MoDL)
7. StarDOM – Transforming Scientific Data into XML
8. Bioinformatic Sequence Markup Language (BSML)
9. BIOpolymer Markup Language (BIOML)
10. Gene Expression Markup Language (GEML)
11. GeneX Gene Expression Markup Language (GeneXML)
12. Genome Annotation Markup Elements (GAME)
13. XML for Multiple Sequence Alignments (MSAML)
14. Systems Biology Markup Language (SBML)
15. Protein Extensible Markup Language (PROXIML).

Relevant website

Chemical Markup Language	http://www.xml-cml.org

Further reading

Bosak J, Bray T (1999) XML and the second generation web. *Scientific American,* May 6 (http://www.sciam.com).

Ezzell C (2001) Hooking up biologists. *Scientific American,* November 19.

Murray-Rust P, Rzepa HS (2002) Scientific publications in XML – towards a global knowledge base. *Data Science.* 1: 84–98.

See also Atomic Coordinate File, MIME Type

Eric Martz is grateful for help from Eric Francoeur, Peter Murray-Rust, Byron Rubin and Henry Rzepa.

Yeast Deletion Project (YDPM)	Yule Process

Concise Encyclopaedia of Bioinformatics and Computational Biology, Second Edition.
Edited by John M. Hancock and Marketa J. Zvelebil.

Y

YEAST DELETION PROJECT (YDPM)

Dan Bolser

A resource containing data for a near complete collection of gene-deletion mutants of the yeast *Saccharomyces cerevisiae*. Each strain carries a precise deletion of one of the genes in the genome, and has been functionally characterized. The fitness contribution of each gene was quantitatively assessed under six different growth conditions. Determining the effect of gene deletion is a fundamental approach to understanding gene function, providing an unparalleled resource for the scientific community.

Related websites

YDPM	http://yeastdeletion.stanford.edu
Wikipedia	http://wikipedia.org/wiki/Yeast_deletion_project
MetaBase	http://metadatabase.org/wiki/YDPM

Further reading

Giaever G, et al. (2002) Functional profiling of the *Saccharomyces cerevisiae* genome. *Nature* 418:387–391.

See also SGD

YULE PROCESS

Matthew He

The Yule process is a birth process with a constant birth rate. The number of births in time interval (0, t), Y (t), has a negative binomial distribution, which is the same as the distribution of the number of new species of a genus produced during (0, t) in Yule's study of evolution.

Z

z-score
Zero Base, *see* Zero Coordinate.
Zero Coordinate (Zero Base, Zero Position)
Zero Position, *see* Zero Coordinate.
Zeta Virtual Dihedral Angle
Zipf's Law, *see* Power Law.

Concise Encyclopaedia of Bioinformatics and Computational Biology, Second Edition.
Edited by John M. Hancock and Marketa J. Zvelebil.

Z

Z-SCORE

Matthew He

z-score or z-value is a mathematical measure to express the divergence of the experimental result x from the most probable result mean μ as a number of standard deviations σ. The larger the value of z, the less probable the experimental result is due to chance.

ZERO BASE, *SEE* ZERO COORDINATE.

ZERO COORDINATE (ZERO BASE, ZERO POSITION)

Thomas D. Schneider

The zero coordinate is the position by which a set of binding sites is aligned. Not having a zero as part of a coordinate system is a disadvantage because it makes computations tricky.

For consistency, one can place the zero coordinate on a binding site according to its symmetry and some simple rules:

- *Asymmetric sites*: at a position of high sequence conservation, or the start of transcription or translation
- *Odd symmetry site*: at the centre of the site
- *Even symmetry site*: for simplicity, the suggested convention is to place the zero base on the side of the axis so that the bases 0 and 1 surround the axis.

See also Binding Site Aymmetry, Alignment, pairwise, Alignment, multiple

ZERO POSITION, *SEE* ZERO COORDINATE.

ZETA VIRTUAL DIHEDRAL ANGLE

Roman Laskowski and Tjaart de Beer

Defined by a protein's four main chain atoms: Cα-N-C′-Cβ Its absolute value lies around 33.9°. A positive value indicates a normal L amino acid, whereas a negative value identifies the very rare D-conformation.

See also Main Chain, Amino Acids.

ZIPF'S LAW, *SEE* POWER LAW.

AUTHOR INDEX

Josep F. Abril (22)
Alternative Splicing, 18
ENCODE (Encyclopedia of DNA Elements), 185
FGENES, 220
GASP (Genome Annotation Assessment Project, E-GASP), 241
GENCODE, 243
Gene Annotation, 244
Gene Annotation, formats, 245
Gene Annotation, hand-curated, 246
Gene Annotation, visualization tools, 247
Gene Prediction, 255
Gene Prediction, ab initio, 257
Gene Prediction, accuracy, 261
Gene Prediction, alternative splicing, 263
Gene Prediction, comparative, 264
Gene Prediction, homology-based (Extrinsic Gene Prediction, Look-Up Gene Prediction), 266
Gene Prediction, NGS-based, 268
Gene Prediction, non-canonical, 269
Gene Prediction, pipelines, 270
GENEID, 274
GENSCAN, 287
Selenoprotein, 661
Translation End Site, 736

Bissan Al-Lazikani (27)
AlogP, 17
Bemis and Murcko Framework (Murcko Framework), 48
Binding Affinity (K_D, K_I, IC_{50}), 52
Bioactivity Database, 56
Chemical Biology, 82
Chemical Hashed Fingerprint, 82
Chemoinformatics, 83
ClogP, 92
Compound Similarity and Similarity Searching, 109
Drug-like, 163
Druggable Genome (Druggable Proteome), 165
Druggability, 164
InChI (International Chemical Identifier), 334
InChi Key, 335
Ligand Efficiency, 378
LogP (ALogP, CLogP), 390
Mol Chemical Representation Format, 443
Oral Bioavailability, 499
Pharmacophore, 522
QSAR (Quantitative Structure Activity Relationship), 583
Rule of Five (Lipinski Rule of Five), 643
SAR (Structure–Activity Relationship), 650
Scaffold, 650
SDF, 655
SMILES, 686
Virtual Library, 764
Virtual Screening, 765

Patrick Aloy (16)
Beta Breaker, 49
Chou & Fasman Prediction Method, 84
Coiled-Coil, 105
DIP, 151
Empirical Pair Potentials, 183
Gene Fusion Method, 253
Gene Neighbourhood, 254
GOR Secondary Structure Prediction Method (Garnier-Osguthorpe-Robson Method), 291
Low Complexity Region, 392
Nearest Neighbor Methods, 475
Phylogenetic Profile, 529
Qindex (Qhelix; Qstrand; Qcoil; Q3), 583
Safe Zone, 649
Secondary Structure Prediction of Protein, 656
SOV, 688
Statistical Mechanics, 696

Concise Encyclopaedia of Bioinformatics and Computational Biology, Second Edition.
Edited by John M. Hancock and Marketa J. Zvelebil.

Rolf Apweiler (5)
Merops, 410
Protein Databases, 565
Protein Family and Domain Signature Databases, 567
Protein Information Resource (PIR), 568
Protein Sequence Cluster Databases, 571

Teresa K. Attwood (23)
Consensus Pattern (Regular Expression, Regex), 112
Consensus Pattern Rule, 113
Diagnostic Power (Diagnostic Performance, Discriminating Power), 149
Domain Family, 160
Fingerprint of Proteins, 221
Gene Family, 252
Helical Wheel, 305
Hydropathy, 324
Hydropathy Profile (Hydrophobicity Plot, Hydrophobic Plot), 326
Indel (Insertion-Deletion Region, Insertion, Deletion, Gap), 335
InterPro, 343
Midnight Zone, 429
Motif, 453
Pfam, 521
PRINTS, 552
ProDom, 554
Profile (Weight Matrix, Position Weight Matrix, Position-Specific Scoring Matrix, PSSM), 554
PROSITE, 561
Regular Expression (Regex), 603
Rule, 642
Scoring Matrix (Substitution Matrix), 652
Twilight Zone, 745
Visualization of Multiple Sequence Alignments –Physicochemical Properties, 768

Jeremy Baum (9)
Amino Acid (Residue), 20
Amino Acid Composition, 21
Cofactor, 104
Disulphide Bridge, 157
Electrostatic Potential, 177
Hydrogen Bond, 324
Hydrophilicity, 327
Hydrophobic Moment, 327
Hydrophobic Scale, 328

Concha Bielza (28)
Affinity Propagation-Based Clustering, 6
Akaike Information Criterion, 10
Bagging, 38
Bayesian Classifier (Naïve Bayes), 42
Bayesian Information Criterion (BIC), 43
Beam Search, 46
Best-First Search, 49
Biclustering Methods, 51
Boosting, 66
Classifiers, Comparison, 91
Decision Surface, 144
Ensemble of Classifiers, 192
Estimation of Distribution Algorithm (EDA), 197
F-Measure, 213
Feature Subset Selection, 218
Finite Mixture Model, 223
Genetic Algorithm (GA), 276
Independent Component Analysis (ICA), 336
K-Medoids, 359
Multilabel Classification, 464
Positive Classification, 546
Random Forest, 595
Random Trees, 595
Regression Tree, 602
Regularization (Ridge, Lasso, Elastic Net, Fused Lasso, Group Lasso), 604
ROC Curve, 636
Rule Induction, 642
Validation Measures For Clustering, 759

M.J. Bishop (7)
Dotter, 163
EMBOSS (The European Molecular Biology Open Software Suite), 181
HMMer, 314
Open Bioinformatics Foundation (OBF), 491
STADEN, 694
VAST (Vector Alignment Search Tool), 762
VecScreen, 763

Enrique Blanco (22)
Alternative Splicing, 18
ENCODE (Encyclopedia of DNA Elements), 185
FGENES, 220
GASP (Genome Annotation Assessment Project, E-GASP), 241
GENCODE, 243
Gene Annotation, 244
Gene Annotation, formats, 245

Gene Annotation, hand-curated, 246
Gene Annotation, visualization tools, 247
Gene Prediction, 255
Gene Prediction, ab initio, 257
Gene Prediction, accuracy, 261
Gene prediction, alternative splicing, 263
Gene Prediction, comparative, 264
Gene Prediction, homology-based (Extrinsic Gene Prediction, Look-Up Gene Prediction), 266
Gene Prediction, NGS-based, 268
Gene prediction, non-canonical, 269
Gene Prediction, pipelines, 270
GENEID, 274
GENSCAN, 287
Selenoprotein, 661
Translation End Site, 736

Dan Bolser (16)

Allen Brain Atlas, 16
DrugBank, 164
e-Mouse Atlas (EMA), 182
Ensembl Plants, 191
Eukaryote Organism and Genome Databases, 199
Flybase, 225
FLYBRAIN, 226
GEISHA, 242
Gramene, 293
Mouse Genome Informatics (MGI, Mouse Genome Database, MGD), 461
PlantsDB, 538
PomBase, 542
The Promoter Database of *Saccharomyces cerevisiae* (SCPD), 560
Solanaceae Genomics Network (SGN), 687
Xenbase, 780
Yeast Deletion Project (YDPM), 785

Stuart Brown (11)

Affymetrix GeneChip™ Oligonucleotide Microarray, 7
Affymetrix Probe Level Analysis, 8
ChIP-Seq, 83
De Novo Assembly in Next Generation Sequencing, 142
Genomics, 285
Microarray, 420
Microarray Image Analysis, 421
Microarray Normalization, 421
Next Generation DNA Sequencing, 483
RNA-seq, 625
Spotted cDNA Microarray, 692

Aidan Budd (26)

Adjacent Group, 4
Bifurcation (in a Phylogenetic Tree), 51
Bootstrapping (Bootstrap Analysis), 66
Branch (of a Phylogenetic Tree) (Edge, Arc), 69
Clade (Monophyletic Group), 89
Cladistics, 89
Clan, 90
Distances Between Trees (Phylogenetic Trees, Distance), 155
Heterotachy, 308
Hypothetical Taxonomic Unit (HTU), 328
Labeled Tree, 373
Mixture Models, 437
Multifurcation (Polytomy), 464
Offspring Branch (Daughter Branch/ Lineage), 493
Parallel Computing in Phylogenetics, 512
Phylogenetic Placement of Short Reads, 529
Phylogenetics, 536
Rate Heterogeneity, 596
Simultaneous Alignment and Tree Building, 683
Sister Group, 684
Split (Bipartition), 692
Subtree, 706
Supermatrix Approach, 708
Taxonomic Unit, 722
Tree Topology, 742
Tree Traversal, 743

Jamie J. Cannone (7)

Covariation Analysis, 123
Phylogenetic Events Analysis (Pattern of Change Analysis), 525
RNA Folding, 624
RNA Structure, 626
RNA Structure Prediction (Comparative Sequence Analysis), 629
RNA Structure Prediction (Energy Minimization), 630
RNA Tertiary Structure Motifs, 631

Liz Carpenter (9)

Asymmetric unit, 32
Electron Density Map, 176

Free R-Factor, 230
R-Factor, 593
Molecular Replacement, 451
NMR (Nuclear Magnetic Resonance), 485
Refinement, 600
Resolution in X-Ray Crystallography, 611
X-Ray Crystallography for Structure Determination, 779

Feng Chen (14)

Association Rule Mining (Frequent Itemset, Association Rule, Support, Confidence, Correlation Analysis), 30
Bayes' Theorem, 41
Data Pre-Processing, 130
Decision Tree, 144
Fuzzy Set (Fuzzy Logic, Possibility Theory), 234
Graph Mining (Frequent Sub-Graph, Frequent Sub-Structure), 294
K-Nearest Neighbor Classification (Lazy Learner, KNN, Instance-Based Learner), 359
Missing Value (Missing Data), 435
Outlier Mining (Outlier), 502
Relational Database, 609
Rough Set, 640
Stream Mining (Time Series, Sequence, Data Stream, Data Flow), 698
Text Mining (Information Retrieval, IR), 723
Transaction Database (Data Warehouse), 728

Yi-Ping Phoebe Chen (14)

Association Rule Mining (Frequent Itemset, Association Rule, Support, Confidence, Correlation Analysis), 30
Bayes' Theorem, 41
Data Pre-Processing, 130
Decision Tree, 144
Fuzzy Set (Fuzzy Logic, Possibility Theory), 234
Graph Mining (Frequent Sub-Graph, Frequent Sub-Structure), 294
K-Nearest Neighbor Classification (Lazy Learner, KNN, Instance-Based Learner), 359
Missing Value (Missing Data), 435
Outlier Mining (Outlier), 502
Relational Database, 609
Rough Set, 640
Stream Mining (Time Series, Sequence, Data Stream, Data Flow), 698
Text Mining (Information Retrieval, IR), 723
Transaction Database (Data Warehouse), 728

Jean-Michel Claverie (14)

Comparative Genomics, 105
Expressed Sequence Tag (EST), 207
Functional Genomics, 231
Functional Signature, 233
Gene Expression Profile, 251
Genome Annotation, 280
Orphan Gene (ORFan), 501
Phylogenomics, 536
Prediction of Gene Function, 549
Protein Array (Protein Microarray), 563
Proteomics, 574
Rosetta Stone Method, 639
SAGE (Serial Analysis of Gene Expression), 649
Structural Genomics, 700
Transcriptome, 732

Andrew Collins (28)

Admixture Mapping (Mapping by Admixture Linkage Disequilibrium), 5
Allele-Sharing Methods (Non-parametric Linkage Analysis), 14
Allelic Association, 15
Association Analysis (Linkage Disequilibrium Analysis), 29
Candidate Gene (Candidate Gene-Based Analysis), 75
Elston-Stewart Algorithm (E-S Algorithm), 178
Exclusion Mapping, 204
Exome Sequencing, 205
Expectation Maximization Algorithm (E-M Algorithm), 206
Extended Tracts of Homozygosity, 209
Family-Based Association Analysis, 215
Gene Diversity, 249
Genome Scans for Linkage (Genome-Wide Scans), 282
Genome-Wide Association Study (GWAS), 283
Genotype Imputation, 286
Haplotype, 301
HapMap Project, 302
Heritability (h2, Degree of Genetic Determination), 307
Homozygosity Mapping, 319
Imprinting, 333
Lander-Green Algorithm (L-G Algorithm), 375
Linkage Analysis, 380
Linkage Disequilibrium (LD, Gametic Phase Disequilibrium, Allelic Association), 381
Linkage Disequilibrium Map, 383
LOD Score (Logarithm of Odds Score), 386

Marker, 400
Mendelian Disease, 409
Thousand Genomes Project, 725

Darren Creek(1)
Metabolomics Databases, 414

Nello Cristianini (18)
Classification in Machine Learning (Discriminant Analysis), 91
Clustering (Cluster Analysis), 96
Cross-Validation (K-Fold Cross-Validation, Leave-One-Out, Jackknife, Bootstrap), 126
Error Measures (Accuracy Measures, Performance Criteria, Predictive Power, Generalization), 196
Feature (Independent Variable, Predictor Variable, Descriptor, Attribute, Observation), 218
Fisher Discriminant Analysis (Linear Discriminant Analysis), 223
Gradient Descent (Steepest Descent Method), 291
Kernel Function, 361
Kernel Method (Kernel Machine, Kernel-Based Learning Method), 362
Label (Labeled Data, Response, Dependent Variable), 373
Machine Learning, 397
Model Selection (Model Order Selection, Complexity Regularization), 438
Neural Network (Artificial Neural Network, Connectionist Network, Backpropagation Network, Multilayer Perceptron), 480
Overfitting (Overtraining), 504
Pattern Analysis, 516
Regression Analysis, 602
Supervised And Unsupervised Learning, 709
Support Vector Machine (SVM, Maximal Margin Classifier), 710

Alison Cuff (1)
Fold, 227

Michael P. Cummings (36)
AWTY (Are We There Yet?), 33
BAMBE (Bayesian Analysis in Molecular Biology and Evolution), 40
BEAGLE, 45
BEAST (Bayesian Evolutionary Analysis by Sampling Trees), 46
BEAUti (Bayesian Evolutionary Analysis Utility), 47
Bio++, 56
DataMonkey, 137
DendroPy, 147
DnaSP, 158
ENCprime/SeqCount, 186
FigTree, 221
GARLI (Genetic Algorithm for Rapid Likelihood Inference), 240
Genealogical Sorting Index (gsi), 273
HyPhy (Hypothesis testing using Phylogenies), 328
IMa2 (Isolation with Migration a2), 332
JELLYFISH, 354
jModelTest, 354
LAMARC, 374
MacClade, 397
MEGA (Molecular Evolutionary Genetics Analysis), 409
Mesquite, 411
MIGRATE-N, 430
Modeltest, 442
MrBayes, 462
PAML (Phylogenetic Analysis by Maximum Likelihood), 510
PAUP* (Phylogenetic Analysis Using Parsimony (and Other Methods)), 516
PHYLIP (PHYLogeny Inference Package), 524
r8s, 593
readseq, 598
Seq-Gen, 663
SITES, 685
Structure, 702
Structurama, 702
Tracer, 727
TreeStat, 741
TreeView X, 743

Tjaart de Beer (33)
Alpha Helix, 18
Amide Bond (Peptide Bond), 19
Amino Acid (Residue), 20
Amphipathic, 23
Aromatic, 27
Backbone (Main chain), 37
Beta Sheet, 50
Beta Strand, 50
Boltzmann Factor, 65
C-α (C-alpha), 75
Catalytic Triad, 77
Coil (Random Coil), 104
Conformation, 111

Dihedral Angle (Torsion Angle), 150
Globular, 289
Kappa Virtual Dihedral Angle, 360
N-Terminus, 473
Peptide, 518
Peptide Bond (Amide Bond), 519
Polar, 540
Polypeptide, 542
Protein Structure, 571
Quaternary Structure, 589
Ramachandran Plot, 594
Root Mean Square Deviation (RMSD), 637
Secondary Structure of Protein, 656
Sequence of A Protein, 670
Sequence Pattern, 671
Side Chain, 679
Supersecondary Structure, 709
TIM-Barrel, 726
Turn, 744
Zeta Virtual Dihedral Angle, 789

Niall Dillon (11)

Codon, 101
Downstream, 163
Enhancer, 189
Exon, 205
Initiator Sequence, 339
Intron, 346
Kozak Sequence, 370
Open Reading Frame (ORF), 496
Promoter, 559
TATA Box, 720
Transcriptional Regulatory Region (Regulatory Sequence), 732
Upstream, 755

Roland Dunbrack (30)

Ab Initio, 3
Anchor Points, 25
CHARMM, 81
CNS, 97
Comparative Modeling (Homology Modeling, Knowledge-Based Modeling), 107
Conformational Energy, 111
Dead-End Elimination Algorithm, 143
Electrostatic Energy, 177
Empirical Potential Energy Function, 184
Energy Minimization, 188
Loop Prediction/Modeling, 391
Molecular Dynamics Simulation, 445
Molecular Mechanics, 450
Monte Carlo Simulation, 452
OPLS, 498
Polarization, 540
Potential of Mean Force, 548
QM/MM Simulations, 583
Rotamer, 640
SCWRL, 654
Self-Consistent Mean Field Algorithm, 661
Side-Chain Prediction, 680
Simulated Annealing, 682
Solvation Free Energy, 688
Statistical Potential Energy, 697
SwissModel, 712
Target, 720
Template (Parent), 723
Whatcheck, 774
WhatIf, 774

Anton Feenstra (1)

Petri Net, 520

Pedro Fernandes (12)

APBIONET (Asia-Pacific Bioinformatics Network), 26
ASBCB (The African Society for Bioinformatics and Computational Biology), 28
The Bioinformatics Organization, Inc (formerly bioinformatics.org), 58
BTN (Bioinformatics Training Network), 70
ELIXIR (Infrastructure for biological information in Europe), 177
EMBnet (The Global Bioinformatics Network) (formerly the European Molecular Biology Network), 181
EUPA (European Proteomics Association), 201
HUGO (The Human Genome Organization), 321
HUPO (Human Proteome Organization), 323
INPPO (International Plant Proteomics Organization), 339
International Society for Computational Biology (ISCB), 342
OBF (The Open Bioinformatics Foundation), 491

James Fickett (4)

Position Weight Matrix of Transcription Factor Binding Sites, 544
Promoter Prediction, 560

Regulatory Region Prediction, 607
Transcriptional Regulatory Region (Regulatory Sequence), 732

Alan Filipski (17)

Ancestral State Reconstruction, 24
Bayesian Phylogenetic Analysis, 44
Branch-Length Estimation, 69
Character (Site), 81
Consensus Tree (Strict Consensus, Majority-Rule Consensus, Supertree), 116
Evolutionary Distance, 203
Maximum Likelihood Phylogeny Reconstruction, 406
Maximum Parsimony Principle (Parsimony, Occam's Razor), 407
Minimum Evolution Principle, 432
Molecular Clock (Evolutionary Clock, Rate of Evolution), 444
Neighbor-Joining Method, 476
Quartets, Phylogenetic, 588
Rooting Phylogenetic Trees, 638
Sequence Distance Measures, 665
Substitution Process, 705
Tree of Life, 740
UPGMA, 754

Juan Antonio Garcia Ranea (1)

Discrete Function Prediction (Function Prediction), 152

Katheleen Gardiner (31)

Base Composition (GC Richness, GC Composition), 41
Centimorgan, 79
Centromere (Primary Constriction), 79
Chromatin, 85
Chromosomal Deletion, 85
Chromosomal Inversion, 85
Chromosomal Translocation, 86
Chromosome, 86
Chromosome Band, 87
CpG Island, 125
Dinucleotide Frequency, 151
Gene Cluster, 248
Gene Distribution, 249
Gene Duplication, 250
Gene Family, 252
Gene Size, 272
Genetic Redundancy, 279
Genome Size (C-Value), 283
Intergenic Sequence, 341
Isochore, 347
Karyotype, 360
Kinetochore, 365
Microsatellite, 428
Minisatellite, 434
Nucleolar Organizer Region (NOR), 488
Sequence Complexity (Sequence Simplicity), 664
Simple DNA Sequence (Simple Repeat, Simple Sequence Repeat), 682
Tandem Repeat, 719
Telomere, 722
Transposable Element (Transposon), 738
Trinucleotide Repeat, 744

Carole Goble (14)

Biological Identifiers, 58
Cross-Reference (Xref), 126
DAS Services, 129
Data Integration, 130
Data Warehouse, 134
Flat File Data Formats, 224
Linked Data, 384
Mark-up Language, 400
Minimum Information Models, 433
RDF, 597
RESTful Web services, 612
Scientific Workflows, 651
Web Services, 773
WSDL/SOAP Web Services, 775

Dov Greenbaum (50)

Affymetrix GeneChip™ Oligonucleotide Microarray, 7
Analog (Analogue), 23
Annotation Transfer (Guilt by Association Annotation), 25
Cluster, 94
Cluster of Orthologous Groups (COG, COGnitor), 95
Concerted Evolution (Coincidental Evolution, Molecular Drive), 110
Consensus Sequence, 114
Conservation (Percentage Conservation), 117
Convergence, 119
Copy Number Variation, 120
Entrez, 192
Epistatic Interactions (Epistasis), 194
Functional Database, 231

Functome, 233
Gene Dispensability, 248
Gene Duplication, 250
Genome-Wide Survey, 284
Horizontal Gene Transfer (HGT), 320
Interactome, 340
Interolog (Interologue), 343
Interspersed Sequence (Long-Term Interspersion, Long-Period Interspersion, Short-Term Interspersion, Short-Period Interspersion, Locus Repeat), 345
Jackknife, 353
Junk DNA, 355
Kingdom, 365
LINE (Long Interspersed Nuclear Element), 378
Maximum Parsimony Principle (Parsimony, Occam's Razor), 407
Metabolome (Metabonome), 413
Molecular Evolutionary Mechanisms, 447
Motif, 453
Ortholog (Orthologue), 501
Paralinear Distance (LogDet), 511
Paralog (Paralogue), 512
Parsimony, 514
Power Law (Zipf's Law), 549
Protein Family, 566
Proteome, 572
Pseudogene, 576
PSI BLAST, 577
Retrotransposon, 613
Robustness, 635
Secretome, 657
Segmental Duplication, 658
Single Nucleotide Polymorphism (SNP), 683
Superfamily, 707
Synteny, 715
Taxonomic Classification (Organismal Classification), 720
Transcriptome, 732
Translatome, 738
Transposable Element (Transposon), 738
Two-Dimensional Gel Electrophoresis (2DE), 746

Malachi Griffith (39)

BioMart, 59
Cancer Gene Census (CGC), 75
CCDS (Consensus Coding Sequence Database), 79
COSMIC (Catalogue of Somatic Mutations in Cancer), 122
dbGAP (The database of Genotypes and Phenotypes), 139
dbSNP, 139
dbVar (Database of genomic structural variation), 141
DDBJ (DNA Databank of Japan), 141
DGV (Database of Genetic Variants), 148
EBI (EMBL-EBI, European Bioinformatics Institute), 174
ENA (European Nucleotide Archive), 185
Ensembl, 190
Entrez Gene (NCBI 'Gene'), 193
GenBank, 243
GOBASE (Organelle Genome Database), 290
GWAS Central, 297
HGMD (Human Gene Mutation Database), 308
HIV Sequence Database, 313
IMGT (International Immunogenetics Database), 333
JASPAR, 353
MGC (Mammalian Gene Collection), 418
Mitelman Database (Chromosome Aberrations and Gene Fusions in Cancer), 436
Model Organism Database (MOD), 438
MPromDB (Mammalian Promoter Database), 461
NCBI (National Center for Biotechnology Information), 474
ODB (Operon DataBase), 493
OMIM (Online Mendelian Inheritance in Man), 495
ORegAnno (Open Regulatory Annotation Database), 499
PAZAR, 517
RefSeq (the Reference Sequence Database), 601
Sequence Read Archive (SRA, Short Read Archive), 671
SGD (Saccharomyces Genome Database), 677
Stanford HIV RT and Protease Sequence Database (HIV RT and Protease Sequence Database), 696
Trace Archive, 727
TRANSFAC, 733
UCSC Genome Browser, 751
Unigene, 753
VEGA (Vertebrate Genome Annotation Database), 763
WormBase, 774

Obi L. Griffith (39)

BioMart, 59
Cancer Gene Census (CGC), 75

CCDS (Consensus Coding Sequence Database), 79
COSMIC (Catalogue of Somatic Mutations In Cancer), 122
dbGAP (The database of Genotypes and Phenotypes), 139
dbSNP, 139
dbVar (Database of genomic structural variation), 141
DDBJ (DNA Databank of Japan), 141
DGV (Database of Genetic Variants), 148
EBI (EMBL-EBI, European Bioinformatics Institute), 174
ENA (European Nucleotide Archive), 185
Ensembl, 190
Entrez Gene (NCBI 'Gene'), 193
GenBank, 243
GOBASE (Organelle Genome Database), 290
GWAS Central, 297
HGMD (Human Gene Mutation Database), 308
HIV Sequence Database, 313
IMGT (International Immunogenetics Database), 333
JASPAR, 353
MGC (Mammalian Gene Collection), 418
Mitelman Database (Chromosome Aberrations and Gene Fusions in Cancer), 436
Model Organism Database (MOD), 438
MPromDB (Mammalian Promoter Database), 461
NCBI (National Center for Biotechnology Information), 474
ODB (Operon DataBase), 493
OMIM (Online Mendelian Inheritance in Man), 495
ORegAnno (Open Regulatory Annotation Database), 499
PAZAR, 517
RefSeq (the Reference Sequence Database), 601
Sequence Read Archive (SRA, Short Read Archive), 671
SGD (Saccharomyces Genome Database), 677
Stanford HIV RT and Protease Sequence Database (HIV RT and Protease Sequence Database), 696
Trace Archive, 727
TRANSFAC, 733
UCSC Genome Browser, 751
Unigene, 753
VEGA (Vertebrate Genome Annotation Database), 763
WormBase, 774

Sam Griffiths-Jones (13)

Folding Free Energy, 228
IsomiR, 348
Mature microRNA, 406
MicroRNA, 423
MicroRNA Discovery, 424
MicroRNA Family, 425
MicroRNA Seed, 426
MicroRNA Target, 426
MicroRNA Target Prediction, 427
miRBase, 434
Mirtron, 435
Rfam, 614
RNA Hairpin, 624

Roderic Guigó (17)

BLASTX, 63
Coding Region (CDS), 98
Coding Region Prediction, 98
Coding Statistics (Coding Measure, Coding Potential, Search by Content), 99
Codon Usage Bias, 102
Gene Prediction, 255
Gene Prediction, ab initio, 257
Gene Prediction, accuracy, 261
Gene Prediction, comparative, 264
Gene Prediction, homology-based (Extrinsic Gene Prediction, Look-Up Gene Prediction), 266
Gene Prediction, pipelines, 270
GENEWISE, 280
GENSCAN, 287
GRAIL, 292
Sequence Motifs: Prediction and Modeling (Search by Signal), 667
Spliced Alignment, 690
Translation Start Site, 737

Robin R. Gutell (10)

Covariation Analysis, 123
Phylogenetic Events Analysis (Pattern of Change Analysis), 525
Ribosomal RNA (rRNA), 615
RNA (General Categories), 621
RNA Folding, 624
RNA Structure, 626

RNA Structure Prediction (Comparative Sequence Analysis), 629
RNA Structure Prediction (Energy Minimization), 630
RNA Tertiary Structure Motifs, 631
Transfer RNA (tRNA), 734

John M. Hancock (59)

Base-Call Confidence Values, 40
Binary Numerals, 52
Binding Affinity (K_D, K_I, IC_{50}), 52
Bioinformatics (Computational Biology), 57
BLAT (BLAST-like Alignment Tool), 64
Complement, 108
Data Standards, 131
Distance Matrix (Similarity Matrix), 154
DNA Sequence, 158
Dotter, 163
EMBOSS (The European Molecular Biology Open Software Suite), 181
Entrez, 192
Euclidean Distance, 198
EuroPhenome, 201
False Discovery Rate Control (False Discovery Rate, FDR), 213
Gene Expression Profile, 251
Gene Symbol, 272
Genetic Code (Universal Genetic Code, Standard Genetic Code), 277
HAVANA (Human and Vertebrate Analysis and Annotation), 305
HMMer, 314
Human Variome Project (HVP), 323
Intron, 346
IUPAC-IUB Codes (Nucleotide Base Codes, Amino Acid Abbreviations), 348
Jaccard Distance (Jaccard Index, Jaccard Similarity Coefficient), 353
Locus-Specific Database (Locus-Specific Mutation Database, LSDB), 386
Markov Chain, 401
Markov Chain Monte Carlo (MCMC, Metropolis-Hastings, Gibbs Sampling), 402
Microfunctionalization, 423
OBO-Edit, 492
OBO Foundry, 492
Omics, 494
Open Source Bioinformatics Organizations, 496
Operational Taxonomic Unit (OTU), 498
PIPMAKER, 538
PRIMER3, 551
Principal Components Analysis (PCA), 552
Protégé, 562
Reference Genome (Type Genome), 600
Regulome, 609
REPEATMASKER, 610
Reverse Complement, 613
Ribosome Binding Site (RBS), 620
Self-Organizing Map (SOM, Kohonen Map), 662
Sequence Tagged Site (STS), 674
SIMPLE (SIMPLE34), 681
Simple DNA Sequence (Simple Repeat, Simple Sequence Repeat), 682
SPARQL (SPARQL Protocol And RDF Query Language), 690
Splicing (RNA Splicing), 692
STADEN, 694
Tanimoto Distance, 719
Transcription, 728
Transcription Start Site (TSS), 732
Translation, 736
VarioML, 761
VAST (Vector Alignment Search Tool), 762
VecScreen, 763
VectorNTI, 763
VISTA, 766
Web Ontology Language (OWL), 773

Andrew Harrison (11)

Contact Map, 118
Fold, 227
Fold Library, 227
Homologous Superfamily, 315
Profile, 3D, 555
Protein Family, 566
Protein Structure, 571
Structural Alignment, 699
Structural Motif, 701
Structure–3D Classification, 703
Superfold, 707

Matthew He (23)

Algorithm, 10
Bayes' Theorem, 41
Data Description Language (DDL, Data Definition Language), 129
Data Manipulation Language (DML), 130
Data Processing, 131
Data Warehouse, 134
Expression Level (of Gene or Protein), 209
Gene Index, 254

Gene Symbol, human, 273
Iteration, 348
Lead Optimization, 377
Multiplex Sequencing, 469
Object-Relational Database, 491
Parameter, 513
Pattern, 515
Recursion, 600
Relational Database Management System (RDBMS), 609
Restriction Map, 612
SQL (Structured Query Language), 693
Stochastic Process, 698
Variation (Genetic), 761
Yule Process, 785
z-score, 789

Jaap Heringa (29)

Alignment (Domain Alignment, Repeats Alignment), 11
Amino Acid Exchange Matrix (Dayhoff matrix, Log Odds Score, PAM (matrix), BLOSUM matrix), 21
BLAST (Maximal Segment Pair, MSP), 61
CLUSTAL, 93
Dot Plot (Dot Matrix), 162
Dynamic Programming, 165
E value, 171
End Gap, 187
FASTA (FASTP), 216
Gap, 239
Gap Penalty, 239
Global Alignment, 289
High-scoring Segment Pair (HSP), 313
Homology, 316
Homology Search, 317
Local Alignment (Local Similarity), 385
Low Complexity Region, 392
Motif Search, 456
Multiple Alignment, 465
Needleman-Wunsch Algorithm, 475
Optimal Alignment, 499
Pairwise Alignment, 509
Phantom Indel (Frame Shift), 522
Phylogenetic Tree (Phylogeny, Phylogeny Reconstruction, Phylogenetic Reconstruction), 530
Profile Searching, 556
Sequence Alignment, 664
Sequence Similarity, 672
Smith-Waterman, 686
Tree-Based Progressive Alignment, 740

A.R. Hoelzel (14)

Allopatric Evolution (Allopatric Speciation), 17
Apomorphy, 26
Autapomorphy, 33
Evolution, 202
Founder Effect, 229
Gene Flow, 253
Hardy-Weinberg Equilibrium, 303
Kin Selection, 363
Natural Selection, 473
Neutral Theory (Nearly Neutral Theory), 482
Overdominance, 503
Plesiomorphy, 539
Population Bottleneck (Bottleneck), 543
Synapomorphy, 714

Simon Hubbard (19)

Database Search Engine (Proteomics) (Peptide Spectrum Match, PSM), 136
Decoy Database, 145
emPAI (Exponentially Modified Protein Abundance Index), 183
False Discovery Rate in Proteomics, 214
Isobaric Tagging, 347
MaxQuant, 408
PeptideAtlas, 519
Peptide Mass Fingerprint, 519
Peptide Spectrum Match (PSM), 519
Posterior Error Probability (PEP), 548
PRIDE, 551
Protein Inference Problem, 568
Proteomics Standards Initiative (PSI), 575
Proteotypic Peptide, 575
Quantitative Proteomics, 586
Selected Reaction Monitoring (SRM), 660
Stable Isotope Labelling with Amino acids in Cell Culture (SILAC), 694
Tandem Mass Spectrometry (MS), 719
Trans-Proteomic Pipeline (TPP), 728

Austin L. Hughes (8)

Exon Shuffling, 206
Homologous Genes, 315
Intron Phase, 346
Paralog (Paralogue), 512

Positive Darwinian Selection (Positive Selection), 547
Purifying Selection (Negative Selection), 578
Retrosequence, 612
Subfunctionalization, 704

David Jones (5)

Accuracy (Of Protein Structure Prediction), 3
CASP, 76
Shuffle Test, 679
Threading, 726
Ungapped Threading Test B (Sippl Test), 752

Pascal Kahlem (2)

Qualitative and Quantitative Databases used in Systems Biology, 584
Systems Biology, 716

Ana Kozomara (3)

IsomiR, 348
MicroRNA Seed, 426
miRBase, 434

Sudhir Kumar (17)

Ancestral State Reconstruction, 24
Bayesian Phylogenetic Analysis, 44
Branch-Length Estimation, 69
Character (Site), 81
Consensus Tree (Strict Consensus, Majority-Rule Consensus, Supertree), 116
Evolutionary Distance, 203
Maximum Likelihood Phylogeny Reconstruction, 406
Maximum Parsimony Principle (Parsimony, Occam's Razor), 407
Minimum Evolution Principle, 432
Molecular Clock (Evolutionary Clock, Rate of Evolution), 444
Neighbor-Joining Method, 476
Quartets, Phylogenetic, 588
Rooting Phylogenetic Trees, 638
Sequence Distance Measures, 665
Substitution Process, 705
Tree of Life, 740
UPGMA, 754

Pedro Larrañaga (28)

Affinity Propagation-Based Clustering, 6
Akaike Information Criterion, 10
Bagging, 38
Bayesian Classifier (Naïve Bayes), 42
Bayesian Information Criterion (BIC), 43
Beam Search, 46
Best-First Search, 49
Biclustering Methods, 51
Boosting, 66
Classifiers, Comparison, 91
Decision Surface, 144
Ensemble of Classifiers, 192
Estimation of Distribution Algorithm (EDA), 197
F-Measure, 213
Feature Subset Selection, 218
Finite Mixture Model, 223
Genetic Algorithm (GA), 276
Independent Component Analysis (ICA), 336
K-Medoids, 359
Multilabel Classification, 464
Positive Classification, 546
Random Forest, 595
Random Trees, 595
Regression Tree, 602
Regularization (Ridge, Lasso, Elastic Net, Fused Lasso, Group Lasso), 604
ROC Curve, 636
Rule Induction, 642
Validation Measures For Clustering, 759

Roman Laskowski (34)

Alpha Helix, 18
Amide Bond (Peptide Bond), 19
Amino Acid (Residue), 20
Amphipathic, 23
Aromatic, 27
Backbone (Main chain), 37
Beta Barrel, 49
Beta Sheet, 50
Beta Strand, 50
Boltzmann Factor, 65
C-α (C-alpha), 75
Catalytic Triad, 77
Coil (Random Coil), 104
Conformation, 111
Dihedral Angle (Torsion Angle), 150
Globular, 289
Kappa Virtual Dihedral Angle, 360
N-terminus (amino terminus), 473
Peptide, 518
Peptide Bond (Amide Bond), 519

Polar, 540
Polypeptide, 542
Protein Structure, 571
Quaternary Structure, 589
Ramachandran Plot, 594
Root Mean Square Deviation (RMSD), 637
Secondary Structure of Protein, 656
Sequence Pattern, 671
Sequence of a Protein, 670
Side Chain, 679
Supersecondary Structure, 709
TIM-Barrel, 726
Turn, 744
Zeta Virtual Dihedral Angle, 789

Antonio Marco (7)
Folding Free Energy, 228
MicroRNA Discovery, 424
MicroRNA Family, 425
MicroRNA Target, 426
MicroRNA Target Prediction, 427
Mirtron, 435
RNA Hairpin, 624

James Marsh (7)
Boolean Logic, 65
Data Structure (Array, Associative Array, Binary Tree, Hash, Linked List, Object, Record, Struct, Vector), 133
Database (NoSQL, Quad Store, RDF Database, Relational Database, Triple Store), 135
Distributed Computing, 156
Operating System, 497
Programming and Scripting Languages, 559
Virtualization, 765

Eric Martz (12)
Atomic Coordinate File (PDB file), 32
Backbone Models, 37
Ball and Stick Models, 39
MIME Types, 431
Modeling, Macromolecular, 439
Models, Molecular, 440
Protein Data Bank (PDB), 564
Schematic (Ribbon, Cartoon) Models, 651
Space-Filling Model, 689
Surface Models, 711
Visualization, Molecular, 766
XML (eXtensible Markup Language), 780

Mark McCarthy (38)
Admixture Mapping (Mapping by Admixture Linkage Disequilibrium), 5
Allele-Sharing Methods (Non-parametric Linkage Analysis), 14
Allelic Association, 15
Association Analysis (Linkage Disequilibrium Analysis), 29
Candidate Gene (Candidate Gene-Based Analysis), 75
Elston-Stewart Algorithm (E-S Algorithm), 178
Exclusion Mapping, 204
Expectation Maximization Algorithm (E-M Algorithm), 206
Family-Based Association Analysis, 215
Gene Diversity, 249
Genome Scans for Linkage (Genome-Wide Scans), 282
Haplotype, 301
Haseman-Elston Regression (HE-SD, HE-SS, HE-CP and HE-COM), 304
Heritability (h2, Degree of Genetic Determination), 307
Homozygosity Mapping, 319
Identical by Descent (Identity by Descent, IBD), 331
Identical by State (Identity by State, IBS), 332
Imprinting, 333
Lander-Green Algorithm (L-G Algorithm), 375
Linkage (Genetic Linkage), 379
Linkage Analysis, 380
Linkage Disequilibrium (LD, Gametic Phase Disequilibrium, Allelic Association), 381
LOD Score (Logarithm of Odds Score), 386
Map Function, 398
Marker, 400
Mendelian Disease, 409
Multifactorial Trait (Complex Trait), 463
Multipoint Linkage Analysis, 469
Oligogenic Inheritance (Oligogenic Effect), 494
Penetrance, 518
Phase (sensu Linkage), 522
Polygenic Inheritance (Polygenic Effect), 540
Polymorphism (Genetic Polymorphism), 541
Positional Candidate Approach, 545
Quantitative Trait (Continuous Trait), 586
Recombination, 598
Segregation Analysis, 659
Variance Components (Components of Variance, VC), 759

Luis Mendoza (3)
Piecewise-Linear Models, 537
Qualitative Differential Equations, 585
Standardized Qualitative Dynamical Systems, 695

Irmtraud Meyer (1)
Hidden Markov Model (HMM, Hidden Semi-Markov Models, Profile Hidden Markov Models, Training of Hidden Markov Models, Dynamic Programming, Pair Hidden Markov Models), 309

Christine Orengo (11)
Contact Map, 118
Fold, 227
Fold Library, 227
Homologous Superfamily, 315
Profile, 3D, 555
Protein Family, 566
Protein Structure, 571
Structural Alignment, 699
Structural Motif, 701
Structure–3D Classification, 703
Superfold, 707

Laszlo Patthy (26)
Alignment Score, 13
BLOSUM (BLOSUM Matrix), 64
Coevolution (Molecular Coevolution), 103
Coevolution of Protein Residues, 103
Dayhoff Amino Acid Substitution Matrix (PAM Matrix, Percent Accepted Mutation Matrix), 137
DNA-Protein Coevolution, 157
Epaktolog (Epaktologue), 194
Exon, 205
Exon Shuffling, 206
Gap Penalty, 239
Homologous Genes, 315
Homology, 316
Modular Protein, 443
Module Shuffling, 443
Multidomain Protein, 463
Non-Synonymous Mutation, 486
PAM Matrix of Nucleotide Substitutions (Point Accepted Mutations), 510
Paralog (Paralogue), 512
Protein Domain, 565
Protein Family, 566
Protein Module, 569
Protein-Protein Coevolution, 569
Pseudoparalog (pseudoparalogue), 576
Sequence Similarity Search, 673
Synonymous Mutation (Silent mutation), 715
Xenolog (Xenologue), 780

Hedi Peterson (2)
Protein-Protein Interaction Network Inference, 570
Regulatory Network Inference, 606

Steve Pettifer (7)
Boolean Logic, 65
Data Structure (Array, Associative Array, Binary Tree, Hash, Linked List, Object, Record, Struct, Vector), 133
Database (NoSQL, Quad Store, RDF Database, Relational Database, Triple Store), 135
Distributed Computing, 156
Operating System, 497
Programming and Scripting Languages, 559
Virtualization, 765

Richard Scheltema (1)
Metabolomics Software, 415

Thomas D. Schneider (47)
After State (After Sphere), 9
Before State (Before Sphere), 47
Binding Site, 53
Binding Site Symmetry, 53
Bit, 60
Box, 68
Channel Capacity (Channel Capacity Theorem), 80
Coding Theory (Code, Coding), 101
Complexity, 108
Consensus Sequence, 114
Coordinate System of Sequences, 120
Core Consensus, 121
DELILA, 146
Delila Instructions, 146
Entropy, 193
Error, 195
Evolution of Biological Information, 203
Gumball Machine, 296
Individual Information, 337
Information, 338
Information Theory, 338
Message, 411
Molecular Efficiency, 446
Molecular Information Theory, 448

Molecular Machine, 449
Molecular Machine Capacity, 449
Molecular Machine Operation, 450
Negentropy (Negative Entropy), 476
Nit, 484
Noise (Noisy Data), 486
Parity Bit, 514
Rfrequency, 614
Ri, 615
Rsequence, 641
Score, 652
Second Law of Thermodynamics, 655
Sequence Logo, 666
Sequence Pattern, 671
Sequence Walker, 674
Shannon Entropy (Shannon Uncertainty), 677
Shannon Sphere, 678
Small Sample Correction, 685
Surprisal, 712
Symmetry Paradox, 713
Thermal Noise, 724
Uncertainty, 752
Zero Coordinate (Zero Base, Zero Position), 789

Rodger Staden (8)

Base-Call Confidence Values, 40
Contig, 119
DNA Sequence, 158
DNA Sequencing, 158
PHRAP, 524
PHRED, 524
Sequence Assembly, 664
Sequence Tagged Site (STS), 674

Alexandros Stamatakis (26)

Adjacent Group, 4
Bifurcation (in a Phylogenetic Tree), 51
Bootstrapping (Bootstrap Analysis), 66
Branch (of a Phylogenetic Tree) (Edge, Arc), 69
Clade (Monophyletic Group), 89
Cladistics, 89
Clan, 90
Distances Between Trees (Phylogenetic Trees, Distance), 155
Heterotachy, 308
Hypothetical Taxonomic Unit (HTU), 328
Labeled Tree, 373
Mixture Models, 437
Multifurcation (Polytomy), 464
Offspring Branch (Daughter Branch/Lineage), 493
Parallel Computing in Phylogenetics, 512
Phylogenetic Placement of Short Reads, 529
Phylogenetics, 536
Rate Heterogeneity, 596
Simultaneous Alignment and Tree Building, 683
Sister Group, 684
Split (Bipartition), 692
Subtree, 706
Supermatrix Approach, 708
Taxonomic Unit, 722
Tree Topology, 742
Tree Traversal, 743

Robert Stevens (35)

Axiom, 34
Binary Relation, 52
Category, 78
Classification, 90
Classifier (Reasoner), 91
Concept, 109
Conceptual Graph, 110
Description Logic (DL), 147
Directed Acyclic Graph (DAG), 152
EcoCyc, 175
Frame-Based Language, 229
Gene Ontology (GO), 255
Gene Ontology Consortium, 255
GRAIL Description Logic, 293
Hierarchy, 312
Individual (Instance), 337
Knowledge, 366
Knowledge Base, 367
Knowledge Interchange Format (KIF), 367
Knowledge Representation Language (KRL), 368
Lattice, 376
Lexicon, 377
Metadata, 417
MGED Ontology, 419
Multiple Hierarchy (Polyhierarchy), 469
Ontology, 495
Protégé, 562
Reasoning, 598
Relationship, 609
Role, 637
Semantic Network, 663
Taxonomy, 722
Term, 723
Terminology, 723
Thesaurus, 725

Guenter Stoesser (20)

dbEST, 138
dbSNP, 139
dbSTS, 140
DDBJ (DNA Databank of Japan), 141
EBI (EMBL-EBI, European Bioinformatics Institute), 174
EMBL Nucleotide Sequence Database (EMBL-Bank, EMBL Database), 180
Ensembl, 190
Flybase, 225
GenBank, 243
GOBASE (Organelle Genome Database), 290
GWAS Central, 297
HIV Sequence Database, 313
IMGT (International Immunogenetics Database), 333
MGD (Mouse Genome Database), 418
NCBI (National Center for Biotechnology Information), 474
Nucleic Acid Database (NDB), 487
Nucleic Acid Sequence Databases, 487
Rat Genome Database (RGD), 596
SRS (Sequence Retrieval System), 693
Stanford HIV RT and Protease Sequence Database (HIV RT and Protease Sequence Database), 696

Neil Swainston (5)

Constraint-Based Modeling (Flux Balance Analysis), 117
Kinetic Modeling, 364
Metabolic Modeling, 412
Metabolic Network, 412
Metabolic Pathway, 413

David Thorne (7)

Boolean Logic, 65
Data Structure (Array, Associative Array, Binary Tree, Hash, Linked List, Object, Record, Struct, Vector), 133
Database (NoSQL, Quad Store, RDF Database, Relational Database, Triple Store), 135
Distributed Computing, 156
Operating System, 497
Programming and Scripting Languages, 559
Virtualization, 765

Denis Thieffry (6)

Genetic Network, 278
Graph Representation of Genetic, Molecular and Metabolic Networks, 294
Logical Modeling of Genetic Networks, 388
Mathematical Modeling (Of Molecular/Metabolic/Genetic Networks), 403
Network (Genetic Network, Molecular Network, Metabolic Network), 477
Regulatory Motifs in Network Biology, 605

Jacques van Helden (15)

Cis-regulatory Element, 87
Cis-regulatory module (CRM), 88
Cis-Regulatory Module Prediction, 88
Motif Discovery, 454
Motif Enrichment Analysis, 455
Motif Search, 456
Phylogenetic Footprint, 527
Phylogenetic Footprint Detection, 528
Position Weight Matrix, 544
Regulatory Region, 607
Software Suites for Regulatory Sequences, 687
Transcription Factor, 729
Transcription Factor Binding Motif (TFBM), 729
Transcription Factor Binding Site, 730
Transcription Factor Database, 731

Juan Antonio Vizcanío (2)

Data Standards in Proteomics, 131
Data Standards in Systems Modeling, 132

Steven Wiltshire (38)

Admixture Mapping (Mapping by Admixture Linkage Disequilibrium), 5
Allele-Sharing Methods (Non-parametric Linkage Analysis), 14
Allelic Association, 15
Association Analysis (Linkage Disequilibrium Analysis), 29
Candidate Gene (Candidate Gene-Based Analysis), 75
Elston-Stewart Algorithm (E-S Algorithm), 178
Exclusion Mapping, 204
Expectation Maximization Algorithm (E-M Algorithm), 206
Family-Based Association Analysis, 215
Gene Diversity, 249

Genome Scans for Linkage (Genome-Wide Scans), 282
Haplotype, 301
Haseman-Elston Regression (HE-SD, HE-SS, HE-CP and HE-COM), 304
Heritability (h2, Degree of Genetic Determination), 307
Homozygosity Mapping, 319
Identical by Descent (Identity by Descent, IBD), 331
Identical by State (Identity by State, IBS), 332
Imprinting, 333
Lander-Green Algorithm (L-G Algorithm), 375
Linkage (Genetic Linkage), 379
Linkage Analysis, 380
Linkage Disequilibrium (LD, Gametic Phase Disequilibrium, Allelic Association), 381
LOD Score (Logarithm of Odds Score), 386
Map Function, 398
Marker, 400
Mendelian Disease, 409
Multifactorial Trait (Complex Trait), 463
Multipoint Linkage Analysis, 469
Oligogenic Inheritance (Oligogenic Effect), 494
Penetrance, 518
Phase (sensu Linkage), 522
Polygenic Inheritance (Polygenic Effect), 540
Polymorphism (Genetic Polymorphism), 541
Positional Candidate Approach, 545
Quantitative Trait (Continuous Trait), 586
Recombination, 598
Segregation Analysis, 659
Variance Components (Components of Variance, VC), 759

Katy Wolstencroft (14)

Biological Identifiers, 58
Cross-Reference (Xref), 126
DAS Services, 129
Data Integration, 130
Data Warehouse, 134
Flat File Data Formats, 224
Linked Data, 384
Mark-up Language, 400
Minimum Information Models, 433
RDF, 597
RESTful Web services, 612
Scientific Workflows, 651
Web Services, 773
WSDL/SOAP Web Services, 775

Marketa J. Zvelebil (9)

Ab Initio, 3
DOCK, 159
Docking, 160
Loop, 390
Proteome Analysis Database (Integr8), 573
QSAR (Quantitative Structure Activity Relationship), 583
Robustness, 635
Structure-Based Drug Design (Rational Drug Design), 704
UniProt, 753